math is/5

Authors

FRANK EBOS, Senior Author
Faculty of Education
University of Toronto

BOB TUCK
Mathematics Consultant
Nipissing Board of Education

Reviewer Consultants

MARY CROWLEY
Department of Education
Dalhousie University
Halifax, Nova Scotia

DENNIS HAMAGUCHI
W.L. Seaton Secondary School
Vernon, British Columbia

GARY HATCHER
Co-ordinator of Mathematics and Science
Avalon Consolidated School District
St. John's, Newfoundland

WILLIAM KORYTOWSKI
Sisler High School
Winnipeg, Manitoba

MARGARET OSBORN
Parkland Composite High School
Edson, Alberta

BILL WEREZAK
Mount Royal Collegiate
Saskatoon, Saskatchewan

NELSON/CANADA

contents

1 concepts and skills: a foundation review

1.1 introduction/1

1.2 language of mathematics/2

1.3 thinking mathematically/6
    inductive thinking/6
    deductive thinking/7
    patterned thinking/7

1.4 working with exponents/10
    applications: exponents and stopping distances/14

1.5 proving laws/15

1.6 skills with monomials/16

1.7 working with polynomials: addition and subtraction/19
    applications: parachuting and sky diving/23

1.8 translating accurately/24

1.9 skills for solving equations/25
    writing equations/27

1.10 applying skills with equations: formulas/28
    interpreting measurement problems/31

1.11 steps for solving problems/32

1.12 a method of proof/36
    function notation $f(x)$/37

skills review: reading accurately/38
problems and practice: a chapter review/39
test for practice/40

math tips/6, 14, 40
problem/matics/9, 23, 31
math is plus: reviewing formulas/41

2 factoring, solving equations and problems

2.1 introduction/42

2.2 using binomials/43
    applications: permissible values/45
    applications: equations and identities/45

2.3 understanding factoring/46
    grouping factors/47
    applications: factoring and calculators/48

2.4 factoring trinomials/49

2.5 writing special factors/52
    factoring incomplete squares/55

2.6 solving quadratic equations by factoring/55
    applications: solving problems/59

2.7 rational expressions: multiplying and dividing/59

2.8 rational expressions: adding and subtracting/61
    solving problems: reading accurately/64

2.9 using the factor theorem/64
    applications: solving cubic equations/68

2.10 solving equations: rational expressions/69

2.11 applying our skills: solving problems/72

skills review: factoring/75
problems and practice: a chapter review/76
test for practice/77

math tips/46, 66, 75
problem/matics/48, 52, 54, 61, 72, 76
math is plus: working with computers/78

3 reals and radicals; solving problems

3.1 symbols: mathematical ideas in shorthand/79

3.2 solving problems: Pythagorean property/80
    applications: television screens/83
    problem-solving: a strategy/84

3.3 real numbers/85
    using graphs and symbols/87
    properties and problem-solving/88

3.4 real numbers and inequations/90
    translating into mathematics/91

3.5 real numbers, radicals and exponents/92

3.6 working with radicals: addition and subtraction/95
    radicals and literal expressions/98

3.7 working with radicals: multiplication/99
    applications: roots and radicals/104

3.8 division with radicals/104

3.9 solving radical equations/108

3.10 solving problems: radical equations/111

3.11 problem-solving: strategy/113

skills review: operations with radicals/114
problems and practice: a chapter review/115
test for practice/116

math tips/83, 94, 96, 108
problem/matics/99, 103, 107, 111, 114
math is plus: absolute value: equations and inequations/117

4 systems of equations, applications, and solving problems

4.1 introduction/120

4.2 graphs of intersecting lines/121
    applications: plotting storms/125

4.3 equivalent linear systems/125
    classifying linear systems/127

4.4 solving systems of equations/128
    applications: using co-ordinates/131

4.5 solving linear systems: addition-or-subtraction method /132

4.6 extending our work: special equations/135
    working with literal coefficients/137
    applications: equations for a computer program/138

4.7 steps for solving problems/139

4.8 organizing the given information/141
    applications: working with gold/144

4.9 solving problems: thinking visually/145
    applications: airline schedules/147

4.10 to solve problems/148

4.11 problem-solving: extraneous information/150

4.12 problems to solve: needing information/152
    applications: sorting information/154

skills review/155
problems and practice: a chapter review/156
test for practice/157

math tips/138, 157
problem/matics/134, 148
math is plus: more linear systems/158

5 ratio, proportion, and variation; solving problems

5.1 working with ratio: skills and concepts/159

5.2 solving problems and applications: ratio/163
    applications: chemicals and ratios/164

5.3 extending our work: ratios and proportions/165
    applications: the fertilizer ratio/168

5.4 working with direct variation/168

5.5 direct squared variation/171
    applications: driving safely/174

5.6 inverse variation and its applications/175

5.7 inverse squared variation/178

5.8 applications with joint variation/180

5.9 partial variation/182

5.10 making decisions: solving problems involving variation/184

skills review: solving equations/185
problems and practice: a chapter review/186
test for practice/187

math tips/165-171
problem/matics/185
looking back: a cumulative review/188

6 concepts and skills: analytic geometry

6.1 introduction / 189

6.2 vocabulary: relations and functions / 190

6.3 the linear function / 194

6.4 working with slope / 197
applications with slope / 200

6.5 slope y-intercept form of the linear equation / 201

6.6 parallel and perpendicular lines / 204

6.7 solving problems: geometric properties / 207

6.8 families of lines / 208

6.9 conditions for writing equations / 211

6.10 combining strategies to solve problems / 214
applications: properties of geometric figures / 217

6.11 systems of inequations and their graphs / 218

6.12 applications: linear programming / 220

skills review: writing equations / 223
problems and practice: a chapter review / 224
test for practice / 225

math tips / 200
problem/matics / 204, 213, 216
math is plus: applications: interpreting / 226

7 analytic geometry: distance, area, and applications

7.1 introduction / 227

7.2 distance between two points / 228
finding a distance formula / 228
applications: solving problems about distance / 230

7.3 choosing strategies for solving problems / 231
applications: distances in Canada / 233

7.4 finding areas of triangles / 234
formula: the area of a triangle / 234
extending our work: area of a polygon / 237

7.5 distance from a point to a line / 238
applications: problems about distance / 241

7.6 satisfying conditions: locus of points / 242
applications: families and areas / 244

7.7 solving problems: analytic geometry / 245

7.8 writing proofs / 248
applications: points of triangle / 250

7.9 circle: analytic properties / 251
extending our work / 254

7.10 regions and inequations: circles / 255

7.11 translations and co-ordinates / 256
equations and mappings / 259

skills review: distance on the plane / 260
problems and practice / 261
test for practice / 262

math tips / 241, 262
problem / matics / 230, 244, 245, 249, 250, 253, 254,
math is plus: applications with calculators / 263

8 statistics, probability and their applications

8.1 introduction: statistics / 264

8.2 statistics and sampling / 265

8.3 sampling: stratified and clustered / 267
applications: hypothesis test / 269

8.4 frequency distributions and histograms / 270

8.5 representatives of data / 273

8.6 predictions, inferences and estimations / 275
applications: predictions and experimental statistics / 276

8.7 measures of dispersion / 277
standard deviation / 278

8.8 decision making and standard deviation / 281

8.9 probability: skills and concepts / 284
charts and sample spaces / 286

8.10 independent and dependent events / 288
dependent events and conditional probability / 289

8.11 applications: probability and statistics / 290

skills review: statistical calculations / 293
problems and practice: a chapter review / 293
test for practice / 294

math tips / 281, 291
problem / matics / 283, 291, 292, 293, 294
math is plus: thinking inductively about circles / 295

9 aspects of geometry: proof, properties, and solving deductions

9.1 language of a deductive system / 297

9.2 congruence in geometry / 299
    congruence assumptions / 299
    writing proofs / 300

9.3 analyzing deductions / 303

9.4 isosceles triangle theorem / 305
    exterior angle theorem / 308

9.5 parallel lines / 309
    sum of the angles of a triangle (ASTT) / 312

9.6 indirect proof and parallel line theorems / 313
    method of indirect proof / 313

9.7 applications: properties of parallelograms / 316
    a method of proof: using constructions / 317

9.8 inequalities for triangles / 318

9.9 properties of circles / 320

geometry from different points of view

9.10 transformational geometry: a deductive system / 322

9.11 translation: properties and proofs / 322
    defining translations / 322
    proving deductions: translations / 325

9.12 reflections: properties and proofs / 326
    proving deductions: reflections / 329

9.13 working with rotations / 330
    proving deductions: rotations / 331

9.14 deductions and analytic geometry / 331

skills review: making conjectures / 333
problems and practice: a chapter review / 333
test for practice / 334

math tips / 303, 332
problem / matics / 305, 313, 318
looking back: cumulative review of chapters 5 to 8 / 335

10 geometry: more concepts and skills

10.1 the nature of mathematics / 336

10.2 area: parallelograms and triangles / 337

10.3 theorems about area / 339
    exercise / 341

10.4 parallel proportion triangle theorem / 344
    exercise / 345

10.5 internal and external division / 346
    working with co-ordinates / 348

10.6 similar triangles AAA ~ / 349
    applications: similar triangles / 351
    applications: proving theorems: 352

10.7 dilatations: similarity transformations / 353

10.8 theorems for similar triangles: SAS~, SSS~ / 355

10.9 areas of similar triangles / 358

10.10 vectors on the plane: another point of view / 360
    geometric vectors and polygons / 361
    applications: vectors and internal division / 363

10.11 proving deductions by vector methods / 364

skills review: solving problems / 365
problems and practice: a chapter review / 366
test for practice / 367

problem / matics / 339, 342, 343, 354, 363, 366, 367
math is plus: matrices / 368

11 elements of trigonometry; concepts, skills, applications

11.1 trigonometry: today and tomorrow / 369

11.2 trigonometry: concepts and skills / 370
   signs of sin θ, cos θ, tan θ / 372

11.3 coterminal and special angles / 373

11.4 trigonometric values: use of tables / 377
   exercise / 378

11.5 working with radian measure / 380
   trigonometric values of $\frac{\pi}{3}, \frac{\pi}{4}, \frac{\pi}{6}$ / 383

11.6 identities in geometry / 383

11.7 drawing graphs: trigonometric functions / 386
   trigonometric function values / 389
   applications: tides / 390

elements of applied trigonometry

11.8 solving right-angled triangles / 391

11.9 solving problems based on right angles / 394

11.10 the law of sines / 396

11.11 solving problems: laws of sines / 399
   applications: golf and the sine law / 402

11.12 law of cosines / 402
   applications: solving problems / 405
   applications: the cosine law and direction / 407

11.13 making decisions to solve problems / 408

11.14 solving 3-dimensional problems / 409

skills review / 410
problems and practice: a chapter review / 410
test for practice / 411

math tips / 376
problem / matics / 373, 390, 393, 399

12 the quadratic function and its applications

12.1 introduction / 412

12.2 quadratic functions and their graphs / 413
   property of a locus / 416
   investigating graphs of quadratic functions / 416
   applications: sports and quadratics / 417

12.3 the method of completing the square / 418
   formulas: for quadratic functions / 421
   translations and congruent parabolas / 422

12.4 applications: maximum and minimum / 424
   applications: problems for Galileo / 426

12.5 solving quadratic equations / 427
   the quadratic formula: roots of a quadratic equation / 430

12.6 solving problems: quadratic equations / 431

12.7 the nature of the roots for quadratic equations / 435
   characteristics of roots / 436

12.8 solving linear quadratic systems / 437
   quadratic regions / 438

skills review: quadratic functions and equations / 438
problems and practice: a chapter review / 439
test for practice / 440

math tips / 421, 432
problem / matics / 418, 422, 427, 429, 434, 438, 439

function plus: a math is plus

f1 the language of functions / 441
   building vocabulary / 442
   inverses of relations and functions / 445

f2 functions and their graphs: techniques for sketching / 447
   investigations / 447
   sketching $y = f(x) + q, y = f(x + p)$ / 447
   sketching $y = af(x), y = f(bx)$ / 448
   sketching $y = f(-x), y = -f(x)$ / 449
   applying skills: sketching graphs / 450

f3 the exponential function / 450

f4 investigating functions / 452

table of trigonometric values / 454
answers / 455
index / 473

1 concepts and skills; a foundation review

review of essential skills, exponents, polynomials, solving equations and formulas; steps for solving problems.

1.1 introduction

Mathematics is the creation of people. Studying mathematics is often like learning about the lasting contributions of famous mathematicians and re-living some of the mathematical ideas and thinking processes which these people developed.

The list of contributors to the development of mathematics goes on and on, and only a partial list is shown here. How many do you know? (They are not listed in any special order.)

Descartes	Euler	Ptolemy
Pythagoras	Abûl Wefâ	Al Karkui
Archimedes	Cardan	Sacrobosco
Fermat	Fibonacci	da Vinci
Leibniz	Newton	Copernicus
Napier	Thales	Recorde
Fourier	Plato	Viète
Pascal	Aristotle	Galileo
DeMoivre	Hipparchus	Appolonius
Cayley	Einstein	

Many of these mathematicians were ordinary people who became interested in someone else's work. As in our own everyday activities, often one mathematician was influenced by the mathematical interests, thinking and thought processes of another. For example, Pythagoras, who has an important theorem named after him, was influenced by the work of Thales, a Greek mathematician.

Pythagoras stated an important relationship among the 3 sides of a right-angled triangle. This law in mathematics advanced the science of surveying and map-making and is used to solve many problems in mathematics.

But not only does mathematics build upon the thoughts of individual people; its collective influence touches every aspect of modern society and, in so doing, it affects people everywhere.

Almost all of the major achievements of mankind involved mathematics in some way.

Many years ago, the development of the formula for the principle of the lever required a knowledge of mathematics.

Archimedes' formula is the basis on which many of the principles of engineering have been developed.

In more recent years, the development of Albert Einstein's famous formula

$$E = mc^2$$

required a knowledge of mathematics.

Albert Einstein's formula relates matter, m, and energy, E. His formula states that a small mass can be converted into a great amount of energy, the principle behind nuclear energy.

One of the main reasons for learning mathematics is to acquire skills and strategies to solve problems. The more problems we solve, the better we will be able to solve problems we have never met before. Throughout our study of mathematics, we, like other people who have developed mathematics, will learn organizational skills that can be applied and used to solve problems.

In our study of mathematics, we will share the methods of many great thinkers. We will see, too, how mathematics, the universal language which crosses time and cultural barriers, helps to solve problems which affect people's lives. Mathematics is created by people and helps to enrich human activity.

1.2 language of mathematics

In the past we have seen how mathematics is used

in food production in business

in industry in space exploration

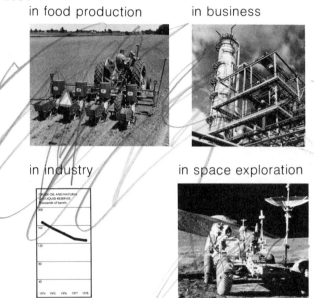

Throughout our study of mathematics we have met and will continue to meet many situations in which we can apply our mathematical skills. Sometimes some of the skills we learn in mathematics cannot be applied in a practical way, as we shall see. It is also important to learn the vocabulary of mathematics in order to learn the skills and work with algebra. (Read the Math Tip in this section.)

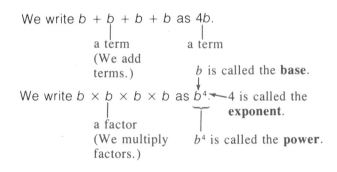

We write $b + b + b + b$ as $4b$.
  a term (We add terms.)   a term
  b is called the **base**.

We write $b \times b \times b \times b$ as b^4. — 4 is called the **exponent**.
  a factor (We multiply factors.)
  b^4 is called the **power**.

For a term we introduce the following vocabulary.

numerical coefficient \ / literal coefficient
$3x$

like terms: $2x, 3x, -4x, 5x$
The literal coefficients are the same.

unlike terms: $3x, -4y, 6x^2, 2xy$
The literal coefficients are different.

To simplify expressions, we may collect like terms.
$$12x + 8y - 3x - 5y$$
$$= 12x - 3x + 8y - 5y$$
$$= 9x + 3y$$

We often use special words in mathematics to describe expressions that consist of sums and differences of terms.

Vocabulary		Example
monomial	← 1 term →	$4x$
binomial	← 2 terms →	$4x - 3y$
trinomial	← 3 terms →	$3x^2 - 2x + 1$

When we want to refer to the above expressions, collectively, we refer to them as **polynomials.**

Our first exposure to the vocabulary of algebra occurred when we studied arithmetic. Whether we study arithmetic or algebra, the meanings of words are consistent no matter which branch of mathematics we are studying.

For example, note the similarity.

tri nomial
$\underbrace{x^2 + 3x + 1}_{3 \text{ terms}}$

tri angle

△ABC

3 angles

In developing the language of mathematics, we often use words that have special meanings in mathematics.

$3x + 4 \qquad x \in \{-1, 0, 1, 2\}$

We refer to x as a **variable**. We call this set the **domain** of the variable x.

Throughout our work with mathematics, we will do many calculations that involve the set of integers.
$$I = \{\ldots, -3, -2, -1, 0, 1, 2, 3, \ldots\}$$

We may evaluate the expression $3x + 4$ for the domain $x \in \{-1, 0, 1\}$.

$x = -1$	$x = 0$	$x = 1$
$3x + 4$	$3x + 4$	$3x + 4$
$= 3(-1) + 4$	$= 3(0) + 4$	$= 3(1) + 4$
$= -3 + 4$	$= 0 + 4$	$= 3 + 4$
$= 1$	$= 4$	$= 7$

When evaluating an expression, we must follow a convention for the order of the operations which has already been established.

$3(8 + 4 - 3) \qquad 3(4^2 - 3)$
$= 3(12 - 3) \qquad = 3(16 - 3)$
$= 3(9) \qquad\quad\; = 3(13)$
$= 27 \qquad\qquad = 39$

$3(8^2 - 2 \times 3) \qquad \dfrac{3^2 - 2^3}{8 \div (24 \div 3)}$
$= 3(64 - 6) \qquad\quad = \dfrac{9 - 8}{8 \div (8)}$
$= 3(58) \qquad\qquad\;\; = \dfrac{1}{1} = 1$
$= 174$

We may evaluate an expression such as $4a + b^2$ for different values of a and b.

Use $a = 3, b = -2.$ Use $a = -2, b = 3.$
$4a + b^2$ $4a + b^2$
$= 4(3) + (-2)^2$ $= 4(-2) + (3)^2$
$= 12 + 4$ $= -8 + 9$
$= 16$ $= 1$

Use brackets to substitute the value.

In algebra, as in mathematics, we define rules to simplify expressions.

$-(a + b) = -a - b$ means the negative of $(a + b)$
$-(a - b) = -a + b$

We may also say the "opposite of $(a - b)$."

Example 1
(a) Simplify
$(2a + b - c) + (a - b) - (a - 3b + c).$
(b) Evaluate the simplified expression in (a) if $a = -1, b = 2, c = -3.$

Solution   Record the given expression.
$(2a + b - c) + (a - b) - (a - 3b + c)$
$= 2a + b - c + a - b - a + 3b - c$
$= 2a + 3b - 2c$

Use $a = -1, b = 2, c = -3.$
$2a + 3b - 2c = 2(-1) + 3(2) - 2(-3)$
$= -2 + 6 + 6$
$= 10$

Symbols are used to write mathematics in a compact form.

Equations	Inequations
$16 \div 4 + 4 = 8$	$3^2 > 2^3$
$5^2 - 4^2 = 9$	$8 \div 4 - 2 < 3(12 - 3)$

We may use the symbol $\neq$ to show "not equal to".

Example 2
Show whether the following are true or false.
A $8 - (-4 + 1) > 5 + 16 \div 4$
B $\dfrac{1}{2}\left(\dfrac{12 + 3}{20 - 5}\right) \neq \dfrac{12 \div 2}{16 - 4}$

Solution means left side

A $LS = 8 - (-4 + 1)$ B $LS = \dfrac{1}{2}\left(\dfrac{12 + 3}{20 - 5}\right)$
$= 8 - (-3)$ $= \dfrac{1}{2}\left(\dfrac{15}{15}\right)$
$= 8 + 3 = 11$ $= \dfrac{1}{2}(1) = \dfrac{1}{2}$

$RS = 5 + 16 \div 4$ $RS = \dfrac{12 \div 2}{16 - 4}$
$= 5 + 4 = 9$ $= \dfrac{6}{12} = \dfrac{1}{2}$

Since $LS > RS$ Since $LS = RS$
then A is true.   then B is false.

When learning mathematics, we need to establish a vocabulary for mathematics and a foundation of useful skills in order to solve problems. We will continue to do this in mathematics.

1.2 exercise

1. For each of the following, write the
   - numerical coefficient
   - literal coefficient

   (a) $3x$ (b) $-4x^2$ (c) $2y^2$
   (d) $-16xy$ (e) $-25mn$ (f) $3m$
   (g) $\dfrac{1}{3}mn$ (h) $25p^2$ (i) $\dfrac{2mn}{3}$

2. Write each of the following as a single term.
   (a) $m + m + m + m$ (b) $x + 3x + 2x$
   (c) $5y - 2y + 3y$ (d) $8y - 11y - 3y$
   (e) $12p - 6p + 6p$ (f) $-3q - 2q - q$
   (g) $p - 3p + 5p$ (h) $-r + 3r + r$

3. What may each of the expressions in the answers to Question 2 be called? Give a reason for your answer.

4. Simplify each of the following.
   (a) $3x + 2x + y + 4y$
   (b) $6x - 3x + x + 2y - y + 3y$
   (c) $3x - 2y + 5x + 5y$
   (d) $3m + p - 2m + p + 3m$
   (e) $p + 3q - 3q + 3p - 2p$

5. Simplify each of the following. Which of the expressions in the answers may be called
   • monomials? • binomials? • trinomials?
   (a) $(x + 2y) + (x + 3y)$
   (b) $(p + 2q) - (p + q)$
   (c) $(3m - 2n) - (m + 3n)$
   (d) $(3u + 2w) + (2u - 3w)$
   (e) $(2p - 3q + 5r) - (3p - 2q - 2r)$
   (f) $(x^2 - 3x) - (x^2 - 2x)$
   (g) $(2x^2 - 3x + 5) - (4x^2 - 3x + 5)$

6. If $a = 0$, $b = 1$, $c = -1$, and $d = 2$, find the value of each expression.
   (a) $a + 2b$
   (b) $b + 3c$
   (c) $3b + 2c - d$
   (d) $9b + cd$
   (e) $2a^2 + b^2 - d^2$
   (f) $ad - bc$
   (g) $a^2 - 2bc - ad$
   (h) $2ab + a^2 - b^2$

7. Find the values of each expression.
   (a) $2x + 5$, $x \in \{-1, 0, 1, 2\}$
   (b) $8 - 3y$, $y \in \{-2, -1, 0, 1\}$
   (c) $3p^2 - 1$, $p \in \{-1, 0, 1\}$
   (d) $\dfrac{m^2 - 6m + 9}{m - 3}$, $m \in \{-1, 0, 1\}$
   (e) $(k + 1)(k - 1)$, $k \in \{-1, 0, 1\}$

8. (a) Simplify $(3m^2 - 2m) - (m^2 + m)$.
   (b) Evaluate the expression in (a) if $m = -1$.

9. Evaluate each expression for the given value of the variable.
   (a) $y^2 + y$, $y = 2$
   (b) $x^2 - 3x$, $x = -1$
   (c) $y^2 - 2y + 1$, $y = -2$
   (d) $3x^2 - 2x + 1$, $x = -3$

10. Evaluate each expression if $a = -3$, $b = 2$, $c = -1$.
    (a) $(3a - 2b) + (2a - 3c) - (b - c)$
    (b) $(2a + c) + (3a - b) - (2c - b)$
    (c) $(3a - b) - (2b - c) - (2c + a)$
    (d) $(a - b - c) - (a + b - c) - (b - a + c)$

11. We may use a chart or table of values to record the values of the variable and the expression. Copy and complete each of the following tables of values.

    (a)
y	$y^2 + 2y$
-1	-1
0	
1	
2	
3	

    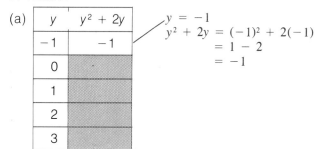

    $y = -1$
    $y^2 + 2y = (-1)^2 + 2(-1)$
    $= 1 - 2$
    $= -1$

    (b)
m	$3m^2 - 3m$
-2	
-1	
0	
1	

x	$x^2 - 3x + 1$
-1	
0	
1	
2	

12. An inequation is given by
    $4(3 - 1) - \dfrac{1}{2} \neq 3^2 - 2(3 - 2)$.
    (a) Calculate the LS (Left Side).
    (b) Calculate the RS (Right Side).
    (c) Is the inequation true or false?

13 An inequation is given by
$\left(\frac{3}{2}\right)\left(\frac{1}{3}\right) + 3 > \left(\frac{15}{3}\right)\left(\frac{3}{2}\right) - 4$.

(a) Calculate the LS (Left Side).
(b) Calculate the RS (Right Side).
(c) Is the inequation true or false?

14 Which of the following are true, (T)? Which are false, (F)?

(a) $3 - (-2 + 1) \leqq 5 + (-1)$
(b) $3 + \frac{4 + 5}{3} = \frac{3 + 5}{4} + 3$
(c) $\frac{1}{4} + \frac{3}{5} < \frac{4}{9}$
(d) $7 + \frac{3}{2} \times \frac{1}{3} \geqq \frac{15}{3} \times \frac{3}{2}$
(e) $\frac{1}{2} + \frac{3}{4} = \frac{3}{2} - \frac{1}{4}$
(f) $\frac{7}{8} \div \frac{3}{4} \geqq \frac{4}{3} \div \frac{8}{7}$
(g) $2 - \frac{5 - 1}{2} \neq \frac{7 - (4 + 3)}{5}$

> ### math tip
> The operation symbols (as well as many others) are intended to simplify the writing and reading of mathematics.
>
> +, − These symbols appeared in print for the first time in 1489, in an arithmetic book.
>
> = This symbol was used for the first time in 1557. The mathematician Robert Recorde used two equal parallel lines (=) "because no things can be more equal".
>
> <, > These symbols, first used by Thomas Harriot (1560-1621), were not accepted as suitable symbols by others at the time.
>
> × The symbol for multiplication was introduced by William Oughtred (1574-1660) but it was not readily adopted since it too closely resembled the variable x. For a time, Leibniz used the symbol ∩ for multiplication.

1.3 thinking mathematically

Once we have learned some mathematics and studied the language of mathematics, we can apply these skills and processes to solve problems. When solving a problem, we must clearly understand two important steps A and B.

A: What we are given.

To use the information given in A and answer the question in B often requires the use of different mathematical skills and strategies, some of which we already know. The more strategies and skills we know or remember, the better we will be able to solve the problem.

B: What we want to know.

Some of the types of thinking that we may use to solve problems and learn new facts are as follows.

inductive thinking

We may investigate properties of figures by conducting experiments.

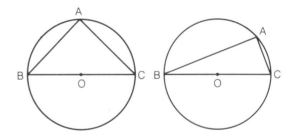

For example, each time we draw an angle as shown in a circle subtended by the diameter of the circle, we find $\angle A = 90°$. Thus we might make the general statement (hypothesize):

The angle at the circumference of a circle subtended by the diameter of that circle is 90°.

When we make a general statement based on the evidence of a number of specific examples, we are *thinking inductively*. However, the general statement we make is only as valid as the results of the specific examples we have explored. As a result, thinking inductively gives us useful information that we may then prove using another type of thinking.

deductive thinking

An example of deductive thinking is shown,

A	We are given a general statement that is true, (or accepted as true).	Banks are closed on a statutory holiday.
B	We are given a specific statement.	Monday is a statutory holiday.
C	We then make a conclusion.	The banks are not open on Monday.

To deduce the conclusion in C, we used the information in steps A and B. In studying mathematics, we deduce many facts based on the above process.

A A general statement is given as true.

The measures of the vertically opposite angles are equal.

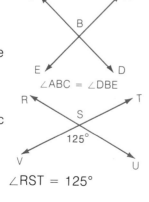

∠ABC = ∠DBE

B We are given a specific piece of information.

C We then deduce a conclusion for the problem.

∠RST = 125°

We shall see that often in solving a problem we use the same steps over and over again—each time thinking deductively.

There are other types of thinking that let us learn new facts, find answers, or solve problems, one of which is patterned thinking.

patterned thinking

For example, how many cannon balls are in the pile?

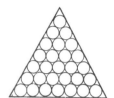

We could answer the question by counting all the cannon balls, but we prefer to look for a pattern in each layer.

4th layer   3rd layer   2nd layer   1st layer

The total in 4 layers of cannon balls is given by
1 + 4 + 9 + 16 = 30.

In the above problem we may recognize that each number is a square, $1^2 + 2^2 + 3^2 + 4^2$. Thus we could answer the question: How many cannon balls in a pile with 10 layers?

Throughout our work in mathematics we will learn many skills and strategies for solving problems. The more we learn, the better we will be at organizing our work when solving a problem we have not met before, regardless of whether it be a mathematical problem or not.

1.3 exercise

A

1 Give reasons why each of the following are examples of inductive thinking.

(a) During the season, the Swedish Team always scored in the last minute of play. The Swedish Team is playing the Canadian

Team. Thus, the Swedish Team will score in the last minute of play.

(b) The last 5 times that Jennifer has gone outside to play, she has come in muddy and dirty. Jennifer just went out to play. Thus when she comes in, she will be muddy and dirty.

How valid or conclusive are the final statements in (a) and (b)?

2. Give reasons why each of the following are examples of deductive thinking.

   (a) All bears are animals. Herman is a bear. Herman is an animal.

   (b) For any rhombus, the diagonals are perpendicular to each other. Figure PQRS is a rhombus. Thus PR ⊥ SQ.

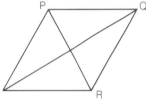

How valid or conclusive are the final statements in (a) and (b)?

3. When Jason squared numbers, he found that certain digits kept recurring. Use patterned thinking to find the possible values for the last digit of the squares of natural numbers.

4. Logs are piled as shown. Use patterned thinking to determine how many logs there are in 10 layers.

5. Refer to the pile of cannon balls shown earlier in this section.

   (a) Use patterned thinking to determine how many cannon balls there are if there are 7 layers.

   (b) How many cannon balls are there for 10 layers?

For Questions 6 to 21, answer the question but identify the type of thinking you used to arrive at your conclusion. Give reasons for your choice.

B

6. Butter becomes soft when warmed. Tracy warmed the butter. Thus, will the butter be soft?

7. To qualify in the finals of the 100-m dash you must run the distance in less than 10.8 s. Susan qualified for the finals of the 100-m dash. What conclusion can you arrive at?

8. Each time Michael touched the wood stove, he burned himself. Yesterday, Michael touched the wood stove. Did Michael burn himself?

9. All sides of an equilateral triangle have equal measures. △DEF is an equilateral triangle. What fact can we deduce about △DEF?

10. Kim is allergic to cats and will break out in a rash on contact with a cat. Kim accidentally held a cat. What conclusion can you come to?

11. In an isosceles triangle the base angles opposite the equal sides have equal measures. △ABC is isosceles. Will ∠B = ∠C?

12. At the class reunion, each person exchanged a memento with each other. If each person kept the memento given to him or her, how many mementos were needed if there were

    (a) 5 people at the reunion?
    (b) 12 people at the reunion?

13. All pelicans are birds. Pepi is a pelican. What conclusion can we come to about Pepi?

14. Central has always beaten Westdale when they play football in the finals. In next week's final game between Central and Westdale, what conclusion can you come to?

15 The law says that it is illegal to cross a street against a red light. Jerome is crossing the street where there is a stoplight. What conclusion can you draw?

16 The first 3 pentagonal numbers are shown.

Write the next two pentagonal numbers.

17 A type of number is defined by the diagram.

(a) What name might be given the number?
(b) Write the next 2 numbers of the above type.

18 Joe and Irene had a party. The first time the doorbell rang 1 person entered. On each successive ring two more people than were in the previous group entered.
How many people were at the party after the

(a) 5th ring?   (b) 25th ring?

19 Susan makes 4 out of 5 free throws during a basketball game. In a game, Susan has made 4 out of 4 free throws. What conclusion can we arrive at about her next free throw?

20 Peter is 16 a old. The movie at the Knox Theatre is restricted to persons 18 a of age or over. Will Peter see the movie?

21 A shape is shown made of cubes. The surface area of the shape is painted (all 6 faces).

How many cubes have

(a) all faces painted?   (b) 1 face painted?
(c) 2 faces painted?   (d) 3 faces painted?

C
22 Use your results to decide how many cubes in the shape have

(a) 1 face painted.
(b) 2 faces painted.
(c) 3 faces painted.

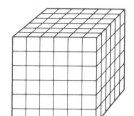

23 (a) What are the missing numbers in the pattern?

  Step 1 $1^3 = 1^2$
  Step 2 $1^3 + 2^3 = 3^2$
  Step 3 $1^3 + 2^3 + 3^3 = (?)^2$
  Step 4 $1^3 + 2^3 + 3^3 + (?)^3 = (?)^2$

(b) What is the missing square number in the 8th step of the above pattern?

24 At a birthday party, 15 relatives attended. If every relative kissed each other, once only, how many kisses in all were given?

problem/matics

Often when solving a problem we try to solve it by using a diagram. Solve this problem:
  A pie is cut into various pieces (using a knife of course). Three cuts were made in a pie as shown, and 7 pieces were obtained. What is the maximum number of pieces you can get, by making 5 cuts in the pie?

1.4 working with exponents

In mathematics, we write expressions compactly. For example we write the sum of terms
$a + a + a + a$ as $4a$.

We write a product of terms
$(a)(a)(a)(a)$ as a^4. — 4 is called the exponent.
The exponent tells *how many* factors to multiply.
a is called the *base*.
factors

Exponents occur frequently in our work in mathematics.

Area A, of a circle, radius r
$$A = \pi r^2$$
We say r squared or the second power of r.

Volume V, of sphere, radius r
$$V = \frac{4}{3}\pi r^3$$
We say r cubed, or the third power of r.

Exponents occur in may subjects.
- The growth of bacteria involve exponents.
- The distance fallen by an object under the influence of gravity is exponential.
- The growth of money is exponential.
And so on.

We use the definition of exponents to simplify expressions involving exponents.

$$a^3 \times a^2 = (a \times a \times a)(a \times a)$$
$$= a^5$$

$$\frac{a^5}{a^2} = \frac{a \times a \times a \times a \times a}{a \times a}$$
$$= a^3$$

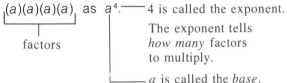

$(a^2)^3 = \overbrace{(a^2)(a^2)(a^2)}^{3 \text{ factors}}$
$= (a \times a)(a \times a)(a \times a)$
$= a^6$

$(ab)^3 = \overbrace{(ab)(ab)(ab)}^{3 \text{ factors}}$
$= (a \times a \times a)(b \times b \times b)$
$= a^3 b^3$

$\left(\dfrac{a}{b}\right)^3 = \left(\dfrac{a}{b}\right)\left(\dfrac{a}{b}\right)\left(\dfrac{a}{b}\right)$
$= \dfrac{a \times a \times a}{b \times b \times b} = \dfrac{a^3}{b^3}$

We may examine other particular cases to summarize the laws of exponents.

Law of Exponents
These *particular cases* suggest the *general cases*.

A $a^4 \times a^2 = a^6$ Multiplication $a^m \times a^n = a^{m+n}$
B $a^5 \div a^2 = a^3$ Division $a^m \div a^n = a^{m-n}$
C $(a^2)^3 = a^6$ Power $(a^m)^n = a^{mn}$
D $(ab)^3 = a^3 b^3$ Power of a Product $(ab)^m = a^m b^m$
E $\left(\dfrac{a}{b}\right)^3 = \dfrac{a^3}{b^3}$ Power of Quotient $\left(\dfrac{a}{b}\right)^m = \dfrac{a^m}{b^m}$

Example 1
Simplify each of the following.
(a) $x^3 \times x^2 \times x$ (b) $y^5 \div y^3$ (c) $(m^2)^3$
(d) $(ab)^3$ (e) $\left(\dfrac{m}{n}\right)^4$

Solution
(a) $x^3 \times x^2 \times x = x^{3+2+1}$ Add the exponent for like bases (A).
$= x^6$

(b) $y^5 \div y^3 = y^{5-3}$ Subtract exponents for like bases (B).
$= y^2$

(c) $(m^2)^3 = m^{2 \times 3}$ Multiply exponents (C).
$= m^6$

(d) $(ab)^3 = a^3 b^3$ Use (D).

(e) $\left(\dfrac{m}{n}\right)^4 = \dfrac{m^4}{n^4}$ Use (E).

We may combine the laws of exponents to simplify expressions.

Example 2

Simplify $\dfrac{a^3 (b^2)^2}{a^2}\left(\dfrac{a}{b}\right)^3$.

Solution

$$\dfrac{a^3 (b^2)^2}{a^2}\left(\dfrac{a}{b}\right)^3 = \dfrac{a^3 b^4}{a^2} \times \dfrac{a^3}{b^3}$$
$$= \dfrac{a^{3+3} b^4}{a^2 b^3}$$
$$= a^{6-2} b^{4-3}$$
$$= a^4 b$$

In mathematics, a skill we learn is applied to do more mathematics. In the previous section we combined simple like terms by adding and subtracting. We now use this skill to simplify expressions that contain exponents.

Example 3

Simplify $\dfrac{m^{3p+q} m^{3q+p}}{(m^p)^2}$.

Solution

Apply the laws of exponents to rewrite the expression

$$\dfrac{m^{3p+q} m^{3q+p}}{(m^p)^2}$$
$$= \dfrac{m^{3p+q} m^{3q+p}}{m^{2p}}$$
$$= \dfrac{m^{4p+4q}}{m^{2p}}$$
$$= m^{4p+4q-2p}$$
$$= m^{2p+4q}$$

$\begin{aligned}& 3p + q + (3q + p) \\ &= 3p + q + 3q + p \\ &= 4p + 4q\end{aligned}$

Some skills permeate the study of mathematics. The need for skills with exponents occurs over and over again in the study of different branches of mathematics, as well as in most scientific work. In Chapter 3 we will extend our study of exponents.

1.4 exercise

A

1 Simplify each of the following. Write with a single base.
(a) $m^3 \times m^2$ (b) $p^4 \times p^3$ (c) $p^2 \times p^3 \times p^4$
(d) $x^3 \times x^2$ (e) $w^3 \times w^5$ (f) 2×2^3
(g) $10^2 \times 10^3$ (h) $(-1)^3(-1)^2$ (i) $(-1)^4(-1)^5$
(j) $(-2)^3(-2)^2$ (k) $(-x)^3(-x)^7$ (l) $(-y)^2(-y)^2$

2 Simplify each of the following. Write with a single base.
(a) $x^6 \div x^3$ (b) $a^5 \times a^2$ (c) $y^{10} \div y^6$
(d) $3^4 \div 3^2$ (e) $2^5 \times 2^3$ (f) $10^{10} \div 10^9$
(g) $(x^{3p})(x^p)$ (h) $y^{2x} \div y^x$ (i) $(m^{3x})(m^{2x})$
(j) $(-1)^5 \div (-1)^3$ (k) $(-1)^8(-1)^2$
(l) $(-1)^7 \div (-1)^4$ (m) $(-m)^3(-m)^2$
(n) $(-a)^4 \div (-a)^2$ (o) $(-a)^5(-a)^3$

3 Simplify each of the following. Write with a single base.
(a) $(3^2)^3$ (b) $(2^3)^2$ (c) $(a^3)^2$
(d) $(y^2)^4$ (e) $[(-1)^3]^2$ (f) $[(-a)^3]^2$
(g) $(my)^2$ (h) $(m^2 y)^2$ (i) $(my^2)^3$
(j) $\left(\dfrac{x}{y}\right)^2$ (k) $\left(\dfrac{a^2}{b}\right)^3$ (l) $\left(\dfrac{b}{a^2}\right)^2$
(m) $y^8 \times y^2 \div y^6$ (n) $x^8 \times x^4 \div x^2$
(o) $(m^3)^4 \div m^2$ (p) $(y^3)^2(y^2)^3$

4 Simplify each of the following.
(a) $2^3 \times 2^4$ (b) $3^8 \div 3^4$ (c) $10^3 \times 10^2$
(d) $3^2 \times 3^4 \times 3$ (e) $2^8 \times 2^2 \div 2^5$ (f) $\left(\dfrac{2}{3}\right)^3$
(g) $y^3 \times y^2$ (h) $m^5 \div m^2$ (i) $(y^2)^3$
(j) $(m^5)(m^4)$ (k) $(xy)^4$ (l) $\left(\dfrac{x}{y}\right)^5$
(m) $(p^3)^5$ (n) $\dfrac{a^9}{a^2}$ (o) $\left(\dfrac{a}{2}\right)^3$
(p) $(2k)^3$ (q) $k^2 \times k^4 \times k^2$ (r) $(mn)^3(mn)^2$

5 Which of the following is greater? By how much?
   (a) 2^3; 3^2 (b) 3^4; 4^3 (c) 2^5; 5^2
   (d) 2^4; 4^2 (e) 5^3; 3^5 (f) 3^6; 6^3

6 Express each of the following as a power of 2.
   (a) $2^2 \times 2^3$ (b) $2^8 \div 2^5$ (c) $2^7 \times 2^5 \div 2^6$
   (d) 4^2 (e) 8^3 (f) $\left(\dfrac{4}{2}\right)^3$
   (g) $(2^m)^3$ (h) $(2^p)^3(2^p)^2$ (i) $32^4 \div 8^3$
   (j) 8^{p+1} (k) $4^{2m} \div 2^m$ (l) $\left(\dfrac{16}{2}\right)^m \div (2^m)^2$
   (m) 4^{2k+1} (n) 16^{2p-1} (o) $16^{4p} \div 4^{2p+1}$

7 Use $m = -2$ and $n = 1$ to evaluate each of the following.
   (a) mn^2 (b) m^2n (c) m^2n^2
   (d) $(mn)^2$ (e) m^3n^3 (f) $(mn)^3$

8 Use $x = 4$ and $y = -2$ to evaluate each of the following.
   (a) $\dfrac{x^3}{y}$ (b) $\dfrac{x}{y^3}$ (c) $\dfrac{x^3}{y^3}$
   (d) $\left(\dfrac{x}{y}\right)^3$ (e) $\dfrac{3x}{y}$ (f) $\dfrac{x}{3y}$

9 (a) Use $x = -3$, $y = 2$ to evaluate each of the following.
      (•) $\dfrac{y^2}{x}\left(\dfrac{x}{y}\right)^3$ (•) $\dfrac{x^2}{y}$
   (b) Why are the answers in (a) the same?
   (c) To evaluate $\dfrac{x^3}{y}\left(\dfrac{y}{x}\right)^2 \div \dfrac{y^3}{x}$ when $x = -4$, $y = 3$, what might be your first step?
   (d) Evaluate the expression in (c).

10 (a) Calculate each of the following.
      $(-1)^2$ $(-1)^3$ $(-1)^4$ $(-1)^5$ $(-1)^6$ $(-1)^7$
   (b) What is the value of $(-1)^n$ when n is odd?
   (c) What is the value of $(-1)^n$ when n is even?

B

11 Indicate whether each of the following is true, (T), or false, (F). Be prepared to justify your answer.
   (a) $3^3 \times 3^3 = 3^6$ (b) $(5^2)^3 = 5^6$
   (c) $9^5 \div 9^4 = 9$ (d) $(-3)^3(-3)^4 = (3)^7$
   (e) $(3a^4)^2 = 6a^8$ (f) $\dfrac{27^3}{9^3} = 3^3$
   (g) $(5^3)^2 = 3125$ (h) $\left(\dfrac{4}{2}\right)^2 = 2^2$
   (i) $\dfrac{27^3}{9^3} = 27$ (j) $(a^2b)^4 = a^8b^4$
   (k) $a^n b^n = (ab)^{2n}$ (l) $x^m y^m = (xy)^{m+n}$
   (m) $a^x a^y a^{x-y} = a^{2x}$ (n) $(ac^d)^{ac} = ac^{acd}$
   (o) $\left(\dfrac{x}{py}\right)^c = \dfrac{x^c}{py^c}$ (p) $(x^m y^n)^2 = x^{2m} y^{2n}$

12 Write
   (a) $3^5 \times 3^2 \times 3^4$ as a power of 3.
   (b) $5^2 \times 5^3 \div 5^2$ as a power of 5.
   (c) $2^2 \times 2^3 \div 2$ as a power of 2.
   (d) $\dfrac{10^4 \times 10^2}{10^5}$ as a power of 10.
   (e) $2^x \times 2^{x+1}$ as a power of 2.
   (f) $3^{y-1} \times 3^{y+2}$ as a power of 3.
   (g) $(4)^3 \times 2^2$ as a power of 2.
   (h) $3^y \times 9^y \div 3^{y+1}$ as a power of 3.

13 Express each of the following with a single base.
   (a) $p^{x+y} p^x$ (b) $y^{2x-p} \div y^p$
   (c) $(p^x)^2(p^3)^x$ (d) $p^{3x-1} \div (p^x)^2$
   (e) $m^x m^y m^{x-y}$ (f) $a^{m+n} a^{m-2n}$
   (g) $y^m \times y^n \div y^{2m}$ (h) $(k^2)^m \times (k^3)^n$
   (i) $(p^3)^m \div p^{2m}$ (j) $\dfrac{a^m}{(a^2)^m}$
   (k) $x^{a+b} x^{b+c} x^{a-c}$ (l) $y^{2a-b} y^{b-2c} \div y^{a+b}$

14 Calculate each of the following if $a = 4, b = -2$.
   (a) $\dfrac{(ab)^2}{ab}$ (b) $\dfrac{3(a^2b)^2}{(a^2)^2}$ (c) $\dfrac{a^2(b^2)^3}{b^3}$
   (d) $(a^2)^3\left(\dfrac{a}{b}\right)^3$ (e) $(a^2b)^3\left(\dfrac{a}{b}\right)^2$ (f) $\dfrac{a^2b^3(ab)^3}{a^2b}$
   (g) $\left(\dfrac{ab}{b}\right)^2(ab)^2$ (h) $\left(\dfrac{b}{a}\right)^2\left(\dfrac{a}{b}\right)^2$ (i) $\dfrac{(ab)^3}{a^2}\left(\dfrac{a}{b}\right)^3$

15 Express each of the following with a single base.
   (a) $\left(\dfrac{a}{b^2}\right)^3(ab^3)^2$ (b) $(mn^6)\left(\dfrac{m}{n^3}\right)^2$
   (c) $(a^k)^2(a^3)^k$ (d) $(x^p)^3 \div (x^p)^2$
   (e) $m^k m^{k+2} m^{k+3}$ (f) $\dfrac{p^k(p^{3k})^2}{p^k}$

16 Evaluate each of the following if $a = -2, k = 1$.
   (a) $(a^3)^k(a^k)^2 a^{2k-1}$
   (b) $a^k \times a^{2k-1} \div a^{3k+1}$
   (c) $(a^k)^2(a^2)^k \div a^{2k+3}$

17 Simplify each of the following.
   (a) $\left(\dfrac{a}{b}\right)^n\left(\dfrac{b}{c}\right)^n\left(\dfrac{c}{a}\right)^n$ (b) $\left(\dfrac{m}{n}\right)^2\left(\dfrac{n}{p}\right)^3\left(\dfrac{p}{m}\right)^4$

Questions 18 to 20 are based on the following formula.

The formula for the average mass of a normal adult male is
$$m = \dfrac{25(h - 100)}{26}$$
where m is his mass in kilograms and h is his height in centimetres.

18 (a) Find the mass of an adult male who is 175 cm tall.
   (b) Jackson has a mass of 65 kg. According to the formula, what should be his height?

19 Medical students are examining the above formula to determine whether it is reasonably accurate. The first person in this experiment has a mass of 74 kg.
   (a) What should be the height of this male if the formula is accurate?
   (b) If the male's height is actually 176 cm, how would you assess the formula based on this single example?

20 The tallest living man is Don Koehler, born in Denton, Montana who stands at 249 cm.
   (a) Use the above relationship between the mass and height of a normal adult male and find Koehler's expected mass.
   (b) His actual mass is 178 kg. Can you account for the difference?

Questions 21 to 23 are based on the following fact.

To clean the exhaust from the furnace of a plant, the exhaust is continuously passed through electronic air cleaners. The amount of pollutants left in the exhaust after passing through the air cleaner n times is given by
$$8\left(\dfrac{1}{2}\right)^n \text{ units.}$$

21 (a) Calculate the amount of pollutants remaining in the exhaust for $n = 1, n = 2$, and $n = 3$.
   (b) What is happening to the amount of pollutant in the exhaust as n increases?

22 One factory requires that the amount of pollutants going into the air should not exceed 0.005 units. Find the required value of n.

23 The exhaust from a blast furnace is passed through the air cleaner 9 times.
   (a) Use the above expression. What is the amount of pollutant that goes into the air?
   (b) What is the value of n?

C

24 For $y = 2$, $k = 1$, which expression has the greater value, A or B?

A $\dfrac{y^{2k}y^{k+2}}{(y^2)^k}$ 	B $\dfrac{(y^k)^3 y^{2k-1}}{(y^k)^3}$

25 Simplify each of the following.

(a) $(b^{p+1})^2 \times b^3 \div (b^p)^3$	(b) $\dfrac{p^{3b+2}(p^{b-1})^2}{(p^{b+1})^3}$

(c) Find the value of each of the above when $b = 2$, $p = -1$.

applications: exponents and stopping distances

When a vehicle is driven, the conditions of the weather affect how far a vehicle travels in coming to a complete stop when the brakes are applied.

On a dry surface
- the speed of a car with a trailer is 20 km/h.
- distance covered in coming to a stop is 12.3 m.

On a wet surface
- the speed of a car with a trailer is 20 km/h.
- distance covered in coming to a stop is 16.8 m.

To calculate the distance covered in coming to a complete stop, we need to calculate expressions involving exponents as shown by the following questions.

Express your answers to 1 decimal place.

26 The distance, in metres, covered by a tandem truck in coming to a stop on a dry paved road is given by $0.02v^2$, where v is the speed of the truck in kilometres per hour.

(a) How far will the truck travel if the brakes are applied at a speed of 30 km/h?

(b) By how much does the required stopping distance increase if the speed is doubled?

27 (a) A car pulling a trailer along a gravel road is able to stop completely in a distance given by $0.015v^2$, in metres. How far will the car travel before coming to a full stop if the brakes are applied at a speed of 40 km/h?

(b) The same vehicle on a paved road requires a distance given by $0.01v^2$, in metres. By how much does the distance decrease for stopping on a paved road at the same speed?

28 (a) A large oil tanker approaches the dock in calm water at 15 km/h. If the braking distance covered in coming to a complete stop is given by $0.095v^2$, in metres, how far from the dock must the tanker begin to stop for a successful docking?

(b) If the speed of the tanker increases by 5 km/h by how much does its stopping distance increase?

29 On a wet slippery road, the stopping distances, in metres are shown for 2 vehicles.

Motorcycle Stopping distance, $0.016v^2$	Car Stopping distance $0.026v^2$

v is the speed of the vehicle in kilometres per hour (km/h).

(a) The motorcycle, travelling at 55 km/h is behind a car which is travelling at 50 km/h. If the driver and the motorcyclist apply their brakes simultaneously, will there be a collision?

(b) If the car and the motorcycle travel at the same speed, and they both apply their brakes simultaneously, will the motorcyclist stop safely?

problem/matics

A sequence of numbers is shown.
  1, 2, 6, 24, 120, 720, ...
Find the next two numbers in the sequence.

1.5 proving laws

Many years ago, Copernicus, the founder of modern astronomy, theorized that the sun is at the centre of the planetary system and that the earth and the other planets revolve about it. His theory was based on his many observations while studying the motions of the planets. He arrived at his conclusions *inductively*.

The rings of Saturn photographed from Pioneer 11. Without a telescope, Copernicus could see Saturn but not the rings.

In mathematics, we may also inductively make statements. For example, if we try values of x for $x^2 + x + 41$ it appears that $x^2 + x + 41$ gives values that are prime numbers.

$x = 1 \quad x^2 + x + 41 = (1)^2 + (1) + 41$
$ = 1 + 1 + 41$
$ = 43$, a prime number

$x = 2 \quad x^2 + x + 41 = (2)^2 + (2) + 41$
$ = 4 + 2 + 41$
$ = 47$, a prime number

However, we cannot conclude that $x^2 + x + 41$ represents only prime numbers. (In fact it does not. Try $x = 41$.)

In mathematics we may deduce facts or laws. To do this, we must understand definitions. At first the laws of exponents were inductively suggested by numerical examples. For example, by using the definition of a^m we may prove one of the laws of exponents, namely, that for m and n, natural numbers,

$$a^m \times a^n = a^{m+n}.$$

Since $a^m = \underbrace{a \times a \times a \times \ldots \times a}_{m \text{ factors}}$

$a^n = \underbrace{a \times a \times a \times \ldots \times a}_{n \text{ factors}}$

then $a^m a^n$
$= \underbrace{(a \times a \times a \times \ldots \times a)}_{m \text{ factors}}\underbrace{(a \times a \times a \times \ldots \times a)}_{n \text{ factors}}$

We use the definition.

$= a^{m+n}$

We use the definition again.

We use the definition.

Thus $a^m a^n = a^{m+n}$.

As we learn more mathematical skills we will be able to prove more and more facts and laws in mathematics.

1.5 exercise

1 Use the definition of a^m to deduce the law of exponents that for all m, n natural numbers, $m > n$,
$$a^m \div a^n = a^{m-n}.$$

2 Prove that $(a^m)^n = a^{mn}$ for all natural numbers m, n.

3 Deduce that for a, b, real numbers
$(ab)^n = a^n b^n$ for all natural numbers, n.

4 Show that for all real numbers a, b
$$\left(\frac{a}{b}\right)^n = \frac{a^n}{b^n}$$
for all natural numbers n.

In mathematics we often ask the same question in different ways. For example, each of these instructions would accomplish the same result.

- Prove that ... .
- Show that ...
- Deduce that ...

5 Show that $\left(\dfrac{ab}{c}\right)^n = \dfrac{a^n b^n}{c^n}$ for all $n \in N$.

6 If $n \in N$, prove that $(abc)^n = a^n b^n c^n$.

7 Prove that $a^n \times a^n \times a^n = a^{3n}$ for all $n \in N$.

8 Prove that for all natural numbers n
$$\frac{a^n a^m}{a^p} = a^{n+m-p}.$$

For Questions 9 and 10, the examples suggest facts we may prove about our results.

9 Calculate each of the following.

(a) $(-1)^{12}(-1)^{11}$ (b) $(-1)^{15} \div (-1)^{12}$

(c) $(-1)^{20}(-1)^3(-1)^2$ (d) $\dfrac{(-1)^{12}(-1)^5}{(-1)^3}$

(e) $\dfrac{(-1)^6(-1)^8}{(-1)^3(-1)^9}$ (f) $\dfrac{(-1)^3(-1)^5}{(-1)^4} \div (-1)^3$

10 Express each of the following with the single base 10.

To evaluate $(-10)^3$ we may think of
$(-10)^3 = [(-1)(10)]^3$
$= (-1)^3(10)^3$
$= -10^3$

(a) $10^8 \times 10^2$ (b) $10^5 \div (-10)^3$

(c) $(-10)^5 \times 10^3 \div 10^3$ (d) $\dfrac{10^2 10^3}{(-10)^4}$

(e) $\dfrac{(-10)^5(10)^3}{(-10)^3}$ (f) $\dfrac{-10^3(-10)^2}{10^3} \div (-10)^2$

(g) $\dfrac{(-10)^2 10^3}{-10^3}$ (h) $\dfrac{-10^2 \times 10^3}{(-10)^3}$

11 Prove that for n an even number, the value of $(-1)^n$ is 1.

12 Prove that for n an odd number, the value of $(-1)^n$ is -1.

13 Prove that for m and n even numbers, $(-1)^m(-1)^n$ is positive.

14 Prove that for consecutive natural numbers m and n, $(-1)^m(-1)^n$ is negative.

1.6 skills with monomials

In mathematics, we often apply one skill to develop additional skills. To multiply monomials, we will apply these skills with exponents.

$a^m \times a^n = a^{m+n}$ $a^m \div a^n = a^{m-n}$

$(a^m)^n = a^{mn}$ $\left(\dfrac{a}{b}\right)^m = \dfrac{a^m}{b^m}$ $(ab)^m = a^m b^m$

To find the product of monomials, which skills with exponents do we need to use?

Example 1
Find the product of $-4a^2b^2$ and $-2ab^3$.
Solution

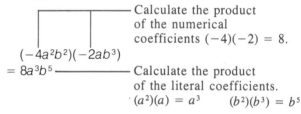

$(-4a^2b^2)(-2ab^3)$ — Calculate the product of the numerical coefficients $(-4)(-2) = 8$.
$= 8a^3b^5$ — Calculate the product of the literal coefficients.
$(a^2)(a) = a^3$ $(b^2)(b^3) = b^5$

To find the quotient of monomials, which skills with exponents do we use?

Example 2
Simplify $(-9x^6y^5) \div (x^2y)^2$.
Solution
$(-9x^6y^5) \div (x^2y)^2$
$= (-9x^6y^5) \div (x^4y^2)$ — Why?
$= -9x^2y^3$

$x^6 \div x^4 = x^{6-4}$
$= x^2$
$y^5 \div y^2 = y^{5-2}$
$= y^3$

To evaluate a monomial, we may first simplify. In Example 2 if $x = -2$, $y = 3$, we may substitute to find the value of the expression.
$-9x^2y^3 = -9(-2)^2(3)^3$ where $x = -2$, $y = 3$
$= -9(4)(27)$
$= -972$

Example 3

Evaluate $\dfrac{m^2n - m^3n^3 - 3mn^2}{-mn}$ if $m = 3$, $n = -2$.

Solution

To evaluate, we simplify first.

$$\dfrac{m^2n - m^3n^3 - 3mn^2}{-mn}$$

$= \dfrac{m^2n}{-mn} - \dfrac{m^3n^3}{-mn} - \dfrac{3mn^2}{-mn}$ We use the same denominator each time.

$= -m + m^2n^2 + 3n$

If $m = 3$, $n = -2$, then
$-m + m^2n^2 + 3n = -(3) + (3)^2(-2)^2 + 3(-2)$
$= -3 + (9)(4) - 6$
$= -3 + 36 - 6$
$= 27$

In Example 3, we used our skills for simplifying monomials *before* we evaluated the expression.

1.6 exercise

A

1. Find each product.
   (a) $(a^2b)(ab^2)$
   (b) $(-3a^3b^2)(-2ab^3)$
   (c) $(x^3y)(xy)$
   (d) $(8x^2y^3)(-2xy^2)$

2. Find each quotient.
   (a) $\dfrac{a^4}{a^2}$
   (b) $\dfrac{8a^5}{2a^3}$
   (c) $\dfrac{-16a^6}{-4a^3}$
   (d) $\dfrac{x^3y^2}{xy}$
   (e) $\dfrac{4x^2y^3}{-2xy^2}$
   (f) $\dfrac{-8x^5y^2}{2x^3y}$

3. (a) Write $(-3x^2y)(-2xy^3)$ as one term.
   (b) Evaluate the expression in (a) if $x = -1$, $y = 2$.

4. (a) Write $(-9x^3y^2) \div (3x^2y)$ as one term.
   (b) Evaluate the expression in (a) if $x = -1$, $y = 2$.

5. (a) Find the square of $-4xy$.
   (b) Evaluate the expression in (a) if $x = 3$, $y = -2$.

6. If $a = 3$, calculate each of the following.
   (a) $-2a^2$
   (b) $-(2a)^2$
   (c) $(-2a)^2$
   (d) $\dfrac{2a^2}{a}$
   (e) $\dfrac{(2a)^2}{a}$
   (f) $\left(\dfrac{2a}{a}\right)^2$

7. If $a = -2$, $b = 3$, calculate each of the following.
   (a) $-2ab^2$
   (b) $-2(ab)^2$
   (c) $(-2ab)^2$
   (d) $\dfrac{2a^2}{b}$
   (e) $\dfrac{(-2a)^2}{b}$
   (f) $\dfrac{-2a^2}{b}$

8. Simplify.
   (a) $(3x)(2x)^2$
   (b) $(-2x^2)(-3x^2)^3$
   (c) $(2xy)(xy)^2$
   (d) $(-3x^2y)(-xy^2)^3$

9. Since $(3x)(2x) = 6x^2$ we may say that $3x$ and $2x$ are factors of $6x^2$. Find the missing factor.
   (a) $(?)(3x^3) = 6x^5$
   (b) $(2xy^2)(?) = -8x^3y^4$

10. (a) Why may $\dfrac{8m^2n + 6mn^2}{2mn}$ be written as $\dfrac{8m^2n}{2mn} + \dfrac{6mn^2}{2mn}$?
    (b) Simplify the expression in (a).

11. (a) Why is the area of rectangle ABCD given by $\left(\dfrac{3x^2y}{2xy^2}\right)\left(\dfrac{12xy^2}{6xy}\right)$?
    (b) Simplify the expression in (a).

12. (a) Why is the area of the square PQRS given by $\left(\dfrac{9a^3b^3}{3ab}\right)^2$?
    (b) Simplify the expression in (a).

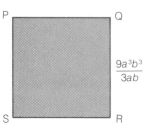

B

13. For each pair of monomials, find the product.
    (a) $-3a^3, 4b^2$
    (b) $-3m^3, 5m^2$
    (c) $-3m^2n, 4mn^2$
    (d) $6p^2q, -3p^2q^3$
    (e) $(mn)^2, -m^3n^2$
    (f) $(p^2q)^2, -4pq^2$

14. Write each product in simplest terms.
    (a) $(3x)(-2y)$
    (b) $(3x)(-4x^2)$
    (c) $(4xy)(2xy)$
    (d) $(-3ab)(2ab)$
    (e) $-2x^2y(-3xy^2)$
    (f) $-p^2q(4p^2q^2)$
    (g) $(xy)(-3xy)^2$
    (h) $(-2ab)(-a)(ab)$
    (i) $(-3mn)(m^2n)(-m)$
    (j) $(-pq)(3pq)(p^2)$

15. Find the square of each of the following.
    (a) $-6a$ (b) $3a^2$ (c) $-6x^3$
    (d) $3mn$ (e) $6a^2b^4$ (f) $-4x^2y$

16. Simplify each quotient.
    (a) $\dfrac{-24x^2y}{6xy}$
    (b) $\dfrac{-18m^3n^2}{-6m^2n}$
    (c) $(-64x^3y^3) \div (8xy^2)$
    (d) $(36p^5q^3) \div (-6p^4q^2)$

17. Find the value of each of the following if $a = 3$, $b = -2$.
    (a) $(3ab)(-2a^2b)$
    (b) $(-12a^2b) \div (4ab)$
    (c) $\dfrac{18a^3b^2}{-6ab^2}$
    (d) $(-2ab^2)(ab)^2$
    (e) $\dfrac{(ab)(a^2b)}{ab^2}$
    (f) $\dfrac{(-3a^2b)(-2ab^2)}{6ab^2}$

18. Find the square of each of the following.
    (a) $(16m^2n) \div (-4mn)$
    (b) $(3x^2y)(-2xy^2)$
    (c) $\dfrac{12a^3b^3}{-3ab^2}$
    (d) $(-9a^3b)(-3ab^2)$

19. Find the missing factor.
    (a) $(-ab)(\ ?\ ) = 3abc$
    (b) $(\ ?\ )(-5m) = -25mn$
    (c) $(9ab)(\ ?\ ) = -18a^2b$
    (d) $(\ ?\ )(-4mn) = 16m^2n$
    (e) $(6x^2y)(\ ?\ ) = -36x^4y^3$

20. Write each of the following in simplest terms.
    (a) $\dfrac{a^2 - ab + ab^2}{a}$
    (b) $\dfrac{8y^2 + 16y}{-4y}$
    (c) $\dfrac{49m^2n - 14m}{-7m}$
    (d) $\dfrac{12ab^2 - 16ab}{-4ab}$
    (e) $\dfrac{28a^3b^2 - 7a^5b^3}{-7a^2b}$
    (f) $\dfrac{p^2q - p^3q - 3pq^2}{-pq}$
    (g) $\dfrac{18m^2n^2 - 12mn^2 - 36m^2n}{-6mn}$
    (h) $\dfrac{10a^2b^3 - 20a^3b^2 + 30a^4b^2}{-10a^2b^2}$

21. Evaluate each expression for the given values.
    (a) $\dfrac{4mn^2 - 8m^2n}{-4mn}$ $m = 2, n = 1$
    (b) $\dfrac{9a^2b^2 - 15a^3b^2}{-3a^2b^2}$ $a = 3, b = -3$
    (c) $\dfrac{25m^3n^3 - 10m^2n^3 - 15m^3n^2}{-5m^2n^2}$ $m = -3, n = -2$

22. Simplify each of the following.
    (a) $(-3a^2b)(5ab)$
    (b) $(-48ab^2) \div (-6ab)$
    (c) $\dfrac{-56m^2n^3}{-7mn^2}$
    (d) $(-2ab)(-3ab) \div (-3a^2b)$
    (e) $\dfrac{9a^3b^5}{-3ab}\left(\dfrac{a^2}{b}\right)^3$
    (f) $(-3x^3y^2)(xy)^2$

(g) $\dfrac{-30a^2b + 5ab}{-5ab}$ (h) $\dfrac{(-3a^2)(6a^2b^3)}{9ab^2}$

(i) $\dfrac{(6a^4)(-2a^2b^3)}{(4ab)^2(-a)^2}$ (j) $\left(\dfrac{ab^2}{b}\right)^2(a^2b)$

(k) $\dfrac{15x^3y - 30x^2y^2 - 10xy^2}{-5xy}$ (l) $\dfrac{36x^4y^5 - 9x^3y^4}{-9x^2y^3}$

23 (a) Simplify $\dfrac{36a^3b^4}{-2ab}\left(\dfrac{a}{b}\right)^3$.

(b) Evaluate the expression in (a) if $a = -2$, $b = 3$.

24 Find an expression for the area of each figure in simplified form.

(a)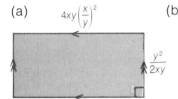

(b)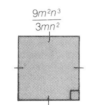

25 Which rectangle has the greater area if $a = 3$, $b = 2$?

(a)

(b)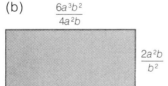

C

26. Which expression, A or B, has the greater value if $a = 1$, $b = -2$?

A $\dfrac{(-8ab)(-6a^2b^2)}{(-4a)(-2b)}$ B $\dfrac{-8ab}{-4a} - \dfrac{-6a^2b^2}{-2b}$

27 What is the maximum area of the rectangle if $a \in \{1, 2, 3\}$ and $b \in \left\{\dfrac{1}{2}, \dfrac{1}{3}, \dfrac{1}{4}\right\}$?

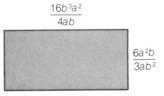

1.7 working with polynomials: addition and subtraction

In studying mathematics, we use algebraic expressions to help us solve problems, and to summarize important results. For example, the expression $4.9t^2$ may be used to calculate the distance, d, in metres, we fall in t seconds.

$t = 1$ $t = 2$ $t = 3$

distance $= 4.9(1)^2$ m

distance $= 4.9(2)^2$ m

distance $= 4.9(3)^2$ m

We often refer to the **degree** of a polynomial.

degree 1 $2x$ { Greatest power of the variable is 1.

degree 2 $3x^2 + 2x + 5$ { Greatest power of the variable is 2.

degree 3 $4x^3 - 3$ { Greatest power of the variable is 3.

What we learn in arithmetic can be used to develop skills in algebra.

4 + 4 may be written as 2 × 4. $y + y$ may be written as $2y$.

In algebra we use our skills to simplify expressions.

Example 1 — Simplify means collect like terms.
Simplify
$(4x + 2y) - (2x - 5y)$.

Solution
$(4x + 2y) - (2x - 5y)$ — The opposite of $2x - 5y$ is written $-(2x - 5y) = -2x + 5y$
$= 4x + 2y - 2x + 5y$
$= 4x - 2x + 2y + 5y$
$= 2x + 7y$ — This step would be done mentally.
$4x - 2x = 2x$
$2y + 5y = 7y$

Since $2x$ and $7y$ are unlike terms, we can not simplify further.

Simplified expressions require fewer calculations when evaluating them. For example, use $x = 2$, $y = 3$.

$(4x + 2y) - (2x - 5y)$
$= (8 + 6) - (4 - 15)$
$= 14 - (-11)$
$= 14 + 11$
$= 25$

$2x + 7y$
$= 2(2) + 7(3)$
$= 4 + 21$
$= 25$

We have shown all the steps.

The two expressions
$(4x + 2y) - (2x - 5y)$ and $2x + 7y$
are said to be *equivalent* since they have the same values for different values of the variables x and y. The word *equivalent* is important to learn for the study of mathematics.

We use the distributive property to simplify expressions, as shown in the next example.

Example 2
Simplify and write the resulting expression in descending powers of x.
$4x(x - 2) - 3(x^2 - 2x + 1) - 3(x - 5)$

Solution — *Write the given expression as the first step of the solution.*

$4x(x - 2) - 3(x^2 - 2x + 1) - 3(x - 5)$
$= 4x^2 - 8x - 3x^2 + 6x - 3 - 3x + 15$
$= 4x^2 - 3x^2 - 8x + 6x - 3x - 3 + 15$
$= x^2 - 5x + 12$

*We often refer to this term as the **constant** term.*

We may often need to simplify expressions that involve more than one set of brackets.

Example 3
(a) Simplify $4 - 2[3x + 5(1 - x)]$.
(b) Evaluate the expression in (a) for $x = -2$.

Solution — *Simplify the inner brackets first.*
(a) $\quad 4 - 2[3x + 5(1 - x)]$
$= 4 - 2[3x + 5 - 5x]$
$= 4 - 2[-2x + 5]$
$= 4 + 4x - 10$
$= 4x - 6$

(b) Use $x = -2$. — *We write brackets to substitute here.*
$4x - 6 = 4(-2) - 6$
$= -8 - 6$
$= -14$

In the above example, it is important that we simplify the expression in (a) correctly or the answer in (b) will definitely be incorrect. Simplify accurately!

Simplifying polynomials is a skill which makes the solution of algebra problems a lot easier.

1.7 exercise

Questions 1 to 13 develop skills for adding and subtracting polynomial expressions.

A

1 Simplify each of the following.
(a) $2x - 3y + 5x$ (b) $3a - 2b + 6a$
(c) $3x^2 - 2x + 5x^2$ (d) $3xy - y^2 + 2xy$

2 Find each product.
(a) $3(x - 5)$ (b) $-2(2y - 1)$
(c) $-3(3a - 5)$ (d) $2(6 - 2y)$
(e) $-3(x - 2y)$ (f) $2(2x + 5y)$
(g) $3(2x^2 - 3x + 5)$ (h) $-3(2a^2 - 5a - 6)$

3 Expand. — *means to find each product Use the distributive law.*
(a) $2x(x - 1)$ (b) $-3a(a - 2)$
(c) $2m(m^2 - 2m)$ (d) $-3y(y^2 + 4y)$
(e) $-2t(t^2 - 2t - 1)$ (f) $-3m(m - 2m^2 - 5)$
(g) $6ab(a^2 + 2ab)$ (h) $-3xy(x^2 + y^2)$

4 (a) Simplify $(2a - 3b) - (3a + b)$.
   (b) Simplify $-(3a - b) + 2a - 5b$.
   (c) Why are the answers to (a) and (b) called equivalent expressions?

5 Simplify. ← means to expand and collect like terms
   (a) $3(x - 2) - 5x + 3$ (b) $2(2y + 5) - 3y$
   (c) $3a - 2b - 2(3a + 2b)$
   (d) $-(3x - 2y) - 3(x - y)$
   (e) $2(m - n) - 3(2m - 3)$

6 (a) Simplify $3(x - 2y) - 2(2x + y)$.
   (b) Simplify $4(x - y) - (5x + 4y)$.
   (c) Why are the answers to (a) and (b) equivalent expressions?

7 Decide what error has been made by a student obtaining each of the following wrong answers.

   A $x^2 + x^2 \stackrel{?}{=} x^4$ D $2x^2 + 3x^2 \stackrel{?}{=} 6x^2$
   B $2x^2 + 3x^2 \stackrel{?}{=} 6x^4$ E $3x + 2y \stackrel{?}{=} 5xy$
   C $x^2 \times x^2 \stackrel{?}{=} 2x^2$ F $x^2 + x^3 \stackrel{?}{=} 2x^5$

8 (a) Simplify the expression inside the square brackets first.
   $$3x - 2[x - 2(x - 3)]$$
   (b) Complete the answer in (a).

9 (a) What might be the first step in simplifying
   $$3[2x - 3(2x - y)] - 3(x - y)?$$
   (b) Simplify the expression in (a).

B

10 An operation has been left out for each. What is the missing operation shown by $*$?
   (a) $x^2 * x^2 = x^4$ (b) $x^2 * x^2 = 2x^2$
   (c) $2x * 2y = 4xy$ (d) $3x * 4x = 12x^2$
   (e) $4y * 2y = 6y$ (f) $3k * 3k = 9k^2$

11 Simplify (collect like terms).
   (a) $x + 2x - 3y - x + 2y$
   (b) $x^2 - 2x - 3 - 2x^2 + x + 4$
   (c) $3 - ab + 2a - 3ab - 4 + 5a$
   (d) $x^2 + 2xy - y^2 - 3x^2 - 6xy - 2y^2$
   (e) $3mn - 2mp + 3np - 2mp - 3mn - 2np$

12 Simplify. Arrange the answers in ascending powers of x.
   (a) $x^2 - 3x + 6 - 2x^2 + 5x + 7$
   (b) $2x^2 - 8 - 6x^2 + x - 3 + 5x$
   (c) $8 - 2x + 6x^2 - 3x - 2x^2 + 16$
   (d) $xy^2 - 2x^2y - 3xy^2 - 2x^2y + 4y^2$

13 Find the sum of
   (a) $a + c$, $2a - 3c$, $3a - 2c$
   (b) $x + y$, $x - 2y$, $y - 3x$
   (c) $3a - b$, $2a + b$, $3b - 2a$
   (d) $x^2 - xy$, $y^2 + xy$, $-x^2 - y^2$
   (e) $6m - 3n$, $-2m + n$, $-3m - 6n$

14 Find an expression for each perimeter.

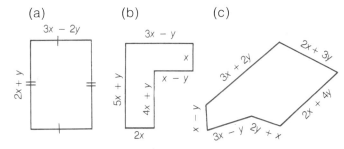

15 Subtract the polynomial in Column B from that in Column A.

Column A	Column B
(a) $2(x - y) - 3x$	$2(y - x) - 4y$
(b) $2(x^2 - 2x + 5)$	$3(x^2 - 3x - 6)$
(c) $-3(a + b) - 2b$	$-2(a - 3b) - 6b$
(d) $2(x^2 - 3x) - 4x^2$	$-3(x^2 - 2x) - 4x$

16 Simplify.
  (a) $3(4x + 2) + 2(x - 1)$
  (b) $2(3a - 2b) - 2(a - b)$
  (c) $3(2x - y) - 3(x - y)$
  (d) $-2(a - 3b + 4) - 6(a - 2b + 5)$
  (e) $3(x - y) - 2(4x - 5y + 6)$
  (f) $-2(x^2 - 2x - 3) - (x^2 - 3x)$
  (g) $3(3a^2 - 4) - 2(a^2 - 2a - 5)$

17 Simplify. Arrange the answers in descending powers of x.
  (a) $3(x^2 - 2x + 5) - 6(x^2 - 6x + 5)$
  (b) $x(2x - 5) - 3x(x - 6)$
  (c) $2(x^2 - x) - 3x(x + 1) - 2(x^2 + 5)$

18 Simplify. Arrange the answers in ascending powers of y.
  (a) $4(y^2 - 2y + 5) - 3(3y^2 - 4y - 6)$
  (b) $y(y - 3) - 2y(y + 6) - 6(y - 3)$
  (c) $3(y^2 - 1) - 2y(y - 5) - 3y(y - 2)$

19 Write a simplified expression which is equivalent to each of the following.
  (a) $3(2a - 3b) - 5(2a + 5b) - 6a$
  (b) $2(y^2 - 6y + 5) - (2y^2 + 5y - 6)$
  (c) $3x(x - 1) - 3(x^2 - 3x) + (x^2 - 2x)$

20 Simplify each of the following. Which expressions are equivalent?
  (a) $-3(a - b) - 2(b - a) - (a + b)$
  (b) $2(3a - 2b) - 3(a - 5b) - 2a$
  (c) $2(a - 5b) - 5(b - a) - 3(a - 2b)$
  (d) $-2(2b - a) + 3(a - 2b) - (4a - 21b)$

21 Simplify each of the following.
  (a) $2x - 4 - 3[-2(x - 2) + 3]$
  (b) $3[(3y - 6) + (-2y - 1)(4)]$
  (c) $b - 2[(1 - 2b) - 5(b + 3)]$

22 Find the value of the expression $x[(5x - 4) - (2 - 3x)]$ if $x = -3$.

23 If $x = -1$, find the value of
  $3(x^2 - 3x - 1) - 2(2x^2 - 3x - 2)$.

24 If $a = -2$ and $b = 3$, by how much does expression A exceed expression B?
  A $3a - 5b - 7[(2a - b) - (b - 2a)]$
  B $9a - (5a - 2b) + 3[1 + 7(a - b) - 4b]$

25 If $x = 3$, $y = 2$ which figure has the greater perimeter, A or B?

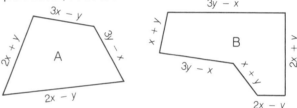

C
26 If $a = -2$, $b = 3$, which expression has the greatest value (maximum value)?
  A $3(a^2 + b^2) - 2ab - 2(a^2 - b^2)$
  B $3(a - b) - 2(b - a) + 3(a - 2b)$
  C $b(a - b) - a(2b - a) - ab$

27 If $m = -1$, $n = 3$, which expression has the least value (minimum value)?
  A $2(m - n) - 3(m - 2n) - (m^2 - n^2)$
  B $3m(m - n) - 2n(m + n) - (m^2 - n^2)$
  C $2n(m - 3n) - (m^2 - 2mn) - 3(n^2 - 2mn)$

28 What is the coefficient of x in the expression
  $ax - ay + bx + by - cx + cy$?

29 Find the coefficient of x and z in each of the following expressions.
  (a) $3z + 6x - ax - 3ax + 2bx - 5bz$
  (b) $-3ax - 4mz - 5az + 3mx - 8z + 9x$

application: parachuting and sky diving

It is important to skydivers to know how high up in the sky they are before a jump.

Skydivers can calculate how high they are above ground using the value of this polynomial,

$A - 4.9t^2$ where A, in metres, is their initial height above ground and t is the time in seconds after the skydivers leave the plane.

If they know how high they will be when they jump and the safe stopping distance for the parachute to bring them to earth, then they will be able to calculate either how long or how far they can free-fall.

30. A skydiving group left the plane at an altitude of 4000 m. Calculate how far above the earth's surface they are after
   (a) 5 s   (b) 10 s   (c) 0.5 min

31. Jean always opens her parachute at 500 m above the earth's surface and glides. If she dropped from a plane 2500 m high, after how many seconds should she open her parachute?

32. Part of the tail section of a 747 Jumbo jet broke off and fell to the ground from an altitude of 3500 m.
   (a) What was the height of the tail section after 25 s?
   (b) Between what two consecutive seconds does the tail section hit the ground?

33. The first stage of a rocket is dropped at an altitude of 10 000 m.
   (a) Calculate the height of the first stage after 40 s.
   (b) How far did the first stage drop in the 41st second?
   (c) If the second stage was dropped from the same height as the first stage after 43 s, how far apart were the two stages after 44 s?

34. Galileo dropped a large cannon ball from the Leaning Tower of Pisa which is 54.55 m tall. One second later he dropped a smaller cannon ball.
   (a) Find the height of the large cannon ball after 3 s.
   (b) How far apart were the two cannon balls after 3 s?
   (c) Were the balls the same distance apart after 2 s?

problem/matics

A locus is a set of points that satisfies one or more conditions. For example, a point P moves so that it is always the same distance from a fixed point. We already know this to be the locus of a circle. However, if the point P moves so that the sum of the distances from two fixed points, A and B, is always the same, then what does the locus of P look like?

$BP + AP$ is a constant.

1.8 translating accurately

An important skill for solving problems is to translate correctly. To do this, we must understand the vocabulary of mathematics.

Example 1
Find the simplified expression if the square of $3y$ is subtracted from the product of $4y$ and $3y - 2$.

Solution
The square of $3y$ is given by $(3y)^2$.
The product of $4y$ and $3y - 2$ is given by $4y(3y - 2)$.
Thus, the required expression is given by
$$4y(3y - 2) - 9y^2$$
$$= 12y^2 - 8y - 9y^2$$
$$= 3y^2 - 8y$$

When a problem has been completed, we may often check the problem by substituting a numerical value for the variable.

To solve problems we must read carefully. Throughout the following exercise, check the solution by choosing appropriate values of the variable.

1.8 exercise

1. Find the sum of $2(y - 3)$, $-3(y + 5)$ and $-5(2 - 3y)$.

2. The product of $3x$ and $x - 3$ is added to the sum of $3x^2 - 2$ and $x + 3$.

3. From the sum of $3x - 2$ and $6x + 5$ subtract $3(x - 2) - 2(x - 1)$.

4. How much less is $3(4 - y) - 2(3y - 1)$ than $2(y + 4) - 6(y - 6)$?

5. The product of $3x$ and $x + y^2$ is added to $3x(y - x) + 3y(y + x)$.

6. From $3(x - 2) - 2(x - 5)$ subtract $3(x - 2) - 2(x - 4)$.

7. How much greater is the sum of $3x - 2$ and $3(4x - 1)$ than the product of 3 and $2x - 3$?

8. By how much does the square of $-3xy$ exceed the sum of $2xy + 3y$ and $-3(xy - y)$?

9. What polynomial increased by $3(x - 5)$ is equal to $-4(2x - 1)$?

10. What polynomial, decreased by $4(2a - 4)$, is equal to $-3(5 - a)$?

11. How much less is $2(y^2 - 3) - 3(4 - y^2)$ than $6(y^2 - 3) - 6y(y - 5)$?

12. From $2(x^2 - 3x - 5)$ subtract $2(x - 3) + 3x(x - 1)$.

13. What polynomial must be added to $3(x - 2) - 2(x + 5)$ to obtain $3x - 6$?

14. Find the sum of the quotient $-25x^3y \div (-5x^2y)$ and the product $5(x - 2)$.

15. From the sum of $3y - 5$ and $6 - 2y$ subtract $2(y - 5)$ and $3(6 - y)$.

16. Find the sum of $3(a^2 - a)$, $-3a(a + 5)$ and $-2(a - 6)$.

17. By how much does $3(a - 2) - 3(a + 1)$ exceed $2(a - 4) - 3(a + 5)$?

18. By how much does the square of mn exceed the quotient $(16m^2n^2) \div (-4mn)$?

19. By how much does the square of $-4y$ exceed the sum of $3(y - 2)$ and $-4(y + 5)$?

20. What polynomial which when decreased by $3x - 2y + z$, is equal to $-4x + 5y - 3z$?

21. How much less is the product of $3x$ and $2x + 5$ than the sum of $3x - 2$ and $2(x + 5)$?

22. Subtract $3(x^2 - 2) - 3(x + 5)$ from $2(x^2 - 6) - 3(x + 5)$.

23. Write the polynomial which when increased by $3(y - 5) - 2(y + 5)$ is equal to $-3(y + 16)$.

1.9 skills for solving equations

We have developed many methods to solve problems.

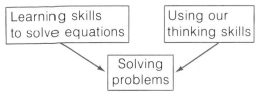

We can build a simple equation to see how to solve the equation. We then apply our skills to solve more advanced equations. Compare each of the following steps.

Building an equation

Begin with
A: $y = 4$
Multiply both sides by 3.
B: $3y = 12$
Subtract 2 from both sides.
C: $3y - 2 = 10$
Multiply both sides by 5.
D: $5(3y - 2) = 50$

Solving an equation

Begin with
D: $5(3y - 2) = 50$
Divide both sides by 5.
C: $3y - 2 = 10$
Add 2 to both sides.
B: $3y = 12$
Divide both sides by 3.
A: $y = 4$

The value 4 for y satisfies the equation $5(3y - 2) = 50$. We call 4 the **root** of the equation.

To *solve* an equation, means to use our skills in algebra to find a set of values of the variable that makes the equation true. Throughout this section, the variables represent real numbers.

To obtain a solution, we use the following properties of equality to obtain equivalent equations.

If $a = b$ then
$a + c = b + c$ $a - c = b - c$
$a(c) = b(c)$ $a \div c = b \div c$

Which of the above properties were used to solve the following equation?

Example 1
Solve $4(2m + 1) - 11 = 2(m + 1) - 3$, $m \in R$.

Solution
$$4(2m + 1) - 11 = 2(m + 1) - 3$$
$$8m + 4 - 11 = 2m + 2 - 3$$
$$8m - 7 = 2m - 1$$
$$8m - 2m = 7 - 1$$
$$6m = 6$$
$$m = 1$$

We refer to R as the *domain* of the equation.

Equivalent equations are equations that have the same root or roots.

We classify an equation according to the degree of the variable. We will study equations of the following types.

Linear	Quadratic	Cubic
$3n + 5 = 9$	$2x^2 + 3x + 1 = 0$	$x^3 + 8 = 0$
first degree	second degree	third degree
1 root	2 roots	3 roots

The set of all roots of an equation is called the **solution set**.

We *verify* our solution to check to see whether we made any error in our calculations.

Example 2
Find the solution set for
$$\frac{1}{3}(4k - 1) = \frac{1}{5}(3 + 3k) + \frac{k}{2}, k \in R.$$
Verify.

Solution
$$\frac{1}{3}(4k - 1) = \frac{1}{5}(3 + 3k) + \frac{k}{2}$$
Multiply both sides by 30.
$$30 \times \frac{1}{3}(4k - 1) = 30 \times \frac{1}{5}(3 + 3k) + 30 \times \frac{k}{2}$$
$$10(4k - 1) = 6(3 + 3k) + 15k$$
$$40k - 10 = 18 + 18k + 15k$$
$$40k - 18k - 15k = 18 + 10$$
$$7k = 28$$
$$k = 4$$
The solution set is $\{4\}$.

We verify in the *original* given equation.

LS $= \frac{1}{3}(4k - 1)$ RS $= \frac{1}{5}(3 + 3k) + \frac{k}{2}$

$= \frac{1}{3}(4 \times 4 - 1)$ $= \frac{1}{5}(3 + 3 \times 4) + \frac{4}{2}$

$= \frac{1}{3}(15)$ $= \frac{1}{5}(15) + 2$

$= 5$ $= 5$

Since LS = RS, then 4 is the root.

The skills for solving equations is important for many subject areas. To *solve any equation*, we apply the properties of equality to obtain equivalent equations. The method and principles we apply in solving many types of equations remain the same. Thus, it is important that we learn the method, as we will apply it over and over again to solve different equations.

1.9 exercise

Throughout our study of solving equations, the domain is the real numbers. For this reason, we do not write the domain each time with the corresponding equation.

Questions 1 to 5 practise skills for solving equations.

A

1 Solve each equation.

(a) $4y + 11 = 6y + 10$ (b) $\frac{4m}{3} + 4 = 8$

(c) $7 - 4k = 5k - 2$ (d) $11 = 8 + \frac{1}{3}m$

2 For each equation, 2 values are given. Which value is the root?

(a) $4(m - 2) - 3(m + 1) = 1 - 3m$ $-3, 3$
(b) $2(y + 4) + 3(y + 2) = -2y$ $2, -2$
(c) $5(4 - 3x) - 14 = 3(x - 4)$ $1, -3$
(d) $2(y + 7) - 4(y + 3) = -2(y - 1)$ $2, 3$

3 (a) Find the solution set for
   $3(p - 1) - 2(2p + 1) = 4(p - 3) - 8$.
   (b) Verify your answer in (a).

4 (a) Why are the following equations equivalent?
   $7(x - 2) = 3(x - 6)$
   $7x - 14 = 3x - 18$
   (b) Find the solution set for the equation in (a).

5 Solve each equation. Which of the following equations are not equivalent to the others?

(a) $3y + 10 = -9 - 2y$ (b) $5y + 8 = 3y + 10$
(c) $3(y - 5) = 2(y - 8)$ (d) $4y - 27 = -2y - 21$

Each of these instructions asks you to solve an equation. The only difference is the method of writing the final answer.
- Solve.
- Find the solution set.
- Find the root.

B

6 Solve.

(a) $3 - (9p - 15) = 4(p - 2)$
(b) $3(y - 7) + 2(y + 3) = 4(y - 1)$
(c) $\frac{1}{3}(y + 5) = \frac{1}{8}(2 - 3y)$
(d) $4(y + 7) - 2(y - 5) = 3(y - 2)$
(e) $2(y + 1) - 3 = 4(2y + 1) - 11$
(f) $\frac{b}{2} - \frac{1}{7}(b + 1) = \frac{1}{3}(2b - 6)$
(g) $15 + 5(m - 20) = 3(m - 1)$

7 Find the root of each of the following.

(a) $8(2y + 4) - 3y = 6(3y + 7)$
(b) $2(p - 3) + 3(p - 5) = 4$
(c) $1 + 3x(x - 6) = x(3x - 1) + 1$
(d) $\frac{1}{2}(3m + 1) = \frac{9m}{4}$
(e) $\frac{1}{9}(11 + 2x) = \frac{2}{3}(x + 1) - \frac{x}{2}$

8 Which of the following equations are not equivalent to the others?
   (a) $4(2y + 3) - 3(y - 1) = 20$
   (b) $2(y - 3) + 5 = 3(y + 2) - 8$
   (c) $3 + 5(2y - 4) = 20y - 7$
   (d) $4(y - 2) - 3(y + 1) = -3y - 7$

9 Solve and verify.
   (a) $4m - 27 = -2m - 9$
   (b) $3(y - 5) + 4(y + 3) = 12$
   (c) $\frac{1}{3}(y + 4) = \frac{1}{2}(y + 1)$

10 Find the solution set and verify.
   (a) $2(y - 3) + 3 = 15 - y$
   (b) $\frac{1}{4}(3y + 7) - \frac{y}{3} = -\frac{1}{3}$

11 Which of the following equations are not equivalent to the others?
   (a) $\frac{1}{7}(2y + 4) = \frac{1}{6}(17 - y)$
   (b) $\frac{3y}{10} + \frac{1}{2} = \frac{1}{3}(16 - 2y)$
   (c) $\frac{1}{3}(4y - 1) = \frac{1}{5}(y + 3) + \frac{y}{2}$
   (d) $\frac{y}{5} + \frac{1}{2}(y + 3) = \frac{1}{7}(6y + 5)$

C

12 For what value of k is 3 a root of the equation $3(y - 2) = 2(y + 5) + k$?

13 Find the value of k if the solution set of the following equation is {4}.
   $\frac{2}{3}(m - 1) = \frac{1}{2}(m + 5) + k$

14 Two equations are given.
   A: $2(y - 3) - 3(y + 2) = 2(y - 5) - 8$
   B: $3(y - 2) + 2(y - k) = 4(y + k) - 4k$
   For what value of k are the two equations equivalent?

writing equations

To apply our skills with equations to solve problems, we first need to practise skills for translating mathematics and writing equations.

Example 1
The expression $4(3 + n)$ is 56 more than the expression $7(n - 5)$. Find the value of n.

Solution
Translate the statement given in the problem to write an equation.
$$4(3 + n) - 56 = 7(n - 5)$$
$$12 + 4n - 56 = 7n - 35$$
$$4n - 44 = 7n - 35$$
$$-44 + 35 = 7n - 4n$$
$$-9 = 3n \quad \text{Or we may write}$$
$$-3 = n \quad n = -3.$$
Thus the value of n that makes $4(3 + n)$ 56 more than $7(n - 5)$ is $n = -3$.

15 For what value of m will $3(m + 2)$ exceed $2(2m - 1)$ by 6?

16 The expression $-4(k - 7)$ is greater than $2k - 2$ by 18. Find the value of k.

17 What value of a will make $6(a + 2)$ greater than $2(2a + 1)$ by 14?

18 What value of y will make the product of -3, and $y + 2$ equal to -57?

19 The sum of $9(b - 5)$ and 50 is equal to $3(b + 3)$. Find the value of b.

20 What value of k will make $6(k - 4)$ exceed $-(k - 1)$ by 38?

21 $5(r - 1)$ is 17 less than $(r + 8)$. What is the value of r?

22 For what value of x will $2(3x - 2)$ equal $(4x + 1)$?

23 What value of x will make $2(-2x + 10y - 15)$ subtracted from $5(3x + 4y - 1)$ equal to 6?

1.10 applying skills with equations: formulas

Formulas are used in many fields of study. For example, the following formula gives the average mass, m kg, of a normal adult male when we are given the height, h cm.

$$m = \frac{25(h - 100)}{26}$$

If we are given values of m, we can determine the corresponding values of h.

m (kg)	72.9	74.5	76.8	78.6	81.3
h (cm)	?	?	?	?	?

To find a value for h, when $m = 72.9$, substitute into the formula.

$$72.9 = \frac{25(h - 100)}{26}$$
$$26(72.9) = 25(h - 100)$$
$$26(72.9) = 25h - 2500$$
$$26(72.9) + 2500 = 25h$$
$$\frac{26(72.9) + 2500}{25} = h$$

Thus, since $h = 175.8$ cm (to 1 decimal place) then we can say that the corresponding height is 175.8 cm when the mass is 72.9 kg.

To find all the corresponding values in the above manner would mean repeating the same steps over and over, thus consuming much time. Thus, if we first solve the formula for the variable, h, in terms of the other variable, then we can substitute the values for m and directly obtain the corresponding values for h.

$$m = \frac{25(h - 100)}{26}$$

Multiply both sides by 26.
$$26m = 25(h - 100)$$
$$26m = 25h - 2500$$
$$26m + 2500 = 25h$$
$$\frac{26m + 2500}{25} = h$$

If $m = 72.9$ then $h = \frac{26(72.9) + 2500}{25}$, and so on.

Example 1
Solve for t in terms of the other variables if $v = u + at$.

Solution To solve for t means to isolate t and express t in terms of the other variables.

$$v = u + at$$
Subtract u from both sides.
$$v - u = u - u + at$$
$$v - u = at$$
Divide both sides by a.
$$\frac{v - u}{a} = \frac{at}{a}$$
Thus $t = \frac{v - u}{a}$.

In working with co-ordinate geometry we often need to express one variable in terms of the other.

Example 2
Express y in terms of x if $2x - 3y = 8$.

Solution
$$2x - 3y = 8$$
Subtract $2x$ from both sides.
$$2x - 2x - 3y = 8 - 2x$$
$$-3y = 8 - 2x$$
Divide both sides by -3.
$$\frac{-3y}{-3} = \frac{8}{-3} - \frac{2x}{-3}$$
$$y = -\frac{8}{3} + \frac{2}{3}x$$
Or we may write
$$y = \frac{2}{3}x - \frac{8}{3}$$

Skill in manipulating formulas is essential for working in many branches of mathematics. This skill is important to know to solve many problems, and to work with computers.

1.10 exercise

Questions 1 to 8 give practice in working with formulas.

A

1. (a) Solve for D in $V = \dfrac{M}{D}$.
   (b) Solve for T in $f = \dfrac{1}{T}$.
   (c) How are the solutions in (a) and (b) alike?

2. (a) Solve for t in $v = u + at$.
   (b) Solve for m in $T = a + sm$.
   (c) How are the solutions in (a) and (b) alike?

3. (a) Solve for a in $y = m(x - a)$.
   (b) Solve for w in $P = 2(l + w)$.
   (c) How are the solutions in (a) and (b) alike?

4. For each given formula, solve for the variable indicated.
   (a) $E = IR$, I
   (b) $I = Prt$, r
   (c) $A = \dfrac{1}{2}bh$, b
   (d) $C = 2\pi r$, r
   (e) $V = \dfrac{M}{D}$, D
   (f) $M = DV$, V
   (g) $PV = RT$, T
   (h) $f = \dfrac{1}{T}$, T
   (i) $s^2 = 2gt$, t
   (j) $P = I^2R$, R

5. (a) Solve the formula $v = u + at$ for a.
   (b) Find the value of a for the following values.

v	u	t	a
46	10	3	?
63	15	8	?
25.7	10.2	5	?

6. (a) Express x in terms of the other variables.
   $y = m(x - a)$

   (b) Find the values of x for the following chart.

m	a	y	x
3	2	9	?
$\dfrac{3}{2}$	-4	12	?
$-\dfrac{1}{3}$	-6	-1	?

7. If $3x - 5y - 10 = 0$, then express
   (a) y in terms of x.
   (b) x in terms of y.

8. (a) What might be the first step in expressing y in terms of x in the equation
   $\dfrac{x}{2} - \dfrac{y}{3} = 1$?
   (b) Express y in terms of x.

B

9. Solve each formula for the variable indicated.
   (a) $t = 3(p + s)$, s
   (b) $P = 2(l + w)$, w
   (c) $u^2 = v^2 - 2vt$, t
   (d) $A = \left(\dfrac{a + b}{2}\right)t$, a
   (e) $s = \dfrac{1}{2}(u + v)t$, u
   (f) $\dfrac{v - u}{a} = t$, v
   (g) $U = \dfrac{-GMm}{R}$, M
   (h) $PV = nRT$, n
   (i) $\dfrac{1}{f} = \dfrac{1}{D_i} + \dfrac{1}{D_o}$, D_o
   (j) $d = vt + \dfrac{1}{2}at^2$, v

10. Solve each of the following for the variable indicated.
    (a) $2x + y = 1$, y
    (b) $3y - 2x = 1$, y
    (c) $x + 2y = 3$, x
    (d) $2x - 5y = 6$, x
    (e) $2x - y - 3 = 0$, y
    (f) $7x - 2y + 2 = 0$, y
    (g) $\dfrac{x}{3} + \dfrac{y}{4} = 1$, x
    (h) $\dfrac{x}{3} - \dfrac{y}{4} = -1$, y

11. For the formula $d = vt + \frac{1}{2}gt^2$ find d,
    (a) if $v = 20$, $g = 10$, $t = 4$.
    (b) if $v = 50$, $g = 8$, $t = 2.5$.

12. (a) To calculate v for different values of the other variables, express v in terms of the other variables.
    $$s = \left(\frac{u + v}{2}\right)t$$
    (b) Complete this chart.

u	s	t	v
12	30	3	?
16	162	4.5	?
12.6	87.1	6.5	?

13. (a) To calculate P for different values of the other variables, solve the formula for P.
    $$I = Prt$$
    (b) Complete this chart.

I	r	t	P
120	0.08	3	?
309.60	0.12	5	?
170.10	0.09	4.5	?
183.75	0.15	3.5	?

14. Solve each equation for x in terms of the other variables.
    (a) $3x + b = c - 2b$
    (b) $\frac{x}{a} - 2b = 2$
    (c) $3ax = 2ax + b$
    (d) $3mx - b = 2b + a$

15. Solve each equation for x.
    (a) $3a + 5x = 4(b - 2x)$
    (b) $6 - (p + x) = 4(3x - p)$
    (c) $a(x + b) - 3ax = 2a(x - b)$
    (d) $4cx - 2(3b - cx) = b - c$

Questions 16 to 18 are based on the following formula.

The effective annual interest rate charged on a loan is given by
$$r = \frac{2NI}{P(n + 1)}$$
where N is the number of payment periods in one year.
I is the amount of interest charged.
P is the principal amount of the loan.
n is the total number of payments.

16. If Bill borrows $1500 to be repaid in monthly installments over 2 a and the total interest charged is $210, find the effective interest rate on the loan.

17. Janet calculated that her loan cost her $350 in interest. The loan was repaid in 30 monthly installments and her effective interest rate was 10.8%. Calculate the amount of the principal of the loan.

18. A loan of $500 is to be repaid in monthly installments. The interest charged was $40. If the effective interest rate was 6.2% per year, find the number of monthly payments required.

Questions 19 to 21 are based on the following formula.

The number of revolutions per minute (r/min) of a metal lathe is given by
$$R = \frac{96c}{\pi d}$$
where c is the recommended cutting speed in metres per minute, and d is the diameter of the shaft in centimetres.

19. Calculate the recommended number of revolutions per minute of the lathe for a shaft 8 cm in diameter if the recommended cutting speed is 45 m/min.

20. A shaft 6 cm in diameter is to be turned at 140 r/min on the lathe. Calculate the recommended cutting speed of the shaft.

21. A steel shaft is to be turned on the lathe at a recommended cutting speed of 20 m/min. If the shaft revolves at a speed of 164 r/min, what can be the maximum diameter of the shaft?

interpreting measurement problems

Often in mathematics and other subjects we need to recall a formula in order to solve a problem. The following questions are based on measurement formulae that are frequently used. All the measurements are in a literal form and are expressed in centimetres.

22. A square piece of metal has sides of length 8 cm. If four holes of diameter $\frac{a}{4}$ cm are drilled in the metal, write an expression to show how much material is left.

23. The water from a rectangular fish tank, a cm by a cm by b cm is transferred to 6 spherical bowls, each of diameter d cm. Write an expression to show how much water is left in the rectangular tank.

24. A basketball of diameter s cm is sold in a box (cube) of sides s cm. Write an expression for the space in the box not occupied by the basketball.

25. A rectangular piece of leather with length s cm and width k cm is to be used to make 6 soccer balls, each of diameter d cm. Write an expression to show how much leather is left over.

26. A chemical container is in the shape of a hemisphere of diameter p cm. How many cone-shaped containers of base diameter $\frac{p}{2}$ cm and height $\frac{p}{3}$ cm can be filled from the hemisphere?

27. A cylindrical tank of diameter k cm and height b cm is full of cement. This cement will be used to make rectangular cement blocks of dimensions s cm by s cm by k cm. Write an expression to show how many such blocks can be made.

28. A cone of base diameter b cm and height h cm is placed in a cylinder of diameter b cm. Write an expression for the volume of the cylinder outside the cone.

29. A cylindrical glass of radius k cm and height b cm is used to fill a semi-circular bowl of radius c cm. Write an expression to show how many glassfuls of water are required to fill the bowl.

30. A circular stop sign, of diameter c cm, is to have a yellow square of sides t cm painted in its centre. Write an expression to show how much of the circle is not to be painted yellow.

31. A triangular piece of metal with base length a cm and height b cm is to have a hole of diameter $\frac{b}{2}$ cm, in its centre. Write an expression to show how much metal is left.

problem/matics

A rectangular box has the following measurements:
- Area of the base is 120 cm².
- Area of the two other sides is 96 cm² and 80 cm².

What are the dimensions of the box?

1.11 steps for solving problems

Being able to use mathematics to solve problems is really what studying mathematics is about. To be able to solve a wide variety of problems, it is necessary to have a plan for problem-solving. In other words, we must establish a sequence of steps to organize the solutions of more advanced problems.

Steps for Solving Problems

Step A Read the problem carefully.
Ask yourself these two questions:
  I What information are we given? (information we know)
  II What information are we asked to find? (information we don't know)
Be sure to understand what it is we are to find, then introduce the variables.

Step B Translate from English to mathematics and write the equations.

Step C Solve the equations.

Step D Check the answers in the original problem.

Step E Write a final statement as the answer to the problem.

In subsequent sections, other skills and strategies will aid us in organizing the given information to solve a problem.

The most important initial step in solving a problem using algebra is to translate from English to mathematics. Each English expression that follows is translated into mathematics using a variable.

English	Translation
a number	n
• the number increased by 4	$n + 4$
• $\frac{1}{4}$ of the number which has been decreased by 3	$\frac{1}{4}(n - 3)$
amount of money invested	$\$m$
• interest paid at 12%	$\$0.12m$
• 15% interest paid on the principal of $\$m$ and $\$20$	$\$0.15(m + 20)$
speed of a 747 jet	s km/h
• the speed of the DC 9 is 120 km/h slower	$(s - 120)$ km/h
• the distance travelled by a 747 jet in 10 h	$10s$ km

We apply the *Steps for Solving Problems* in Examples 1 and 2.

Example 1
Roberta is deciding between building a den (an extra room) on the house or installing a pool. The pool costs $1500 less than the den. If she were to build both, the cost would be $15 500. Find the cost of the pool.

Solution

Step A Let $\$p$ represent the cost of the pool.

Step B The cost of the den is $\$(p + 1500)$.
$p + (p + 1500) = 15\,500$

Step C $p + p + 1500 = 15\,500$
$2p + 1500 = 15\,500$
$2p = 14\,000$
$p = 7000$

Step D Check
Cost of pool   7 000
Cost of den 7000 + 1500 = 8 500
                Total Cost 15 500 ✓
                                 checks

Step E The cost of the pool is $7000.
Always write a final statement when you solve a problem.

We often need to review our earlier skills to solve problems about time, distance, and rate. Remember that $d = vt$.
We may also write the above formula as
$$\frac{d}{v} = t, \qquad \frac{d}{t} = v$$

Example 2
Jay and Alan went on a bicycle trip. On the first day they averaged 15 km/h. On the second day they averaged 20 km/h and rode 1 h less than the first day. If they rode a distance of 190 km in all, how far did they travel each day?

Solution
Step A   Let t represent the number of hours they rode on the first day.

We may use a chart to record the information given in the problem.

Step B

	Time (h)	Speed (km/h)	Distance (km)
First Day	t	15	$15t$
Second Day	$t - 1$	20	$20(t - 1)$

From the chart
$$15t + 20(t - 1) = 190$$
(Do not write units in the equation.)

↑ distance first day   ↑ distance second day   ↑ total distance

Step C
$$15t + 20t - 20 = 190$$
$$35t = 210$$
$$t = 6$$

Step D   Check

	Time (h)	Rate (km/h)	Distance (km)
First Day	6 h	15 km/h	90 km
Second Day	5 h	20 km/h	100 km
		Total	190 km ✓ checks

Step E   Thus, they rode 90 km the first day, and 100 km the second day.

As we become more experienced in solving problems, we will probably devise a plan of our own. The *Steps for Solving Problems* is suggested as a sequence of steps that we will find useful in organizing our work in solving a problem.

1.11 exercise

Questions 1 to 6 practise essential skills for solving problems.

Use the variable given and translate from English into mathematics.

A
1   Let h cm represent John's height. If Ron is 2 cm shorter than John, express the following in terms of h.
   (a) Ron's height   (b) their combined heights
   (c) 3 times Ron's height
   (d) John's height last year if he has grown 3 cm in 1 a
   (e) the sum of John's and Ron's heights last year if Ron has grown only 1 cm since then

2   Darlene is m a old today. Express the following in terms of m.
   (a) Darlene's age 5 a from now
   (b) Darlene's age 3 a from now
   (c) twice Darlene's age now
   (d) half Darlene's age now
   (e) 4 a less than 3 times her age now
   (f) 2 a more than $\frac{1}{2}$ her age now

3   If t a represents Jay's age and Dave is 2 a older than Jay, express the following in terms of t.
   (a) Dave's age
   (b) the sum of their present ages
   (c) Jay's age 3 a from now
   (d) Dave's age 5 a from now
   (e) Dave's age 4 a ago

4 Find the interest, in 1 a, on each of the following investments.
   (a) $200m invested at 15%
   (b) $m invested at 10%
   (c) $(m + 2) invested at 9%

5 Ann jogs 2 km/h faster than Lynn. If Lynn's speed is s km/h, express the following in terms of s.
   (a) Ann's speed
   (b) the time Ann takes to jog 17 km
   (c) the time Lynn takes to jog 15 km
   (d) the distance Ann jogs in 6 h
   (e) the distance apart the girls are if they start at the same time and jog in the same direction for t min

6 Jennifer can travel at n km/h on her one-speed bicycle, but goes twice as fast on her 10-speed bicycle. Express each of the following in terms of n.
   (a) the speed of her 10-speed
   (b) the distance she goes in 2 h on her 10-speed
   (c) the time it takes her to go 12 km on her one-speed
   (d) the time it takes her to go 20 km on her 10-speed

In the following problems, we may organize our work using the *Steps for Solving Problems*. Read carefully.

B

7 In a skate-a-thon to raise money for handicapped children, Nicole skated 50 more laps than Paul and Sam skated 10 more than twice the number of laps Paul skated. If they skated a total of 860 laps, how many laps did each of them skate?

8 A wrestling mat is rectangular in shape. The length of the mat exceeds its width by 1.2 m. If the perimeter of the mat is 21.6 m, find the area of the mat.

9 A square pool of sides 7 m in length is enclosed by a rectangular fence. The fence is 30 m longer than it is wide. If the perimeter of the pool is $\frac{1}{5}$ that of the fencing, find the area enclosed by the fence.

10 Eleanor read a 300-page thriller in 3 d. The second day she read 3 times as many pages as on the first day but on the third day she read only twice as many pages as on the first day. How many pages did she read each day?

11 Each year the school student council donates $1000 to three different charities. This year they donated twice as much to the Knee Fund as to the Bone Fund but $200 less to the Foot Fund than to the Knee Fund. How much did they donate to each fund?

12 Sarah works from Wednesday to Saturday as a receptionist at a health-spa. Each day she works 2 h more than the previous day. If she works a total of 24 h during the 4-d period, how many hours does she work each day?

13 The office staff has a weekly jackpot draw. This week's jackpot has one more $2-bill than $1-bills. The number of $2-bills is 1 more than twice the number of $5-bills. If the jackpot is worth $35, how many of each bill are there?

14 It takes Steve 9 h, spread over 3 d to type a 4000-word English essay. The second day he types 1 h longer than the first day but on the third day he types twice as long as on the first day. How many hours does he type each day?

15. In a survey done on the types of vehicles passing a school crossing during morning rush-hour, it was found that twice as many compact cars as sports cars and 8 less vans than sports cars, pass the intersection. How many of each type pass the crossing if a total of 64 vehicles are counted?

16. San Francisco had 3 major quakes between 1865 and 1906. The total damage from these quakes was $524 000 000. The first quake caused $150 000 more damage than the second, but the third and more devastating quake caused $24 million more damage than 8 times the first. What was the dollar value of the damage from each quake?

17. The length of a tennis court is 2 m longer than twice its width. If the perimeter of the court is 88 m, find its area.

18. On a 21-d trip through the Latin countries of Europe, Mary and Pat spent twice as many days in Italy as in France but 1 d less in France than in Spain. For how many days were they in each country?

19. Richard invested the winnings of a sweepstake in two bank accounts for one year. One was a term deposit paying $10\frac{1}{2}$% while the other was a savings account paying 6%. He invested twice as much and earned $750 more in the term deposit than in the savings account. How much did he invest at 6%?

20. Rosemary spends $149 each month on gas, hydro, and water. The charge for gas is $7 more than four times the water charge. If hydro costs $2 less than three times the water, find the cost of each utility.

21. A survey taken of the amount of time a student spends in front of a television indicated that on the average, the student spends twice as much time watching comedy as music-shows and 10 min less watching sports than comedy. If television is watched for 140 min, how much time is spent on each category?

22. How would you invest $8000, part at 10% and part at 12%, so that in the first year the interest from the 10% investment will be $120 more than twice the interest from the 12% investment?

23. "The Tiger" received 244 min in penalties in one season due to tripping, elbowing (each 2 min) and major misconducts (5 min). He had twice as many tripping as elbowing penalties and 10 fewer major penalties than tripping penalties. How many elbowing penalties did he receive in that season?

24. Annoyed by a $10 parking ticket, Petra decides to pay the fine using pennies, nickels, and dimes. The number of nickels is 2 more than four times the number of dimes and the number of pennies is 6 more than six times the number of nickels. How many pennies, nickels, and dimes are there?

25. Keith and his friends spend 2 weeks each year fishing in an isolated lake near Brandon. To get there they must drive by car, fly by plane, and then row the remaining distance. They drive 200 km more than they fly and fly 60 km more than three times the distance they row. If their trip is 600 km altogether, how far do they fly?

1.12 a method of proof

Often there is more than one way of solving a problem. The more problems we solve, the more methods we will acquire of solving problems that we have not met before.

It is possible to obtain more than 1 value of the variable to satisfy the following equation.

$$7(3x + 2) - 4(2x + 5) = 5(3x - 2) + 2(2 - x)$$

Try $x = 2$.

LS	RS
$= 7(3x + 2) - 4(2x + 5)$	$= 5(3x - 2) + 2(2 - x)$
$= 7(3 \times 2 + 2) - 4(2 \times 2 + 5)$	$= 5(3 \times 2 - 2) + 2(2 - 2)$
$= 7(8) - 4(9)$	$= 5(4) + 0$
$= 56 - 36$	$= 20$
$= 20$	

Thus, LS = RS when $x = 2$.

Try $x = 3$.

LS	RS
$= 7(3x + 2) - 4(2x + 5)$	$= 5(3x - 2) + 2(2 - x)$
$= 7(3 \times 3 + 2) - 4(2 \times 3 + 5)$	$= 5(3 \times 3 - 2) + 2(2 - 3)$
$= 7(11) - 4(11)$	$= 5(7) + 2(-1)$
$= 77 - 44$	$= 35 - 2$
$= 33$	$= 33$

Thus, LS = RS when $x = 3$.

We may try any value of x for the domain of real numbers and find that the value we choose *satisfies the equation*. An equation that is true for all values of the domain is called an **identity**.

We cannot possibly test all the values of the domain to prove that an equation is an identity, since it would take forever. We use our algebraic skills to prove that an equation is an identity.

Example 1
Prove that the equation
$2(x - 3) + 4(2x + 5) = 3(5x - 3) - 5(x - 4) + 3$
is an identity.

Solution
We simplify each side of the equation to obtain equivalent expressions.
$$\begin{aligned}\text{LS} &= 2(x - 3) + 4(2x + 5) \\ &= 2x - 6 + 8x + 20 \\ &= 10x + 14 \\ \text{RS} &= 3(5x - 3) - 5(x - 4) + 3 \\ &= 15x - 9 - 5x + 20 + 3 \\ &= 10x + 14\end{aligned}$$
Since LS = RS = $10x + 14$, then LS = RS for all values of x. Thus, the equation is an identity.

We may use our skills with substitution to prove statements, as Example 2 shows.

Example 2
If $x = a + b$, prove that for all a, b, and x
$$a(x - b) + b(x - b) = ax.$$

Solution
Use $x = a + b$.
$$\begin{aligned}\text{LS} &= a(x - b) + b(x - b) \\ &= a(a + b - b) + b(a + b - b) \\ &= a(a) + b(a) \\ &= a^2 + ab\end{aligned}$$

$$\begin{aligned}\text{RS} &= ax \\ &= a(a + b) \\ &= a^2 + ab\end{aligned}$$

Thus LS = RS

The need for the above skill of proving statements occurs throughout the study of the different branches of mathematics.

1.12 exercise

Questions 1 to 7 practise skills for working with identities and proving statements.

A
1. Simplify each expression.

   (a) $2(x - 3) + 4(2x + 5)$

(b) $5 - 3(4m - 6) + 4(2 - 3m)$
(c) $3(6y + 4) + 8 - 6(5y + 2)$
(d) $2(9 - 8k) + 6(2 - k) - (2k - 1)$
(e) $3(k - 3) - 2(2k + 1) - 3k$

2. Simplify. Which of the following expressions are equivalent?
   (a) $3(x - 2) - 2(x + 3)$
   (b) $4(2x + 1) - (7x + 16)$
   (c) $2(x - 1) - 3(x - 3) - (19 - 2x)$
   (d) $3(x - 1) - 2(2x - 3)$

3. If $x = a - b$, find an expression for $2x - 3a + b$.

4. If $y = 2a - b$, find an expression for $2y + 3(a - y)$.

5. If $x = a - b$, simplify each expression.
   (a) $\dfrac{x - 2b}{6a} + \dfrac{1}{a}$ (b) $\dfrac{2x + 2b - a}{a}$
   (c) $\dfrac{3x - (a - b)}{2}$

6. What value of k will make the following expression equal to 5?
   $2(3m - k) - 3(2m + k)$

7. Find the value of k that will make
   $2(m + k) - 3(k - 2) - 2m$
   equal to 4.

B

8. Prove that $\dfrac{1}{2}(2y - 6) - 3y = 2 - (2y + 5)$ is an identity.

9. Prove that
   $2(x - 3) + 4(2x + 5) = 3(5x - 3) - 5(x - 4) + 3$
   for all values of x.

10. Prove that
    $2(9 - 8y) + 6(2 - y) - 2(y - 1)$
    $= 6 - 3(4y - 6) + 4(2 - 3y)$
    is an identity.

11. Show that the expression
    $8 + 3(6m + 4) - 6(5m + 2)$
    is equivalent to the expression
    $10(3m + 2) - 6(7m + 2)$
    for all values of m.

12. If $x = 2a - b$ show that
    $2a(x - 2a) = b(x - 2a) - bx$.

13. Show that $2b = \dfrac{b(b + x)}{a} + a - x$ if $x = a - b$.

14. Prove that if $x = 2a - 3b$, then
    $\dfrac{x - 2a + 3b}{a} = a - \dfrac{x + 3b}{2}$.

C

15. Find the value of k that will make the following equation an identity.
    $3(4x - 6) + 4(2 - 4x) = k - 3(2x - 5) + 2(x - 15)$

16. The following equation is an identity.
    $2(x - 5) + 3(2x + 8) = 5(2x + k) - 2(x + 8)$
    Find the value of k.

function notation: f(x)

In order to speak a language we must learn the vocabulary. To do mathematics, we also need to learn the vocabulary. We use the symbol $f(x)$, $g(x)$ and so on to represent expressions in the variable x.

$f(x) = x + 3$ $g(x) = 2x^2 - 3x + 5$
We read f at x equals $x + 3$
We read g at x equals $2x^2 - 3x + 5$.

We may then evaluate $f(x)$ and $g(x)$ for values of the variable.

If $x = 2$, then   If $x = -3$
$f(x) = x + 3$ $g(x) = 2x^2 - 3x + 5$
$f(2) = 2 + 3$ $g(-3) = 2(-3)^2 - 3(-3) + 5$
$ = 5$ $ = 18 + 9 + 5$
    $ = 32$

You will find many symbols used in the same way such as $h(x)$, $H(x)$, $K(x)$, $P(x)$, and so on.

17 If $f(x) = 2x + 5$, find
   (a) $f(0)$ (b) $f(-1)$ (c) $f(3)$

18 If $f(x) = 3x^2 - 6x + 5$, find
   (a) $f(3)$ (b) $f(-3)$ (c) $f(0)$

19 If $H(y) = \dfrac{y^2 - 2}{4}$ find
   (a) $H(2)$ (b) $H(0)$ (c) $H(-2)$

20 If $g(x) = 3x^2 - 5x$ find
   (a) $g(0)$ (b) $g(-3)$
   (c) $g(3)$ (d) $g(-3) - g(3)$
   (e) $g(-3) + g(3)$ (f) $g(3) - g(-3)$
   Find $g(-3)$, $g(3)$. Then find the sum.

We may use the notation $f^2(x)$ to mean $f(x) \times f(x)$.

---

21 If $H(x) = 2x^2 - 3$, then find
   (a) $H(3) + H(-3)$ (b) $H^2(-3)$ (c) $H^2(3)$

We may now use our earlier skills with equations to solve the following problems.

22 If $f(x) = 3x - 8$, find the value of x that makes $f(x)$ equal to
   (a) 7   (b) -2 (c) 0

23 If $g(x) = 16 - 3(x - 5)$, find the value of x that makes $g(x)$ equal to
   (a) 34   (b) 22   (c) -5

24 If $f(x) = 3x + 2$ and $g(x) = 2x - 3$, for what value of x is $f(x) = g(x)$?

25 If $H(x) = 2(x - 5) + 3$ and $F(x) = 3(x - 1) - 3$ for what value of x is $F(x) = H(x)$?

skills review: reading accurately

To do mathematics well we need to practise skills in mathematics regularly. The sections called Skills Review will review not only skills you develop in each chapter, but skills you need for developing more mathematics. You may wish to refer back to this Skills Review more than once to review these skills.

For Questions 1 to 12 write your answers in simplest terms. Read carefully.

1 Write the square of $-3ab^2$.

2 Find the quotient when $-3xy^2 + 9x^2y$ is divided by $-3xy$.

3 What is the resulting numerical coefficient when $-3xy^3$ is multiplied by $-6x^3y^4$?

4 Find the sum of $3x - 2y$ and $2(3x - 2y)$ decreased by the product of 2 and $-3x + 2y$.

5 What is the exponent of b when the product of $6a^2b$ and $-5a^3b^3$ is written in simplest terms?

6 Find the missing factor of $-18x^3y^5$ if the other factor is $-6xy^3$.

7 Decrease the sum of $6x - 12y$ and $-8x + 5y$ by the product of -3 and $2x - 6y$.

8 By how much does $(12x^2) \div (-4xy)$ exceed $2(3x - 5y)$?

9 One factor of $16a^3b^2$ is $-4ab^2$. Find the other factor.

10 What is the exponent of q when the quotient of $-16p^2q^9$ divided by $-4pq^3$ is written in simplest terms?

11 What is the numerical coefficient of the square of $(-3x^2y)^3$?

12 What is the greater exponent when the square of $-2x^3y^2$ is divided by $2xy^2$?

problems and practice: a chapter review

At the end of each chapter, this section will provide you with additional questions to check your skills and understanding of the topics dealt with in this chapter. An important step for problem-solving is to decide which skills to use. For this reason, these questions are not placed in any special order. When you have finished the review, you might try the *Test for Practice* that follows.

1. Which are true (T)? Which are false (F)?
   (a) $-4 - (3 - 1) < 8 \div (-2 + 1)$
   (b) $-3(-1)^2 - (6 - 3) \neq \dfrac{18 - 2}{(-2)^2}$
   (c) $-3 + \dfrac{-6 + 3}{-3} > -8 \div \dfrac{4 \div 2}{-6 \div 3}$

2. What type of thinking is involved in answering the following question? Give reasons for your choice.

   To qualify for the Boston Marathon, you must previously have run another Marathon in less than 2.5 h. Tom is running in the Boston Marathon. What conclusion can you make?

3. Write each of the following in simplest terms.
   (a) $(-3mn)(2m^2n)$
   (b) $(3m^2n)(-2mn)$
   (c) $(-18x^4y^3) \div (-9x^2y^2)$
   (d) $\dfrac{27a^5b^3}{-9ab^2}$
   (e) $\dfrac{3(-2mn^2)(-mn)^2}{-6m^2n^2}$
   (f) $\dfrac{-8pq^2 + 4p^2q}{-2pq}$

4. Simplify each of the following.
   (a) $2(x + 3y) - [x + 2(x - y)]$
   (b) $-a^2 - 4a[2(a - 2) - 3(5a + 4)]$

5. Find the coefficient of y in the expression
   $-2ax + 6ay - 2bx + 3by + 2cx + 3cy$.

6. From the sum of $2x - 6$ and $3 - 2x$, subtract the product of -3 and $4 - 6x$.

7. A polynomial, increased by $2(y^2 - 6y + 11)$, is equal to $4y^2 - 8y + 5$. Write the polynomial.

8. If $a = -9$, $b = 3$, find the value of each of the following.
   (a) ab^3
   (b) a^3b
   (c) a^3b^3
   (d) $(ab)^3$
   (e) $\dfrac{a^3}{b}$
   (f) $\left(\dfrac{a}{b}\right)^3$

9. Write the following expression with a single base.
   $$\dfrac{(x^m)^2(x^n)^2(x^p)^2}{x^{m+n}x^{n+p}x^{p+m}}$$

10. The conductor on a train travelling at 90 km/h spotted a cow 260 m ahead, standing on the track and immediately applied the brakes. If the distance required to stop is given by the expression $0.04v^2$ m where v is the speed of the train in kilometres per hour, determine whether the cow is saved.

11. The most successful wrestler ever was Strangler Lewis who won all but 33 of 6200 bouts. Lewis wrestled in the 77-kg class. If the mass, m, in kilograms, of a normal adult male is given by
    $$m = \dfrac{25(h - 100)}{26}$$
    where h is the height in centimetres, determine the Strangler's height.

12. Solve and verify.
    (a) $5(2m + 1) - 4(2 - m) = 5m + 6$
    (b) $7 + \dfrac{1}{2}(k + 1) = \dfrac{2}{3}(2k + 5)$

13. Adult tickets cost $9.00 each and senior citizen tickets cost $7.50 at a soccer game. At one game, gate receipts totalled $292 500. If this had been a playoff game, ticket prices would have increased by 50¢ each and the same crowd would have paid $309 000. How many adult tickets were sold?

test for practice

Try this test. Each test will be based on the mathematics you have learned in the chapter. Try this test later in the year as a review. Keep a record of those questions that you were not successful with and review them periodically.

1. Evaluate.
   (a) $-3(-2)^2(-1)^3$
   (b) $(-16) \div (-4) + 3$

2. Simplify.
   (a) $\dfrac{(xy)^2}{xy}$
   (b) $\dfrac{6(x^2y)^2}{(x^2)^2}$
   (c) Find the value of each expression if $x = -1$, $y = 3$.

3. Simplify each of the following. Write a single base.
   (a) $(p^b)^3(p^3)^b(p^3)^b$
   (b) $\dfrac{p^k p^{k-1}}{(p^3)^k}$

4. Write each of the following in simplest terms.
   (a) $(3x^2y)(2xy^3)$
   (b) $(-3a^2b)(-4ab^2)$
   (c) $-(-mn)(2m^2n)$
   (d) $(-16x^3y) \div (-4xy)$
   (e) $\dfrac{6(3mn^2)(-mn)^2}{-9mn}$
   (f) $\dfrac{3x^2y - 9xy^3}{-3xy}$

5. Simplify each of the following.
   (a) $-3(a + b) - 2(2a - b) - (a - b)$
   (b) $-4(x^2 - 5x + 1) + 2(3x^2 - 4x + 5)$
   (c) $x(4x - 1) - [x^2 + 3(x - 4) - 12]$

6. For the formula
   $$d = vt + \dfrac{1}{2}gt^2$$
   find v if $d = 93$, $g = 4$, and $t = 3$.

7. On a dry road the distance a car takes to stop once the brakes are applied is given by $0.025v^2$ m, where v is the speed in metres per second. Find the distance a car takes to stop if its speed is 30 m/s.

8. Solve and verify.
   (a) $4(x + 8) - 3(x + 5) = 2x$
   (b) $\dfrac{1}{2}(m + 1) = \dfrac{1}{3}(m + 4)$

9. What value(s) of x will make the sum of $2(x + 4)$ and $6(x - 1)$ equal to 22?

10. Show that the following is true for all values of x.
    $2(x - 5) + 3(2x + 8) = 5(2x + 6) - 2(x + 8)$

11. What type of thinking is needed to answer each question? Give reasons for your choice.
    (a) All horses have four legs. Northern Dancer is a horse. What conclusion can you make?
    (b) Based on consumer reports, every fifth new car has a problem of some sort. Laurier bought a new car. What conclusion can you make?

12. Use the definition of x^m. Prove that $(x^m)^n = x^{mn}$.

13. The parking rates in a downtown Winnipeg lot are $3/d for cars and $5/d for trucks. The day before rates were due to increase, the parking lot attendant collected $250. The next day, when rates went up by 50¢, the same lot of cars and trucks would have earned $285. How many cars and trucks were in the lot?

math tip

An important mathematicial skill is learning methods of checking our work. One method is to substitute convenient values for the variable (e.g. $a = 1$, $b = 1$) in both the original expression and the simplified expression. The answers should be the same! Try this check in the above test.

math is plus reviewing formulas

There are many common formulas that we have used in our earlier work in mathematics to calculate perimeters (P) and areas (A).

square   rectangle   circle

$P = 4s$ $P = 2(L + w)$ $C = 2\pi r$, $\pi \doteq 3.14$
$A = s^2$ $A = Lw$ $A = \pi r^2$

trapezoid   parallelogram   triangle

$A = \frac{1}{2}h(a+b)$ $A = bh$ $A = \frac{1}{2}bh$

Solve the following problems to review your skills. Use $\pi \doteq 3.14$ and round your answers to 1 decimal place.

1. Find the area of each shaded region.
   $a = 15$ cm, $b = 6$ cm, $h = 5$ cm.

   (a), (b), (c), (d), (e), (f)

   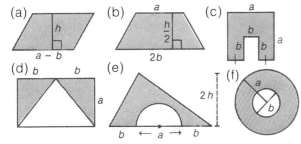

2. Two circles have radii 8 cm and 6 cm. What is the length of the diameter of a circle whose area is the sum of the areas of the two given circles?

3. A wheel has a radius of 28 cm.
   (a) How many turns will the wheel make in rolling, without slipping, 1 km?
   (b) If the radius is increased by 2 cm, how many turns less will be obtained as those in (a)?

4. A sector of a circle is shown by the shaded region. The area of the sector is given by
   $A = \frac{n}{360} \times \pi r^2$.

   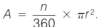
   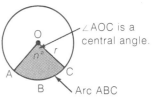
   ∠AOC is a central angle. Arc ABC

   (a) Find the area of a sector if the central angle is 108° and the radius of the circle is 2.1 m.
   (b) Find the central angle, to the nearest degree, if the sector area is 38.2 cm² and the radius is 5 cm.

Often in mathematics we may derive a different formula if the information is given in another way as shown by the next question.

5. For the circle, the length of the arc ABC is a units. The area, A, of the sector, with radius r, is given by $A = \frac{1}{2}ar$ units².

   (a) Find the area of a sector if the arc length is 9.2 cm and the radius is 8.1 cm.
   (b) Find the length of the arc of a sector if the sector area is 10.2 cm² and the radius is 3.2 cm.
   (c) The length of the arc of a sector is 12.1 cm and the area of the sector is 27.8 cm². What is the radius?

6. For each of the following diagrams, calculate the missing part.

   (a)
   $n° = 165°$, $r = 12$ cm, $A = ?$ cm²

   (b)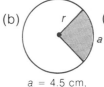
   $a = 4.5$ cm, $r = ?$ cm, $A = 7.2$ cm²

   (c)
   $a = ?$ cm, $r = 4.5$ cm, $A = 14.4$ cm²

   (d)
   $r = 4.2$ cm, $d = 1.3$ cm, $n° = 80°$, $A = ?$ cm²

   (e)
   $r = 4.5$ cm, $d = 5.5$ cm, $n° = ?°$, $A = 4.53$ cm²

2 factoring, solving equations and problems

understanding factoring, factoring skills, binomials, trinomials, solving quadratic equations, and solving problems, operations with rational expressions.

2.1 introduction

In studying mathematics, the skills and concepts we learn are applied to solving problems we use

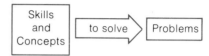

Skills and concepts we learn are extended to learn new skills and concepts, which in turn provide us with other methods and strategies for solving problems.

As we learn new skills and concepts we increase our vocabulary, which enables us to make further progress in mathematics. New words lead to the use of new symbols.
Many symbols in our everyday activities are universal.

)

The symbols used to study mathematics are universal. In fact, we may choose a mathematics book in any language and we will notice that we have seen much of the material before.

II Ἰδιώτης. Ἐὰν πολλαπλασιάσωμε ἢ διαιρέσωμε καὶ τὰ δύο μέλη μιᾶς ἐξισώσεως μὲ τὸν αὐτὸν μὴ μηδενικὸν ἀριθμόν, αἱ λύσεις τῆς ἐξισώσεως δὲν μεταβάλλονται, ἡ νέα ἐξίσωσις εἶναι ἰσοδύναμος μὲ τὴν ἀρχικήν.

Π.χ. ἡ ἐξίσωσις $\dfrac{7x-3}{5} = \dfrac{4x}{3}$

εἶναι ἰσοδύναμος μὲ τὴν

$15 \cdot \dfrac{7x-3}{5} = 15 \cdot \dfrac{4x}{3}$, ἤτοι τὴν $3(7x-3) = 5 \cdot 4x$,

καὶ ἡ ἐξίσωσις

$$8x = 24x$$

εἶναι ἰσοδύναμος μὲ τὴν

$\dfrac{8y}{8} = \dfrac{24x}{8}$, ἤτοι τὴν $y = 3x$.

2.2 using binomials

We have used the distributive property to find products.

Multiplication distributes over addition.

$$3(x + y) = 3(x + y) \qquad (a + b)a = (a + b)a$$
$$= 3x + 3y \qquad\qquad\qquad = a^2 + ab$$

We may then use the distributive property to learn what the result is when we multiply binomials.

$$(a + b)(a + b) = (a + b)(a + b)$$
$$= (a + b)a + (a + b)b$$
$$= a^2 + ab + ab + b^2$$
$$= a^2 + 2ab + b^2$$

Thus
$$(a + b)(a + b) = a^2 + 2ab + b^2$$

We may write $(a + b)(a + b) = (a + b)^2$.

The Math Tip in this section shows a shortcut, called FOIL, which can be used to multiply binomial expressions.

We use the word **expand** to describe the following procedure.

$$(a + b)^2 = a^2 + 2ab + b^2$$

$$\boxed{\text{EXPAND} \Rightarrow}$$

For example, we may interpret the significance of the square of the binomial $(a + b)^2$ by using this diagram.

	a	b
a	a^2	ab
b	ab	b^2

Area of large square
$$= (a + b)(a + b)$$
$$= (a + b)^2$$

Area of parts of square
$$= a^2 + ab + ab + b^2$$
$$= a^2 + 2ab + b^2$$

Thus
$$(a + b)^2 = a^2 + 2ab + b^2$$

Once we understand how the distributive property enables us to obtain the product of binomials, we then establish more efficient methods. An example follows.

$$(a + 5b)(a - 3b) = a^2 + 2ab - 15b^2$$

with a^2, $-15b^2$, $5ab$, $-3ab$ shown.

Mentally find the result.
$$5ab - 3ab = 2ab$$

Example 1
Simplify $(a - 3b)(2a + 3b) - 4(a - b)^2$.

Solution Here, we use the distributive property again.

$$(a - 3b)(2a + 3b) - 4(a - b)^2$$
$$= (2a^2 - 3ab - 9b^2) - 4(a^2 - 2ab + b^2)$$
$$= 2a^2 - 3ab - 9b^2 - 4a^2 + 8ab - 4b^2$$
$$= -2a^2 + 5ab - 13b^2$$

Simplify by collecting like terms.

Use our earlier methods to check this result. See page 4.

We now use the skill of finding products of binomials to solve equations.

Example 2
Solve $(2a - 1)(3a + 2) - (3a - 1)^2 = -3a(a + 1)$.

Solution

Always begin by writing the original equation.

$$(2a - 1)(3a + 2) - (3a - 1)^2 = -3a(a + 1)$$
$$(6a^2 + a - 2) - (9a^2 - 6a + 1) = -3a^2 - 3a$$
$$6a^2 + a - 2 - 9a^2 + 6a - 1 = -3a^2 - 3a$$
$$6a^2 - 9a^2 + 3a^2 + a + 6a + 3a = 2 + 1$$
$$10a = 3$$
$$a = \frac{3}{10}$$

Do this step mentally.

Use earlier skills to check the result. Refer to page 4.

Once we have mastered the skill of solving these equations, we may then apply our skill to solving word problems.

2.2 exercise

Questions 1 to 9 develop your skills in working with products of binomials.

1. Find the products.
   (a) $(x + 3)(x + 5)$
   (b) $(x - 6)(2x + 1)$
   (c) $(3a - 5)(2a + 1)$
   (d) $(2y - 3)(3y + 2)$
   (e) $(2x + 3y)^2$
   (f) $(3a - b)^2$
   (g) $(1 - 2x)^2$
   (h) $(2a - b)(2a + b)$
   (i) $(a - 3b)(a - 3b)$
   (j) $(xy - 2)(xy - 9)$
   (k) $(6 - p)(7 + p)$
   (l) $(x - 4y)(x - 3y)$
   (m) $(y^2 - 2)(y^2 + 8)$
   (n) $\left(3x - \frac{1}{2}y\right)(x + 4y)$

2. Expand.
   (a) $(3y - 1)(3y + 1)$
   (b) $(x^2 - 3)(x^2 + 3)$
   (c) $(4x - 3y)(4x + 2y)$
   (d) $(8 - 2y)(3 - y)$
   (e) $(3k - 2)(3k - 3)$
   (f) $(3y - 2)^2$
   (g) $(1 - 3x)^2$
   (h) $(m - 3y)(m - 2y)$
   (i) $(3y^2 - 1)(2y^2 + 1)$
   (j) $(ab - 2)(ab + 6)$
   (k) $\left(2x - \frac{3}{2}\right)^2$
   (l) $(xy - 6)(xy + 5)$

3. Simplify.
   (a) $2(y - 1)(y - 4)$
   (b) $3(y + 6)(y - 5)$
   (c) $2(3x - 1)^2$
   (d) $-3(x - y)^2$
   (e) $3x(x - y)(x + y)$
   (f) $-(3m - n)(3m + 6n)$
   (g) $\frac{1}{2}(4x - 3)(2x + 1)$
   (h) $-3m(m - n)(m + n)$
   (i) $-2(3a - 2b)^2$
   (j) $4h(2x - y)(x + 2y)$

4. (a) Simplify $3(a - 2b)^2$.
   (b) Simplify $(3a - 2b)^2$.
   (c) Why do the answers in (a) and (b) differ?

5. (a) Simplify $(a + b)^2 + (a - 3b)^2$
   (b) Simplify $(a + b)^2 - (a - 3b)^2$.
   (c) Why do the answers in (a) and (b) differ?

6. (a) Simplify $3(x - 3y)^2 + 2(x + 6y)(x - 6y)$.
   (b) Simplify $3(x - 3y)^2 - 2(x + 6y)(x - 6y)$.
   (c) Why do the answers in (a) and (b) differ?

7. (a) Simplify $2(y - 3)(y + 3) - 2(y - 1)^2 + y^2$.
   (b) If $y = 1$, find the value of the expression in (a).
   (c) Use $y = 1$ to evaluate $y^2 + 4y - 20$.
   (d) Why are the answers in (b) and (c) the same?

8. (a) Simplify $(4x - 1)^2 - 6(3x + 1)(2x - 3)$.
   (b) Use $x = 1$ to check the answer in (a).

9. (a) Simplify $2(2x - y)(3x + 4y) - 2(x - 2y)^2$.
   (b) Use $x = 1$ and $y = 1$ to check the answer in (a).

10. Simplify.
    (a) $(y - 3)(y + 9) - 3(y - 3)(y + 3)$
    (b) $2(x + 6)(x - 2) + 3(x - 4)(x + 6)$
    (c) $(3x - y)(2x + y) - 2(x - y)^2$
    (d) $-3(a - 4b)^2 - 2(a + 5b)^2$
    (e) $(m + 4n)(m - 4n) - 3(m - n)^2$
    (f) $2(3c - d)(2c + d) - 5(c + d)(c - 3d)$
    (g) $3x(x - y)^2 - 2y(x + 2y)^2$

11. Simplify. Check your answers.
    (a) $(x - 1)^2 - (x + 2)^2 - (x - 3)^2$
    (b) $2(y - 3)^2 - 3(y - 1)(y + 3)$
    (c) $3m(m - 5) - 2(m + 6)(m - 3)$
    (d) $2(x - y)(x + y) - 3(x - y)^2$
    (e) $3(a - b)^2 + 3(2a - b)(a + b) + b(a - b)$

12. Subtract $(x + 3)^2 - (x - 4)^2$ from the sum of $2(x - 1)^2$ and $-3(2x + 2)$.

13. By how much does $2(x - 3)^2 - 5$ exceed $-3(x + 1)^2 - (x - 1)^2$?

14. Find the sum of $(y - x)^2 + 3(x - y)^2$ and $3(x + y)^2 - 2(x + y)^2$.

applications: permissible values

Often the variables we use in some problems have certain restrictions. For example, when we use the equation to find the length of the hypotenuse, we place a restriction on x, namely $x > 0$, since distance is always positive.

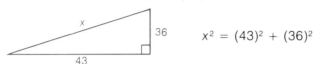

$x^2 = (43)^2 + (36)^2$

In the next diagram, the sides are given as variable expressions.

Restrictions:
$2y - 4 > 0$ or $y > 2$
$3x - 6 > 0$ or $x > 2$

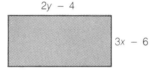

Restrictions will occur in studying many topics in mathematics.

15 (a) Find an expression for the area of the shaded region.
    (b) Find the shaded region if $x = 6$.
    (c) Why is the value $x = 1$ not permissible?

16 Find an expression for the area of each shaded region. Write the expression in simplest terms.
    (a)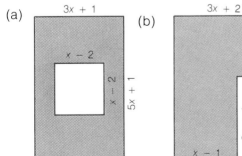
    (b)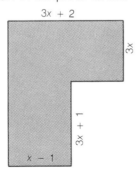

17 The area of a region is given by the expression
    $A(n) = 3(2n - 1)^2 - 2(n + 1)^2$, $n \in N$.
    Calculate the area for
    (a) $n = 2$ (b) $n = 3$ (c) $n = 4$
    (d) Why is $n = 1$ not a permissible value?

18 For a computer program, the area of a region is given by
    $A(n) = 2(2n + 1)^2 - (3n - 1)^2$, $n \in N$.
    Calculate
    (a) $A(1)$ (b) $A(2)$ (c) $A(3)$ (d) $A(4)$
    (e) Use the pattern in your answers (a) to (d) to predict the value of $A(5)$, $A(6)$, $A(7)$.

19 If $x = -2$, $y = 3$, which expression has the greatest value?
    (a) $3(x + 2y)(x - y) - (2x - y)(x + 3y)$
    (b) $2(x - y)^2 - 3(x - 2y)(x + 2y)$
    (c) $3(2x - y)^2 + (x + y)(x + 2y)$
    (d) $(x + 2y)^2 - 2(x - y)^2$

applications: equations and identities

20 A prime number, p, may be expressed by the following formula in terms of the whole number before it, namely $p - 1$, and the whole number following it, namely $p + 1$.
    $$p = \frac{(p + 1)^2}{4} - \frac{(p - 1)^2}{4}$$
    (a) Check the formula for the prime numbers 5, 47, and 101.
    (b) Show why the above formula is true for all prime numbers, p. (Hint: use your earlier work with identities.)

21 (a) Show that $(y + 3)(y - 2) = y(y + 4)$ is true when $y = -2$.
    (b) Is the equation in (a) true for any other values of y?

22. (a) Show that
$$y(y + 4) - (y + 2)^2 = 4(2y + 1) - 8(y + 1)$$
is true for $y = -3$.
(b) Is the equation in (a) true for any other values of y?

23. Solve.
(a) $(y + 3)^2 = (y - 1)^2 + 40$
(b) $(m + 3)^2 + (2m - 1)^2 = 5m^2 + 8$
(c) $(3x + 1)^2 - 6x^2 = 3(x^2 + 1) - 2$
(d) $(2y + 3)^2 - 4y(y + 1) = 5$
(e) $(5 - 2y)^2 + (y - 3)^2 = 6y^2 - 4 - (y + 4)^2$
(f) $2x^2 - (7 - 2x)^2 = (x - 3)(4 - 2x) - 37$

24. Show that each of the following are identities.
(a) $x(x - 4) - 1 = (x - 1)^2 - 2(x + 1)$
(b) $(m + 5)^2 - m(m + 6) = 35 + 2(2m - 5)$
(c) $(a^2 - 1)^2 + 3(a^2 - 1) = a^2(a^2 + 3) - 2(a^2 + 1)$
(d) $4x(x + 4) + 7 = (2x + 3)^2 + 2(2x - 1)$

25. Solve. Which of the following are identities?
(a) $3(n + 3)(2n - 1) = (3n + 3)^2 - 3(n^2 + 3)$
(b) $(y + 2)(y - 4) = (y + 3)(y - 4) - 3$
(c) $(x + 3)^2 + 2x(x - 5) = (3x + 1)(x - 1) + 6$
(d) $(x - 9)^2 + 3(x + 2) = x(x - 15) + 87$
(e) $(4a + 1)^2 + 2(2a - 5)^2 = 24a^2 + 35$

math tip

In arithmetic, we have often learned a shortcut for computing mentally. The following shortcut will help us to multiply binomials mentally.

$$(2a + 3b)(4a - b) = 8a^2 - 2ab + 12ab - 3b^2$$

First terms, Last terms, Inner terms, Outer terms

F — First
O — Outer
I — Inner
L — Last

2.3 understanding factoring

We use the distributive property to expand expressions.

To Expand: The final expression is written as a *sum* or *difference* of terms.

$a(x + y) = ax + ay$

$3y(y - 2) = 3y^2 - 6y$

To Factor: The final expression is written as a *product* of factors.

$ax + ay = a(x + y)$
  a is a factor of each term.

$3y^2 - 6y = 3y(y - 2)$
  $3y$ is a factor of each term.

Thus, factoring and expanding may be thought of as inverse operations. When factoring expressions the first thing we must do is obtain the *greatest common factor* of the terms.

Example 1
Write the greatest common factor of each expression.
(a) $6x^2 - 12y^2$
(b) $-2a^2x - 4ax - 6x$

Solution — 6 is the greatest common factor of each term.

(a) $6x^2 - 12y^2 = 6(x^2 - 2y^2)$
(b) $-2a^2x - 4ax - 6x = -2x(a^2 + 2a + 3)$
  — $-2x$ is the greatest common factor of each term.

We may check the factors by expanding.

Example 2
Factor $4a^2b + 8ab^2 - 12abc$.

Solution
The greatest common factor of each term is $4ab$. Thus
$$4a^2b + 8ab^2 - 12abc = 4ab(a + 2b - 3c).$$

Skills we acquire at one stage of learning mathematics are often utilized to develop additional skills. For example, compare the following expressions.

$$ax + ay = a(x + y)$$ — The common factor is a monomial.
$$(a + b)x + (a + b)y = (a + b)(x + y)$$ — The common factor is a binomial.

In the next example, there is no monomial that is a common factor of all terms. To obtain a common factor that is a binomial, we group the terms that have a common factor, as shown.

Example 3

Factor $2ax + bx + 2ay + by$.

Solution

$$2ax + bx + 2ay + by = x(2a + b) + y(2a + b)$$
$$= (2a + b)(x + y)$$

Common factor is a binomial.

To factor, we write a sum or difference of terms as a product.

2.3 exercise

Questions 1 to 5 develop essential skills in factoring.

1. Find the missing factor.
   (a) $6ab = (3a)(\ ?\ )$ $2b$
   (b) $-16xy = (\ ?\ )(-8y)$ $2x$
   (c) $-8y^4 = (-4y^2)(\ ?\ )$ $2y^2$
   (d) $36ab^2c = (\ ?\ )(6b^2)$ $6ac$
   (e) $-6a^3b^2 = (-3ab)(\ ?\ )$ $2a^2b$

2. Find the greatest common factor of each of the following.
   (a) a^2, ab a
   (b) $-24ab$, $8ab$ $8ab$
   (c) ay^2, $-by$ y^2
   (d) $-6x^3$, $2x^4$, $4x^5$ $2x$
   (e) $3a^4$, $-3a^3b$, $6a^2b^2$ $3a^2$
   (f) $9a^2$, $-12ab$, $-6b^2$ $3ab$

3. (a) Factor the expression $2x^3y - 8x^2y^2 + 6xy^3$.
   (b) How would you check your factors is (a)?

4. Find the missing factor.
   (a) $2mx + 2my = (\ ?\ )(x + y)$
   (b) $a^2 - ab = (\ ?\ )(a - b)$
   (c) $3x^2 - 12xy = (\ ?\ )(x - 4y)$
   (d) $-2m^2n + 2mn^2 - 6mnp = (\ ?\ )(m - n + 3p)$
   (e) $2\pi mR - 2\pi mr + 2\pi hm = (\ ?\ )(R - r + h)$

5. Find the greatest common factor of each of the following expressions.
   (a) $-13ab - 39ac$
   (b) $-15xy + 25xy^2$
   (c) $56m^2 - 4mn$
   (d) $18x^2 - 9x + 3$
   (e) $x^3 - x^2y + xy^2$
   (f) $2axy - 4bxy + 6cxy$
   (g) $4a^2b^3 - 6a^2b^2 + 2ab^3$
   (h) $6m^2n^2 + 3m^3n^2 - 9m^2n^2$

grouping factors

6. (a) What might be your first step in factoring the expression $x(a - b) + 2y(a - b)$?
   (b) Factor the expression in (a).

7. Factor each of the following.
   (a) $2x(a + b) + y(a + b)$
   (b) $3m(x - y) - k(x - y)$
   (c) $3y(m - n) - 2(m - n)$
   (d) $2m(y + 3) + n(y + 3)$
   (e) $3x(3m - 2n) + 2y(3m - 2n)$
   (f) $4m(4a - 2b) - 3n(4a - 2b)$

8 (a) Find the factors of $a(x + y) + b(x + y)$.
   (b) Find the factors of $x(a + b) + y(a + b)$.
   (c) Why are the answers in (a) and (b) the same?

   > Question 8 shows that the terms may be grouped in more than one way to obtain the same factors.

9 Factor each of the following.
   (a) $ax + bx + ay + by$.
   (b) $2ax - 2bx - ay + by$.
   (c) $a^2 - ab + ac - bc$.
   (d) $3bn - 2bm - 9an + 6am$.
   (e) $a^2c^2 + acd + abc + bd$
   (f) $y^4 + y^3 + 2y + 2$

10 Factor.
   (a) $4ax - 8bx$ (b) $18y^4 - 27y^3$
   (c) $49xy - 14x^2y^2$ (d) $25ab - 10ab^2$
   (e) $3y^3 - y^2 + y$ (f) $-6m^2 - 3m^3 - 9m$
   (g) $8a^3 + 2a^4 + 8a^5$ (h) $a^3 - a^2b + ab^2$
   (i) $2x^2y^3 - 6x^2y^2 + 2xy^3$ (j) $9x^4 - 6x^3y + 12x^2y^2$
   (k) $x^2 + y - xy - x$ (l) $a^2 - ac + ab - bc$
   (m) $3mxy - 6mx - 3nxy + 6nx$

problem/matics

A tape is placed snugly around the earth's equator. Of course, if 2 m more of tape are added, the tape will no longer be snug. Suppose the tape is placed a uniform distance, d, in centimetres, from the earth's surface; then which of the following would you agree with?

The distance, d, is
A less than 1 cm.
B 1 cm.
C greater than 1 cm.

applications: factoring and calculators

We may use our factoring skills to write expressions in a formula in a more convenient form.

	Formula	Factored Form
Physics	$s = \dfrac{ut}{2} + \dfrac{vt}{2}$	$s = \dfrac{1}{2}(u + v)t$
Geometry	$y = mx - ma$	$y = m(x - a)$

We may use the formula $A = \dfrac{1}{2}(a + b)h$ more easily with a calculator.

11 Rewrite each of the following formula in factored form.
   (a) $P = 2l + 2w$ (b) $S = n^2 + n$
   (c) $S = 4n^2 + 3n$ (d) $E = Ir + IR$
   (e) $S = 180n - 360$ (f) $S = \dfrac{n^2}{2} + \dfrac{n}{2}$
   (g) $A = 2\pi rh + 2\pi r^2$ (h) $S_n = \dfrac{3n^2}{2} - \dfrac{n}{2}$

12 The area of a trapezoid is given by the formula, $A = \dfrac{1}{2}ah + \dfrac{1}{2}bh$.
   (a) Use $h = 4$, $a = 3$, $b = 6$ to calculate A.
   (b) Factor the expression given above for A. Calculate the value of A using $h = 4$, $a = 3$, and $b = 6$.
   (c) Which answer, (a) or (b), requires fewer steps?

13 The formula for the sum of n even numbers
   $$\underbrace{2 + 4 + 6 + 8 + \ldots}_{n \text{ terms}}$$
   is $S_n = n^2 + n$ (We read "S subscript n". The n indicates that the formula gives the sum of n terms.)

   (a) Calculate the sum of the first 20 even numbers. Use $S_n = n^2 + n$.
   (b) Factor the formula in (a). Use the factored form to find the sum of the first 20 even numbers.

14 The surface area, S, of a cylinder is given by the formula $S = 2\pi r^2 + 2\pi rh$.

(a) Calculate the surface area of a cylinder with height 10 cm and radius 12 cm.
(b) By how much does the area of the cylinder in (a) increase if the radius is increased by 1 cm and the height by 2 cm?

Use the following formula in Questions 15 to 18. The sum of the cubes of the natural numbers to n terms

$$\underbrace{1^3 + 2^3 + 3^3 + 4^3 + \ldots + n^3}_{n \text{ terms}}$$

is given by the formula $S_n = \left(\dfrac{n^2 + n}{2}\right)^2$.

15 Use the above formula for S_n.
 (a) Calculate S_6.
 (b) Calculate $1^3 + 2^3 + 3^3 + 4^3 + 5^3 + 6^3$, without using the formula.
 (c) Are the answers in (a) and (b) the same?

16 Calculate.
 (a) S_7 (b) S_8 (c) S_{10}

17 (a) Calculate S_{15} to obtain the sum $1^3 + 2^3 + 3^3 + \ldots + 14^3 + 15^3$.
 (b) Calculate S_{10} to obtain the sum $1^3 + 2^3 + 3^3 + \ldots + 9^3 + 10^3$.
 (c) Explain why $S_{15} - S_{10}$ represents the sum $11^3 + 12^3 + 13^3 + 14^3 + 15^3$.

18 Use the above formula. Calculate each of the following sums.
 (a) $12^3 + 13^3 + 14^3$
 (b) $20^3 + 21^3 + 22^3 + 23^3 + 24^3$
 (c) $3^3 + 4^3 + 5^3 + \ldots + 9^3 + 10^3$

2.4 factoring trinomials

In studying mathematics, we may often use "what we know" to help us develop a strategy for finding "what we do not know".

known information

How is each term of the trinomial related to the original factors?

$(y - 2)(y - 5) = y^2 - 7y + 10$

How are the factors related to the trinomial?
Which numbers have a sum of -7
and a product of $+10$?

$(y - 2)(y + 5) = y^2 + 3y - 10$

Which numbers have a sum of $+3$
and a product of -10?

We use the above known information to develop a method.

unknown information

Factor
$y^2 + 7y + 12$ $\quad (y + 3)(y + 4)$

Which numbers have a sum of $+7$ and a product of $+12$? → $+3, +4$

Example 1
Factor $y^2 - 4y - 45$.

Solution
We ask this important question:
What two numbers have
• a sum of -4 and • a product of -45?

$(-9)(+5) = -45$
$(-9) + (+5) = -4$

We may then use the numbers, -9 and $+5$ to write the required factors.
$y^2 - 4y - 45 = (y - 9)(y + 5)$

$y^2 + 5y - 9y + 45$
$= y^2 - 4y - 45$

To factor a trinomial $ax^2 + bx + c$ where $a \neq 1$, we again use

Notice how the following coefficients of the trinomial are related to the original factors. For the original factors, note that

For the trinomial below, how are 36 and 20 related coefficients?

$(6)(1)(2)(3) = 36 \qquad (12)(3) = 36$
$(6y + 1)(2y + 3) = 12y^2 + 20y + 3$
$+18 \qquad 18 + 2 = 20$

Thus, to find the factors of $12y^2 + 20y + 3$, we may ask this question:

Which two numbers have
- a product of +36 and • a sum of +20?

We use the answers +18 and +2 to decompose the given trinomial.

$12y^2 + 20y + 3$

$ \underbrace{}_{20y}$
$= 12y^2 + 18y + 2y + 3$
$= 6y(2y + 3) + (2y + 3) \quad$ We use our earlier skills
$= (2y + 3)(6y + 1) \quad$ in grouping.

We call the above procedure the **method of decomposition**. For this method, we may decompose and write either of the following for the above example:
$20y = 18y + 2y \quad$ or $\quad 20y = 2y + 18y$

Thus we have used *what we know* to develop a strategy for finding *what we do not know*.

Example 2
Factor $6y^2 + 5y - 25$.

Solution
We ask the important question:
What two numbers have
- a product of -150 and • a sum of $+5$?

$(+15)(-10) = -150 \qquad (+15) + (-10) = +5$

Thus, we may decompose the trinomial.

$6y^2 + 5y - 25 = 6y^2 + 15y - 10y - 25$
$ = 3y(2y + 5) - 5(2y + 5)$
$ = (2y + 5)(3y - 5)$

Check your answer by multiplying the factors.

As a first step in factoring any trinomial, always check for a common factor first. For example,

$3m^2 - 6m - 24 = 3(m^2 - 2m - 8)$
$ = 3(m - 4)(m + 2)$

2.4 exercise

Questions 1 to 10 develop skills for factoring trinomials.

1 Find the missing factor.

(a) $x^2 + 8x + 15 = (\ ?\ )(x + 3)$
(b) $y^2 - 12y + 20 = (y - 10)(\ ?\ )$
(c) $a^2 + 4a - 12 = (\ ?\ )(a + 6)$
(d) $m^2 - 17m + 42 = (m - 14)(\ ?\ )$

2 Find the numbers that have the properties shown in the following table. Then write the factors of each trinomial.

	Trinomial	What two numbers have	
		a sum of	a product of
(a)	$b^2 + 6b + 8$	+6?	+8?
(b)	$y^2 - 12y + 20$	-12?	+20?
(c)	$a^2 - 2a - 8$	-2?	-8?
(d)	$m^2 + 7m - 30$	+7?	-30?

3. (a) How are the coefficients A, B, and C used in the *method of decomposition* to find the factors of the trinomial $6x^2 - 7x - 3$?
   (b) Factor the trinomial.

   $\uparrow \quad \uparrow \quad \uparrow$
   A B C

4. (a) For the trinomial $3y^2 + 10y + 3$, what two integers have a sum of +10 and a product of +9?
   (b) Use the information in (a) to write the factors of $3y^2 + 10y + 3$.

5. (a) For the trinomial $3x^2 - 19x - 14$, what two integers have a sum of -19 and a product of -42?
   (b) Use the integers in (a) to factor the trinomial.

6. The *method of decomposition* was used to write each trinomial in the following form. Find the factors.
   (a) $2y(y - 1) + (y - 1)$.
   (b) $3x(x - 3) - 2(x - 3)$
   (c) $m(2n - 1) - 3(2n - 1)$
   (d) $2x(x - y) - 3y(x - y)$

7. Factor. Look for a common factor first.
   (a) $x^2 + 13x + 42$ (b) $3m^2 + 7m + 2$
   (c) $x^2 - x - 2$ (d) $y^2 + 2y - 3$
   (e) $3m^2 + 9m + 6$ (f) $5m^2 + 11m + 2$
   (g) $4y^2 - 4y - 3$ (h) $2a^2 + 8a + 8$

8. To find the factors of the following trinomial, Peter used a trial and error method to discover the positive factors as shown.

   $15x^2 - 28x + 5$

   Factors of 15   Factors of 5

   15   1   ⑤   3 ⟶ 5   ①
   1   15 ③   5 ⟶ 1   ⑤

   (a) Use the circled factors to complete the factors $(5x \quad 1)(3x \quad 5)$.
   (b) Why are the signs placed within the factors as shown? $(5x - 1)(3x - 5)$

9. A perfect trinomial square may be written as $x^2 - 2x + 1 = (x - 1)^2$.

   Find the missing term in each of the following if each is a perfect trinomial square.
   (a) $x^2 - 2xy + (\ ?\ )$ (b) $(\ ?\ ) + 4a + 1$
   (c) $(\ ?\ ) - 12x + 4$ (d) $4m^2 - 4m + (\ ?\ )$
   (e) $4x^2 + (\ ?\ ) + 81$ (f) $9a^2 - (\ ?\ ) + 10$

10. Write each of the following as a perfect trinomial square.
    (a) $m^2 + 6m + 9$ (b) $a^2 - 8a + 16$
    (c) $4x^2 + 4x + 1$ (d) $9y^2 - 12y + 4$
    (e) $25a^2 - 40ab + 16b^2$ (f) $4 - 20m + 25m^2$

11. Each of the following trinomials is to be a perfect square. Find k, $k > 0$.
    (a) $x^2 - 6x + k$ (b) $x^2 - kx + 121$
    (c) $x^2 - 12x + k$ (d) $4x^2 - 4x + k$
    (e) $4x^2 + kx + 9$ (f) $4x^2 - kx + 16$

12. Factor completely.
    (a) $a^2 + 3a + 2$ (b) $a^2 + 4a - 12$
    (c) $x^2 + x - 56$ (d) $6x^2 + x - 7$
    (e) $50 - 20x + 2x^2$ (f) $20x^2 - 19x - 6$
    (g) $16m^2 - 8mn + n^2$ (h) $x^2 - 24xy + 144y^2$
    (i) $x^3 - 6x^2 + 9x$ (j) $3x^2 - 2x - 1$
    (k) $10 + 43x + 28x^2$ (l) $3a^2 + 18a + 27$
    (m) $3x^2 + 21x + 36$ (n) $5x^2 - 29x - 6$

13. Express each trinomial in factored form.
    (a) $x^2 + 8x + 15$ (b) $a^2 - a - 72$
    (c) $b^2 + 6b + 8$ (d) $4x^2 - 16x + 7$
    (e) $4x^2 - 4x - 3$ (f) $a^2 + 8a + 16$
    (g) $25t^2 - 10t + 1$ (h) $2x^2 + 8x - 10$
    (i) $6x^2 - 23x + 21$ (j) $3x^2 - 7x - 6$
    (k) $10 + 21x + 2x^2$ (l) $3x^2 - 11x + 10$

14 For each trinomial find the values of k so that two binomials factors are obtained, $0 < k \leq 20$.

(a) $x^2 + 6x + k$ (b) $x^2 + 7x - k$
(c) $x^2 + kx - 15$ (d) $a^2 + kab + 6b^2$
(e) $x^2 + 5xy + ky^2$ (f) $x^2 - 8xy + ky^2$

applications: restrictions on the variable

When we write a rational expression there are certain restrictions on the variables.

$\dfrac{a+b}{a}$ ← Why is $a \neq 0$? $\dfrac{y^2 + 4y + 4}{y - 3}$ ← Why is $y \neq 3$?

The denominator of each rational expression must be non-zero. Thus $a \neq 0$ and $y \neq 3$ for the above expressions.

15 For each expression
- indicate the restrictions on the variables.
- simplify the rational expression.

(a) $\dfrac{x^2 + 15x + 50}{x + 5}$ (b) $\dfrac{a^2 + 2a - 35}{a + 7}$

(c) $\dfrac{y^2 - y - 20}{y - 5}$ (d) $\dfrac{m^2 - 9m - 90}{m - 15}$

(e) $\dfrac{x^2 - 4xy - 32y^2}{x - 8y}$ (f) $\dfrac{t^2 + 7t + 12}{t + 4}$

(g) $\dfrac{a^2 - 5ab - 24b^2}{a + 3b}$ (h) $\dfrac{x^2 - 2xy - 8y}{x + 2y}$

(i) $\dfrac{2x^2 - 5x - 3}{x - 3}$ (j) $\dfrac{5k^2 + 22k - 48}{k + 6}$

(k) $\dfrac{6m^2 + mn - 35n^2}{3m - 7n}$ (l) $\dfrac{6x^2 - 107x + 35}{3x - 1}$

(m) $\dfrac{2x^2 + 5xy - 12y^2}{2x - 3y}$ (n) $\dfrac{18x^2 + 9xy - 5y^2}{3x - y}$

problem/matics

Refer to page 23. This time the point P moves so that the difference $BP - AP$ is a constant. What is the locus of P?

2.5 writing special factors

When we were learning to factor trinomials we used patterns to help us find methods of factoring.

Common Factor
$mx + my = m(x + y)$
$2x^3 - 2x^2 + 2x = 2x(x^2 - x + 1)$

Trinomials
$a^2 + 3ab + 2b^2 = (a + b)(a + 2b)$
$6x^2 - 11xy + 3y^2 = (3x - y)(2x - 3y)$

Perfect Squares
$m^2 - 2mn + n^2 = (m - n)(m - n)$
$\quad = (m - n)^2$
$9p^2 + 12pq + 4q^2 = (3p + 2q)(3p + 2q)$
$\quad = (3p + 2q)^2$

Difference of squares

If we study the following products we will notice a pattern that will help us to devise a more efficient method of factoring special polynomials which we call a **difference of squares**.

Difference of squares

$(x + 3)(x - 3) = x^2 - 9$ $(x)^2 - (3)^2$
$(2m - n)(2m + n) = 4m^2 - n^2$ $(2m)^2 - (n)^2$
$(p - 3q)(p + 3q) = p^2 - 9q^2$ $(p)^2 - (3q)^2$

Example 1
Factor $9m^2 - n^2$.   Our first step is to identify the squares. This may eventually be a mental step.

Solution
$9m^2 - n^2 = (3m)^2 - (n)^2$
$\quad = \underbrace{(3m + n)}_{\text{Sum of the two terms.}}\underbrace{(3m - n)}_{\text{Difference of the terms.}}$

We call the above procedure: **factoring a difference of squares.**

When we factor any polynomial, we must
- remove the greatest common factor.
- check the remaining factors to see if any of them can be factored again.

Example 2
Factor completely $3a^4 - 48b^4$.

Solution
$3a^4 - 48b^4 = 3(a^4 - 16b^4)$
- check: for a common factor.
- check: Can we still factor?

$= 3[(a^2)^2 - (4b^2)^2]$
$= 3(a^2 + 4b^2)(a^2 - 4b^2)$

check: Can we still factor?

$= 3(a^2 + 4b^2)(a + 2b)(a - 2b)$

None of the factors can be factored further.

Example 3
Factor $x^2 + 6xy + 9y^2 - 25x^2$.

Solution
$x^2 + 6xy + 9y^2 - 25x^2$
$= (x + 3y)^2 - (5x)^2$
$= (x + 3y + 5x)(x + 3y - 5x)$
$= (6x + 3y)(-4x + 3y)$

We recognize that $x^2 + 6xy + 9y^2$ is a perfect square.

In Example 3, we *grouped* the terms of the polynomials so that we could rewrite the polynomial as a difference of squares.

2.5 exercise

Questions 1 to 10 develop some essential skills using the method of factoring a difference of squares.

A

1. Expand each of the following. What property do they have in common?

   (a) $(a + 5)(a - 5)$ (b) $(2x - 3)(2x + 3)$
   (c) $(x + 3y)(x - 3y)$ (d) $(3x + b)(3x - b)$
   (e) $(a^2 - 2b^2)(a^2 + 2b^2)$ (f) $(xy - 1)(xy + 1)$
   (g) $(x + y + z)(x + y - z)$
   (h) $(a + 2b - y)(a + 2b + y)$
   (i) $(2x - 3y + 5z)(2x - 3y - 5z)$

2. We must be able to identify a difference of squares before we can factor it. Complete each of the following.

   (a) $9a^2 - b^2 = (\ ?\ )^2 - b^2$
   (b) $m^2 - 16n^2 = m^2 - (\ ?\ )^2$
   (c) $25x^2 - 49y^2 = (\ ?\ )^2 - (\ ?\ )^2$
   (d) $(x - y)^2 - 16a^2 = (x - y)^2 - (\ ?\ )^2$
   (e) $100m^2 - (2a + b)^2 = (\ ?\ )^2 - (2a + b)^2$

3. For each polynomial, group the terms and write a difference of squares. The first one has been done for you.

   (a) $a^2 + 2ab + b^2 - c^2$ ← $a^2 + 2ab + b^2 - c^2$
   $= (a + b)^2 - c^2$

   (b) $4x^2 + 4xy + y^2 - c^2$ (c) $x^2 + 6x + 9 - 4y^2$
   (d) $4x^2 - a^2 - 2ab - b^2$ (e) $9y^2 - 4a^2 + 4ab - b^2$
   (f) $16x^2 - m^2 - 6mn - 9n^2$
   (g) $25y^2 - 1 + 6x - 9x^2$
   (h) $a^2 - 2ab + b^2 - x^2 - 2xy - y^2$

4. Look at the polynomials $k^2 - x^2$ and $(a + b)^2 - x^2$.
   (a) How are they alike? How are they different?
   (b) Write them in factor form.

5. Look at the polynomials $4k^2 - x^2$ and $4(a + b)^2 - x^2$.
   (a) How are they alike? How are they different?
   (b) Write them in factor form.

6. Look at the polynomials $p^2 - 9k^2$ and $p^2 - 9(q + r)^2$.
   (a) How are they alike? How are they different?
   (b) Write them in factor form.

7. Write the factors of each of the following.
   Remember: regroup to obtain a difference of squares.

   (a) $9a^2 - 6ab + b^2 - x^2$ (b) $1 - 2y + y^2 - 4x^2$
   (c) $9a^2 + 6a + 1 - 9b^2$ (d) $100x^2 - m^2 - 2m - 1$
   (e) $49x^2 - 4m^2 - 4m - 1$ (f) $25y^2 - 1 + 6x - 9x^2$
   (g) $x^2 + 2xy + y^2 - a^2 + 2ab - b^2$
   (h) $9x^2 - 6x + 1 - 16x^2 + 8xy - y^2$

8 (a) Write the factors of $m^4 - 1$.
   (b) Check the factors you wrote in (a). Are they factorable?

9 (a) Write the factors of the trinomial
       $x^4 - 5x^2 + 4$.
   (b) Check the factors in (a). Write the factors of those that are factorable.

10 A square with sides x units is drawn and then a square with sides y units is removed from the original square. Use the diagram to show why $x^2 - y^2 = (x + y)(x - y)$.

B
11 Factor fully.
   (a) $y^2 - 16$ (b) $m^2 - 100$ (c) $36 - x^2$
   (d) $18y^2 - 2$ (e) $9 - 4k^2$ (f) $36x^2 - y^2$
   (g) $25m^2 - 16a^2$ (h) $1 - 25y^2$ (i) $a^2b^2 - 1$
   (j) $m - 16m^3$ (k) $9m^2 - 121$ (l) $98a - 72a^3$
   (m) $4a^2 - 9b^4$ (n) $16m^4 - 9n^2$ (o) $x^2y^2 - 4$

12 Factor fully.
   (a) $n^2 - 36$ (b) $9h^2 - 49m^2$
   (c) $225m^2 - 1$ (d) $x^2y^2z^2 - 9$
   (e) $-1 + 36m^2$ (f) $(x + y)^2 - 9$
   (g) $(x + y)^2 - m^2$ (h) $y^4 - 81$
   (i) $8x^2 - 50y^2$ (j) $y^2 - (a + h)^2$
   (k) $(x - 3y)^2 - 4a^2$ (l) $9m^2 - 4(x - y)^2$
   (m) $\frac{1}{4}x^2 - \frac{1}{9}y^2$ (n) $a^4 - b^4$
   (o) $y^2 - 0.16x^2$ (p) $2(x - 2y)^2 - \frac{1}{2}m^2$
   (q) $(x - y)^2 - 4(p - q)^2$ (r) $(x + y - 3z)^2 - 4k^2$
   (s) $(x + y)^2 - 9(a + b - c)^2$

13 Write each of the following in factor form.
   (a) $a^2 - 2ab + b^2 - m^2$ (b) $k^2 - m^2 + 2mn - n^2$
   (c) $a^2 - 2ay + y^2 - 4x^2$ (d) $a^2 - 2ab + b^2 - c^4$
   (e) $x^2 - y^2 - 2yz - z^2$ (f) $m^4 - x^2 + 2xy - y^2$
   (g) $x^2 + 2xy + y^2 - a^2 - 2ab - b^2$
   (h) $x^2 + y^2 - a^2 - b^2 + 2xy + 2ab$
   (i) $x^2 + 2xw + w^2 - y^2 - 2yz - z^2$

14 Factor each of the following completely. Which two cannot be factored?
   (a) $16x^2 - y^2$ (b) $x^2 - 49y^2$
   (c) $4x^2 - 1$ (d) $4x^2 - 25$
   (e) $(x + 3)^2 - 25$ (f) $x^2 + 6x + 9 - y^2$
   (g) $2x^2 - 18$ (h) $-75 + 3x^2$
   (i) $a^2 + 4bc - 4c^2 - b^2$ (j) $49 - (x - 2)^2$
   (k) $6x^2 - 54y^2$ (l) $27x^2 - 3y^2$
   (m) $9a^2 - 6a + 1 - 9m^2$ (n) $x^2 - 18$
   (o) $m^4 - 17m^2 + 16$ (p) $(x + 2)^2 - (y - 5)^2$
   (q) $25x^2 + 16y^2$ (r) $\frac{x^2}{16} - \frac{y^2}{36}$
   (s) $4a^4 - 37a^2 + 9$ (t) $81m^8n^4 - y^8$
   (u) $a^2 - 2ab + b^2 - 9x^2 - 18xy - 9y^2$
   (v) $m^2 + n^2 - p^2 - q^2 - 2mn - 2pq$

problem/matics

Historically, many people have used the above method to develop important achievements, not only in mathematics, but also in medicine, science, and astronomy. Blaise Pascal used what he knew to develop what he did not know.

$(x + 1)^0 = 1$
$(x + 1) = x + 1$
$(x + 1)^2 = x^2 + 2x + 1$
$(x + 1)^3 = x^3 + 3x^2 + 3x + 1$
$(x + 1)^4 = $?
$(x + 1)^5 = $?

```
        1
       1 1
      1 2 1
     1 3 3 1
```
Can you continue the next two rows?

factoring incomplete squares

To factor a trinomial such as $x^4 + 4x^2 + 16$, we combine our factoring skills. Notice that these two trinomials are the same, except for the term containing x^2.

$$x^4 + \underbrace{4x^2}_{\text{We refer to this expression as an } \textit{incomplete square.}} + 16 \qquad x^4 + 8x^2 + 16 = (x^2 + 4)^2$$

To factor the trinomial, $x^4 + 4x^2 + 16$, we rewrite its form as follows.

$$x^4 + \underbrace{4x^2}_{} + 16$$
$$= x^4 + 8x^2 - 4x^2 + 16 \qquad \text{Write } 4x^2 \text{ in a}$$
$$= x^4 + 8x^2 + 16 - 4x^2 \qquad \text{form to obtain}$$
$$= (x^2 + 4)^2 - (2x)^2 \qquad \text{a perfect square.}$$
$$= (x^2 + 4 + 2x)(x^2 + 4 - 2x)$$

15. For what value of m is each of the following a perfect square?
   (a) $x^4 + mx^2 + 144$
   (b) $9a^4 + ma^2 + 1$
   (c) $16k^4 - mk^2 + 9$
   (d) $y^4 - my^2 + 25$
   (e) $25x^4 - mx^2y^2 + 16y^4$
   (f) $9 + mp^2 + 36p^4$

16. Write each of the following as a difference of squares. Then factor fully.
   (a) $x^4 + x^2 + 1 = x^4 + 2x^2 + 1 - x^2$
   (b) $a^4 + 7a^2 + 16 = a^4 + 8a^2 + 16 - a^2$
   (c) $a^4 + 4 = a^4 + 4a^2 + 4 - 4a^2$
   (d) $m^4 - 11m^2n^2 + n^4 = m^4 - 2m^2n^2 - n^4 - 9m^2n^2$
   (e) $4y^4 - 13y^2 + 1 = 4y^4 - 9y^2 + 1 - 4y^2$
   (f) $a^4 + 2a^2b^2 + 9b^4 = a^4 + 6a^2b^2 + 9b^4 - 4a^2b^2$

17. Factor each of the following by completing the square.
   (a) $y^4 + y^2 + 25$
   (b) $x^4 + 2x^2 + 9$
   (c) $a^4 + 3a^2 + 4$
   (d) $y^4 - 23x^2y^2 + x^4$
   (e) $16x^4 - 12x^2 + 1$
   (f) $9m^4 + 2m^2 + 1$
   (g) $9a^4 - 52a^2b^2 + 64b^4$
   (h) $4m^4 - 16m^2 + 9$

2.6 solving quadratic equations by factoring

We apply our skills in factoring to solve equations such as:

$$x^2 - 6x + 8 = 0 \qquad 2x^2 - 3x + 1 = 0$$
$$x^2 - 9 = 0$$

Each of the above equations is of *degree 2* and is called a quadratic equation.

Since we are already able to solve linear equations, we try to transform a quadratic equation into equivalent linear equations.

To solve a quadratic equation we use the principle:

If $ab = 0$ then either $a = 0$ or $b = 0$.

Example 1

Solve $x^2 - 4x + 3 = 0$.

Solution

$x^2 - 4x + 3 = 0$ We factor the trinomial expression.
$(x - 1)(x - 3) = 0$

$x - 1 = 0$ or $x - 3 = 0$ We replace the
$x = 1$ $x = 3$ quadratic equation by two other equivalent equations.

We must verify *both* roots in the original equation.

For $x = 1$ | For $x = 3$
RS = 0 | RS = 0
LS = $x^2 - 4x + 3$ | LS = $x^2 - 4x + 3$
$= (1)^2 - 4(1) + 3$ | $= (3)^2 - 4(3) + 3$
$= 1 - 4 + 3 = 0$ | $= 9 - 12 + 3 = 0$
LS = RS | LS = RS

The roots of the equation are $\{1, 3\}$.

In order to solve for the variable, using the same principle, the quadratic equation in Example 2 needs to be rewritten so that one side of the equation is equal to zero.

Example 2

Find the solution set of $(3x - 1)(2x + 3) = -5$.

Solution

$(3x - 1)(2x + 3) = -5$ ← In this form we are not able to solve the equation.
$6x^2 + 7x - 3 = -5$
$6x^2 + 7x + 2 = 0$
$(3x + 2)(2x + 1) = 0$ ← We may equate each factor to zero.

$3x + 2 = 0$ or $2x + 1 = 0$
$3x = -2$ | $2x = -1$
$x = -\frac{2}{3}$ | $x = -\frac{1}{2}$

Thus the solution set is $\left\{-\frac{2}{3}, -\frac{1}{2}\right\}$.

To solve the following equation we first divide both sides by the constant 3.

$3x^2 - 3x - 36 = 0$

Divide by 3.

$x^2 - x - 12 = 0$
$(x - 4)(x + 3) = 0$
$x - 4 = 0$ or $x + 3 = 0$
$x = 4$ or $x = -3$

The roots are 4, −3.

We may multiply or divide both sides of one equation by a constant and obtain equations that are equivalent to the original equation. However, this is not the case when dividing or multiplying both sides by an expression containing a variable. For example, the following equation has two roots.

$x^2 - 5x = 0$
Thus $x(x - 5) = 0$
$x = 0$ or $x - 5 = 0$
$x = 5$

The equation has two roots i.e. 0 and 5.

However, if we were to divide both sides by x, *in error*, we would obtain only one root given by $x = 5$. We may divide only by a numerical constant.

Consider the equation $x = 5$.

Multiply both sides by $x - 2$.
$x(x - 2) = 5(x - 2)$
$x^2 - 2x = 5x - 10$
$x^2 - 7x + 10 = 0$
$(x - 5)(x - 2) = 0$
$x - 5 = 0$ or $x - 2 = 0$
$x = 5$ | $x = 2$

Thus, by multiplying both sides by $x - 2$, we have produced an equation that is not equivalent to the original equation. The root of the original equation is 5. The roots of the equation we made are 2 and 5. The root, 2, does not satisfy the original equation. We call 2 an **extraneous root** since it satisfies the new equation but not the original equation.

Example 3

Solve $12a^2 - 5a = 3$.

Solution

In the present form, we may not use the principle: if $pq = 0$ then $p = 0$ or $q = 0$. Rewrite the equation so that one side of the equation is equal to zero.

$12a^2 - 5a = 3$
$12a^2 - 5a - 3 = 0$
$(3a + 1)(4a - 3) = 0$
$3a + 1 = 0$ or $4a - 3 = 0$
$3a = -1$ | $4a = 3$
$a = -\frac{1}{3}$ | $a = \frac{3}{4}$

Thus the roots are $-\frac{1}{3}$ and $\frac{3}{4}$

If we know what the roots of a quadratic equation are we can then write the equation. For example, if the roots are −3 and 2, then we may introduce a variable, say x, and write the following using a reverse procedure.

$x = -3$ or $x = 2$
$x + 3 = 0$ | $x - 2 = 0$
Then $(x + 3)(x - 2) = 0$
$x^2 + x - 6 = 0$

A quadratic equation with roots −3 and 2 is $x^2 + x - 6 = 0$.

applications: solving problems

26. Dave and Jim entered the annual spring canoe race last year. The number of times they paddled the course during practice was one less than the number of kilometres in the course. If they paddled a total of 72 km during the practice, how long was the course?

27. The product of two whole numbers is 45. If the sum of the numbers is 14, find the numbers.

28. The sides of one square are 1 cm shorter than twice the length of the sides of a smaller square. If their combined area is 65 cm², find the lengths of the sides of each square.

29. The corner drugstore ordered a small quantity of cashews. The cost per kilogram was $2 more than the number of kilograms bought. If the total cost was $35, how many kilograms of cashews did the store order?

30. Find two consecutive positive numbers such that their product is 156.

31. One square silicone computer chip has a side 2 mm longer than a smaller square chip. Two of the smaller chips have a combined area which is 8 mm² smaller than the area of one of the bigger chips. Find the length of the side of the smaller chip.

32. A polygon of n sides has $\frac{1}{2}(n)(n-3)$ diagonals. A certain polygon has 65 diagonals. How many sides does this polygon have?

33. A rectangular medal is to be mounted on a block of wood 16 cm wide and 20 cm long in such a way that the area of the medal will be three fifths of the area of the block. Find the dimensions of the medal if the border around the metal is of uniform width.

2.7 rational expressions: multiplying and dividing

We can apply our factoring skills to simplify rational expressions. For the following products of rational expressions, we must be aware of the restrictions as the denominators $\neq 0$.

Example 1

Calculate $\dfrac{1-3y}{2y+1} \times \dfrac{4y^2-1}{1-9y^2}$ if $y = 5$.

Solution

$\dfrac{1-3y}{2y+1} \times \dfrac{4y^2-1}{1-9y^2}$ First express the products in simplest terms.

$= \dfrac{1-3y}{2y+1} \times \dfrac{(2y-1)(2y+1)}{(1-3y)(1+3y)}$

$= \dfrac{\cancel{1-3y}}{\cancel{2y+1}} \times \dfrac{(2y-1)\cancel{(2y+1)}}{\cancel{(1-3y)}(1+3y)}$

$= \dfrac{2y-1}{1+3y}$ Remember the restriction that is understood here. $1 + 3y \neq 0$

Use $y = 5$. Then $\dfrac{2y-1}{1+3y} = \dfrac{2(5)-1}{1+3(5)}$

$= \dfrac{10-1}{1+15}$

$= \dfrac{9}{16}$

The skills we learned when working with rational numbers extend to our work with rational expressions.

Example 2

Simplify $\dfrac{2x^2-7x-4}{x^2-7x+12} \div \dfrac{4x^2-1}{2x-6}$.

Solution

$\dfrac{2x^2-7x-4}{x^2-7x+12} \div \dfrac{4x^2-1}{2x-6}$ Remember: to divide, multiply by the reciprocal.

$= \dfrac{2x^2-7x-4}{x^2-7x+12} \times \dfrac{2x-6}{4x^2-1}$

$$= \frac{(2x+1)(x-4)}{(x-3)(x-4)} \times \frac{2(x-3)}{(2x-1)(2x+1)}$$

$$= \frac{\overset{1}{\cancel{(2x+1)}}\overset{1}{\cancel{(x-4)}}}{\underset{1}{\cancel{(x-3)}}\underset{1}{\cancel{(x-4)}}} \times \frac{2\overset{1}{\cancel{(x-3)}}}{(2x-1)\underset{1}{\cancel{(2x+1)}}}$$

$$= \frac{2}{2x-1}$$

We will not write the restrictions each time, but the restrictions are understood for rational expressions.

2.7 exercise

Questions 1 to 6 review some earlier skills related to multiplying and dividing rational expressions.

A

1 Which of the following are equivalent to $\frac{y-x}{x+y}$?

(a) $-\frac{x-y}{x+y}$ (b) $-\frac{y-x}{x+y}$ (c) $-\frac{-(y-x)}{x+y}$ (d) $-\frac{x-y}{y+x}$

2 Find the value of each of the following.

(a) $\frac{a+b}{b+a}$ (b) $-\frac{(b-a)}{b-a}$ (c) $\frac{a-b}{-(b-a)}$

(d) $\frac{-(b+a)}{-(b+a)}$ (e) $\frac{-(b-a)}{a-b}$ (f) $\frac{-(b-a)}{-(a-b)}$

3 Simplify.

(a) $\frac{a^2-b^2}{b-a}$ (b) $-\frac{4x^2-y^2}{y-2x}$

(c) $\frac{x^2-2x+1}{1-x}$ (d) $\frac{a^2-10a+25}{25-a^2}$

4 Simplify each of the following.

(a) $\frac{-3a}{10b} \times \frac{20b}{-24a}$ (b) $\frac{a}{2y} \times \frac{-4by}{-10ax}$

(c) $\frac{3t^2}{-2y} \div \frac{6t^3}{4y^2}$ (d) $\frac{3x}{-4y^3} \div \frac{-32}{15x^2y}$

(e) $\frac{4km}{9m} \times \frac{-3m^3k}{6k^2}$ (f) $\frac{a}{-2b} \div \frac{-a}{4b}$

(g) $\frac{6x^3}{-x^2} \times \frac{x^5}{18x} \div \frac{8}{6x^6}$ (h) $\frac{36x^3y^5}{-x} \div \frac{12x^2y}{10xy^2} \times \frac{-x^4}{15y^4}$

5 Simplify.

(a) $\frac{y-1}{y+3} \times \frac{2y+6}{1-y}$ (b) $\frac{x-3}{2-x} \div \frac{2x-6}{x-2}$

(c) $\frac{x-1}{3x} \times \frac{x}{x^2-1}$ (d) $\frac{y+1}{y^2-1} \div \frac{y-1}{y}$

6 Simplify each expression. Which values of the variable are not permissible?

(a) $\frac{x^2-2x-15}{4x-20}$ (b) $\frac{y^2-25}{y^2-3y-10}$

(c) $\frac{y-y^2}{y^3-y}$ (d) $\frac{y^2-8y+15}{y^2-y-6}$

B

7 Simplify.

(a) $\frac{x^2+7x}{x^2-1} \times \frac{x^2+3x+2}{x^2+14x+49}$

(b) $\frac{(x^2-9)}{x^2+5x+4} \div \frac{x^2-4x+3}{x^2+5x+4}$

(c) $\frac{x^2-3x+2}{x^2+6x+9} \times \frac{x^2-9}{2x-4}$

(d) $\frac{4a^2+8ab+4b^2}{8} \times \frac{3a-3b}{a^2-b^2}$

(e) $\frac{3m^2-17m+20}{3m^2+7m-20} \times \frac{m^2+9m+20}{m^2+m-20}$

(f) $\frac{2x^2-14x+24}{4x^2-64} \div \frac{x^2-8x+15}{2x^2-11x+5}$

8 If $x = -3$, find the value of each of the following.

(a) $\frac{x^2+9}{x+9}$ (b) $\frac{x^2+3x+2}{4-x^2}$

(c) $\frac{x^2-9}{2x+1} \div \frac{x^2-6x+9}{2x-6}$

(d) $\frac{x^2+4x+4}{x^2-4} \times \frac{2x-4}{x^2-3x-10} \div \frac{4x+16}{x^2-25}$

9 Write each of the following as a rational expression in lowest terms.

(a) $\dfrac{m^2 + 7m + 12}{m^2 + m - 6} \div \dfrac{2m + 8}{m^2 + 9m + 20}$

(b) $\dfrac{x^2 - 4}{2x^2 + 11x + 5} \times \dfrac{x^2 + 2x - 15}{x^2 - x - 6}$

(c) $\dfrac{a^2 - 4a - 21}{a^2 - 3a - 28} \div \dfrac{a^2 - 9}{a^2 - a - 20}$

(d) $\dfrac{x^4 - y^4}{x^4 + 2x^2y^2 + y^4} \times \dfrac{x^2 + y^2}{-x^2 + y^2}$

(e) $\dfrac{a + 1}{a - 1} \times \dfrac{a + 3}{1 - a^2} \div \dfrac{(a + 3)^2}{1 - a}$

(f) $\dfrac{x^2 - 5x + 6}{2x} \div \dfrac{x - 3}{4x^2} \times \dfrac{4x^2 - 4x - 8}{8x - 16}$

10 How does the value of $\dfrac{2x^2 - 18}{x^2 - 6x + 9} \div \dfrac{3x + 6}{x^2 + 2x}$ change if x increases from 1 to 2?

11 How does the value of $\dfrac{28x - 4x^2}{16x^2 + 4x^3} \times \dfrac{x^2 + 7x + 12}{x^2 - 9x + 14}$ change if x decreases from 4 to 3?

12 If $f(m) = \dfrac{m^2 - 3m + 2}{2m^2 + 5m + 2} \times \dfrac{2m^2 + 11m + 5}{m^2 + 4m - 5}$ find

(a) $f(-1)$ (b) $f(2)$ (c) $f(0)$
(d) $f(3) - f(-3)$ (e) $3f(-1) - f(-3)$

problem/matics

Three freighters leave St. John's, Newfoundland, for Montreal, Quebec, at the same time. One ship takes 20 d to make the round trip, another ship takes 16 d, and the final ship takes 12 d. If these three ships continuously make this round trip between St. John's and Montreal, how many trips would each freighter have to make so that all three ships are again in port at the same time?

2.8 rational expressions: adding and subtracting

Skills for factoring we learned earlier are extended to adding and subtracting rational expressions. How are these examples the same? How are they different?

Simplify.

$\dfrac{1}{x - 1} - \dfrac{2}{x^2 - 1}$

$= \dfrac{1}{x - 1} - \dfrac{2}{(x - 1)(x + 1)}$

$= \dfrac{(x + 1)}{(x - 1)(x + 1)} - \dfrac{2}{(x - 1)(x + 1)}$

$= \dfrac{x + 1 - 2}{(x - 1)(x + 1)}$

$= \dfrac{x - 1}{(x - 1)(x + 1)}$

$= \dfrac{\cancel{(x - 1)}^1}{_1\cancel{(x - 1)}(x + 1)}$

$= \dfrac{1}{x + 1}$ ← Remember the restriction that is understood. $x + 1 \neq 0$ or $x \neq -1$.

Simplify.

$\dfrac{1}{x} + \dfrac{3}{xy}$

$= \dfrac{y}{xy} + \dfrac{3}{xy}$

$= \dfrac{y + 3}{xy}$

Remember the restriction that is understood. $x, y \neq 0$.

Throughout our work with rational expressions we should always be aware of the restrictions that occur. We use our skills with least common multiples to find a suitable denominator.

Example 1

Simplify $\dfrac{\dfrac{a}{1 + a} + \dfrac{1 - a}{a}}{\dfrac{a}{1 + a} - \dfrac{1 - a}{a}}$

We refer to this expression as a complex fraction.

Solution

Simplify complex expressions one step at a time so that we reduce the risk of major errors.

$$\frac{\frac{a}{1+a} + \frac{1-a}{a}}{\frac{a}{1+a} - \frac{1-a}{a}}$$

$$= \left(\frac{a}{1+a} + \frac{1-a}{a}\right) \div \left(\frac{a}{1+a} - \frac{1-a}{a}\right)$$

$$= \left(\frac{a^2 + (1+a)(1-a)}{a(a+1)}\right) \div \left(\frac{a^2 - (1-a)(1+a)}{a(1+a)}\right)$$

$$= \left(\frac{a^2 + 1 - a^2}{a(a+1)}\right) \div \left(\frac{a^2 - 1 + a^2}{a(a+1)}\right)$$

$$= \frac{1}{a(a+1)} \times \frac{a(a+1)}{2a^2 - 1}$$

$$= \frac{1}{\cancel{a(a+1)}} \times \frac{\cancel{a(a+1)}}{2a^2 - 1}$$

$$= \frac{1}{2a^2 - 1}$$

Now it is easier to work with a simplified expression which is *equivalent* to the original expression.

In solving some problems in the scientific world and in economics we will encounter complex expressions that may be simplified significantly using our algebraic skills. We must understand the language of algebra to use algebra to solve problems.

2.8 exercise

Questions 1 to 7 practise essential skills needed to add and subtract rational expressions.

A

1 Express each of the following in simplest terms.

(a) $\dfrac{x^2 - y^2}{x^2 - 2xy + y^2}$
(b) $\dfrac{16 - b^2}{b^2 - 8b + 16}$
(c) $\dfrac{a^3 - a^2}{3 - 3a^2}$
(d) $\dfrac{2k^2 + k - 1}{2k^2 - 3k + 1}$

2 Remember that $y - x = -(x - y)$. Simplify each of the following.

(a) $\dfrac{3 - x}{x^2 - 9}$
(b) $\dfrac{2x + 1}{2x^2 - 5x - 3}$
(c) $\dfrac{16 - y^2}{y + 4}$
(d) $\dfrac{2a^2 - a - 1}{1 - a^2}$
(e) $\dfrac{x^3 + x^2 + x}{x^2 + x + 1}$
(f) $\dfrac{(y - x)^2}{x - y}$
(g) $\dfrac{m^2 - 5m}{25 - m^2}$
(h) $\dfrac{25 + 20a + 4a^2}{4a^2 - 25}$

3 Copy and complete.

(a) $\dfrac{x}{x - 1} = \dfrac{?}{x^2 - 1}$
(b) $\dfrac{3}{a + 2} = \dfrac{?}{a^2 + 4a + 4}$
(c) $\dfrac{m - 2}{m + 2} = \dfrac{?}{4 - m^2}$
(d) $\dfrac{y - 1}{y + 3} = \dfrac{?}{2y^2 + 5y - 3}$
(e) $\dfrac{q}{3p - q} = \dfrac{?}{6p^2 + 13pq - 5q^2}$
(f) $\dfrac{m - 3}{3m + 4} = \dfrac{?}{16 - 9m^2}$

4 Find the least common multiple for each of the following.

(a) $2x - 5$, $4x^2 - 20x + 25$
(b) $y^2 + y - 2$, $y^2 + 3y + 2$
(c) $2m^2 + 5m - 3$, $3m^2 + 5m - 12$
(d) $a^2 - 25$, $a + 5$, $a - 5$
(e) $12p^2 + 3p - 42$, $12p^3 + 30p^2 + 12p$

5 (a) Simplify each term of the rational expression
$$\dfrac{3x - 3}{x^2 - 1} - \dfrac{2x + 2}{(x + 1)^2}.$$
(b) Find a common denominator for the expression in (a) and simplify.

6 First simplify each term of the expression
$$\dfrac{3y^2 - 9y}{y^2 - 6y + 9} - \dfrac{y^2 + 3y}{y^2 - 9}.$$
Then simplify the resulting expression.

7 Find the lowest common denominators of each of the following.
   (a) $\dfrac{2}{x^2 - xy} - \dfrac{3}{xy - y^2}$
   (b) $\dfrac{6a}{a^2 - 9} - \dfrac{2}{a - 3} + \dfrac{3}{a + 3}$
   (c) $\dfrac{6}{2y - 1} - \dfrac{3}{2y + 1} - \dfrac{8y - 2}{1 - 4y^2}$
   (d) $\dfrac{3x}{(x + 1)(x + 2)} - \dfrac{4x}{(x + 2)(x + 3)} + \dfrac{5x}{(x + 3)(x + 1)}$

B

8 Simplify each of the following.
   (a) $\dfrac{y + 1}{5} - \dfrac{y}{4}$
   (b) $\dfrac{6a}{6} - \dfrac{3a}{9} + \dfrac{a}{12}$
   (c) $\dfrac{1}{x} - \dfrac{1}{3x}$
   (d) $\dfrac{1}{y^2} - \dfrac{3}{2y}$
   (e) $\dfrac{4}{a + 1} - \dfrac{1}{a}$
   (f) $\dfrac{3}{p} - \dfrac{2}{p(p + 1)}$
   (g) $\dfrac{3}{x - 2} - \dfrac{4}{x + 1}$
   (h) $\dfrac{x - 2}{3} - \dfrac{x + 1}{4}$
   (i) $\dfrac{2a - 3b}{a} - \dfrac{3a + 2b}{b}$
   (j) $\dfrac{a}{2a - 3b} - \dfrac{b}{3a + 2b}$
   (k) $\dfrac{m + x}{m} + \dfrac{m^2 - x^2}{mx} - \dfrac{m - x}{x}$

9 Simplify.
   (a) $\dfrac{x - y}{x + y} - \dfrac{2xy}{x^2 - y^2}$
   (b) $\dfrac{3a - 1}{a^2 - 9} - \dfrac{5}{a - 3}$
   (c) $\dfrac{a - b}{a + b} - \dfrac{a + b}{a - b}$
   (d) $\dfrac{3}{x^2 - xy} - \dfrac{3}{xy - y^2}$
   (e) $\dfrac{x + y}{x - y} - \dfrac{x - y}{x - 2y}$
   (f) $\dfrac{x - 4}{4} + \dfrac{4}{x - 4} - \dfrac{1}{4x - 16}$

10 (a) Simplify $\dfrac{2y - 3}{1 - y} - \dfrac{2y + 7}{y - 1}$.
   (b) Find the value of the expression in (a) when $y = -1$.

11 (a) Simplify $\dfrac{5}{x^2 - 3x - 18} - \dfrac{2}{x^2 + x - 6}$.
   (b) Find the value of the expression in (a) if $x = -\dfrac{2}{3}$.

12 Find the value of each expression for $y = -1$.
   (a) $\dfrac{y}{y^2 + 3y} + \dfrac{1}{y + 3}$
   (b) $\dfrac{y + 2}{y^2 - 4} - \dfrac{3}{2 - y}$

13 Write each of the following as a single rational expression.
   (a) $\dfrac{2x}{3x - 3} + \dfrac{3x^2}{5x + 5}$
   (b) $\dfrac{4}{3x + 2} + \dfrac{3x - 1}{3x - 2}$
   (c) $\dfrac{2a}{a^2 + a - 20} - \dfrac{3}{a^2 + 6a + 5}$
   (d) $\dfrac{3}{2m - 5} - \dfrac{4}{4m^2 - 20m + 25}$
   (e) $3(y + 3)\left(\dfrac{2y}{y + 3} - \dfrac{3}{4}\right)$
   (f) $\dfrac{2y^2 + x^2}{x^4 - y^4} - \dfrac{1}{x^2 - y^2} + \dfrac{1}{y^2 + x^2}$
   (g) $\dfrac{3}{(m + 1)^2} - \dfrac{4}{m^2 - 1} - \dfrac{1}{m^2 - 2m + 1}$
   (h) $6x(x - 5)\left(\dfrac{x}{x - 5} - \dfrac{6}{x}\right)$

14 If $g(x) = \dfrac{2}{x + 4} - \dfrac{12 - 3x}{x^2 - 16}$, how does the value of $g(x)$ change if x increases from -2 to -1?

15 If $f(y) = \dfrac{3 - y}{y + 2} - \dfrac{2y^2 + 7}{2 - y - y^2}$ then how does the value of $f(y)$ change if y increases from 3 to 4?

16 Simplify each of the following.
   (a) $\dfrac{\dfrac{2}{3} - \dfrac{1}{6y}}{\dfrac{1}{5y} + \dfrac{2}{3}}$
   (b) $\dfrac{3 - \dfrac{2}{3y}}{5 - \dfrac{1}{2}y}$

solving problems: reading accurately

In solving problems, we must read the problem accurately. Questions 17 to 26 provide practice in reviewing language and in reading accurately to interpret a problem and translate it correctly into algebra.

17 Subtract $\dfrac{1}{1 - 9x^2}$ from $\dfrac{2}{1 + 3x}$.

18 Add $\dfrac{3}{x + 2}$ to the sum of $\dfrac{2}{x}$ and $\dfrac{2}{x - 2}$.

19 Decrease the sum of $\dfrac{3x}{x^2 - 9}$ and 1 by $\dfrac{x}{x - 3}$.

20 By how much does $\dfrac{x^2 + 2}{x^2 - 16}$ exceed the sum of $\dfrac{x}{x + 4}$ and $\dfrac{x}{x - 4}$?

21 Subtract the sum of $\dfrac{2}{x + 3}$ and $\dfrac{5}{x - 3}$ from $\dfrac{6}{x^2 - 9}$.

22 Find the result of $\dfrac{x - 3}{x - 3}$ subtracted from $\dfrac{x + 3}{x - 3}$.

23 The product of $x - 1$ and $\dfrac{1}{x + 1}$ is subtracted from $\dfrac{2x}{x^2 + 7x + 6}$.

24 Find the product of $\dfrac{x}{x^2 + 7x + 10}$ and the sum of $\dfrac{x^2 + 5x}{x - 3}$ and $-\dfrac{x^2 - x - 6}{x}$.

25 $4x^2 - 25$ is divided by $2x - 5$ and then increased by $\dfrac{2}{2x + 5}$.

26 By how much does $x + 1$ divided by $3x^2 + 6x + 3$ exceed $\dfrac{1}{x^2 + 9x + 8}$?

2.9 using the factor theorem

Often in mathematics as we solve problems or practise skills we may make observations that allow us to develop other useful mathematical ideas. Often we may interpret our results of problems in a different way and thus develop additional strategies to solve problems.
For example, we are able to factor the trinomial using previously acquired skills.

$$\text{Let } f(x) = x^2 - 3x + 2$$
$$= (x - 1)(x - 2)$$

We note that

$$f(1) = (1 - 1)(1 + 2) \qquad f(2) = (2 - 1)(2 - 2)$$
$$= 0 \qquad\qquad\qquad\qquad = 0$$

It seems that if

$f(1) = 0$ then $x - 1$ is a factor.
$f(2) = 0$ then $x - 2$ is a factor.

Using the above observation, we may test other examples.

$$f(y) = 2y^2 - 5y + 2$$
$$f(2) = 2(2)^2 - 5(2) + 2$$
$$= 8 - 10 + 2$$
$$= 0$$

Is $y - 2$ a factor of the expression $2y^2 - 5y + 2$?

By factoring, we find that
$$2y^2 - 5y + 2 = (y - 2)(2y - 1).$$

Thus $f(2) = 0$ and $y - 2$ is a factor. Therefore it seems we may use this method to find the factor of a polynomial.

The nature of mathematics is such that we ought to determine whether the above procedure may be extended to obtain the factors of a cubic expression.

Let $g(x) = x^3 - 5x^2 + 7x - 3$.

Test $x = 1$

$$g(1) = (1)^3 - 5(1)^2 + 7(1) - 3$$
$$= 1 - 5 + 7 - 3$$
$$= 0$$

Thus it seems that $x - 1$ is a factor of $g(x)$.

We then divide to test our result.

$$\begin{array}{r} x^2 - 4x + 3 \\ x - 1 \overline{\smash{\big)}\, x^3 - 5x^2 + 7x - 3} \\ \underline{x^3 - x^2} \\ -4x^2 + 7x \\ \underline{-4x^2 + 4x} \\ 3x - 3 \\ \underline{3x - 3} \\ 0 \end{array}$$

Thus
$(x - 1)(x^2 - 4x + 3)$
$= x^3 - 5x^2 + 7x - 3$

The above results are stated as the following theorem.

Factor Theorem
$x - a$ is a factor of $f(x)$ if and only if $f(a) = 0$.
 A: If $f(a) = 0$ then $x - a$ is a factor.
 B: If $x - a$ is a factor, then $f(a) = 0$.

Example 1
Factor $y^3 - 4y^2 + y + 6$.

Solution
Use the factor theorem. Test values of y:

$y = 1 \quad f(y) = y^3 - 4y^2 + y + 6$
$ f(1) = (1)^3 - 4(1)^2 + (1) + 6$
$ = 1 - 4 + 1 + 6$
$ = 4 \neq 0$

Thus, $y - 1$ is not a factor.

$y = -1 \quad f(-1) = (-1)^3 - 4(-1)^2 + (-1) + 6$
$ = -1 - 4 - 1 + 6 = 0$

Thus, $y + 1$ is a factor.
By dividing we find that
$y^3 - 4y^2 + y + 6 = (y + 1)(y^2 - 5y + 6)$
$ = (y + 1)(y - 3)(y - 2)$

$$\begin{array}{r} y^2 - 5y + 6 \\ y + 1 \overline{\smash{\big)}\, y^3 - 4y^2 + y + 6} \\ \underline{y^3 + y^2} \\ -5y^2 + y \\ \underline{-5y^2 - 5y} \\ 6y + 6 \\ \underline{6y + 6} \\ 0 \end{array}$$

The factor theorem states that if $x - a$ is a factor then $f(a) = 0$. We may use this result to complete the next example.

Example 2
If $2x^3 + kx^2 - 3x - 6$ is divisible by $x + 2$, find the value of k.

Solution
Since $(x + 2)$ is a factor of $f(x)$ then $f(-2) = 0$.

$f(x) = 2x^3 + kx^2 - 3x - 6$
$f(-2) = 2(-2)^3 + k(-2)^2 - 3(-2) - 6$
$ = 2(-8) + 4k + 6 - 6$
$ = -16 + 4k$

Use $f(-2) = 0$. Then
$-16 + 4k = 0$
$4k = 16$
$k = 4$

To check whether $k = 4$ is correct divide
$2x^3 + 4x^2 - 3x - 6$
by $x + 2$.

sum and difference of cubes

We may use the factor theorem to develop other factoring rules. For example

$f(x) = x^3 - 27 \qquad f(x) = x^3 + 27$
$f(3) = (3)^3 - 27 \qquad f(-3) = (-3)^3 + 27$
$ = 0 \qquad\qquad\quad = 0$
$x - 3$ is a factor $\qquad x + 3$ is a factor
of $x^3 - 27$. $\qquad\qquad$ of $x^3 + 27$.

Divide to find the other factor.
$x^3 - 27$
$= (x - 3)(x^2 + 3x + 9)$

Divide to find the other factor.
$x^3 + 27$
$= (x + 3)(x^2 - 3x + 9)$

└── Cannot be factored ──┘
further

In general, we write
$x^3 + y^3 = (x + y)(x^2 - xy + y^2)$
$x^3 - y^3 = (x - y)(x^2 + xy + y^2)$

We refer to the above results as factoring a *sum or difference of cubes*. Note that a sum (or difference) of cubes of two numbers is divisible by the sum (or difference) of the numbers themselves.

2.9 exercise

Questions 1 to 8 examine skills needed to apply the factor theorem.

A

1. (a) Find the value of $f(1)$ if $f(x) = x^2 - 9x + 8$.
   (b) What is one factor of $x^2 - 9x + 8$?

2. (a) Find the value of $f(-2)$ if $f(x) = x^3 - 5x^2 - 2x + 24$.
   (b) What is one factor of $f(x)$?
   (c) Write the other factor of $f(x)$.

3. (a) Find the value of $g(2)$ if $g(x) = x^3 + x^2 - 12$.
   (b) What is one factor of $x^3 + x^2 - 12$?

4. Use $(x - 3)(x - 1)(x + 2) = x^3 - 2x^2 - 5x + 6$.
   (a) How are 3, 1, and -2 related to the polynomial?
   (b) How may we use the information in (a) to find the factors of $x^3 - 2x^2 - 5x + 6$?

5. A polynomial is given by
   $f(x) = x^3 - 16x^2 - 19x + 34$.
   (a) What are the possible combinations of factors of 34?
   (b) How may the information in (a) be used to find the factors of $f(x)$?
   (c) Find the factors of $f(x)$.

6. Use the factors of the constant term to help write the factors of each polynomial.
   (a) $x^3 - 6x^2 + 11x - 6$
   (b) $y^3 - y^2 - 9y - 12$
   (c) $27m^3 + 9m^2 - 3m - 10$

7. Without dividing show that the factors of $x^3 - 4x^2 + x + 6$ are $x + 1, x - 2, x - 3$.

8. (a) If $f(-3) = 0$, then what is one factor of $f(x)$?
   (b) If $f(a) = 0$, then what is one factor of $f(x)$?
   (c) If $x + 3$ is a factor of $f(x)$, what is the value of $f(-3)$?
   (d) If $f(0) = 0$, what is a factor of $f(x)$?

To apply the factor theorem, we must also be able to divide polynomials.

9. Find the remainder when
   (a) $3x^2 + 5x - 4$ is divided by $x + 2$.
   (b) $x^3 - 9x^2 + 26x - 38$ is divided by $x - 2$.
   (c) $y^3 - 6y^2 + 11y - 6$ is divided by $y - 3$.

10. Find the quotient for each of the following.
    (a) $(y^2 + 5y + 6) \div (y + 2)$
    (b) $(2x^2 - 9x - 5) \div (x - 5)$
    (c) $(y^3 - 3y^2 + 3y - 1) \div (y - 1)$
    (d) $(27m^3 + 9m^2 - 3m - 10) \div (3m - 2)$
    (e) $(4a^3 + 7a^2 - 3a - 15) \div (4a - 5)$
    (f) $(a^3 - 6a - 9) \div (a - 3)$

11. Divide.
    (a) $y^3 + 6y^2 + 11y + 6 \div (y + 2)$
    (b) $3m^3 + 4m^2 + 5m + 18 \div (m + 2)$
    (c) $y^3 - 7y + 6 \div (y - 1)$

    Write in descending powers of m.
    (d) $m^3 + 7m + 5m^2 + 3 \div (m + 3)$
    (e) $y^3 - 9y - 4y^2 + 36 \div (y - 4)$
    (f) $6k^3 + 15k - 17k^2 - 4 \div (3k - 4)$
    (g) $x^3 - 8 \div (x - 2)$
    (h) $y^3 + 27 \div (y + 3)$

> ## math tip
> The remainder theorem states:
> > If $f(x)$ is divided by $x - m$, the remainder is $f(m)$.
>
> Use this theorem to check the remainders for our previous work.
> > Why is the factor theorem a special case of the remainder theorem?

B

12. If $f(-2) = 0$, factor fully
    $f(x) = x^3 - 16x^2 - 19x + 34$.

13. If $g(2) = 0$, factor fully
    $g(x) = x^3 - 3x^2 + 4x - 4$.

14. (a) Why is $x + 1$ not a factor of
    $x^3 - 6x^2 + 11x - 6$?
    (b) Find the factors of $x^3 - 6x^2 + 11x - 6$.

15. Without factoring, determine if $x + 2$ is a factor of $f(x) = 3x^2 + 4x - 4$.

16. Find a common factor of each of the four polynomials. $x^2 - 9$, $x^2 + 6x + 9$, $x^2 + 2x - 3$, $2x^2 + x - 15$.

17. Show that $x + a$ is a factor of $x^3 + a^3$.

18. (a) Use the factor theorem to determine whether $x - 5$ is a factor of the polynomial
    $x^4 + x^3 - 31x^2 - x + 30$.
    (b) Find the other factors.

19. A cubic polynomial $f(x)$ has zero value when $x = -3$, $x = 2$ and $x = 5$. Find an expression for $f(x)$.

20. A polynomial $h(y)$ vanishes (has zero value) when $y = -1$, $y = 1$, $y = 2$. Find an expression for $h(y)$.

21. A polynomial $g(x)$ has the properties $g(-3) = 0$, $g(3) = 0$, $g(-1) = 0$. Find $g(x)$.

22. If $x - a$ is a factor of $f(x)$, find the value of $(f(a) - a)^2$.

23. (a) Find a factor of
    $P(x) = 2x^4 - 3x^3 + x^2 + 85x + 3$.
    (b) What are the other factors?

24. Factor fully.
    (a) $x^3 - x^2 - 9x - 12$
    (b) $2m^3 - 3m^2 - 17m - 12$
    (c) $x^3 + 5x^2 + 7x + 3$
    (d) $3x^3 - 4x - 7x^2 - 76$
    (e) $8 - 6y^3 - y^2 - 14y$

25. Find k so that
    (a) $x^3 + kx + 2x + 1$ is divisible by $x + 1$.
    (b) $kx^3 - 3x^2 + x - 3$ is divisible by $x - 3$.
    (c) $2x^3 + kx^2 - 3x - 6$ is divisible by $x + 2$.

26. Find k so that
    (a) $x^3 - kx + 6$ is divisible by $x - 1$.
    (b) $x^4 + kx^3 - kx - 1$ is divisible by $x - 1$.
    (c) $kx^4 - 11x^3 + 6kx^2 + x - 4$ is divisible by $x - 4$.

27. $f(x) = kx^3 - 8x^2 - x + 3k + 1$ is divisible by $x - 2$.
    (a) Find the value of k.
    (b) Find the other factors.

28. $g(-3) = -80$ and $g(x) = kx^3 + kx^2 + 5x - 1$
    (a) Find the value of k.
    (b) Find the other factors of $g(x)$.

29. (a) Substitute $-y$ for x in $x^3 + y^3$. What result should you expect?
    (b) Write the factors of $x^3 + y^3$.

30. Find the factors of each of the following.
    (a) $y^3 + 1$ (b) $x^3 + 8$ (c) $y^3 - 1$
    (d) $x^3 - 8$ (e) $8x^3 + 1$ (f) $8y^3 - 1$
    (g) $x^3 + 64$ (h) $x^3 - a^3$ (i) $x^3 + y^3$
    (j) $27x^3 + y^3$ (k) $1 - 27y^3$ (l) $\dfrac{y^3}{8} + 1$

31 Factor fully. Look for common factors first.
(a) $a^3 + 64b^3$ (b) $2x^3 - 2$
(c) $2 - 54x^3$ (d) $y^4 + y$
(e) $x^6 - y^6$ (f) $27a^3 - 125$
(g) $216m^3 - 27$ (h) $1000m^3 - y^3$

If $f(m) = 0$ then we say that m is a **zero** of $f(x)$.
For example.
$f(x) = x^3 - 3x^2 + 4x - 4$
$f(2) = (2)^3 - 3(2)^2 + 4(2) - 4$
$= 0$
Thus, we say 2 is a zero of $f(x)$.

C
32 (a) Use the factor theorem to determine a zero of $f(x) = x^2 - 8x + 7$.
(b) What is another zero of $f(x)$?

33 Use the factor theorem to determine which of the three values $-2, 1, \frac{1}{2}$, is a zero of $g(x) = 2x^3 + 5x^2 + x - 2$.

34 Find a zero of each of the following.
(a) $g(x) = x^2 + x - 12$
(b) $h(y) = y^3 + 5y^2 + 7y + 3$
(c) $f(x) = x^3 + x^2 - 9x - 9$
(d) $g(x) = 2x^3 - 16$

35 Find a common zero of each of the polynomials.
$x^2 - 25$, $x^2 - 10x + 25$, $x^2 - 4x - 5$, $3x^2 - 13x - 10$.

applications: solving cubic equations

To solve a quadratic equation, we found the factors.
$$x^2 + 3x + 2 = 0$$
$$(x + 1)(x + 2) = 0$$
$$x + 1 = 0 \quad \text{or} \quad x + 2 = 0$$
$$x = -1 \quad \text{or} \quad x = -2$$
To solve a cubic equation, we extend the same principle.

$x^3 - 2x^2 - 5x + 6 = 0$
$(x - 1)(x - 3)(x + 2) = 0$

Use the factor theorem $f(1) = 0$, and so on. Thus the factors are $(x - 1)(x - 3)(x + 2)$.

Thus
$x - 1 = 0$ or $x - 3 = 0$ or $x + 2 = 0$
$x = 1$ $x = 3$ $x = -2$
Thus the solution set of the cubic equation is $\{1, 3, -2\}$.
The roots are $1, 3, -2$.

Finding roots of cubic equations is an important skill which will help us to draw graphs of cubic functions in our later work.

36 (a) Find the factors of $x^3 + 2x^2 - 9x - 18$.
(b) Solve the cubic equation $x^3 + 2x^2 - 9x - 18 = 0$.

37 (a) Show that 3 is a root of the equation $x^3 - 2x^2 - 5x + 6 = 0$.
(b) Find the other roots.

38 Solve and verify
$2x^3 - x^2 - 3x + 2 = 0$, where $x \in Q$.

39 Solve for x, $x \in Q$.
(a) $x^3 - 3x^2 - x + 3 = 0$
(b) $x^3 + 5x^2 + 7x + 3 = 0$
(c) $x^3 - 4x^2 - 9x + 36 = 0$
(d) $2x^3 + 5x^2 - 4x - 3 = 0$

40 (a) What is the first step in solving $x(1 - 4x^2) + 12x^3 = 3$?
(b) Solve the equation in (a).

41 Solve each equation, $x \in Q$.
(a) $x^3 - 2x^2 - x + 2 = 0$
(b) $x^3 - x = 2(1 - x^2)$ (c) $3(x^2 - 4) = 4x - x^3$
(d) $5x^2 - x^3 - 2 - 2x = 0$
(e) $2(x^3 - 1) = x^2 + 5x$

2.10 solving equations: rational expressions

Our later work in mathematics is often based on skills we have learned earlier. We use the following strategy a great deal in our work.

> What we have learned
> is applied
> to develop new skills.

For example, to solve the following equation involving **rational numbers**, we multiply by the least common multiple of the denominators, to simplify our calculations.

$$\frac{y + 6}{2} = \frac{y - 2}{3} + 5$$

Multiply both sides by the least common denominator, namely, 6.

$$6\left(\frac{y + 6}{2}\right) = 6\left(\frac{y - 2}{3}\right) + 6(5)$$
$$3(y + 6) = 2(y - 2) + 30$$
$$3y + 18 = 2y - 4 + 30$$
$$3y - 2y = 8$$
$$y = 8$$

[We multiply all terms occurring in the equation by 6.]

We refer to equations involving rational expressions as **fractional equations.** In these equations, sometimes the variables occur in the denominator. As we have seen earlier, certain restrictions occur when variables occur in the denominator. We may extend our earlier work in solving equations to learn to solve fractional equations.

To solve a fractional equation that involves **rational expressions,** we multiply by the least common multiple of the denominator.

To solve the fractional equation

$$\frac{y + 6}{y} = \frac{y + 2}{y - 2} - 2$$

Compare the following solution to the one above.

multiply both sides by the least common denominator, namely $y(y - 2)$.
Since we can only multiply by non-zero expressions then $y(y - 2) \neq 0$ or $y \neq 0$ and $y \neq 2$.

$$y(y - 2)\frac{(y + 6)}{y} = y(y - 2)\frac{y + 2}{y - 2} - 2(y)(y - 2)$$
$$(y - 2)(y + 6) = y(y + 2) - 2(y)(y - 2)$$
$$y^2 + 4y - 12 = y^2 + 2y - 2y^2 + 4y$$
$$2y^2 - 2y - 12 = 0$$
$$y^2 - y - 6 = 0$$
$$(y - 3)(y + 2) = 0$$
$$y - 3 = 0 \quad \text{or} \quad y + 2 = 0$$
$$y = 3 \qquad\qquad y = -2$$

Check the roots that you obtain in the *original* equation.

Thus, the roots are 3, and −2.

Example 1

Solve $\dfrac{2m - 3}{m + 1} = -3$.

Solution

Multiply both sides by $m + 1$, $m \neq -1$.

$$\frac{2m - 3}{m + 1} = -3$$
$$(m + 1)\frac{2m - 3}{m + 1} = -3(m + 1)$$
$$2m - 3 = -3m - 3$$
$$5m = 0$$
$$m = 0$$

You may verify the root quickly in the original equation.

Thus, the root of the equation is 0.

In Example 1, the restriction we placed on the variable m was that $m \neq -1$. To solve the equation in Example 2, we need again to multiply by the least common multiple.

Example 2

Solve and verify $\dfrac{y}{y-2} = \dfrac{y^2+3y}{y^2-4} - \dfrac{3}{y+2}$.

Solution

Multiply both sides of the equation by the least common multiple of the denominator, namely $(y-2)(y+2)$ where $y \neq 2$, $y \neq -2$.

$(y-2)(y+2)\dfrac{y}{y-2}$
$= (y-2)(y+2)\dfrac{y^2+3y}{y^2-4} - (y-2)(y+2)\dfrac{3}{y+2}$

$y(y+2) = y^2 + 3y - 3(y-2)$
$y^2 + 2y = y^2 + 3y - 3y + 6$
$2y = 6$
$y = 3$

Verify in the original equation.

Record the expression for the original left and right side.

$\text{LS} = \dfrac{y}{y-2}$ $\text{RS} = \dfrac{y^2+3y}{y^2-4} - \dfrac{3}{y+2}$

$= \dfrac{3}{3-2}$ $= \dfrac{(3)^2+3(3)}{(3)^2-4} - \dfrac{3}{3+2}$

$= 3$ $= \dfrac{9+9}{9-4} - \dfrac{3}{5}$

$= \dfrac{18}{5} - \dfrac{3}{5} = 3$

LS = RS   checks

Thus, the required root of the equation is 3.

In learning to solve word problems
- we learn skills in algebra.
- we use these skills to solve equations.
- then we apply our skills with equations to solve problems.

We learn to solve equations in this section and then use these skills in the next section to solve problems.

2.10 exercise

Questions 1 to 9 develop skills in solving equations involving rational expressions.

A

1. Write the least common multiple of each of the following.
   (a) $x+1$, x^2-1 (b) $3y$, $3y+5$
   (c) $m+6$, m^2-36 (d) $m-3$, m^2-6m+9
   (e) k, $k-1$, k^2-k (f) x^2-3x+2, x^2-1
   (g) y^2-9, y^2-5y+6

2. (a) What might be the first step in solving
   $\dfrac{2}{y-1} = \dfrac{2}{y+1} - \dfrac{1}{y}$?
   (b) Solve the equation in (a).

3. (a) Solve the equation $\dfrac{x+2}{3} = \dfrac{x+6}{4}$.
   (b) Solve the equation $\dfrac{x+2}{x+3} = \dfrac{x+6}{x+4}$.
   (c) How are the solutions in (a) and (b) alike? How are they different?

4. (a) Solve the equation
   $\dfrac{y+18}{4} = y + 2 - \dfrac{y-5}{2}$.
   (b) Solve the equation
   $\dfrac{y+18}{y} = y + 2 - \dfrac{y-5}{y-2}$.
   (c) How are the solutions in (a) and (b) alike? How are they different?

5. (a) Solve the following equation by finding the least common denominator of the terms.
   $\dfrac{6}{a-3} - \dfrac{2}{a-2} = 5$.
   (b) Solve the equation in (a) by multiplying both sides by $(a-3)(a-2)$.
   (c) Which method do you prefer, (a) or (b)?

6 For each of the following equations,
   - write the least common multiple of the denominators, before you solve any of the equations.
   - solve the equations.

   (a) $\dfrac{y-5}{y-3} = \dfrac{2}{y} + 1$ (b) $\dfrac{a-2}{a} + \dfrac{4}{5a} = \dfrac{1}{5}$

   (c) $\dfrac{5-m}{2+m} = \dfrac{2m-2}{m+2} - 2$ (d) $\dfrac{3}{5} = \dfrac{2s+3}{s-1} - 2$

   (e) $2 - \dfrac{y-5}{y+2} = \dfrac{y-2}{y+1}$ (f) $\dfrac{5m}{3(m-4)} = \dfrac{m}{m-4} + 5$

7 For each of the following equations,
   - indicate clearly the restrictions on the variables.
   - solve the equations.

   (a) $\dfrac{2y-1}{3y} = 1$ (b) $2 + \dfrac{3}{m} = \dfrac{2m+3}{m-1}$

   (c) $\dfrac{s-1}{2s+1} = 6$ (d) $\dfrac{k+1}{k+3} = \dfrac{k-1}{k}$

   (e) $4 + \dfrac{3m}{m^2-1} = \dfrac{m}{m+1}$

   (f) $\dfrac{5}{y(y-1)} + \dfrac{2}{y} = \dfrac{y+4}{y-1}$

8 (a) Solve for y.
   $$\dfrac{5}{y-2} = \dfrac{3}{y+6} + \dfrac{4}{y}$$
   (b) How would you check your work in (a)?

9 Solve for m and verify.
   $$\dfrac{7}{m-3} = \dfrac{1}{2} + \dfrac{3}{m-4}$$

B

10 Solve and verify.

   (a) $\dfrac{3m+1}{2m} = 2$ (b) $\dfrac{3}{a+5} = \dfrac{2}{a+4}$

   (c) $\dfrac{x}{3x+2} = 2$ (d) $\dfrac{y+2}{y-3} = \dfrac{y}{y+5}$

   (e) $\dfrac{p-8}{p} = \dfrac{p+7}{p-15}$ (f) $\dfrac{m-2}{m-1} = \dfrac{m+4}{m+7}$

11 Find the solution set of each of the following.

   (a) $\dfrac{5x+7}{3x+2} = x - 1$ (b) $\dfrac{3m+2}{3m+4} = \dfrac{m-1}{m+1}$

   (c) $\dfrac{3k}{k-3} + 2 = \dfrac{3k-1}{k+3}$ (d) $\dfrac{m+5}{m-3} = \dfrac{m+3}{m-4}$

   (e) $\dfrac{x^2+x}{x-1} - \dfrac{x+1}{x} = \dfrac{x+1}{x^2-x}$

   (f) $2 - \dfrac{m+2}{m-1} = \dfrac{m-4}{m+1}$

12 Solve.

   (a) $y - \dfrac{3}{2} = \dfrac{7}{y}$ (b) $\dfrac{m}{2} - \dfrac{1}{2} = 3 - \dfrac{2m}{3}$

   (c) $\dfrac{2}{p} - \dfrac{1}{2} = 2 - \dfrac{p}{2}$ (d) $\dfrac{2}{3k} + \dfrac{3}{5} = \dfrac{1}{4k}$

   (e) $\dfrac{y+1}{y-2} = \dfrac{y+3}{y-4}$ (f) $\dfrac{2}{y} + \dfrac{5}{y(y-1)} = \dfrac{y+4}{y-1}$

   (g) $\dfrac{3m}{m-3} + 2 = \dfrac{3m-1}{m+3}$

   (h) $\dfrac{m+8}{m+3} = \dfrac{m}{m-3}$

   (i) $\dfrac{5}{y+2} + \dfrac{y(y+3)}{y^2-4} = \dfrac{y}{y-2}$

   (j) $\dfrac{3m}{m-3} + 2 = \dfrac{3m-1}{m+3}$

   (k) $\dfrac{2k}{k+3} - \dfrac{3k-4}{2(k+3)} = \dfrac{5}{k+3}$

   (l) $\dfrac{y}{y-2} = \dfrac{5}{y+2} + \dfrac{y^2+3y}{y^2-4}$

   (m) $\dfrac{x-2}{x^2-1} + \dfrac{x-5}{x^2+x-2} = \dfrac{x}{x-1}$

13 Solve and verify.
   $$\dfrac{3}{m-5} + \dfrac{1}{5-m} = \dfrac{8}{7-m} + \dfrac{5}{m-7}$$

   In solving the above equations, you were asked to verify the roots.

14. (a) Write the restrictions for the following equation.
$$\frac{y+2}{2} + \frac{y^2+5}{y-5} = \frac{11y^2 - 23y - 10}{y^2 - 5y}$$
(b) Solve the equation in (a). Why does the equation have no roots?

15. For each equation,
   - write the restrictions first
   - then, find the solution set.

   (a) $\dfrac{x^2+2x}{x+3} + \dfrac{x+2}{x} = \dfrac{3x+6}{x^2+3x}$

   (b) $\dfrac{x^2+1}{x-2} - \dfrac{x}{x+2} = \dfrac{2x^2+3x-2}{x^2-4}$

   (c) $\dfrac{x^2-x}{x+1} + \dfrac{x-1}{x+1} = \dfrac{x^2-1}{x^2+x}$

   (d) $\dfrac{x^2+1}{x-2} - \dfrac{x}{x+2} = \dfrac{5x+2}{x^2-4}$

16. (a) To solve the following equation, write each side with a common denominator.
$$\frac{1}{k-7} - \frac{1}{k-2} = \frac{1}{k-10} - \frac{1}{k-5}$$
(b) Solve the equation in (a).

17. Solve $\dfrac{m+1}{m-1} + \dfrac{m+2}{m-2} = \dfrac{m+5}{m-2} - 1$.

18. Solve $\dfrac{a+2}{a+3} - \dfrac{a+3}{a+4} = \dfrac{a+5}{a+6} - \dfrac{a+6}{a+7}$.

---

problem/matics

To get a hole-in-one at the mini golf course you have to "bank" the ball.

Bank the ball once to get a hole-in-one.   Bank the ball twice to get a hole-in-one.

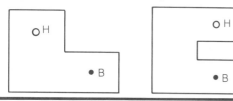

2.11 applying our skills: solving problems

To solve a problem we must not only organize our work carefully, but also answer these two important questions.

- What do we know? (What information is given in the problem?)
- What are we asked to find?

To solve more difficult problems, we use the *Steps for Solving Problems* developed in Chapter 1, page 32. The methods we learn for solving problems may be applied over and over again.

As an aid in solving problems, we may use a chart to sort the given information in a problem. Once the given information is sorted, we may then use our skills with equations to solve the problem. In solving problems about motion we need to know that

$$d = vt, \quad \frac{d}{v} = t, \quad \frac{d}{t} = v \quad \text{where } d = \text{distance}$$
$$v = \text{speed}$$
$$t = \text{time}$$

Example 1
The Olympic World record for the women's 600-m speed skating is 42.76 s. At the Olympic trials Jean knew that if she skated 3 m/s faster she could cut 10 s from her time.

(a) What is Jean's normal speed?
(b) If she skates at the faster rate, will she have a new Olympic record?

Solution
Step A Let Jean's normal speed be v m/s.
Then $(v + 3)$ m/s is the rate 3 m/s faster, $v > 0$. We use a chart to sort the given information.

	Distance (m)	Speed (m/s)	Time (s)
Normal speed	600	v	$\frac{600}{v}$
Faster speed	600	$v + 3$	$\frac{600}{v+3}$

We use $\frac{d}{v} = t$.

We use the fact that she made the trip 10 s faster.

Step B $\quad \dfrac{600}{v + 3} + 10 = \dfrac{600}{v}$

Step C Simplify the equation by multiplying by $v(v + 3)$ $v \neq 0, v \neq -3$.

$600v + 10v(v + 3) = 600(v + 3)$

$600v + 10v^2 + 30v = 600v + 1800$

$10v^2 + 30v - 1800 = 0$

$v^2 + 3v - 180 = 0$

$(v - 12)(v + 15) = 0$

$v - 12 = 0 \quad$ or $\quad v + 15 = 0$

$v = 12 \quad$ or $\quad v = -15$

This value is inadmissible, since the normal speed of Jean is positive.

Step D We check the answer.

Normal speed is 12 m/s.

Time taken $\dfrac{600}{12}$ s or 50 s.

Faster speed is 15 m/s.

Time taken $\dfrac{600}{15}$ s = 40 s.

Difference 10 s   Checks

Step E Make a final statement.
(a) Jean's normal speed is 12 m/s.
(b) The Olympic record is 42.76 s. At the faster rate Jean will make a new Olympic record of 40 s.

In Example 1, we needed to interpret our results carefully and identify the inadmissible value. In solving problems, we will also obtain values that do not satisfy the original problem. For example, to solve the problem,

Find two consecutive integers such that, if twice the larger is added to 3 times the square of the smaller, the sum will be 58.

we write the following equation.

$$2x + 3(x - 1)^2 = 58$$
$$2x + 3(x^2 - 2x + 1) = 58$$
$$2x + 3x^2 - 6x + 3 - 58 = 0$$
$$3x^2 - 4x - 55 = 0$$
$$(3x + 11)(x - 5) = 0$$

Thus $3x + 11 = 0 \quad$ or $\quad x - 5 = 0$

$3x = -11 \quad$ or $\quad x = 5$

$x = -\dfrac{11}{3}$

Since $-\dfrac{11}{3}$ is not an integer then $x = -\dfrac{11}{3}$ *does not* satisfy the original problem.

We refer to this root as an **extraneous root**. Thus, we *must* check the answers we obtain in the original problem.

The skills we develop in solving equations and applying them to the solution of problems are important foundation skills needed to solve more advanced real-life problems.

2.11 exercise

1 To solve problems we need to translate accurately. Let n represent the number. Write an expression for each of the following in terms of n.

(a) the number increased by 4
(b) the number decreased by 7
(c) 4 times the number
(d) $\dfrac{1}{4}$ of the number, increased by 2
(e) the square of the number

(f) 5 less than $\frac{1}{2}$ the number squared

(g) twice the square of the number

(h) twice the square of the number, decreased by 3

2 One number exceeds another number by 5. If the sum of their squares is 433, find the two numbers.

3 The sum of two numbers exceeds twice the smaller number by 7. The sum of their squares is 709. Find the two numbers.

4 The smaller number is 8 less than the larger. The sum of the square of twice the smaller and the square of the larger is 76. Find the two numbers.

5 If a car travels 30 km/h slower, it will take 3 h more to travel 700 km. How fast is the car travelling now?

(a) Complete the chart.

	Distance (km)	Speed (km/h)	Time (h)
present rate	700	s	?
slower rate	700	s − 30	?

(b) Write a complete solution to the problem using the chart as an aid.

6 If Jan could increase her speed by 1 m in 10 s she could cut 20 s off her time for a 60-m swim. Find her present swimming speed.

(a) Copy and complete the chart.

	Distance (km)	Speed (m/s)	Time (s)
present rate	60	s	?
slower rate	60	s + 0.1	?

(b) Use the chart as an aid in solving the problem.

7 The wind is picking up on the lake. If the sailboat increases its speed by 2 km/h, a 30-km course will take 30 min less. Find the speed of the sailboat.

(a) Complete the chart.

	Distance (km)	Speed (km/h)	Time (h)
present rate	30	s	?
faster rate	30	s + 2	?

(b) Use the chart to help you write a complete solution to the problem.

For each of the following problems, use steps A, B, C, D, E, to help you organize your solutions.

8 A Greyhound bus travelled 540 km. Reducing its speed by 15 km/h would increase the journey time by 72 min. Find the speed of the bus.

9 Jim, a cross-country runner, trains all year round and finds that in the winter with slippery road conditions he slows down by 2 km/h, so that it takes him 2 h longer to run the 48-km course. How fast does he run

(a) in the winter?   (b) in the summer?

10 Supplies at the campsite became low so one of the campers had to go to the outfitters to pick up more food. If 3 campers had paddled instead of 2 they would have increased their speed by 2 km/h and saved 1 h. How long did it take the canoe to travel the 40-km distance to the outfitters?

11 Lesley is travelling from Edmonton to Winnipeg to visit friends. She can leave today by bus or she can leave tomorrow by express train which travels 25 km/h faster, saving her 2.8 h. Determine how long each trip would take her if the distance between Edmonton and Winnipeg is 1400 km.

12. Arnold and Sheila enter a 330-km bicycle race. Sheila rides 5 km/h faster than Arnold, but she gets a flat which takes her half an hour to repair. They both finish the race in a tie.
    (a) What are the speeds of Arnold and Sheila?
    (b) How long did each person take to cover the distance?

13. Two planes are 2500 km from an airport. It takes one plane flying 50 km/h faster 1 h 40 min less than the other plane to reach the airport. Find the speeds of the two planes.

14. Motorcycle racing is gruelling. One lap of a course is 650 m. At the start of the race Jason leaves 4 s after Gerard, but goes 5 m/s faster and finishes the lap 2.5 s sooner.
    (a) Determine their speeds.
    (b) Determine the time taken by the winner to cover the distance.

15. Samuel paddled a rubber dinghy 12 km down the French River. Paddling up the river took 20 min longer because he paddled 6 km/h slower than before. Find his speed down the river.

16. Greg, who lives in Montréal, is going to visit the famous Colosseum, which is 7200 km away in Rome. He can either leave now on an L-1011 or wait 5 h at the airport and take an Al'Italia 747 which travels 120 km/h faster than the L-1011. How long will it take him to get from Montréal to Rome if the length of time of the 747 flight is 3 h less than the L-1011 flight?

math tip

These patterns often occur in studying mathematics. Learn them!
$(a + b)^2 = a^2 + 2ab + b^2$
$(a - b)^2 = a^2 - 2ab + b^2$
$(a + b)(a - b) = a^2 - b^2$

skills review: factoring

Factor with rational factors, where possible, each of the following expressions. Look out for expressions that cannot be factored.

1. $4x^3 + x^2$
2. $3x - 9xy$
3. $6x^4 - 12x$
4. $9x^4 - 16y^2$
5. $x^2 + 9x + 20$
6. $5x^2 - 5y^2$
7. $y^2 - 13y + 42$
8. $6x^2 - 13x - 5$
9. $x^2 + xy - 12y^2$
10. $9x^2 + 27x + 8$
11. $-4 + 25x^2$
12. $y^3 - y - 6$
13. $3y^2 - 9y^3$
14. $5 + 6x + x^2$
15. $18x^2 - 25x - 3$
16. $9x^2 + 1$
17. $6x^2 - 28x - 10$
18. $x^4 - 64$
19. $m^4 + 3m^2 + 4$
20. $-(9x^2 - y^4)$
21. $x^4 - 3x^2 - 4$
22. $10x^2 + 7x + 1$
23. $4x^6 - y^6$
24. $x^2 + 4x - 21$
25. $2x^4 - 3x^2 - 2$
26. $y^4 + 2y^2 + 9$
27. $2x^2 - 4xy + 8x$
28. $x^2 - 121$
29. $2x^2 - 2x - 28$
30. $(2x + y)^2 - z^2$
31. $m^4 - 6m^2 - 27$
32. $4y^4 - 16y^2 + 9$
33. $x^4 + 2x^2 - 15$
34. $8y^3 + 1$
35. $9x^{10} - 4$
36. $2y^2 + 24y + 40$
37. $5x^2 - 20$
38. $3x^2 - 27x + 54$
39. $x^4 - 225y^2$
40. $4x^2 - 28x - 32$
41. $x^6 - y^6$
42. $3x^5 + 15x^3 + 12x$
43. $-75 + 12x^4$
44. $3x^2yz^3 + 18xy^2$
45. $y^4 - 17y^2 + 16$
46. $4(x - y)^2 - (x + y)^2$
47. $(x - y)^2 - 9(2x + y)^2$
48. $m^2 + 6m + 9 - 4n^2$
49. $16y^2 - a^2 - 6ab - 9b^2$
50. $x^3 - 2x^2 - 9x + 18$

problems and practice: a chapter review

This section, found at the end of each chapter, will provide you with additional questions to check your skills with and understanding of the topics developed in the chapter. An important step for problem-solving is to decide which skills to use. For this reason, these questions are not placed in any special order. When you have finished the review, you might try the *Test for Practice* that follows.

1. (a) Simplify $(3x + 1)^2 - 2(x - 3)(2x + 6)$.
   (b) Use $x = 1$ to check your answer in (a).

2. Solve. Which of the following is an identity? Give reason(s) why.
   (a) $(a + 2)(4a - 6) - (2a - 1)^2 = 17$
   (b) $a(a + 4) + 8(a + 1) = (a + 2)^2 + 4(2a + 1)$
   (c) $(a + 5)^2 + 3a^3 - (2a + 1)(2a + 11) = 0$

3. Factor each of the following.
   (a) $2y + by + 2b + b^2$
   (b) $6m^2 + 3mn - 2qm - qn$

4. Each trinomial has 2 binomial factors. Find suitable values for f if $0 < f \leq 20$.
   (a) $x^2 - fx + 12$ (b) $x^2 - 7xy - fy^2$

5. Solve each of the following.
   (a) $3x^2 + 5x - 2 = 0$ (b) $1 - 5x = 6x^2$

6. Write the following in factored form.
   $9m^2 - 6m + 1 - k^2 - 8ky - 16y^2$

7. Ned jogged 15 km from his home at a steady pace and then jogged back 2 km/h slower. Find his average speed on the first part of his route if his time was 2 h 45 min.

8. Simplify.
   (a) $\dfrac{a + 1}{a - 1} \times \dfrac{a + 3}{1 - a^2} \div \dfrac{(a + 3)^2}{1 - a}$ where $a \neq \pm 1$
   (b) $\dfrac{2}{(x - 1)(x + 1)} - \dfrac{3}{(x - 1)^2}$ where $x \neq \pm 1$

9. Simplify.
   $\dfrac{15x^2}{x^2 + 3x + 2} \times \dfrac{x^2 - 4x - 5}{5x^2 - 25x} \div \dfrac{x^2 - 16}{x^2 + 8x + 16}$

10. Solve and verify.
    $\dfrac{5y}{y + 2} - 2 = \dfrac{y}{y - 2} - \dfrac{3y + 1}{y^2 - 4}$, $y \neq \pm 2$

11. Find k so that $kx^3 - 19x - 30$ is divisible by $x + 2$.

12. If $(x - 1)$ and $(x - 2)$ are factors of $x^4 + ax^3 - 8x^2 - bx + 16$, find a and b.

13. Rick and Charlie spent the weekend fishing at a cottage 300 km from home. Going home, Rick started out 30 min before Charlie and travelled 20 km/h slower than Charlie. If they arrived home at the same time, find their speeds.

problem/matics

If $\dfrac{a}{b} = \dfrac{a}{d}$, then $b = d$ was used to solve the following equation. What went wrong?

$$3 - \dfrac{12}{x + 1} = \dfrac{3x - 9}{x - 2}$$

$$\dfrac{3x + 3 - 12}{x + 1} = \dfrac{3x - 9}{x - 2}$$

$$\dfrac{3x - 9}{x + 1} = \dfrac{3x - 9}{x - 2} \quad \text{Equal numerators.}$$

$$x + 1 = x - 2 \qquad x + 1 = x - 2$$

$$1 = -2$$

test for practice

Try this test. Each *Test for Practice* is based on the mathematics you have learned in the preceding chapter. Try this test later in the year as a review. Keep a record of those questions that you were not successful with, get help in obtaining solutions and review them periodically.

1. Simplify each of the following.
   - (a) $(3x - 1)(2x + 5) - (3x + 5)(2x - 1)$
   - (b) $(2a - 3b)^2 - 3(4a + b)(5a - 3b)$

2. Solve each of the following. Which is an identity?
   - (a) $(y + 5)^2 + 5(2y - 1) = y(y + 10) + 10(y + 2)$
   - (b) $2(g + 2)(g + 1) = (2g + 5)(g + 1) - 2$

3. Find the missing factor.
   - (a) $-8x^6 = (\ ?\ )(-2x)$
   - (b) $-6m^3 + 3m^2n = -3m^2(\ ?\ )$
   - (c) $3x^3 - 6x^2 + 18x = (\ ?\ )(x^2 - 2x + 6)$

4. Find the missing factor.
   - (a) $2x^2 + 7x + 3 = (\ ?\ )(x + 3)$
   - (b) $3x^2 + 5x - 2 = (3x - 1)(\ ?\ )$

5. Factor fully.
   - (a) $36x^2 - y^2$
   - (b) $9 - 4k^2$
   - (c) $2a^2 - 17a + 21$
   - (d) $5x^2 + 22xy - 48y^2$
   - (e) $(x - y)^2 - (x + 2y)$

6. For the trinomial find the values of m so that two binomial factors are obtained.
   $x^2 - mx - 6$, $0 < m \leq 20$, $m \in W$

7. Write the following in factored form.
   $m^2 + n^2 + 2mn - x^2 - y^2 - 2xy$

8. Solve each of the following.
   - (a) $12x^2 - 31x - 15 = 0$
   - (b) $19x = 3x^2 - 14$
   - (c) $2x^2 - 18x + 40 = 0$
   - (d) $2(m^2 + 1) = 5m$

9. One number is 14 greater than another number. The sum of their squares is 260. Find the two numbers.

10. Simplify.
    - (a) $\dfrac{1 - 3y}{2y + 1} \div \dfrac{1 - 9y^2}{4y^2 - 1}$
    - (b) $\dfrac{4 - x}{x^2 - 4} \div \dfrac{x^2 - 16}{x^2 + 2x - 8}$
    - (c) $\dfrac{4k^2 - 36k + 72}{k^2 - 36}$
    - (d) $\dfrac{1}{y + 3} - \dfrac{3y - 1}{y^2 - 9} + \dfrac{4}{3 - y}$

11. Use the factor theorem to find k given that $x^3 - kx^2 - (kx + 1)$ is divisible by $x - 3$.

12. Factor fully $x^3 - 5x^2 - 2x + 24$.

13. Solve and verify $\dfrac{3p}{p - 1} + 4 = \dfrac{p}{p + 1}$.

14. Find an expression in simplified form if you decrease the sum of $\dfrac{4}{x^2 - xy}$ and $\dfrac{1}{x - y}$ by $\dfrac{1}{xy - y^2}$.

15. The area of a triangular bandage is 2400 cm². If the length of the bandage is 20 cm longer than the width, find the dimensions of the bandage.

16. Derek ran a 3000-m race at Boyd Conservation Area. Later he rode his bike along the same course, averaging 4 m/s faster than his running speed. The bike ride took 3 min 20 s less. How fast was Derek running in the race?

working with computers

We often need to do complex operations in subjects such as business, scientific experiments, weather forecasting, to name but a few. Fortunately, the computer has been invented, but before the advent of the computer, many computing devices had been invented to reduce the drudgery of tedious calculations.

- John Napier (1550-1617) invented Napiers Bones, and later logarithms.
- Pascal, in 1642, invented the first calculating machine. This machine had a limited number of operations.
- Leibniz, in 1671, and Morland, in 1673, invented calculating machines that could multiply.

Many other people contributed to the development of computing devices.

The early computers were bulky. For example, one of the forerunners of today's computers, ENIAC, was invented in 1945, occupied a room 9 m × 9 m, and had a mass of 27 000 kg. The development of electronic technology has resulted in new generations of machines that are faster and more reliable. For example, in 1946 it took about 70 h to calculate π to 2037 places; in 1961 it took 8.7 h to calculate π to 100 265 decimal places. Today's computers can perform *36 million operations in one second*.

Although a computer can do many things, it cannot think or plan for itself. We must program it to do things, and the computer, having been programmed, will do exactly as it is told. When we prepare instructions for a computer in some computer language we are *writing a program*. There are different computer languages for different families of computers. The one we will use is called BASIC.

These symbols are used in computer language.

+ add   ↑ exponential   ≠ not equal to
− subtract   > greater than   < less than
* multiply   >= greater than or equal to   <= less than or equal to

The following program is written in BASIC. Each statement is numbered. The computer will do the steps involved in each statement in order. The statement numbers are not written consecutively in case we may add other steps later. These statements are then coded in computer language on cards, or typed at a terminal which tells the computer what to do.

The following program, in Basic, finds the area of a trapezoid given by $A = \frac{1}{2}(a + b)h$.

```
10  PRINT "TRAPEZOID AREA"
11  INPUT A, B, H
20  LET S = (A + B)*H/2
30  PRINT A, B, H, S;
31  IF S = 0 THEN 50
40  GO TO 11
50  END
```

10 — The computer will print the information between " and ".

11 — Provide this information for the computer, called the INPUT.

30 — The computer provides the answers, called the OUTPUT.

40 — You may repeat the steps.

50 — The computer will stop. All computer programs must have an end.

In the above, only a brief introduction to computers and a computer language, BASIC, have been provided. To study computers as well as BASIC, obtain further information from the many books written about them. In the subsequent pages we will find other programs written in the computer language BASIC.

3 reals and radicals; solving problems

symbols in mathematics, real numbers, operations with radicals, radical equations, solving problems.

3.1 symbols: mathematical ideas in shorthand

Symbols, like any other human invention, are introduced to serve specific needs. For example, the invention and acceptance of hieroglyphics, alphabets and other systems of written language came about only after thousands of years of human thought and non-written language. As the need to communicate became more complex and sophisticated, written language became standardized, accepted and convenient.

Similarly, although the concept of radical numbers was known to Ancient Greeks, it was not until late in the fifteenth century that the symbol $\sqrt{}$, as we know it today, appeared to provide a convenient way of showing a radical number.

Imagine how difficult it would be to talk about radical numbers without the mathematical symbol $\sqrt{}$. Imagine how difficult (or impossible) it would be to develop or use any mathematical ideas without mathematical symbols.

Mathematics is a type of shorthand. It uses symbols to describe and use "big" ideas simply, precisely, and in ways that anyone who knows the language of mathematics can understand. In this chapter, we'll look at the language of mathematics.

3.2 solving problems: Pythagorean Property

Many ideas in mathematics, including those related to radical numbers, were developed as a result of the work of great thinkers in mathematics. Many years ago, Pythagoras, who was a Greek mathematician, discovered an important property about right-angled triangles. The theorem that eventually resulted now bears his name.

In a right-angled triangle, the following relationship holds among the sides of the triangle.

$a^2 + b^2 = c^2$

c represents the number of units of length of the hypotenuse.

For example, in △ABC if $a = 6$, $b = 8$, then we may calculate c.

$c^2 = a^2 + b^2$, $c > 0$
$= 6^2 + 8^2$
$= 36 + 64 = 100$

Then $c = 10$.

A number like 100 has 2 square roots.
$(-10)^2 = 100$ $(10)^2 = 100$
-10, 10 are the square roots of 100.

In problems involving distances, such as those above, we may use the symbol $\sqrt{100}$ to indicate the positive or principal square root of 100.

$\sqrt{100} = 10$ — 100 is called the **radicand**
$\sqrt{}$, the **radical sign**.

Example 1
In △PQR, $\angle Q = 90°$, PQ = 6 cm, QR = 9 cm. Calculate the length of the hypotenuse, RP, to 1 decimal place.

Solution
Sketch a diagram to record the given information.

Since RP is opposite $\angle Q$, we often use q to represent the side opposite $\angle Q$

$p = 9$, $r = 6$, $q > 0$

$q^2 = p^2 + r^2$
$= 9^2 + 6^2$
$= 81 + 36$
$= 117$
$q = \sqrt{117} = 10.8$ (to 1 decimal place)

Thus the length of the hypotenuse to 1 decimal place is 10.8 cm.

In answering a problem, be sure to make a final statement.

We may solve many problems that require the use of the Pythagorean property of a right-angled triangle.

Example 2
For a change, Jessica decides to swim diagonals instead of lengths in the square pool. If the pool is 6 m by 6 m, how far does she swim along a diagonal? (Express your answer to 1 decimal place.)

Solution
Let d represent the number of metres in the length of the diagonal.

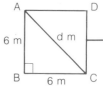

Sketch a diagram to organize the information given on the problem.

△ABC is a right-angled triangle.
AC² = AB² + BC²
$d^2 = 6^2 + 6^2$
$= 36 + 36$
$= 72$
$d = \sqrt{72} = 8.5$ (to 1 decimal place)

Thus Jessica swims 8.5 m along a diagonal.

We can solve many different problems involving distance by using the Pythagorean property of a right-angled triangle.

Example 3

A cruiser travels due west from a port at 24 km/h. A steamship leaves at the same time and heads due south at 16 km/h. After 2 h how far apart are the ships? Express your answer to the nearest kilometre.

Solution

Let d represent the number of kilometres between the two ships after the ships have sailed for 2 h.

	Time	Speed	Distance
Cruiser	2 h	24 km/h	48 km
Steamship	2 h	16 km/h	32 km

We use a chart to organize the information.

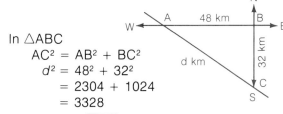

In $\triangle ABC$
$AC^2 = AB^2 + BC^2$
$d^2 = 48^2 + 32^2$
$= 2304 + 1024$
$= 3328$
$d = \sqrt{3328} = 57.69$ (to 2 decimal places)

Thus, the ships are 58 km apart, to the nearest kilometre, after 2 h.

3.2 exercise

Questions 1 to 10 develop essential skills for working with the Pythagorean property of right-angled triangles.

A

1. Solve for x. $x > 0$.
   (a) $x^2 + 16 = 25$
   (b) $x^2 = 10^2 + 24^2$
   (c) $x^2 = 7^2 + 24^2$
   (d) $15^2 + x^2 = 39^2$
   (e) $36 + x^2 = 100$
   (f) $x^2 + 14^2 = 50^2$
   (g) $144 + x^2 = 400$
   (h) $20^2 + 48^2 = x^2$
   (i) $x^2 + 21^2 = 75^2$
   (j) $78^2 = x^2 + 72^2$
   (k) $x^2 + 40^2 = 41^2$
   (l) $289 = 225 + x^2$

2. Solve each equation, $x > 0$.
   (a) $x^2 = 14^2 + 8^2$
   (b) $38^2 = x^2 + 12^2$
   (c) $14^2 + x^2 = 30^2$
   (d) $36^2 + 14^2 = x^2$
   (e) $x^2 + 18^2 = 28^2$
   (f) $39^2 + x^2 = 53^2$
   (g) $26^2 = x^2 + 9^2$
   (h) $x^2 + 18^2 = 26^2$

3. Find the value of d in each of the following.

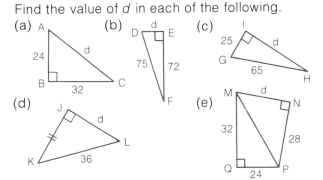

4. For each figure find the length of a diagonal.

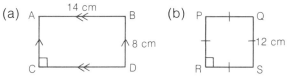

5. The altitude, h, of a triangle is shown.

Calculate the altitude for each triangle.

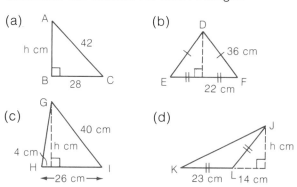

6. A square has sides that measure 8 cm. Find the length of the diagonal.

7. A rectangle has sides that measure 10 cm and 5 cm. Find the length of the diagonal.

8. The diagonal of a rectangle is 16 cm. If the width of the rectangle is 6 cm, find the length.

9. The diagonal of a square is 12 cm. Find the measure of the sides.

10. An equilateral triangle has sides that measure 12 cm. Find the altitude of the triangle.

B

11. Jason is painting the eavestrough on his house. How far from the wall must the foot of a 4-m ladder be placed for the ladder to reach exactly 3 m up the wall?

12. (a) A sign is supported by a guy wire as shown. Calculate the length of the supporting wire BC.

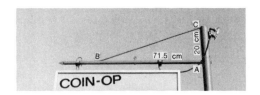

(b) If the sign is made 20 cm longer, by how much will the supporting wire have to be increased?

13. Michael decides to swim diagonals instead of lengths in his backyard rectangular pool. If the pool is 8 m by 6 m, how many diagonals must he swim to complete 1 km?

14. A room at the museum is to hold two displays from the same period in history. The curator decides to divide the 12 m by 9 m room diagonally with the use of a cord. How long should the cord be?

15. Jennifer is late for school and decides to walk across the diagonal of a rectangular field rather than walk around the two adjacent sides. What distance does she save if the field is 200 m by 250 m?

16. Can a circular table top, 2.3 m in diameter, be carried through a doorway 1.5 m × 2 m?

17. Two speedboats leave a dock at the same time. One boat travels at 32 km/h and travels due south. The other travels at the same speed and heads due west. How far apart are the boats after 3 h?

18. How far should the base of a ladder be placed from a wall so that a 10-m ladder will reach 8.7 m up the wall?

19. Two persons carry a length of pipe along a corridor, 1.7 m wide, and around a corner. What is the longest pipe which they can carry if the pipe is to be carried horizontally?

Questions 20 to 23 are based on the cube.

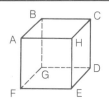

20. (a) If the sides of the above cube measure 8 cm, calculate the distance BE.
    (b) The sides are increased by 4 cm. By how much does the length of the diagonal BE increase?

21. If the edges of a cube are 5 cm long, how long is the diagonal?

22. How far is the centre of a 20-cm cube from each of its vertices?

23. The volume of a cube is 243 cm³. Find the length of the diagonal.

In mathematics we may often get surprising results in solving a problem. For example, a railway track is constructed 2000 m in length. No spaces are provided for the track to expand. On a hot sunny day the track expands 2 m and buckles. To illustrate this expansion, a right-angled triangle is drawn as follows.

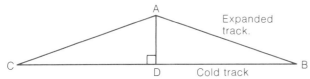

C
24 How high do you think the lift, AD, is in the diagram? Before you solve the problem, choose one of the following answers (A, B, C or D) to provide your estimate of how "high" the track lifts.

A: Track does not lift at all.
B: Track lifts much less than 2 m.
C: Track lifts 2 m.
D: Track lifts much more than 2 m.

25 (a) What is the length of AC?
(b) What is the length of CD?
(c) Calculate the height AD.

26 Using your results in Question 25, how high does the 2000-m track lift if the lift is represented by the above diagram after the track expands 2 m?

27 If the above track on another day expanded only 1 m, what would be the height shown by AD in the diagram?

math tip

Numbers were invented to aid the solution of equations. To solve $x^2 = 5$, radical numbers were invented. $x = \pm \sqrt{5}$. However, radical numbers were used for many years before the symbol $\sqrt{}$ was invented. The Latin word for root is *radix* and from it the word *radical* was derived. The symbol $\sqrt{}$ was used for the first time in 1525.

applications: television screens

Manufacturers mass-produce standard picture tubes for television sets. Over the years, certain sizes have become popular while others have become obsolete. A television tube is always greater than the *viewable area* of the television picture. In order to advertise TV sets manufacturers refer to the term *viewable diagonal*, given by AB in the diagram. For the viewable area of a TV set, the dimensions are given by the ratio
$$\frac{\text{width}}{\text{height}} = \frac{4}{3}.$$

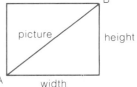

Example
The viewable diagonal of a TV set is 67 cm. Calculate the approximate dimensions of the viewable area to the nearest centimetre.

Solution
Let the width, in centimetres, be 4x. Then the height, in centimetres, is 3x.

In $\triangle ABC$
$BC^2 + AC^2 = AB^2$
$(3x)^2 + (4x)^2 = 67^2$
$9x^2 + 16x^2 = 4489$
$25x^2 = 4489$
$x^2 = 179.6$ (to 1 decimal place)
$x = 13.4$ (to 1 decimal place)

Thus, the dimensions of the viewable area to the nearest centimetre are 40 cm and 54 cm.

Once we have calculated the above dimensions, we are able to calculate the viewable area, VA.
VA = 40 × 54
   = 2160
The viewable area is 2160 cm².

28 Calculate the dimensions of the viewable area if the length of the viewable diagonal is given by
   (a) 33 cm   (b) 42 cm   (c) 49 cm

29 A popular TV set has a viewable diagonal that measures 51 cm.
   (a) Calculate the dimensions of the viewable area to the nearest centimetre.
   (b) Calculate the viewable area.
   (c) The TV set given in the above worked example is also a popular size. How much greater is its viewable area than the one given above in (a)?

30 The viewable diagonal of a TV set is 42 cm. If its diagonal is increased by 5 cm, by how much does the viewable area increase?

31 How much wider is the TV screen on a set with a viewable diagonal of 59 cm as compared to one with a viewable diagonal of 42 cm? Express your answer to 1 decimal place.

32 The smallest popular TV set available has a viewable diagonal of 37 cm. How many times greater is the viewable area of a TV set that has a 67-cm viewable diagonal? Express your answer to 1 decimal place.

problem-solving: a strategy

The Pythagorean property of a right-angled triangle is shown in the diagram. The square on the hypotenuse is equal to the sum of the squares on the other two sides.

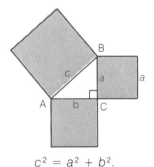

$c^2 = a^2 + b^2$.

In the previous section we calculated distances by using the Pythagorean Theorem *without actually proving the theorem*. We may now apply our skills with algebra to **prove** the result:

In $\triangle HDG$, a, b, and c represent the sides of a right triangle. ABCD is a square. Another square EFGH is constructed as shown with sides a, b, c marked.

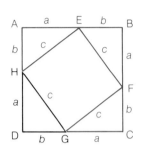

From the diagram,

$\square ABCD = \square HEFG + \triangle AEH + \triangle EBF + \triangle FCG + \triangle HGD$

$(a + b)^2 = c^2 + \frac{1}{2}ab + \frac{1}{2}ab + \frac{1}{2}ab + \frac{1}{2}ab$

$a^2 + 2ab + b^2 = c^2 + 2ab$

$a^2 + b^2 = c^2$

area of $\triangle FCG$
$= \frac{1}{2}$ (base)(height)
$= \frac{1}{2} ab$

Thus, for $\triangle HDG$ we have proved the Pythagorean Theorem.

33 Prove that the area of a circle drawn on the hypotenuse as diameter, is equal to the sum of the areas of the circles drawn on the other two sides as diameter.

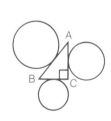

34 In the diagram prove that the area of the hexagon drawn on the hypotenuse is equal to the sum of the areas of the hexagons drawn on the other two sides.

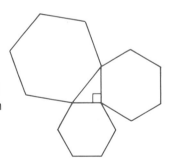

3.3 the real numbers

We have already worked with many of these numbers and have often used the mathematical ideas and symbols associated with them.

Natural Numbers $N = \{1, 2, 3, 4, \ldots\}$
Whole Numbers $W = \{0, 1, 2, 3, \ldots\}$
Integers $I = \{\ldots, -2, -1, 0, 1, 2, \ldots\}$

We may graph these sets of numbers on a number line.

$x < 3, x \in I$

$x \leq 3, x \in I$

The above examples illustrate the difference in meaning between $<$ and $\leq$.

We may also construct another set of numbers, the set of *rational numbers* Q, based on the set I as shown.

$Q = \left\{\dfrac{a}{b} \,\middle|\, a, b, \in I \; b \neq 0\right\}$

If $a = -6$, $b = 5$, then the rational number is $\dfrac{-6}{5}$.

Each rational may be written in an equivalent decimal form as shown.

Terminating
$\dfrac{1}{2} = 0.5$
$\dfrac{1}{4} = 0.25$

Periodic
$\dfrac{1}{3} = 0.3333\ldots$ or $0.\overline{3}$
$\dfrac{31}{99} = 0.313131\ldots$ or $0.\overline{31}$

$0.\overline{31}$ is a compact form for the decimal $0.313131\ldots$

We introduce new vocabulary to refer to periodic decimals:
$\dfrac{31}{99} = 0.31313131\ldots$

The **length** of the period is 2.
31 is the **period**. The period repeats.

Conversely, if we are given a periodic decimal we may find its corresponding rational number. Consider the following.

Example 1
Find the rational number equivalent to each periodic decimal.

A: $0.\overline{13}$ B: $0.1\overline{3}$

Solution

A
Let x represent the periodic decimal.
Then $x = 0.\overline{13}$ or
$ 0.131313\ldots$
$100x = 13.131313\ldots$
$x = 0.131313\ldots$
$\overline{}$
$99x = 13$
$x = \dfrac{13}{99}$

Thus $0.\overline{13} = \dfrac{13}{99}$.

B
Let y represent the periodic decimal.
Then $y = 0.1\overline{3}$ or
$ 0.133333\ldots$
$100y = 13.33333\ldots$
$10y = 1.33333\ldots$
$\overline{}$
$90y = 12$
$y = \dfrac{12}{90} = \dfrac{4}{30}$

Thus $0.1\overline{3} = \dfrac{4}{30}$.

For the set of rational numbers, Q,
- the equivalent decimal for any rational number is either a terminating decimal or a periodic decimal.
- any terminating decimal or periodic decimal represents a rational number.

We may invent decimals that are non-terminating and non-periodic. They represent numbers that are *not rational* such as
$0.393993999399993\ldots$
We refer to these numbers as *irrational* numbers.
- The decimal for π is non-periodic and non-terminating.
- The decimal for numbers given by $\sqrt{2}$, $\sqrt{3}$, $\sqrt[3]{2}$, and so on, are non-periodic and non-terminating.
- The number represented by the following decimal is an irrational number also. Even though there is a pattern in the decimal, the decimal is not periodic.
$0.13133133313333\ldots$

The set of all irrational numbers is shown as $\bar{Q}$. All rationals Q and irrational numbers $\bar{Q}$ are referred to as the **real numbers**, **R**.

$R = Q \cup \bar{Q}$ periodic decimals (terminating and non-terminating)
           non-periodic, non-terminating decimals

Locating integers on the number line is easy but where would we mark $\sqrt{2}$ or $-\sqrt{2}$?

To show $\sqrt{2}$ on the number line, we may use the following construction based on our knowledge of the Pythagorean Theorem.

$OB^2 = BC^2 + OC^2$
$= 1 + 1 = 2$
$OB = \sqrt{2}$

Once we have located $\sqrt{2}$, we may then locate $-\sqrt{2}$.

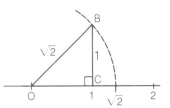

3.3 exercise

Questions 1 to 11 examine skills needed to study real numbers.

A

1. Which of the following numbers are
   - rational?
   - irrational?

   (a) $\frac{2}{3}$ (b) $\sqrt{2}$ (c) $-\frac{6}{7}$ (d) 2π (e) $\sqrt{7}$

   (f) 3 (g) $1 + \sqrt{2}$ (h) 0.25 (i) $\sqrt{3} + 1$ (j) $-\frac{3}{2}$

   (k) $\frac{\sqrt{3}}{2}$ (l) -3π (m) $\sqrt{19}$ (n) $\sqrt{4}$ (o) $\sqrt{\frac{25}{16}}$

2. Express each rational number as a decimal.

   (a) $\frac{1}{4}$ (b) $\frac{3}{8}$ (c) $-\frac{2}{5}$ (d) $\frac{23}{50}$ (e) $-\frac{1}{11}$

   (f) $\frac{2}{9}$ (g) $\frac{11}{9}$ (h) $-2\frac{1}{8}$ (i) $\frac{53}{100}$ (j) $\frac{-7}{15}$

   (k) $\frac{3}{7}$ (l) $1\frac{2}{3}$ (m) $-3\frac{1}{8}$ (n) $\frac{11}{31}$ (o) $-\frac{7}{13}$

3. (a) Write a decimal for each rational $\frac{11}{30}, \frac{9}{26}$.
   (b) Which rational in (a) is greater?

4. For each pair, which is the greater number?
   (a) $\frac{1}{6}, \frac{6}{29}$ (b) $-\frac{3}{13}, -\frac{2}{7}$ (c) $\frac{2}{13}, \frac{11}{57}$

5. Write the period and length of each periodic decimal.
   (a) $0.\overline{6}$ (b) $0.1\overline{6}$ (c) $2.\overline{36}$
   (d) $-4.3\overline{8}$ (e) $2.9\overline{16}$ (f) $-0.\overline{162}$
   (g) $6.\overline{382}$ (h) $0.00\overline{45}$ (i) $0.\overline{142\,857}$

6. (a) Describe the decimal equivalent of a rational number.
   (b) Explain why 0.2356 is the decimal equivalent of a rational number.
   (c) Explain why $0.\overline{2356}$ is the decimal equivalent of a rational number.

7. (a) A decimal is given by 0.431431431431.... Is the number represented by the decimal rational or irrational? Can you explain why?
   (b) A decimal is given by 0.431433143331.... Is the number represented by the decimal rational or irrational? Can you explain why?

8. Use the Pythagorean Theorem to show the length of AB in each triangle.

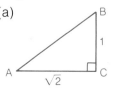

9. To show $\sqrt{5}$ on the number line, the following construction was done. Show why $OE = \sqrt{5}$.

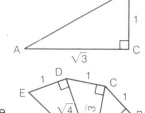

10 Use the above construction to mark each number on the number line.
   (a) $-\sqrt{2}$ (b) $\sqrt{6}$ (c) $\sqrt{7}$
   (d) $1 + \sqrt{2}$ (e) $\sqrt{2} - 1$ (f) $1 + \sqrt{3}$

11 Draw the graph of each of the following.
   (a) $\{x \mid x < 3, x \in I\}$ (b) $\{x \mid -2 > x, x \in R\}$
   (c) $\{x \mid 5 \leq x, x \in R\}$ (d) $\{x \mid x \geq -3, x \in I\}$
   (e) $\{x \mid x < 2, x \in I\}$ (f) $\{x \mid 0 > x, x \in R\}$
   (g) $\{x \mid 5 \geq x, x \in I\}$ (h) $\{x \mid x \leq 2, x \in R\}$

B

12 Write the rational number represented by each decimal.
   (a) 0.25 (b) $0.\overline{7}$ (c) $-0.\overline{25}$
   (d) $0.\overline{13}$ (e) $2.1\overline{3}$ (f) $0.0\overline{13}$
   (g) $-3.6\overline{3}$ (h) $2.8\overline{56}$ (i) $0.4\overline{9}$

13 Express each of the following as an equivalent rational number.
   (a) 0.18 (b) $0.\overline{18}$ (c) $0.1\overline{8}$
   (d) $0.\overline{9}$ (e) $0.1\overline{9}$ (f) $0.11\overline{9}$

14 Construct the decimal for an irrational number between each of the following.
   (a) $\frac{1}{8}, \frac{2}{15}$ (b) $-\frac{4}{5}, -\frac{7}{10}$
   (c) 0.25, 0.26 (d) 0.01, 0.012
   (e) 0.7, $0.7\overline{5}$ (f) 0.63, $0.6\overline{3}$
   (g) $0.\overline{46}, 0.\overline{47}$ (h) $-2.3\overline{15}, -2.31\overline{5}$

15 Determine the equivalent rational number for
   (a) 0.5 (b) $0.4\overline{9}$
   (c) What do you notice about your answers in (a) and (b)?
   (d) Use your answers above to predict the rational number given by $0.74\overline{9}$. Check your answer by finding the rational number represented by $0.74\overline{9}$.

16 Use your results in Question 15. Write 2 different decimal equivalents for each of the following.
   (a) $\frac{1}{4}$ (b) $\frac{3}{8}$ (c) $-1\frac{1}{8}$ (d) $\frac{3}{20}$ (e) $\frac{5}{8}$ (f) $-\frac{6}{25}$

17 A periodic decimal such as $0.\overline{7}$ may be written as $0.\overline{7} = 0.7 + 0.07 + 0.007 + 0.0007 + \ldots$. Find the equivalent rational number for each decimal.
   (a) $0.2 + 0.02 + 0.002 + \ldots$
   (b) $0.06 + 0.006 + 0.0006 + \ldots$
   (c) $0.4 + 0.04 + 0.004 + 0.0004 + \ldots$
   (d) $0.25 + 0.0025 + 0.000025 + \ldots$

18 The circumference, C, of a circle is given by $C = 2\pi r$, with radius r, and $C = \pi d$, with diameter d.
   Develop a method of showing how the irrational number π may be marked on the number line.

C

19 Mark each of the following on a number line.
   (a) $\sqrt{3} + \sqrt{2}$ (b) $\sqrt{3} - \sqrt{2}$

20 Show that $\sqrt{3} + \sqrt{2} \neq \sqrt{5}$.
   Use your constructions on a number line to aid you.

using graphs and symbols

We can show the union and intersection of sets using number lines.

Intersection

$A = \{x \mid x \leq 3, x \in R\}$

$B = \{x \mid 0 < x, x \in R\}$

$A \cap B$

The intersection of sets A and B.

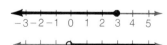

We may use a compact form to show the set of numbers:

$A \cap B = \{x \mid 0 < x \leq 3, x \in R\}$ or $\{x \in R \mid 0 < x \leq 3\}$

Union

$C = \{x \mid x > 2, x \in R\}$

$D = \{x \mid x \leq -2, x, x \in R\}$

$C \cup D$

The union of sets C and D

We may use a compact form to show the set of numbers:

$C \cup D = \{x \mid x > 2 \text{ or } x \leq -2, x \in R\}$
or $\{x \in R \mid x > 2 \text{ or } x \leq -2\}$

Why is $C \cap D = \phi$? — the empty set

21. Draw the graph of each set.
    (a) $\{x \in I \mid -3 < x\}$
    (b) $\{x \in R \mid -1 \geq x\}$
    (c) $\{x \in I \mid x > 0\}$
    (d) $\{x \in R \mid 2 < x\}$
    (e) $\{x \in R \mid 3 \leq x\}$
    (f) $\{x \in R \mid x \geq -1\}$

22. Draw the graph of each set.
    (a) $\{x \in R \mid -3 \leq x \leq 2\}$
    (b) $\{x \in I \mid -1 < x \leq 3\}$
    (c) $\{x \in R \mid -1 \geq x \text{ or } x \geq 2\}$
    (d) $\{x \in R \mid 0 \leq x < 2\}$
    (e) $\{x \in R \mid x > 0 \text{ or } x \leq -3\}$
    (f) $\{x \in I \mid -2 < x < 2\}$

23. For the sets given by
    $E = \{x \in R \mid x \geq 2\}$ $G = \{x \in R \mid x > 1\}$
    $F = \{x \in R \mid x \leq -2\}$ $H = \{x \in R \mid x < -1\}$
    draw the graph of each of the following.
    (a) $E \cup F$
    (b) $E \cap G$
    (c) $E \cup H$
    (d) $F \cup G$
    (e) $E \cap F$
    (f) $F \cap H$

24. If $x \in R$, draw the graph of each of the following.
    (a) $\{x \mid x > 2\} \cup \{x \mid x \leq -3\}$
    (b) $\{x \mid x \leq 4\} \cap \{x \mid x \geq -1\}$
    (c) $\{x \mid x > 0\} \cup \{x \mid x \leq 2\}$
    (d) $\{x \mid -3 \leq x \leq 1\} \cap \{x \mid 0 < x \leq 3\}$
    (e) $\{x \mid -1 \leq x \leq 2\} \cup \{x \mid 2 < x \leq 5\}$
    (f) $\{x \mid -3 \leq x \leq 3\} \cap \{x \mid x < 0\}$

properties and problem-solving

Often in mathematics to show something is true requires more skill than to show something is not true. For example, to prove that the statement

"The set of integers is closed with respect to division",

is false we need only to find one counter example, or one case where the set of integers is not closed with respect to division.

These examples show the statement may be true.

$\dfrac{-8}{4} = -2 \in I$

$\dfrac{-6}{-2} = 3 \in I$

One example shows the statement is not true.

$9 \div (-2) = -4\dfrac{1}{2}$

Since $-4\dfrac{1}{2}$ is not an integer, then the set of integers is not closed under division.

Thus, to show that a certain statement is not true we need only to find *one* example that disproves the statement.

On the other hand, to show that a certain property is true in all cases, we need to *prove* the property. Even though we may test a great many cases, this does not prove that the property or statement is true *for all cases*. For example, we may prove the sum of two even numbers is always even, for *all* even numbers.

Let $2m$ and $2n$ be any even numbers, $m, n \in I$.

Since $2n$ has a factor 2, then $2n$ is an even number.

Write $2m + 2n = 2(m + n)$.

$2(m + n)$ is an even number since it has a factor 2.

Thus, the sum of any two even numbers is always even.

25. Which of the following are true? Which are false? $a, b, c \in R$.
    (a) If $a < b$, then $b - a < 0$.
    (b) If $b > a$, then $b - a > 0$.
    (c) If $b > a$, and $c < 0$, then $b - c > a$.
    (d) If $b > a$, then $-b > -a$.

26. Which of the following are true for all $a, b, c \in R$? Use examples to test each one.
    (a) If $a < b$ and $b < c$ then $a < c$.
    (b) If $a \leq b$ and $b \leq c$ then $a \leq c$.
    (c) If $a > b$ and $c < 0$ then $a + c < b + c$.
    (d) If $a > b$ and $c < 0$ then $ac < bc$.
    (e) If $a < b$ and $c > 0$ then $a + c < b + c$.
    (f) If $a < b$ and $c < 0$ then $ac > bc$.

27. Prove each of the following.
    (a) The sum of two odd numbers is an even number.
    (b) The difference of two odd numbers is an even number.
    (c) The difference of two even numbers is an even number.
    (d) The product of two even numbers is an even number.

28. (a) Prove that if a number is divisible by 3 and also divisible by 2 then it must be divisible by 6.
    (b) Prove that if a number is divisible by any two numbers, then it must be divisible by the product of the two numbers.

29. Which of the following properties is true? Use numerical examples to test each one.
    (a) Every integer has an additive inverse.
    (b) The set of integers is closed under division.
    (c) For each real number, a, there is another real number x so that $a + x = 0$.
    (d) Every rational has a multiplicative inverse.
    (e) The set N is dense.
    (f) The set I is dense.

30. The following chart may be used to summarize the properties of the set of real numbers and its important subsets.

    N, Natural numbers $\{1, 2, 3, \ldots\}$
    W, Whole numbers $\{0, 1, 2, 3, \ldots\}$
    O, Odd numbers $\{1, 3, 5, 7, \ldots\}$
    E, Even numbers $\{2, 4, 6, 8, \ldots\}$
    I, Integers $\{\ldots, -3, -2, -1, 0, 1, 2, 3, \ldots\}$
    R, Reals
    Q, Rationals
    $\overline{Q}$, Irrationals

Set	Closed with respect to				Commutative				Associative				Multiplication distributes over				There is an Additive Identity.	There is a Multiplicative Identity.	Each element has an Additive Inverse.	Each element has a Multiplicative Inverse.
	+	−	×	÷	+	−	×	÷	+	−	×	÷	+	−	×	÷				
N	no																			
W																				
E									yes											
O																			no	
I		no											yes						no	
Q																				
$\overline{Q}$																				
R																				

3.4 real numbers and inequations

The skills we have used to solve equations extend to solving inequations.
To solve an inequation we use our skills in algebra to simplify the inequation so that we may write its solution set.

The vocabulary and symbols that we used to solve equations are used in solving inequations.

Example 1
Solve. Draw the graph of the solution set, if $2(y + 4) + 3(y + 2) \geq -11, y \in R$.

Solution
$$2(y + 4) + 3(y + 2) \geq -11$$
$$2y + 8 + 3y + 6 \geq -11$$
$$5y + 14 \geq -11$$
$$5y \geq -25$$
$$y \geq -5, y \in R$$

Thus, the solution set is given by
$$\{y \mid y \geq -5, y \in R\}$$
The graph of the solution set is shown.

We may find the solution set for compound sentences such as the following.
$$2y - 3 \leq 3y - 2 < 6 + y$$

Example 2
Solve and graph the solution set if
$2y - 3 \leq 3y - 2 < 6 + y$.

Solution
Two inequations are given by
$2y - 3 \leq 3y - 2 < 6 + y$
$2y - 3 \leq 3y - 2$ and $3y - 2 < 6 + y$
$-1 \leq y$ $2y < 8$
                     $y < 4$
Thus $-1 \leq y < 4$
The graph of the solution set is drawn.
$\{y \mid -1 \leq y < 4, y \in R\}$

3.4 exercise

Questions 1 to 7 examine skills needed to solve inequations.
Throughout the exercise, the domain of the variable is the set of real numbers.

A

1 Write each inequation in a simpler form.
   (a) $\frac{1}{2}y - 6 < 8$ (b) $8 > 5 + 3m$
   (c) $3m - 2 \geq 16$ (d) $4k \leq 35 + 3k$
   (e) $4y - y < 50 - 7y$ (f) $5m - 3 > 2m + 9$
   (g) $7p + 3 \leq 23 + 2p$ (h) $\frac{4p}{3} - 8 \geq -4$

2 Two values are given. Which values belong to the solution set?
   (a) $2(y + 4) < 3y + 7$ 4, −3
   (b) $3m + 2(m + 1) \geq 2(2m + 4)$ −3, 8

3 (a) Solve $4(x + 3) - 3 < 5 + 3x$.
   (b) Draw a graph of the solution set.

4 (a) Why are the following inequations equivalent?
      A: $3(x - 1) \leq 2(x + 1) - 4$
      B: $4 - 3(x + 2) \geq x - 3(1 + x)$
   (b) Draw the graph of the solution set for each inequation.

5 (a) Find the solution set for $2(y - 1) \geq 3y - 4$.
   (b) Find the solution set for
       $3 + 2(y - 1) < 3(y + 1)$.
   (c) Graph the solution set for $2(y - 1) \geq 3y - 4$
       and $3 + 2(y - 1) < 3(y + 1), y \in R$.

6 (a) Find the solution set of $3y + 4 \leq 2y - 1$.
   (b) Find the solution set of
       $5(y - 1) > 3(y - 2) + 9$.
   (c) Graph the solution set that satisfies
       $3y + 4 \leq 2y - 1$ or
       $5(y - 1) > 3(y - 2) + 9, y \in R$.

7 (a) Write 2 inequations given by
    $y - 3 < 2y + 1 \leq y + 3$.
   (b) Find the solution set in (a).

B

8 Solve each of the following.
   (a) $2(x - 3) \leq 4 - 3(x - 5)$
   (b) $15 - 3(y - 1) > 5(20 - y)$
   (c) $11 - 4(2m + 1) < 3 - 2(m + 1)$.

9 Graph the solution set for each of the following.
   (a) $y - 3(y + 1) > 2(y - 5) - 1$.
   (b) $3(2m - 1) - 5(1 - m) \leq 3(1 + m)$
   (c) $2(2x^2 + x - 6) + 8 \geq 4(x^2 + x + 7)$

10 Which inequation is not equivalent to the others?
   (a) $3(2 + x) - 1 > 4(x + 3) - 3$
   (b) $2x + 3(x - 6) < 9(x - 1) + 3$
   (c) $(x + 1) - 2(x^2 + x + 3) < 13 - 2x^2 + 3(x - 2)$

11 Solve each of the following.
   (a) $\frac{1}{4}(2 - y) \leq -\frac{1}{6}(y + 2)$
   (b) $\frac{1}{3}(6 - 2m) + \frac{m}{2} > \frac{1}{7}(1 - m)$

12 (a) Find the solution set A for
    $6x + 2(x - 7) \leq 5(x + 1) + 7(x - 1)$.
   (b) Find the solution set B for
    $2(x + 3) + 2x \leq 5 - 3(x - 5)$.
   (c) Draw the graph of $A \cap B$.

13 Graph the intersection of the solution sets of
   A: $5 + 2(4 - x) \geq x + 2(x + 4)$
   B: $4x + 2(x - 2) > 3(x - 2) - 2(x - 1)$

14 Solve.
   (a) $6y^2 + 3(y^2 + 1) < (3y + 1)^2 + 2$
   (b) $(2y + 3)^2 - 5 \geq 4y(y + 1)$
   (c) $2(y - 5) + y^2 \leq (y + 2)(y - 1)$

15 Find the solution set for each of the following.
   (a) $3(y - 2) \leq 5(y + 1) - 3$ and $11 > 8 + \frac{1}{3}y$
   (b) $4p + 11 < 6p + 10$ and $\frac{4p}{3} + 4 < 8$
   (c) $7 - 4k < 5k - 2$ or $2(k - 1) < (1 - k) - 2$

16 Solve.
   (a) $2y - 3 < 3(y - 1) \leq 4y - 3(y - 1)$
   (b) $2(m - 3) \leq 2m + 2(m - 1) < 3(m + 2) - 3$
   (c) Graph the solution sets in (a) and (b).

translating into mathematics

To solve each of the following problems, we need to translate them into symbols by writing an inequation. For example:

For what values of m does $4(m + 2)$ exceed $5(m + 1) + 2(m - 2)$?
We write the inequation
$4(m + 2) > 5(m + 1) + 2(m - 2)$.

We use $>$ to translate the significance of *exceed*.

17 Find the values of m that satisfy the above inequation.

18 Find the integers that make $3(x + 1) - (2x + 3)$ at least equal to $2(x + 2)$.

19 For what values of m is $3(m + 4) - 4(m - 1)$ no greater than $6(m + 5)$?

20 Find the values of b so that $3(b - 1) - 2(1 - 2b)$ is less than 3?

21 The sum of $2(x - 1)$ and $3(x + 1)$ is less than $2(x - 2)$. For what values of x is this true?

22 Find the values of k so that $12(k + 1) - 8(k - 1)$ is at most equal to $5(2k - 1)$.

23 If the value of $5(1 - y) + 3(1 - 5y)$ is to exceed the value of $2(1 - 4y)$ by more than 2, then find the range of values of y.

3.5 real numbers, radicals and exponents

In mathematics we frequently look for relationships. For example, in our earlier work, we established that for the real numbers a, where $m, n \in N$
$$a^m a^n = a^{m+n}$$
For example, we may investigate the meaning of a^0:

> We may then ask, Does the law of exponents extend to other numbers, not in set N?

$$a^m a^0 = a^{m+0} = a^m$$

In order that we may extend the laws of exponents to include the use of a^0 in
$$a^m a^0 = a^m,$$
we must define $a^0 = 1$.
We may then further extend our work with exponents for the following to be true:
$$a^m a^{-m} = a^{m+(-m)}$$
We note that $a^{m+(-m)} = a^0 = 1$. Thus, we use
$$a^m a^{-m} = 1.$$
Thus, to extend our work with exponents, we

define: $a^{-m} = \dfrac{1}{a^m}$, and $a^m = \dfrac{1}{a^{-m}}$, $a \neq 0$

Thus, $2^0 = 1 \qquad 3^0 = 1 \qquad 5^0 = 1$
$\qquad 2^{-1} = \dfrac{1}{2^1} = \dfrac{1}{2} \qquad 2^{-2} = \dfrac{1}{2^2} = \dfrac{1}{4}$

Example 1
Find the value of $4^0 + 2^{-2} - 3^0 + 2^{-1}$.

Solution
$$4^0 + 2^{-2} - 3^0 + 2^{-1} = 1 + \dfrac{1}{2^2} - 1 + \dfrac{1}{2^1}$$
$$= 1 + \dfrac{1}{4} - 1 + \dfrac{1}{2} = \dfrac{3}{4}$$

Is it possible to extend our work with exponents to fractional exponents? For example, what meaning may we give to $5^{\frac{1}{2}}$ or $8^{-\frac{2}{3}}$? We notice the following patterns.
$$\sqrt{5} \times \sqrt{5} = 5 \qquad 5^{\frac{1}{2}} \times 5^{\frac{1}{2}} = 5$$
Examples such as the above suggest that we use $\sqrt{5} = 5^{\frac{1}{2}}$ or in general
$$\sqrt{a} = a^{\frac{1}{2}}, \ a \geq 0.$$
In defining the second order radical $\sqrt{a}$, we may state that
$$\sqrt{a} = b, \text{ if } b^2 = a, \ a \geq 0. \quad \text{We call } b \text{ the principal square root of } a.$$

To extend our work to n^{th} order radicals, we state that
$$\sqrt[n]{a} = b \text{ if } b^n = a. \quad \text{Where, if } n \text{ is even, then } a \geq 0.$$

We call b the principal n^{th} root of a. Our vocabulary includes the following new symbols:

$\sqrt[n]{a}$ where $\sqrt{}$ is the radical sign
$\qquad a$, the radicand
$\qquad n$, the index (exponent)

If we use the above relationship for radicals and exponents we may simplify expressions as follows.

$$\sqrt[3]{8} = (8)^{\frac{1}{3}} = (2^3)^{\frac{1}{3}} = 2^{3(\frac{1}{3})} = 2^1 = 2$$

Thus we may extend the law of exponents $(a^m)^n = a^{mn}$ to include rational numbers.

Example 2
Simplify
(a) $\sqrt[5]{32^2} - \sqrt[3]{8}$ \qquad (b) $32^{-\frac{3}{5}}$

Solution
(a) We may write the terms using exponents.
$$\sqrt[5]{32^2} - \sqrt[3]{8} = (32^2)^{\frac{1}{5}} - (8)^{\frac{1}{3}}$$
$$= [(2^5)^2]^{\frac{1}{5}} - (2^3)^{\frac{1}{3}}$$
$$= (2^{10})^{\frac{1}{5}} - 2 = 2$$

(b) $32^{-\frac{3}{5}} = \dfrac{1}{32^{\frac{3}{5}}} = \dfrac{1}{(2^5)^{\frac{3}{5}}} = \dfrac{1}{2^3} \text{ or } \dfrac{1}{8}$

In solving problems in mathematics, there is often more than one way of looking at the problem and obtaining the answer. The more strategies we have for looking at problems in different ways, the more problems we will be able to solve. For this reason, in studying mathematics we often look at skills in different ways and relate them. In the exercise that follows, remember the useful relationships

$a^0 = 1 \quad \sqrt{a} = a^{\frac{1}{2}} \quad \sqrt[n]{a} = a^{\frac{1}{n}}$
$\sqrt[n]{a^m} = a^{\frac{m}{n}} \quad \text{and} \quad (\sqrt[n]{a})^m = a^{\frac{m}{n}}$
$a^{-n} = \dfrac{1}{a^n} \quad \text{where} \quad a \neq 0.$

3.5 exercise

Questions 1 to 8 examine skills for simplifying expressions containing radicals and exponents.

A

1. Find the value of each of the following.
   (a) 4^0
   (b) $4^{\frac{1}{2}}$
   (c) $4^{-\frac{1}{2}}$
   (d) $27^{\frac{1}{3}}$
   (e) $27^{-\frac{1}{3}}$
   (f) 27^0
   (g) -100^0
   (h) $-100^{\frac{1}{2}}$
   (i) $-100^{-\frac{1}{2}}$
   (j) $(0.04)^0$
   (k) $(0.04)^{-\frac{1}{2}}$
   (l) $(0.04)^{\frac{1}{2}}$

2. Find the value of each of the following.
   (a) $25^{\frac{1}{2}}$
   (b) $\sqrt{25}$
   (c) $25^{-\frac{1}{2}}$
   (d) 4^2
   (e) $(\sqrt{4})^{-1}$
   (f) 4^{-2}
   (g) $16^{\frac{1}{4}}$
   (h) $\sqrt[4]{16}$
   (i) $-\sqrt[4]{16}$
   (j) $27^{\frac{1}{3}}$
   (k) $27^{-\frac{1}{3}}$
   (l) $-\sqrt[3]{27}$

3. Find the value of each of the following.
   (a) $32^{\frac{1}{5}}$
   (b) $32^{\frac{2}{5}}$
   (c) $32^{\frac{3}{5}}$
   (d) $\sqrt[5]{32}$
   (e) $\sqrt[5]{32^2}$
   (f) $\sqrt[5]{32^3}$

4. Find the value of each of the following.
   (a) $25^{\frac{1}{2}}$
   (b) $4^{\frac{1}{2}}$
   (c) $100^{\frac{1}{2}}$
   (d) $16^{\frac{1}{4}}$
   (e) $-27^{\frac{1}{3}}$
   (f) $256^{\frac{1}{4}}$
   (g) $8^{\frac{2}{3}}$
   (h) $64^{\frac{2}{3}}$
   (i) $-4^{\frac{3}{2}}$
   (j) $-32^{\frac{1}{5}}$
   (k) $125^{\frac{2}{3}}$
   (l) $-27^{\frac{2}{3}}$

5. Since $2^3 = 8$ we may write the radical form $2 = \sqrt[3]{8}$. Write each of the following in a radical form.
   (a) $3^3 = 27$
   (b) $3^5 = 243$
   (c) $(-1)^5 = -1$
   (d) $2^4 = 16$
   (e) $(-2)^5 = -32$
   (f) $4^3 = 64$
   (g) $5^4 = 625$
   (h) $7^4 = 2401$

6. Which of the following represent rational numbers? Give reasons for your answers.
   (a) $\sqrt[3]{-27}$
   (b) $\sqrt{-4}$
   (c) $(4)^{\frac{1}{2}}$
   (d) 10^0
   (e) $\sqrt[4]{-16}$
   (f) $(-8)^{\frac{1}{2}}$

7. Simplify each of the following. Write with a single base.
   (a) $m^3 \times m^2 \times m^{-1}$
   (b) $x^4 \times x^5 \times x^0$
   (c) $p^{-2} \times p^3 \times p^4$
   (d) $2^3 \times 2^{-2} \times 2^0$
   (e) $10^3 \times 10^{-1} \times 10^0$
   (f) $(-3)^2(-3)^{-2}(-3)^0$
   (g) $(3^{-2})^3$
   (h) $(2^{-3})^2$
   (i) $[(-a)^3]^{-2}$
   (j) $\left(\dfrac{b}{a^{-2}}\right)^2$
   (k) $\left(\dfrac{x^{-1}}{y}\right)^{-5}$
   (l) $\left(\dfrac{a^2}{2^2}\right)^{-3}$

8. Simplify.
   (a) $16^{\frac{1}{4}}$
   (b) $-125^{\frac{1}{3}}$
   (c) $\left(\dfrac{1}{4}\right)^{\frac{1}{2}}$
   (d) $\sqrt[4]{16}$
   (e) $\left(\dfrac{25}{36}\right)^{\frac{1}{2}}$
   (f) $-\sqrt[3]{\dfrac{27}{125}}$
   (g) $16^{-\frac{5}{4}}$
   (h) $\left(\dfrac{1}{4}\right)^{\frac{1}{2}}$
   (i) $-1000^{\frac{1}{3}}$
   (j) $-1.44^{\frac{1}{2}}$
   (k) $\dfrac{1}{(27)^{\frac{2}{3}}}$
   (l) $-125^{\frac{2}{3}}$
   (m) $\sqrt[3]{8^2}$
   (n) $-\sqrt[4]{16^{-2}}$
   (o) $\left(\dfrac{32}{50}\right)^{\frac{1}{2}}$
   (p) $(0.008)^{\frac{2}{3}}$
   (q) $(0.008)^{-\frac{2}{3}}$
   (r) $16^{\frac{5}{4}}$
   (s) $\sqrt[3]{\dfrac{1}{8}}$
   (t) $\dfrac{1}{16^{-\frac{3}{4}}}$
   (u) $-\sqrt[5]{243}$

B

9 Evaluate each of the following and express your answer with denominator 1.

(a) $4^{-\frac{1}{2}}$ (b) $-4^{\frac{3}{2}}$ (c) $16^{-\frac{3}{4}}$
(d) $(\sqrt{16})^{-1}$ (e) $25^{-\frac{3}{2}}$ (f) $\sqrt[3]{-8}$
(g) $(\sqrt[4]{16})^{-1}$ (h) $(0.008)^{\frac{2}{3}}$ (i) $-\sqrt[5]{-32}$
(j) $(-\sqrt[3]{27})^{-1}$ (k) $\left(\frac{36}{49}\right)^{\frac{1}{2}}$ (l) $(-\sqrt{144})^2$
(m) $\left(\frac{100}{196}\right)^{\frac{1}{2}}$ (n) $\left(0.25^{\frac{1}{2}}\right)^0$ (o) $(\sqrt[3]{-8})^3$

10 Simplify each of the following.

(a) $81^{\frac{1}{2}} + \sqrt[3]{8} - 32^{\frac{3}{5}} + 32^{-\frac{1}{5}}$
(b) $9^{\frac{1}{2}} - \sqrt[4]{16} + 81^{\frac{1}{4}} - 3(3^{-2})$
(c) $\left(\frac{1}{8}\right)^{\frac{1}{3}} - \sqrt[3]{\frac{27}{125}} + 2(16^{-\frac{3}{4}}) - (\sqrt{0.01})^0$
(d) $-\sqrt{49} + \left(-\frac{1}{27}\right)^{\frac{1}{3}} + (-8)^{-\frac{2}{3}}$
(e) $(-125)^{\frac{1}{3}} - 32^{-\frac{3}{5}} + \frac{1}{25^{\frac{1}{2}}} + \sqrt[3]{8^{-1}}$

11 Simplify each of the following.

(a) $\frac{3^{-1}}{2^{-1}}$ (b) $\left(9^{-\frac{1}{2}}\right) \div \left(16^{-\frac{3}{4}}\right)$
(c) $3^{-1} + 2^{-1}$ (d) $(3^{-2}) \div \left(27^{-\frac{2}{3}}\right)$
(e) $\frac{2^{-1} + 3^{-1}}{6^{-1}}$ (f) $\frac{2^{-1}}{2^{-2} - 2^{-3}}$
(g) $\frac{2^{-1} - 2^{-3}}{3^{-1} - 2^{-1}}$ (h) $\frac{4^{-\frac{1}{2}} + 9^{-\frac{1}{2}}}{27^{-\frac{2}{3}}}$

12 Express each of the following with a single base.

(a) $\left(\frac{a}{b}\right)^{-3}(ab^3)^2$ (b) $(mn^6)\left(\frac{m}{n^{-3}}\right)^{-2}$
(c) $(x^k)^{-2}(-a^{-3})^k$ (d) $(x^p)^{-3} \div (x^{-p})^2$
(e) $\left(\frac{-b}{a}\right)^{-2}\left(\frac{a}{-b}\right)^2$ (f) $\frac{(ab^{-1})^3}{a^{-2}}\left(\frac{a}{b}\right)^{-3}$
(g) $\frac{p^{-k}(p^{3k})^{-2}}{p^0}$ (h) $\frac{(a^3)^{-k}(a^{-k})^2}{a^{-5k}}$

13 Find the value of each of the following if $a = 1$, $b = 2$, $k = 2$.

(a) ab^k (b) $a^k b$ (c) $(ab)^k$
(d) $(ab)^{-k}$ (e) $(-ab)^k$ (f) $(a^{-1}b^{-2})^k$
(g) $(a^{-1}b^{-2})^{-k}$ (h) $(a^k b^k)^k$ (i) $(a^3 b^2)^k$

14 Express each of the following with positive exponents.

(a) $a^3 b^{-2}$ (b) $\frac{3a^{-1}b}{c^{-1}}$ (c) $\frac{3ab^{-2}}{4m^{-2}}$
(d) $\frac{3^{-1}a^2b}{c^{-2}}$ (e) $\frac{(5a)^{-1}b}{c^{-2}}$ (f) $\frac{a^{-3}b^{-2}}{m^{-1}}$

C

15 Each of the following represents a real number. What are the restrictions on the variables?

(a) $\sqrt{3x + 1}$ (b) $\sqrt{2y - 3}$
(c) $\sqrt{m^2 - 25}$ (d) $\sqrt{k + 1}$
(e) $\sqrt[3]{3 - p}$ (f) $\sqrt[4]{y + 1}$

16 Find the smallest positive value of the variable so that each of the following represents an integer.

(a) $\sqrt{x + 1}$ (b) $\sqrt{3p - 1}$
(c) $\sqrt{16 - y}$ (d) $\sqrt{\frac{y + 5}{2}}$
(e) $\sqrt{\frac{2(x - 5)}{3}}$ (f) $\sqrt{x^2 - 1}$

17 For $p = 2$, $k = -1$, which expression has the greater value?

A: $\dfrac{p^{-2k} \, p^{-k+2}}{(p^{-2})^k}$ B: $\dfrac{(p^k)^{-3} \, p^{-(1-2k)}}{(p^{-k})^3}$

---

math tip

Remember: a very serious, but common, error when working with inequalities is to forget to do the following:

$-x < 3$ $-x < 3$
$(-1)(-x) > (-1)(3)$ $\dfrac{-x}{-1} > \dfrac{3}{-1}$
$x > -3$ $x > -3$

3.6 working with radicals: addition and subtraction

In solving a problem, we often obtain terms that involve radicals.

$BC^2 + AC^2 = AB^2$
$x^2 + 4 = 9$
$x^2 = 5$
$x = \sqrt{5}$

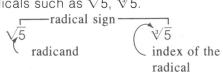

We use precise vocabulary when referring to radicals such as $\sqrt{5}$, $\sqrt[3]{5}$.

$\sqrt{5}$ — radical sign, radicand
$\sqrt[3]{5}$ — index of the radical

For the radical $\sqrt{5}$, we do not write the index but we understand it to be 2. Namely, we write $\sqrt[2]{5}$ as $\sqrt{5}$.
$\sqrt{5}$ is a radical of **order 2**, $\sqrt[3]{5}$ is a radical of **order 3**, and so on.

The vocabulary used with polynomials is extended to our work with radicals.

Like Terms
$k, -4k, -\frac{1}{2}k$

Like Radicals
$\sqrt{5}, -4\sqrt{5}, -\frac{1}{2}\sqrt{5}$

Unlike Terms
$k, 2p, -3q$

Unlike Radicals
$\sqrt{5}, 2\sqrt{3}$ — The radicands are unlike.
$\sqrt[3]{5}, \sqrt{5}$ — The indices are unlike.

Entire Radicals
$\sqrt{8}, \sqrt{12}, \sqrt{20}$

Mixed Radicals
$2\sqrt{2}, 2\sqrt{3}, 2\sqrt{5}$

To add or subtract **like radicals**, we may add or subtract the coefficients of each radical. The following example illustrates this.

Example 1
Simplify A: $4\sqrt{3} - 2\sqrt{5} + 6\sqrt{3} + 5\sqrt{5}$
B: $2\sqrt{8} + \sqrt{18} - \sqrt{32} - 3\sqrt{50}$

Solution
A: All the terms are expressed in the simplest form. We may add or subtract numerical coefficients.

$4\sqrt{3} - 2\sqrt{5} + 6\sqrt{3} + 5\sqrt{5}$
$= \underbrace{4\sqrt{3} + 6\sqrt{3}}_{\text{like radicals}} \underbrace{- 2\sqrt{5} + 5\sqrt{5}}_{\text{like radicals}}$
$= 10\sqrt{3} + 3\sqrt{5}$ ← To simplify, we add or subtract the numerical coefficients.

B: The terms are not expressed in simplest form as mixed radicals.
$2\sqrt{8} + \sqrt{18} - \sqrt{32} - 3\sqrt{50}$
$= 2(2\sqrt{2}) + (3\sqrt{2}) - (4\sqrt{2}) - 3(5\sqrt{2})$
$= 4\sqrt{2} + 3\sqrt{2} - 4\sqrt{2} - 15\sqrt{2}$
$= -12\sqrt{2}$ Expressed in simplest form as a mixed radical.

$\sqrt{50}$
$= \sqrt{25 \times 2}$
$= 5\sqrt{2}$

3.6 exercise

A

1. Write each of the following as a mixed radical in simplest form.
   (a) $\sqrt{32}$
   (b) $\sqrt{48}$
   (c) $-\sqrt{27}$
   (d) $-3\sqrt{32}$
   (e) $2\sqrt{50}$
   (f) $\frac{1}{2}\sqrt{32}$
   (g) $-\frac{3}{2}\sqrt{24}$
   (h) $-6\sqrt{150}$
   (i) $-3\sqrt{98}$
   (j) $-\frac{1}{5}\sqrt{125}$
   (k) $-4\sqrt{27}$
   (l) $2\sqrt{75}$
   (m) $2\sqrt[3]{2}$
   (n) $-4\sqrt[3]{3}$
   (o) $2\sqrt[4]{3}$
   (p) $-3\sqrt{5}$
   (q) $2\sqrt[5]{3}$
   (r) $-3\sqrt{8}$

2. Write each of the following as an entire radical.
   (a) $3\sqrt{2}$
   (b) $-4\sqrt{3}$
   (c) $2\sqrt{5}$
   (d) $5\sqrt{27}$
   (e) $6\sqrt{8}$
   (f) $-2\sqrt{27}$
   (g) $2\sqrt[3]{3}$
   (h) $-3\sqrt[3]{2}$
   (i) $2\sqrt[4]{27}$

3. Which of the following radicals are equivalent to $\sqrt{72}$?
   A: $6\sqrt{2}$ B: $3\sqrt{8}$ C: $6\sqrt{12}$ D: $2\sqrt{18}$

4. Which of the following radicals are equivalent to $\sqrt[3]{128}$?
   A: $8\sqrt[3]{2}$ B: $2\sqrt[3]{16}$ C: $4\sqrt[3]{2}$ D: $4\sqrt[3]{8}$

5. (a) Write $5\sqrt{2}$ as an entire radical.
   (b) Write $4\sqrt{3}$ as an entire radical.
   (c) Use your answers in (a) and (b) to show why $5\sqrt{2} > 4\sqrt{3}$.

6. (a) Write each of the following as an entire radical.
      A: $6\sqrt{2}$ B: $5\sqrt{3}$
   (b) Which is greater, A or B?

7. (a) Write each term as a mixed radical in simplest form.
      $6\sqrt{8} + 2\sqrt{18} - \sqrt{72}$
   (b) Simplify the radical expression in (a).

8. (a) Write each term as a mixed radical in simplest form.
      $6\sqrt{48} + \sqrt{75} - 2\sqrt{28} + 3\sqrt{63}$
   (b) Simplify the radical expression in (a).

9. Simplify each of the following.
   (a) $3\sqrt{2} - 4\sqrt{2} + 5\sqrt{2} - 3\sqrt{2}$
   (b) $5\sqrt{2} - 3\sqrt{3} - 6\sqrt{2} + 5\sqrt{3}$
   (c) $3\sqrt{5} - 4\sqrt{3} - 3\sqrt{5} + 6\sqrt{3}$
   (d) $4\sqrt{3} - 2\sqrt{7} + 3\sqrt{7} - 3\sqrt{3} - 2\sqrt{3}$

10. Simplify. Which of the following, A, B, C, are equivalent?
    A: $2\sqrt{2} - 8\sqrt{3} + \sqrt{2} + 3\sqrt{3}$
    B: $8\sqrt{2} - 3\sqrt{3} - 5\sqrt{2} + 2\sqrt{3}$
    C: $4\sqrt{2} + \sqrt{3} - 6\sqrt{3} - \sqrt{2}$

11. Simplify each of the following.
    (a) $2(3 - 3\sqrt{3})$ (b) $3(3\sqrt{2} - 5)$
    (c) $-2(2\sqrt{3} - 4\sqrt{2})$ (d) $5(3\sqrt{5} + 5\sqrt{6})$
    (e) $-\dfrac{1}{2}(2\sqrt{3} + 4\sqrt[3]{2})$ (f) $3(2\sqrt[3]{2} - 3\sqrt{2})$

B

12. Simplify each of the following.
    (a) $2\sqrt{12} - 5\sqrt{27} + 3\sqrt{48}$
    (b) $-3\sqrt{8} - 2\sqrt{18} + 5\sqrt{72}$
    (c) $\sqrt{20} - 3\sqrt{245} - 2\sqrt{20}$
    (d) $2\sqrt[3]{16} + 3\sqrt[3]{54} - 2\sqrt[3]{128}$
    (e) $2\sqrt{8} - 3\sqrt{98} - 2\sqrt{200}$
    (f) $-3\sqrt{50} - \sqrt{32} + 5\sqrt{200}$

    Why may we refer to each of the answers as a monomial?

13. Simplify each of the following.
    (a) $6\sqrt{8} - 2\sqrt{27} - 3\sqrt{18} + 2\sqrt{3}$
    (b) $2\sqrt{32} - 3\sqrt{20} - 3\sqrt{50} + \sqrt{80}$
    (c) $-2\sqrt{72} + 3\sqrt{28} - 2\sqrt{102} - 6\sqrt{98}$
    (d) $3\sqrt{3} - 2\sqrt{48} - 6\sqrt{20} - 2\sqrt{45}$
    (e) $3\sqrt[3]{81} + \dfrac{1}{2}\sqrt[3]{128} - 3\sqrt[3]{192} + 4\sqrt[3]{54}$

    Why may we refer to each of the answers as a binomial?

---

math tip

The absolute value of a real number, r, is defined as

		Example
$\|r\| = r$	if $r > 0$	$\|5\| = 5$
$\|r\| = 0$	if $r = 0$	$\|0\| = 0$
$\|r\| = -r$	if $r < 0$	$\|-5\| = -(-5) = 5$

Read this as: The absolute value of -5 is 5.

In the *Math Is Plus* at the end of this chapter (pages 117 to 119) we will examine its properties as well as solve equations involving absolute value.

14. Simplify.
    (a) $-3\sqrt{12} + 5\sqrt{27} - 6\sqrt{48} + 2\sqrt{75}$ Which answers are monomials? binomials?
    (b) $12\sqrt{8} - 2\sqrt{27} - 2\sqrt{18} + 6\sqrt{3}$
    (c) $2\sqrt{20} - 3\sqrt{245} - 2\sqrt{20} + \sqrt{125}$
    (d) $2\sqrt{45} - 4\sqrt{63} - 2\sqrt{125} - 9\sqrt{112}$
    (e) $3\sqrt{175} - 2\sqrt{28} + 3\sqrt{63} - \sqrt{112}$
    (f) $-2\sqrt[3]{40} - 3\sqrt[3]{135} + 5\sqrt[3]{320} + 8\sqrt[3]{5}$
    (g) $\sqrt{108} - 2\sqrt{27} - \sqrt{40} - 5\sqrt{160}$
    (h) $4\sqrt[3]{54} - 6\sqrt[3]{81} - 4\sqrt[3]{16} + 3\sqrt[3]{24}$

15. Which radical expression has the greater value?
    A: $4\sqrt{12} - 3\sqrt{27} + 5\sqrt{48} - 3\sqrt{75}$
    B: $2\sqrt{27} - 3\sqrt{48} + \sqrt{108} - 2\sqrt{192}$

16. Which radical expression has the lesser value?
    A: $2\sqrt{12} - \sqrt{12} + 2\sqrt{27} - 2\sqrt{75}$
    B: $2\sqrt{18} - \sqrt{8} - 2\sqrt{8} - \sqrt{18}$

17. Which radical expressions, A, B, C, are equivalent?
    A: $2\sqrt{20} - 2\sqrt{45} + 2\sqrt{80} + \sqrt{20}$
    B: $4\sqrt{20} - 2\sqrt{45} - 2\sqrt{125} + 2\sqrt{20}$
    C: $2\sqrt{45} - 4\sqrt{20} + 3\sqrt{80} - \sqrt{20}$
    Give reasons for your answer.

We may apply the distributive property to simplify radical expressions. Compare

$$2(3a - 2b) \qquad\qquad 2(3\sqrt{2} - 4\sqrt{3})$$
$$= 6a - 4b \qquad\qquad = 6\sqrt{2} - 8\sqrt{3}$$

18. (a) Simplify $3(2\sqrt{8} - 3\sqrt{125})$.
    (b) Simplify $3(4\sqrt{2} - 15\sqrt{5})$.
    (c) Why are your answers in (a) and (b) the same?

19. Simplify.
    (a) $-2(5\sqrt{2} + 5) + 5(6 - 3\sqrt{2})$
    (b) $3(3\sqrt{2} - 3\sqrt{3}) - 2(4\sqrt{2} - 2\sqrt{3})$
    (c) $-2(2\sqrt{12} - \sqrt{18}) - 5(3\sqrt{32} - \sqrt{27})$
    (d) $3(3\sqrt[3]{40} - \sqrt[3]{135}) + 4(\sqrt[3]{320} - \sqrt[3]{10})$

20. Express the value of each expression in simplest form if $a = 3\sqrt{2} - 4\sqrt{3}$, $b = 2\sqrt{3} - 2\sqrt{2}$.
    (a) $a + b$ (b) $2a + b$ (c) $3b - 2a$
    (d) $3a + 2b$ (e) $4b + 6a$ (f) $\frac{1}{2}(2a - 3b)$

21. Express the value of each expression in simplest form if $a = 2\sqrt{50} - 3\sqrt{32}$, $b = 2\sqrt{75} - 2\sqrt{27}$.
    (a) $2a + b$ (b) $a - 2b$ (c) $3b + a$
    (d) $a + 3b$ (e) $3b + 2a$ (f) $3a - 3b$

22. If $a = 2\sqrt[3]{16} - \sqrt[3]{54}$, $b = 3\sqrt[3]{81} - 2\sqrt[3]{24}$, express each of the following in simplest form.
    (a) $3a + b$ (b) $2b - 3a$ (c) $3a + 6b$
    (d) $4a - 2b$ (e) $3a + 2b$ (f) $\frac{2}{3}(2a - 3b)$

23. Simplify each of the following.   Remember $\frac{3\sqrt{8}}{2} = \frac{3}{2}\sqrt{8}$.
    (a) $\frac{3}{2}\sqrt{8} - \frac{4}{3}\sqrt{27} + 6\sqrt{50}$
    (b) $\frac{2\sqrt{27}}{3} - 3\sqrt{48} + \frac{4\sqrt{50}}{5} - \frac{4\sqrt{18}}{3}$
    (c) $-\frac{1}{2}\sqrt{80} - \frac{3}{4}\sqrt{64} + \frac{\sqrt{20}}{2} - 3\sqrt{16}$
    (d) $\frac{5\sqrt{100}}{2} - \frac{2\sqrt{99}}{3} - 4\sqrt{125} + \frac{6\sqrt{49}}{2}$
    (e) $\frac{2}{3}\sqrt[3]{81} - \frac{1}{2}\sqrt[3]{24} + \frac{2\sqrt[3]{135}}{3} - \frac{3\sqrt[3]{40}}{2}$

radicals and literal expressions

We may extend our previously acquired skills to add and subtract radical expressions. If $a \neq b$ then $\sqrt{a}$ and $\sqrt{b}$ are *unlike* radicals. For $\sqrt{b}$, we restrict the values of b so that $b \geq 0$, but for $\sqrt[3]{b}$, b may be positive or negative. To simplify expressions, we may write mixed radicals as shown.

$$\sqrt{9a} - \sqrt{4b} + \sqrt{36a} + \sqrt{25b}$$
$$= 3\sqrt{a} - 2\sqrt{b} + 6\sqrt{a} + 5\sqrt{b}$$
$$= 9\sqrt{a} + 3\sqrt{b}$$

Once we have simplified the expression, we may evaluate for given values of a and b.
If $a = 9$, $b = 25$, then
$$9\sqrt{a} + 3\sqrt{b} = 9\sqrt{9} + 3\sqrt{25}$$
$$= 9(3) + 3(5)$$
$$= 27 + 15$$
$$= 42$$

If the variables occur as radicands we may need to use our skills with absolute value to simplify expressions and write the following.
$$\sqrt{36y^2} = 6|y|$$

In the above example if $y = -2$,

LS $= \sqrt{36y^2}$ RS $= 6|y|$
$= \sqrt{36(-2)^2}$ $= 6|-2|$
$= \sqrt{144}$ $= 6(2)$
$= 12$ $= 12$

However, note the answer if $y = -2$ is used to calculate the following expression.
$$6y = 6(-2) = -12$$

Thus $\sqrt{36y^2} \neq 6y$ but rather $\sqrt{36y^2} = 6|y|$.

24 Simplify each of the following.
(a) $3a\sqrt{8} - 2b\sqrt{32} + 3b\sqrt{50} - 2a\sqrt{72}$
(b) $-m\sqrt{27} + 3n\sqrt{12} - 2m\sqrt{75} - 3n\sqrt{48}$
(c) $y\sqrt{20} + 3x\sqrt{80} + 2y\sqrt{45} - 3x\sqrt{125}$
(d) $-2m\sqrt{8} - 3m\sqrt{27} + 6m\sqrt{18} - 5m\sqrt{48}$

25 Find the coefficient of $\sqrt{2}$ in the following expression.
$$6m\sqrt{8} - 3m\sqrt{18} + 2m\sqrt{32} + 6m\sqrt{2}$$

26 What is the coefficient of $\sqrt{5}$ in the following expression?
$$3x\sqrt{20} - 2y\sqrt{45} + 5x\sqrt{80} - 6y\sqrt{125}$$

27 What is the coefficient of $\sqrt{3}$ in the following expression?
$$3a\sqrt{18} - 3a\sqrt{27} - 2b\sqrt{72} - 6b\sqrt{48}$$

28 What is the coefficient of $\sqrt{5}$ in the following expression?
$$-3x\sqrt{20} + 2y\sqrt{24} + 5x\sqrt{45} - 3y\sqrt{54}$$

29 (a) Simplify:
$$6a\sqrt{8} - 2b\sqrt{50} - 46b\sqrt{32} + 3a\sqrt{98}$$
(b) Find the value of the expression in (a) if $a = -3$ and $b = 2$.

30 (a) Evaluate each expression if $p = 4$, $q = -3$.
A: $\sqrt{36pq^2}$ B: $6|q|\sqrt{p}$ C: $6q\sqrt{p}$
(b) Why are the expressions A and B equivalent?
(c) Why are the expressions A and C *not* equivalent?

31 (a) Evaluate each expression if $a = -2$, $y = -1$.
A: $\sqrt{a^2y^4}$ B: $|ay^2|$ C: $y^2|a|$
(b) Why are the expressions A, B, and C equivalent?

32 Express each of the following as mixed radicals in simplest form.
(a) $\sqrt{8x^2}$ (b) $\sqrt{98x^3}$ (c) $\sqrt{18x^4}$
(d) $\sqrt{36a^2b^3}$ (e) $\sqrt{27a^4b^3}$ (f) $\sqrt{72x^5y^2}$
(g) $\sqrt{80m^4n^3}$ (h) $\sqrt{64x^2yz^4}$ (i) $\sqrt{112a^3b^4}$
(j) $\sqrt[3]{16a^3b^3}$ (k) $\sqrt[3]{-54x^4b^3}$ (l) $\sqrt[3]{128a^5y^3}$
(m) $\sqrt[4]{81x^4y^2}$ (n) $\sqrt[3]{24p^3q^5}$ (o) $\sqrt[4]{256s^4t^5}$

33 Simplify.
(a) $3\sqrt{x^2} - 2\sqrt{x^3} - 5\sqrt{x^4}$
(b) $4\sqrt{b^3} - 3\sqrt{b^5} - 2\sqrt{b^3}$
(c) $3\sqrt{x^2y} - 2\sqrt{x^4y^3}$
(d) $-3\sqrt{a^3b^5} - 2\sqrt{a^5b^3} + 5\sqrt{a^7b^3}$
(e) $3\sqrt[3]{x^4y^7} - 6\sqrt[3]{x^7y^4} + 2\sqrt[3]{x^7y^{10}}$

34 Find the value of the following expression if $a = 3, b = 2$.
$\sqrt{4a} + a\sqrt{a^2b} + \sqrt{b^2a} + b\sqrt{9a}$

35 If $a = 1, b = 3$, find the value of
$b\sqrt{b} - \sqrt{16b} + \sqrt{9ab^2} - \sqrt{25a^2b}$.

36 (a) Simplify the following expression.
$\sqrt{16b} + 3\sqrt{9a} + 3a\sqrt{b^3} - 2\sqrt{a^3}$
(b) If $a \geq 0, b \geq 0$ in the expression in (a), how would you write your answer?

37 Simplify each of the following if $a \geq 0$ and $b \geq 0$.
(a) $\sqrt{a^2b} - 2\sqrt{ab^2} - \sqrt{9b}$
(b) $\sqrt{ab} - \sqrt{ab^3} - \sqrt{9a^3b^3} - \sqrt{a^3b}$
(c) $\sqrt{4b} - 2\sqrt{25a^2b} - 3\sqrt{a^2b^3} + 7\sqrt{b}$
(d) $6\sqrt{a^2b} - \sqrt{4ab^2} - \sqrt{9b} + \sqrt{64a}$

38 Simplify each of the following.
(a) $7\sqrt{b^3} + \sqrt{4a^2b} - \sqrt{4b}$
(b) $8\sqrt{49b} - 7\sqrt{9b^3} + \sqrt{4a} + \sqrt{a^3}$
(c) $7\sqrt{a^3b} + 9\sqrt{9a^3b^3} - \sqrt{a^3b^3} + 2\sqrt{4ab}$

> **problem/matics**
> Is the following statement true or false?
> In a right-angled triangle, the area of the semi-circle drawn on the hypotenuse as the diameter is equal to the sum of the areas of the semi-circles drawn on the other sides as diameters.

3.7 working with radicals: multiplication

In the study of mathematics, a mathematician will often use a numerical example to see whether a general mathematical statement is true.
For example, is $\sqrt{4} \times \sqrt{9} = \sqrt{36}$?

$$\begin{aligned} \text{LS} &= \sqrt{4} \times \sqrt{9} & \text{RS} &= \sqrt{36} \\ &= 2 \times 3 & &= 6 \\ &= 6 \end{aligned}$$

Since LS = RS, then $\sqrt{4} \times \sqrt{9} = \sqrt{36}$.
The above example suggests that the following statement might be true:

$$\sqrt{m} \times \sqrt{n} = \sqrt{mn}, m \geq 0, n \geq 0$$

To prove the statement for all m and n, let us recall that by the definition of $\sqrt{m}$, it follows that $\sqrt{m} \times \sqrt{m} = m$.

Let $x = \sqrt{m} \times \sqrt{n}$. Thus $x \geq 0$.
Then $x^2 = (\sqrt{m} \times \sqrt{n})^2$ (Why?)
$x^2 = (\sqrt{m})^2 (\sqrt{n})^2$
$x^2 = mn$
Thus $x = \pm \sqrt{mn}$
Since $x \geq 0$, then $x = +\sqrt{mn}$, or $\sqrt{mn}$. ①
But we let $x = \sqrt{m} \times \sqrt{n}$. ②
Thus $\sqrt{m} \times \sqrt{n} = \sqrt{mn}$.
   ①   ②

In our earlier work in mathematics we often accepted rules without proof and used them. The more mathematics we learn, the more we are able to prove statements in general and then use the results.

In studying mathematics, we may also apply the skills we have developed earlier. We now apply them to multiplying radicals.

Multiplying Monomials   Multiplying Radicals
$(2a)(-4b)$ $(2\sqrt{2})(-4\sqrt{3})$
$= -8ab$ $= -8\sqrt{6}$

Example 1
Simplify $(3\sqrt{2})(-2\sqrt{3}) - (5\sqrt{2})(-3\sqrt{3})$.

Solution
$(3\sqrt{2})(-2\sqrt{3}) - (5\sqrt{2})(-3\sqrt{3})$
$= -6\sqrt{6} + 15\sqrt{6}$
$= 9\sqrt{6}$

To multiply monomials, we may first wish to write the terms in simplest form. Compare these two methods.

$(3\sqrt{8})(2\sqrt{12})$	$(3\sqrt{8})(2\sqrt{12})$
$= (6\sqrt{2})(4\sqrt{3})$ This step often	$= 6\sqrt{96}$
$= 24\sqrt{6}$ —— simplifies subsequent calculations.	$= 6\sqrt{16 \times 6}$
	$= 24\sqrt{6}$

We apply our skills for finding the product of binomials in the next example.

Remember that $(\sqrt{5})(\sqrt{5}) = 5$ or $(\sqrt{5})^2 = 5$.

With practice, each step that is shown may be done mentally. Practise completing the steps mentally when simplifying binomials involving radicals.

$(2\sqrt{5} - 3)(3\sqrt{5} + 2)$

a: $(2\sqrt{5})(3\sqrt{5}) = 30$
b: $(-3)(2) = -6$
c: $(-3)(3\sqrt{5}) = -9\sqrt{5}$
d: $(2\sqrt{5})(2) = 4\sqrt{5}$

With practice we may do these steps mentally.

$= 30 + 4\sqrt{5} - 9\sqrt{5} - 6$
$= 30 - 5\sqrt{5} - 6$
$= 24 - 5\sqrt{5}$

Example 2
Simplify $(2\sqrt{5} - 3)(3\sqrt{5} + 2) - 3(3\sqrt{5} - 1)^2$.

Solution
$(2\sqrt{5} - 3)(3\sqrt{5} + 2) - 3(3\sqrt{5} - 1)^2$
$= 30 + 4\sqrt{5} - 9\sqrt{5} - 6 - 3(45 - 6\sqrt{5} + 1)$
$= 24 - 5\sqrt{5} - 3(46 - 6\sqrt{5})$
$= 24 - 5\sqrt{5} - 138 + 18\sqrt{5}$
$= -114 + 13\sqrt{5}$

There are some skills that occur in almost every branch of mathematics. The skill of substituting is one of those skills that occur frequently and is important to know. This skill is used in the next example.

Example 3
If $m = 2\sqrt{3} - 3\sqrt{2}$ and $n = 2\sqrt{3} + 3\sqrt{2}$, find the value of each of the following in simplest form.
  A mn B $(m - n)(m + n)$ C $m^2 + mn$

Solution
A Recall that $(x + y)(x - y) = x^2 - y^2$.
Thus $(2\sqrt{3} - 3\sqrt{2})(2\sqrt{3} + 3\sqrt{2})$
$= (2\sqrt{3})^2 - (3\sqrt{2})^2$
$= 12 - 18$
$= -6$

B We first simplify $(m - n)$ and $(m + n)$ separately.
$m - n = 2\sqrt{3} - 3\sqrt{2} - (2\sqrt{3} + 3\sqrt{2})$
$= 2\sqrt{3} - 3\sqrt{2} - 2\sqrt{3} - 3\sqrt{2}$
$= -6\sqrt{2}$
$m + n = 2\sqrt{3} - 3\sqrt{2} + 2\sqrt{3} + 3\sqrt{2}$
$= 4\sqrt{3}$
$(m - n)(m + n) = (-6\sqrt{2})(4\sqrt{3})$
$= -24\sqrt{6}$

C We use our factoring skills to simplify the calculations.
$m^2 + mn = m(m + n)$
$= (2\sqrt{3} - 3\sqrt{2})(4\sqrt{3})$
$= 24 - 12\sqrt{6}$

To develop our skills with radicals, we will again apply many skills we have learned earlier: substituting, factoring, collecting like terms, to name a few.

3.7 exercise

Questions 1 to 11 examine skills needed to simplify expressions which contain radicals.

A

1. (a) Find each product.
   A: $(3\sqrt{2})(-2\sqrt{3})$ B: $(3\sqrt{6})(2\sqrt{3})$
   (b) Which product may be written in simpler form?

2. Find each product.
   (a) $(6\sqrt{2})(-12\sqrt{3})$ (b) $(3\sqrt{8})(-3\sqrt{48})$
   (c) Why are the answers the same? Write each answer in simplest form.

3. (a) To find the product $(-3\sqrt{75})(-2\sqrt{48})$, what first step may be taken?
   (b) Find the product in (a).

4. To find the product $(-3\sqrt{48})(2\sqrt{50})$, compare the following.

Method A	Method B
$(-3\sqrt{48})(2\sqrt{50})$	$(-3\sqrt{48})(2\sqrt{50})$
$= (-12\sqrt{3})(10\sqrt{2})$	$= -6\sqrt{2400}$
$= -120\sqrt{6}$	$= -6\sqrt{400 \times 6}$
	$= -120\sqrt{6}$

   Which method do you prefer?

5. Write each product in simplest form.
   (a) $(2\sqrt{3})(3\sqrt{2})$ (b) $(4\sqrt{6})(-2\sqrt{5})$
   (c) $(3\sqrt{5})(5\sqrt{3})$ (d) $(2\sqrt{3})\left(-\frac{1}{2}\sqrt{5}\right)$
   (e) $(-3\sqrt{3})(-5\sqrt{5})$ (f) $(-\sqrt{10})(-2\sqrt{2})$
   (g) $(3\sqrt{2})(5\sqrt{15})$ (h) $\left(-\frac{3}{4}\sqrt{8}\right)(2\sqrt{2})$
   (i) $(6\sqrt{8})(-3\sqrt{2})$ (j) $\left(\frac{2}{5}\sqrt{10}\right)(-3\sqrt{15})$

6. Simplify. Write each product in simplest form.
   (a) $(-3\sqrt{3})(4\sqrt{6})(3\sqrt{7})$
   (b) $(-4\sqrt{5})(2\sqrt{3})(4\sqrt{6})$
   (c) $(2\sqrt{8})(4\sqrt{3})(-3\sqrt{2})$
   (d) $(3\sqrt{48})\left(\frac{1}{2}\sqrt{50}\right)(-3\sqrt{2})$
   (e) $\left(\frac{2}{3}\sqrt{75}\right)(-2\sqrt{12})(3\sqrt{8})$
   (f) $(-3\sqrt{20})\left(\frac{1}{2}\sqrt{48}\right)(5\sqrt{32})$

7. Simplify each of the following.
   (a) $3\sqrt{5}(2\sqrt{2} - 3\sqrt{3})$ (b) $-3\sqrt{3}(3\sqrt{6} - 3\sqrt{2})$
   (c) $-4\sqrt{3}(2\sqrt{6} - 2\sqrt{5})$ (d) $3\sqrt{3}(2\sqrt{2} - 4)$
   (e) $-3\sqrt{5}(3\sqrt{2} - 4)$ (f) $2\sqrt{3}(5 - 3\sqrt{3})$
   (g) $-6\sqrt{2}(3\sqrt{2} - 3\sqrt{3})$ (h) $2\sqrt{5}(3\sqrt{3} - 5\sqrt{2})$

8. (a) To find the following product, what first step may be taken?
   $2\sqrt{3}(2\sqrt{48} - 3\sqrt{72})$
   (b) Find the product in (a).

9. Find each product.
   (a) $3\sqrt{3}(3\sqrt{8} - 2\sqrt{18})$
   (b) $-2\sqrt{2}(3\sqrt{12} - 5\sqrt{27})$
   (c) $-3\sqrt{5}(4\sqrt{20} - 2\sqrt{45})$
   (d) $2\sqrt{6}(-3\sqrt{24} - 5\sqrt{54})$
   (e) $\frac{1}{2}\sqrt{3}(2\sqrt{48} - 3\sqrt{32})$ (f) $\frac{3}{2}\sqrt{2}(2\sqrt{18} - 3\sqrt{48})$
   (g) $2\sqrt{5}(\sqrt{125} - \sqrt{108})$ (h) $2\sqrt{5}(2\sqrt{80} - 3\sqrt{98})$

10. (a) Find the product.
    $(a + 2b)(3a - b)$
    (b) Use the steps in (a) to help you find the product $(\sqrt{2} + 2\sqrt{3})(3\sqrt{2} - \sqrt{3})$.

11 Use the statement that $(a + b)^2 = a^2 + 2ab + b^2$ to find the square of each binomial.
   (a) $(\sqrt{2} + 3\sqrt{3})$ (b) $(3\sqrt{6} - 2\sqrt{3})$

B

12 Simplify each product.
   (a) $(\sqrt{3} + \sqrt{2})(\sqrt{3} - 2\sqrt{2})$
   (b) $(2\sqrt{2} - \sqrt{3})(3\sqrt{2} - 3\sqrt{3})$
   (c) $(2\sqrt{5} - 2\sqrt{2})(3\sqrt{5} - 3\sqrt{2})$
   (d) $(3\sqrt{3} - 2\sqrt{5})(5\sqrt{5} - 3\sqrt{3})$
   (e) $(3\sqrt{5} - 3\sqrt{3})(3\sqrt{3} - 2\sqrt{5})$

Did you write your answers in simplest form?

13 Simplify each of the following.
   (a) $(\sqrt{50} - \sqrt{75})(\sqrt{32} - \sqrt{48})$
   (b) $(\sqrt{125} - \sqrt{75})(\sqrt{80} - \sqrt{48})$
   (c) $(3\sqrt{18} - 3\sqrt{27})(2\sqrt{8} - 2\sqrt{12})$
   (d) $(\sqrt{80} - 2\sqrt{27})(-3\sqrt{20} - 3\sqrt{12})$

14 The measures of the rectangles are given. Find a radical expression for their areas.

   (a)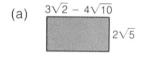
   (b)
   (c)
   (d)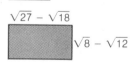

15 Find the square of each binomial.
   (a) $(\sqrt{5} + \sqrt{3})$ (b) $(2\sqrt{3} + 2\sqrt{2})$
   (c) $(3\sqrt{2} - 2\sqrt{5})$ (d) $(4\sqrt{2} - 3\sqrt{8})$
   (e) $(3\sqrt{3} - \sqrt{2})$ (f) $(5\sqrt{7} - 3\sqrt{10})$
   (g) $(2\sqrt{3} - \sqrt{6})$ (h) $(5\sqrt{3} + 5)$
   (i) $8 - 3\sqrt{5}$ (j) $2\sqrt{6} - 2\sqrt{3}$
   (k) $5\sqrt{5} - 3\sqrt{10}$ (l) $2\sqrt{6} - 3\sqrt{2}$

16 Simplify each of the following.
   (a) $(\sqrt{3} + \sqrt{5})^2$ (b) $(\sqrt{3} - \sqrt{5})^2$
   (c) $(8 - \sqrt{3})^2$ (d) $(8 + \sqrt{3})^2$
   (e) $(2\sqrt{3} - \sqrt{2})^2$ (f) $(5\sqrt{3} + \sqrt{5})^2$
   (g) $(2\sqrt{3} - \sqrt{6})^2$ (h) $(3\sqrt{2} - \sqrt{10})^2$

17 The measures of the sides of each square are given. Find a radical expression for each area.
   (a) $2\sqrt{3} - \sqrt{2}$ (b) $\sqrt{32} - \sqrt{18}$
   (c) $2\sqrt{50} - 2\sqrt{12}$ (d) $\sqrt{45} - 2\sqrt{8}$

18 Simplify each of the following.
   (a) $(2\sqrt{5} - \sqrt{3})(\sqrt{5} + 2\sqrt{3}) - (3\sqrt{5} - 2\sqrt{3})^2$
   (b) $(3\sqrt{2} - 8)^2 + (5\sqrt{2} + 4)(-2\sqrt{2} - 5)$
   (c) $3(3\sqrt{2} - 3\sqrt{3})^2 - (4\sqrt{2} - 3\sqrt{3})(3\sqrt{2} - 2\sqrt{3})$
   (d) $2(2\sqrt{5} - 3\sqrt{3})(3\sqrt{5} + \sqrt{3}) - 4(\sqrt{5} + \sqrt{3})^2$
   (e) $4(3\sqrt{6} - 2\sqrt{2})^2 - 8(5\sqrt{6} - 2\sqrt{2})^2$

19 (a) What first step may be taken to simplify
       $(3\sqrt{8} - 2\sqrt{2})^2 - (5\sqrt{8} - 3\sqrt{2})^2$?
   (b) Simplify the expression in (a).

20 Simplify each of the following.
   (Look for ways to simplify your calculations.)
   (a) $(\sqrt{32} - \sqrt{50})(\sqrt{18} - \sqrt{2}) - 2(\sqrt{72} - \sqrt{98})^2$
   (b) $3(\sqrt{12} - \sqrt{27})^2 + (\sqrt{48} - \sqrt{75})(-3\sqrt{48} - 2\sqrt{75})$
   (c) $(\sqrt{32} - \sqrt{48})^2 - (\sqrt{50} - \sqrt{75})^2$
   (d) $3(\sqrt{125} - 2\sqrt{48})^2 - 2(3\sqrt{20} - 2\sqrt{27})^2$

21 Find each product.
   (a) $(\sqrt{3} - \sqrt{5})(\sqrt{3} + \sqrt{5})$

(b) $(2\sqrt{3} + \sqrt{2})(2\sqrt{3} - \sqrt{2})$
(c) $(3\sqrt{5} - \sqrt{6})(3\sqrt{5} + \sqrt{6})$
(d) $(2\sqrt{3} + 4\sqrt{2})(2\sqrt{3} - 4\sqrt{2})$
(e) What do you notice about your answers in (a) to (d)?

In general, the expressions $(a\sqrt{b} - c\sqrt{d})$ and $(a\sqrt{b} + c\sqrt{d})$ are called **conjugates**. The product of a pair of conjugates is a rational number.

---

22 Find the product of each binomial and its corresponding conjugate.
(a) $\sqrt{5} - \sqrt{2}$
(b) $3\sqrt{5} + 2\sqrt{2}$
(c) $3\sqrt{2} - 2\sqrt{5}$
(d) $5\sqrt{7} + 2\sqrt{10}$
(e) $2\sqrt{3} - 8$
(f) $9 + 2\sqrt{5}$
(g) $3\sqrt{5} - 2\sqrt{10}$
(h) $3\sqrt{6} + \sqrt{8}$
(i) $\sqrt{18} - \sqrt{27}$
(j) $\sqrt{50} - 2\sqrt{80}$
(k) $3\sqrt{12} - 2\sqrt{32}$
(l) $4\sqrt{98} - 3\sqrt{45}$

23 Find each product.
(Look for conjugates to simplify your work.)
(a) $(\sqrt{3} - \sqrt{2})(3\sqrt{3} - \sqrt{2})(\sqrt{3} + \sqrt{2})$
(b) $(2\sqrt{5} - 3\sqrt{2})(2\sqrt{5} + 3\sqrt{2})(3\sqrt{5} - \sqrt{2})$
(c) $(4\sqrt{6} - 3\sqrt{2})(4\sqrt{6} + 3\sqrt{2}) - (2\sqrt{6} - \sqrt{8})^2$
(d) $3(2\sqrt{27} - 3)^2 - (2\sqrt{3} - 8)(\sqrt{12} + 8)$
(e) $(5\sqrt{6} - 2\sqrt{3})(3\sqrt{2} - \sqrt{3})(3\sqrt{2} + \sqrt{3})$
(f) $(3\sqrt{2} + 5\sqrt{3})(3\sqrt{2} - 5\sqrt{3}) - 4(\sqrt{18} - \sqrt{27})^2$

24 Simplify.
(a) $(2\sqrt{32} - 4\sqrt{48} + 2\sqrt{18})(2\sqrt{27} - 2\sqrt{50} + 5\sqrt{75})$
(b) $(4\sqrt{20} - 3\sqrt{27} + 3\sqrt{80})(3\sqrt{48} - 2\sqrt{125} - 4\sqrt{45})$
(c) $(3\sqrt{48} - 2\sqrt{27} - \sqrt{12})(\sqrt{50} - 3\sqrt{32} - 5\sqrt{8})$

25 If $a = 3\sqrt{2}$, $b = -2\sqrt{3}$, simplify each of the following.
(a) ab
(b) $-3ab$
(c) $\frac{1}{2}ab$
(d) $a^2 + b^2$
(e) $a^2 - b^2$
(f) $a^2 - 2b^2$
(g) $a^2 + 2ab$
(h) $b^2 - 2ab$

26 If $m = 3\sqrt{5} - 2$, $n = 2\sqrt{5} + 5$, find the value of
(a) $2mn$
(b) $-3mn$
(c) $m^2 - n^2$

C

27 Find a radical expression for the shaded area.

$3\sqrt{50} - 2\sqrt{32} + \sqrt{12}$

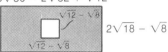

$2\sqrt{18} - \sqrt{8}$

28 Prove that the product of any pair of conjugate binomials is always a rational number.

29 (a) For all $a, b \in R$ prove that
$\sqrt[3]{a}\sqrt[3]{b} = \sqrt[3]{ab}$.
(b) Use a numerical example that suggests the generalization in (a).

30 (a) For all $a, b \in R$, prove that
$\sqrt[4]{x}\sqrt[4]{y} = \sqrt[4]{xy}$.
(b) Use a numerical example that suggests the generalization in (a).

31 For all real numbers p, q prove that
$\sqrt[n]{p}\sqrt[n]{q} = \sqrt[n]{pq}$ n, a natural number.

problem/matics

The floor of a gym is 12 m × 30 m and is 12 m high. A spider sits 1 m above the floor on the centre line of the east wall. A sleeping bug is 1 m from the ceiling on the west wall. What is the shortest distance from the spider to the bug? (Remember spiders don't fly.)

applications: roots and radicals

We have learned that a root of an equation satisfies the equation.

Example
Show that $2\sqrt{2} + 3$ is a root of the quadratic equation $x^2 - 6x + 1 = 0$

Solution
Verify the root in both sides of the equation.
$$LS = x^2 - 6x + 1 \qquad RS = 0$$
$$= (2\sqrt{2} + 3)^2 - 6(2\sqrt{2} + 3) + 1$$
$$= (8 + 12\sqrt{2} + 9) - 12\sqrt{2} - 18 + 1$$
$$= 12\sqrt{2} - 12\sqrt{2} + 17 - 17 = 0$$
$$LS = RS$$
Thus, $2\sqrt{2} + 3$ is a root of the equation $x^2 - 6x + 1 = 0$.

32 Show that $\sqrt{3} - 1$ is a root of the quadratic equation $x^2 + 2x - 2 = 0$.

33 Show that $7 + \sqrt{5}$ is a root of the equation $x^2 - 14x + 44 = 0$.

34 (a) Show that $\sqrt{7} - \sqrt{5}$ is a root of the equation $x^2 - 2\sqrt{7}x + 2 = 0$.
    (b) Test whether $\sqrt{7} + \sqrt{5}$ is a root of the above equation.

35 (a) Show that $6 - \sqrt{3}$ is root of $x^2 - 12x + 33 = 0$.
    (b) Test whether $6 + \sqrt{3}$ is also a root of the above equation.

36 Show that $7 - \sqrt{3}$ is not a root of $x^2 - 14x + 45 = 0$.

37 Show that $3 - \sqrt{3}$ is not a root of $x^2 - 3x - 2 = 0$.

38 Show that $\sqrt{6} + \sqrt{3}$ is a root of the equation $2x^2 - \sqrt{3}x + 3 = x^2 + \sqrt{3}x + 6$.

39 Show that $\sqrt{3} - \sqrt{2}$ is a root of the equation $x(2x + 5\sqrt{2}) - 4 = x^2 + 3(\sqrt{2}x - 1)$.

3.8 division with radicals

To multiply second order radicals, we proved the general result that for $a, b \in R$, $a \geq 0, b \geq 0$
$$\sqrt{a} \times \sqrt{b} = \sqrt{ab}$$
The above general example was suggested by numerical examples, such as
$$\sqrt{4} \times \sqrt{9} = \sqrt{36} \text{ or } \sqrt{9} \times \sqrt{16} = \sqrt{144}.$$
For n^{th} order radicals, we may write $a, b \in R$, $n \in N$ then
$$\sqrt[n]{a} \times \sqrt[n]{b} = \sqrt[n]{ab}.$$
In a similar way, numerical examples suggest a rule for dividing radicals.

Is $\dfrac{\sqrt{36}}{\sqrt{4}} \stackrel{?}{=} \sqrt{\dfrac{36}{4}}$ 　　$LS = \dfrac{\sqrt{36}}{\sqrt{4}}$ 　　$RS = \sqrt{\dfrac{36}{4}}$

$$= \dfrac{6}{2} = 3 \qquad = \sqrt{9} = 3$$

Since $LS = RS$ then $\dfrac{\sqrt{36}}{\sqrt{4}} = \sqrt{\dfrac{36}{4}}$.

In general, we may prove that
$$\dfrac{\sqrt{a}}{\sqrt{b}} = \sqrt{\dfrac{a}{b}}, a, b \in R, a \geq 0, b > 0$$

Let $x = \dfrac{\sqrt{a}}{\sqrt{b}}$ 　　① 　　Thus $x \geq 0$. 　　Why?

Then $x^2 = \left(\dfrac{\sqrt{a}}{\sqrt{b}}\right)^2 = \dfrac{(\sqrt{a})^2}{(\sqrt{b})^2} = \dfrac{a}{b}$

Thus $x = \sqrt{\dfrac{a}{b}}$ 　　② 　　since $x \geq 0$.

From ① and ② we may write
$$\dfrac{\sqrt{a}}{\sqrt{b}} = \sqrt{\dfrac{a}{b}}, a, b \in R, a \geq 0, b > 0.$$

Our skills with algebra extend to simplifying radical expressions, as shown in the next example.

Example 1

Simplify $\dfrac{2\sqrt{10} + 3\sqrt{30}}{\sqrt{5}} + \dfrac{3\sqrt{6} - 4\sqrt{18}}{\sqrt{3}}$.

Solution

$\dfrac{2\sqrt{10} + 3\sqrt{30}}{\sqrt{5}} + \dfrac{3\sqrt{6} - 4\sqrt{18}}{\sqrt{3}}$
$= \dfrac{2\sqrt{10}}{\sqrt{5}} + \dfrac{3\sqrt{30}}{\sqrt{5}} + \dfrac{3\sqrt{6}}{\sqrt{3}} - \dfrac{4\sqrt{18}}{\sqrt{3}}$
$= 2\sqrt{2} + 3\sqrt{6} + 3\sqrt{2} - 4\sqrt{6}$
$= 5\sqrt{2} - \sqrt{6}$

To find the value of some radical expressions we may use our knowledge of conjugate radicals.

Example 2

(a) Simplify $\dfrac{3\sqrt{3} - 2\sqrt{2}}{\sqrt{3} - \sqrt{2}}$.

(b) Find the value of the expression above to 2 decimal places.

Solution

(a) To obtain a denominator that is a rational number use a conjugate radical expression as shown.
$\dfrac{3\sqrt{3} - 2\sqrt{2}}{\sqrt{3} - \sqrt{2}} = \dfrac{(3\sqrt{3} - 2\sqrt{2})(\sqrt{3} + \sqrt{2})}{(\sqrt{3} - \sqrt{2})(\sqrt{3} + \sqrt{2})}$
$= \dfrac{5 + \sqrt{6}}{1} = 5 + \sqrt{6}$

(b) Use $\sqrt{6} \doteq 2.449$
Then $5 + \sqrt{6} \doteq 5 + 2.449 \doteq 7.449$

Thus $5 + \sqrt{6} \doteq 7.45$ (to 2 decimal places)

For n^{th} order radicals, we may use
$\dfrac{\sqrt[n]{a}}{\sqrt[n]{b}} = \sqrt[n]{\dfrac{a}{b}}$
to simplify expressions.

Example 3

Simplify $\dfrac{2\sqrt[3]{15} - 3\sqrt[3]{10}}{\sqrt[3]{5}}$.

Solution

$\dfrac{2\sqrt[3]{15} - 3\sqrt[3]{10}}{\sqrt[3]{5}} = \dfrac{2\sqrt[3]{15}}{\sqrt[3]{5}} - \dfrac{3\sqrt[3]{10}}{\sqrt[3]{5}}$
$= 2\sqrt[3]{\dfrac{15}{5}} - 3\sqrt[3]{\dfrac{10}{5}}$
$= 2\sqrt[3]{3} - 3\sqrt[3]{2}$

Often in order to compare radicals we write them in simplest form.

	Written in simplest form		Not in simplest form	
A:	$2\sqrt{3}$	$3\sqrt{5}$	$\sqrt{12}$	$\sqrt{45}$
B:	$\dfrac{\sqrt{6}}{2}$	$\dfrac{\sqrt{15}}{3}$	$\dfrac{\sqrt{3}}{\sqrt{2}}$	$\dfrac{\sqrt{60}}{6}$

3.8 exercise

Questions 1 to 7 examine skills needed to divide expressions that contain radicals.

A

1. Which of the following radical expressions are *not* in simplest form? Simplify

(a) $\dfrac{4\sqrt{5}}{3}$ (b) $\dfrac{\sqrt{33}}{\sqrt{11}}$ (c) $\dfrac{8\sqrt{2}}{16}$ (d) $3\sqrt{33}$

(e) $\dfrac{6\sqrt{3}}{2}$ (f) $\dfrac{-\sqrt{48}}{\sqrt{27}}$ (g) $3\sqrt{27}$ (h) $\sqrt{\dfrac{2}{3}}$

(i) $\dfrac{-3\sqrt{27}}{6}$ (j) $\dfrac{\sqrt[3]{16}}{2}$ (k) $\dfrac{\sqrt[3]{9}}{3}$ (l) $\sqrt[4]{\dfrac{1}{8}}$

2. (a) Simplify each of the following.

A: $\dfrac{\sqrt{120}}{2}$ B: $\dfrac{\sqrt{120}}{\sqrt{2}}$

(b) Why do your answers differ?

3. Which two expressions do not have the same answer when written in simplest form?

(a) $\dfrac{3}{\sqrt{10}}$ (b) $\dfrac{\sqrt{9}}{\sqrt{10}}$ (c) $\dfrac{9}{\sqrt{30}}$ (d) $\dfrac{\sqrt{18}}{\sqrt{20}}$

(e) $\dfrac{6}{\sqrt{40}}$ (f) $\dfrac{3\sqrt{3}}{\sqrt{30}}$ (g) $\dfrac{2\sqrt{3}}{\sqrt{30}}$ (h) $\dfrac{\sqrt{18}}{\sqrt{20}}$

4. Simplify.

(a) $\dfrac{6\sqrt{10}}{\sqrt{5}}$ (b) $\dfrac{2\sqrt{3}}{\sqrt{5}}$ (c) $\sqrt{\dfrac{2}{3}}$

(d) $8\sqrt{\dfrac{3}{4}}$ (e) $\dfrac{-6}{\sqrt{2}}$ (f) $-\dfrac{2}{\sqrt{12}}$

(g) $\dfrac{-2\sqrt{5}}{3\sqrt{2}}$ (h) $\dfrac{-3}{\sqrt{18}}$ (i) $\dfrac{30}{\sqrt{5}}$

(j) $-\dfrac{2\sqrt{6}}{\sqrt{2}}$ (k) $-\dfrac{2}{3}\sqrt{\dfrac{3}{5}}$ (l) $-\dfrac{1}{2\sqrt{5}}$

5. Simplify.

(a) $\dfrac{8}{\sqrt[3]{2}}$ (b) $\dfrac{12\sqrt[3]{6}}{\sqrt[3]{2}}$ (c) $\sqrt[3]{\dfrac{7}{2}}$

(d) $\dfrac{40}{\sqrt[4]{4}}$ (e) $\sqrt[4]{\dfrac{32}{4}}$ (f) $\sqrt[5]{\dfrac{1}{32}}$

6. Simplify. You may need to divide or multiply.

(a) $1 \div \sqrt{24}$ (b) $8 \times \sqrt{8}$

(c) $\dfrac{2\sqrt{84}}{-\sqrt{12}}$ (d) $(2\sqrt{75}) \div \sqrt{15}$

(e) $3\sqrt{3} \times \sqrt{27}$ (f) $\left(\dfrac{\sqrt{72}}{2\sqrt{8}}\right)(\sqrt{2})$

(g) $\dfrac{5\sqrt{25}}{\sqrt{75}} \div \sqrt{5}$ (h) $(2\sqrt{3})^2 \div \sqrt{3}$

(i) $\dfrac{2\sqrt{3}}{-4} \div \sqrt{27}$ (j) $\left(\dfrac{8\sqrt{5}}{\sqrt{2}}\right) \div \left(\dfrac{6\sqrt{5}}{-\sqrt{10}}\right)$

(k) $\dfrac{25\sqrt[3]{48}}{5\sqrt[3]{6}}$ (l) $\dfrac{\sqrt[4]{125}}{\sqrt[4]{80}} \div (\sqrt[4]{5})$

7. To simplify radical expressions, we may need to apply our work with conjugate radicals. Multiply each of the following by its conjugate.

(a) $\sqrt{3} - \sqrt{2}$ (b) $3\sqrt{2} + 2\sqrt{3}$

(c) $\sqrt{5} + 3\sqrt{2}$ (d) $3\sqrt{5} - 4\sqrt{2}$

(e) $3\sqrt{6} - 2\sqrt{5}$ (f) $3\sqrt{8} + 2\sqrt{18}$

(g) $2\sqrt{50} - 3\sqrt{27}$ (h) $\sqrt{98} + \sqrt{125}$

B

8. Write each of the following in simplest form.

(a) $\dfrac{-12\sqrt{22}}{4\sqrt{11}}$ (b) $3\sqrt{28} \div (-4\sqrt{7})$

(c) $\dfrac{-5\sqrt{3}}{\sqrt{6}}$ (d) $\dfrac{-4}{2\sqrt{8}}$

(e) $\left(\dfrac{\sqrt{8}}{\sqrt{27}}\right)\left(-\dfrac{\sqrt{25}}{\sqrt{5}}\right)$ (f) $\left(\dfrac{3\sqrt{2}}{2\sqrt{4}}\right) \div 3\sqrt{2}$

(g) $\dfrac{6\sqrt{7}}{3\sqrt{6}} \div \dfrac{4\sqrt{21}}{2\sqrt{3}}$ (h) $\left(\dfrac{-8\sqrt{50}}{2\sqrt{2}}\right)\left(\dfrac{\sqrt{8}}{4}\right)$

(i) $5\sqrt[3]{100} \div 25\sqrt[3]{5}$ (j) $(-3\sqrt[3]{20})(\sqrt[3]{50})$

9. Rationalize the denominator of each expression. Write your answer in simplest form.

(a) $\dfrac{\sqrt{3} + \sqrt{5}}{\sqrt{2}}$ (b) $\dfrac{2\sqrt{3} - 3\sqrt{2}}{\sqrt{2}}$

(c) $\dfrac{4\sqrt{3} + 3\sqrt{2}}{2\sqrt{3}}$ (d) $\dfrac{3\sqrt{5} - \sqrt{2}}{2\sqrt{2}}$

10. Simplify. Express each answer with a rational denominator.

(a) $\dfrac{-4\sqrt{6} - 2\sqrt{18}}{\sqrt{3}}$ (b) $\dfrac{\sqrt{20} - 3\sqrt{10}}{4\sqrt{5}}$

(c) $\dfrac{36 - \sqrt{6}}{\sqrt{6}}$ (d) $\dfrac{3\sqrt{2} - 3\sqrt{3}}{2\sqrt{2}}$

(e) $\dfrac{2\sqrt{5} - 3\sqrt{2}}{-3\sqrt{2}}$ (f) $\dfrac{5\sqrt{5} - 2\sqrt{2}}{8\sqrt{2}}$

11. Simplify. Express each answer in simplest form.

(a) $\dfrac{3}{\sqrt{5} - \sqrt{2}}$

(b) $\dfrac{2\sqrt{5}}{2\sqrt{5} + 3\sqrt{2}}$

(c) $\dfrac{\sqrt{3} - \sqrt{2}}{\sqrt{3} + \sqrt{2}}$

(d) $\dfrac{2\sqrt{5} - 8}{2\sqrt{5} + 3}$

(e) $\dfrac{2\sqrt{3} - \sqrt{2}}{5\sqrt{2} + \sqrt{3}}$

(f) $\dfrac{3\sqrt{3} - 2\sqrt{2}}{3\sqrt{3} + 2\sqrt{2}}$

12. (a) Rationalize the denominator of $\dfrac{8\sqrt{2}}{\sqrt{20} - \sqrt{18}}$.

(b) Rationalize the denominator of $\dfrac{8\sqrt{2}}{2\sqrt{5} - 3\sqrt{2}}$.

(c) Why are your answers in (a) and (b) the same?

13. Express each of the following in simplest form.

(a) $\dfrac{3}{3\sqrt{12} - \sqrt{2}}$

(b) $\dfrac{2\sqrt{2}}{2\sqrt{3} - \sqrt{8}}$

(c) $\dfrac{8}{2\sqrt{75} - 3\sqrt{50}}$

(d) $\dfrac{2\sqrt{6}}{2\sqrt{27} - \sqrt{8}}$

(e) $\dfrac{2\sqrt{2}}{\sqrt{16} - \sqrt{12}}$

(f) $\dfrac{3}{2\sqrt{80} - \sqrt{45}}$

(g) $\dfrac{3\sqrt{2} + 2\sqrt{3}}{\sqrt{12} - \sqrt{8}}$

(h) $\dfrac{3\sqrt{5}}{4\sqrt{3} - 5\sqrt{2}}$

(i) $\dfrac{2\sqrt{3} - \sqrt{2}}{\sqrt{12} + \sqrt{8}}$

(j) $\dfrac{\sqrt{18} + \sqrt{12}}{\sqrt{18} - 3\sqrt{12}}$

problem/matics

To check whether a result is true or not we may often use a numerical example. Use a numerical example to check which of the following seem true and which seem false.

A $\sqrt{x^2} + \sqrt{y^2} \stackrel{?}{=} |x| + |y|$

B $\sqrt{x^2} - \sqrt{y^2} \stackrel{?}{=} |x| - |y|$

C $\sqrt{x^2} \times \sqrt{y^2} \stackrel{?}{=} |xy|$

D $\sqrt{x^2} \div \sqrt{y^2} = |x| \div |y|$

To find the decimal value of a radical expression we may use our skills to first write each expression in simplest form and then use the following decimal equivalents for radicals, shown to 3 decimal places.

$\sqrt{2} \doteq 1.414 \quad \sqrt{5} \doteq 2.236 \quad \sqrt{7} \doteq 2.646$
$\sqrt{3} \doteq 1.732 \quad \sqrt{6} \doteq 2.449 \quad \sqrt{10} \doteq 3.162$

14. Find the value of each of the following to 2 decimal places.

(a) $\dfrac{3}{\sqrt{2}}$

(b) $\dfrac{-2}{\sqrt{3}}$

(c) $\dfrac{3\sqrt{2}}{\sqrt{3}}$

(d) $\dfrac{-4\sqrt{2}}{3\sqrt{3}}$

(e) $\dfrac{-5\sqrt{2}}{\sqrt{8}}$

(f) $\dfrac{\sqrt{18}}{\sqrt{3}}$

(g) $\dfrac{2\sqrt{2} - \sqrt{3}}{\sqrt{2}}$

(h) $\dfrac{5\sqrt{3} - 2\sqrt{2}}{\sqrt{3}}$

15. Evaluate each radical expression to 1 decimal place.

(a) $\dfrac{8}{\sqrt{3} + \sqrt{2}}$

(b) $\dfrac{-3\sqrt{2}}{\sqrt{3} - \sqrt{2}}$

(c) $\dfrac{5}{6 - \sqrt{2}}$

(d) $\dfrac{9}{2\sqrt{2} - \sqrt{3}}$

(e) $\dfrac{\sqrt{3} + 1}{\sqrt{3} - 1}$

(f) $\dfrac{1}{2\sqrt{3} - \sqrt{2}}$

16. If $m = 2\sqrt{3} + \sqrt{5}$ and $n = 2\sqrt{3} - \sqrt{5}$, find the value of the following expression in simplest form.

(a) $\dfrac{mn + m^2}{m}$

(b) $\dfrac{m^2 - n^2}{m + n}$

17. If $m = \sqrt{2} - 3\sqrt{3}$ and $n = 2\sqrt{2} + \sqrt{3}$, write each of the following in simplest form.

(a) $\dfrac{m^2 + 3mn}{m}$

(b) $\dfrac{n^2 - 2mn}{n^2}$

(c) $\dfrac{m^2 + 2mn + n^2}{m + n}$

(d) $\dfrac{m^2 - n^2}{mn + m^2}$

18 Evaluate the area of each triangle in simplest form.
(a)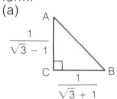
(b)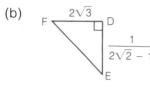

19 Evaluate the area of each rectangle in simplest form.
(a)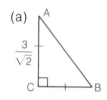
(b)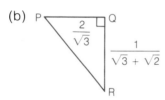

20 Express the hypotenuse of each triangle in simplest form.

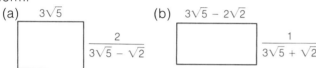

21 Prove that
$$\frac{\sqrt[3]{a}}{\sqrt[3]{b}} = \sqrt[3]{\frac{a}{b}} \quad a, b, \in R, b \neq 0.$$

22 Prove that
$$\frac{\sqrt[n]{a}}{\sqrt[n]{b}} = \sqrt[n]{\frac{a}{b}} \quad a, b, \in R, n \in N, b \neq 0.$$

23 Without using decimal equivalents, prove that $2 + \frac{1}{\sqrt{2}}$ is greater than $3 - \frac{1}{\sqrt{3}}$.

math tip
Use a calculator to estimate each expression to 1 decimal place.

A $\dfrac{1}{2\sqrt{3} - \sqrt{2}}$ B $\dfrac{2\sqrt{3} + \sqrt{2}}{10}$

What do you notice about your answers?

3.9 solving radical equations

The principles we have developed to solve equations apply to solving any type of equation. In each of the following equations, the variables occur in the radicand.

$$\sqrt{y} = 4 \qquad \sqrt{3m + 1} = 7$$
$$\sqrt{x + 8} - \sqrt{x} = 2$$

We refer to the above equations as **radical** equations since the variable occurs in the radicand.

To solve an equation, we find the values of the variable that satisfy the equation. Compare the steps needed to solve the following equation with the steps needed to solve equations in earlier work.

Example 1
Solve $2\sqrt{x} - 3 = 9 - \sqrt{x}$.

Solution
$$2\sqrt{x} - 3 = 9 - \sqrt{x}$$ — Collect the terms with the variable to isolate x.
$$2\sqrt{x} + \sqrt{x} = 9 + 3$$
$$3\sqrt{x} = 12$$
$$\sqrt{x} = 4$$
$$(\sqrt{x})^2 = (4)^2$$
$$x = 16$$

We may note by inspection that $x = 16$ or we may square both sides.

Verify $x = 16$.
LS $= 2\sqrt{x} - 3$ RS $= 9 - \sqrt{x}$
$= 2\sqrt{16} - 3$ $= 9 - \sqrt{16}$
$= 2(4) - 3$ $= 9 - 4$
$= 8 - 3$ $= 5$
$= 5$
LS = RS

Thus, the root of the equation is 16.

To solve a radical equation, we have introduced the need to square the radical containing the variable. However, in squaring the left and right members of an equation we may introduce values of the variable that are not roots.

Consider the following.
A: The solution of $x - 1 = 4$ is $x = 5$.
B: Square both sides of the above equation.
$$(x - 1)^2 = 4^2$$
$$x^2 - 2x + 1 = 16$$
$$x^2 - 2x - 15 = 0$$
$$(x + 3)(x - 5) = 0$$
$$x + 3 = 0 \quad \text{or} \quad x - 5 = 0$$
$$x = -3 \quad \text{or} \quad x = 5$$

From A
- we know the root of the original equation is 5.
- we know that -3 as obtained in B is not a root of the original equation.

Thus, the method of squaring both sides of an equation does not always result in obtaining an equivalent equation. Squaring may introduce "extra" values of the variable, which may not be a root. We refer to these values as **extraneous roots**. Thus, we *must* verify all values obtained because the squaring process often does not give equivalent equations as shown in the next example.

Example 2
Solve $y - 2 = \sqrt{y - 1} + 1$.
Solution   Isolate the
$$y - 2 = \sqrt{y - 1} + 1$$ term with the
$$y - 3 = \sqrt{y - 1}$$ radical.
Square both sides of the equation.
$$y^2 - 6y + 9 = y - 1$$
$$y^2 - 7y + 10 = 0 \qquad \text{We use our earlier skills}$$
$$(y - 5)(y - 2) = 0 \qquad \text{in algebra to factor.}$$
$$y - 5 = 0 \quad \text{or} \quad y - 2 = 0$$
$$y = 5 \qquad\qquad y = 2$$
Verify
$y = 5$ LS $= y - 2$ and   RS $= \sqrt{y - 1} + 1$
               $= 5 - 2$ $= \sqrt{5 - 1} + 1$
               $= 3$ $= \sqrt{4} + 1 = 3$
LS = RS
5 is a root.

$y = 2$ LS $= y - 2$ RS $= \sqrt{y - 1} + 1$
            $= 2 - 2$ $= \sqrt{2 - 1} + 1$
            $= 0$ $= 1 + 1 = 2$
LS $\neq$ RS
2 is not a root. (It is an *extraneous value* obtained by squaring in the original equation.)

To simplify the solution of a radical equation, we often use a technique by which we square the original equation in two steps:

Example 3
Solve $\sqrt{2m + 1} + \sqrt{m} = 5$.
Solution   To simplify the
$$\sqrt{2m + 1} + \sqrt{m} = 5$$ solution, rewrite the
$$\sqrt{2m + 1} = 5 - \sqrt{m}$$ equation as shown.
Square both sides of the equation.
$$2m + 1 = 25 - 10\sqrt{m} + m$$
$$10\sqrt{m} = 24 - m \qquad \text{Isolate the term}$$
$$\qquad\qquad\qquad\qquad \text{with the radical.}$$
Square both sides of the equation.
$$100m = 576 - 48m + m^2$$
$$m^2 - 148m + 576 = 0$$
$$(m - 144)(m - 4) = 0$$
$$m = 144 \quad \text{or} \quad m = 4$$

Check   You must verify to show that 144 is extraneous.

The root of the equation is 4.

When we add, subtract, multiply or divide numerical coefficients in an equation, we obtain equivalent equations. However, when we square terms of an equation we may obtain equations that are not equivalent and therefore we must verify all values obtained for the variable.

3.9 exercise

Questions 1 to 6 examine skills needed to solve radical equations.

A

1. Simplify each of the following.
   (a) $(\sqrt{2x-1})^2$
   (b) $(5 + \sqrt{x})^2$
   (c) $(3\sqrt{x} - 1)^2$
   (d) $(8 - 2\sqrt{m})^2$
   (e) $(2 + \sqrt{x-5})^2$
   (f) $(\sqrt{y-3} + 5)^2$
   (g) $(3\sqrt{m-2} - 2)^2$
   (h) $(8 - 3\sqrt{y-1})^2$

2. If $y = 3$, find the value of each expression.
   (a) $\sqrt{3y} + 3$
   (b) $\sqrt{3y - 5}$
   (c) $\sqrt{3y} - 5$
   (d) $\sqrt{3y-5}$
   (e) $\sqrt{15 - 2y}$
   (f) $3\sqrt{3y} - 2$
   (g) $\sqrt{2y-5} + \sqrt{3y}$
   (h) $5\sqrt{4-y} - \sqrt{3y-5}$

3. Calculate the value of each radical expression for the value given.
   (a) $\sqrt{2x+1}$, $x = 144$
   (b) $2 + \sqrt{4y-1}$, $y = \frac{1}{2}$
   (c) $\sqrt{4m+6} - 2m$, $m = -\frac{1}{2}$
   (d) $\sqrt{2k+1} - \sqrt{k}$, $k = 144$
   (e) $\sqrt{2-m} + \sqrt{11+m}$, $m = -2$
   (f) $\sqrt{2p+3} + \sqrt{3-p}$, $p = 3$

4. Which of the following are radical equations, (Yes, Y)? Which are not (No, N)? Give reasons for your answer.
   (a) $\sqrt{3x} + 5 = \sqrt{2}$
   (b) $2\sqrt{x} - 3 = \sqrt{x}$
   (c) $\sqrt{2x-3} = \sqrt{x+1}$
   (d) $\sqrt{3}(x-1) + 2 = 5(3x-1)$
   (e) $\sqrt{x-1} - 5 = 3$
   (f) $\sqrt{x} - \sqrt{x-1} = 2$

5. Two values are given for the variable in each equation. Use verification to indicate whether both, one, or none of them is a root.
   (a) $3 + \sqrt{3x+1} = 10$, 16, 5
   (b) $\sqrt{y+2} = y - 1$, 2, 7
   (c) $\sqrt{5x-6} = x$, 2, 3
   (d) $2 + \sqrt{3m-5} = \sqrt{4m+1}$, 2, 42
   (e) $\sqrt{x-5} - 6 = \sqrt{x}$, 9, 21

6. Solve each equation.
   (a) $\sqrt{x} = 6$
   (b) $\sqrt{m} = 3$
   (c) $\sqrt{k} = -2$
   (d) $\sqrt{x+1} = 8$
   (e) $\sqrt{y-1} = 3$
   (f) $3\sqrt{y+1} = 9$

Throughout these questions, we must remember to verify the values obtained for the variable.

B

7. Solve.
   (a) $\sqrt{y} - 1 = 4$
   (b) $\sqrt{y-1} = 4$
   (c) $\sqrt{2x-1} = 3$
   (d) $\sqrt{2x} - 1 = 3$
   (e) $8 - \sqrt{2m} = 0$
   (f) $3\sqrt{x} - 1 = 6$
   (g) $3\sqrt{y} - 2 = 2\sqrt{y} + 4$
   (h) $6(\sqrt{y} - 2) = 3(\sqrt{y} + 3)$

8. Two values 6, 14 for the variable are given for $7 - \sqrt{2y+4} = \sqrt{2y-3}$. Which value is a root?

9. Solve.
   (a) $\sqrt{10y+16} = 3y$
   (b) $\frac{1}{2}x = \sqrt{2x-4}$
   (c) $\sqrt{5p+4} = 5 - 2p$
   (d) $p - 1 = \sqrt{p^2-7}$
   (e) $2t + 4 = 1 + \sqrt{4t+6}$

10 Solve.
 (a) $\sqrt{3x+1} = \sqrt{5x+1}$
 (b) $\sqrt{2y+1} + \sqrt{y} = 5$
 (c) $\sqrt{p-9} + \sqrt{p+1} = 1$
 (d) $\sqrt{p+5} + \sqrt{p-3} = 4$
 (e) $\sqrt{2y+5} - \sqrt{y-2} = 3$

11 Solve each equation. How many roots does each equation have?
 (a) $\sqrt{3-x} + \sqrt{2x+3} = 3$
 (b) $\sqrt{2y+1} - \sqrt{y} = 5$ (c) $\sqrt{y-5} = \sqrt{y} + 6$

12 Solve.
 (a) $\dfrac{\sqrt{2x+3}}{\sqrt{5x+9}} = 1$ (b) $1 + \dfrac{\sqrt{y+4}}{\sqrt{y-3}} = \dfrac{7}{\sqrt{y-3}}$
 (c) $\sqrt{5+m} + \sqrt{m} = \dfrac{3}{\sqrt{m}}$

C

13 Solve each of the following equations.
 (a) $\sqrt[3]{2x+1} = 3$ (b) $\sqrt[3]{1-3x} - 4 = 0$
 (c) $\sqrt[3]{x^2 - 11} = 5$ (d) $\sqrt[3]{2x^2 - 5} = 3$

---

problem/matics

The following BASIC program may be used to apply the factor theorem for
$ax^3 + bx^2 + cx + d$ for $-10 \le x \le 10$.

```
10 INPUT A, B, C, D
20 FOR X = -10 TO 10
30 LET Y = A*X↑3 + B*X↑2 + C*X + D
40 IF Y = 0 THEN 60
50 GO TO 70
60 PRINT "ONE FACTOR IS X =", X
70 NEXT X
80 END
```

3.10 solving problems: radical equations

We must never overlook the fact that one of the main reasons we study mathematics and acquire skills and strategies is to solve problems. We often use a combination of our skills to solve a problem.

Example 1
Two rectangles have dimensions in metres expressed as radical expressions.

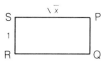

If the difference in the lengths of their diagonals is 2, find the dimensions of the rectangles.

Solution
The length of diagonal AC, in metres, is given by
$AC^2 = (\sqrt{2x+1})^2 + (\sqrt{x})^2$
$ = 2x + 1 + x = 3x + 1$
$AC = \sqrt{3x+1}$
The length of diagonal QS, in metres, is given by
$QS^2 = (\sqrt{x})^2 + (1)^2 = x + 1$
$QS = \sqrt{x+1}$
Thus, we obtain an equation.   *The difference of diagonals is 2.*
$\sqrt{3x+1} - \sqrt{x+1} = 2$
$\sqrt{3x+1} = 2 + \sqrt{x+1}$
$3x + 1 = 4 + 4\sqrt{x+1} + x + 1$
$2x - 4 = 4\sqrt{x+1}$
$x - 2 = 2\sqrt{x+1}$ *Simplify the equation before squaring again.*
$x^2 - 4x + 4 = 4(x+1)$
$x^2 - 4x + 4 = 4x + 4$
$x^2 - 8x = 0$
$x(x - 8) = 0$
$x = 0$ or $x - 8 = 0$ *0 is extraneous.*
$\phantom{x = 0 \text{ or }} x = 8$ *8 is the root.*

Thus, using x = 8, the dimensions of ABCD are 2√2 m × √17 m and of PQRS are 1 m × 2√2 m.

Check our results in the original problem.
For rectangle ABCD
$$BD^2 = (\sqrt{17})^2 + (\sqrt{8})^2$$
$$= 17 + 8 = 25$$
$$BD = 5 \leftarrow \text{Length of diagonal is 5 m.}$$

For rectangle PQRS
$$QS^2 = (\sqrt{8})^2 + (1)^2$$
$$= 8 + 1 = 9 \quad \text{The length of the}$$
$$QS = 3 \leftarrow \text{diagonal is 3 m.}$$

Thus the difference in the lengths of the diagonals is 2 m, and our solution is verified.

3.10 exercise

For each problem:
- decide on the skills needed to solve the problem.
- write the appropriate radical equation.
- solve the problem.

1 Two rectangles have sides shown. If the diagonal of the larger rectangle is 2 units longer than that of the smaller rectangle, find the length of each diagonal.

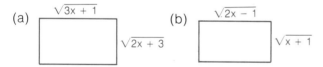

2 The difference in the lengths of the diagonals in the rectangles below is 4 units. Find the area of each rectangle, if the diagonal of rectangle A is the greater.

3 If the difference in length of the hypotenuses of the triangles below is 3, find the length of each hypotenuse, AB > DF.

 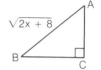

4 The difference in lengths of the hypotenuses below is 2. Find the length of each hypotenuse, if ST > YW.

 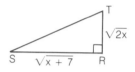

5 The lengths of the hypotenuses in the triangles below differ by 3. Find the area of each triangle, if UT > MR.

 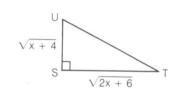

6 Find a number that has the property that the sum of the square root of the number and the square root of 3 less than the number is 3.

7 A number has the property that the square root of 3 less than the number, increased by the square root of 4 more than the number is 7. Find the number.

8 The square root of five times a number increased by 34, diminished by the square root of five times the number increased by six gives a result of 2. What is the number?

9 The sum of the square root of three times a number increased by 1 and the square root of 4 less than the number is 5. Find the number.

3.11 problem-solving: strategy

Often the skills and strategies we acquire to solve a particular problem can be used in different ways to solve a problem we have not met before. For example, we are given

$$\sqrt{y + 12} + \sqrt{y} = 6 \quad \text{①}$$

and we are asked to find the value of

$$\sqrt{y + 12} - \sqrt{y} \quad \text{②}$$

From equation ①, we could directly solve for y and then use this value of y to substitute in ② and eventually find the value of the expression ②.

However, before we directly involve ourselves with calculations, we may ask the question

How is [A: What we are given.]
related to [B: What we are asked to find.] ?

Namely, the question we ask is:

How is [A: $\sqrt{y + 12} + \sqrt{y} = 6$]
related to [B: $\sqrt{y + 12} - \sqrt{y}$] ?

A and B are conjugates! Thus, we may use this information to simplify our work.
Since
$$\sqrt{y + 12} + \sqrt{y} = 6$$
then multiply both sides by $\sqrt{y + 12} - \sqrt{y} \, (\neq 0)$
$$(\sqrt{y + 12} - \sqrt{y})(\sqrt{y + 12} + \sqrt{y})$$
$$= 6(\sqrt{y + 12} - \sqrt{y})$$
$$(y + 12 - y) = 6(\sqrt{y + 12} - \sqrt{y})$$
$$12 = 6(\sqrt{y + 12} - \sqrt{y})$$

Thus,
$$\sqrt{y + 12} - \sqrt{y} = 2.$$

By first relating the information we are given to the information we are asked to find, we have avoided many tedious calculations.

3.11 exercise

1. (a) Simplify
    $(\sqrt{2x - 7} - \sqrt{2x})(\sqrt{2x - 7} + \sqrt{2x})$.
   (b) Use your results in (a) to find the value of $\sqrt{2x - 7} + \sqrt{2x}$ if $\sqrt{2x - 7} - \sqrt{2x} = 1$.

2. (a) Simplify
    $(\sqrt{3x + 1} + \sqrt{3x})(\sqrt{3x + 1} - \sqrt{3x})$.
   (b) Show that when $\sqrt{3x + 1} + \sqrt{3x} = 1$, then $\sqrt{3x + 1} - \sqrt{3x} = 1$.

3. (a) Solve the equation $\sqrt{x + 4} + \sqrt{x} = 8$.
   (b) Use your results in (a) to calculate $\sqrt{x + 4} - \sqrt{x}$.
   (c) What other method can be used to find the value of $\sqrt{x + 4} - \sqrt{x}$ *without* solving the equation in (a)?

4. (a) If the value of the expression $\sqrt{7y + 8} + \sqrt{7y}$ is 8, then find the value of $\sqrt{7y + 8} - \sqrt{7y}$.
   (b) Find the value of $\sqrt{x + 9} + \sqrt{x}$ if we know that $\sqrt{x + 9} - \sqrt{x} = 3$.
   (c) What is the value of $\sqrt{x - 9} - \sqrt{x}$ if we know that $\sqrt{x - 9} + \sqrt{x} = 9$?

5. (a) If $\sqrt{2x - 3} + \sqrt{2x} = 3$, then find the value of $\sqrt{2x - 3} - \sqrt{2x}$.
   (b) Given $\sqrt{9 - x} - \sqrt{4 - x} = 1$, then what is the value of $\sqrt{9 - x} + \sqrt{4 - x}$?

6. (a) If the value of $\sqrt{4x + 3} + \sqrt{4x}$ is 3, what is the value of $\sqrt{4x + 3} - \sqrt{4x}$?
   (b) Determine the value of $\sqrt{3x - 5} - \sqrt{3x}$ if $\sqrt{3x - 5} + \sqrt{3x} = 5$.

7 Use the given information as follows to calculate the value of each of the following expressions.

A: $\sqrt{5x+11} + \sqrt{5x} = 7$
B: $\sqrt{3x+6} - \sqrt{3x} = \dfrac{2}{3}$
C: $\sqrt{x+8} + \sqrt{x} = 4$
D: $\sqrt{x-3} + \sqrt{x} = -1$

(a) $\sqrt{3x} + \sqrt{3x+6}$ (b) $\sqrt{x} - \sqrt{x+8}$
(c) $-(\sqrt{x} - \sqrt{x-3})$ (d) $\sqrt{5x+11} - \sqrt{5x}$

8 (a) Show that $4\sqrt{3} < 5\sqrt{2}$.
   (b) Show that $4\sqrt{5} > 5\sqrt{3}$.

9 Which figure has the greater perimeter?

(a) 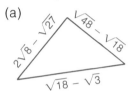 (b)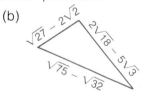

10 Which figure has the greater perimeter?

(a) 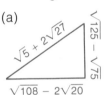 (b)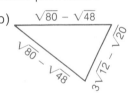

11 Show that $4\sqrt{7} - 5\sqrt{3} < 6\sqrt{3} - 2\sqrt{7}$.
12 Show that $4\sqrt{7} - 3\sqrt{5} > 4\sqrt{5} - 2\sqrt{7}$.
13 Which figure has the greater perimeter?

(a) 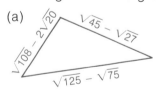 (b)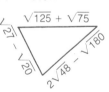

math tip
Estimate which of the following pipe systems carry more water?
A 5 pipes, each with a diameter of 10 cm.
B 10 pipes, each with a diameter of 5 cm.

skills review: operations with radicals

1 Simplify.
(a) $6\sqrt{72} - 8\sqrt{50} + 3\sqrt{18}$
(b) $-4\sqrt{20} - 5\sqrt{45} - 6\sqrt{5}$
(c) $3\sqrt{8} + 5\sqrt{12} - 2\sqrt{18} - 6\sqrt{48}$
(d) $5\sqrt{32} - 3\sqrt{48} + 5\sqrt{50} - 3\sqrt{108}$
(e) $2\sqrt{6} - 3\sqrt{5} + \sqrt{96} - 3\sqrt{125}$
(f) $6\sqrt[3]{16} - 8\sqrt[3]{54}$
(g) $-3\sqrt[3]{7} - 2\sqrt[3]{56} + 5\sqrt[3]{189}$
(h) $2\sqrt[4]{16} - 3\sqrt[4]{64} + 5\sqrt[4]{256}$

2 Simplify.
(a) $3a\sqrt{9a} - 4b\sqrt{25a} + 16\sqrt{36a}$
(b) $2\sqrt{x} - 3\sqrt{16x} + 5\sqrt{25x}$
(c) $3\sqrt{x^3y} + 5\sqrt{xy} - 6\sqrt{x^3y^3}$

3 Multiply.
(a) $(8 - \sqrt{3})(12 + \sqrt{3})$
(b) $(2\sqrt{2} - 4)(\sqrt{2} + 6)$
(c) $(3\sqrt{2} - \sqrt{3})(5\sqrt{2} - 2\sqrt{3})$
(d) $(3\sqrt{2} - 2\sqrt{5})(3\sqrt{2} + 2\sqrt{5})$
(e) $(3\sqrt{5} - \sqrt{2})^2 - 2(4\sqrt{5} + \sqrt{2})^2$
(f) $5(2\sqrt{3} + \sqrt{2})^2 - 3(2\sqrt{3} - 3\sqrt{2})^2$

4 Rationalize the denominator and write in simplest form.
(a) $\dfrac{\sqrt{3} - \sqrt{5}}{\sqrt{2}}$ (b) $\dfrac{2\sqrt{6} - 3\sqrt{12}}{\sqrt{3}}$
(c) $\dfrac{2}{\sqrt{3} - \sqrt{2}}$ (d) $\dfrac{5\sqrt{3}}{2\sqrt{5} - \sqrt{3}}$

problems and practice: a chapter review

This section at the end of each chapter will provide you with additional questions to check your skills and understanding of the topics dealt with in the chapter. An important step for problem-solving is to decide which skills to use. For this reason, these questions are not placed in any special order. When you have finished the review, you might try the *Test for Practice* that follows.

1. If $m + 2$, $m + 3$ and $m + 4$ are the measures of the sides of a right-angled triangle, find the value of m.

2. A lazy housefly decides to walk diagonally across a rectangular table instead of walking along the sides. How much shorter, to the nearest centimetre, is the route if the table is 2 m by 1 m?

3. Express in lowest terms each of the following in the form
   $$\frac{a}{b}, \ a, b, \in I.$$
   (a) 0.375 (b) $0.\overline{8}$ (c) $0.\overline{12}$
   (d) $0.3\overline{4}$ (e) $0.3\overline{9}$ (f) $0.0\overline{48}$

4. Solve each of the following. Draw the graph of the solution set. Variables are real numbers.
   (a) $2(2m - 1) - 8 > -39 - 3(m - 5)$
   (b) $(y - 3)^2 \leq (y - 1)(y - 2) + 1$
   (c) $\frac{1}{2}(2x + 4) \geq \frac{1}{5}(4x - 1)$

5. Simplify each of the following.
   (a) $8^{-\frac{1}{3}} + 10^0 + 4^{\frac{1}{2}}$
   (b) $2^{-4} + 5^0 - 3(-2)^{-4}$
   (c) $16^{\frac{3}{4}} - (3)^2 + 8^{-\frac{2}{3}} - 3^4 \times 3^{-5}$

6. A triangle has vertices at A(1, 5), B(−2, −4) and C(4, −2). What is the perimeter of the triangle? Irrational numbers may be used.

7. Simplify each of the following.
   (a) $2a\sqrt{8} - 3a\sqrt{18} + 5a\sqrt{32}$
   (b) $2y\sqrt{20} - 3y\sqrt{45} - 3y\sqrt{125}$

8. Simplify.
   (a) $3(2\sqrt{2} - 1)(3\sqrt{2} + 2) - 3(\sqrt{2} - 1)$
   (b) $(2\sqrt{5} - \sqrt{2})^2 - 3(\sqrt{5} - \sqrt{2})(\sqrt{5} + \sqrt{2})$

9. Find a radical expression for the shaded area.

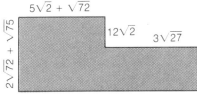

10. Show that
    (a) $7 - \sqrt{5}$ is a root of $x^2 - 14x + 44 = 0$.
    (b) $\sqrt{3} - \sqrt{2}$ is a root of $x^2 + 2\sqrt{2}x - 1 = 0$.

11. Solve.
    (a) $\sqrt{1 - 2x} - 1 = 0$ (b) $\sqrt{2m - 1} = \sqrt{m + 3}$
    (c) $3\sqrt{y - 1} = 1 - 2y$ (d) $\sqrt{5x + 4} + 2x = 5$

12. Simplify.
    (a) $(2\sqrt[3]{2})(3\sqrt[3]{4})$ (b) $\sqrt[3]{20} \times \sqrt[3]{50}$
    (c) $(-3\sqrt[3]{4})(2\sqrt[3]{4})$ (d) $(-5\sqrt[3]{3})(2\sqrt[3]{9})$

13. Solve.
    (a) $\sqrt[3]{x^2 - 26x} = 3$ (b) $6 - \sqrt[3]{x^2 - 30x} = 0$

14. For what values of x is the sum of $2(2x + 1) + 3(x - 5)$ and $5(x - 2) - (x + 5)$ positive?

test for practice

Try this test. Each *Test for Practice* will be based on the mathematics you have learned in the chapter. Try this test later in the year as a review. Keep a record of those questions that you were not successful with, find out how to answer them, and review them periodically.

1. Find the value of y if the sides of a right-angled triangle measure y, $y + 49$, and $y + 50$.

2. A 42-cm television screen is one with a diagonal that measures 42 cm. If the screen is 30 cm high, find the area of the screen, to the nearest square centimetre.

3. Find the equivalent rational number represented by each decimal, in lowest terms.
   (a) 0.36 (b) $0.\overline{36}$ (c) $0.3\overline{6}$

4. (a) Solve $4(y + 3) - 3(2 + y) < 2$, $y \in R$.
   (b) Draw the graph of the solution set in (a).

5. Solve each of the following. Variables are real numbers.
   (a) $\frac{1}{3}(m - 4) > \frac{1}{4}(m - 2)$
   (b) $3(m - 1)^2 - 3(m^2 - 1) \leq m - 15$

6. Simplify.
   (a) $\sqrt{12} + 5\sqrt{27} - \sqrt{48}$ (b) $3\sqrt{8} + \sqrt{18} - 5\sqrt{72}$
   (c) $13\sqrt{8} + \sqrt{27} - 2\sqrt{18} + \sqrt{3}$
   (d) $6\sqrt{27} - \sqrt{396} - 8\sqrt{108} - 7\sqrt{99}$

7. Which figure has the greater perimeter?

8. Simplify each expression. Write as mixed radicals in simplest form.
   (a) $\sqrt{25x^3yz^3}$ (b) $\sqrt[3]{-27x^3y^4}$

9. Simplify $3\sqrt{a^4b} - 2a^2\sqrt{b} + 6\sqrt{a^4b^5}$.

10. Show that $3\sqrt{5} + 1$ is a root of $x^2 - 2x - 44 = 0$.

11. Simplify.
    (a) $(3\sqrt{2} - 3\sqrt{3})(2\sqrt{2} - 5\sqrt{3})$
    (b) $(2\sqrt{5} - 3)^2$
    (c) $(5\sqrt{3} - 2\sqrt{2})^2 - 3(\sqrt{3} - \sqrt{2})(\sqrt{3} + \sqrt{2})$

12. Solve.
    (a) $\sqrt{3x + 1} = 7$ (b) $\sqrt{x - 1} = x - 3$
    (c) $\sqrt{y + 11} + \sqrt{2 - y} = 5$

13. Given that $\sqrt{x + 5} + \sqrt{x} = 5$, find the value of $\sqrt{x + 5} - \sqrt{x}$.

14. Simplify each of the following.
    (a) $8^{\frac{2}{3}}$ (b) $27^{-\frac{1}{3}}$ (c) $3^0 - 16^{\frac{3}{4}} + 2^{-2}$

15. For what values of x is $3(2x + 1) - 2(x - 1)$ non-negative?

16. Prove that the product of an even number and an odd number is an even number.

17. Find the perimeter of a rectangle if the length is given by $2\sqrt{3} + \sqrt{2}$ and the area is 10.

18. A number b has the property that the square root of 3 more than the number is equal to 5 diminished by the square root of 2 less than the number. Calculate b.

absolute value: equations and inequations

definition of absolute value

Mathematics is a language that uses many symbols to represent important concepts. The concepts dealt with here illustrate how mathematics grows.

The development of absolute value that follows shows that once a definition in mathematics is stated, it can be applied, along with our earlier skills, to the development of more mathematics. For example, it doesn't matter if we walk 15 km east from a starting point, or 15 km west from the same starting point, we are the same distance from the starting point in each case. Thus, absolute value symbols were invented so that we could write the result only when we are interested in *how far* and not *in what direction*.

$$|+15| = 15 \qquad |-15| = 15$$

In mathematics, the definitions we use are precise. For any real number, $a \in R$, we define its absolute value as

$$|a| = a \text{ if } a > 0, \quad |a| = 0 \text{ if } a = 0,$$
$$|a| = -a, a < 0.$$

The definition covers all cases. *There are no exceptions*. You may often use the above definition in the following form. For $a \in R$,

$$|a| = a \text{ if } a \geq 0, \quad |a| = -a \text{ if } a < 0.$$

Once a concept is introduced, we must be accurate in applying our skills based on it.

1. Find the value of each of the following.
   (a) $|-2|$ (b) $4|-3|$ (c) $3|-8|$
   (d) $|-4| + 3|-8|$ (e) $|8| - 3|-2|$
   (f) $|-3| + 2|-2| - |-3| + |5| - |8|$
   (g) $2|-3| - 3|-2| + 2|6| - 3|5|$

2. Simplify.
   (a) $3|-3 + 2| - 2|4 - 3| - 3|8 - 6|$
   (b) $|9 - 5| - |6 - 8| - 3|8 - 5|$
   (c) $|9 - 9| - 3|7 - 7| - 2|-6 + 6|$

3. If $a = -3$, $b = -2$, $c = 2$, find the value of each of the following.
   (a) $2|a| - 3|b|$ (b) $|a| - |b|$ (c) $|a - b|$
   (d) $2a - 3|b|$ (e) $2|a| - 3b$ (f) $3|a - b|$
   (g) $|ab| - |bc|$ (h) $a|b + c| - 3|a^2| - b^2$

Once we have mastered the fundamental properties, we may examine the general properties.

4. (a) Use numerical examples to show that for all $a, b \in R$, $|a||b| = |ab|$.
   (b) Prove that $|a||b| = |ab|$ for all $a, b \in R$.

5. (a) Use numerical examples to show that for all $a, b \in R$, $\frac{|a|}{|b|} = |\frac{a}{b}|$.
   (b) Prove that $\frac{|a|}{|b|} = |\frac{a}{b}|$ for all $a, b \in R$.

6. Use numerical values for a and b to test which of the following are likely to be true.
   (a) $|a + b| = |a| + |b|$ (b) $|a + b| = a + b$
   (c) $|a + b| > 2|a|$ (d) $|a| - |b| = |a - b|$
   (e) $|a - b| > 2|b|$ (f) $|a| + |b| \geq |a + b|$
   (g) $|a| - |b| \geq |a - b|$

We may use absolute value symbols to write inequations in a compact form. For example,

$-3 \leq x$ and $x \leq 3$ is written compactly as $|x| \leq 3$.

$x \geq 3$ or $x \leq -3$ is written compactly as $|x| \geq 3$.

We may draw a graph of each of the above inequations.

$|x| \leq 3$

$|x| \geq 3$

7. Use absolute value symbols to write each of the following in a compact form.
   (a) $-6 \leq x \leq 6$ (b) $-2 \leq x \leq 2$
   (c) $x \geq 4$ or $x \leq -4$ (d) $x \leq -3$ or $x \geq 3$

8. Draw a graph of each of the following.
   (a) $|x| \leq 3$ (b) $|y| \geq 4$ (c) $|x| > 4$
   (d) $2 < |y|$ (e) $|x| < 2$ (f) $|x| \geq \frac{1}{2}$

solving equations

We may solve equations involving absolute value.

Solve $|x| = 4$.
Since $|4| = 4$ and $|-4| = 4$, thus the solution set is $\{4, -4\}$.

We may apply our earlier skills with equations to solve the following equation.

$5|x| + 8 = 3|x| + 26$
$5|x| - 3|x| = 18$
$2|x| = 18$
$|x| = 9$ Solution set
Thus $x = 9$ or $x = -9$.   is $\{9, -9\}$.

To solve advanced equations involving absolute value we must clearly *understand* and apply the definition of absolute value. To solve an equation involving an absolute value such as $|a|$, we need to consider these cases.

Case 1   or   Case 2
$|a| = a$ if $a \geq = 0$ $|a| = -a$ if $a < 0$

Example 1
Solve $|x - 2| = 7 - 3x$, $x \in R$.

Solution
Case 1
Solve the above equation for the condition $x - 2 \geq 0$. Find x that satisfy
$x > 2 \geq 0$ ① and $|x - 2| = 7 - 3x$ ②

Since $x - 2 \geq 0$, then we may write $|x - 2|$ as $x - 2$.

$x - 2 = 7 - 3x$
$4x = 9$
$x = \frac{9}{4}$.

Thus, $\frac{9}{4}$ is a solution since $\frac{9}{4}$ satisfies conditions ① and ②.

Case 2
Solve the above equation for the remaining condition $x - 2 < 0$. Find x that satisfy
$x - 2 < 0$ ① and $|x - 2| = 7 - 3x$ ②

Since $x - 2 < 0$, then we may write $|x - 2|$ as $-(x - 2)$.

$-(x - 2) = 7 - 3x$
$-x + 2 = 7 - 3x$
$2x = 5$
$x = \frac{5}{2}$.

Since $\frac{5}{2}$ does not satisfy condition ①, then $\frac{5}{2}$ is not a root.

Thus, the solution set for $|x - 2| = 7 - 3x$ is $\{\frac{9}{4}\}$.

As we have seen many times in our study of mathematics, as long as we understand and carefully apply our definitions, we should be able to solve almost any problem. In other words, we solve equations or problems by applying first principles.

9 Solve.
(a) $|x| = 6$
(b) $|x| + 3 = 8$
(c) $2|x| - 1 = 7$
(d) $3|x| = |x| + 4$
(e) $5|x| - 3 = |x| + 2$
(f) $6|m| - 14 = 4|m| + 6$
(g) $3|y| - 8 = 5|y| - 26$

10 Solve and verify.
(a) $|x - 5| = 1$
(b) $|2x + 3| = 0$
(c) $4|x + 2| = 3$
(d) $2|x - 3| + 1 = 4$
(e) $|x - 4| = 5x$
(f) $|2x - 3| = 4x - 2$

11 Solve.
(a) $|2x - 5| = -3x$
(b) $\frac{1}{3}|x + 3| + 5x = 0$
(c) $|x + 4| = -2x$
(d) $|\frac{x - 6}{3}| + 4x = 0$
(e) $|2x + 1| = |x + 5|$
(f) $|3x - 2| = |2x + 1|$
(g) $|2x - 5| = |x - 1|$
(h) $|x + 5| = |2x - 5|$

solving inequations

We may apply the definition for absolute value when solving inequations.

Example 2
Solve and draw the graph of $|x - 5| \leq 3$, $x \in R$.

Solution
Case 1 $x - 5 \geq 0$ and $|x - 5| \leq 3$ If $x - 5 \geq 0$, then $|x - 5| = x - 5$.
$x \geq 5$ $x - 5 \leq 3$
$x \geq 5$ $x \leq 8$
$5 \leq x \leq 8$ ←— written compactly

Case 2 $x - 5 < 0$ and $|x - 5| \leq 3$ If $x - 5 < 0$, then $|x - 5| = -(x - 5)$.
$x < 5$ $-(x - 5) \leq 3$
$x < 5$ $-x + 5 \leq 3$
$x < 5$ $2 \leq x$
$2 \leq x < 5$ ←— written compactly

Thus x satisfies $2 \leq x < 5$ or $5 \leq x \leq 8$ which is written compactly as $2 \leq x \leq 8$.

The graph of the solution set is shown. $|x - 5| \leq 3 \Leftrightarrow 2 \leq x \leq 8$

We may again apply the definition for absolute value and solve an inequation of the following type.

Example 3
Solve $|6 - x| \geq 2$, $x \in R$.

Solution
Case 1 $6 - x \geq 0$ and $|6 - x| \geq 2$ If $6 - x \geq 0$, then $|6 - x| = 6 - x$.
$6 \geq x$ $6 - x \geq 2$
$6 \geq x$ $4 \geq x$
Which simplifies to $4 \geq x$.

Case 2 $6 - x < 0$ and $|6 - x| \geq 2$ If $6 - x < 0$, then $|6 - x| = -(6 - x)$.
$6 < x$ $-(6 - x) \geq 2$
$6 < x$ $-6 + x \geq 2$
$6 < x$ $x \geq 8$
Which simplifies to $x \geq 8$.

Thus $\{x \mid |6 - x| \geq 2, x \in R\} = \{x \mid x \leq 4 \text{ or } x \geq 8\}$.

The graph of the solution set is shown.

12 Solve and draw the graph.
(a) $|x - 2| \leq 4$
(b) $|x - 2| \geq 4$
(c) $|3 - x| > 6$
(d) $|3 - x| < 6$

13 Solve.
(a) $|3 - x| < 8$
(b) $|x + 2| - 3 > 0$
(c) $|2x + 1| < 4$
(d) $|2x - 1| \geq 3$
(e) $|4x + 5| < 13$
(f) $|2 + 3x| \geq 5$
(g) $|3 - 2x| > 5$
(h) $|2x - 3| \geq 0$

14 Solve and verify.
(a) $|3y + 2| \geq -3y$
(b) $|2m| < m + 3$
(c) $|2x - 1| \geq x$
(d) $|3k - 5| \leq k$
(e) $|1 - 3y| < y + 5$
(f) $|3a + 1| > 3a + 5$

4 systems of equations; applications and solving problems

graphing, equivalent linear systems, solving problems, steps for solving problems; strategies for problem-solving.

4.1 introduction

Every place in Canada, and for that matter, every place on our planet, can be pinpointed as to location using latitude and longitude. For example, the following table lists the locations of several major Canadian cities.

Latitude and Longitude of Canadian Cities

	Latitude North (degrees, minutes, seconds)			Longitude West (degrees, minutes, seconds)		
Vancouver	49	16	30	123	07	30
Edmonton	53	32	45	113	29	15
Regina	50	27	02	104	36	30
Winnipeg	49	53	56	97	08	20
Toronto	43	39	12	79	23	00
Quebec City	46	48	46	71	12	20
Fredericton	45	57	40	66	38	30
Charlottetown	46	14	00	63	07	45
Halifax	44	38	39	63	34	34
St. Johns	47	34	00	52	43	30

In air navigation, geological surveying, mapmaking, weather forecasting and in many other ways, being able to locate the position of a landmark accurately depends on the mathematics-based concept of latitude and longitude.

Similarly it is possible to determine the position of a point on the co-ordinate plane if we know either of two facts about the point:

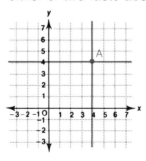

For the point A, we know the value of the x-co-ordinate and y-co-ordinate.

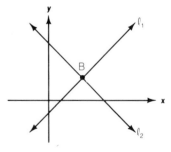

For the point B, we know two lines ℓ_1 and ℓ_2 intersect at the point B.

Although the mathematics involved is far more complex, the same basic principle of determining a point of intersection has been used to successfully guide unmanned satellites towards the most distant planets in our solar system.

Scientists used mathematics to calculate a course for the spacecraft that would *intersect* the orbit for Saturn at exactly the right moment.

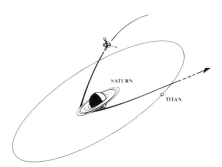

The study of intersecting lines and intersecting curves is very important in the study of mathematics. In this chapter, we will learn to find the co-ordinates of a point of intersection if we know the equations of the two intersecting lines. The equations we will deal with in our later work will vary, but the principles we will learn about finding the points of intersection of intersecting lines may be applied when we want to find the points of intersection of more complex curves.

4.2 graphs of intersecting lines

René Descartes was one of the great thinkers of the seventeenth and eighteenth centuries. He probably developed his ideas about the Cartesian plane so that he could use it to study numbers.

We can now use the Cartesian plane to show relations between pairs of numbers that satisfy the following conditions.

> The sum of one integer and twice another integer is always 6. What are the integers?

We use algebra to translate this question into mathematics.

> Find x, y so that $x + 2y = 6$, $x, y \in I$.

By inspection we find ordered pairs that satisfy the equation. For example, if $x = 2$, then

$$x + 2y = 6$$
$$2 + 2y = 6$$
$$2y = 4$$
$$y = 2$$

Substitute 2 for x in the equation.

Thus, one ordered pair that satisfies $x + 2y = 6$ is $(x, y) = (2, 2)$.

We may write other ordered pairs.

$x + 2y = 6$
(x, y)
$(6, 0)$
$(4, 1)$
$(0, 3)$
$(-2, 4)$
$(-4, 5)$

Using the co-ordinate plane, we plot these ordered pairs.

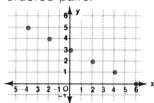

By plotting these points we have drawn the graph of the equation given by

$\underbrace{x + 2y = 6, \; x, y \in I}$

This is the *defining equation* of the above graph.

Of course, we cannot list all the possible ordered pairs (x, y) that could satisfy the equation x + 2y = 6. Similarly, we are not able to show all the points on the graph. For this reason what we draw is a *partial* graph.

If we use the domain and the range of the equation as R we write x + 2y = 6, x, y ∈ R. The graph of the equation with domain and range R is then drawn and appears as a solid line.

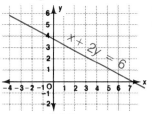

The equation x + 2y = 6 defines this relation among numbers: *the sum of a number and twice another number is always 6.* We may also use algebra to write this relation, using set builder notation.

$$S = \{(x, y) \mid x + 2y = 6, x, y \in R\}$$

| The set of all ordered pairs (x, y) | such that they satisfy the equation x + 2y = 6 | where x and y are real numbers. |

Since the graph of the above set, S, is a straight line, we refer to x + 2y = 6 as a *linear equation.*

Two important concepts which are fundamental to our study of graphs are,

I The co-ordinates of any *point on the* graph satisfy the equation of the graph.

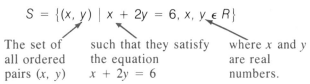

The values x = −8 and y = 7 *satisfy* the equation of the graph defined by x + 2y = 6.

II Any ordered pair (x, y) which satisfies the equation of the graph represents a point on the graph.

The values x = 8 and y = −1 satisfy the equation of the graph defined by x + 2y = 6.

The point with co-ordinates (8, −1) lies *on the graph.*

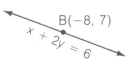

Example 1
(a) Draw the graph of 3x − y = 9, x, y ∈ R.
(b) Find the value of the missing co-ordinate in each of (x, 9) and (−2, y)

Solution
(a) Since the domain and the range are both R then we need find only two points to draw the line given by the equation 3x − y = 9. We find the co-ordinates of a third point on the graph as a check on our accuracy.

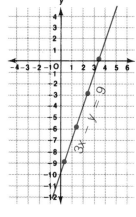

x	y
0	−9
3	0
1	−6
2	−3

Use x = 0
Then 3x − y = 9
0 − y = 9
y = −9

(b) The partial graph we have drawn does not include the points of the graph we need to find x and y for (x, 9) and (−2, y). Thus, we must find the missing co-ordinates, using algebra.
To find x for (x, 9), we use the value y = 9 in the defining equation.

$$3x - y = 9$$
$$3x - 9 = 9$$
$$3x = 18$$
$$x = 6$$

Thus the point (6, 9) is on the line 3x − y = 9.

To find y for (−2, y), we use the value x = −2 in the defining equation.

$$3x - y = 9$$
$$3(-2) - y = 9$$
$$-6 - y = 9$$
$$-y = 15$$
$$y = -15$$

The co-ordinates of another point on the graph would be (−2, −15).

Example 2
(a) Draw the graphs defined by $x + y = 4$, $x, y \in R$ and $x - 2y = 10$, $x, y \in R$.
(b) Write the co-ordinates of their intersection point.

Solution
(a) Construct a table of values for each equation. Then plot the points and draw their graphs.

$x + y = 4$ $x - 2y = 10$

x	y
0	4
4	0
1	3
2	2

x	y
0	-5
2	-4
4	-3
10	0

We use values of x (or y) that simplify our calculations.

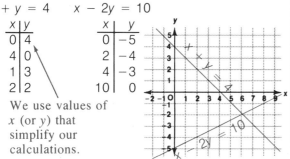

(b) From the graph, the co-ordinates of the intersection point of the lines is $(6, -2)$. This is the only point which lies on both lines.

Since the ordered pair $(6, -2)$ in the above example satisfies both $x + y = 4$ and $x - 2y = 10$, we say that $(6, -2)$ is a *solution* of the *linear system of equations*. To solve a linear system of equations means to find values of the variables that *satisfy* all equations of the system. We may write the solution as $(x, y) = (6, -2)$. We can check or verify that we have obtained the correct solution to the above system.

Use $x = 6$, $y = -2$

For $x + y = 4$	For $x - 2y = 10$
LS $= x + y$ RS $= 4$	LS $= x - 2y$ RS $= 10$
$= (6) + (-2)$	$= 6 - 2(-2)$
$= 6 - 2 = 4$	$= 6 + 4 = 10$
LS = RS	LS = RS
Solution checks ✓	Solution checks ✓

Another ordered pair such as $(x, y) = (5, -1)$ may satisfy one equation but *not* the other. Thus, for non-parallel lines the solution of a linear system is unique and may be seen graphically as the intersection point of the lines.

4.2 exercise

Throughout this exercise, the variable will represent real numbers. Thus, we will write the equation as $x + 2y = 6$ with the understanding that all $x, y \in R$.

A

1 The graph of the equation $3x - 2y = 6$ is drawn. Which of the following points satisfy the equation? Which do not?

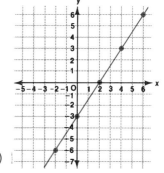

(a) (6, 6)   (b) (3, 4)
(c) (−2, 6) (d) (2, 0)
(e) (−3, 0) (f) (−6, −1)

2 The equation of a graph is given by $2x + y = 10$. Indicate which of the following points lie on the graph.

(a) (6, −2)   (b) (2, 8)   (c) (2, 6)
(d) (−3, 16) (e) (−6, 18) (f) (−6, 22)

3 For each equation, indicate which points lie on the graph of the equation. Do not graph the points.

(a) $x - y = 8$ A(10, 2)   B(−2, −10) C(4, 4)
(b) $y = 3x - 6$ D(0, −6)   E(3, 15)   F(2, 0)
(c) $2x - y = 12$ G(7, −2)   H(6, 0)   I(−4, −20)

4 (a) Find x if $(x, -3)$ is a point on the graph given by $3x - y = -6$.
  (b) Find y if $(-4, y)$ is a point on the graph given in (a).

5 A linear relation is given by the equation $y = -3x + 7$ where $x, y \in R$.
Find the missing co-ordinates if each of the following satisfies the above relation.

(a) $(x, -2)$ (b) $(0, y)$ (c) $(x, -5)$
(d) $(-1, y)$ (e) $(6, y)$ (f) $(x, -8)$

B 6. (a) Draw the graph of the equation $2x - y = 12$.
   (b) Use your graph in (a) to decide which of the following points are on the graph.
   A $(0, -12)$ B $(2, -6)$ C $(5, -2)$

7. Draw the graph of each of the following equations using the same axes.
   (a) $y = 3x + 3$
   (b) $2x - y = -3$
   (c) $2x = 3 - y$
   (d) What property do the above graphs have in common?

8. Draw the graph of each of the following equation using the same axes.
   (a) $4x - 2y = 5$
   (b) $2x - y = -8$
   (c) $y = 2x + 5$
   (d) What property do the above graphs have in common?

9. A co-ordinate system is used to determine the position of an island at $(-4, 5)$. The following equations define the paths of various recent hurricanes. Which hurricanes crossed the island?
   (a) Hurricane Arnold; $4x + 2y = -6$
   (b) Hurricane David; $\frac{x}{2} - y = 7$
   (c) Hurricane Hazel; $3x - y = -17$
   (d) Hurricane Ernie; $2x = 3y + 10$

10. The position of a sinking freighter is $(6, -8)$. Four rescue planes travel paths defined by the following equations. Which of the planes will make the rescue?

    A Piper Cub; $x - 3y = 20$
    B Cherokee; $y = 4 - 2x$
    C Wingstar; $30 + 2y = 3x$
    D Starcraft; $y + 10 = \frac{x}{3}$

11. A weather station uses a co-ordinate system to locate places in Alberta, as shown. A major storm is travelling along the path given by the equation $x + 4y = -18$.

    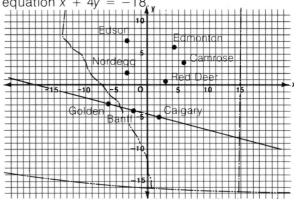

    Which of the following places are in the path of the storm?
    (a) Edson $(-3, 7)$
    (b) Golden $(-6, -3)$
    (c) Nordegg $(-3, 2)$
    (d) Red Deer $\left(3, \frac{1}{2}\right)$
    (e) Calgary $(2, -5)$
    (f) Edmonton $\left(\frac{9}{2}, 6\right)$
    (g) Banff $(-2, -4)$
    (h) Camrose $\left(6, \frac{7}{2}\right)$

12. Which places in Question 11 will be hit by a storm travelling along the path given by $y = \frac{11}{2}x - 16$?

13. Use a graphical method to solve each of the following linear systems of equations.
    (a) $3x - y = 7$
        $x + 2y = 7$
    (b) $y = 3x + 13$
        $2x + y = -2$
    (c) $x - 3y = 13$
        $2x + 3y = -1$
    (d) $y = 2x + 10$
        $x + y + 8 = 0$

14. Write the co-ordinates of the point of intersection of lines defined by the following equations.
    (a) $x + y = -2$
        $5x - 3y = -2$
    (b) $4x + y = 0$
        $y = 2x + 6$

applications: plotting storms

The weather office collects data in order to predict the weather. The principles used in plotting the path of a storm are similar to those used in the following questions.

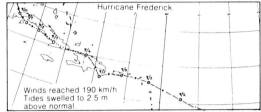

15. A cruiser is on a course given by the equation $4x - 3y = 6$, while a tugboat follows the course given by $2x + 3y = 12$.

    (a) Find the co-ordinates of the point where their paths will cross.
    (b) A buoy is located at $(-3, 6)$. Which ship will need to alter its course?
    (c) Which ship will come closest to a shoal located at the point $\left(\frac{1}{2}, -2\right)$?

16. In Canada's Arctic, a co-ordinate system is one way of locating position. Two surveying crews trek along the following paths.
    A: Survey crew Hollander: $6x + y = 0$
    B: Survey crew Williams: $2x - y = -8$
    At which point might the two crews meet?

17. A tornado is following a path plotted by the weather office as given by $x - 2y = -8$. At the same time the centre of a thunderstorm is on the path given by $y = 7 - x$. Indicate which towns with the given co-ordinates will experience:
    I a thunder storm
    II a tornado
    III a thunder storm and a tornado

    (a) Delhi $(-4, 2)$ (b) Marysville $(7, 0)$
    (c) Everett $(11, 4)$ (d) Norwich $(2, 5)$
    (e) Walton $(-1, 8)$ (f) Vernon $(8, 8)$

4.3 equivalent linear systems

The skills and concepts we study in one part of mathematics are usually further developed when we study another area of mathematics. For example, to solve equations in 1 variable, we used our knowledge of equivalent equations. To solve a system of linear equations, we use the concept of equivalent equations again. *Two linear systems are equivalent if they have the same solution.* For example, each of the following systems of equations are equivalent.

A: $x = 2y$ B: $x + 4y = -12$
$3x + y = -14$ $x = -4$

C: $2x - y = -6$ D: $x = -4$
$y = -2$ $y = -2$

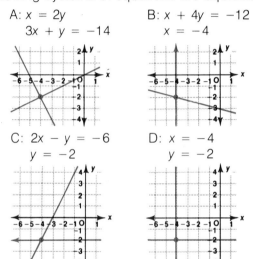

We can easily obtain the solution by inspection of system D. For solving a linear system of equations, we will learn another method, called the *addition- or- subtraction method*. This alternative method enables us to simplify a system of equations (like system A above) such that we can then obtain its solution by inspection (as in system D above). In other words:

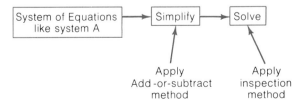

4.3 exercise

Questions 1 to 11 develop some essential skills needed to understand why we are able to use the algebraic add-or-subtract method (in the next section) to solve linear systems.

A

1. (a) Draw the graph of each system of equations using the same axes. Label each line clearly.
   A: $2x - y = 8$ B: $3x + y = 7$
       $x + y = 1$ $3x - 2y = 13$
   (b) Explain why systems A and B are equivalent.

2. (a) The solution of two systems, C and D, is given as $(x, y) = (3, 0)$.
   C: $3x - 2y = 9$ D: $x - y - 3 = 0$
      $x + y = 3$ $2x - y = 6$
   Verify the given solution.
   (b) Explain why systems C and D are equivalent.

3. (a) Which of the following systems have $(x, y) = (-2, 3)$ as the solution?
   P: $x + y = 1$ Q: $2y - x = -4$
      $2x - y = -7$ $y = 3x + 3$

   R: $x = 7 - 2y$ S: $x = 2y - 8$
      $2x - 3y = 0$ $3 + y = -3x$
   (b) Which systems in (a) are equivalent?
   (c) Find the solution for the other equations not named in (b).

4. The following lines pass through the same intersection point.

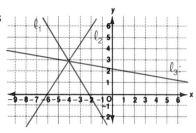

Write 3 different linear systems that are equivalent.

5. (a) From the following graph, write pairs of equations that give equivalent systems.

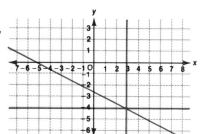

   (b) Which linear system equivalent to those in (a) has the simplest form?

6. (a) Draw the graph given by $3x + y = 6$.
   (b) Draw the graph given by $6x + 2y = 12$.
   (c) What do you notice about the graphs in (a) and (b)?
   (d) Explain how these equations are related.
   A: $3x + y = 6$ B: $6x + 2y = 12$

B

7. Based on your results in Question 6, describe how each of the graphs in Column A is related to the corresponding graph in Column B.

Column A	Column B
(a) $x - y = 3$	$2x - 2y = 6$
(b) $-x + 2y = 6$	$x - 2y = -6$
(c) $x + 5y = 1$	$3x + 15y = 3$
(d) $x + y = 2$	$\frac{1}{2}x + \frac{1}{2}y = 1$

8. (a) Draw the graphs of the lines given by
   A: $x + y = 3$ B: $2x - 3y = 1$
   (b) Draw the graph of equation C obtained as follows.
      A: $x + y = 3$ Both sides of equations A
      B: $2x - 3y = 1$ and B are added to
      C: $3x - 2y = 4$ obtain equation C.
   (c) Draw the graph of equation D obtained as follows.
      A: $x + y = 3$ Both sides of equations A
      B: $2x - 3y = 1$ and B are subtracted to
      D: $-x + 4y = 2$ obtain equation D.
   (d) What do you notice about the intersection point of the graphs of equations A, B, C, and D?

9. Repeat the same procedures of steps outlined in the following equations.
   A: $x - y = -7$ B: $3x + y = -9$

10. Use your results in Questions 8 and 9. For each of the following systems, write another equation that passes through the same intersection point.
    (a) $x - y = 7$
    $x + 2y = -2$
    (b) $x - 2y = 0$
    $2x + y = -5$
    (c) $2x + y = 1$
    $x - 2y = -17$
    (d) $3x + 2y = -8$
    $2x - y = 4$

11. For each system, another equation is given. Explain why the other equation passes through the same intersection point.
    (a) $\left.\begin{array}{l}2x + y = 1 \\ x + y = 2\end{array}\right\}$ A $3x + 2y = 3$
    (b) $\left.\begin{array}{l}3x - y = 9 \\ x - 2y = 8\end{array}\right\}$ B $2x + y = 1$
    (c) $\left.\begin{array}{l}x - 2y = -1 \\ x + 2y = -5\end{array}\right\}$ C $2x = -6$

classifying linear systems

This linear system
P ℓ_1: $x + y = 4$
    ℓ_2: $x - 2y = -2$
has one intersection point. Thus, the system has exactly one solution. We classify this system as a **consistent system**.

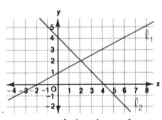

This linear system
Q ℓ_3: $2x - 3y = 6$
    ℓ_4: $4x - 6y = 12$
has more than one solution. This system is also consistent, but since one equation may be derived from the other, we further define Q as a **consistent** but **dependent system**. Since system P above

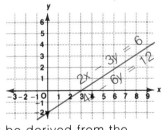

has exactly one solution we refer to it as a **consistent** but **independent system**.

This linear system
R ℓ_5: $x - 2y = -6$
    ℓ_6: $x - 2y = 6$
has no solution, since the graph is a pair of parallel lines which, of course, do not intersect. We classify this linear system as an **inconsistent system**.

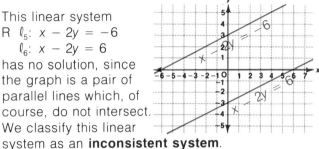

12. (a) Draw the graph of each equation.
    A: $2x - y = 8$ B: $4x - 2y = 16$
    (b) Use your results in (a) to explain why the following system is dependent.
    $2x - y = 8$ $4x - 2y = 16$

13. (a) Draw the graph of each equation.
    C: $x + 2y = 6$ D: $x + 2y = 10$
    (b) Use your results in (a) to explain why the following system is inconsistent.
    $x + 2y = 6$ $x + 2y = 10$

14. The graphs of equations are drawn. Write 1 linear system that is
    (a) consistent
    (b) inconsistent
    (c) consistent but dependent

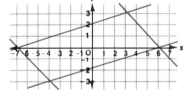

15. Classify each of the following as
    • Consistent   • Inconsistent
    • Consistent but Dependent
    (a) $y = 3x - 5$
    $y = 3x + 8$
    (b) $x - 2y = 1$
    $x + y = 4$
    (c) $2x - 3y = 3$
    $6 + 6y = 4x$
    (d) $x + 3y = 9$
    $x + 3y = -2$
    (e) $x + y = -1$
    $x - y = -3$
    (f) $4x + y = 8$
    $4 - \frac{1}{2}y = 2x$

4.4 solving systems of equations

In mathematics, we learn many skills and strategies for solving problems. We can often solve a problem in more than one way. An important skill in solving problems is deciding which skill is the most suitable in solving equations and problems.

For example, to solve this system of 2 equations in 2 variables, we may decide to use our skills in co-ordinate geometry.

System of 2 equations in 2 variables
$$x + y = 1$$
$$4x - 8y = 49$$

Graph of the system of 2 equations in 2 variables.

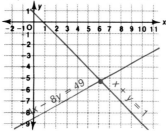

From the graph, we can only estimate the solution of the above system of equations.

Then we need to develop a more exact method of solving a system of 2 equations in 2 variables.

Substitution method

To solve the following system, we may substitute the given value of x into the remaining equation to find the value of y.

$$x = 5 \qquad ①$$
$$x + y = 7 \qquad ②$$

From equation ① use the value of x in equation ②.

$$x + y = 7$$
$$5 + y = 7$$
$$y = 2$$

Thus, $x = 5$ and $y = 2$.

From the graph of the above system we see that the lines given by $x = 5$ and $x + y = 7$ intersect at (5, 2).

In Example 1, we substitute the expression for y given in terms of x into the remaining equation.

Example 1

Solve $3x + y = 2$
$2x + 5y = 23$

Solution

$$3x + y = 2 \qquad ①$$
$$2x + 5y = 23 \qquad ②$$

From equation ① we obtain an expression for y in terms of x.

$$3x + y = 2 \qquad ①$$
$$y = 2 - 3x \qquad ③$$

Substitute the expression for y in equation ③ into equation ②.

$$2x + 5y = 23$$
$$2x + 5(2 - 3x) = 23 \qquad ④$$

Equation ④ is in 1 variable, which is solved for x.

We might also have written the next equation as
$$10 - 13x = 23$$
$$10 - 23 = 13x$$
$$-13 = 13x$$
$$-1 = x$$

$$2x + 5(2 - 3x) = 23$$
$$2x + 10 - 15x = 23$$
$$10 - 13x = 23$$
$$-13x = 13$$
$$x = -1 \qquad ⑤$$

Now substitute $x = -1$ in equation ①.

$$3x + y = 2$$
$$3(-1) + y = 2$$
$$-3 + y = 2$$
$$y = 5$$

Remember: to avoid errors, use brackets when substituting.

Thus, the solution is $x = -1, y = 5$.
Check the solution.

For equation $3x + y = 2$
LS $= 3x + y$ RS $= 2$
$= 3(-1) + 5$
$= -3 + 5$
$= 2$
LS = RS
checks ✓

For equation $2x + 5y = 23$
LS $= 2x + 5y$ RS $= 23$
$= 2(-1) + 5(5)$
$= -2 + 25$
$= 23$
LS = RS
checks ✓

In the above method we use one equation to express one variable in terms of the other variable and then substitute into the remaining equation. We call this algebraic method of solving a system of 2 equations in 2 variables the **substitution method**.

Comparison method
We may extend the method of substitution and from each equation obtain an expression for one of the variables. For example, to solve this system
$x - 2y = -3$ ①
$x + 4y = 3$ ②
we obtain
from equation ① $x - 2y = -3$
$x = -3 + 2y$ ③
from equation ② $x + 4y = 3$
$x = 3 - 4y$ ④

Equate the two expressions for x from equations ③ and ④ and then solve the resulting equation in 1 variable.
$-3 + 2y = 3 - 4y$
$2y + 4y = 3 + 3$
$6y = 6$
$y = 1$

Use the value of $y = 1$ in equation ① (or ②) to obtain a value of x.
$x - 2y = -3$
$x - 2(1) = -3$
$x - 2 = -3$
$x = -1$
Thus we obtain $x = -1$, $y = 1$ as the solution of the system.

In the above method, we obtain expressions for one of the variables from both equations and then *compare* them to obtain 1 equation in 1 variable. We often refer to the above method as the **comparison method**.

To solve problems such as the following, we need to practise and develop our skills in solving linear systems of 2 equations in 2 variables. Do you know the answer to this problem?

> 300 people applied for a rebate on their cars under a new financial plan. The plan provides a rebate of $100 for small cars and $45 for full-size cars. If a total of $25 325 was paid out, how many applicants owned full-size cars?

4.4 exercise

The skills in Questions 1 to 3 are essential when using either the method of *substitution* or *comparison*.

A
1 Express y in terms of x.
(a) $3x + y = 4$ (b) $5x + y = 8$
(c) $y - 3x = 8$ (d) $y - 2x = -6$
(e) $2x = y + 4$ (f) $3x = 4 - y$
(g) $2x + y - 7 = 0$ (h) $2x - y + 5 = 0$
(i) $4x = 3y$ (j) $5x + 2y = 0$

2 Express x in terms of y.
(a) $x + 3y = 8$ (b) $x - 2y = 6$
(c) $x + y - 5 = 0$ (d) $y - x + 30 = 0$
(e) $3y - x = 0$ (f) $4x + 3y = 0$

3 For each equation, solve for the variable indicated.
(a) $2m - n = 6$, n (b) $3y = 8 + x$, x
(c) $4a - b - 6 = 0$, b (d) $3x = y - 8$, y
(e) $3y - x = 8$, x (f) $5a = 8 - b$, b
(g) $2m + 3n - 5 = 0$, m (h) $4x + 3y = 0$, y

B
4 (a) Use the *method of substitution* to solve this system.
$x = 2y - 1$ $2x + y = 3$
(b) Verify your solution in (a).

5 Use the *method of substitution* to solve each system. (Did you check your answers?)

(a) $x = 3$
$3x - 2y = 11$

(b) $y = 5$
$2x + 3y = 11$

(c) $3a - b = -6$
$a = -2$

(d) $2b - 5a = 14$
$b = -3$

(e) $m = 3n - 8$
$2n - m = 5$

(f) $3a - 2b = 1$
$a = b + 1$

6 (a) Use the *method of comparison* to solve this system.
$a = b - 1,\quad 3a + b = 3$

(b) Verify your solution in (a).

7 Use the *method of comparison* to solve each system.

(a) $m = 1 - n$
$m = 1 - 2n$

(b) $-3 - 2y = x$
$1 + 2y = x$

(c) $n = 2m - 8$
$n = 1 - m$

(d) $a = -2b + 3$
$a = 3b - 7$

(e) $m = 3 - 2n$
$n + 3 = m$

(f) $n = m - 1$
$2 - 2m = n$

(g) $x = 2y + 3$
$x + y = 0$

(h) $b + a = 0$
$2b = 3 + a$

8 (a) Solve this system using the *method of substitution*.
$3x - 2y = 9 \qquad x + y = -2$

(b) Solve the system in (a) using the *method of comparison*.

(c) Which method do you prefer, (a) or (b)? Why?

9 (a) Use the *method of comparison* to solve this linear system.
$3(x + 1) + y = 7 \qquad 3(2x - 5) - y = 26$

(b) Use the *method of substitution* to solve the system in (a).

(c) Which method do you prefer, (a) or (b)? Why?

10 For each of the following systems, *first* decide which method to use to solve each system.
- Method of Substitution
- Method of Comparison

Then, solve each system.

(a) $b = 8 - 3a$
$-b = 7 - 2a$

(b) $x = 3y - 10$
$y - 4 = x$

(c) $13 - 9x = y$
$3x - 5y = -33$

(d) $x = 2y - 2$
$9y - 6x = 11$

(e) $a - b = 2$
$a + 3b = -10$

(f) $x + 3y = 6$
$2x - y = 5$

(g) $3x - y = -9$
$y - 2x = 7$

(h) $3m - n = 4$
$m = 2n + 3$

11 (a) Draw a graph to find the co-ordinates of the point of intersection of the lines given by
$x - y = -1 \qquad 2x - y = -1$

(b) How can you check your answer in (a) by using an algebraic method?

12 Find the co-ordinates of the point of intersection of each of the following systems.

(a) $2x - 3y = -9$
$x + y = -2$

(b) $x - y = 1$
$3x + y = 7$

(c) $y - x + 1 = 0$
$x - 2y = 1$

(d) $x + 3y = 4$
$2x - y = 8$

13 Find the co-ordinates of the point that lies on both lines given by the following systems.

(a) $x = 2y + 1$
$y - x = -1$

(b) $3x - 2y = -5$
$x = y - 2$

(c) $x - y = -1$
$2x - y = -1$

(d) $y - 3x = -4$
$x - y = 2$

(e) $3x + 2y - 11 = 0$
$x = y - 3$

(f) $2x + 3y = 0$
$2x + y = -4$

14 (a) What might be your first step in solving the system given by
$3(x - 2) - 2(y - 3) = 4 \qquad 3y - x = -6$?

(b) Solve the system in (a).

15. Solve each system.
    (a) $x + y = 2$
        $2(x - 1) - y = -7$
    (b) $3x - y = 16$
        $2(x + 3) = y + 16$
    (c) $x - 2y = 2$
        $2(x - 3) - 3(y - 1) = 0$
    (d) $x + y = -3$
        $3(x + 1) - 2(y - 1) = -4$
    (e) $2x - y = 0$
        $3(x - 1) = 2(y + 1) - 6$

16. To solve the system of equations given by
    A: $y + \frac{3}{4}x = -1 \qquad x - \frac{1}{4}y = 5$
    Jean wrote the following:
    B: $4y + 3x = -4$ and $4x - y = 20$
    (a) Show why the equations in A are equivalent to those in B.
    (b) Solve the system.

17. Solve each system.
    (a) $\frac{1}{3}x - y = 1$
        $\frac{2}{3}x + y = 5$
    (b) $2a + \frac{1}{2}b = 8$
        $a - \frac{1}{2}b = 1$
    (c) $\frac{1}{3}a - b = -5$
        $a - \frac{2}{3}b = -1$
    (d) $\frac{m}{6} + \frac{n}{2} = -\frac{1}{2}$
        $\frac{m}{3} - 3n = 7$

applications: using co-ordinates

Often to rescue people, the pilot of the helicopter or the rescue crew must know co-ordinates exactly.

The line of sight of a forest fire is given from one observation post as $x - y = 1$. From another observation post the line of sight is given by the equation $x + 3y = 21$.

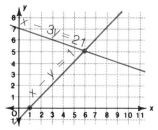

The fire is located at the co-ordinates of the point of intersection of the two lines of sight. In real life, more advanced methods are used to determine the positions of wrecked ships, forest fires, survivors, but the principles needed to calculate their position are the same.

18. For the above forest fire:
    (a) Solve the equations.
        $x - y = 1 \qquad x - 3y = 21$
    (b) Where is the fire located?

19. The lines of sight for a forest fire are given.
    From Observation Deck A: $3x + y - 9 = 0$
    From Observation Deck B: $2x + 3y - 13 = 0$
    What are the co-ordinates of the forest fire?

20. The position of a wrecked oil tanker is given at the intersection of
    $2x - y = 2$ and $3(2x - 1) - 2(y - 1) = 9$.
    Find the co-ordinates of its position.

21. A piper cub crashed in the desert and was located by the lines given by $x - y = -2$ and $3y - 4 = x$. Find the co-ordinates of its position.

22. The coasts of a triangular island are given by the following equations.

    (a) Find the co-ordinates of the northern tip.
    (b) Find the co-ordinates of the farthest west point.

4.5 solving linear systems: addition-or-subtraction method

When mathematicians find that their methods are limited, they invent better methods of overcoming their difficulties. The graphical method of solving a linear system had shortcomings. Thus the substitution method of eliminating one of the variables of the linear system was developed.

However to solve the following linear system, the *method of substitution* is not efficient, since the solution of

$$2x + 3y = 21 \qquad 4x - 2y = 2$$

may involve tedious calculations.

Thus, another method of solving a linear system was invented based on the following properties of equivalent systems which we studied in Section 4.3.

Property I

Equations $x + y = 4$ and $2x + 2y = 8$ are equivalent. This property means that we can perform operations on an equation such that the resulting equation and the original equation are equivalent.

Property II

In any linear system, adding or subtracting the equations results in an equation that has the same intersection point.

$$\begin{array}{l} 3x + 2y = 12 \\ 2x + 3y = 13 \\ \hline 5x + 5y = 25 \end{array}$$
Add

$$\begin{array}{l} 3x + 2y = 12 \\ 2x + 3y = 13 \\ \hline x - y = -1 \end{array}$$
Subtract

The equations $5x + 5y = 25$ and $x - y = -1$ have the same intersection point as the original system.

We refer to the method based on the above properties as the **addition-or-subtraction method** for solving a system of 2 linear equations in 2 variables.

Example 1

Solve the following system. Use the *addition-or-subtraction method*.
$$x + 2y = 11 \qquad 3x - 2y = 9$$

Solution

We add or subtract the two equations in order to eliminate *one* of the variables. A new equation in one variable is obtained. This new equation should be easy to solve.

Details of the Solution   Reasons

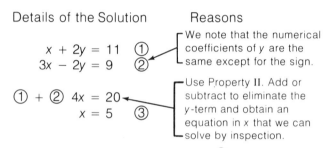

$$x + 2y = 11 \quad ① $$
$$3x - 2y = 9 \quad ② $$

We note that the numerical coefficients of y are the same except for the sign.

$$① + ② \quad 4x = 20$$
$$x = 5 \quad ③$$

Use Property II. Add or subtract to eliminate the y-term and obtain an equation in x that we can solve by inspection.

Use $x = 5$ in ① to obtain a value for y.

From ③ we know the value of x by inspection.

$$\begin{aligned} x + 2y &= 11 \\ 5 + 2y &= 11 \\ 2y &= 6 \\ y &= 3 \end{aligned}$$

Substitute this value for x into one of the original equations to obtain the value of the remaining variable, y.

Thus $(x, y) = (5, 3)$

Use $x = 5$, $y = 3$

We check our work to verify our solution.

$$\begin{array}{ll} x + 2y = 11 & 3x - 2y = 9 \\ LS = x + 2y \quad RS = 11 & LS = 3x - 2y \quad RS = 9 \\ \quad = 5 + 2(3) & \quad = 3(5) - 2(3) \\ \quad = 5 + 6 & \quad = 15 - 6 \\ \quad = 11 & \quad = 9 \\ LS = RS & LS = RS \\ \text{checks } \checkmark & \text{checks } \checkmark \end{array}$$

To use the *addition-or-subtraction method* of solving a system of equations the coefficients of one of the variables must be the same or the same except for the sign. In Example 2, we use Property I to change the form of one of the equations so that we end up with a set of equations in which one of the variables has the same numerical coefficient.

Example 2

Solve $3x - 2y = 13$ and $x + y = 1$. Use the *addition-or-subtraction method*.

Solution

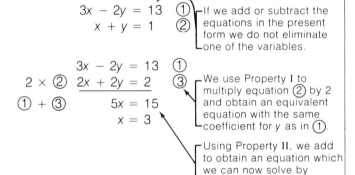

Reasons

If we add or subtract the equations in the present form we do not eliminate one of the variables.

We use Property I to multiply equation ② by 2 and obtain an equivalent equation with the same coefficient for y as in ①.

Using Property II, we add to obtain an equation which we can now solve by inspection.

Use $x = 3$ in ①.

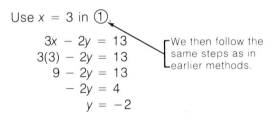

We then follow the same steps as in earlier methods.

Thus $(x, y) = (3, -2)$ You may check the solution.

The principles of the methods we use to solve a linear system are basically the same. However, one method may be more convenient to apply than another, as we shall see in the exercise.

4.5 exercise

Questions 1 to 7 develop some essential skills required to apply the *add-or-subtract method* of solving a linear system.

A

1. How are the equations in Column B obtained from those in Column A?

	Column A	Column B
(a)	$x + 3y = 6$	$\longrightarrow 2x + 6y = 12$
(b)	$2p - 3q = 12$	$\longrightarrow 6p - 9q = 36$
(c)	$6a + 10b = 12$	$\longrightarrow 3a + 5b = 6$
(d)	$2y - 3x = 8$	$\longrightarrow 6x - 4y = -16$
(e)	$-m + 4n = -11$	$\longrightarrow 3m - 12n = 33$

    Why are the equations in columns A and B equivalent?

2. For each system, how is equation ③ obtained from equations ① and ②?

    (a) $a + 3b = 6$ ①
    $2a - b = 5$ ②
    $2a + 6b = 12$ ③

    (b) $5x - 3y = 14$ ①
    $2x - y = 6$ ②
    $6x - 3y = 18$ ③

3. (a) Which variable of the following system is more easily eliminated?
    $2x - y = -1$ $3x + y = -9$

    (b) Solve the system in (a).

4. (a) Which variable of the following system is more easily eliminated?
    $3a - 5b = 10$ $a - 2b = 4$

    (b) Solve the system in (a).

5. • Decide which variable is more easily eliminated in each of the following systems.
    • Solve the system.

    (a) $2m + n = -1$
    $3m - n = -4$

    (b) $x + 3y = 2$
    $x - 2y = 2$

    (c) $a + 2b = -7$
    $3a - 2b = -5$

    (d) $3p - q = 14$
    $3p + 2q = 8$

    (e) $3x + 2y = -15$
    $x - 3y = 6$

    (f) $4m - 3n = 23$
    $2m + 5n = 5$

B 6
- First decide which method to use to solve each of the following systems.
- Then solve each system.

(a) $y = 2x + 5$
$2x - 3y = -11$

(b) $3m - 2n = 8$
$7m + 2n = 32$

(c) $2a + 3b = -6$
$5a - 2b = -15$

(d) $2p - q = 7$
$3p + 5q = 17$

7 (a) What might be your first step in solving this system?
$3y - 2x = 4$ $\qquad$ $2x + 5y = -4$

(b) Solve the system in (a).

8 Solve.
In solving systems, be sure to look ahead and decide *first* which method is preferable. Save yourself tedious calculations.

(a) $a + 4b = 6$
$2a - b = 3$

(b) $2m = 6 + n$
$3m + n = 9$

(c) $2p - 3q = -1$
$2p + 6q - 8 = 0$

(d) $b - 3a = 0$
$2a + 5b = 17$

(e) $2m = 3 - 3n$
$6n + 10m + 3 = 0$

(f) $5p = 2q$
$2p = 5q$

(g) $6y = 21 - 9x$
$3y - 2x = 4$

(h) $2x - 18y = 0$
$3x - 5y = 22$

(i) $3x - 2y - 5 = 0$
$4x + 14y - 15 = 0$

(j) $2m + 14 = 3n$
$48 = 3m + 7n$

9 (a) What might be your first step in solving this system? Why?
$\frac{x}{2} - y = 5$ $\qquad$ $2x - \frac{y}{3} = 9$

(b) Solve the system in (a).

10 (a) A system is given by
$3(a - 2) - 3(b + 6) = -9$ $\qquad$ $2a + 3b = -5$
What might be your first step in solving the system?

(b) Solve the system. Then verify your solution.

11 Solve.

(a) $\frac{a}{2} + b = -4$
$a - \frac{2b}{3} = 8$

(b) $x - \frac{3}{4}y = -5$
$x - \frac{y}{4} = -1$

(c) $2(x + 1) - 3y = 5$
$2x + y = 7$

(d) $3x - y = 7$
$5x - 2(y - 3) = 18$

(e) $\frac{1}{2}a - b = 8$
$a + \frac{1}{3}b = 2$

(f) $p + 3q = 2$
$3(p - 2q) = 1$

(g) $2a = 4(2b + 2)$
$3(a - 3b) + 2 = 17$

(h) $2(m - 1) + n = 1$
$m - 3n = 5$

12 Two lines are defined by the following equations.
$2x - 3y = -25$ $\qquad$ $3x + y = -21$
Find the co-ordinates of the point of intersection.

13 Find the co-ordinates of the intersection for each of the following pairs of lines.

(a) $2x + 5y = 19$
$3x - y = 3$

(b) $3x - 2y = 13$
$2x + y = 4$

(c) $7x = 5 + 3y$
$2y + 4 = 5x$

(d) $x - 3 = 2y$
$5x + 4y - 8 = 0$

14 Find the co-ordinates of the vertices of the triangle shown.

problem/matics

In how many ways can the 4 × 4 square be cut into congruent parts?

4 × 4 square $\qquad$ This is one way

4.6 extending our work: special equations

The skills we learn in solving one type of equation are applied again as we learn to solve new types of equations and problems. Earlier, in solving the following type of equation, we first simplified the equation so that it did not involve fractions.

$$\frac{m-5}{3} - \frac{2m+6}{2} = -4$$

To solve a system of equations involving fractions, we apply the above principles. Simplify the equation by clearing fractions as shown in the following example.

Example 1
Solve the linear system given by
$$\frac{x-2}{3} - \frac{y+5}{2} = -3 \qquad x + y = 8$$

Solution
Simplify the first equation by multiplying it by the least common multiple of 3 and 2 which is 6.

$$\frac{x-2}{3} - \frac{y+5}{2} = -3 \qquad \text{①}$$

$$6\left(\frac{x-2}{3}\right) - 6\left(\frac{y+5}{2}\right) = 6(-3)$$

$${}^2\cancel{6}\left(\frac{x-2}{\cancel{3}_1}\right) - {}^3\cancel{6}\left(\frac{y+5}{\cancel{2}_1}\right) = 6(-3)$$

$$2(x-2) - 3(y+5) = -18$$
$$2x - 4 - 3y - 15 = -18$$
$$2x - 3y = 1 \qquad \text{②}$$

We now solve the simplified system.
$$2x - 3y = 1 \qquad \text{②}$$
$$x + y = 8 \qquad \text{③}$$

Substitute $y = 8 - x$ from ③ into ②.
Thus
$$2x - 3y = 1$$
$$2x - 3(8 - x) = 1$$
$$2x - 24 + 3x = 1$$
$$5x = 25$$
$$x = 5$$

Use $x = 5$ in the *original* ③ to obtain the value of y.
$$x + y = 8 \qquad \text{To verify our solution}$$
$$5 + y = 8 \qquad \text{use the original equation ①.}$$
$$y = 3 \qquad \text{This is left for us to try.}$$
Thus the solution is $(x, y) = (5, 3)$.

Mathematicians often introduce new steps to save tedious calculations. This is illustrated in Example 2.

Example 2
Solve the linear system.
$$\frac{3}{x} + \frac{4}{y} = 10 \qquad \frac{2}{x} - \frac{1}{y} = 3 \qquad x, y \neq 0$$

Solution
Use the intermediate step $\frac{1}{x} = a$ and $\frac{1}{y} = b$.
The equations are then rewritten as
$$\frac{3}{x} + \frac{4}{y} = 10 \longrightarrow 3a + 4b = 10 \qquad \text{①}$$
$$\frac{2}{x} - \frac{1}{y} = 3 \longrightarrow 2a - b = 3 \qquad \text{②}$$

$$\begin{array}{r} 3a + 4b = 10 \qquad \text{①} \\ 4 \times \text{②} \quad \underline{8a - 4b = 12} \qquad \text{③} \\ \text{① + ③} \quad 11a = 22 \\ a = 2 \qquad \text{④} \end{array}$$

Use $a = 2$ in equation ②.
$$2a - b = 3$$
$$2(2) - b = 3$$
$$4 - b = 3$$
$$-b = -1 \text{ or } b = 1$$
Thus the solution is $(a, b) = (2, 1)$
or $a = 2, b = 1$.
Since we introduced an intermediate step, we need to return to the solution of the original equations. Thus
$$a = 2, b = 1 \text{ means } \frac{1}{x} = 2, \frac{1}{y} = 1$$
$$\text{or } x = \frac{1}{2}, y = 1.$$

The solution of the original system is $(x, y) = \left(\frac{1}{2}, 1\right)$.

4.6 exercise

Questions 1 to 3 develop our skills in solving special types of equations.

A 1 Explain how the equations in Column B are obtained from the corresponding equations in Column A.

Column A	Column B
(a) $3x - \frac{1}{2}y = 4$	$6x - y = 8$
(b) $\frac{2}{3}x + y = -3$	$2x + 3y = -9$
(c) $3(x - 1) + y = -2$	$3x + y = 1$
(d) $y - 2(x + 5) = 6$	$y - 2x - 10 = 6$
(e) $\frac{x - 1}{2} + y = 4$	$x - 1 + 2y = 8$

2. (a) What might be your first step in solving the following linear system?

$$\frac{x-1}{2} + y = 5 \qquad x + y = 8$$

(b) Solve the system in (a).

3. (a) How is the least common multiple of 3 and 4 used to simplify the following system?

$$\frac{(2x+1)}{3} - \frac{(y+5)}{4} = -3$$
$$2x + y = -1$$

(b) Solve the system in (a).

B 4 Solve each of the following systems.

(a) $\frac{x+1}{5} = 2y - 3$
   $x + \frac{y+7}{4} = -4$

(b) $\frac{4x-5}{3} - 3y = 22$
   $3x - \frac{(y+1)}{6} = 7$

(c) $5x - \frac{(5y-2)}{3} + 11 = 0$
   $\frac{x+3}{2} + 4y = -8$

(d) $2x + \frac{(4y+3)}{2} = 2$
   $\frac{3x+1}{3} = y + \frac{5}{4}$

(e) $\frac{2(x-1)}{3} + y = 3$
   $x - 3y = -8$

5. (a) Solve the following system.

$$\frac{x-1}{2} + y = -2 \qquad 3y + \frac{2x+11}{5} = 1$$

(b) Verify the answers you obtained in (a).

6. Solve and verify.

(a) $\frac{x+1}{2} + \frac{3y-3}{8} = 0$
   $\frac{2x+3}{3} + 6y = -1$

(b) $\frac{x}{7} - \frac{2y+1}{5} = 0$
   $\frac{x-2}{9} + \frac{y}{3} + 2 = 0$

7. (a) What is the first step needed to solve the following system?

$$\frac{1}{x} + \frac{4}{y} = 3 \qquad \frac{2}{x} - \frac{2}{y} = 1$$

(b) Solve the system in (a).

8. Solve each of the following systems.

(a) $\frac{2}{x} + \frac{8}{y} = 0$
   $\frac{3}{x} + \frac{4}{y} = -2$

(b) $\frac{1}{x} - \frac{3}{y} = \frac{1}{2}$
   $\frac{4}{x} + 4 = \frac{9}{y}$

(c) $\frac{10}{x} + \frac{25}{y} = 0$
   $\frac{2}{x} = 2 + \frac{5}{y}$

(d) $\frac{7}{x} = 15 - \frac{1}{y}$
   $\frac{14}{x} - \frac{2}{y} = -14$

9. (a) Solve the following system.

$$\frac{2}{x} + \frac{4}{y} = 10 \qquad \frac{4}{x} + \frac{6}{y} = 23$$

(b) Verify the answers you obtain in (a).

10. Solve and verify.

(a) $\frac{6}{y} = 5 + \frac{2}{x}$
   $\frac{3}{y} = 15 - \frac{4}{x}$

(b) $\frac{2}{x} - \frac{1}{y} - 13 = 0$
   $\frac{3}{x} + \frac{2}{y} = 2$

(c) $\frac{1}{x} = \frac{2}{y} + 12$
   $\frac{3}{y} = 2 - \frac{1}{x}$

(d) $\frac{3}{x} - \frac{4}{y} - 15 = 0$
   $\frac{3}{y} + \frac{4}{x} = \frac{15}{2}$

working with literal coefficients

The skills we learn in solving equations involving numerical coefficients may be extended to solving equations involving literal coefficients. In mathematics, we often wish to study a general situation rather than a particular situation. For example, using a computer to solve all numerical examples of linear systems requires the skill of solving a linear system that has literal coefficients. In the following applications we can use a computer program to solve all specific linear systems. Compare the solutions of these two systems at each step.

Solve for x and y.
$$3x + 2y = 12 \quad \text{①}$$
$$2x + 3y = 13 \quad \text{②}$$
Use the *addition-or-subtraction method*.
$$3 \times \text{①} \quad 9x + 6y = 36 \quad \text{③}$$
$$2 \times \text{②} \quad 4x + 6y = 26 \quad \text{④}$$
$$\text{③} - \text{④} \quad \overline{5x = 10}$$
$$x = 2$$

Solve for x and y.
$$ax + by = c \quad \text{①}$$
$$dx + ey = f \quad \text{②}$$
Use the *addition-or-subtraction method*.
$$e \times \text{①} \quad aex + bey = ce \quad \text{③}$$
$$b \times \text{②} \quad bdx + bey = bf \quad \text{④}$$
$$\text{③} - \text{④} \quad \overline{aex - bdx = ce - bf}$$
$$x(ae - bd) = ce - bf$$
$$x = \frac{ce - bf}{ae - bd}$$
$$ae - bd \neq 0$$

For the system with literal coefficients, we have solved x in terms of other variables a, b, c, d, e, f, called **parameters**.

11 Compare the solutions of the above system which uses the *addition-or-subtraction method*.

(a) How are ths solutions of the systems alike?
(b) How are they different?

12 To answer these questions refer to the solution of the system with literal coefficients.

(a) Why do we need to indicate that $ae - bd \neq 0$? Why is this step not necessary in the solution of the other system involving numerical coefficients?

(b) Substitute $x = \dfrac{ce - bf}{ae - bd}$, into equation ① to find the value of y.

(c) Using equations ① and ②, solve for y, from first principles using the *addition-or-subtraction method*.

13 Verify that the solution of the system $x + y = 2b$ and $4x - 5y = 17b$ is $(x, y) = (3b, -b)$.

14 Match the solutions A, B, C, D, with the corresponding system of equations.

Solutions

(a) $x - 2y = 7a$
$5x + 3y = -17a$ A $(x, y) = (a, 3b)$

(b) $x + y = 2a$
$x - y = 2b$ B $(x, y) = (a + 2b, a + b)$

(c) $x + a = 2y$
$2x - 3y = b - a$ C $(x, y) = (a + b, a - b)$

(d) $x - y = a - 3b$
$3x + 2y = 3a + 6b$ D $(x, y) = (-a, -4a)$

15 (a) Solve the system for x and y.
$$3x - y = a \quad \text{and} \quad 2x + 4y = 10a$$
(b) Verify your solution in (a).

16 (a) Solve the system for x and y.
$$ax + y = 2, \quad bx - y = 4$$
(b) Verify your solution in (a).

17 Solve for x and y.

(a) $x - y = 5a$ (b) $2x - 4y = 0$
$2x - 3y = 13a$ $3x - a = 2y$

(c) $2x + y = b - 4a$ (d) $2x - 5y = 6a - 20b$
$3x - 2y = -(6a + 2b)$ $3x - 4y = 9a - 16b$

(e) $2ax + y = -2$ (f) $3x + ay = 4$
$bx - y = 3$ $ax + 2y = 6$

18 Solve for x and y.
$$ax + by = e \qquad cx + dy = f$$

19 Use your results in Question 18 and write the solution for each of the following systems. *Do not solve these systems* from first principles.

(a) $x + y = -1$
$x - y = -3$

(b) $2x + y = 1$
$x - y = 2$

(c) $x - 2y = 1$
$x + y = 4$

(d) $x - y = -2$
$x - 3y = -4$

applications: equation for a computer program

Often, in solving problems in business and science, we need to use systems of equations. Many complex problems involve systems of equations that contain many variables and equations. To save time, a computer is used to solve complex systems of equations. However in order to use a computer we first need to write a program for the computer.

System of Equations
$Ax + By = C$
$Dx + Ey = F$

Solutions
$x = \dfrac{CE - BF}{AE - BD}$, $AE \neq BD$
$y = \dfrac{CD - AF}{BD - AE}$, $BD \neq AE$

We thus write the computer program in BASIC to solve all systems in 2 equations in 2 variables.

Computer Program in BASIC
```
10  Print    SOLVE TWO EQUATIONS IN
             TWO VARIABLES.
20  Input A
21  Input B
22  Input C
23  Input D
24  Input E
25  Input F
30  If (B * D) = (A * E), then 70
40  Let Y = (C * D − A * F)/(B * D − A * E)
50  Let X = (C * E − B * F)/(A * E − B * D)
60  Print X; Y;
70  Stop
80  End
```

Means multiplication in BASIC language. (refers to *)

Means division in BASIC language. (refers to /)

20 A system of equations is given by
$3x - 4y = 5$ $x + 5y = 8$

(a) What are the values of A, B, C, D, E, and F?
(b) Use the program to calculate the value of x and y.
(c) Solve the above system to verify your answers in (b).

21 For each system of equations, use the computer program to obtain the values of x and y.

(a) $4x + 5y = 7$
$3x + 4y = 6$

(b) $8x - 9y = 41$
$4x + 3y = 3$

(c) $2x - y = -7$
$y + 3x = -8$

(d) $3x = 9 - 4y$
$4 + 8y = 5x$

(e) $3x = 12 - y$
$2x + 5y = 21$

(f) $3x + 4 = 2y$
$6x = 3y$

22 For each system of equations, make the appropriate changes in the computer program to obtain the values of x and y.

(a) $2x + ay = -2$
$bx - 3y = 5$

(b) $ax - by = e$
$cx - dy = f$

23 A system is given by
$\dfrac{x + 7}{5} + \dfrac{y + 13}{3} = 7$ $\dfrac{1}{2}x + \dfrac{3}{4}y = 3$

Change the form of the above equations so that the computer program may be used. Then use the program to obtain the values of x and y.

math tip

To solve some problems, we have translated the given information and have written 1 equation in 1 variable. For example,
  h, for the height of a building
  t, for the thickness of a gym mat.
To solve problems involving a linear system in the following sections, use the same procedure.
  r, for the distance to the rink
  p, for the distance to the pool.

4.7 steps for solving problems

Previously, when solving a problem using 1 variable, we used the following plan to organize our work and then solved the problem.

> Steps for Solving Problems

Step A Read the problem carefully.
Answer these two important questions:
 I What information are we given (information we know)?
 II What information are we asked to find (information we don't know)?
Be sure to understand what it is we are to find, then introduce the variables.

Step B Translate from English to mathematics and write the equations.

Step C Solve the equations.

Step D Check the answers in the original problem.

Step E Write a final statement as the answer to problem.

The above plan may be used over and over again to solve problems.

Some problems require 2 variables. In these cases we require 2 equations in 2 variables to solve the problem. To solve any problem, the most important first step is to translate correctly from English to mathematics.

English expression	Variable expression
The value of m quarters and n dimes in cents	$25m + 10n$
The denominator of the fraction $\frac{a}{b}$ is decreased by 3 and the numerator increased by 6.	$\frac{a + 6}{b - 3}$
The total interest paid per year on $\$m$ invested at 12% and $\$n$ invested at 15%.	$0.12m + 0.15n$

In order to be successful in solving problems we must organize our work by using the *Steps for Solving Problems*. We *must* be sure that we see how each step of the solution is arrived at.

Example 1
Bob and Kevin bought a lottery ticket and won $100 000. Since they paid different amounts for the winning ticket, Bob received $10 000 more than twice what Kevin received. How much did each person receive?

Solution

Step A Let $\$b$ represent Bob's share of the lottery and $\$k$ Kevin's share.

Step B Bob's share, plus Kevin's share is $100 000.
$$b + k = 100\,000 \quad ①$$
Bob's share is $10 000 more than twice Kevin's share.
$$b = 10\,000 + 2k \quad ②$$

Step C
$$b + k = 100\,000 \quad ①$$
$$b - 2k = 10\,000 \quad ③$$
$$3k = 90\,000 \quad ①-③$$
$$k = 30\,000$$
Use $k = 30\,000$ in equation ①.
$$b + k = 100\,000$$
$$b + 30\,000 = 100\,000$$
$$b = 70\,000$$

Step D Check in the original problem.
Amount received by Bob $ 70 000
Amount received by Kevin $ 30 000
                   Total $100 000   checks ✓

Twice Kevin's share $ 60 000   Bob's
+ $10 000 $ 10 000   share.
                          $ 70 000   checks ✓

Step E Thus Bob received $70 000 and Kevin received $30 000 of the lottery winnings.

To illustrate the important steps in solving a problem, a straightforward example was chosen. We will need lots of practice with these types before attempting more complex ones.

4.7 exercise

Questions 1 to 5 develop some essential skills related to the steps in solving problems.

A 1. Find the value in cents of each of the following.
   (a) 6 dimes
   (b) t dimes
   (c) $2m$ dimes
   (d) $(k + 2)$ quarters

2. Find the value of each of the following in cents.
   (a) t dimes, k quarters
   (b) s quarters, t dimes
   (c) a nickels, b quarters
   (d) d pennies, e quarters

3. Find the amount of interest paid on each of the following investments. (The rate of interest is yearly and the money is invested for a year.)
   (a) $\$s$ at 8% and $\$y$ at 9%
   (b) $\$k$ at 15% and $\$y$ at 10%
   (c) $\$(y - 1)$ at 10% and $\$(m + 1)$ at 12%

4. Find the total distance travelled.
   (a) Lori walked for 14 h at y km/h.
   (b) A jet travelled 12 h at m km/h.
   (c) A truck travelled at 90 km/h for x min.
   (d) A car travelled for y h at 40 km/h at p h at 100 km/h.
   (e) A motorbike travelled for 6 h at 50 km/h.

5. For each of the following statements
   - use two variables of your own choice.
   - write an equation in 2 variables.
   (a) A total of 960 tennis and badminton racquets were ordered.
   (b) The total number of marks Joanne received in Physics and Economics was 185.
   (c) The sum of Barb's age and three times Frank's age is 160 a.
   (d) The interest on an amount of money invested at 9% exceeds the interest on another amount of money invested at 12% by $190.
   (e) When twice the width of a rectangle is added to half the length the sum is 126 m.
   (f) A car travelled for y h at 40 km/h and x h at 65 km/h and covered a total of 480 km.
   (g) A number is 9 greater than another number.
   (h) The total cost of the seeds at $3.86/kg and the kernels at $2.90/kg is $12.90.
   (i) Four times Cy's age increased by Elaine's age is 196 a.
   (j) The width of a rectangle is 36 m less than the length.
   (k) The total interest on an amount of money invested at 9% and on another amount invested at 12% is $900.

To solve each of the following problems, use the steps on page 139 to help organize your work.

B 6. There were 52 more girls than boys at the last school dance at Vernon High. Total attendance at the dance was 420. How many boys and girls attended?

7. Two numbers differ by 25. If 3 times the smaller exceeds the larger by 3, find the two numbers.

8. Alex and Sam won $2000 in a television game show. If Sam received twice as much as Alex less $100, how did they share their prize?

9. 1400 students attended a concert at Stoney Creek High School. The cost for students with activity cards was $2.00 and $2.50 for those without activity cards. If $3025 was collected, how many students had activity cards?

10. The Book Store cash register had $30 in $1- and $2-bills. If there were 3 times as many $1-bills as $2-bills, how many of each were there?

11. A hockey team bought a total of 36 pucks and sticks. The number of pucks was 4 less than the number of sticks. How many sticks and pucks were purchased?

12 The sum of two numbers is 88. If the greater number is 2 less than 5 times the smaller, find the numbers.

13 The sum of two numbers exceeds twice the smaller number by 7. One half the larger number is 3 less than the smaller. Find the two numbers.

14 In the playoffs, the Leafs played 13 games before being eliminated. If they won 3 more games than they lost, how many games were won and lost?

15 Find two numbers such that their sum is 4 times the smaller and whose difference is 20.

16 The average of two numbers is 36. If the larger number is twice the smaller, find the two numbers.

17 Maple Sports is having a closing-out sale. 4% is lost on all equipment and 7% is lost on all clothing. In 1d sales were $2300. If the total loss was $125, what portion of sales were for clothing?

18 15 uniforms were ordered for the badminton team. The number of medium sized uniforms exceeds 3 times the number of small uniforms by 3. How many uniforms of each size were ordered?

19 200 people attended an acrobatics show. Adults paid $2.00 each and children paid 75¢ each. If total receipts were $337.50, how many adults and children attended the show?

20 Lori is 5 cm shorter than Lesley. Michael's height is 170 cm. Twice Lesley's height is 3 cm more than the combined height of Lori and Michael. How tall is Lesley?

21 A 10-speed bicycle cost $40 more than a 3-speed model and $55 more than a standard model. Two 3-speed and 2 standard bikes can be purchased for $80 less than 3 10-speed bikes. How much does a 10-speed bicycle cost?

4.8 organizing the given information

Already we know several ways of solving the same problem.

Choosing Variables

We may use variables that remind us of given information.

Problem	Choice of Variables
• The sum of two numbers is 170 and their difference is 26. Find the two numbers.	Let the numbers be represented by m and n.
• There were 200 Rams and Jasons at a weekend meet. If the Rams outnumbered the Jasons by 18, how many Rams were at the meet?	Let r represent the number of Rams and j, the number of Jasons.
• Trevor and Matt were going to a baseball game when Matt realized he did not have enough money. Since Trevor had more than enough, he lent Matt $1.30. Now they each had the same amount and enough for the game. If they had $9.50 altogether how much did each have originally?	Let t cents represent the amount of money Trevor has and m cents the amount Matt has.

Listing Information

We may organize the information given in a problem by using these headings.

Facts We are Given	What We Are Asked to Find

Using a Chart

Charts may also be used to organize the information given in a problem as the following example shows.

Example 1

From the Yearbook receipts the student council deposited part of the $1200 in a savings account receiving 9% interest, and the remainder in a chequing accounting receiving 4% interest. If the total interest received for a year is $88.00, how much money was deposited in each account?

Solution

Step A Let $s represent the amount of money deposited in the savings account and $c in the chequing account.

	Amount of money deposited	Rate of interest	Interest paid
Savings Account	$s	9%	$0.09s
Chequing Account	$c	4%	$0.04c

Add to obtain total amount deposited.
Add to obtain total interest.

Step B $s + c = 1200$ ①
$0.09s + 0.04c = 88$ ②

Step C $100 \times$ ② $9s + 4c = 8800$ ③
$4 \times$ ① $\underline{4s + 4c = 4800}$ ④
③ − ④ $5s = 4000$
$s = 800$

Use $s = 800$ to find c in ①
$s + c = 1200$
$(800) + c = 1200$
$c = 400$

Step D Check
I Total amount $800 + $400 = $1200
II 9% of $800 = $72
    4% of $400 = $\underline{$16}$ checks ✓
    Total Interest $88 checks ✓

Step E Thus $800 was deposited in the savings account and $400 was deposited in the chequing account.

4.8 exercise

A

1. Find the distances Tony and Ned ran
   - if the total distance they ran was 1500 m.
   - if, had they each run 1000 m farther, Ned would have run 1200 km less than twice Tony's distance.

	Distance run (m)	Distance plus 1000 m
Ned	n	
Tony	t	

2. Find Henry's age and Janice's age now if
   - the difference in their ages is 6 a.
   - in 7 a, Henry will be twice as old as Janice was 17 a ago.

	Present age (a)	Age in 7 a	Age 17 a ago
Henry	n	h + 7	
Janice	j		j − 17

3. Janet wishes to invest $660 so that the interest from a 10% investment is equal to the income from a 12% bond. How much should be invested at each rate?

Money invested ($)	Interest rate	Income earned ($)
i	10%	
b	12%	

4. Simco Gas sells gas at 23¢/L and premium gas at 25¢/L. It also sells a mixture of the regular gas and the premium gas at a medium price of 24.2¢/L. If a 10 000-L tanker is to be filled with the medium price gas, how much regular and premium gas should be used?

Gas Type	Amount of gas (L)	Value of gas ($)
Regular	x	0.23x
Premium	y	0.25y
Medium	x + y	

5. A chemical firm produces an 80% iron alloy by combining 90% iron ore and 60% iron ore by mass. If the company wishes to fill an order for 150 kg of this alloy, how much of each type of ore must be used?

Ore Type	Amount of ore (kg)	Amount of iron (kg)
90% iron	x	0.90x
60% iron	y	0.60y
80% iron		

In Questions 6 to 25, use a chart where possible to help organize the information given in the problem. You may not need to use a chart to solve every problem.

B

6. Susan invested part of her $1800 savings at 9% and the remainder at 6%. In one year, the 9% investment earned $102 more than the 6% investment. How much did she invest at each rate?

7. Steven won $10 000 in a lottery. He invested his winnings, part at 10% and part at 11%. If the total interest earned was $1060, how much was invested at each rate?

8. How much gasoline should be added to 36 L of a 10% oil mixture to decrease the oil concentration to 5%?

9. The difference between two numbers is 20. The greater exceeds three times the smaller by 10. Find the two numbers.

10. A nut mixture containing 10% cashews is mixed with another nut mixture containing 30% cashews. The result is 10 kg of a nut mixture containing 24% cashews. How much of each nut mixture was used?

11. Joan wishes to invest $3600 in two types of bonds, one yielding 10% and the other 8%. How should she split her investments so that each bond earns the same amount of interest?

12. Premium gasoline sells at 23¢/L. Regular gasoline sells at 21.5¢/L. To boost sales, a middle octane gasoline is formed by mixing premium and regular. If 1000 L of this middle octane gasoline is produced and sold at 22.1¢/L, how much of each type of gasoline was used?

13. Chris uses a mixture of gasoline and oil in the boat's motor. How much oil should be added to a 20-L solution to increase the oil concentration from 10% to 25%?

14. Jim is always thinking about food. He had bacon and eggs for breakfast and noticed there were twice as many pieces of bacon as eggs. If he had eaten 1 more egg, he would have had two-thirds the number of pieces of bacon. How many eggs and pieces of bacon were on his plate?

15. Michael uses a solution of vinegar and water to clean his car windows. How much vinegar should be added to increase the strength in a 900-mL tub of solution from 25% to 40% vinegar?

16. Sarah uses a 40% milk mixture (other ingredients include eggs, flour) to make pancakes. How much milk must be added to a 2-L mixture in order to increase the milk concentration to 60%?

17. How much alcohol must be evaporated from a 50-mL solution to reduce the concentration of alcohol from 30% to 20%?

18. Total sales on Tuesday at Balmoral Pharmacy were $4200. A profit of 4% was realized on part of these sales and a profit of 5% on the rest. If the profit from the 4% part exceeded the profit on the 5% part by $51, how much of sales were realized at each of the two profit rates?

19. A 20% dye solution is mixed with a 50% dye solution to form a 44% dye solution. How much of each dye solution is required to make 250 L of the 44% dye solution?

20. Dandelion weed killer sells at $3.50/L and the crabgrass weedkiller sells at $2.80/L. Rob decides to make an all-purpose weed killer by mixing the two types of weed killer and selling it at $3.08/L. If he makes 50 L of this mixture, how much of each type of weed killer does he use?

applications: working with gold

Since Antiquity, gold has been one of our precious metals. Many of the ancient relics found by archeologists are pure gold. The purest gold nugget ever found was discovered in Victoria, Australia and was 98.59% pure gold. To show the purity of jewellery in terms of the amount of gold in it,

Gold bars in Bank of Canada

we use the term carat (k). 1 carat represents one 24th part. Pure gold is described as 24 k gold. For example, to say a bracelet is 10 k gold means the bracelet is made of an alloy consisting of

    10 parts gold   14 parts of another metal or alloy.

Thus 14 k gold means 14 parts gold and 10 parts alloy. In North America we usually deal in 10 k, 14 k, or 18 k gold. In Great Britain we could find jewellery that is 9 k.

We may use our work with 2 equations in 2 variables to solve problems about gold. To solve the following problems we may use a chart to organize our work.

21. A 14 k gold statue is found at an auction by an antique dealer. In order to maximize profits from this statue, it is melted down with cheap 6 k gold trinkets to produce an 8 k gold antique replica. If the replica has a mass of 3.2 kg, what were the masses of the statue and trinkets used?

22. Bing Crosby won an Oscar Award in 1951, for his role in the movie "Going My Way". This award is 0.5 k gold (bronze with a gold plate). If this award were melted together with raw ore of 5 k gold, a cheap costume jewellery with 2 k gold could be made. Find the mass of the Oscar if 5.25 kg of this new structure is formed.

23. Alan wants to have a pin made for his girlfriend. For sentimental reasons he decides to have a jeweller melt some gold coins from his collection to make the pin. One coin was 18 k gold and the other was 11 k gold. If the pin is 14 k and has a mass of 17.5 g, what was the mass of each coin?

24. The largest gold nugget was the Holtermann Nugget found in Hill End, Australia in 1872. The nugget was 23 k. This nugget was melted with 10 k gold alloy to produce jewellery of 14 k gold. If 702 kg of this alloy was melted, find the mass of the nugget.

25. Chains and Things Limited sold a 12 k gold chain. After many complaints, the owner realized that the chain was too soft. It was decided to melt it down with a bar of 6 k gold and make a new harder chain of 10 k gold. If the new chain has a mass of 31.5 g, find the mass of

(a) the original chain.
(b) the amount of 6 k gold used.

4.9 solving problems: thinking visually

Another strategy which helps us solve problems is using a diagram to record the information and visually interpret the problem. For example, we may draw a diagram to illustrate this problem and use the diagram to help us complete the chart.

> A cleanser containing 20% ammonia is mixed with another cleanser containing 10% ammonia to dilute its strength. If 100 L of this 17% ammonia mixture is obtained, how much of each cleanser was used?

The following diagram helps us better understand the problem.

Original Solutions | Mixture Solution
20% solution | 10% solution

Use the diagrams to help complete the information in the chart.

	20% solution	0% solution	mixture
Amount of liquid	x L	y L	100 L
Concentration	20%	10%	17%
Amount of ammonia	0.20x L	0.10y L	0.17(100) = 17 L

From the above information we may construct the equations, and solve the problem.

$$x + y = 100$$
$$0.20x + 0.10y = 17$$

To solve a problem involving time, distance and rate, we *must* understand this formula and its different forms.

$$d = vt \qquad \frac{d}{v} = t \qquad \frac{d}{t} = v \qquad \begin{array}{l} d,\text{ distance} \\ t,\text{ time} \\ v,\text{ speed} \end{array}$$

To understand the following problem involving the time, distance, or rate, it is helpful to visualize the problem and draw a diagram.

Problem:

> Sue drove her car from Pembroke to North Bay in 3 h while Peter drove his truck for 2 h in the opposite direction to Ottawa. Sue drove 20 km/h faster than Peter. If the distance between North Bay and Ottawa is 400 km, find the speed of the car and truck.

Diagram:

North Bay ← Pembroke → Ottawa
Sue | Peter

	Sue	Peter
speed (km/h)	s	p
time (h)	3	2
distance (km)	$3s$	$2p$

Let s km/h represent Sue's speed and p km/h Peter's speed.

From our sketch we may then complete the chart we have used to record the information.

	Distance (km)	Speed (km/h)	Time (h)
Sue	$3s$	s	3
Peter	$2p$	p	2

total distance
$3s + 2p$

From the chart and the problem, we write the required equations.

$$3s + 2p = 400 \quad \text{— The total distance is 400 km.}$$
$$s = p + 20 \quad \text{— Sue drove 20 km/h faster than Peter.}$$

Now we are able to solve the equations and answer the problem.

4.9 exercise

Questions 1 to 7 develop skills for using diagrams to aid in solving problems. Use diagram and chart to complete the solution for each problem.

For Questions 1 to 4,
- draw a diagram to illustrate the physical situation stated in the problem.
- then construct a chart, record the information and solve the problem.

A

1. A 50% lime solution is mixed with a 30% lime solution to produce 300 L of a 46% solution. How much of each solution was used?

2. A nut mixture contains 20% peanuts by mass and another contains 40%. The two mixtures are combined to obtain 100 kg which contains 36% peanuts by mass. What quantities of each of the original mixtures were used?

3. Terry made 10 L of orange juice using 70% water. How much orange concentrate should be added to bring the water concentration down to 50%?

4. The coffee trucks sell 250 mL of coffee containing 2% cream. How much hot water should be added to each portion of coffee to decrese the cream concentration to 1.5%?

Questions 5 to 7 illustrate some basic diagrams for problems involving distance, speed, and time.
- Use the diagram to aid you in drawing a chart to organize the information.
- Then solve the problem.

5. Grant and John jog every Saturday. Grant runs 0.2 m/s faster than John. After 20 min, Grant got a stitch and stopped for rest for 5 min while John kept running. At the end of the rest period John was 1200 m ahead of Grant. How fast was each person running?

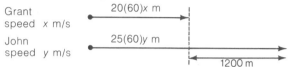

6. Two trains are 750 km apart and start towards each other (on different lines) at the same time. One train travels 24 km/h faster than the other. If they passed each other after 3 h how fast was each train travelling?

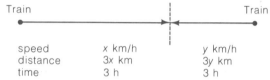

	Train	Train
speed	x km/h	y km/h
distance	$3x$ km	$3y$ km
time	3 h	3 h

7. A glider takes 2 h to travel 480 km with the wind, but takes 3 h to make the same trip against the wind. Find the speed of the plane and the speed of the wind.

time (h)	speed (km/h)	distance (km)
2	$g + w$	$2(g + w)$
3	$g - w$	$3(g - w)$

Solve each of the following problems. Use as necessary, charts and diagrams to aid you in solving them.

B

8. Two trucks leave Edmonton with one going north and the other heading south. The truck going north had a heavier load so it drove 10 km/h slower. In 8 h the trucks were 1840 km apart. Find the speeds of the trucks.

9. Two birds fly towards each other from two trees which are 240 m apart. One bird flies 2 m/s faster than the other. If they meet in 12 s, how fast was each bird flying?

10. Two planes are observed on a radar system to be 2550 km apart and heading towards each other. One plane is flying 50 km/h faster than the other and it is estimated that the planes will collide in 3 h unless their courses are altered. Find the speed of the planes.

11. Justin invested $10 000 in two stocks yielding 9% and 10% respectively. If he earned $430 more from the 10% stock, how much did he invest in each stock?

12. Two cars leave two cities at the same time and travel towards each other. One car travels at 100 km/h and the other at 120 km/h. How long will it take before the cars pass each other if the cities are 550 km apart?

13. Torpedoes are shot from ships 1000 m towards each other. One torpedo travels 4 m/s faster than the other. If the torpedoes collide and explode after 20 s, find the speed of each torpedo.

14. Two sailboats leave their clubhouses and sail toward each other. One boat sails 1 km/h faster than the other and they pass each other in 4 h. If the clubhouses are 164 km apart, find the speed of each sailboat.

15. How much sugar should be added to 100 mL of tea to increase its sugar concentration from 7% to 10%?

16. Two nut mixtures were not selling. To increase sales, a retailer mixed one type selling at $2.20/kg with another selling at $2.40/kg. If 100 kg were mixed and sold for $2.28/kg, how much of each type of nut mixture was used?

17. How much windshield fluid should Rick add to 5 L of a 20% fluid solution to increase its concentration to 75%?

applications: airline schedules

If you look at an airline schedule you will see the times of departure and arrival of flights across Canada. However, very often a flight time is different than the one in the schedule because of the air currents.

From Edmonton

To Ottawa/Hull (cont./suite)
Via Toronto
01:50 08:55 X
07:45 14:55 X   AC158-AC442
12:45 19:55 X   AC106-AC464
12:45 20:55 X   AC126-AC464
Via Winnipeg
15:45 23:20 X   AC126-AC466

Québec EDT/HAE
Via Montréal
09:50 18:25 X   AC760-AC180
Via Toronto-Montréal
07:45 17:10 X   AC104-AC524
12:45 23:05 X   AC106-AC414-AC518

Regina/Moose Jaw CST/HNC
10:15 11:25   AC126-AC424-AC532
21:10 22:15   TZ600
                    AC274

Time of departure.
Time of arrival.
Flight numbers.

Speed of airplane in still air	Tailwind or headwind	Speed of airplane with respect to the ground
450 km/h	tailwind 120 km/h	570 km/h
450 km/h	headwind 90 km/h	360 km/h
V km/h	tailwind W km/h	$(V + W)$ km/h
V km/h	headwind H km/h	$(V - H)$ km/h

In a similar way, the river current helps you downstream, but works against you upstream.

Use the portions of the above airline schedule to solve each of the following problems.

18. The Edmonton Eskimos football team flies to Regina leaving Edmonton at 10:15.
    (a) What is the flight number of the aircraft?
    (b) What is the estimated time of arrival?
    (c) On the above flight, the team arrives in Regina at 11:15 because of a tailwind. The return trip into the same wind takes 1 h 15 min. If the distance between Regina and Edmonton is 500 km, find the speed of the tailwind.

19. A piper cub plane leaves Prince Albert for Estevan, flying into the wind. The trip takes 2.5 h. The return trip, with this tailwind, takes only 2 h. If the distance between the two cities is 350 km, what is the speed of the wind?

20 A traffic helicopter pilot finds that with a tailwind a 120-km distance takes 45 min. but that the return trip, into the wind, takes 1 h. What is the speed of the helicopter? What is the speed of the wind?

21 A twin engine plane flies into the wind from Windsor to the Quebec Winter Carnival in 5 h 20 min. The return flight with the wind is done in 4 h 24 min. Windsor and Quebec City are 1760 km apart. What is the rate of the wind and the rate of the plane?

22 Flying into a headwind, a 747 jet takes 3.5 h to travel 1890 km. On the return flight, the 747 jet took 3 h under the same weather conditions. Find the speed of

   (a) the plane in still air. (b) the wind.

23 A canoeist takes 5 h to deliver a telegram to an outpost 85 km downstream. Returning upstream with the reply the canoeist takes 1 h to travel 10 km. What is the rate of the canoeist in still water? What is the rate of the current?

24 On her way up the St. Lawrence River for shore leave, the oil tanker Southern Belle averaged 12 km/h. Going back out to sea she averaged 22 km/h. At what rate is the St. Lawrence River flowing? At what rate does the tanker move in still water?

25 At the Pan-Am Games, the Canadian kayak team averaged 40 km/h downstream but only 6 km/h upstream, in the white water obstacle course. What was the rate of the kayak in still water? What was the rate of the current?

problem/matics

Before going on a spare, Michael looked at his watch. After the spare, he noted that the hour hand and the minute hand had exchanged places. How long was the spare?

4.10 to solve problems

To solve problems we need to be organized. In solving all problems we must carefully answer these two important questions:

A: What information is given in the problem?
B: What information in the problem are we asked to find?

Checking in the *original* problems ensures that any errors may be identified. To solve a problem we need to successfully complete all 5 steps on page 139. To aid us in solving problems, we may use the various strategies we have dealt with:

- Using a chart to organize the information in some problems.
- Using a diagram to understand visually the information given in some problems.

The variety of questions that occur in the exercise will enable us to apply our skills and make decisions as to which skills we need to use to solve the problems.

4.10 exercise

Solve each of the following questions.

1 In an effort to make inexpensive peach yogurt, Jeff mixed plain yogurt worth $1.75/kg with substandard peaches worth $1.10/kg. If he mixed 100 kg of peach yogurt in all and the cost was $155.50, how much plain yogurt and peaches did he use?

2 Tomkin Publishing charges $150 for each page of advertising and $90 for each half-page of advertising. In their Sunday newspaper, 16 pages were used for advertising. The revenue from this advertising was $2520. How many full pages and half-pages of advertising were sold?

3. Dick left a 500-mL pot of water boiling on the stove. Originally the pot had contained 2% salt, but now contains 5% salt. How much water had evaporated?

4. A solution of phosphorus and water is used as a plant nutrient. How much phosphorus should be added to 25 L of solution to decrease the water concentration from 90% to 80%?

5. Sheila earned $360 in dividends from two stocks. The income from her investment in 8% yield stock is twice the income from her investment in 6% yield stock. How much was invested in each stock?

6. When a plane flies into the wind it can travel 2400 km in 11 h. With the wind behind it, the plane can travel a third again as far in 11 h. What is the windspeed and the ground speed of the plane?

7. The Walker family borrowed money for their vacation from a bank and a trust company. They borrowed $100 more from the bank at 10% interest than from the trust company at 11% interest. If their interest payment in the first year totalled $136, how much was borrowed from each institution?

8. A crop-dusting plane flies over a 1000-m field spraying weed killer. Flying east, against the wind, takes 40 s but flying west takes only 20 s. Find the rate of the plane and the rate of the wind.

9. Chris caught 7 trout and pickerel while fishing in Little Joe Lake. The catch was 11 kg. If the trout averaged 1.8 kg each and the pickerel 1 kg each, how many of each fish did he catch?

10. A student typing service charges $1.25/page for the body of an essay and $2.00/page for the bibliography. If a 24-page essay costs $33.75, how many pages of bibliography are in the essay?

11. Fodder or forage crops are terms used to describe foods used for feeding animals for meat or milk production.

Milk production has increased in Canada 4% in the last year. Some of the Forage Crops such as clovers and lucerne contain 75-90 per cent water, and are important for milk production.

Two types of seed were ordered for forage crops. Clover seed costs $8.50 per sack, while lucerne costs $6.00 per sack. The total cost of 100 sacks was $762.50. How much of each type of seed was bought?

12. The attendance at Music Night in Rock Bend High School was 100 people. Those persons seated in the first 3 rows paid $4 each and those in the remaining seats paid $2.50 each. If the total receipts was $280, how many people were seated in the first 3 rows?

13. Last week, Joanne sold real estate valued at $240 000. On some part of this, she earned 3.5% commission and on the remainder she earned 5%. If her total commission was $10 500, on what part was 5% commission earned?

14. The Stampeders have won 4 more games than the Ticats and 2 fewer games than the Roughriders. Three times the number of games the Ticats won is 6 more than the total number of games won by the Stampeders and the Roughriders. How many games did the Stampeders win?

15. The height of a triangular banner exceeds its width by 50 cm. If the height is increased by 10 cm, the area of the banner is increased by 500 cm². Find the dimensions of the original banner.

16. Jerry invested part of a $10 000 lottery winning in a bonus savings account earning 11% and the remainder in a chequing account yielding 6%. If the total interest on the two accounts is $1000, how much was invested in each account?

17. Members of the Birchmount Tennis Club pay a yearly membership fee and hourly court fees. In her first year, Cathy played 100 h and paid $1000. In the second year, she became very serious and played 500 h, at a cost of $3400. Find the yearly membership fee and the hourly court fee.

18. Wildewood Golf and Country Club sells new and used golf balls. When Tom played the front nine holes, he bought 2 new and 6 used balls because of the water traps and paid $3.60. The next day he purchased 1 new and 2 used balls for the back nine. If this purchase cost $1.45, what is the cost of new and used balls?

19. A quantity of 50% silver alloy is melted together with a quantity of 70% silver alloy to produce 500 g of alloy containing 65% silver. How much of each alloy was used?

20. A Boeing 747 and a DC10 are 7000 km apart and flying towards each other on different air lanes. The 747 flies at 1000 km/h.

    (a) If the DC10 flies at 750 km/h, when will the planes pass each other?
    (b) How far has the DC10 travelled when they meet?

4.11 problem-solving: extraneous information

In solving many real life problems, we often need a room full of people.

To calculate the path that a satellite will take on its journey back to earth requires many people to sort all the information. Using information obtained from all parts of the world the staff at the manned space centre at Houston decide on which pieces of information are needed to solve the problem of directing the satellite.

An important strategy in problem-solving is to be able to sort out information and identify what is not needed to solve the problem. We refer to the unneeded information as **extraneous** information.

Example 1
(a) Read the following problem and decide which information is extraneous.

   Jennifer invested her latest lottery winnings of $5000, one part at 8%/a and the rest at 9%/a. For the past 5 a, she has spent $865 on lottery tickets. If Jennifer receives $420 as the total interest on her investment at the end of the year, how much money did she invest at each rate?

(b) Solve the problem.

Solution
(a) The information that is extraneous (not needed) to solve the problem is *"For the past 5 a, she has spent $865 on lottery tickets."*

(b) Let $x represent the amount invested at 8% and $y represent the amount at 9%

The sum of both amounts is $5000.
$$x + y = 5000 \quad \text{①}$$
$$0.08x + 0.09y = 420 \quad \text{②}$$
The total interest is $420.

Simplify equation ② by multiplying both sides by 100.
$$100(0.08x + 0.09y) = 100(420)$$
$$8x + 9y = 42\,000 \quad \text{③}$$

Multiply equation ① by 8.
$$8x + 8y = 40\,000 \quad \text{④}$$

Use equations ③ and ④ to obtain 1 equation in 1 variable.
$$8x + 9y = 42\,000 \quad \text{③}$$
$$\underline{8x + 8y = 40\,000} \quad \text{④}$$
Subtract $\quad y = 2\,000$

Substitute $y = 2000$ in equation ①.
$$x + 2000 = 5000$$
$$x = 3000$$

Thus $3000 was invested at 8% and $2000 was invested at 9%.

Check
8% of $3000 = $240
9% of $2000 = $180
Total interest $420   checks ✓

— Be sure to make a concluding statement for all problems.

In solving a problem two important questions need to be answered:

A: | What information is given? |

B: | What information are we asked to find? |

In solving a real-life problem, we are often given too much information. An important skill in problem-solving is deciding what information is not needed in order to solve the problem. The problems that follow have only 1 item of extra information. They will provide practice in this important skill.

4.11 exercise

For each problem
- decide which information is extraneous.
- solve the problem.

A

1. Samuel works for a landscaping company planting trees and shrubs. Last week he planted 56 trees and shrubs. Each tree is twice the size of each shrub. If Samuel planted 8 more trees than twice the number of shrubs, how many of each did he plant?

2. Michelle is paying interest on a loan at the rate of 12%. If the interest rate had been 10% she could have borrowed $200 more and still paid the same annual amount in interest. Michelle also pays a $50 000 mortgage at 12%. How much did Michelle borrow on the loan?

3. At the World Population Conference, there were 168 more biologists than sociologists. However, there were 268 engineers. If there was a total of 1134 biologists and sociologists, how many of each were there?

4. Mary has a rectangular garden plot with perimeter 11 m. Her garden is surrounded by a fence 1 m high. If the length of the plot is 49 cm shorter than twice the width, find the dimensions of the plot.

5. Twice the larger of two numbers is equal to three times the smaller. If the greater number is increased by 2, it will be a prime number. The difference of the numbers is 5. Find the two numbers.

B

6. On the shelf, only 40% and 60% acid solutions are available. For the experiment, the acid must have a concentration of 45%. For the experiment yesterday, the acid had to be 63%. If 200 L of the acid solution is required, how many litres of each acid solution should be mixed?

7. Mary paid a $5 parking ticket using only nickels and dimes. She used 10 more nickels than dimes and still had 4 dimes in her purse for a cup of coffee. How many nickels and dimes did she use in paying her fine?

8. Alan bought a used car privately and paid $25 for a safety inspection. An engine overhaul costs $\frac{3}{4}$ of the price of the car. New tires will cost at least $185. If the cost of the car and the inspection together was $100 more than the cost of overhaul, find the cost of the overhaul and the price of the car.

9. In hockey, a player earns a point for each goal and each assist. In the last season, Maurice accumulated 75 points and 312 min in penalties. If he had two-thirds as many assists as goals, how many goals did he score and how many assists did he receive?

10. In football, a field goal is worth 3 points and a converted touchdown 7 points.

Last year, 2 890 000 fans watched football games in the Canadian Football League (CFL). Each year, a team from the Eastern Division plays a team from the Western Division to win the coveted Grey Cup.

In their last game the Astros scored 33 points. The Astros had one more field goal than converted touchdowns. If 2 field goal attempts were blocked and no other plays scored points, how many touchdowns and field goals were successful?

4.12 problems to solve: needing information

Often in solving a real-life problem, we need some information that does not occur in the problem.

At the Annual Demolition Derby tickets were sold at $3.75 and $2.75 per ticket. The total receipts collected for the Annual Demolition Derby was $1525.00. How many people paid $3.75 per ticket?

To solve the problem, we need to know *how many people in all attended the Annual Demolition Derby*.

Missing Information: 500 people attended the Annual Demolition Derby.

With the above information we are able to write the equations and solve the problem.

Example 1
Which of the following facts is needed to solve the problem that follows?

Facts
A) The driver earns $327.00 a week
B) The courier delivered 10 pieces of mail.
C) The truck used diesel fuel which costs 17.8¢/L.

Problem
A courier service charges 2¢/km to deliver special letters and 5¢/km to deliver parcels. The charge for one delivery over a 6-km distance was $1.56. How many letters and how many parcels did the courier take on this delivery?

Solution
- The missing fact we need to know is how many pieces of mail were delivered.
  Missing Fact: The courier delivered 10 pieces of mail.
- Now solve the problem.

Let s represent the number of special letters and p the number of parcels. Thus for 6 km, the cost of the special letters was $12s$¢ and the parcels $30p$¢.

$s + p = 10$ ①

$12s + 30p = 156$ ②

From equation ①
$30s + 30p = 300$ ③
$12s + 30p = 156$ ②

Subtract $18s = 144$
$s = 8$
From ① $s + p = 10$
$8 + p = 10$
$p = 2$

Thus the number of special letters is 8 and parcels is 2.

Check   Cost of special letters 8×12¢ $= 96$¢
        Cost of parcels 2×30¢ $= 60$¢
        Total Cost $1.56

The answer checks √ with original problem.

4.12 exercise

A
1 Greg is an avid tennis player who plays whenever he can. Court fees are $8.00/h on off-hours and $12.00/h at peak hours (of which he pays half while his opponent pays the other half). This month his court fees totalled $74.00. How many hours did he play during off-hours?

(a) Why are you not able to solve the problem?
(b) Which of the following facts is needed to solve the problem?
   A: He played a total of 14 h this month.
   B: His tennis racquet costs $90.00.
   C: He pays $6.00 for a tube of balls.
(c) Use your choice in (b) to solve the problem.

2 Shirley won $1500 in a lottery and decided to open two new accounts; a savings account and a chequing-savings account, each of which offers different interest rates. After a year she earned a total of $131.25 in interest. How much did she put into each account?

(a) Why are you not able to solve the problem?
(b) Which of the following facts is needed to solve the problem?
   A: The savings account pays $9\frac{1}{2}$%.
   B: The chequing-savings account pays 5%.
   C: The savings account pays $9\frac{1}{2}$% while the savings-chequing account pays 5%.
(c) Use your choice in (b) to solve the problem.

For Questions 3 to 8, choose one of the facts A to C to solve each problem.

B
3 Ticket prices were $4 per adult and $2.50 per student. The receipts were $920 for the Saturday night performance. How many adults attended that night?

Facts
A: 465 people attended the performance last Wednesday.
B: The receipts for the previous night was $1840.
C: 305 people went Saturday night.

4 All-Cover Insurance pays their sales people expenses for driving. The company pays 12¢/km for calls during regular hours and 15¢/km for calls made after hours. Sam Bradie travelled 1150 km in all this week. How much did he receive in expenses for his after-hour calls?

Facts
A: He earned $145.50 for expenses.
B: His base salary is $230 per week.
C: His car gets 8 km/L.

5 An engineering consultant is paid 18¢/km for in-city trips and 16¢/km for long distance trips. At the end of the month the consultant received $131.40 for trips. How many kilometres did the consultant travel on long distance trips?

   Facts
   A: Two new tires cost him $185.00.
   B: He travelled a total of 780 km.
   C: A tune-up costs $98.50.

6 For a certain cloth, a 34% final dye is needed (the concentration of the dye solution must be 34%). A 30% dye solution is mixed with a 40% dye solution. How much of each solution must be used to obtain the required 34% concentration?

   Facts
   A: To dye the cloth, 60 L of dye solution is needed.
   B: 100 L of the final solution is required.
   C: 300 L of the 40% dye solution was available.

7 Three years ago, three times John's age added to twice Tom's age totalled 30 a. How old is each boy today?

   Facts
   A: The sum of their present ages is 19 a.
   B: John and Tom are brothers.
   C: Tom is older than John.

8 Two ocean front docks in the Atlantic Ocean are 396 km apart. A sailboat leaves the northern dock while an empty fishing boat leaves the southern dock and head towards each other. If they meet after 6 h, how fast does the fishing boat go?

   Facts
   A: With a wind the sailboat needs 2 h less time.
   B: With a full load the fishing boat travels 10 km/h faster.
   C: The sailboat goes 18 km/h faster than the fishing boat.

applications: sorting information

Each of Questions 9 to 17 does not contain enough information to solve the problem. Each question requires a fact to be chosen from the following list. (There are more facts than problems.)

A A total of 28 racquets were bought.
B The wholesaer sold 110 kg of fresh potatoes and onions.
C Parcel rates are 35¢/kg.
D There are 580 members.
E Shag rug costs $25.75/m^2.
F Sandra bought a total of 5.5 m of fabric.
G The amount of carpet bought was 24 m^2.
H John made a deposit of $9.75.
I The flight from Montreal to New York takes 45 min.
J John's net pay was $92.73.
K The rates of interest she charged were 2% and 1.5% per month.
L She worked a total of 27 h.
M The pub serves 3 different kinds of imported cheeses.
N The mixture costs $2.62/kg.
O Total plane fare was $1667.
P 8 bottles of wine were bought.
Q Bob travelled a total of 5.5 h.
R The office used 19 plastic cups.
S The cheapest brand of coffee costs $3.20 for a large jar.
T The bus travels 20 km/h slower than the car.
U They used 12 bottles of soft drinks.

9 The cook at Mac's Restaurant buys fresh potatoes and onions from a vegetable wholesaler. The potatoes cost 12¢/kg and the onions cost 16¢/kg. If this month's total bill was $14.80, how many kilograms of potatoes did the cook buy?

10. The Munson family is to have a family reunion in Halifax. Fifteen members of the family fly from Montreal for the reunion. If fares are $137.00 return for adults and $40.00 return for children, how many adults went to the reunion?

11. Sandra decides to sew herself a fancy dress for her cousin's wedding. The satin costs $12/m and lining costs $3.00/m. How much satin did she buy if her bill for the fabric (without notions) was $39.00?

12. The school team decided to buy new badminton and tennis racquets this year. They paid $7.50 for badminton racquets and $9.50 for tennis racquets. How many of each did they buy if the total bill was $242.00?

13. Leslie works part-time for a large grocery store and makes $3.50/h for parcelling groceries and $4.50/h for stocking shelves at night. Last week his pay was $106.50. How many hours did he spend parcelling?

14. Betty lent $500 in all for a month to two different friends, charging each a different interest rate. If she received a total of $9.00 in interest, how much did she lend to each of her friends?

15. The membership fees at the Racquet Club are $250 a year for men and $225 for women. This year the club collected $139 250 in membership fees. How many men belong to the club?

16. Jay was in charge of purchasing wine for the staff party. Red wine costs $3.20 a bottle, while white wine costs $3.80 a bottle. How many bottles of red wine did he buy if he spent a total of $28.60?

17. The health-food store sells a sunflower seed and raisin mixture. Sunflower seeds cost $2.20/kg and raisins cost $3.60/kg. How much of each did the store use to make 50 kg of the mixture?

skills review

To do mathematics well we need to practise skills in mathematics regularly. This section, called Skills Review, will review not only skills you develop in each chapter, but skills you need for developing more mathematics. You may wish to refer back to this Skills Review more than once to sharpen your skills.

Solve each of the following systems.

(a) $y + 4 = 3x$
    $3 + 2y = x$

(b) $3a - 5b = 7$
    $b + 3 = a$

(c) $3x + y = 50$
    $x + y = 26$

(d) $2m = 4 + 5n$
    $2m + 2 = 8n$

(e) $2x - y = 14$
    $3x + 4y = -23$

(f) $m + 5n = -80$
    $2m - n = 5$

(g) $2m - n = 0$
    $m - 2n = -30$

(h) $4a + 12b = 7$
    $4a - 8b = -9$

(i) $2x + 3y = -3$
    $5x - y = 35$

(j) $3x - 12 = -2y$
    $2x - y = 1$

(k) $10 - y = 4x$
    $y = 2 - 2x$

(l) $3m + n - 11 = 0$
    $2m = -1 + n$

(m) $\dfrac{a}{6} = 3 + \dfrac{b}{2}$
    $\dfrac{5a}{6} - (b + 3) = 6$

(n) $\dfrac{3}{a} + \dfrac{1}{b} = 2$
    $\dfrac{4}{a} - \dfrac{1}{b} = 12$

(o) $a - \dfrac{2}{3}b = -2$
    $b - 4a = -5$

(p) $\dfrac{1}{m} = \dfrac{3}{n} - 12$
    $\dfrac{4}{m} - 27 = -\dfrac{3}{n}$

(q) $2(x - 1) = 3(1 - y) - 7$ $x - 5y = -1$

(r) $2(x + 2) + 3(y - 1) = 8$
    $3(5 - x) - (y + 2) = 6$

(s) $2(3x - 2) + 3(y - 1) = 1$ $6x + 5 + 4(y - 3) = 5$

(t) $\dfrac{x + y}{3} + \dfrac{x - y}{5} = -2$ $x + 4y = 0$

(u) $\dfrac{x + 2y}{3} + \dfrac{x + 2y}{2} = -5$ $\dfrac{2x - y}{7} - \dfrac{3x + 2y}{2} = 6$

problems and practice: a chapter review

At the end of each chapter, this section will provide you with additional questions to check your skills and understanding of the topics dealt with in the chapter. An important step in problem-solving is to decide which skills to use. For this reason, these questions are not placed in any special order. When you have finished this review, you might try the *Test for Practice* that follows.

1. Find the co-ordinates of a rendezvous for two submarines on the course shown.
   Sub A: $y = 10 - 4x$ Sub B: $2x = y - 7$

2. (a) Solve the system.
   $x + y = 344$ $x - y = 108$
   (b) Why have you used an algebraic method rather than a graphical method to obtain your answers in (a)?

3. Solve and verify.
   (a) $\dfrac{5(x + 1)}{3} - \dfrac{3(y - 1)}{2} = 8$ $y - x = -3$
   (b) $4(x + 2) - 5(3y - 1) = -46$
   $5(2x - 1) + 4(y + 2) = 63$

4. Remember this problem from Section 4.4? Now you may use your skills with equations and solve it.

   300 people applied for a rebate on their cars under a new financial plan. The plan provides a rebate of $100 for small cars and $45 for full-size cars. If a total of $25 325 was paid out, how many applicants owned full-size cars?

5. Solve each system for x and y.
   (a) $x + y = m + n$
   $mx - ny = m^2 - n^2$
   (b) $px + qy = 1 + q^2$
   $p^2x + y = p + q$

6. A flour mill mixes a 40% whole wheat flour with a 70% whole wheat flour to obtain a 60% whole wheat flour. If 45 kg of the final product was obtained, how much of each type of flour was used?

Flour type	Amount of flour (kg)	Amount of whole wheat (kg)
40% Whole Wheat	x	$0.40x$
70% Whole Wheat	y	$0.70y$
60% Whole Wheat	$x + y$	$0.60(x + y)$

7. Dave wants to increase his concentration of antifreeze in his coolant from 30% to 50%. How much water must evaporate from the coolant for this to happen if he has 10 L of solution?

8. A mailing house in Saskatoon processed 1200 advertisements. The cost of delivering them locally was 10¢ each, but out-of-province deliveries cost 14¢ each. If the total budget was $152, how many were delivered out of Saskatchewan?

9. Ron is conducting a survey on television viewing. 500 homes were chosen at random. Some surveys require 19¢ stamps while others require 23¢ stamps. If the cost of stamps is $106.20, how many of each type of stamp is required for the survey?

10. To increase sales, vegetable seeds selling at $1.40/kg were mixed with seeds selling at 80¢/kg. 400 g of seeds were mixed and sold at $1.22/kg. How much of each type of seed were mixed?

11. Two joggers running in opposite directions pass each other. Sarah jogs at 3 m/s while Janet runs at 5 m/s. How long will it take them to be 1000 m apart?

test for practice

Try this test. Each test will be based on the mathematics you have learned in the chapter. Try this test later in the year as a review. Keep a record of those questions that you were not successful with and review them periodically.

1. The graph of an equation is given by
   $3x - 2y = 12$.
   Which of the following points do not lie on the line?
   (a) $(2, -3)$ (b) $(6, 2)$ (c) $(4, 0)$

2. (a) Draw the graph of the following equations using the same axes.
   $3x + 2y + 10 = 0$, $x + \frac{3}{2} = \frac{1}{4}y$, $7y = 3x - 8$
   (b) What do the lines have in common?

3. (a) Solve this system,
   $4x + y = 8$ $2x + 3y = 4$
   (b) Verify your answer.

4. Write the co-ordinates of the intersection point for each of the following systems.
   (a) $\frac{x + y}{2} - \frac{2y}{3} = \frac{5}{2}$ $\frac{3}{2}x + 2y = 0$
   (b) $12(x - 2) - (2y - 1) = 14$
   $5(x - 1) + 2(1 - 2y) - 14 = 0$

5. Solve each system.
   (a) $2x + 3y = -3$ (b) $a = 3 + 2b$
   $3x - y = -10$ $5a + 4b - 8 = 0$

6. Solve and verify.
   (a) $\frac{3}{a} - \frac{4}{b} - 15 = 0$ $\frac{3}{a} + \frac{4}{b} = \frac{15}{2}$
   (b) $\frac{3(a + 5)}{5} - \frac{2(b - 3)}{4} = 4$ $2a + b = -15$

7. Solve.
   (a) $3x = 6b - 2y$ (b) $ax - by = 1$
   $4x - 3y = 25b$ $4x + 9y = 4$

8. Jeremy earns 10¢ per week for each newspaper subscription he delivers and 6¢ per week for each business flyer he delivers. If Jeremy earned $24 in one week for delivering 320 items, how many business flyers did he deliver?

9. Two rare herbs are mixed together to form a cold remedy. One herb costs $5.50/kg and the other herb costs $7.00/kg. 600 kg of this mixture is produced and sold at $6.25/kg. How much of each herb was used?

10. A showboat travelled up the Mississippi River and back to its dock in 25 h. Travelling upstream, the boat could only manage a speed of 20 km/h but gained 10 km/h downstream. How far upstream did the boat travel?

11. Rick runs at 4 m/s over flat land and 3 m/s over rough terrain. On Saturday, he ran 3000 m in 13 min over flat and rough terrain. What distance did he run over flat land?

12. A 60% salt solution was mixed with an 80% salt solution to obtain a 68% salt solution. If a total of 100 L of this solution was mixed, how many litres of each solution were used?

> **math tip**
>
> Remember: to solve a word problem requiring an equation
> - use a variable to represent what you are to find.
> - use a chart, table, etc., to help you organize the given information.
> - write an equation.
> - check your answer in the *original* problem.
> - write a *final* statement to answer what you were asked to find.

math is plus — more linear systems

We must extend the skills we have learned in mathematics to study more mathematics. For example, to draw a line we have used a co-ordinate system with two axes, in two dimensions. In three dimensions, we introduce another co-ordinate axis. To show a point in space we use the following co-ordinates.

$P(x, y, z) = P(4, 5, 3)$
   on the x axis
     on the y axis
       on the z axis

A line is given by an equation such as $2x - y = 5$ using the x-y co-ordinate system. When we solved 2 equations in 2 variables, we found the co-ordinates of the intersection, namely a point.

A plane is given by an equation such as $2x + y + 6z = 4$. In this section, we will solve a linear system of 3 equations in 3 variables. The equations we solve here represent planes that intersect in a point. The skills we have developed for solving 2 equations in 2 variables may be applied to the solution of 3 equations in 3 variables. *The same principles apply.* Look for them.

Solve
  $x + y + z = 2$ ①
  $3x + y - 2z = 5$ ②
  $x - y + z = 6$ ③

Subtract equation ② from ①.
  $-2x + 3z = -3$ ④

Add equations ② and ③.
  $4x - z = 11$ ⑤

Now we may solve ④ and ⑤ in the same way as we learned to do earlier.

$2 \times$ ④ $-4x + 6z = -6$ ⑥
        $4x - z = 11$ ⑤

⑥ + ⑤ $5z = 5$
         $z = 1$

Use the value $z = 1$ in equation ⑤.
  $4x - z = 11$
  $4x - 1 = 11$
  $4x = 12$
  $x = 3$

Use the values $x = 3$ and $z = 1$ in ①.
  $x + y + z = 2$
  $3 + y + 1 = 2$
  $y = -2$

Thus we obtain the solution
  $(x, y, z) = (3, -2, 1)$

As before, be sure to check your solution in the *original* equations.

Now try these questions.

1. For each plane given in the form $ax + by + c = d$, a point is given. Check to see whether the point is in the plane.

    (a) $2x + y - z = 4$ $(1, 3, 1)$
    (b) $3x - y - 2z = -3$ $(0, -1, -2)$
    (c) $x - y - z = -1$ $(-1, -1, 3)$
    (d) $3x - 2y + z = -8$ $(-2, 3, 4)$

2. (a) Solve the following system of 3 equations in 3 variables.
    $x - y + 2z = 1$
    $2x + 3y - 2z = -6$
    $3x - y + z = -1$

    (c) Draw a graph of your solution in (a).

3. Solve each of the following for x, y, and z.

    (a) $x - 2y - z = -6$ (b) $x - 3y + 2z = 5$
        $x - y + z = 2$ $2x + y - z = -5$
        $2x + y - z = 1$ $x - y + 2z = 5$

    (c) $x + y - z = 1$ (d) $3x - 2y + z = -5$
        $2x - y - 2z = 8$ $2x - 4y - z = -7$
        $x + y + 3z = 1$ $-x + 2y + 3z = 6$

4. You may again extend your skills to the solution of a linear system of 4 equations in 4 variables.
    Solve. $a + 2b + 3c + d = 9$
              $a - b + c - d = -5$
              $a + b - c + 2d = 6$
              $3a - b + 2c - d = -6$

5 ratio, proportion and variation; solving problems

ratio skills, proportion, direct and inverse variation, joint and partial variation, solving problems, decisions for solving problems.

5.1 working with ratio: skills and concepts

We often compare numbers.

(a) In sports we compare scores.

> Blue Bombers beat Stampeders
> 36 to 24

(b) In business we compare amounts of money.
profits to *losses* 3 to 2

A comparison of numbers is a **ratio**.
For (a) we write 36:24.   For (b) we write 3:2.

36 is compared to 24 or   3 is compared to 2 or
more compactly 36 to 24   more compactly 3 to 2

We may compare more than two numbers. For example
  wins to *losses* to *ties*   8:3:2
is a 3-term ratio.

  2-term ratio $a:b$ 3-term ratio $a:b:c$

  a, b, c, are called **terms** of the ratio.

For a 2-term ratio we may use the fraction notation $a:b = \dfrac{a}{b}$, if $b \neq 0$.

When we state that two ratios are equivalent we are writing a proportion.

$$\underbrace{36:24 = 3:2}_{\text{proportion}} \qquad \underbrace{a:b = c:d}_{\text{proportion}}$$

Again the fraction form may be written for 2-term ratios.

$$\dfrac{36}{24} = \dfrac{3}{2} \qquad \text{or} \qquad \dfrac{a}{b} = \dfrac{c}{d}$$

For the proportion $a:b = c:d$,
- a, b, c, d are the first, second, third, and fourth terms in the proportion, and as such are called the first, second, third, and fourth proportionals.
- a, d are called the **extremes** of the proportion.
- b, c are called the **means**.

If $\frac{a}{b} = \frac{c}{d}$ $b \neq 0, d \neq 0$

then $bd \times \frac{a}{b} = bd \times \frac{c}{d}$

$ad = bc$

Thus the product of the extremes equals the product of the means.

- If $\frac{a}{b} = \frac{b}{c}$, we call b the mean proportional between a and c.
- A ratio may be reduced to lower terms if all of the terms of the ratio contain the same factor.

$\frac{36}{24} = \frac{12 \times 3}{12 \times 2} = \frac{3}{2}$ $15:10:5$
$= (5 \times 3):(5 \times 2):(5 \times 1)$
$= 3:2:1$

Example 1
(a) Write $(x^2 - y^2):(ax + ay)$ in lowest terms.
(b) Find the fourth proportional for 4, 8, 24.
(c) Find the mean proportional between 3 and 75.

Solution
(a) Factor the terms of the ratio. Thus
$(x^2 - y^2):(ax + ay)$
$= (x - y)(x + y):a(x + y)$
$= (x - y):a$ provided $x + y \neq 0$.
Thus $(x^2 - y^2):(ax + ay) = (x - y):a$

(b) Let x represent the fourth proportional.
Then $\frac{4}{8} = \frac{24}{x}$.
Multiply by $8x$.   You may do
$8x\left(\frac{4}{8}\right) = 8x\left(\frac{24}{x}\right)$ ← this step mentally.
$4x = 192$
$x = 48$
The fourth proportional is 48.

(c) Let x represent a mean proportional.
Then $\frac{3}{x} = \frac{x}{75}$
$x^2 = 225$
$x = \pm 15$
Thus, the mean proportional is 15 or −15.

To solve a problem we may introduce a **parameter**. A parameter is a constant represented by a letter, say k. Its value varies from relation to relation, but for one given relation it does not change.

Example 2
If $x:y = a:b$ prove that $\frac{x + a}{y + b} = \frac{a}{b}$.

Solution
Since $x:y = a:b$ then $\frac{x}{y} = \frac{a}{b}$ $y, b \neq 0$.

Write $\frac{x}{a} = \frac{y}{b} = k$, $k \neq 0, k \neq -1$.
Then $x = ak, y = bk$.

By substitution $\frac{x + a}{y + b} = \frac{ak + a}{bk + b}$
$= \frac{a(k + 1)}{b(k + 1)}$
$= \frac{a}{b}$ as required.

Thus $\frac{x + a}{y + b} = \frac{a}{b}$.

When we learn new skills in mathematics, we often combine them with skills we have learned earlier to solve problems, as shown in Example 3.

Example 3
Find $x:y$ if
$(x + 9)(y - 2) = (x + 3)(y - 6)$.

Solution
Expand the products.
$xy + 9y - 2x - 18 = xy + 3y - 6x - 18$
$9y - 2x = 3y - 6x$
$4x = -6y$
Divide by $4y$. $\frac{4x}{4y} = \frac{-6y}{4y}$
$\frac{x}{y} = \frac{-3}{2}$ or $x:y = -3:2$

The skills we learn in this section to use with ratios and proportions are applied to solve problems in the next section.

5.1 exercise

1. Write each ratio in lowest terms.
   (a) 120 cm to 20 cm
   (b) 2 m to 25 cm
   (c) 1.5 km to 200 m
   (d) 320 km/h to 80 km/h
   (e) 4 h to 30 min
   (f) 2 min to 30 s
   (g) $5 to 25¢
   (h) $1.75 to 60¢

2. Express each of the following in lowest terms.
   (a) $36a : 72a$
   (b) $28ab : 14b$
   (c) $30a^2b : 60ab^2$
   (d) $-16xy : 32y$
   (e) $8k - 2k : 2k$
   (f) $4p : 6p + 2p$
   (g) $\frac{3}{p} : \frac{q}{6}$
   (h) $\frac{x}{3} : \frac{9}{y}$
   (i) $(2x + 4y) : 2x$
   (j) $30a : (10a + 20b)$
   (k) $(ap + aq) : (bp + bq)$
   (l) $(p^2 - q^2) : (p + q)$
   (m) $(a^2 - b^2) : (a - b)$
   (n) $(ma^2 + mb^2) : (ma + mb)$

3. If $\frac{x}{y} = \frac{m}{n}$ then find an equivalent expression for each of the following.
   (a) $\frac{y}{x}$
   (b) my
   (c) $\frac{x}{m}$

4. If $\frac{m}{n} = \frac{2}{3}$, what is the value of each of the following?
   (a) $\frac{m-n}{m}$
   (b) $\frac{m-n}{n}$
   (c) $\frac{m-n}{m+n}$
   (d) $\frac{m+n}{m-n}$

5. Find the missing value (?) for each of the following.
   (a) If $\frac{x}{y} = \frac{2}{3}$ then $\frac{y}{x} = ?$
   (b) If $\frac{3}{b} = \frac{2}{a}$ then $\frac{b}{a} = ?$
   (c) If $\frac{m}{n} = \frac{4}{5}$ then $\frac{m+1}{n} = ?$
   (d) If $7t = 4s$ then $\frac{s}{t} = ?$
   (e) If $\frac{a}{b} = \frac{4}{5}$ then $\frac{a+1}{b-1} = ?$

6. Find the missing terms for each proportion.
   (a) $x : 7 = 5 : 2$
   (b) $2 : 5 = 4 : x$
   (c) $4 : 3x = 45 : 63$
   (d) $3s : 36 = 9 : 16$
   (e) $8 : 4 = 20 : 4x$
   (f) $6 : 2x = 7 : 4$

7. Find the value of the variables that satisfy the proportions.
   (a) $\frac{y}{45} = \frac{2}{3}$
   (b) $\frac{12}{y} = \frac{16}{9}$
   (c) $\frac{6}{12} = \frac{3y}{12}$
   (d) $\frac{m}{24} = \frac{3}{10}$
   (e) $\frac{3}{7p} = \frac{1}{2}$
   (f) $\frac{3}{k} = \frac{18}{24}$

8. If $x = 3k$ and $y = 4k$, find $x : y$.

9. If $p = 5k$ and $q = -2k$, find $q : p$.

10. Find the ratio $x : y$ for each of the following.
    (a) $2x = 3y$
    (b) $4x - 5y = 0$
    (c) $\frac{3x}{2} = \frac{4y}{3}$
    (d) $\frac{2x}{3} - \frac{y}{2} = 0$
    (e) $2x + 3y = 3x$
    (f) $2x - y = 3y - x$
    (g) $\frac{x-y}{x+y} = \frac{2}{3}$
    (h) $\frac{y+2x}{x-y} = \frac{5}{4}$

11. If $\frac{m}{15} = \frac{n}{8} = 2$, find the values of m and n.

12. Find the values of the variables that satisfy the proportions.
    (a) $\frac{x}{2} = \frac{y}{5} = 3$
    (b) $\frac{m}{3} = \frac{4}{n} = 5$
    (c) $\frac{30}{p} = \frac{25}{q} = \frac{5}{4}$
    (d) $\frac{25}{r} = \frac{s}{2} = \frac{5}{2}$

B

13. If $x : y = a : b$, $x, y, a, b \neq 0$, prove that
    (a) $\frac{x}{a} = \frac{y}{b}$
    (b) $bx = ay$
    (c) $bx - ay = 0$
    (d) $\frac{y}{x} = \frac{b}{a}$
    (e) $\frac{a}{x} = \frac{b}{y}$

161

14. Find the value of each variable that satisfies the proportion.
    (a) $\dfrac{4}{y+1} = \dfrac{2}{3}$
    (b) $\dfrac{2}{3} = \dfrac{3}{x-3}$
    (c) $5:7 = 2:(p+3)$
    (d) $10:(s-2) = -5:2$

15. Solve each proportion.
    (a) $\dfrac{x+3}{x+2} = \dfrac{2}{3}$
    (b) $\dfrac{y+3}{y-2} = \dfrac{2}{3}$
    (c) $(2+y):2 = (2-y):3$

16. Solve for each variable.
    (a) $\dfrac{y+6}{6} = \dfrac{y+2}{2}$
    (b) $\dfrac{a+5}{6} = \dfrac{a+6}{7}$
    (c) $(b-7):(b+5) = 5:3$

17. Solve.
    (a) $\dfrac{3}{5m-9} = \dfrac{-7}{21}$
    (b) $\dfrac{x+7}{9} = \dfrac{21+3x}{4}$
    (c) $\dfrac{2x-7}{2x+7} = \dfrac{7}{5}$
    (d) $\dfrac{2p-8}{6} = \dfrac{p+3}{24}$

18. Find the value of the variable which satisfies the proportion.
    (a) $(2x-3):3 = 3:1$
    (b) $(p-1):1 = (p+1):3$
    (c) $(4y+3):5 = y:2$
    (d) $(y+3):2 = (2-y):3$
    (e) $(x-3):2 = (x+2):3$
    (f) $4:(3-y) = 3:4$
    (g) $3:(2s+3) = 2:(3s+2)$
    (h) $4a:5 = (a+3):2$

19. Solve for the ratio $x:y$ in each of the following.
    (a) $(x+4)(y-4) = (x-8)(y+2)$
    (b) $(x-9)(y-4) = (x-12)(y-3)$
    (c) $(4x+3)(y+8) = (2x+6)(2y+4)$
    (d) $(3x+3)(2y+2) = (2x-3)(3y-2)$

20. Solve for the ratio $x:y$ in each of the following.
    (a) $x^2 - 4xy + 3y^2 = 0$
    (b) $x^2 + 3xy - 10y^2 = 0$
    (c) $2x^2 - 7xy + 3y^2 = 0$
    (d) $10x^2 - 9xy + 2y^2 = 0$

21. Find the mean proportionals between
    (a) 4, 9
    (b) 3, 27
    (c) m, mn^2
    (d) $3m, 12m^3$
    (e) $2ab^2, 8a$
    (f) $4p^2, 16p^4$
    (g) $\sqrt{3}, 3\sqrt{3}$
    (h) $2\sqrt{2}a^2, \sqrt{2}a$

22. Find the fourth proportional for each of the following.
    (a) 3, 4, 12
    (b) 5, 4, 25
    (c) 3, 8, 12
    (d) $\dfrac{1}{2}, \dfrac{2}{3}, \dfrac{3}{4}$

23. If $x:y = 2:1$, find the value of each of the following.
    (a) $\dfrac{2x}{3y}$
    (b) $\dfrac{x+y}{x-y}$
    (c) $\dfrac{x^2}{y^2}$
    (d) $\dfrac{x^2 + 2y^2}{x^2}$
    (e) $\dfrac{2x^2 - y^2}{y^2}$
    (f) $\dfrac{x^2 - y^2}{x^2 + y^2}$
    (g) $\dfrac{x^2 - y^2}{(x+y)^2}$

24. If $a:b = 3:2$, find the value of each of the following.
    (a) $\dfrac{3a}{2b}$
    (b) $\dfrac{a+b}{2a}$
    (c) $\dfrac{a-3b}{2a+b}$
    (d) $\dfrac{a-b}{a+b}$

25. If $\dfrac{x}{y} = \dfrac{a}{b}$, $x, y, a, b, \neq 0$, prove that
    (a) $\dfrac{x+y}{y} = \dfrac{a+b}{b}$
    (b) $\dfrac{x-y}{y} = \dfrac{a-b}{b}$
    (c) $\dfrac{x+a}{a} = \dfrac{y+b}{b}$
    (d) $\dfrac{x+a}{y+b} = \dfrac{a}{b}$

26. If $\dfrac{x}{a} = \dfrac{y}{b}$, $x, y, a, b, \neq 0$, prove that
    (a) $\dfrac{x+a}{x-a} = \dfrac{y+b}{y-b}$
    (b) $\dfrac{x+y}{x-y} = \dfrac{a+b}{a-b}$

C

27. If $m:n = -3:5$, $n:p = 6:7$, and $p:q = -10:21$ then find the ratio of $m:q$.

28. If $a:b = 6:5$, $b:c = -2:9$, and $c:d = 10:3$ then find the ratio $a:d$.

5.2 solving problems and applications: ratio

The skills we have learned about ratio in the previous section can now be combined with our earlier skills for solving problems. As well, with skills for solving problems we may also explore applications that involve ratios.

Example 1
Find the two positive numbers, where
(a) they are in the ratio 5:4.
(b) the sum of their squares is 1025.

Solution
Let the numbers be represented by x and y.
From (a), $\frac{x}{y} = \frac{5}{4}$ or $\frac{x}{5} = \frac{y}{4}$.
Let $\frac{x}{5} = \frac{y}{4} = k$, $k > 0$. Then $x = 5k$, $y = 4k$.
Thus using (b), we may write
$$(5k)^2 + (4k)^2 = 1025$$
$$25k^2 + 16k^2 = 1025$$
$$41k^2 = 1025$$
$$k^2 = 25$$
$$k = \pm 5$$
Since the numbers are positive, then $k = 5$. The numbers are 20 and 25.

Check: $\frac{25}{20} = \frac{5}{4}$ checks ✓
$20^2 + 25^2 = 1025$ checks ✓

We may use our skills with proportions to solve the next problem.

Example 2
In carrying the Olympic Torch, Samuel, Jerome, and Donna ran a total of 6000 m. The ratio of Samuel's distance to Jerome's distance was 5:4 and Donna's distance to Jerome's distance was 3:4. How far did each person run?

Solution
Let s, j, and d represent the number of kilometres run respectively by Samuel, Jerome and Donna.

Then $\frac{d}{j} = \frac{3}{4}$ $\quad\quad$ $\frac{s}{j} = \frac{5}{4}$
$d = \frac{3}{4}j$ $\quad\quad$ $s = \frac{5}{4}j$

The total distance travelled was 6000 m.
Thus $j + \frac{3}{4}j + \frac{5}{4}j = 6000$ — Distance run by Samuel.
— Distance run by Donna.
$$4j + 3j + 5j = 24\,000$$
$$12j = 24\,000$$
$$j = 2000$$
Thus, $d = \frac{3}{4}j$ $\quad\quad$ $s = \frac{5}{4}j$
$= 1500$ $\quad\quad$ $= 2500$

Thus, the distances run by Samuel, Jerome and Donna are respectively 2500 m, 2000 m and 1500 m.

You may check the answers!

5.2 exercise

A

1. A picture frame is esthetically pleasing if the ratio of the width to length is 5:8. What should be the width of a picture frame whose length is 56 cm?

2. The ratio of the number of strikeouts to walks of a baseball pitcher is 5:4. If the pitcher has given up 60 walks, how many strikeouts should he have?

3. John and Tim invest in a company in the ratio 4:3. If John's profit in 1 a from the company is $28 360, how much will Tim receive?

4. In creating a wood filler substitute, clean sawdust and white glue is mixed in the ratio 5:2 by volume. How much glue must be added to 22 mL of sawdust to form this substitute?

5   The ratio of the circumference of a circle to its diameter is 3.14:1. What is the circumference of a bicycle which has a 65-cm diameter?

6   One season the Montreal Canadiens outscored their opponents by a ratio of 8:5. If the Canadiens scored 328 goals, how many goals did the opponents score?

For 30 a, the Montreal Canadiens have maintained a top position in Canadian Hockey. They have won the Stanley Cup more than 20 times.

7   Two numbers, in the ratio 2:3, have a sum of 140. What are the numbers?

B

8   In a basketball game, Tony, Colin and André scored a total of 72 points. The ratio of Tony's points to Colin's points was 2:7 and Colin's points to André's points was 7:3. How many points did each score?

9   Jane, Karen and Rose share the rent on their townhouse. Due to differences in room sizes, the ratio of Jane's rent to Karen's rent is 4:3 and Jane's rent to Rose's rent is 2:1. Find each girl's rent if the monthly rent is $540.

10 The ratio of two positive numbers is 2:3. Their product is 96. Find the numbers.

11 The ratio of two positive numbers is 3:5. Twice their product when added to the sum of their squares is 256. Find the numbers.

12 The ratio of two numbers is 5:4. If the sum of their squares is 1025, find the numbers.

13 The difference of the squares of two numbers is 180. If the numbers are in the ratio 7:2, find the numbers.

14 The square of one number exceeds the square of a second number by 612. Find the numbers if their ratio is 13:4.

15 A ratio is given by 2:9. What number added to each term of the ratio will result in a ratio equivalent to 3:4?

16 Find the number such that when it is doubled and then added to each term of the ratio 3:10, the resulting ratio is equivalent to 3:4.

17 Bill, Chris and Ann work at the Pizza Palace. They share 50 h of counter work so that
    • Ann's share:Bill's share is 3:5
    • Bill's share:Chris' share is 10:9
    How long did each person work at the counter?

18 A will divides 300 shares of IBM stock among 3 brothers as follows
    • Bill's legacy:Brad's legacy is 3:2
    • Bill's legacy:Bob's legacy is 6:5
    How many shares did each brother receive?

C

19 If $a:b = b:c$, prove that $(ac + b^2):b = 2b:1$.

20 If b is a mean proportional for a and c, prove that $(a - b^2):b = (b - bc):c$.

applications: chemicals and ratios

When pure substances combine in a chemical reaction they do so in certain ratios. For example, dry ice is solid carbon dioxide. In the substance called carbon dioxide, carbon has combined with oxygen in the ratio 3:8 by mass. For example, 3 g of carbon combine with 8 g of oxygen to produce (3 + 8) g or 11 g of carbon dioxide.

21 Carbon dioxide consists of carbon and oxygen which have combined in the ratio 3:8 by mass. What mass of oxygen is needed to produce 55 g of carbon dioxide?

2. Laughing gas consists of nitrogen and oxygen combined in the ratio 28:16 by mass. How much oxygen can be obtained from 22 g of gas?

3. Heavy water or deuterium consists of hydrogen and oxygen which have combined in the ratio 1:4 by mass. What mass of oxygen can be obtained from 60 g of heavy water?

4. Silicon and oxygen combine in the ratio 7:8 by mass to produce ordinary sand. What mass of silicon can be extracted from 45 g of sand?

5. Common salt consists of sodium and chlorine in the ratio 23:35.5 by mass. What mass of chlorine can be obtained from 117 g of salt?

6. Ammonia consists of nitrogen and hydrogen combined in the ratio 14:3 by mass. How many grams of ammonia were used to obtain 56 g of nitrogen?

7. Moth balls consist of carbon and hydrogen which have combined in the ratio 15:1 by mass. If the carbon from 48 g of moth balls could be obtained, what would be its mass?

8. (a) Sand is composed of silicon and oxygen in the ratio 7:8 by mass. What mass of oxygen could be obtained from 90 g of sand?
   (b) Dry ice consists of carbon and oxygen in the ratio 3:8 by mass. Using the oxygen obtained in part (a) above, what mass of dry ice would be created if sufficient carbon were available?

---

math tip

The proportion $x:y = a:b$ or $\dfrac{x}{y} = \dfrac{a}{b}$ may be written in other useful equivalent forms.
$$\dfrac{x}{a} = \dfrac{y}{b}, \quad bx = ay, \quad \dfrac{y}{x} = \dfrac{b}{a}, \quad \dfrac{a}{x} = \dfrac{b}{y}$$
Can you show that they are indeed equivalent forms?

5.3 extending our work: ratios and proportions

As we develop skills and learn concepts in mathematics, we apply them to solve problems.

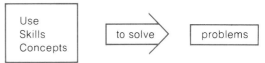

We also extend what we have learned to learn new skills and concepts

If we are given the following proportions.
$$\dfrac{a}{2} = \dfrac{b}{3} \qquad \dfrac{b}{3} = \dfrac{c}{4}$$
we may write them in a compact form, as
$$\dfrac{a}{2} = \dfrac{b}{3} = \dfrac{c}{4} \quad \text{or} \quad a:b:c = 2:3:4$$

The skills we learned to solve proportions for 2-term ratios may be applied to solve proportions involving 3-term ratios as shown in the next example.

Example 1
Find the values of k and m.
$20:k:4 = m:5:1$

Solution
Write an equivalent form.
$$\dfrac{20}{m} = \dfrac{k}{5} = \dfrac{4}{1}$$
Use the proportion to write each equation.

$\dfrac{20}{m} = \dfrac{4}{1}$ $\qquad\qquad$ $\dfrac{k}{5} = \dfrac{4}{1}$

$20 = 4m$ $\qquad\qquad\quad$ $k = 20$

$5 = m$

Check: $20:20:4 = 5:5:1$
$$\dfrac{20}{5} = \dfrac{20}{5} = \dfrac{4}{1} \quad \text{checks } \checkmark$$

The principles we used to solve the above proportion are the same principles we use to solve a more advanced one as in Example 2.

Example 2
Solve for a and b if
$(a + 2):2:3 = 7:-14:(2b - 3)$.

Solution
Write
$$\frac{a + 2}{7} = \frac{2}{-14} = \frac{3}{2b - 3}$$
From the proportion

$\frac{a + 2}{7} = \frac{2}{-14}$ | $\frac{3}{2b - 3} = \frac{2}{-14}$

$\frac{a + 2}{7} = -\frac{1}{7}$ | $\frac{3}{2b - 3} = -\frac{1}{7}$

$7(a + 2) = -7$ | $21 = -(2b - 3)$

$7a + 14 = -7$ | $21 = -2b + 3$

$7a = -21$ | $2b = -18$

$a = -3$ | $b = -9$

We may, as with 2-term ratios, introduce a parameter k, to evaluate expressions.

Example 3
Three positive numbers are in the ratio $2:3:4$. If the sum of their squares is 116, find the numbers.

Solution
Let the numbers be x, y, z.
Thus $x:y:z = 2:3:4$ or $\frac{x}{2} = \frac{y}{3} = \frac{z}{4}$.
Let $\frac{x}{2} = \frac{y}{3} = \frac{z}{4} = k$, $k > 0$.
Then $x = 2k, y = 3k, z = 4k$.
$(2k)^2 + (3k)^2 + (4k)^2 = 116$
$4k^2 + 9k^2 + 16k^2 = 116$
$29k^2 = 116$
$k^2 = 4$
$k = \pm 2$
Since the numbers are positive, use $k = 2$, $2k = 4, 3k = 6, 4k = 8$. The required numbers are 4, 6, and 8.

5.3 exercise

Express your answers as required to 1 decimal place.

A

1. Express each ratio in lowest terms.
   (a) $8:12:16$
   (b) $15:-5:10$
   (c) $x^3:x^2y:xy^2$
   (d) $3a^2b:6ab^2:9ab$
   (e) $(a^2 - b^2):(a - b):(a^2 - 2ab + b^2)$
   (f) $(x^2 + 2xy + y^2):(x + y):(x^2 - y^2)$

2. Solve for a and b.
   (a) $3:a:12 = 6:4:b$
   (b) $a:3:b = 5:15:10$
   (c) $7:5:a = 14:b:4$
   (d) $a:10:16 = 3:b:8$
   (e) $4:a:6 = 10:20:b$
   (f) $a:4:1 = 6:b:8$
   (g) $3:a:b = 6:20:24$
   (h) $a:4:10 = 9:6:b$

3. (a) If $x = 6k, y = 2k, z = 8k$, find $x:y:z$.
   (b) If $a = 9k, b = -3k, c = -12k$, find $a:b:c$.

4. If $7:(a + 3):2 = 14:10:(b - 1)$, then find the values of a and b.

5. Find the values of a and b if
   $(a + 2):12:8 = 3:(2b + 1):6$.

6. Calculate the values of a and b if
   $(3a + 2):3:7 = 8:12:(3b + 1)$.

7. If $3:6:(4a + 1) = -2:(b + 1):10$, find the values of a and b.

8. Solve each of the following for m and n.
   (a) $\dfrac{2m + 2}{n + 1} = \dfrac{1 - 6m}{n - 2} = \dfrac{2m + 9}{3n + 2}$
   (b) $\dfrac{3m - 1}{2n + 1} = \dfrac{6m + 3}{4n + 9} = \dfrac{15m}{15n - 3}$
   (c) $\dfrac{3m - 1}{4n - 1} = \dfrac{6m}{8n + 2} = \dfrac{3 - 6m}{1 - 4n}$

9. A football team had a win-loss-tie record in the ratio of 4:2:3. In that season, they played 18 games. How many games were won, lost and tied?

10. For a triangle the measures of the angles are in the ratio 2:2:5. Find the measure of each angle.

11. Three numbers are in the ratio 1:3:5. If the sum of the numbers is 252, find the numbers.

12. The analysis of skim milk powder reveals that the ratio of protein to carbohydrates to water is 7:10:1 by mass. How many grams of each is contained in a 500-g box of powder?

13. Jenny, Lori and Alex have formed a pool to buy lottery tickets. Respectively, they contribute $5, $7.50, and $12.50. How should they divide the $100 000 cash prize?

14. The sum of three numbers is 230 and they are in the ratio 2:3:5. What are the numbers?

B

15. The sum of the squares of 3 positive numbers is 1368. If the ratio of the numbers is 2:3:5, find the numbers.

16. Three positive numbers are in the ratio of 3:4:5. When the square of the middle number is subtracted from the product of the other 2 numbers, the answer is −9. Find the numbers.

17. Silver, tin, and lead are used in the ratio 10:8:12 to forge an alloy for costume jewellery. How much silver is required to forge a bracelet which has mass 40 g?

18. A sidewalk is built using gravel, cement and sand in the ratio 14:4:8 by mass. How much gravel is needed to produce 42 kg of this mixture?

19. Three positive numbers have the ratio 2:5:7. The sum of their squares is 296 more than the square of the smallest number. Find the numbers.

20. The ratio of three positive numbers is 3:4:6. The sum of the squares of the first 2 numbers is 44 less than the square of the third number. Find the numbers.

21. Sugar consists of carbon, hydrogen and oxygen combined in the ratio 72:11:88. What mass of hydrogen can be obtained from 1 kg of sugar?

22. Nitroglycerine is a compound of carbon, hydrogen, nitrogen and oxygen in the ratio 36:5:42:48. What mass of each of oxygen, carbon, and hydrogen will react with 6 g of nitrogen to form this compound?

23. Nitric acid is formed by the reaction of hydrogen, nitrogen and oxygen combining in the ratio 1:14:48. How much nitrogen and oxygen is required to combine with 2.5 g of hydrogen to form this acid?

24. Glucose or dextrose is a compound of carbon, hydrogen and oxygen combined in the ratio 84:12:96. How much of each of carbon and oxygen is required to react with 3 g of hydrogen?

25. If $x:y:z = 1:2:3$, find the values of each of the following.
    (a) $\dfrac{x - 2y}{2x - z}$
    (b) $\dfrac{x^2 - y^2}{y^2 + z^2}$

26. If $x:y:z = 2:5:3$, find the values of
    (a) $\dfrac{x + y + z}{x + z}$
    (b) $\dfrac{x^2 + y^2 + z^2}{x^2 + y^2 - z^2}$

27. If $x:y:z = 2:5:7$, find the value of each of the following.
    (a) $\dfrac{x + y}{x - y}$
    (b) $\dfrac{(x + y + z)^2}{x^2 + y^2 + z^2}$
    (c) $(x + 2y - 3z)^2 \div y^2$

applications: the fertilizer ratio

Each of the terms of the ratio as shown on the bag of fertilizer has a special significance. The ratio indicates the relative amounts of nitrogen, phosphoric acid, and potash in the fertilizer.

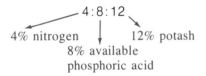

4% nitrogen   8% available phosphoric acid   12% potash

28. A plant fertilizer contains nitrogen, phosphoric acid and potash in the ratio 14:14:12. If a bag of this fertilizer has 21 kg of nitrogen how much of the other two ingredients does it contain?

29. Lawn Booster gives lawns an early feeding and provides vigorous growth. If a bag contains nitrogen, phosphoric acid and potash in the ratio 20:5:5, how much nitrogen and phosphoric acid is in such a bag containing 2 kg potash?

30. All-Purpose Fertilizer is applied in the late spring when lower nitrogen content is needed. This fertilizer contains nitrogen, phosphoric acid, and potash in the ratio 14:7:7. If the bag contains 2 kg of nitrogen, how much potash does it contain?

31. 4-12-8 Garden Special has a high proportion of phosphoric acid and potash for large, healthy flowers and fruits. A bag of this fertilizer contains nitrogen, phosphoric acid and potash in the ratio 4:12:8. How much potash and nitrogen is contained in a bag which has 4 kg of phosphoric acid?

5.4 working with direct variation

Each year, swimmers test their skills to swim across the English Channel.

During one try Cindy swam at an average rate of 3 km/h. The distance she travelled is shown in the chart.

t Time to swim (h)	v Rate (km/h)	d Distance (km)
1	3	3
2	3	6
3	3	9
4	3	12

We use the formula to calculate the values in the chart.
$d = vt$
When $v = 3$
$d = 3t$

The relationship between the distance travelled (d) and the time taken (t), is given by
$d = 3t$.
The graph of $d = 3t$ is shown at the right. From the equation, we may write
$\dfrac{d}{t} = 3$. ← A constant

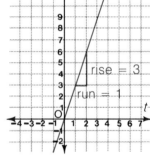

From the equation we note that
- if d increases then t increases.
- if d decreases then t decreases.

Thus, we say that d varies directly with t or simply d varies with t, and we write in symbols that
$d \propto t$
↖ varies directly with

In general, if y varies directly with x, then we may write the direct variation as $y \propto x$, or $\dfrac{y}{x} = k$ where k is a constant, ($k \neq 0$), called the **constant of variation**. If (x_1, y_1) and (x_2, y_2) satisfy the direct variation, then we may write the following, where k is a constant.

$$\frac{y_1}{x_1} = k \quad \text{and} \quad \frac{y_2}{x_2} = k$$

From the above, we may then write the following proportion for a direct variation.

$$\frac{y_1}{x_1} = \frac{y_2}{x_2} = k \qquad \text{We also refer to the constant } k \text{ as the } \textit{constant of proportionality}.$$

Example 1

If y varies directly with x and $y = 384$ when $x = 192$, find the value of y when x is 48.

Solution

To find the answer, we first need to find k, the constant of variation.

Write $y = kx$, where k is constant, $k \neq 0$.

Step 1 Use $y = 384$, $x = 192$.
$$y = kx$$
$$384 = k(192)$$
$$2 = k$$

The equation for the direct variation is then $y = 2x$.

Step 2 Solve for y if $x = 48$.
$$y = 2x$$
$$y = 2(48)$$
$$= 96$$

Thus $y = 96$ when $x = 48$.

To solve the above problem, we could have used the proportion

$$\frac{y_1}{x_1} = \frac{y_2}{x_2}$$

From the given information we write,
$y_1 = 384$, $x_1 = 192$, $y_2 = ?$, $x_2 = 48$.
Thus, to solve the problem we use the proportion and the given information.

$$\frac{y_1}{x_1} = \frac{y_2}{x_2} \qquad \frac{384}{192} = \frac{y_2}{48}$$
$$\frac{48(384)}{192} = y_2$$
$$96 = y_2$$

Thus $y = 96$ when $x = 48$.

An important skill we used in solving word problems was translating the problem into mathematics. This same skill is again needed to solve problems involving direct variation.

Example 2

The amount of bend of a diving board varies directly with the mass of the diver. If a 64-kg diver causes the board to bend 4.0 cm as shown, how much will the diving board bend if the diver has a mass of 72 kg? Express your answer to 1 decimal place.

Solution

Let D represent the mass of the diver in kilograms, and B the amount of bend, in centimetres.
Thus $D \propto B$ or $D = kB$ where k is a constant.

Step 1 Find k for $D = kB$, $D = 64$ (kg), $B = 4.0$ (cm).
$$64 = k(4.0)$$
$$16 = k$$
$$D = 16B.$$

Step 2 Use $D = 16B$, $D = 72$ (kg).
Thus $72 = 16B$
$$4.5 = B$$

Thus, the amount of bend in the diving board is 4.5 cm.

5.4 exercise

Questions 1 to 7 develop skills for solving problems involving direct variation.

A

1. (a) If A varies directly with W, then write an equation to express the direct variation.
   (b) If $A = 720$ when $W = 12$, find A when $W = 15$.

2 (a) If H is directly proportional to A, then write an equation to express the direct variation.
   (b) If $H = 40.3$ when $A = 6.5$, find H when $A = 9.5$.

3 Determine the constant of variation for each direct variation.
   (a) m varies directly with n and $m = 12$ when $n = 72$.
   (b) p varies directly with q and $p = 36$ when $q = 3$.
   (c) L is directly proportional to T and $T = 225$ when $T = 15$.
   (d) y varies directly with x and $y = 300$ when $x = 30$.
   (e) m varies directly with n and $m = 28$ when $n = 14$.

4 Which ordered pairs belong to each direct variation?
   (a) $P = 4T$ (T, P); $(1, 16), (8, 34), (3, 12)$
   (b) $S = 6P$ (P, S); $(3, 18), (1, 48), (10, 68)$
   (c) $y = 15x$ (x, y); $(5, 70), (10, 150), (2, 40)$
   (d) $\dfrac{d}{t} = 12$ (t, d); $(5, 60), (2, 20), (10, 120)$

5 For each of the following, calculate the constant of variation first.
   (a) A direct variation is given by $P = kV$, k is constant. Find the value of V when $P = 1500$ if $P = 600$, when $V = 3$.
   (b) If d varies directly with t and $d = 325$ when $t = 6.5$, find d when $t = 12$.
   (c) h is directly proportional to t. If $t = 8$ when $h = 288$, find h when $t = 20$.
   (d) A direct variation is given by $y = kx$, k is constant. If $x = 12.2$ when $y = 183$, find x when $y = 300$.
   (e) H is directly proportional to t. Find t_2 when $H_1 = 27.5$, $t_1 = 2.2$, $H_2 = 77.5$.

6 For each direct variation, find the missing value.
   (a) $d_1 = 375$ $t_1 = 5$ $d_2 = ?$ $t_2 = 9$
   (b) $h_1 = ?$ $t_1 = 4$ $h_2 = 240$ $t_2 = 15$
   (c) $P_1 = 750$ $V_1 = ?$ $P_2 = 1250$ $V_2 = 10$
   (d) $P_1 = 1560$ $T_1 = ?$ $P_2 = 1586$ $T_2 = 12.2$
   (e) $W_1 = ?$ $t_1 = 129.2$ $W_2 = 3.5$ $t_2 = 8.5$
   (f) $h_1 = ?$ $t_1 = 3.2$ $h_2 = 209.1$ $t_2 = 8.2$

7 Introduce suitable variables and express each of the following as a direct variation in the form $y = kx$, k is constant.
   (a) The perimeter of a square varies directly with the length of a side.
   (b) The distance travelled varies directly with the time (if the speed is constant).
   (c) In any tire the amount of pressure on the tire wall is directly proportional to the amount of air in the tire.
   (d) The force of gravitation between the earth and a body on its surface varies directly with the mass of the body.
   (e) The crop yield of a corn field is directly proportional to the amount of rainfall during the season.
   (f) The mass of a metal ball varies directly with the cube of its diameter.

B

8 The mass of a cylinder rod is directly proportional to its length.
   (a) Write an equation to express the direct variation.
   (b) The mass of the cylinder is 100 kg when its length is 6 m. Find the mass of a cylinder 15 m in length.

9 The amount of stretch of a spring is directly proportional to the mass added.
   (a) The spring stretched 2 cm when 6 g was added. How many grams must be added to stretch the spring 7 cm?

(b) What will be the stretch on the spring when 10 g are added?

10. Sam can do 70 situps in 1 min.
    (a) How many situps can he do in 2.5 min?
    (b) How long will it take him to do 14 situps?
    (c) What assumptions do you make in finding your answers in (a) and (b)?

11. The distance a car travels varies directly with volume of fuel the tank holds. If a car can travel 500 km on 30 L of gas, how far can the car travel if there is 20 L of gas in the tank?

12. The maximum speed of a car varies directly with the size of its engine. If a car with a 300-cm³ engine has a maximum speed of 120 km/h, what would be the maximum speed of a car with a 2200-cm³ engine?

13. Parking fees in a downtown lot vary directly with the length of time parked. If Harry paid $3.90 for 6.5 h, what would be the charge for 2 h?

14. The annual interest earned on a savings account varies directly with the amount on deposit. If Janet earns $13 on a deposit of $200, how much should she earn on a $500 deposit?

15. Under constant speed, the amount of gas consumed varies dirrectly with the distance travelled by the car. When Tom travelled from Prince George to Kamloops (375 km), his car used 12.5 L. At the same speed, how many litres of gas would his car consume travelling 450 km?

---

math tip

These equivalent forms are useful to know.

If $\dfrac{a}{b} = \dfrac{c}{d}$ then

$$\dfrac{a+b}{b} = \dfrac{c+d}{d} \quad \text{or} \quad \dfrac{a-b}{b} = \dfrac{c-d}{d}.$$

Can you prove these results?

5.5 direct squared variation

There are many phenomena that involve a squared relationship.

- The area A of a circle varies directly with the square of the radius r.
  $$A \propto r^2$$
- The distance, d, that a falling body travels varies directly with the square of the time, t, of falling.
  $$d \propto t^2$$
- The value, V, of a diamond, varies directly with the square of its mass, M.
  $$V \propto M^2$$

We may make the following comparison.

direct variation	direct squared variation
• $y \propto x$	• $y \propto x^2$
• $y = kx$, k constant	• $y = kx^2$, k, constant
• y varies directly with x.	• y varies directly with the square of x.
• y is directly proportional to x.	• y is directly proportional to the square of x.

Example

The volume of water rushing from a hose in a unit of time varies directly with the square of the diameter of the hose (if the water pressure is constant). From a fire hose, with a diameter of 8 cm, 400 kL of water were obtained. For the same amount of time, how much water could be obtained from a garden hose 2 cm in diameter?

Solution 1   You may use one of two methods to solve problems involving variation as follows.

Let V represent the volume of water, in kilolitres, obtained in a unit of time. Let d represent the diameter of the hose, in centimetres.

Then
$V = kd^2$ where k is a constant, ($k \neq 0$).

V (kL)	d (cm)
400	8
?	2

← Record the information given in the problem in a chart.

Step 1 Use $V = kd^2$.
$400 = k(8)^2$ ← From the chart.
$\frac{400}{8^2} = k$ or $k = \frac{25}{4}$

Thus, the variation equation is given by $V = \frac{25}{4} d^2$.

Step 2 Solve for V, using $d = 2$.
$V = \frac{25}{4} d^2$
$V = \frac{25}{4} (2)^2$ ← From the chart.
$= 25$

Thus, the volume of water obtained from the garden hose is 25 kL.

Be sure to make a final statement.

Solution 2

Let V represent the volume of water, in kilolitres obtained in a unit of time.
Let d represent the diameter of the hose, in centimetres
Then,
$\frac{V_1}{d_1^2} = \frac{V_2}{d_2^2}$

Given Information
$V_1 = 400 \quad d_1 = 8$
$V_2 = ? \quad d_2 = 2$

Use the given information to solve for V_2.
$\frac{400}{(8)^2} = \frac{V_2}{(2)^2}$
$\frac{400}{64} = \frac{V_2}{4}$
$\frac{4(400)}{64} = V_2$
$25 = V_2$ ← Or we may write $V_2 = 25$.

Thus the volume of water obtained from the garden hose is 25 kL.

In Solution 1 (which we may refer to as the k-method) we wrote a variation equation.
$V = kd^2$
↑
└── constant of variation

In Solution 2 (which we may refer to as the proportion method) we wrote a proportion.
$\frac{V_1}{d_1^2} = \frac{V_2}{d_2^2}$

5.5 exercise

A

1 If b is directly proportional to c^2, then find the constant of variation if $b = 72$ when $c = 12$.

2 (a) If s varies directly as p^2, then write a variation equation to express the relationship.
   (b) If $s = 9$ when $p = 6$, then find s when $p = 10$.

3 If R is directly proportional to the square of V and $R = 50$ when $V = 15$, then find V when $R = 200$.

4 If air resistance is neglected the distance a meteorite travels through the atmosphere varies directly with the square of the time of falling.
   (a) Find the constant of variation if a meteor takes 12 s to travel 2304 m.
   (b) How long does it take a meteor to travel 6400 m?

5 The mass of a china plate is directly proportional to the square of the diameter of the plate.
   (a) Write a proportion to express the above variation.
   (b) A plate, 24 cm in diameter, has a mass of 64 g. What is the mass of a plate that is 30 cm in diameter?

B

6 When a vehicle moves, the resistance of air to it varies directly with the square of its speed. For a car travelling at 10 km/h, the resistance to it is 4 units. Find the resistance to the car when the speed is 25 km/h.

7 The surface area of a sphere varies directly with the square of its radius. The surface area of a basketball, with radius 15 cm, is 900π cm². What would be the surface area of a ball with a radius 12 cm?

8 The distance a car rolls down a hill from a standstill is directly proportional to the square of the time it rolls. If a car rolls 135 m in 6 s, how far will it roll in 10 s?

9 (a) The value, in dollars, of a diamond is directly proportional to the square of its mass. If a diamond, worth $6300, is 200 mg, then find the mass of a diamond that is worth $25 200.
(b) Calculate the value of the diamond (28 mg).

10 The number of bacteria in a culture is directly proportional to the square of the time the bacteria have been growing. A culture growing for 16 min has 600 bacteria. How long will it take to grow 15 000 bacteria?

11 If the height of a cone is constant, then the volume of the cone varies directly with the square of the radius of its base. The volume of a cone is 2464 cm³ when the radius is 14 cm. What is the volume when the radius is 21 cm?

12 The amount of leather needed to cover a ball is directly proportional to the square of the radius. A ball of radius 9 cm requires 162 cm² of leather. What is the radius of a ball if the amount of leather used is 144.5 cm²?

13 On an ice surface, the distance, in metres, travelled by a frozen puck before it stops is directly proportional to the square of its initial speed, in metres per second.
(a) Find the constant of variation if a puck with initial speed of 8 m/s could travel 320 m.
(b) The initial speed of a puck in some National Hockey League games may be as high as 42 m/s. How far would a puck go before it stops? Express your answer to the nearest kilometre.

C For Questions 14 to 16 use the following.
The distance an object falls from a height varies directly with the square of the time the object has been falling.

14 (a) Write a proportion statement for the above variation.
(b) A ball falls 312 m in 8 s. How far would a ball fall in 10 s?

15 Calculate the distance a ball would fall in
(a) 3 s   (b) 4 s
(c) Use the information in (a) and (b) to determine the distance fallen during the 4th second.

16 A parcel is dropped in the Arctic from an airplane, but its parachute does not open.
(a) How far will the parcel drop in the 4th second?
(b) How far will the parcel drop in the 5th second?
(c) How much farther will the parcel drop in the 5th second than in the 4th?

applications: driving safely

We may use our work with direct squared variation to solve problems about driving safely. The distance, d, in metres, a car travels after braking is directly proportional to the square of its speed, v, in kilometres per hour, at the time of braking.

The two graphs show how the distance for braking increases as the speed of a car increases, on different surfaces.

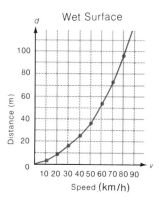
Wet Surface

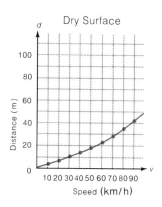
Dry Surface

17. The distance, d, in metres a truck needs to come to a full stop varies directly with the square of its speed, v km/h. A truck requires 8 m to come to a full stop when travelling at 20 km/h. At what distance will the truck come to a stop travelling at 30 km/h?

18. The distance, d, in metres a car travels to a full stop is directly proportional to the square of its speed, v km/h, at the time of braking. At an initial speed of 60 km/h a car travels 36 m before stopping. If the car was travelling at 90 km/h, how far would it travel before stopping?

19. On a wet surface the distance a car travels upon braking is directly proportional to the square of its speed at the time of braking. A car at an initial speed of 80 km/h travels 90 m before stopping. If the car was travelling at 60 km/h, how far would it travel before stopping? Express your answer to 1 decimal place.

20. The distance required for a truck to come to a full stop is directly proportional to the mass of the truck. At a speed of 60 km/h, a 10-t truck required 55 m to come to a full stop safely. A truck enters an intersection and a car comes directly into its path, at a distance of 38.5 m. If the truck has a mass of 7.5 t, will there be a collision? Give reasons for your answer.

21. The conditions of a road greatly affect the distances a car travels upon braking. On an icy road the stopping distance is directly proportional to the cube of the car's speed at the time of braking.
    (a) At 20 m/s a car travels 120 m before stopping. What is the stopping distance of a car travelling at 10 m/s upon braking?
    (b) What are the speeds in (a) in terms of kilometres per hour?

22. The stopping distance (in metres) of a car is directly proportional to the reaction time of the driver (in seconds). For a reaction time of 0.5 s a car travels 36 m at 60 km/h. At the same speed, calculate to 2 decimal places the maximum reaction time if the driver is to avoid hitting a cyclist who has fallen 20 m ahead of the car.

23. The stopping distance of a car is directly proportional to the square of the car's speed upon braking. At 16 m/s a car travels 40 m. A car is 24 m behind a bus and both are travelling at 12 m/s. If the bus suddenly comes to a dead stop, can the driver stop safely? Give reasons.

5.6 inverse variation and its applications

The basketball team marked out a course of 16 km for their workouts. The formula is used to complete the chart.

Speed, v	Time, t	
16 km/h	1 h	←Bicycle
8 km/h	2 h	←Jogging
4 km/h	4 h	←Walking
1 km/h	16 h	

$vt = 16$
speed ↙ ↓time ↘course
         taken   is 16 km

From the above chart, note that
- if the speed is increased the time taken to travel 16 km is decreased.
- if the speed is decreased, the time taken to travel 16 km is increased.

We may draw a graph of $vt = 16$ to show the graphical relationship. The domain for $vt = 16$ is the real numbers, $t \in R, t \geq 0$.
From $vt = 16$, we may write $v = \dfrac{16}{t}$.

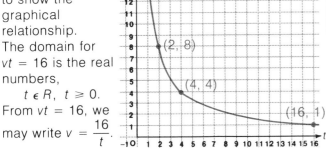

We say that
- v varies inversely with t. We write $v \propto \dfrac{1}{t}$.
- v is inversely proportional to t.

In general, an inverse variation is given as
$xy = k$
or $y = \dfrac{k}{x}$
Where k is the constant of variation, $k \neq 0$.

From the above variation equation, we may obtain a proportion as follows.
$x_1 y_1 = k \qquad x_2 y_2 = k$
Thus $x_1 y_1 = x_2 y_2$.

We may also rewrite the above equation as
$\dfrac{x_1}{x_2} = \dfrac{y_2}{y_1}$.

Example 1
If s varies inversely with t and $s = 30$ when $t = 25$, then find the value of s when t is 150.

Solution
Write the variation equation $s = \dfrac{k}{t}$, k constant.
When $s = 30$, then $t = 25$.
$s = \dfrac{k}{t}$
$30 = \dfrac{k}{25}$
$k = 750$
The variation equation is $s = \dfrac{750}{t}$.
Use $t = 150$ to solve for s.
$s = \dfrac{750}{t} = \dfrac{750}{150} = 5$
Thus $s = 5$ when $t = 150$.

To solve the above problem we may also use the proportion, $s_1 t_1 = s_2 t_2$.
Given information: $s_1 = 30$, $t_1 = 25$, $s_2 = ?$, $t_2 = 150$.
Thus $30(25) = s_2(150)$
$\dfrac{750}{150} = s_2$
$5 = s_2$
Thus $s = 5$ when $t = 150$. We use a proportion to solve the following problem.

Example 2
The number of chairs on a lift at a ski resort varies inversely with the distance between them. When they are 16 m apart the ski lift can accommodate 45 chairs. If the distance between chairs is 12 m, how many more chairs can be placed on the ski lift?

Solution
Let n represent the number of chairs. Let d represent the distance between the chairs,

in metres. Then, we may write the proportion.

Proportion   Information from the given problem.

$n_1 d_1 = n_2 d_2$ $n_1 = 45$ $d_1 = 16$
                $n_2 = ?$ $d_2 = 12$

Use the given information and substitute.

$$n_1 d_1 = n_2 d_2$$
$$(45)(16) = n_2(12)$$
$$n_2 = \frac{(45)(16)}{12} = 60$$

Thus, the number of chairs on the ski lift is 60 when placed 12 m apart.

Thus the lift can accommodate 15 more chairs when the chairs are placed 12 m apart.

5.6 exercise

A Questions 1 to 14 develop skills for working with inverse variation.

1   Calculate the constant of variation for each of the following inverse variations.

(a) m varies inversely with n and $m = 25$ when $n = 2$.

(b) P is inversely proportional to T and $T = 16$ when $P = 4$.

(c) x varies inversely with y and $x = 8$ when $y = 12$.

(d) S is inversely proportional to T and $S = 36$ when $T = 8$.

2   Which ordered pairs in columns A, B, C, satisfy each inverse variation?

                                 A   B   C

(a) $sp = 60$, (s, p);   (4, 15)   (3, 30)   (2, 30)

(b) $t = \frac{100}{q}$, (q, t);   (2, 50)   (10, 10)   (4, 30)

(c) $P = \frac{72}{T}$, (T, P);   (2, 36)   (3, 24)   (4, 18)

(d) $xy = 144$, (x, y);   (6, 18)   (24, 6)   (12, 12)

3   If P varies inversely with T and $P = 108$ when $T = 3$, then find

(a) the constant of variaton.
(b) T if $P = 18$.   (c) P if $T = 54$.

4   H is inversely proportional to S.

(a) Write a proportion to express the inverse variation.

(b) If $H = 9$ when $S = 140$, then find S when $H = 60$.

5   Use the proportion $x_1 y_1 = x_2 y_2$ to solve each of the following.

(a) $S_1 = 10$, $T_1 = 100$, $S_2 = ?$, $T_2 = 40$
(b) $x_1 = 10$, $x_2 = ?$, $y_1 = 20$, $y_2 = 20$
(c) $P_1 = 6$, $Q_1 = ?$, $P_2 = 24$, $Q_2 = 2$

6   T is inversely proportional to P. If $T = 50$ then $P = 2$. Find P when $T = 20$.

7   If x varies indirectly with y, and $x = 30$ when $y = 15$, then find x when $y = 90$.

8   If $ab = $ constant and $a = 3$ when $b = 32$, then find a when $b = 12$.

9   An inverse variation is given by $VT = k$, when k is constant. If $V = 75$ when $T = 36$, then find V when $T = 50$.

To solve any problem you must read carefully. The following questions include examples of direct and inverse variation. Be sure you read carefully!

10   An inverse variation is given by $PT = k$, k is constant. Find P_1 when $T_1 = 50$, $P_2 = 125$ and $T_2 = 5$.

11   A direct variation is given by $\frac{S}{t} = k$, k constant. Find s_1 when $t_1 = 48$, $s_2 = 25$, and $t_2 = 100$.

2. y is inversely proportional to x. If $y = 12$ when $x = 45$, then find x when $y = 30$.

3. P varies directly with Q. If $P = 36$ when $Q = 9$, then find P when $Q = 16$.

4. Write a proportion for each of the following variations. Introduce appropriate variables.

   (a) The value of a car varies inversely with the age of the car.
   (b) The volume of a gas kept at constant pressure varies directly with the temperature.
   (c) The frequency of sound waves is inversely proportional to the wave length of the sound waves.
   (d) At a constant speed, an automobile's gasoline consumption varies directly with the distance travelled.
   (e) If the temperature of a gas is constant, then the volume of the gas is inversely proportional to the pressure.

Solve each problem. Be sure to identify carefully whether the problem requires skills for direct variation or inverse variation.

5. The time required to cook a roast varies inversely with the oven temperature. If a 3-kg roast takes 3 h to cook at 175°C, how long will it take the same roast to cook at 210°C?

6. Boyle's Law states that at a fixed temperature the volume of the gas is inversely proportional to the pressure. If the volume of the gas is 30 L when the pressure is 8 units, determine the pressure when the gas expands to 80 L.

7. The amount of chlorine needed for a pool is directly proportional to the size of the pool. If 27 units of chlorine are used for 15 kL of water, how much chlorine is required for 25 kL of water?

18. A baseball pitcher throws a fastball at a speed of 15 m/s. If it takes 1.5 s for the ball to cross the plate, how long will it take to return the ball if the catcher has a sore elbow and can only throw the ball at a speed of 10 m/s?

In Major League Baseball the speed at which the ball crosses the home plate is often 125 km/h. Do you understand why the batter is wearing a helmet?

19. A Concorde Jet travelling at 2400 km/h takes 3 h to make a New York-to-London crossing. If a 747 jet makes the same trip in 9 h, how fast does the 747 jet fly?

20. The effect of a sunlamp is inversely proportional to the distance from the source. When a person is 2 m from the source the tanning power is 40 units. What is the tanning power at 0.5 m?

21. The amount of interest earned in a year in a savings account is directly proportional to the amount of money in the account. If $36.75 interest is earned on $350, how much interest will be earned on $420?

22. Richard rode his "pedal power" bicycle to visit a friend. The trip takes 3 h pedalling at 8 km/h. How much time would he have saved if he had taken his 10-speed bicycle which travels at 12 km/h?

23. A noise pollution researcher found that the sound level of heavy traffic varies inversely with the distance from the road. At a distance of 5 m the sound level is 120 dB. At what distance would the sound level be a soft 30 dB?

5.7 inverse squared variation

When you show a film on a screen, the intensity, I, of the beam decreases quickly as you move the projector away from the screen. In fact, the intensity, I, varies *inversely as the square* of the distance, d.

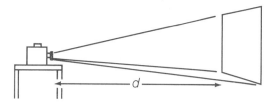

We may write the *inverse squared variation*

$$I \propto \frac{1}{d^2} \text{ or } I = \frac{k}{d^2} \quad \text{Where } k \text{ is the constant variation.}$$

The skills we have learned in working with inverse variation extend to learning skills with inverse squared variation. Compare the following.

Inverse Variation	Inverse Squared Variation
$y = \dfrac{k}{x}$	$y = \dfrac{k}{x^2}$
• y varies inversely as x.	• y varies inversely as x^2.
• y is inversely proportional to x.	• y is inversely proportional to x^2.

Example 1
T is inversely proportional to the square of s. If $T = 150$ when $s = 6$, then find the value of T when $s = 30$.

Solution
$T = \dfrac{k}{s^2}$ where k is the constant of variation.

Step 1 Find k when $T = 150$ and $s = 6$.
$$T = \frac{k}{s^2}$$
$$150 = \frac{k}{6^2}$$
$$5400 = k \text{ or } k = 5400$$

Thus, the variation equation is $T = \dfrac{5400}{s^2}$.

Step 2 Find T when $s = 30$.
$$T = \frac{5400}{s^2}$$
$$T = \frac{5400}{(30)^2}$$
$$= 6$$
Thus, $T = 6$ when $s = 30$.

We could also have solved the problem in Example 1 by using a proportion of the form
$$T_1 s_1^2 = T_2 s_2^2$$
In Example 2, a proportion is used to solve the problem.

Example 2
The force of gravity between two planets in the solar system is inversely proportional to the square of the distance between them. The moon is now about 400 000 km from the earth. If the moon were only 100 000 km from the earth, what would be the change in the force of gravity between the earth and the moon?

Solution
Let the present force of gravity between the earth and the moon be represented by F units. Let d represent the distance, in kilometres, between the planets.

Write the proportion statement as follows.
$$F_1 d_1^2 = F_2 d_2^2 \quad F_1 = F \quad d_1 = 400\,000$$
$$F_2 = ? \quad d_2 = 100\,000$$
Thus, $F(400\,000)^2 = F_2(100\,000)^2$
Solve for F_2.
$$\frac{F(400\,000)^2}{(100\,000)^2} = F_2 \quad \text{Original force between the planets.}$$
$$16F = F_2 \text{ or } F_2 = 16F$$
Since the original force was F units, then the force between the planets has increased by a factor of 16, (namely $16F$).

5.7 exercise

A

1. Calculate the constant of variation if P varies inversely with the square of T and $P = 6$ when $T = 8$.

2. Calculate the constant of variation if y is inversely proportional to the square of x, and $y = 0.25$ when $x = 4$.

3. R varies inversely with the square of S. If $R = 24$ when $S = 15$, then find
   (a) the constant of variation.
   (b) R when $S = 5$. (c) S when $R = 150$.

4. An inverse squared variation is given by $a_1 b_1^2 = a_2 b_2^2$. If $a_2 = 6$ when $b_2 = 30$, then find a_1 when $b_1 = 6$.

5. Write the missing value for each inverse squared variation $P \propto \dfrac{1}{Q^2}$.

	P_1	Q_1	P_2	Q_2
(a)	36	25	225	?
(b)	?	12	16	30

6. A variation is given by $ST^2 = k$, k is constant. If $S = 36$ when $T = 2$ then find T when $S = 16$.

In solving any problem we must read carefully. The following includes not only inverse squared variations also other types of variations. Be sure to read carefully.

7. An inverse squared variation is given by $rs^2 = k$, k is constant. If $r_1 = 15$ when $s_1 = 9$ then find r_2 when $s_2 = 6$.

8. P varies directly with V^2. If $P = 16$ when $V = 10$, find P when $V = 4$.

9. Q varies inversely with the square of T. If $Q = \dfrac{7}{2}$ when $T = 5$, then find T when $Q = 14$.

10. If $\dfrac{P}{V} = k$, k is constant, and we know $P_1 = 15$, $P_2 = 25$, and $V_2 = 125$, find the value of V_1.

11. Write an equation using a constant k to express each variation.
    (a) The intensity, I, of a light source varies inversely with the square of the distance, d, from the source.
    (b) The surface area, A, of a sphere varies directly with the square of the diameter, d, of the sphere.
    (c) The frequency, f, of sound waves is inversely proportional to the wavelength, λ, of the sound waves.
    (d) The force of gravity, g, between any two bodies is inversely proportional to the square of the distance, d, between them.
    (e) The mass, M, of a substance varies directly with its volume, V.

Solve each problem. Be sure you correctly identify the type of variation that is involved.

B

12. The temperature of a location on the earth's surface varies inversely with the distance from the equator. On a given day the temperature is 30°C at a place 5250 km from the equator. What is the temperature 7500 km from the equator?

13. The gas efficiency of a car varies inversely with the mass of the car. If a 900-kg car obtains a gas efficiency of 9 km/L, what will the gas efficiency be for a 1500-kg car?

14. The volume of a cylinder is directly proportional to the square of its radius. If a cylinder of radius 3 cm holds 49.5 L, what is the radius of the cylinder that would hold 137.5 L?

15 The price of a diamond ring varies directly with the square of the mass of the stone. A stone, whose mass is 40 mg, costs $1000. What is the mass of a stone which costs $1500?

16 The base of a triangle varies inversely with the altitude, if the area is constant. If a triangle with base 6 cm has an altitude of 16 cm, what base length corresponds to an altitude of 4 cm?

17 The mass of commercial soup cans varies directly with the square of the radius. A large can with radius 3 cm has a mass of 300 g. What is the mass of a similar can with radius 5 cm?

18 The time needed to fill the gas tank of a car varies inversely with the square of the diameter of the hose. If a hose of diameter 2 cm takes 10 min to fill the tank, how long will it take to fill the same tank if the hose has a diameter of 5 cm?

19 The exposure time required for a photograph is inversely proportional to the square of the diameter of the camera lens. If the lens diameter is 5 cm, the exposure time needed is $\frac{1}{625}$ s. Find the diameter of the lens if an exposure time of $\frac{1}{3600}$ s is needed.

20 The value of fine china is inversely proportional to the square of the thickness of the porcelain. Calculate the value of a plate 4 mm in thickness if a plate 1.5-mm thick, costs $50.00.

21 If the height of a right circular cone is kept constant, the volume varies directly with the square of the radius of the base. If a cone with a base radius 6 cm has a volume of 189 cm³, what is the volume of a cone with a base radius 4 cm?

5.8 applications with joint variation

To help us understand joint variation, we may refer to the area of a rectangle.

If the width of a rectangle is constant, the area, A, varies directly with the length, L; or simply the greater the length, the greater the area. Thus if L, W, and A represent the number of consistent units of length, width, and area respectively,
$A \propto L$ when W is constant.

Similarly
$A \propto W$ when L is constant.

Thus, the area of a rectangle varies directly with its length *and* its width. We may say that the area of the rectangle varies *jointly* with its length and its width. We write the variation relation as
$A \propto L \times W$ We may also write $A \propto LW$.
and $A = kL \times W$ where k is a real number, called the constant of variation.
We may also write $\frac{A_1}{A_2} = \frac{L_1 W_1}{L_2 W_2}$
and if the area is constant $L_1 W_1 = L_2 W_2$.

Example 1
The cost of publishing a sailing magazine varies directly with the number of pages and inversely with the number of advertisements in the magazine. A magazine with 50 advertisements and 125 pages costs $2.00 to publish. How many advertisements are necessary for a 250-page magazine so that it costs $2.50 to publish?

Solution
Let C represent the cost of the magazine, in dollars; p the number of pages, and n, the

number of advertisements. Then
$$C \propto p \times \frac{1}{n} \text{ or } C \propto \frac{p}{n}.$$

Step 1 Write the proportion as follows.
$$\frac{C_1 n_1}{p_1} = \frac{C_2 n_2}{p_2}$$

Step 2 Record the information given.

C_1, C_2 expressed in dollars. $C_1 = 2$, $n_1 = 50$, $p_1 = 125$
$C_2 = 2.5$, $n_2 = ?$, $p_2 = 250$

Substitute the given information into the proportion. — Solve the equation for n_2.
$$\frac{2(50)}{125} = \frac{2.5 n_2}{250}$$
$$80 = n_2$$

Always make a final statement to answer the original problem.

Thus, the number of advertisements is 80.

Many other situations occur that involve joint variation.

- The interest paid on a sum of money varies jointly with the principal, the rate of interest, and the length of time.
- The resistance of a wire to an electric current varies directly as the length of the wire and inversely as the cross sectional area of the wire.

To solve problems involving joint variation, we apply skills that we have developed for direct and inverse variation.

5.8 exercise

Questions 1 to 9 develop skills for solving problems involving joint variation.

A

1. Write a variation statement involving a constant $k \neq 0$ for each of the following joint variations.
   (a) V varies directly with T and Q.
   (b) R varies directly with T and inversely with P.
   (c) U varies inversely with S and directly with the square of T.
   (d) E varies directly with M and the square of V.
   (e) Q varies directly with M and inversely with the square of P.

2. Write a proportion to express each of the following joint variations.
   (a) J varies directly with T and inversely with P.
   (b) P varies directly with Q and inversely with the square of R.
   (c) A varies directly with B and the square of C.
   (d) R varies inversely with the square of S and directly with G.
   (e) V varies directly with R and inversely with the square of T.

3. A varies directly with T and inversely with M^2. If $A = 75$ and $T = 15$, then $M = 2$.
   (a) Find the constant of variation.
   (b) Use the value found in (a) to find a value of T if $A = 100$ when $M = 5$.

4. R varies directly with L and inversely with D^2. If $R = 14$ and $D = 3$, then $L = 0.8$.
   (a) Write a proportion.
   (b) Use the proportion obtained in (a). Find R if $D = 6$ when $L = 0.4$.

5. A joint variation is given by $P \propto \frac{S}{R}$. Which of the following are equivalent variation statements for the above?
   (a) $\frac{P_1 R_1}{S_1} = \frac{P_2 R_2}{S_2}$
   (b) $\frac{P_2}{P_1} = \frac{S_1}{R_1} \times \frac{S_2}{R_2}$
   (c) $\frac{P_1}{P_2} = \frac{S_1 S_2}{R_1 R_2}$
   (d) $\frac{P_1}{P_2} = \frac{S_1}{R_1} \times \frac{R_2}{S_2}$

6. Q varies directly with R and V. If $Q = 192$ and $R = 3$ when $V = 16$, then find R when $Q = 96$ and $V = 4$.

7. E varies directly with F and inversely with the square of G. If $E = 75$ and $F = 60$ when $G = 12$ then find F when $E = 12$ and $G = 15$.

8 For the joint variation $Q \propto RV$, find the missing value if $Q_1 = 840$, $R_1 = 2$, $V_1 = 30$, $R_2 = 12$, $V_2 = 15$.

9 For $R \propto \dfrac{Q}{T^2}$ we know that $R = 20$ and $Q = 15$ when $T = 21$. Find R when $Q = 60$ and $T = 14$.

Solve the following problems involving joint variation.

B

10 The area of a triangle varies directly with its base and altitude. The area of a triangle with base 8 cm and altitude 12 cm is 48 cm². Find the area when the base is 11 cm and the altitude is 22 cm.

11 The resistance of a wire to an electric current is directly proportional to the length of the wire and inversely proportional to the square of the wire's diameter. If 3.2 km of wire with a diameter of 6 mm has a resistance of 14 units, then find the resistance for 8.4 km of wire with a diameter of 7 mm.

12 The energy of a moving billiard ball is directly proportional to the mass of the ball and the square of its speed. A 150-g billiard ball moving at 2 m/s has an energy of 300 units. Calculate the speed of a 140-g ball that has an energy of 630 units.

13 The distance a car travels from rest is directly proportional to the acceleration and the square of the time elapsed. If the car travels 400 m in 4 s with an acceleration of 50 m/s², find the acceleration of a car that travels 1080 m in 6 s.

14 The destructive force of a falling object varies directly with the mass of the object and the height from which the object is dropped. From a height of 75 m, a 20-kg rock has a destructive force of 15 000 units. Calculate the destructive force of a 40-kg rock dropped from a height 25 m less than in the first case.

5.9 partial variation

Each of these situations has similar characteristics.

- The total cost, C, of placing an advertisement in the newspaper partly depends on a fixed cost, and partly varies directly with the number of words in the ad.
- The expenses of a basketball tournament are partly constant and partly vary with the number of players that attend the tournament.

In each of the above situations, the total amount is based on two parts:

- a part that is fixed
- a part that varies.

The total cost of placing an advertisement in the newspaper depends on one part that is constant and on another part that varies. Each of the above situations illustrates an example of **partial variation**.

Example

The total cost of running an advertisement in the newspaper is partly constant and partly varies directly with the number of words in the advertisement. An advertisement with 20 words costs $7.00 and an advertisement with 30 words costs $9.50. How many words would be in an advertisement that costs $13.50?

Solution

Let the total cost, T, in cents, be represented by a constant part, C cents, and vary directly with the number of words, n.
Then $T = C + kn$ where k is a constant of variation.

Then $700 = C + k(20)$ ①
$950 = C + k(30)$ ②

Solve equations ① and ②.
Subtract ② − ①.
$250 = 10k$
$25 = k$

Use $k = 25$ in ①.
$$700 = C + (25)(20)$$
$$700 = C + 500$$
$$200 = C$$
Thus the partial variation is given by the equation
$$T = 200 + 25n$$
Use $T = 1350$. Then
$$1350 = 200 + 25n$$
$$1150 = 25n$$
$$46 = n$$
Thus the number of words in an advertisement that costs $13.50 is 46.

In the following exercise, we will work with many examples of partial variation.

5.9 exercise

A

1. A partial variation is given by $y = p + mx$. If $y = 22$ when $x = 3$ and $y = 64$ when $x = 17$, find m and p.

2. A partial variation is given by $C = K + pb$. If $C = 200$ when $b = 8$ and $C = 140$ when $b = 5$, find K and p.

For Questions 3 to 5 use the following partial variation.

The total expenses, E, in cents, of a basketball tournament are partly constant, C, in cents, and partly vary directly with the number of players, n.

3. Use the variation statement
$E = C + kn$, k is constant.
Find the constant of variation for the above tournament if the fixed cost is $60 and the total expenses are $240 when 120 players attend.

4. Find the total expenses of the above tournament if the number of players is 230.

5. Find the number of players that attended the above tournament if the total expenses were $511.50.

B

6. The cost of producing handbills is partly constant and partly varies directly with the number printed. To produce 50 copies, the total cost is $53.60 and the variable cost is $4.80. Find the cost of producing 1000 copies.

7. The toll charge over the Humber Skyway is partly constant and partly varies directly with the number of people in the vehicle. The Walker family paid $2.50 with 4 persons in the car, while the Driver family in the next car had 6 people and was charged $3.00. How many persons were in the next vehicle if the charge was $3.75?

8. The value of a car is partly constant and partly varies inversely with the age, in years, of the car. A 2-a-old car is worth $3800, but after 4 a, it is worth $2500. What would be the value of the car after 8 a?

9. The cost of auto repairs at Lethbridge Auto is partly constant and partly varies directly with the length of time of repair. Melissa's compact car was in for an oil change and other services which took 45 min. The bill was $19.25. Later an 8-h transmission overhaul cost $171.50. Calculate the hourly rate in the shop.

C

10. Cooking time for roast beef is partly constant and partly varies directly with the mass of the roast. If it takes Betty 3 h to cook a 3-kg roast, and 4 h to cook a 5-kg roast, how soon before dinner is to be served should Betty start cooking a 6-kg roast?

11. The expenses of a basketball tournament are partly constant and partly vary directly with the number of players that attend the tournament. If 12 players attend, each player will pay $15.00. For 24 players, the price will be $11.00 each. What will it cost each player if 30 players attend?

5.10 making decisions: solving problems involving variation

In the previous sections you have solved each of the following types of problems involving variation.

- direct variation
- direct squared variation
- joint variation
- inverse variation
- inverse squared variation
- partial variation

To solve a problem involving variation, you must recognize which type of variation is involved. Once the type of variation is noted then you need only apply the steps you have learned to solve the problem.

5.10 exercise

For each of the following problems,
- first identify the type of variation.
- second, solve the problem.

Read carefully and accurately!

1. The owner of a concession booth at a stadium has discovered that the amount of sales varies directly with the attendance at football games. If he sold $5540 worth of food and soft drinks at a game attended by 22 160 fans, how much can he expect to sell at the next game if the expected attendance is 30 000?

2. The number of chairs in a row varies inversely as the distance between them. If 45 seats are needed when they are 6 cm apart, what would be the distance between the seats if only 27 are needed?

3. The cost of operating a boat varies directly as the cube of the speed of the boat. If a trip of 300 km at 50 km/h costs $150, find the cost of the trip at 60 km/h.

4. When Todd sits 2 m away from the TV screen he notices that the intensity bothers his eyes. By what factor is the intensity decreased when he sits twice as far away if intensity varies inversely as the square of the distance from the source?

5. At a given speed the distance that a car travels varies directly as the amount of gasoline consumed. A driver notices that on a certain day she drove 216 km and used 54 L of gas. How much gas will she need for a night run of 296 km at the same speed on roads that have no gas stations open?

6. The spring factor of car shock absorbers varies inversely as the square root of the distance the car has travelled. After 2500 km of travel, the spring factor is 9 units. How far can the car go before getting new shock absorbers if a spring factor of less than 2 is considered unsafe?

7. To play a hockey tournament in another city the expenses are partly constant and partly vary directly with the number of players that go to the tournament. If 10 hockey players were to go, they would each have to pay $14. However, if 20 players were to go, then each player would need to pay $10.25. What would it cost each player if 40 players were to go?

8. The amount of rubber needed to make a racquet-sport ball varies directly as the square of the radius of the ball. A squash ball of radius 2 cm requires 20 cm² of rubber. How much rubber is required for a racquet ball which has a radius of 4 cm?

9. After a fixed length of time, the interest generated by a movie varies inversely as the square of time the movie continues. For an additional 6 min the interest level is 25 units. How much time beyond the fixed length does the movie continue to be interesting if 4 units indicates boredom setting in?

10. On a 289-cm swing Jason takes 3.4 s for one swing. He then goes to a swing that is 625 cm in length. If the time for one swing is proportional to the square root of the length, find the time for one swing on the longer swing.

11. A stereo turntable spins at a constant speed so that the number of revolutions varies directly with time. If the turntable makes 666 revolutions during the 20 min of one side of a record, how many revolutions will it take to play a song 6 min in length?

12. Donna discovers that when she shines her flashlight at an object the intensity of illumination varies inversely as the square of the distance from that object. When standing 4 m from an object she notices the intensity is 3 units. At what distance from the object will the intensity be doubled?

13. The sales of coffee on a given day are inversely proportional to the price per kilogram. A 1-kg jar costs $2.50 on Tuesday and 2000 jars were sold. If the price was increased by $1.50 on Friday, how many jars would you expect to be sold?

14. The amount of water a humidifier releases into the air varies directly as the square of the radius of the wheel in the humidifier. A wheel of radius 20 cm releases 5 L of water over a fixed period of time. How many litres will be released during the same period if the radius of the wheel is 25 cm?

problem/matics

Prove that if a varies directly with b when c is constant and if a varies directly with c when b is constant then a varies directly with bc when both b and c vary.

skills review: solving equations

To solve problems about ratio, we need to solve equations and proportions. Solve each of the following.

1. Find the missing terms for each of the following.
   (a) $5:x = 7:3$
   (b) $45:25 = 9:x$
   (c) $15:y = 3:2$
   (d) $2:3 = k:45$
   (e) $s:3 = 24:18$
   (f) $4:8 = y:10$
   (g) $3:x = 5:8$
   (h) $8:4 = 5:x$

2. Solve for a and b.
   (a) $2:3:a = 14:b:49$
   (b) $a:2:6 = 6:18:b$
   (c) $4:a:30 = b:3:5$
   (d) $20:a:b = 4:3:7$

3. Solve each of the following.
   (a) $\dfrac{x+2}{5} = \dfrac{4-x}{7}$
   (b) $\dfrac{x-1}{3} = \dfrac{5x-1}{5}$
   (c) $\dfrac{8-x}{5} = \dfrac{x+16}{10}$
   (d) $\dfrac{x+7}{x+17} = \dfrac{-3}{7}$
   (e) $\dfrac{3-x}{2} = \dfrac{x+12}{4}$
   (f) $\dfrac{x+3}{x-3} = \dfrac{7}{6}$

4. Solve for m and n.
   (a) $6:(m+n):12 = 2:5:m$
   (b) $4:2:1 = (m+3):(n+2):1$
   (c) $(m+n):3:8 = 2:6:m$

5. Solve for a and b.
   (a) $\dfrac{2}{a} = \dfrac{b}{12} = \dfrac{7}{21}$
   (b) $\dfrac{2a}{12} = \dfrac{3b}{1} = \dfrac{7}{3}$

6. Solve for a, b, and c, if
   $\dfrac{a}{6} = \dfrac{b}{2} = \dfrac{c+1}{3} = 3$.

7. Solve for m and n.
   $\dfrac{m+n}{3n+2} = \dfrac{2m}{2n+3} = \dfrac{4m-1}{5+4n}$

problems and practice: a chapter review

This section at the end of each chapter will provide you with additional questions to check your skills and understanding of the topics dealt with in the chapter. An important step for problem-solving is to decide which skills to use. For this reason, these questions are not placed in any special order. When you have finished the review, you might try the *Test for Practice* that follows.

1. Solve.
   (a) $3:(x - 3) = 1:3$ (b) $(x + 2):(x - 1) = 5:6$

2. Solve for the ratio $x:y$ in each of the following.
   (a) $x^2 - xy - 2y^2 = 0$
   (b) $10x^2 - 19xy - 15y^2 = 0$

3. Calculate x and y.
   (a) $-3:x:8 = y:-16:28$
   (b) $2:(2x + 1):8 = (y - 1):15:20$
   (c) $\dfrac{3x + 4}{2y - 9} = \dfrac{-3x - 2}{-2y + 4} = \dfrac{5x}{3y + 4}$

4. Find the number that must be subtracted from each term of the ratio $5:8$ to make the ratio equivalent to $2:15$.

5. The difference of the squares of two numbers is 19. If the ratio of the numbers is $10:9$, find the numbers.

6. If $x:y:z = -3:4:-12$, find the values of
   (a) $\dfrac{x + 2y + 3z}{3x + 2y + z}$ (b) $\dfrac{x^2 - y^2}{y^2 - z^2}$

7. If the measures of angles of a triangle are in the ratio $1:2:6$, find the angles.

8. Susan, Marilyn, and Nancy drove 1530 km on their vacation. If they shared the driving according to the ratio $3:2:5$, how far did each girl drive?

9. If $a:b = b:c$ prove that $(abc^2 - b^2):(b^2c^2) = (bc - 1):c^2$.

10. The speed at which a bird can fly is directly proportional to the square root of the surface area of its wings.
    (a) Express this relationship with a variation statement.
    (b) If a robin whose wing surface area is 49 cm² flies at a speed of 10 m/s, how fast does a crow fly whose wings have a surface area of 196 cm²?
    (c) Determine the surface area of the wings of a bluejay if it flies at 15 m/s.

11. The membership fee of a local golf club is partly constant and partly varies inversely with the number of members. For 500 members, the fee is $800 per member, but if there are 700 members, the fee would be only $700. What would be the membership fee if there were only 350 members?

12. At a given speed, the distance that a car travels varies directly as the amount of gasoline consumed. A driver notices that on a certain day he drove 200 km at a certain speed and used 40 L of gasoline. How much gasoline will he need for a night run of 300 km at the same speed on roads that have no gas stations open?

13. Boyle's Law states that at the same temperature the volume of a gas is inversely proportional to the pressure. If the volume of the gas is 30 L when the pressure is 8 units, find the pressure when the gas expands to 80 L.

test for practice

Try this test. Each *Test for Practice* will be based on the mathematics you have learned in the chapter. Try this test later in the year as a review. Keep a record of those questions that you were not successful with, find out how to answer them and review them periodically.

1. Solve for x.
   (a) $(4 - x):8 = 3:5$
   (b) $(x + 6):(x - 6) = 1:2$

2. If $x:y = 3:2$, find the value of $\dfrac{2x + 3y}{x - y}$.

3. Find the ratio $x:y$ for each of the following.
   (a) $3x = 2y$
   (b) $3y - 2x = 5y$
   (c) $\dfrac{2y}{3} = \dfrac{2x}{5}$
   (d) $\dfrac{3x - 2y}{5x + y} = \dfrac{4}{3}$
   (e) $(x + 7)(y + 4) = (x + 14)(y + 2)$

4. Two numbers have the property that
   • the ratio of one number to the other is $4:5$.
   • the sum of the numbers is 243.
   Find the numbers.

5. If 1296 is added to the square of one number, the result is the square of the second number. If the numbers are in the ratio $13:5$, find the numbers.

6. Terry makes a special batch of oatmeal cookies which contains oatmeal, raisins, and water in the ratio $18:6:3$. How much oatmeal is required to make up a 90-g batch of cookies?

7. A football team was purchased by Smith, Jones, and Wong who invested $50 000, $40 000, and $60 000 respectively. In the first season, the team made a profit of $30 000. How was this divided among the owners?

8. Calculate x and y.
   (a) $x:-4:14 = -2:\dfrac{1}{2}:y$
   (b) $18:(2y - 5):42 = 6:-5:(3x - 4)$
   (c) $\dfrac{4x + 9}{-2y - 5} = \dfrac{11x + 3}{-10y} = \dfrac{-2x + 3}{y - 10}$

9. Write an equation for each of the following variations.
   (a) The volume of a circular cylinder varies directly as the square of its radius.
   (b) The force of attraction between two magnets is inversely proportional to the square of the distance between them.
   (c) The distance a car rolls down a hill is directly proportional to the square of the time.

10. The speed of a sailboat varies directly as the square of the size of its sail. A boat with a sail of area 5 m² travels at 3 km/h.
    (a) Write a variation statement to express the relationship.
    (b) How big a sail is required to go at a speed of 12 km/h?
    (c) Suppose the sailboat in (b) hoisted a spinnaker of area 5 m² in addition to its sail. How fast would the boat go?

11. At a given speed the distance a car travels is directly proportional to the amount of gas consumed. At an average speed of 50 km/h a car travels 315 km and uses 30 L of gas. At the same speed, what is the maximum distance the car could travel on a 50-L capacity fuel tank?

12. The cost per student of taking a class on a geography field trip is partly constant and partly varies inversely as the number of students. If 16 students go each pays $5.00. If 24 students go, they each pay $4.00. What would it cost to take 32 students?

looking back: a cumulative review

(Chapters 1 to 4)

1. Solve and verify.
   (a) $\frac{1}{4}(2m + 2) + \frac{3}{5}(3m - 4) = \frac{5m}{2}$
   (b) $(a + 3)^2 - (a + 2)^2 = a + 3$
   (c) $(2m - 1)(3m + 2) + 3m(m + 1) = (3m - 1)^2$
   (d) $\frac{y^2}{y - 1} + 5 = \frac{1}{y - 1}$

2. Show that the following is true for all values of y.
   $3(y - 5) + 4(3 - 6y) - (y - 1) = 2(3y - 5) + 7(1 - 4y) + 1$

3. Simplify.
   $\frac{3m + 9}{m^2 - 4m - 21} - \frac{12 - 3m}{2m^2 - 7m - 4}$

4. George sailed his kite 320 m on his first flight. On his second flight the wind eased up and his speed dropped 4 m/s. If he flew the same distance and took 4 s longer, find his speed on his first flight.

5. Factor.
   (a) $4 - 9x^4$
   (b) $x^4 + 4y^4$
   (c) $3y^3 + 5y^2 + 3y + 5$
   (d) $3x^2 - 2xy - y^2$
   (e) $2mx^2 + 3mxy - 2nxy - 3ny^2$

6. Find k so that $x^4 - kx^2 + 36$ is divisible by $x - 2$.

7. Solve $\sqrt{2 - x} + \sqrt{11 + x} = 5$.

8. Which radical expression has the greater value?
   A: $\sqrt{8} - 2\sqrt{512} - 5\sqrt{32}$
   B: $7\sqrt{27} - 4\sqrt{3} - 5\sqrt{75}$

9. If $p = 3\sqrt{2} + 2\sqrt{3}$, $q = 3\sqrt{3} - 2\sqrt{2}$, simplify each of the following.
   (a) $p + q$
   (b) $p - 2q$
   (c) $3p + 2q$
   (d) pq
   (e) $2pq$
   (f) $-3pq$

10. (a) Find the perimeter of a rectangle if its area is $11\sqrt{5} - 22$ and its length is $2\sqrt{5} - 3$.
    (b) Find the perimeter of a square if the length is given by $\sqrt{3} - \sqrt{2}$ and the area is $5 - 2\sqrt{6}$.

11. Solve.
    (a) $2x = y - 2$, $y = -4x + 5$
    (b) $2x + 3y + 9 = 0$, $x - 2y - 6 = 0$
    (c) $\frac{3p}{2} - \frac{q}{8} = -1$, $4p + \frac{3q}{4} = -7$

12. The greatest hoard of gold, valued at $30 000, was recorded off the coast of Florida and dates back to a Spanish bullion fleet of 1715. This treasure contained 20 k (20 karat) gold coins and 12 k gold coins. How much of each coin would have to be melted and struck to produce a 24-g coin of 18 k gold?

13. James travelled 50 km from camp in search of a lost friend. He used skis on one part of the journey averaging 10 km/h and then used snowshoes for the rest of the trip averaging 4 km/h. If the journey took 8 h, how far did he travel on snowshoes?

14. Liz gives her animals 3 buckets of liquid feed solution daily. The normal feed concentration is 30%. How much water must be added to each bucket to reduce the concentration to 25% if each bucket contains 2 L of the stronger solution?

6 concepts and skills; analytic geometry

vocabulary, relations and functions, linear function, slope, writing equations, solving problems, systems of inequations, applications, linear programming

6.1 introduction

The concepts and skills we learn in mathematics are used over and over again to develop new skills, as well as to develop further concepts. By using mathematics we may often relate certain phenomena that otherwise seem unrelated. For example, to the mathematician

the relationship between the amount of candle burned and the period of time the candle burns	the relationship between the circumference of a circle and the corresponding diameter.
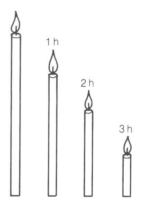	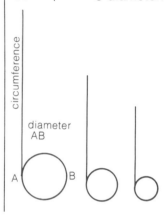

is similar

Through mathematics we learn some properties of various shapes and employ these properties in many ways.

6.2 vocabulary: relations and functions

Using mathematics, we may study patterns and relationships that are important in order to understand the many phenomena about us. We may use a table to show how numbers of one set are associated with another set.

length of spring (cm)	mass added (g)
12	4.5
12.8	4.8
13.6	5.1
14.4	5.4
15.2	5.7

We may show a relationship among numbers by using the following diagram.

The number 0 is mapped into 2. In symbols, we write $0 \rightarrow 2$. We refer to the following as a mapping diagram.

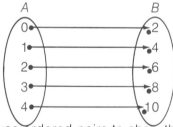

We may use ordered pairs to show the relation between the numbers in sets A and B. A relation may be a set of **ordered pairs**. Here we have ordered pairs (0, 2), (1, 4), etc. We may write the above relation as

$$S = \{(0, 2), (1, 4), (2, 6), (3, 8), (4, 10)\}$$

The following words are used in working with relations.

Domain: The set of all first elements of the ordered pairs of a relation is called the domain of the relation. The domain for the relation S above is given by,
domain = $\{0, 1, 2, 3, 4\}$

Range: The set of all second elements of the ordered pairs of the relation is called the range of the relation. The range for the relation S above is given by,
range = $\{2, 4, 6, 8, 10\}$

A relation may also be defined as a set involving an equation. For example,

$$P = \{\underbrace{(x, y)}_{\text{ordered pairs of relation}} | \underbrace{y = 2x + 2}_{\text{defining equation of the relation}}, \underbrace{-2 \leq x \leq 3, x \in R}_{\text{domain of the relation}}\}$$

We may draw a partial graph for the above relation P by constructing a table of values given by the equation,
$y = 2x + 2$, $-2 \leq x \leq 3$

We choose integral values for x.

Table of Values

x	y
-2	-2
-1	0
0	2
1	4
2	6
3	8

Draw the graph for the points in the table of values.

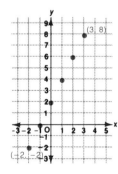

Draw the complete graph as shown.

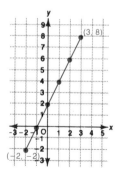

Since the domain of P is $\{x | -2 \leq x \leq 3, x \in R\}$, the partial graph of P is a line segment with end points $(-2, -2)$, $(3, 8)$. The graph of P is the straight line containing $(-2, -2)$ and $(3, 8)$. From the completed graphs, we see that the interval of values for y is given by $-2 \leq y \leq 8$. The graph of relation P is a straight line. Thus we often refer to a relationship such as P as a **linear** relation.

We may follow the same procedure when drawing the graph of **non-linear** relations.

Example 1

Draw the graphs defined by
A: $y = x^2 - 3$ B: $x = y^2 - 3$

Solution

Construct a table of values for each of A and B. Plot the corresponding points showing the ordered pair associated with each point. Draw a smooth continuous curve through the points.

A: Graph of the relation defined by $y = x^2 - 3$. Choose representative values for x.

x	y
-3	6
-2	1
-1	-2
0	-3
1	-2
2	1
3	6

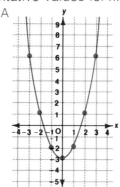

Note: The bold dots are plotted to locate the position of the graph.

The graph is called a **parabola**, and we have shown only that part of the graph which turns about. The point $(0, 3)$ is called the **turning point**. As a rule, we plot only that part of the parabola which contains the turning point. Is it possible to construct the complete graph?

B: Graph of the relation defined by $x = y^2 - 3$. Choose representative values for y.

x	y
6	-3
1	-2
-2	-1
-3	0
-2	1
1	2
6	3

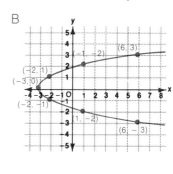

- In graph A, for every x co-ordinate there is *only one* corresponding y co-ordinate.
- In graph B, for some x co-ordinates there is more than one y co-ordinate.

A relation is a function such that for every x co-ordinate there is only one corresponding y co-ordinate. In other words, no two ordered pairs in the function have the same first element. Thus the relation shown in A is a function. The relation shown in B is *not* a function.

To determine whether a relation is a function or not we may draw a line parallel to the y-axis to intersect the graph.

- If the vertical line cuts the graph at most once, the relation is a function.
- If the vertical line cuts the graph more than once, the relation is not a function.

The above test is called the **vertical line test** and is used as follows.

A: Graph for the relation
$\{(x, y) | x^2 + y^2 = 9, x \in R\}$

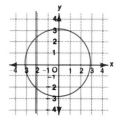

We see that by the vertical line test the above relation is not a function.

B: Graph for the relation
$\{(x, y) | y = (x - 1)^2 - 2, x \in R\}$
We see that by the vertical line test the above relation is a function.

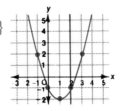

We may write the domain and range for the above.
A Domain $\{x | -3 \leq x \leq 3, x \in R\}$
   Range $\{y | -3 \leq y \leq 3, y \in R\}$
B Domain = R ← The domain is all the real numbers.
   Range = $\{y | y \geq -2, y \in R\}$

A function, or relation, may also be denoted by using an arrow or mapping notation. For example, an equivalent form of the function defined by $y = 2x - 2$, $x \in R$ is

$f : x \longrightarrow 2x - 2$, $x \in R$

We find it is useful to have different ways of denoting a function. For example, these two forms are also equivalent.

$f : x \longrightarrow 2x - 2$, $x \in R$ ← We read this as f maps x into $2x - 2$.
$f(x) = 2x - 2$, $x \in R$
We read this as "f at x".
$f(-1)$ means the value of $f(x)$ when $x = -1$.

$f(x) = 2x - 2$
$f(-1) = 2(-1) - 2 = -2 - 2 = -4$

Thus, for two sets A and B, a function or relation, shown by f, may be thought of as a rule or correspondence shown as follows.

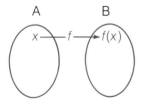

We say that f maps x onto its image $f(x)$ and that x is the pre-image of $f(x)$.

f associates each element x in A, (the domain), with an element $f(x)$ in B (the range).

6.2 exercise

A

1. For the relation shown by the graph, write the
   (a) domain   (b) range
   (c) Is the relation a function? Give reasons for your answer.

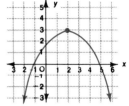

2. For the relation shown by the graph, write the
   (a) domain
   (b) range
   (c) Is the relation a function? Give reasons for your answer.

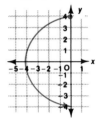

3. (a) With which axis is the domain associated?
   (b) With which axis is the range associated?

4. For the following:
   A use the vertical line test to determine which represent a function; which do not.
   B write the domain and range.

(a) 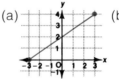   (b)   (c)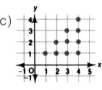

(d) 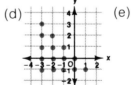   (e) 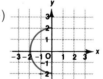   (f)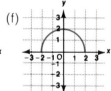

(g) 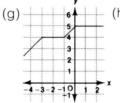   (h) 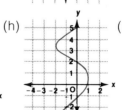   (i)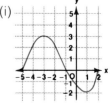

5   For each relation given by the mapping diagram,
    A write the domain and range.
    B which relations are functions? Why or why not?

(a) 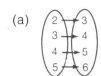 (b)   (c)

(d)   (e)   (f)

6   A relation is given by

x	−1	0	1	2	0	1	2	1	2
y	−2	−2	−2	−2	−1	−1	−1	0	0

(a) Write the domain; range.
(b) Draw a graph of the relation. Is the relation a function? Give reasons for your answer.

7   (a) Draw a mapping diagram for the relation given by the ordered pairs (0, 1), (1, 2), (2, 3), (3, 4).
    (b) Write the domain; range.
    (c) Is the relation a function? Why or why not?

8   For the relation given by ordered pairs, determine which are functions and which are not. Give reasons for your answer.

(a) (4, 0) (3, 0) (−2, 0) (4, 1) (2, 1)
(b) (1, 3) (6, 8) (5, 7)   (4, 6) (0, 2)
(c) (−1, 4) (0, 4) (1, 3)   (−1, 5) (2, 2)
(d) (8, 5) (5, 2) (3, 0)   (−1, −4) (−3, −6)

9   A relation is defined by
    $y = 3x - 1$, $x \in \{0, 1, 2, 3, 4\}$. the domain
    (a) Find the ordered pairs of the relation.
    (b) Write the range of the relation.
    (c) Is the relation a function? Why or why not?

10 A relation is defined by $y^2 = x$, $x \in R$.
    (a) Write the range of the relation.
    (b) Is the relation a function? Why or why not?

11 A function is given by $f : x \longrightarrow 2x + 5$, $x \in R$.
    Write three other ways of representing the function.

B

12 For each function, the domain is $\{-2, -1, 0, 3\}$.
    Write the function values for each of the following.

(a) $f(x) = 3x + 1$ (b) $H(u) = 1 - 2u$
(c) $f(m) = m^2$ (d) $g(k) = k^2 - 3$
(e) $k(t) = \frac{1}{t}$ (f) $F(x) = \frac{1}{3x - 1}$

13 (a) Draw the graph of the relation defined by $f : x \longrightarrow 3x^2 + 1$, $x \in R$.
    (b) Write the domain and range of the relation.
    (c) Is f a function? Why or why not?

14 (a) Draw the graph of the relation defined by $y = \frac{1}{x}$, $x \in R$.
    (b) Write the domain and range of the relation.
    (c) Is the relation a function? Why or why not?

15 (a) Draw the graph of the relation defined by $y^2 - 3 = x$, $x \in R$.
    (b) Write the domain and range of the relation.
    (c) Is the relation a function? Why or why not?

16 Draw the graph of each of the following relations. $x, y \in R$.
(a) $y = 3x + 2$ (b) $x^2 - 1 = y$ (c) $y = \frac{1}{x^2}$
(d) $f(x) = x(x - 1)$ (e) $h : x \longrightarrow x^2 + 3$
(f) $f : x \longrightarrow x(x - 3)$ (g) $g(x) = 2x^2 - 3$
(h) $y = 3x - 3$, $-2 \leq x \leq 5$ (i) $x^2 + y^2 = 25$
(j) $f(x) = 2x^2 + 1$, $0 \leq x \leq 2$
(k) $f : x \longrightarrow x(x + 1)$, $-1 \leq x \leq 3$

17 (a) Draw the graph of $y = 2x + 1$, $-3 \leq x \leq 2$, $x \in R$.
    (b) Write the range of the relation in (a).

18 A function is given by $G(p) = p^2 + 2p$. Find each of the following.
   (a) $G(0)$ (b) $G(1)$ (c) $G(2)$
   (d) $G(-2)$ (e) $G(3)$ (f) $G\left(\frac{1}{2}\right)$

19 For each function, find the values as shown.
   (a) $f:x \longrightarrow 5x - 2$, $f(2), f(-3)$
   (b) $g:x \longrightarrow 5 - 2x$, $g(-2), g(-3)$
   (c) $F:x \longrightarrow 2x^2 - 1$, $F(-2), F(2)$

20 Given that $f(s) = s^2 - 3s + 1$, calculate
   (a) $f(2)$ (b) $3f(2)$ (c) $-2f(3)$
   (d) $\frac{f(2)}{2}$ (e) $\frac{f(-2)}{-2}$ (f) $[f(-1)]^2$

21 If $f(x) = 4 - x^2$, $g(x) = 2x^2 - 1$, find
   (a) $f(3)$ (b) $g(3)$ (c) $f(3) + g(3)$
   (d) $f(-1)$ (e) $g(-1)$ (f) $f(-1) + g(-1)$
   (g) $3[f(2) - g(2)]$ (h) $\frac{1}{2}[g(-2) - f(2)]$

22 A function is given by $g:x \longrightarrow 3x - 1$. Find a value of k so that
   (a) $g(k) = 8$ (b) $g(k) = -4$

Two functions are given by
   $f(x) = 2x^2 - 1$ and $g(x) = 3x - 1$.
We may calculate $f[g(2)]$
   Find $g(2)$ first.   Then find $f[g(2)]$ since
   $g(2) = 3(2) - 1$ $f[g(2)] = f(5) \longleftarrow g(2) = 5$
   $= 6 - 1 = 5$ $= 2(5)^2 - 1 = 49$

C
23 If $f(x) = 2x^2 - 1$ and $g(x) = 3x - 1$, calculate
   (a) $g(2)$ (b) $f(2)$ (c) $g[f(2)]$ (d) $f[g(2)]$
   (e) $g(-3)$ (f) $f(-3)$ (g) $g[f(-3)]$ (h) $f[g(-3)]$

24 Use your result in Question 23 to prove that $f[g(x)] \neq g[f(x)]$ for all x.

25 Use the functions in Question 23 to find a value, k, such that $f[g(k)] = g[f(k)]$.

6.3 the linear function

There are many phenomena that may be associated with an equation of the form $y = kx$ where k is some constant.

- The distance travelled by a jet varies directly with the time in flight. If the jet travels at 800 km/h,
  $d = 800t$.   ($k = 800$)
- The circumference of a circle varies directly with its diameter.
  $C = \pi d$ ($k = \pi$)
- The cost of printing books is partly constant and partly varies with the number of books. Thus we might have
  $C = 5000 + 4.5n$.   ($k = 4.5$)

The graph of each of the above relations is a straight line or ray. Thus each relation is a linear function. The method of mathematics is to study the linear function in general. Using specific examples from travel, science, economics and so on, we find many cases where a relation may be defined by the equation $Ax + By + C = 0$ where A, B, C are constants. Each corresponding graph is linear.

For example, if
   $A = 3, B = -2, C = -6$
the equation is
   $3x - 2y - 6 = 0$.

Graph of $3x - 2y = 6$.

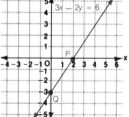

The concept of intercept is associated with the graphs of functions. For the linear function at right, the intercepts are given as follows.

x-intercept: the directed distance from the origin to the point where the graph crosses the *x*-axis. The directed distance, OP, is 2 units.

   The *x*-intercept of $3x - 2y = 6$ is 2.

y-intercept: the directed distance from the origin to the point where the graph crosses the *y*-axis. The directed distance, OQ, is -3 units.

The y-intercept of $3x - 2y = 6$ is -3.

We may also use our skills in algebra to calculate the intercepts.

Example 1
Find the x- and y-intercepts of the graph of the linear function given by $2x - 6 = y$, $x, y \in R$.

Solution

To find the x-intercept, use $y = 0$.	To find the y-intercept, use $x = 0$.
$2x - 6 = y$	$2x - 6 = y$
$2x - 6 = 0$	$2(0) - 6 = y$
$2x = 6$	$-6 = y$
$x = 3$	The y-intercept is -6.
The x-intercept is 3.	

To draw the graph of a line, we need to plot 2 points. A third point is used to check our work. Thus, for the above equation the x-intercept is 3. A(3, 0) is a point on the graph.
The y-intercept is -6.
B(0, -6) is a point on the graph. Draw the graph of $2x - 6 = y$.

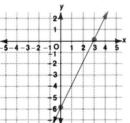

Example 2
Find the value of k if the linear functions have equal y-intercepts.
A: $3x - y = -4$ B: $2y = x - k$

Solution
To find the y-intercept, use $x = 0$ in A.
$3x - y = -4$
$3(0) - y = -4$
$y = 4$
Thus, the y-intercept is 4.

From B the y-intercept is given by
$2y = 0 - k = -k$
$y = -\dfrac{k}{2}$

For equal y-intercepts then $-\dfrac{k}{2} = 4$
$k = -8$

6.3 exercise

1 Which of the following are linear functions?
(a) $2x - y = 6$
(b) $x = \dfrac{3}{y}$
(c) $y - 2x = 4$
(d) $f(x) = x - 5$
(e) $g : x \longrightarrow \dfrac{2x - 1}{3}$
(f) $y = x^2$
(g) $f(x) = 2x^2 - 1$
(h) $y = \sqrt{x}$, $x \geq 0$

2 Draw the graph of each of the following. $x \in R$.
(a) $y = 2x - 1$, $-1 \leq x \leq 3$
(b) $2x - y = 3$, $-3 \leq x \leq 0$
(c) $y = -8x + 5$, $-2 \leq x \leq 2$
(d) $3x - y = 2$, $-3 \leq x \leq -1$

What is the range for each of the above?

3 Write the equation of a line parallel to the y-axis that passes through
(a) $(-3, 1)$
(b) $(5, -3)$
(c) $(2, 0)$

4 Write the equation of a line parallel to the x-axis that passes through
(a) $(-2, 1)$
(b) $(3, -4)$
(c) $(0, -3)$

5 A linear relation is given by the equation
$y = mx + 5$, $m \in R$. —m is a parameter.
Using the same axes, draw the graph of the above relation for
(a) $m = 1$
(b) $m = 3$
(c) $m = -1$

What do you notice about your results?

6 A linear relation is given by
$y = 3x + b$, $b \in R$, b is a parameter
Using the same axes, draw the graph of the above relation for
(a) $b = -3$
(b) $b = 2$
(c) $b = -1$

What do you notice about your results?

B

7 Write each of the following defining equations in the form $Ax + By + C = 0$.

(a) $y - 3 = 2(x - 1)$ (b) $\dfrac{x}{3} - y = 5$

(c) $y = \dfrac{2x - 1}{3}$ (d) $\dfrac{x - 3}{2} - y = 6$

(e) $\dfrac{x}{2} - \dfrac{y}{3} = 6$ (f) $2(x - 1) - 3(y + 1) = 1$

(g) $\dfrac{y - 1}{3} - \dfrac{x + 1}{2} = 6$ (h) $\dfrac{x + 1}{2} - \dfrac{3y - 1}{4} = 3$

8 Find the x- and y-intercepts for each of the following.

(a) $3x - y = 2$ (b) $2x - 5y = 10$

(c) $3x = \dfrac{y - 1}{2}$ (d) $\dfrac{y - 2x}{3} = 4$

(e) $f(x) = \dfrac{2x - 1}{3}$ (f) $x = 5$

(g) $y = -6$ (h) $2y = -\dfrac{3x - 1}{2}$

(i) $f : x \longrightarrow 3x - 1$ (j) $g : x \longrightarrow \dfrac{2x - 1}{3}$

9 Find the range for each of the following.

(a) $3x + y = -6$, $0 \leq x \leq 3$
(b) $x - 2y = 8$, $-2 \leq x \leq 2$
(c) $y = \dfrac{-x + 1}{3}$, $-3 \leq x \leq 0$
(d) $3x = \dfrac{2y - 1}{2}$, $-5 \leq x \leq -2$

10 Find the 2 intercepts of the graph for each linear function and draw the graph.

(a) $3x + y - 9 = 0$ (b) $y = \dfrac{2}{3}x - 4$

(c) $3x + 5y - 15 = 0$ (d) $x = 3(y + 3)$

(e) $2x - 4 = x + y$ (f) $y = \dfrac{3x - 1}{2}$

(g) $f(x) = 2x - 3$ (h) $g : x \longrightarrow \dfrac{2x - 3}{3}$

11 Find the value of k if the given point lies on the graph defined by each equation.

(a) $3x + ky = 2$ $(1, -1)$
(b) $y + kx = 8$ $(-3, 2)$
(c) $2x - 3y = k$ $(-1, 5)$
(d) $2x - 4 = ky$ $(2, 0)$
(e) $ky = \dfrac{x - 3}{2}$ $(5, 1)$

12 Find the value of k if the lines have equal y-intercepts.
$3x + y - 4 = 0$, $x + ky + 2 = 0$

13 Find the value of k if the lines have equal x-intercepts.
$3x - y = 6$, $kx - 2y = 8$

14 A linear function is given by $x + ky = 6$. Find the value of k if the x-intercept is 5 more than the y-intercept.

15 For the linear function given by $2x - ky = 3$, find the value of k if the x-intercept exceeds the y-intercept by 3.

C

16 For the linear function defined by
$Ax + By + C = 0$, $A, B, C \in R$, $A, B \neq 0$.

(a) prove that the x-intercept is given by $-\dfrac{C}{A}$.

(b) prove that the y-intercept is given by $-\dfrac{C}{B}$.

17 Use the results of Question 16 to write the x- and y-intercepts for each of the following.

(a) $3x + 6y - 9 = 0$ (b) $2x - y + 4 = 0$
(c) $3x - y - 6 = 0$ (d) $2x + y = 8$
(e) $3x - y = -6$ (f) $y - 3x + 5 = 0$

18 For the linear function given by $px + qy = 6$, prove that $p = \dfrac{3q}{3 + q}$ if the x-intercept exceeds the y-intercept by 2.

6.4 working with slope

We use our work with ratios to compare the steepness of two ramps.

Ramp A 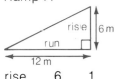   Ramp B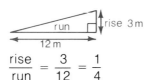

$$\frac{\text{rise}}{\text{run}} = \frac{6}{12} = \frac{1}{2} \qquad \frac{\text{rise}}{\text{run}} = \frac{3}{12} = \frac{1}{4}$$

We define $\text{Slope} = \dfrac{\text{rise}}{\text{run}}$

From the calculations, we may say that the *steepness* or *slope* of ramp A is greater than that of ramp B.

To calculate the slope of a line on the co-ordinate plane, choose any two points to determine the rise and run.

$$\text{Slope} = \frac{\text{rise}}{\text{run}} = \frac{\text{vertical change}}{\text{horizontal change}}$$
$$= \frac{\Delta y}{\Delta x} \leftarrow \text{change in } y$$
$$\phantom{= \frac{\Delta y}{\Delta x}} \leftarrow \text{corresponding change in } x$$

Δy is read as delta y.

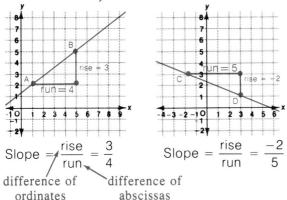

$$\text{Slope} = \frac{\text{rise}}{\text{run}} = \frac{3}{4} \qquad \text{Slope} = \frac{\text{rise}}{\text{run}} = \frac{-2}{5}$$

difference of ordinates   difference of abscissas

In general, if $P(x_1, y_1)$ and $Q(x_2, y_2)$ are points on a line then

$$m_{PQ} = \frac{y_2 - y_1}{x_2 - x_1} \quad \text{Slope of the line segment PQ.}$$

Example 1

Calculate the slope of the line given by the equation $2x + y = 6$.

Solution

Points on the line are shown.

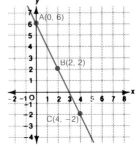

$$\text{Slope}_{AC} = \frac{\Delta y}{\Delta x}$$
$$= \frac{-2 - 6}{4 - 0}$$
$$= \frac{-8}{4} = -2$$

The slope of $2x + y = 6$ is -2.

From the diagram, note that

$$\text{Slope}_{AB} = \frac{\Delta y}{\Delta x} \qquad \text{Slope}_{BC} = \frac{\Delta y}{\Delta x}$$
$$= \frac{2 - 6}{2 - 0} \qquad \qquad = \frac{-2 - 2}{4 - 2}$$
$$= \frac{-4}{2} \qquad \qquad = \frac{-4}{2}$$
$$= -2 \qquad \qquad = -2$$

From the result

$$\text{Slope}_{AB} = \text{Slope}_{BC} = \text{Slope}_{AC}$$

we see two important results.

Result 1

- The calculation of the slope of a line is independent of the choice of the points used to calculate the slope. *The slope of a line is constant* and is called the **constant slope property** of a line. We use the symbol m, to represent the slope. We may write
$$m_{AB} = m_{BC} = m_{AC}$$

Result 2

- Since $m_{AB} = m_{BC}$ then the points A, B, C are on the same line. Three points are said to be **collinear** if they lie on the same straight line. We may use the fact $m_{AB} = m_{BC}$ to show points are collinear.

Example 2

Show that the points P(3, 2), Q(−3, −2), R(6, 4) are collinear.

Solution

$$m_{PQ} = \frac{\Delta y}{\Delta x}$$
$$= \frac{-2 - 2}{-3 - 3}$$
$$= \frac{-4}{-6} = \frac{2}{3}$$

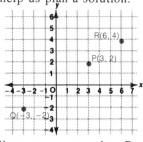

A diagram is drawn to help us plan a solution.

$$m_{PR} = \frac{\Delta y}{\Delta x}$$
$$= \frac{4 - 2}{6 - 3} = \frac{2}{3}$$

Since $m_{PQ} = m_{PR}$ then P, Q, R, are collinear points.

We may apply the skills we have developed for solving equations to solve problems about slope.

Example 3

A line with slope 1 passes through the points M(1, k + 2) and N(−8, 2). Find the value of k.

Solution

$$\text{Slope}_{MN} = \frac{\Delta y}{\Delta x} = \frac{2 - (k + 2)}{-8 - 1}$$

Since $\text{Slope}_{MN} = 1$, write the equation

$$\frac{2 - (k + 2)}{-8 - 1} = 1$$
$$\frac{2 - k - 2}{-9} = 1$$
$$-k = -9$$
$$k = 9$$

Check: Co-ordinates are M(1, 11), N(−8, 2).
$$\text{Slope}_{MN} = \frac{2 - 11}{-8 - 1} = \frac{-9}{-9} = 1 \quad \text{checks } \checkmark$$

Skills with slope are needed to solve many problems in analytic geometry, as we shall see in subsequent work.

6.4 exercise

A

1 Calculate the slope of each line segment.

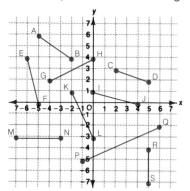

2 For each of the following, calculate the slope. Then sketch the line segment containing the two points.

(a) (−9, 4), (−3, 1)   (b) (−8, −6), (−6, −2)
(c) (−2, −5), (4, −5)   (d) (0, 6), (8, 2)
(e) (7, −5), (10, 1)   (f) (−6, −5), (3, −1)
(g) (−4, 7), (0, 2)   (h) (6, 1), (6, 6)
(i) (−2, 3), (4, 3)   (j) (−6, 1), (−6, 5)

3 Based on the results in Questions 1 and 2, what conclusion can you make about the graph of line segments that have

(a) positive slopes?   (b) negative slopes?
(c) zero slopes?   (d) undefined slopes?

4 A line passes through the indicated points. Calculate the slope of each line.

(a) (3, −4), (9, 2)   (b) (14, −3), (8, −9)
(c) (−5, 1), (5, 3)   (d) (2, −1), (−8, −3)

5 From the points
A(−6, 3)   B(−2, 2)   C(−2, −3)
D(3, 4)   E(10, 1)   F(−4, 1)
calculate the slope of each line segment.

(a) AC   (b) EF   (c) BD   (d) DE   (e) BA   (f) CB

6. Calculate the slope of each side of each figure.

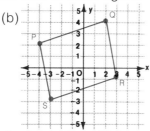

7. Use your work with slope to determine which sets of points are collinear.

   (a) (3, −1)   (5, 4)   (1, −6)
   (b) (−3, −1)   (3, −2)   (−7, 1)
   (c) (−1, 2)   (1, 5)   (−3, 3)
   (d) (3, 0)   (−1, 4)   (6, −3)

8. For each line given by the equation, calculate its slope.

   (a) $x + y = 8$
   (b) $x = y + 6$
   (c) $2x + y = 3$
   (d) $2y = x - 6$
   (e) $3x + 2y - 6 = 0$
   (f) $y - 2x = 3$
   (g) $3y - 2x - 6 = 0$
   (h) $\frac{x}{3} - y = 4$
   (i) $3 - 2x = y$
   (j) $y = 6$
   (k) $x = 3$
   (l) $y = 0$

9. To draw a line through P(−3, 1), with slope $\frac{2}{5}$, we plot the point P(−3, 1). Since the slope of the line is $\frac{2}{5}$ we measure 5 units to the right and 2 units up to locate a point Q. Draw a line through P and Q. In each of the following, sketch a line that passes through the given point and has the given slope.

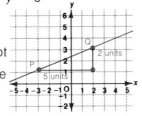

   (a) (2, 3), slope 4
   (b) (−2, 4), slope −3
   (c) (1, −6), slope 0
   (d) (−6, 2), slope $\frac{1}{2}$
   (e) (−3, −2), slope $-\frac{2}{3}$
   (f) (3, −4), slope $\frac{-6}{5}$

10. Calculate the slope of each line segment.

    (a) (2.5, 1.3), (4.2, 3.1)
    (b) $(\sqrt{3}, -\sqrt{2}), (2\sqrt{3}, 3\sqrt{2})$
    (c) (2a, 3b), (−4a, 6b)
    (d) $(x_1, y_1), (x_2, y_2)$
    (e) (ax, −by), (−ay, bx)
    (f) $(x^2, -y), (y^2, x)$

B

11. The vertices of △DEF are
    D(8, 3), E(−1, −5), F(−1, 3).
    Calculate the slope of each side of △DEF.

12. The vertices of a figure are
    P(−2, 0), Q(6, 5), R(1, −2), S(−7, −7).

    (a) Calculate the slope of each side.
    (b) What do you notice about the slopes of the sides?
    (c) What type of polygon is figure PQRS?

13. The figure PQRS is a parallelogram with vertices P(8, 1), Q(5, −3), R(−5, −5), S(−2, −1).

    (a) Calculate the slope of each side.
    (b) What do you notice about your answers in (a)?
    (c) Which of the following figures gives a parallelogram?
       I M(4, −1) P(4, −8) R(−8, 3) S(0, 3)
       II S(−3, −5) T(−3, 0) Q(9, 3)   R(9, −2)

14. A line has slope $\frac{2}{3}$ and passes through the point (0, 2). Which of the following are points on the line?

    (a) (−1, 3)   (b) (−3, 0)   (c) (3, −4)

15. A line has slope $-\frac{3}{2}$ and passes through the point (−2, 1). Which of the following are points on the line?

    (a) (4, −8)   (b) (10, −16)   (c) (6, −11)

16. Two points A(−2, 3b) and B(b, 4) determine a line. If the slope of the line is $\frac{1}{3}$, what are the co-ordinates of A and B?

17. Find the missing co-ordinates for each point if
    (a) M(k, 7), N(2, 0) lie on a line with slope 7.
    (b) P(0, 3), Q(2, k) lie on a line with slope $\frac{1}{2}$.
    (c) R(6, 5), S(−3, k) lie on a line with slope $\frac{2}{3}$.
    (d) A(−2, 3k), B(k, 4) lie on a line with slope $\frac{1}{3}$.

18. Two vertices of a triangle are C(1, 3) and B(5, 15). If D(d, 6) is a point on BC, what is the value of d?

19. A line with slope −2 passes through the points M(k, 7) and N(2k, 3). What is the value of k?

20. The following set of points are collinear. What is the value of k?
    (a) (−2, k), (2, −2), (4, −1)
    (b) (0, −3), (k + 1, −1), (3, 3)
    (c) (2, 6), (0, k + 1), (−5, −9)

21. In △PQR, PQ has slope 2. Calculate k if the co-ordinates are $P(\frac{1}{2}k + 3, -2)$ and Q(k + 3, 3).

C
22. The following points are collinear.
    A(−1, 1)   B(1, p + 4)   C(−1 − p, −4)
    Calculate the value(s) of p.

23. The slope of the line through AB is 1. Calculate the value of k for the given co-ordinates.
    $A(k + 1, \frac{1}{3}k - 1)$ $B(0, \frac{1}{2}k - 1)$

24. Square PQRS is given by P(7, 0), Q(5, −8), R(−3, −6), S(−1, 2). A point M(2, −3) is inside the square. Show that M lies on PR and QS.

applications with slope

A good intermediate skier should be able to ski on a hill with a slope of 0.90 or less.

25. Jennifer skied Renegade Run which has a vertical drop of 300 m.
    (a) If the AB in the diagram is 440 m long, calculate the slope of the hill, to 2 decimal places.
    (b) How would you rate Jennifer's skiing ability?

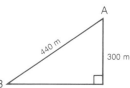

26. Peter skied Pine Trail which has a vertical drop of 250 m and he fell 8 times.
    (a) If the length of the trail, AB, is 420 m long, calculate the slope of the hill.
    (b) How would you rate Peter's skiing ability?

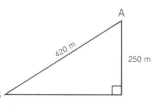

math tip

A linear function may be defined in different forms as follows:

- by an equation $y = 2x + 1$
- by ordered pairs $(x, 2x + 1)$
- by a mapping notation $f : x \longrightarrow 2x + 1$
- by function notation $f(x) = 2x + 1$

Be sure to recognize the linear function that may be written in these different ways.

6.5 slope y-intercept form of the linear equation

If we draw the graphs of different lines we should notice the relationship between the equation, the slope, and the y-intercept, of the line.

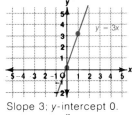
Slope 3; y-intercept 0.

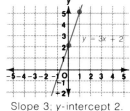
Slope 3; y-intercept 2.

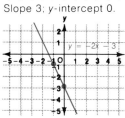
Slope −2; y-intercept −3.

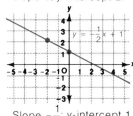
Slope $-\frac{1}{2}$; y-intercept 1.

Equation	Slope	y-intercept
$y = 3x$	3	0
$y = 3x + 2$	3	2
$y = -2x - 3$	−2	−3
$y = -\frac{1}{2}x + 1$	$-\frac{1}{2}$	1

From the above examples and others that we may graph, we write the **slope y-intercept** form of a linear equation as

$$y = mx + b$$

where m is the slope and b is the y-intercept.

Thus, from this special form of the equation, we can obtain the slope and y-intercept of a line directly from its equation when written in the above form. The ordinate of the point of intersection of the line with the y-axis is the y-intercept, given as b.

Example 1

Determine the slope and y-intercept of the line given by $3x - 4y = 12$.

Solution

Write the given equation in the *slope y-intercept* form.
$$3x - 4y = 12$$
$$-4y = 12 - 3x$$
$$y = -3 + \frac{3}{4}x$$

Divide both sides of the equation by −4.

or
$$y = \frac{3}{4}x - 3$$

Compare this equation with $y = mx + b$.

Thus, the slope is $\frac{3}{4}$, y-intercept is −3.

We may use the *slope y-intercept* form of the equation to help us draw the graph of the line.

We apply our algebraic skills to solve problems about slope as shown in the next example.

Example 2

The lines given by the equations
$y + 2 = 2(x - 3) + kx$ and $3(x + 2) = 3 + y$
have equal slopes. Find the value of k.

Solution

Write each equation in the *slope y-intercept* form.

$y + 2 = 2(x - 3) + kx$	$3(x + 2) = 3 + y$
$y + 2 = 2x - 6 + kx$	$3x + 6 = 3 + y$
$y = 2x + kx - 8$	$3x + 3 = y$
$ = (2 + k)x - 8$	$y = 3x + 3$

Compare the slopes:

For equal slopes $2 + k = 3$
$k = 1$

In the following exercise remember that a linear equation written in the form
$$Ax + By + C = 0$$
is said to be in **standard form**.

6.5 exercise

Questions 1 to 10 develop skills with equations in the *slope y-intercept* form.

A

1. What are the slope and the y-intercept for each of the following?

   (a) $y = 2x - 4$ (b) $y = \frac{3}{2}x + 4$ (c) $y = 2x + 1$

   (d) $y = -3x + \frac{5}{2}$ (e) $y = 3x$ (f) $y = 2x - \frac{1}{2}$

   (g) $y = \frac{2}{3}x$ (h) $y = \frac{1}{2}x - \frac{3}{2}$ (i) $y = \frac{5}{2}x - 3$

2. For each of the following, write the equation of the line for the given slope, *m*, and y-intercept, *b*.

   (a) $m = 3, b = 5$ (b) $m = -2, b = 3$

   (c) $m = \frac{1}{2}, b = 6$ (d) $m = -\frac{3}{2}, b = 0$

   (e) $m = \frac{5}{2}, b = -\frac{2}{3}$ (f) $m = -\frac{2}{3}, b = 0$

3. For each line, write the value of the slope, y-intercept and the equation of the line.

   (a) 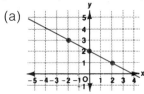 (b)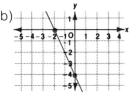

   (c) 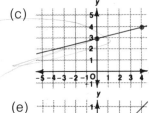 (d)

   (e) 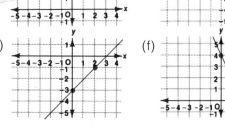 (f)

4. Draw each line on squared paper using the given slope and y-intercept.

   (a) $m = -2, b = 1$ (b) $m = 3, b = -\frac{1}{2}$

   (c) $m = -\frac{1}{4}, b = 3$ (d) $m = \frac{2}{3}, b = 2$

5. Draw the graph given by each equation. Use the *slope y-intercept* form of the equation to help you.

   (a) $x + y = 6$ (b) $x - y = 3$ (c) $x - 2y = 8$
   (d) $x = 2y - 6$ (e) $3x - 2y = 12$ (f) $2x + 3y = 6$
   (g) $y - 2x = 8$ (h) $3x - 5y = 0$ (i) $3x + 4y = 9$

6. Copy and complete the chart.

Slope y-intercept form	Standard form	Slope	y-intercept
$y = \frac{3}{2}x - 5$	?	?	?
?	$3x - 2y - 8 = 0$	?	?
?	?	$\frac{2}{3}$	-4

7. Write the slope and y-intercept of each line.

   (a) $4x - y = 3$ (b) $2x + y - 9 = 0$
   (c) $3x = 2y - 6$ (d) $2(x - 1) = y - 2$
   (e) $\frac{x - y}{2} = 6$ (f) $\frac{1}{2}x - y = 9$
   (g) $\frac{3x}{2} - 4y = 9$ (h) $y - 9 = 0$
   (i) $f : x \longrightarrow x - 3$ (j) $g : x \longrightarrow \frac{2x - 1}{2}$

8. (a) Write each line in the *slope y-intercept* form.
      $y = 2(x - 2)$, $4x - 2y = -5$, $2x - y = 6$
   (b) What characteristic, slope or y-intercept, do the lines in (a) have in common?

9. (a) Draw the graph of each of the following.
      $3x - y = -5$, $x + y = 5$, $2x - y = -5$
   (b) What characteristic do the lines in (a) have in common?
   (c) Write the lines in (a) in the *slope y-intercept* form. What do you notice?

10 Draw the graph of each of the following lines.
   (a) $y = 2x - 1$, $y = 2x + 5$, $y = 2x - 4$
   (b) $y = -3x + 2$, $y = -3x - 1$, $y = -3x + 5$
   (c) Use your results in (a) and (b). What characteristic is common to the lines in (a) and (b)?
   (d) From the above results, suggest a relationship among the slopes of parallel lines.

B

11 Draw the graph of the line given by each of the following.
   (a) slope 3, y-intercept −2
   (b) y-intercept 3, slope $-\frac{3}{2}$
   (c) x-intercept 2, slope 4
   (d) x-intercept 3, y-intercept −2

12 Write the co-ordinates of the points where each line crosses the x- and y-axis.
   (a) $3x - 4y = 12$
   (b) $3x - y = \frac{1}{2}$
   (c) $\frac{1}{2}x - \frac{1}{3}y = 8$
   (d) $2(x - 1) = 3(y - 2)$
   (e) $f : x \longrightarrow 3x + 6$
   (f) $h : x \longrightarrow \frac{3x - 1}{3}$

13 Write the x- and y-intercept of each of the following lines.
   (a) $2x + y = 6$
   (b) $3x - y - 2 = 0$
   (c) $\frac{1}{2}x - y = 8$
   (d) $2(2x - y) = 3$
   (e) $3(x + 1) = y + 2$
   (f) $\frac{x}{2} - \frac{y}{3} = 6$

14 Parallel lines have equal slopes. Which of the following are parallel lines?
   A: $2x - y = 3$
   B: $2y - x = 3$
   C: $3y = 2(9 - x)$
   D: $2x - 3y - 15 = 0$
   E: $2y + x = 12$
   F: $x - 2y = 5$
   G: $y - 2x - 5 = 0$
   H: $y + 2x = 5$

15 For the lines, ℓ_1, ℓ_2, the slopes are given.
   $\text{slope}_{\ell_1} = \frac{k + 5}{6}$ $\text{slope}_{\ell_2} = \frac{k + 6}{7}$
   If the lines are parallel, find the value of k.

16 The slopes of pairs of parallel lines are given. Calculate the value of k.
   (a) $\frac{k + 6}{2}$, $\frac{k - 6}{4}$
   (b) $\frac{k - 7}{5}$, $\frac{k + 5}{3}$
   (c) $2(2k + 3)$, $3(3k + 2)$
   (d) $3(k + 3)$, $2(2 - k)$

17 The lines given by $x - 2y = 2$ and $4 + 2y = kx$ are parallel. Find the value of k.

18 The lines $2x - y = 1$ and $4x - 2y + k = 0$ have equal y-intercepts. Find the value of k.

19 A line has slope $\frac{1}{2}$ and y-intercept 3. If $(k, 2)$ is a point on the line, what is the value of k?

20 A line with slope $\frac{2}{3}$ passes through the point $(0, -2)$. If $(3, k)$ is a point on the line, what is the value of k?

21 Two lines have slopes $2k^2$ and $5 - 9k$ respectively. Find the value of k if the lines are parallel.

C

22 Two aircraft are designated by the flight paths
   F_1: $2x - y = 1$ F_2: $2y - k = 4x$.
   If their paths cross on the y-axis, find the value of k.

23 The paths of 2 hovercraft are given by $3x - y = 8$ and $y + 5 = 4x - k$. If the paths intersect on the y-axis, find the value of k.

24 The sides of a quadrilateral are given by
   $3x - y + 1 = 0$, $x + 3y - 13 = 0$,
   $3x - y - 9 = 0$, $x + 3y - 3 = 0$
   Use the *slope y-intercept* form of an equation to determine whether the quadrilateral is a parallelogram. Give reasons for your answer.

25 The lines defined by $x - 2y + 4 = 0$, $3x - y + 7 = 0$, $x - 2y - 6 = 0$ and $2x + y - 7 = 0$ intersect to form a quadrilateral. Establish whether the quadrilateral is a parallelogram. Give reasons for your answer.

26 (a) Write the equation $Ax + By + C = 0$ in the *slope y-intercept* form.
(b) From your answer in (a), what is an expression for the
 • slope of the line? • y-intercept of the line?

formulas for $Ax + By + C = 0$

The equation $Ax + By + C = 0$, written in the *slope y-intercept* form as follows gives a formula to calculate the slope, m, and y-intercept, b, of any line. $B \neq 0$.

$$y = -\frac{A}{B}x - \frac{C}{B} \qquad m = -\frac{A}{B} \qquad b = -\frac{C}{B}$$

27 Use the above formulas. Calculate the slope and y-intercept for each line.
(a) $2x + y - 3 = 0$ (b) $3x - y + 5 = 0$
(c) $4x - 3y = 0$ (d) $3x - 2y - 6 = 0$
(e) $3x - 2 = y$ (f) $3(x + 1) = 2(y - 3)$
(g) $\frac{x + y}{3} = 2$ (h) $\frac{x}{2} - \frac{3y}{4} = 1$

28 Find the value of k, if the line given by $3y + kx - 2 = 0$ has slope -1.

29 A line with slope $\frac{3}{2}$ is given by the equation $3x - 2ky + 2 = 0$. Find the value of k.

problem/matics

Cy, famous for his corny riddles, replied, when asked his age and the age of his son: "If you subtract 2 from my age and divide this by 5, you will have my son's age. But in 3 a, he will be $\frac{1}{4}$ of my age." How old is Cy?

6.6 parallel and perpendicular lines

We can use the *slope y-intercept* form to decide whether or not lines are parallel.

Standard Form   Slope y-intercept Form
$Ax + By + C = 0$ $y = mx + b$ Since the slope
$2x - y - 5 = 0 \longrightarrow y = 2x - 5$ of each line is
$4x - 2y + 7 = 0 \longrightarrow y = 2x + \frac{7}{2}$ 2, the lines are parallel.

Perpendicular lines are drawn on the plane. The equations are written in the *slope y-intercept* form.

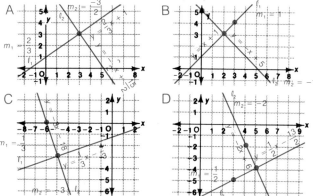

From the pairs of perpendicular lines, we notice a relationship between their slopes.

A	B	C	D
$m_1 = \frac{2}{3}$	$m_1 = 1$	$m_1 = \frac{1}{3}$	$m_1 = \frac{1}{2}$
$m_2 = -\frac{3}{2}$	$m_2 = -1$	$m_2 = -3$	$m_2 = -2$
$(m_1)(m_2) = -1$	$(m_1)(m_2) = -1$	$(m_1)(m_2) = -1$	$(m_1)(m_2) = -1$

We may also say that the slopes of perpendicular lines are negative reciprocals. $m_1 = -\frac{1}{m_2}$

The following two results are very important in studying analytic geometry. Two lines ℓ_1 and ℓ_2 have slopes m_1 and m_2 respectively.

| Parallel lines | $\ell_1 \| \ell_2$ if and only if $m_1 = m_2$. |

| Perpendicular lines | $\ell_1 \perp \ell_2$ if and only if $m_1 = -\frac{1}{m_2}$ or $m_1 m_2 = -1$. |

Example 1
Which pairs of lines are parallel?
Which pairs of lines are perpendicular?
ℓ_1 $x - 2y = 6$ ℓ_2 $2(x - 3) = y$
ℓ_3 $4x - 2y = -11$ ℓ_4 $2x + y = 3$

Solution
To compare the slopes, write each line in the *slope y-intercept* form.

ℓ_1 $y = \frac{1}{2}x - 3$ ℓ_2 $y = 2x - 6$
Slope, $m_1 = \frac{1}{2}$ Slope, $m_2 = 2$

ℓ_3 $y = 2x + \frac{11}{2}$ ℓ_4 $y = -2x + 3$
Slope, $m_3 = 2$ Slope, $m_4 = -2$

For parallel lines, the slopes are equal.
Since $m_2 = m_3$, then $\ell_2 \parallel \ell_3$.

For perpendicular lines, the slopes are negative reciprocals (or the product of the slopes is -1).
Since $m_1 = \frac{1}{2}$, $m_4 = -2$
then $(m_1)(m_4) = \left(\frac{1}{2}\right)(-2) = -1$.
Since the slopes, m_1 and m_4 are negative reciprocals, then $\ell_1 \perp \ell_4$.

We may apply our work with slopes to determine properties of various figures.

Example 2
The vertices of $\triangle ABC$ are A(-3, 2), B(2, 3), C(3, -2). Determine whether $\triangle ABC$ is right-angled or not. Give reasons for your answer.

Solution
Calculate the slopes of the sides of $\triangle ABC$.

It is often helpful to sketch a diagram to visualize the problem.

$m_{AB} = \frac{3-2}{2-(-3)}$ $m_{BC} = \frac{-2-3}{3-2}$ $m_{AC} = \frac{-2-2}{3-(-3)}$

$= \frac{1}{2+3}$ $= \frac{-5}{1}$ $= \frac{-4}{6}$

$= \frac{1}{5}$ $= -5$ $= -\frac{2}{3}$

For perpendicular lines, the slopes of lines are negative reciprocals.
Since $(m_{AB})(m_{BC}) = \left(\frac{1}{5}\right)(-5) = -1$
then $AB \perp BC$.
Thus $\triangle ABC$ is right-angled with $\angle B = 90°$.

Skills with slopes for parallel and perpendicular lines will enable us to prove properties about many geometric figures.

6.6 exercise

Questions 1 to 7 practise skills for parallel and perpendicular lines.

A
1. The slopes m_1 and m_2 are given for lines ℓ_1 and ℓ_2. Which lines are parallel, perpendicular, or neither?

(a) $m_1 = \frac{1}{2}$, $m_2 = \frac{1}{2}$ (b) $m_1 = -3$, $m_2 = \frac{1}{3}$

(c) $m_1 = 4$, $m_2 = \frac{1}{4}$ (d) $m_1 = -\frac{2}{3}$, $m_2 = \frac{3}{2}$

(e) $m_1 = \frac{2}{3}$, $m_2 = \frac{4}{6}$ (f) $m_1 = \frac{3}{4}$, $m_2 = \frac{4}{3}$

(g) $m_1 = -\frac{1}{2}$, $m_2 = \frac{1}{2}$ (h) $m_1 = \frac{3}{4}$, $m_2 = -\frac{8}{6}$

(i) $m_1 = 1$, $m_2 = -1$ (j) $m_1 = -4$, $m_2 = \frac{2}{4}$

2. The slope of a line m is given. What is the slope of a line that is
 • parallel to the line? • perpendicular to the line?

(a) $m = 3$ (b) $m = \frac{2}{3}$ (c) $m = -\frac{1}{2}$

(d) $m = -1$ (e) $m = \frac{5}{4}$ (f) $m = 0$

3 Find the slope of a line perpendicular to each of the following lines.

(a) $y = 3x - 4$
(b) $y = -\frac{2}{3}x + 5$
(c) $2x - y = 6$
(d) $x + 2y = 8$
(e) $y = 8$
(f) $x = -6$

B

4 For the following points, pairs of line segments are given. Determine whether the pairs of line segments are parallel (||), perpendicular (⊥) or neither (N).

$A(-5, 1)$ $B(0, 3)$ $C(-1, 1)$ $D(-3, -3)$
$E(1, -3)$ $F(2, 1)$ $G(1, -1)$ $H(4, 3)$
$I(7, 3)$ $J(6, 6)$ $K(1, 3)$ $L(1, 8)$

(a) AB, FI
(b) CE, DG
(c) HJ, CK
(d) LH, KJ
(e) CH, FI
(f) GI, KJ

5 Write each line in the *slope y-intercept* form, then select pairs of lines that are
  • parallel.
  • perpendicular.

(a) $2x - y = 5$
(b) $x + 3y = -9$
(c) $3x - 2y = -16$
(d) $2x - 3y = 15$
(e) $2x - 4y = 11$
(f) $6x - 3y = -11$
(g) $3x - y = -2$
(h) $2x + 3y = -9$
(i) $3x - 2y = 16$
(j) $x - 2y = -5$

6 Show that

(a) the line through (1, 4) and (5, 5) is parallel to the line through (3, −4) and (7, −3).
(b) the line through (−1, 7) and (3, 5) is perpendicular to the line through (−4, 1) and (−1, 7).

7 The vertices of triangles are given. Determine which of the following triangles are right-angled. Give reasons why.

(a) $M(-2, 3)$ $N(2, 1)$ $P(-4, -1)$
(b) $A(-1, -1)$ $B(2, 3)$ $C(5, -1)$
(c) $R(-3, 4)$ $S(2, 6)$ $T(4, 1)$

(d) $D(7, -8)$ $E(10, -4)$ $F(-2, 5)$
(e) $K(-1, 8)$ $L(1, 3)$ $R(-4, 1)$

8 Two lines, ℓ_1 and ℓ_2, are given by the following pairs of equations. Find the value of k if $\ell_1 \| \ell_2$.

(a) $x - 4y = 24$, $2kx - y = -4$
(b) $y + kx = 7x - 2$, $y + x = 3(kx - 1)$

9 Two lines, ℓ_1 and ℓ_2, are given by the following pairs of equations. Find the value of k if $\ell_1 \perp \ell_2$.

(a) $y = 2kx + 8$, $3y = -x + 9$
(b) $x - 2y = -8$, $kx - y = -3(x + 1)$

10 The line passing through the points (2, k) and (0, 5) is parallel to the line passing through (2, 6) and (−2, −2). Find the value of k.

11 The line containing one side of a square passes through the points (8, 3) and (−6, 4). Find the value of k if the opposite side of the square passes through the points (3, 4) and (k, 2).

12 The vertices of the base of a triangle are given by (6, 0) and (−2, 4). If the altitude to the base passes through the points (k, 5) and (4, 3), then find the value of k.

13 In △CDE, find the slope of the altitude from C(2, 6) to the side determined by the vertices E(−2, 4) and D(3, −2).

C

14 The line passing through (7, k) and (4, 3) is parallel to the line joining (k, 5) and (−1, 1). Find two values of k.

15 The base of △ABC is given by the vertices A(−5, 2) and B(k, 6). The altitude to the base passes through the points (k, 3) and (−6, 6). Find the value(s) of k.

6.7 solving problems: geometric properties

In developing mathematics, we may use our skills to develop additional methods of solving problems and to extend our knowledge about the properties of various geometric figures. We can show some properties about geometric figures by using our skills with parallel and perpendicular lines.

Example 1
Show that the quadrilateral given by the vertices P(−1, 0), Q(3, 2), R(4, −6), and S(0, −8) is a parallelogram.

Solution
Drawing a diagram is often helpful in organizing the solution to a problem.
To show that quadrilateral PQRS is a parallelogram, show that PQ||SR and QR||PS.

$m_{PQ} = \dfrac{2 - 0}{3 - (-1)}$ $\qquad$ $m_{SR} = \dfrac{-6 - (-8)}{4 - 0}$

$\qquad = \dfrac{2}{4} = \dfrac{1}{2}$ $\qquad\qquad = \dfrac{-6 + 8}{4} = \dfrac{2}{4}$ or $\dfrac{1}{2}$

Since $m_{PQ} = m_{SR}$, PQ||SR.

$m_{QR} = \dfrac{-6 - 2}{4 - 3}$ $\qquad$ $m_{PS} = \dfrac{-8 - 0}{0 - (-1)}$

$\qquad = \dfrac{-8}{1} = -8$ $\qquad\qquad = \dfrac{-8}{1} = -8$

Since $m_{QR} = m_{PS}$, then QR||PS. Since PQ||SR and QR||PS then quadrilateral PQRS is a parallelogram.

Example 2
The sides of △ABC are defined by the equations as shown. Is the triangle right-angled? Give reasons for your answer.

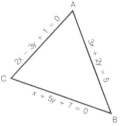

Solution
Express each equation in the *slope y-intercept* form.

AC: $y = \dfrac{2}{3}x + \dfrac{1}{3}$ $\qquad$ AB: $y = -\dfrac{3}{2}x + \dfrac{5}{2}$

BC: $y = -\dfrac{1}{5}x - \dfrac{7}{5}$

$m_{AC} = \dfrac{2}{3}$ $\qquad m_{AB} = -\dfrac{3}{2}$ $\qquad m_{BC} = -\dfrac{1}{5}$

The slopes of perpendicular lines are negative reciprocals.

Since $(m_{AC})(m_{AB}) = \left(\dfrac{2}{3}\right)\left(-\dfrac{3}{2}\right) = -1$ then ← Slopes are negative reciprocals.

△ABC is right-angled with a right angle at A.

In the exercise that follows we will use our knowledge of analytic geometry to establish certain properties of geometric figures.

6.7 exercise

A
1. The vertices of a triangle are given by
   A(−5, −4) B(4, −1) C(−5, 4).
   Which of the following lines are perpendicular to the base AB?

   (a) $3x + y = -8$ $\qquad$ (b) $3y - x = 4$
   (c) $3x - y - 3 = 0$ $\qquad$ (d) $6x = 9 - 2y$

2. The vertices of a square are P(2, 7), Q(−6, 3), R(−2, −5), and S(6, −1). Which of the following lines are parallel to the diagonal QS?

   (a) $x - 2 = 3y$ $\qquad$ (b) $3y + x = 8$
   (c) $5 - 2x = 6y$ $\qquad$ (d) $3x - y - 3 = 0$

B
3. Show that the quadrilateral DEFG given by
   D(−1, 3) E(6, 4) F(4, −1) G(−3, −2)
   is a parallelogram.

4 Show that the quadrilateral MPRS given by
   M(−5, 1) P(3, 3) R(4, −1) S(−4, −3)
   is a rectangle.

5 The vertices of quadrilateral PQRS are P(0, −5), Q(−9, 2), R(−5, 8), and S(4, 2). Show that the figure is not a rectangle.

6 The vertices of square RSTU are R(−2, 3), S(2, 1), T(0, −3), and U(−4, −1). Show that the diagonals RT and SU are perpendicular to each other.

7 The vertices of a rhombus JKLM are given by J(−5, 2), K(−1, 3), L(−2, −1), and M(−6, −2). Show that the diagonals JL and KM are perpendicular to each other.

8 The positions of 3 islands are at the points of intersection of the lines as shown. Determine whether △ABC is right-angled or not. Give reasons for your answer.

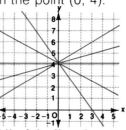

9 Three forest fire stations, F_1, F_2, and F_3, occur at the intersection of the lines $3x + 4y = 25$, $7x + y = 25$, and $4x − 3y = 25$. Is $\triangle F_1F_2F_3$ right-angled? Give reasons for your answer.

10 The roads bordering a tract of land are defined by the equations. Use your work with slopes to decide whether the tract of land is a parallelogram or not.

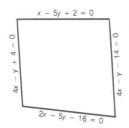

11 The sides of a quadrilateral are defined by the equations
    $3x − y + 13 = 0$ $7y = x + 31$
    $y = 3x − 7$ $x + 3y + 11 = 0$
    Use your work with slopes to decide whether the quadrilateral is a parallelogram.

6.8 families of lines

Each of these sets of lines has a common property.

The lines are concurrent in the point (0, 4).

The lines are parallel.

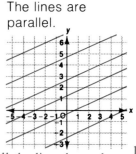

All of the lines have y-intercept 4.

All the lines have slope $\frac{1}{2}$.

Since these lines have a common property, we call them a **family of lines**. The defining equation for each of the above families is shown.

$y = mx + 4$ $y = \frac{1}{2}x + b$

Each member of the family has y-intercept 4.

Each member of the family has slope $\frac{1}{2}$.

In each equation, we use a **parameter**, which is an arbitrary constant.

$y = mx + 4$ parameter, m, for the slope

$y = \frac{1}{2}x + b$ parameter, b, for the y-intercept

We use these arbitrary constants, b and m, when we wish to show a common characteristic of the family of lines. Parameters, such as b and m, represent constants or variables other than the co-ordinate variables x and y.

To write a particular line (member) of the family we give the parameter a value.

$m = 3$, $y = mx + 4$ Equation of the family of lines

$y = 3x + 4$ Equation of the member of the family when $m = 3$ (i.e. the slope is 3)

$b = −3$, $y = \frac{1}{2}x + b$ Equation of the family of lines

$y = \frac{1}{2}x - 3$ ← Equation of the member of the family when $b = -3$ (i.e. the y-intercept is -3).

Each value assigned to the parameter m or b above determines a unique member of the family. If we know a point through which a member of the family passes we may use the co-ordinates of that point to determine the specific value of the parameter.

Example 1
A family of lines is given by $y = mx - 8$. Find the equation of the member of the family passing through the point $A(-3, 1)$.

Solution
Since the line passes through the point $A(-3, 1)$, the co-ordinates of $A(-3, 1)$ satisfy the equation of the line.
Use $y = mx - 8$ where $x = -3, y = 1$.

$1 = m(-3) - 8$ Thus the slope of the member
$1 = -3m - 8$ of the family of lines is -3.
$9 = -3m$ The equation is $y = -3x - 8$
$-3 = m$ or $3x + y + 8 = 0$
in standard form.

Throughout our work in analytic geometry, we use this important concept about locus to solve problems.

If a point is on a line or curve, the co-ordinates of that point satisfy the equation of the line or curve.

We use another important concept to write the family of lines passing through any point.

The slope of a line is constant.
The line passing through $A(-3, 2)$ is shown. Let $P(x, y)$ be another point on the line.

$\text{Slope}_{PA} = \frac{y - 2}{x + 3}$

If the slope of the line is given by the parameter, m, then, we may write
$$\frac{y - 2}{x + 3} = m.$$
This equation is called the **point slope form** of the equation of a line. The set of all lines through the point $(-3, 2)$ are a family. They are **concurrent** in the point $(-3, 2)$.

In general, we write
$$\frac{y - y_1}{x - x_1} = m$$
or $y - y_1 = m(x - x_1)$

Example 2
(a) Write the defining equation of the family of lines passing through the point $(3, -4)$.
(b) Write the equation of the member of the family of lines in (a) passing through $(5, -8)$.

Solution
A sketch is often helpful in visualizing the problem.
(a) The equation of the family of lines passing through $(3, -4)$ is given by
$\frac{y + 4}{x - 3} = m$, parameter m.

(b) A member of the family of lines passes through the point $(5, -8)$. Thus co-ordinates $(5, -8)$ satisfy the equation of the family of lines.

$\frac{y + 4}{x - 3} = m$ Use $x = 5, y = -8$.
$\frac{-8 + 4}{5 - 3} = m$
$\frac{-4}{2} = m$ or $m = -2$.

Thus the equation of the member of the family of lines passing through $(5, -8)$ is
$\frac{y + 4}{x - 3} = -2$.

We may write the equation in standard form.

$y + 4 = -2(x - 3)$
$y + 4 = -2x + 6$
$2x + y - 2 = 0$

The use of families of lines or curves, occurs throughout the study of mathematics. It is a powerful strategy for solving problems.

6.8 exercise

A

1. A family of lines is given by $y = mx + 6$.
   (a) What common property do the members of the family have?
   (b) Sketch the lines of the family with the values of the parameter $m = 2, 1, 0, -1, -2$.

2. A family of lines is given by $y = -\frac{1}{2}x + b$.
   (a) What common property do the members of the family have?
   (b) Sketch the lines of the family with the values of the parameter $b = -3, -1, 1, 3$.

3. Use each of the following slopes to write the equation of the family of lines in the *slope y-intercept* form.
   (a) $m = 2$ (b) $m = -3$ (c) $m = \frac{3}{2}$
   (d) $m = 0$ (e) $m = -\frac{4}{3}$ (f) slope undefined

4. Use the *slope y-intercept* form. Write a family of lines for each of the following y-intercepts.
   (a) $b = 3$ (b) $b = -4$ (c) $b = \frac{1}{3}$
   (d) $b = -\frac{3}{2}$ (e) $b = \frac{4}{3}$ (f) $b = 0$

5. Write the defining equation of the family of lines passing through each point.
   (a) $(3, 2)$ (b) $(-2, 5)$ (c) $(-1, -3)$
   (d) $(0, 8)$ (e) $(4, -5)$ (f) $(-6, 0)$
   (g) (a, b) (h) $(0, b)$ (i) $(a, 0)$

6. Each equation is the defining equation of a family of lines. What common property does each family have?

   (a) $y = mx + 6$ (b) $\frac{y-4}{x-6} = m$
   (c) $y = -\frac{1}{2}x + b$ (d) $y = m(x + 2)$

7. (a) What is the slope of the x-axis? y-axis?
   (b) Write the defining equation of the family of lines parallel to the x-axis; y-axis.

8. Write the equation of the family of lines having the common property
   (a) y-intercept -6.
   (b) passing through the point $(4, -3)$.
   (c) slope -8. (d) parallel to the y-axis.
   (e) slope $-\frac{2}{3}$. (f) y-intercept $\frac{2}{3}$.
   (g) concurrent in the point $(3, -4)$.
   (h) parallel to the x-axis. (i) x-intercept -6.

B

9. The defining equation of a family of lines with slope -2 is given by $y = -2x + b$, where b is a parameter. Find the member of the family
   (a) with y-intercept 3.
   (b) passing through the point $(-3, 4)$.
   (c) with x-intercept -4.

10. The defining equation of a family of lines with y-intercept 6 is given by $y = mx + 6$ where m is the parameter. Find the member of the family
    (a) with slope $\frac{2}{3}$.
    (b) passing through the point $(3, -5)$.
    (c) with the x-intercept 3.

11. (a) Write the equation of the family of lines with x-intercept -3.
    (b) What is the equation of member of the family in (a) passing through the point $(3, 2)$?

12. (a) Write the defining equation of the family of lines concurrent in the point $(3, -4)$.

(b) Write the equation of the member of the family in (a) that also passes through the point $(-3, 1)$.

13. (a) Write the defining equation of the family of lines perpendicular to the x-axis.
    (b) What is the equation of the member of the family in (a) passing through $(3, -2)$?

4. (a) Write the defining equation of the family of lines parallel to the x-axis.
   (b) What is the equation of the member of the family in (a) passing through $(-3, 8)$?

5. (a) Write a defining equation of the family of lines passing through the origin.
   (b) What is the equation of the member of the family passing through $(-2, 5)$?

6. Each equation defines a family of lines. Write the equation, in *standard form*, of the member of the family passing through the point $(-3, 5)$.
   (a) $y = \frac{2}{3}x + b$
   (b) $(y - 3) = m(x - 2)$
   (c) $y = m(x + 4)$
   (d) $y = mx - 6$
   (e) $y = k$
   (f) $x = p$
   (g) $3x - 2y + k = 0$
   (h) $mx + y = 5$

7. (a) The point $(-3, 4)$ is on the line $y = -4x + k$. Find k.
   (b) A line is given by $y = 3x - k$. If $(3, 3)$ is a point on the line, find the value of k.

8. The line $y = (2k + 7)x + 8$ passes through the point $(2, 14)$. Find k.

9. The line with equation $(2k - 5)y = (2 - 2k)x + 3$ contains the point $\left(\frac{1}{4}, 2\right)$. Find k.

10. Find the value of k if the point $(-4, -7)$ is on the line $(2k^2 - 1)y = (k + 1)x + k$.

6.9 conditions for writing equations

A family of lines is a set of lines with a common property. To find the equation for a specific member of the family, we must be given additional information as shown in Example 1.

In solving problems, we often need to combine skills to obtain an answer.

Example 1
A family of lines is parallel to the line $x - 5y = 5$. Find the equation of the member passing through the intersection of the lines $x + 3y = 21$ and $x - 3y = -9$.

Solution
Step 1 Find the defining equation of the family of lines.
From $x - 5y = 5$
$-5y = -x + 5$
$y = \frac{1}{5}x - 1$ — slope $\frac{1}{5}$

The equation of the family of lines is $y = \frac{1}{5}x + k$.

Step 2 Find the equation of a specific line.
Solve $x + 3y = 21$ ①   Solve the
$x - 3y = -9$ ②   system of
Equation ① − ②   equations to
$6y = 30$ obtain the point
$y = 5$ through which
Use $y = 5$ in ①.   the required
$x + 3y = 21$ equation
$x + 15 = 21$ passes.
$x = 6$

Thus the required member of the family passes through the point $(6, 5)$.
$y = \frac{1}{5}x + k$ $x = 6, y = 5$
$5 = \frac{1}{5}(6) + k$

Thus $k = \frac{19}{5}$.

Thus the required equation is

$y = \frac{1}{5}x + \frac{19}{5}$ ← Slope y-intercept form

or

$x - 5y + 19 = 0$ ← Standard form.

6.9 exercise

A

1. Write the defining equation for the family of lines parallel to each of the following.
   (a) $4x + 2y = -3$ (b) $2y - x = 7$
   (c) $3x + 2y - 6 = 0$ (d) $x = 8$

2. Write the defining equation for the family of lines perpendicular to each of the following.
   (a) $3x + y = 8$ (b) $3x - 2y = 6$
   (c) $2y - 3x - 16 = 0$ (d) $2 - y = 3x$

3. Write an equation for each family of lines with the common property as shown.
   (a) y-intercept 2
   (b) concurrent with point $(-3, 8)$
   (c) slope $\frac{4}{3}$
   (d) x-intercept -3
   (e) passing through the point $(2, -6)$
   (f) parallel to the y-axis

4. (a) Write the *slope y-intercept* form of the family of lines parallel to the line given by $2x - y = 6$.
   (b) Write the equation of the member in (a) passing through the point $(-3, 1)$.
   (c) Write the equation of the member in (a) that has x-intercept -2.

5. A family of lines passes through the point $(0, 3)$.
   (a) Use the *point-slope* form to find the equation of the member with slope 2.
   (b) use the *slope y-intercept* form to find the equation of the member with slope 2.
   (c) Which method do you prefer, (a) or (b), to find the equation of the member? Give reasons why.

6. (a) A family of lines passes through the point $(-3, 2)$. Write the equation of the family.
   (b) What is the equation of the member in (a) for each property?
   A: slope 2   B: y-intercept -3
   C: x-intercept -1 D: parallel to the y-axis

7. To write the equation for each of the following, decide which form is more suitable to find the answer.
   * *slope y-intercept* form
   * *point slope* form
   Then find the equation.
   (a) x-intercept 3, passes through $(0, -2)$
   (b) slope $-\frac{4}{3}$, y-intercept 2
   (c) passing through $(-1, 3)$ and $(-2, 6)$
   (d) x-intercept $\frac{2}{5}$, slope 4
   (e) slope 0, passes through $(-3, -4)$
   (f) perpendicular to $y = 2x$, x-intercept -3
   (g) slope 3, y-intercept 2
   (h) parallel to $3x + 3y = 1$, x-intercept -2
   (i) x-intercept -4, slope -2
   (j) y-intercept -2, slope $\frac{1}{2}$
   (k) passes through $(-2, 1)$, parallel to the x-axis
   (l) perpendicular to $3x - y = 6$, passes through $(-8, 0)$
   (m) parallel to $2x + y = 4$, y-intercept 7

B

8. A line passes through the point $(-2, 3)$. Write its equation in *standard form* if its slope is the same as for $2x + y = 9$.

9. Write the equation of a line in *standard form* if it passes through the point $(-2, -6)$ and its *y*-intercept is equal to that for $x - 2y = 6$.

10. A line is parallel to $2x + 3y = 10$ and passes through $(-2, 0)$. Find its equation in *standard form*.

11. The *y*-intercept of a line is 3 more than the *y*-intercept of $2x - y = 6$. Find its equation in *standard form* if it passes through $(3, -3)$.

12. Write the equation for each side of the rhombus.

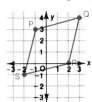

13. A line is perpendicular to $3x + y = 2$ and has *x*-intercept -3.
    (a) Find the equation of the line.
    (b) Which of the following points are also on the line?
    P$(-6, -1)$ Q$(9, 3)$ R$(3, 2)$

14. Three points are given.
    T$(-2, 4)$ Q$(-6, -2)$ S$(0, -4)$
    (a) Find the equation of the line through T parallel to QS.
    (b) Find the equation of the line through Q perpendicular to TS.

5. Find the equation of each altitude of △EFG given the co-ordinates
    E$(3, 4)$ F$(1, -3)$ G$(6, 0)$

6. A line passes through the point $(4, 3)$ and has *y*-intercept -2. Find its *x*-intercept.

7. Write the equation of the line with the same *y*-intercept as $3x + y = 6$ and the same slope as $2x - y = 8$.

18. What is the *y*-intercept of a line that has slope $\frac{3}{4}$ and passes through the point $(-2, 3)$?

19. Write the equation of a line through $(-1, 3)$ with the same *y*-intercept as $4x - y = 8$.

20. Find the *y*-intercept of a line that has slope $\frac{2}{3}$ and *x*-intercept -4.

21. Write the equation of the line through $(-8, 2)$ with equal *x*- and *y*-intercepts.

22. A line passes through the intersection of $2x - 5y = 18$ and $x + 7y = -29$ and through the point $(5, 4)$. Find its equation.

23. A line passes through a point, P, on the *y*-axis and through $(8, 2)$ so that it is parallel to the line given by $2x - 4y + 10 = 0$. Find the co-ordinates of P.

C
24. The line $x - 3y + 9 = 0$ is perpendicular to the line formed by joining P$(4, -3)$ and a point on the *x*-axis. Find the co-ordinates of this point.

problem/matics

The computer may be used to perform calculations in *any* branch of mathematics. For example, in Chapters 6 and 7 we will study analytic geometry. The following program finds the co-ordinates of the midpoint of a line segment if we know the co-ordinates of the end points.
(x_1, y_1) (m_1, m_2) (x_2, y_2)

```
10 INPUT X1, Y1, X2, Y2
20 LET M1 = (X1 + X2)/2
30 LET M2 = (Y1 + Y2)/2
40 PRINT "MIDPOINT IS" M1, M2
50 END
```

6.10 combining strategies to solve problems

We may combine the skills and strategies for co-ordinate geometry to solve problems, as well as learn additional properties about geometric figures. One of the most important skills in solving problems is to decide *which* skill may be used to solve a problem. We have developed these skills for working on the co-ordinate plane.

- finding the slope of a line.
- finding the equation of a line.
- finding the intercepts of a line.

We also know

- when lines are parallel.
- when lines are perpendicular.
- when points are collinear.

As we learn new skills in co-ordinate geometry we may apply these skills to solve additional problems. For example, these examples suggest a general method for finding the midpoint of a line segment.

A(2, 3)   B(6, 3)
Midpoint M(4, 3)
$$4 = \frac{2 + 6}{2}$$
i.e. 4 is the average between 2 and 6.

C(2, 3)   D(2, −1)
Midpoint M(2, 1)
$$1 = \frac{3 - 1}{2}$$

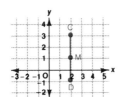

E(1, 1)   F(5, 3)
Midpoint M(3, 2)
$$3 = \frac{1 + 5}{2}$$
$$2 = \frac{1 + 3}{2}$$

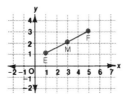

midpoint of line segment

From the above examples, it seems that for any two points $P_1(x_1, y_1)$ and $P_2(x_2, y_2)$ the midpoint M has co-ordinates
$$\left(\frac{x_1 + x_2}{2}, \frac{y_1 + y_2}{2}\right)$$

We may now combine our earlier skills with the skill of finding the co-ordinates of the midpoint and apply them to writing the equation of a line, as shown in the following example.

Example 1

Find the equation of the line perpendicular to the line defined by $2x - y = 6$ and passing through the midpoint of the segment with end points A(−8, 4), B(6, −6).

Solution

Denote the required line by ℓ. The midpoint M(m, n) of AB is given by

$$m = \frac{x_1 + x_2}{2} \qquad n = \frac{y_1 + y_2}{2}$$
$$= \frac{-8 + 6}{2} \qquad = \frac{4 + (-6)}{2}$$
$$= \frac{-2}{2} = -1 \qquad = \frac{-2}{2} = -1$$

Then M is (−1, −1).

To find the slope of ℓ, write $2x - y = 6$ in the *slope y-intercept* form.
$$2x - y = 6$$
$$-y = -2x + 6$$
$$y = 2x - 6$$
The slope of $2x - y = 6$ is 2.
Thus the slope of ℓ is $-\frac{1}{2}$.

Thus the equation of the required line, ℓ, is given by
$$\frac{y - (-1)}{x - (-1)} = -\frac{1}{2}$$
$$\frac{y + 1}{x + 1} = -\frac{1}{2}$$
$$2y + 2 = -x - 1 \text{ or } x + 2y = -3$$

To solve any problem, we must clearly be able to answer these two questions.

A: What information is given in the problem? (What do we know?)

B: What are we asked to find? (What don't we know?)

Once we have the answers to Questions A and B,
- we may use our skills and strategies to proceed *step by step* from what we know to the solution of the problem.
- we may record the solution so that other persons may understand how the problem was solved, as shown above.

6.10 exercise

Questions 1 to 7 provide practice with some essential skills required to solve problems using co-ordinate geometry.

A

1. Find the midpoint of each line segment given by the co-ordinates of the end points.

   (a) (−1, −2), (−7, 10) (b) (6, 4), (0, 0)
   (c) (5, −1), (−2, 9)   (d) (0, 0), (6, 4)
   (e) (4, −5), (9, −6)   (f) (0, −4), (12, 0)
   (g) (−2, 3), (3, 5)   (h) (5, 0), (−8, −3)

2. △PQR is given by the vertices
   P(−6, 5)   Q(−6, −9)   R(8, 5)

   (a) Find the co-ordinates of the midpoint M of PQ.
   (b) Find the co-ordinates of the midpoint N of RP.
   (c) Find the slope of MN, QR.
   (d) What do you notice about the slopes in (c)?

3. Find the slopes of each segment, given its end points.

   (a) (−2, 6), (6, −4)   (b) (7, −5), (−1, 3)
   (c) (3, −4), (2, 0)   (d) (8, −6), (3, −2)

4. For the following points,
   P(−4, 0)   Q(4, −5)   R(−4, −6)
   S(−2, −9) T(4, 1)   K(2, 4)
   show that
   (a) KP∥SQ (b) KT⊥PK (c) KT∥RS (d) SQ⊥RS

5. Which sides of the figure, given by the co-ordinates A(−3, −2), B(3, 4), C(9, −2), and D(3, −8), are parallel?

6. Write the equation of the line
   (a) through (2, −1) and having the same x-intercept as the line $y = 2x + 4$.
   (b) parallel to $y = \frac{1}{2}x + 3$ and having the same y-intercept as the line $2x + 3y - 6 = 0$.
   (c) with x-intercept the same as the line $3x + y - 6 = 0$ and parallel to the line $y = \frac{2}{3}x + 4$.
   (d) with the same y-intercept as the line $2x + 4y - 8 = 0$ and perpendicular to the line $y = 2x - 1$.

7. Determine the equation of the line through (4, −2) parallel to
   (a) the x-axis.   (b) the y-axis.

B

8. Find the value of t if the following lines have the same x-intercept.
   $2x - 3y + 4 = 0$ $x + ty + 2 = 0$

9. If the points P(−1, s), Q(−2, 3) and R(4, −2) are collinear, find the value of s.

10. AB is perpendicular to the line defined by $8 + y = 3x$. If A is (m, 0) and B is (3, −2), find the value of m.

11. The end point of segment AB is A(−2, 4). If the co-ordinates of the midpoint are (−1, 7), find the co-ordinates of B.

12 Find the co-ordinates of the point A on the y-axis, so that AB is perpendicular to the line defined by $3x - 2y - 6 = 0$ where B is $(-1, 4)$.

13 If the midpoint of a segment is $(-1, -8)$ and one endpoint is $(7, -9)$, find the co-ordinates of the remaining endpoint.

14 A diameter of a circle has end points $A(9, -4)$ and $B(3, -2)$. Find the centre of the circle.

15 The end points of AB are $A(\sqrt{72}, -\sqrt{12})$ and $B(\sqrt{32}, -\sqrt{48})$. Find the midpoint.

16 Find the equation of the line passing through the midpoint of the segment CD where C is $(-4, 5)$, and D is $(-6, 3)$ and parallel to the line $3x + 2y = 6$.

17 Find the equation of the perpendicular bisector of EF where E is $(-2, 10)$, and F is $(-12, 4)$.

18 Write the equation of the line that meets the segment AB at right angles at its midpoint where A is $(-2, -1)$ and B is $(4, 3)$.

19 Two lines, ℓ_1 and ℓ_2, intersect in the point $(0, 2)$. The slope of ℓ_2 is equal to $\frac{3}{2}$ the slope of ℓ_1. Find the equation of ℓ_2 if the equation of ℓ_1 is $2x - 5y + 10 = 0$.

20 Find the equation of the line through the origin which bisects the part of the line $3x + y = 6$ which is contained between the axes.

21 Given the two points $A(-6, 2)$ and $B(-3, 7)$. If the line $5x + py + k = 0$ is parallel to the line through AB and passes through the point $C(-7, 6)$, find the value of p and k.

22 Given the points $A(-6, k)$ and $B(k + 2, 1)$. If the line $2x - 7y - 41 = 0$ is parallel to AB, find k.

23 A fighter plane landed at a base with map co-ordinates $(-5, -2)$ while a cargo plane was loaded at an airport with map co-ordinates $(2, -3)$. If the fighter plane leaves the base on a straight line path with slope 2 and the cargo plane flies on a line with slope -3, at what co-ordinates will their flight paths cross?

24 One vertex of a parallelogram is determined by the intersection of the lines $7x - 2y + 29 = 0$ and $x - 6y + 27 = 0$.
   (a) If the other two sides intersect at $(1, -2)$, find the equation of the diagonal of the parallelogram through $(1, -2)$.
   (b) Find the equations of the other two sides.
   (c) Find the equation of the other diagonal.

25 One vertex of a rectangle is determined by the intersection of the lines $3y = x + 10$ and $3x + y + 10 = 0$. If the other two sides intersect at $(1, -3)$, then find the equations of the other two sides.

C

26 Given two points $A(2, 5)$ and $B(6, 1)$. If a line parallel to AB has an x-intercept $p + 4$, and slope equal to the reciprocal of $p - 4$, find an equation of the line.

27 Three sides of a parallelogram are formed by the lines, $x + 8y = 11$, $2y + 3x + 11 = 0$, $8y + x = -11$. Find the equation of the line forming the fourth side if the diagonals of the parallelogram intersect at the point $(0, 0)$.

problem/matics

Steven, Brian and Jason are married to Barbara, Jackie and Gale, but not in this order. On Saturday night they played bridge, but the wives and husbands were with different partners. Jackie played with Jason, and Brian played with Steven's wife. If Brian and Gale played together, then who is married to whom?

applications: properties of geometric figures

We may use our strategies and skills from co-ordinate geometry to explore the properties of various geometric figures. To do so, review the following vocabulary.

In △ABC, the **median** is the line segment joining vertex A to the midpoint of BC.

If three points are in a straight line, we say the points are **collinear**.

If three lines meet in a common point, we say the lines are **concurrent**.

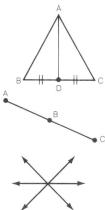

28 △PQR is given by the vertices
P(1, 4)   Q(−5, 2)   R(−1, −4).
(a) Write the equation of the medians of △PQR.
(b) Find the co-ordinates of G, the point of intersection of any two of the medians.
(c) Show that the third median contains G.

29 △MNQ is given by the vertices
M(−2, 2)   N(2, −6)   Q(6, 0).
(a) Write the equations of the right bisectors of each side of △MNQ.
(b) Show that the right bisectors of △MNQ are concurrent at a point B. Find the co-ordinates of B.

30 △EFG is given by the vertices
E(5, 3)   F(−1, 1)   G(3, −3).
(a) Write the equation of the altitudes of △EFG.
(b) Show that the altitudes of △EFG are concurrent in a point C and give its co-ordinates.

31 △PQR has vertices
P(−1, 4)   Q(1, 8)   R(−9, −4).
Show that the line joining the midpoints of two sides is parallel to the third side.

32 The vertices of △PQR are given by
P(3, 4)   Q(−1, −8)   R(11, −2).
(a) Write the co-ordinates of the midpoint S of PQ.
(b) Write the equation of the line parallel to QR through S.
(c) Find the co-ordinates of the intersection of RP and the line in (b).
(d) Show that the point in (c) is the midpoint of RP.

33 The vertices of parallelogram ABCD have co-ordinates A(−2, −4), B(10, 2), C(12, 7), and D(0, 1).
(a) Find the midpoint of the diagonal AC.
(b) Find the midpoint of the diagonal BD.
(c) Use your results in (a) and (b) to show that the diagonals bisect each other.

34 (a) Show that the diagonals of parallelogram with vertices A(0, −1), B(−2, −5), C(10, 1), and D(12, 5) bisect each other.
(b) Use the results in (a) to show that quadrilateral PQRS with vertices P(−2, 3), Q(5, 1), R(7, −3), and S(0, −2) is *not* a parallelogram.

35 A quadrilateral ABCD is given by the vertices
A(0, 1)   B(3, 0)   C(5, −4)   D(−1, −2).
If M, N, P, Q are the midpoints of the sides of ABCD taken in order, prove that the figure MNPQ is a parallelogram.

36 A rectangle EFGH has vertices E(−11, 6), F(7, 0), G(3, −12), and H(−15, −6). Show that the diagonals bisect each other.

37 Show that the quadrilateral formed by joining the midpoints of the sides of square ABCD where A(9, 8), B(−3, 4), C(1, −8), and D(13, −4), is also a square.

6.11 systems of inequations and their graphs

The skills we have learned to work with equations may be now extended to work with linear inequations.
The graph of a linear equation such as
$y = x + 2$
separates the Cartesian plane into 3 regions, shown by A, B, and C.

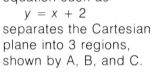

A: points that are above the line.
B: points that are on the line.
C: points that are below the line.
From the diagram, P(4, 6) is on the line. Every point on the plane must satisfy one of the following conditions:

- Points on the line satisfy the equation $y = x + 2$.

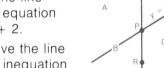

- Points above the line satisfy the inequation $y > x + 2$.

The y co-ordinate of Q is greater than the y co-ordinate of P.

- Points below the line satisfy the inequation $y < x + 2$.

The y co-ordinate of R is less than the y co-ordinate of P.

We may draw the graph of each inequation as shown.

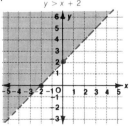 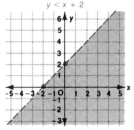

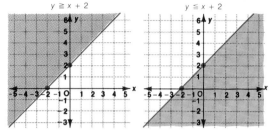

We may define a region by a system of linear inequations.
Figure A shows all (x, y) that satisfy $2x + y \geq 4$.

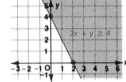

Figure B shows all (x, y) that satisfy $x - y \geq -1$.

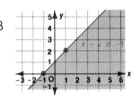

All (x, y) that satisfy the system
$2x + y \geq 4$
$x - y \geq -1$
are shown by the intersection of the regions A and B.

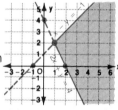

We may draw a region defined by a system of more than 2 inequations as shown in Example 1.

Example 1
Draw the region defined by $5x - 6y \geq 3$, $5x + 2y < 19$, $y \geq -3$.

Solution

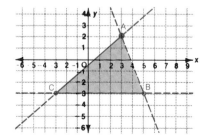

To find the solution of the linear inequations $5x - 6y \geq 3$, $5x + 2y < 19$ and $y \geq -3$ means to find all (x, y) that satisfy all the above inequations and equations. To show the solution set, we draw the triangular region as shown by $\triangle ABC$ in Example 1.

Drawing regions defined by a system of linear inequations is important in solving real-life problems that occur in linear programming in the next section.

6.11 exercise

Questions 1 to 6 practise skills required to for drawing regions defined by a system of linear inequations.

A

1. Which of the following equations or inequations describes the region as shown?

   A $2x + 3y = 6$
   B $2x + 3y > 6$
   C $2x + 3y \leq 6$
   D $2x + 3y < 6$
   E $2x + 3y \geq 6$

2. A region is shown. Which of the following points satisfy $3x + 4y \leq 12$?

   (a) $(1, 1)$ (b) $(1, 3)$
   (c) $(6, -2)$ (d) $(7, -2)$

3. A region defined by $3x - 4y \leq 12$ is drawn. Which of the following points may be used to check the accuracy of the region that has been shaded?
   A$(1, 1)$ B$(-1, 1)$ C$(0, 0)$

4. For each region, the corresponding equation is given to indicate the boundary of the region. Write the inequation that defines the region.
   (a) 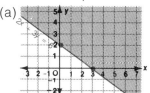 (b)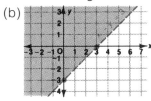

5. For each region, write the inequation that defines it.
   (a) 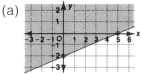 (b)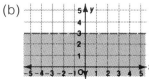
   (c) 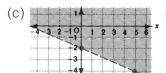 (d)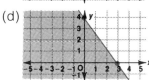

6. Draw the region defined by each of the following inequations.
   (a) $x \leq 3$ (b) $x > -2$ (c) $y < 2$
   (d) $y \geq -3$ (e) $x + y \leq 4$ (f) $x - y > 2$
   (g) $2x + y < 8$ (h) $x - 2y \geq 4$ (i) $5x - 2y > 10$

B

7. For each region the corresponding equations are given. Write the system of linear inequations that define the region.
   (a) 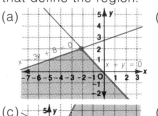 (b)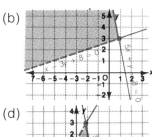
   (c) 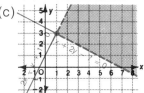 (d)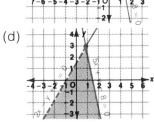

8 Draw a region to show the solution for each system of linear inequations.
   (a) $x + y > 9$ and $x - y \leq 3$.
   (b) $3x - y \leq -2$ and $x - y \geq -6$.
   (c) $3x - y < 4$ and $x - 2y > 3$.
   (d) $2x - 3y < -12$ and $2x + 3y \geq 0$.

9 Draw the region defined by each system of inequations.
   (a) $2x - y \geq 3$ and $x + y \leq 3$
   (b) $x + 4y < 16$ and $4x - 3y \geq -12$
   (c) $2x - y < 8$ and $x - y \geq 3$
   (d) $x + 2y < 4$ and $x - y > 1$
   (e) $x > y - 1$ and $3y - x \leq 1$

10 For each region, write the defining system of inequations.

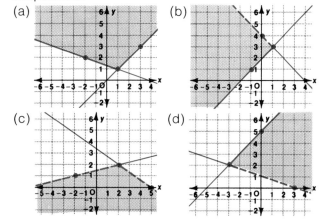

11 Draw the graph of each region defined by the system of inequations.
   (a) $x \geq 0$, $y \geq 0$, $2x - 3y \leq 6$
   (b) $x + y \geq 6$, $x - y \leq 4$, $y \leq 2$
   (c) $x \leq 3$, $y < 2$, $x + y < 4$
   (d) $2x - 3y > 0$, $x - 6y \leq -5$, $x > 2$
   (e) $y \geq 2x + 2$, $x - 2 > -y$, $x \geq 0$
   (f) $y - x \leq 4$, $y - 3x \geq 0$, $x + y > 1$
   (g) $y + 1 > x$, $6 < 2x + 3y$, $x > 0$

6.12 applications: linear programming

The skills we have learned for graphing regions defined by systems of inequations may be used to solve many problems in industry.

A small company is interested in maximizing its profit and also minimizing its expenses. For example, a company manufactures a pair of ping-pong bats at a $3.00 profit and a pair of badminton racquets at a profit of $5.00. In 1 min the number of
• sets of ping-pong bats manufactured is x.
• sets of badminton racquets manufactured is y.
The profit P, in dollars, for each minute is then given by $P = 3x + 5y$.

From the graph, the value of P is increased as shown. However, since there are certain conditions at the plant that impose restrictions on x and y, the company finds the maximum value of P that satisfies certain conditions as shown in Example 1. We refer to these restrictions as the **constraints** of the business problems.

Example 1
The constraints on P, where $P = 3x + 5y$ is given by the region $2x + y \leq 12$, $x + 4y \leq 13$, $y \geq 0$, and $x \geq 0$. Find the maximum value of P for the given region.

Solution
The region given by the system of inequations is shown. The family of lines given by
  $P = 3x + 5y$
is shown for different values of P. From the above diagram, the value of P is a maximum if $x = 5$ and $y = 2$.

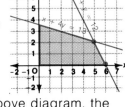

Thus, the company, for the given constraints, would manufacture 5 sets of ping-pong bats and 2 sets of badminton racquets to maximize the profit P.

The above graphical method of drawing a region defined by a system of inequations and solving a problem based on certain constraints is referred to as **linear programming**. The next example illustrates how to solve a problem in business using this method which combines the skills of algebra and co-ordinate geometry.

Example 2

The Pro Shop makes golf hats and visors. Each golf hat requires 4 min on the cutting machine and 3 min on the stitching machine. Each visor requires 3 min on the cutting machine and 1 min on the stitching machine. The cutting machine is available 2 h/d while the stitching machine is only available 1 h/d. If the profit on a golf hat is $1.10 and the profit on a visor is $0.60, calculate how many of each should be made to realize a maximum profit.

Solution

Let x represent the number of golf hats made. Thus $x \geq 0$.
Let y represent the number of visors made. Thus $y \geq 0$.
We may use a chart to organize the information.

	Number of each	Cutting time (min)	Stitching time (min)	Profit ($)
golf hats	x	$4x$	$3x$	$1.10x$
visors	y	$3y$	y	$0.60y$
total		$4x + 3y$	$3x + y$	$1.10x + 0.60y$

The cutting machine is available 2 h or 120 min. Thus, $4x + 3y \leq 120$. The stitching machine is available 1 h or 60 min. Thus, $3x + y \leq 60$. Let P represent the maximum profit.
Thus $P = 1.10x + 0.60y$ or $P = 1.1x + 0.6y$.

The region that satisfies the following system of inequations is drawn.
$$x \geq 0$$
$$y \geq 0$$
$$4x + 3y \leq 120$$
$$3x + y \leq 60$$

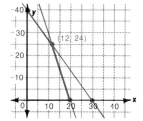

From the graph, P is a maximum at the point of the region, M(12, 24). Thus for maximum profit
- the number of golf hats made is 12/h.
- the number of visors made is 24/h.
- the maximum profit is $27.60/h.

To solve a business problem in industry, more than 2 variables are often used. Since many constraints are needed, a computer is programmed to solve the problem. The problems that follow involve only 2 variables and may be solved using the graphical method of drawing regions. However, to solve advanced problems in linear programming the same principles are used.

6.12 exercise

Questions 1 to 10 develop essential skills for solving linear programming problems.

A

1 (a) Draw the region given by the system of inequations $x \geq 0$, $y \geq 0$, $x + 19 \geq 3y$, $3x + 2y \leq 31$.
   (b) Name the co-ordinates of the vertices of the region in (a).

2 (a) Draw the region given by the system of inequations $x \geq 0$, $y \geq 0$, $5x + 4y \geq 32$, $x + 2y \geq 10$.
   (b) Name the co-ordinates of the vertices of the region given in (a).

3 For the region in Question 1, find
   (a) the maximum value of P, given by $P = 2x + 4y$.
   (b) The maximum value of R, given by $R = 3y + x$.

4 For the region in Question 2, find
   (a) the minimum value of P, given by $P = 3x + 2y$.
   (b) the minimum value of S, given by $S = 3y + 2x$.

5 Find the maximum value of P if $P = x + y$ over the region as shown.

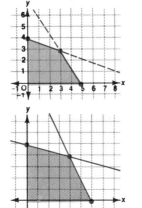

6 The profit P, of making two types of dishes is given by $P = 2x + 4y$. Find the maximum profit for the given region.

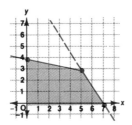

7 The expenses, E, for manufacturing two types of dog food are given by $E = 3x + 6y$. Find the minimum value of E.

8 Find the minimum value of P given by $P = 3x + 4y$ for the region given by $x \geq 0, y \geq 0, 2x + y \geq 8, x + 3y \geq 9, x + y \geq 7$

9 The constraints for manufacturing two types of hockey skates are given by the following region.
   $x \geq 0, y \geq 0, x + 4y \leq 41, y \geq 2x - 10,$
   $7x + 64 \geq 11y.$
   Find the maximum value of Q over the region if $Q = 3x + 5y$.

10 The constraints for making basketballs and soccer balls are given by the region
   $x \geq 0, y \geq 0, 2x + y \leq 260, x + 5y \leq 400.$
   Find the minimum value of the expenses, E, given by $E = 5x + 4y$ over the above region.

Solve each problem using the method of linear programming.

B
11 Burlington Runners repairs sports shoes, particularly tennis and jogging shoes. Two operations are required on each shoe and the times required on each operation are shown.

	Strip	Re-sew
Tennis	16 min	12 min
Jogging	8 min	16 min

If two people repair these shoes, each working 8 h/d and the profits on tennis and jogging shoes are $3 and $5 respectively, how many pairs of each shoe should be repaired daily to maximize the profits?

12 It is recommended that cattle be supplemented with 19 g of iron and 12 g of riboflavin in their diet. Two feeds are available which contain both nutrients in different amounts as shown.

Feed	Iron	Riboflavin
Husky	5%	2%
Vibrant	2%	3%

Husky Feed sells for $25/kg and Vibrant Feed sells for $32/kg.

(a) How many kilograms of each feed are required to feed 100 cattle as economically as possible?

(b) What is the cost of these quantities in (a)?

13 A firm manufactures 2-bulb bedroom lamps and 4-bulb living room lamps. Each day they get a consignment of 480 bulbs and 180 shades. Determine how many of each lamp they should manufacture to yield a maximum profit if their profit on a 2-bulb lamp is $20 and on a 4-bulb lamp is $35. What is the maximum profit?

14. The Sundial Watch Company manufactures two types of watches — a self-winding and an automatic model. In any day there are 3 h of machine time and 7 h of jeweller time available. The self-winding model requires 1.5 h of machine time and 1 h of jeweller time, while the automatic model requires 30 min of machine time and 2 h of jeweller time. If the profit on the self-winding model is $25 and the profit on the automatic model is $18, how many of each model should be manufactured daily for a maximum profit? What is the maximum profit?

5. John and his son service electrical appliances such as toasters and kettles. John can service a toaster in 15 min and a kettle in 12 min, while his son can service the same appliances in 10 min and 20 min respectively. If John is paid $12/h and his son earns $8/h, how long should each work on an order of 50 toasters and 40 kettles so as to minimize the cost of labour?

6. To produce a top quality yield of carrots a vegetable farmer must spray his crop with at least 5.3 L of nutrient A and 3.2 L of nutrient B per field. Two suppliers provide the nutrients in a special blend but the amounts of nutrients A and B vary in each litre they provide as follows:

Supplier	Amount of nutrient A	Amount of nutrient B
Top Grade Brand	30%	70%
Quality Brand	80%	20%

If the price for Top Grade Brand is $3.20/L and for Quality Brand is $3.80/L, then

(a) determine how many litres of each brand must be bought from each supplier to provide the required amounts of each nutrient per field at a minimum cost.
(b) determine what the minimum cost will be for the nutrients bought.

skills review: writing equations

In each of the following questions leave the answer in the form $Ax + By + C = 0$, with A, B, C in lowest terms.

1. (a) Write an equation for the family of lines parallel to the line $y = 3x - 6$.
   (b) What is the equation of a line passing through the point $(-3, 2)$?

2. Write the equation of the line which
   (a) passes through $(-2, 6)$ and $(2, 8)$.
   (b) has slope 4 and passes through $(2, 6)$.
   (c) passes through $(5, -3)$ and has x-intercept -3.
   (d) passes through $(0, -9)$ and $(5, -3)$.
   (e) has x- and y-intercepts -5 and 3 respectively.
   (f) has y-intercept -3 and passes through $(5, 4)$.
   (g) has slope $-\frac{2}{3}$ and x-intercept 3.
   (h) is perpendicular to the x-axis and passes through $(-3, 2)$.
   (i) has x-intercept -2 and passes through $(2, 5)$.
   (j) has y-intercept $\frac{3}{2}$ and slope -4.
   (k) has x- and y-intercepts 4 and -5 respectively.
   (l) passes through $(5, 4)$ and $(-2, -2)$.

3. A line is perpendicular to $y = 5x - 1$ and passes through $(4, -2)$. Find its equation.

4. A line is parallel to $3x + y = 5$ and passes through $(-2, 3)$. Find its equation.

5. Determine an equation of the line having the same y-intercept as $2x + 3y = -24$ and the same slope as $x - 3y = 5$.

problems and practice: a chapter review

At the end of each chapter, this section will provide you with additional questions to check your skills and understanding of the topics dealt with in the chapter. An important step for problem-solving is to decide which skills to use. For this reason, these questions are not placed in any special order. When you have finished the review, you might try the *Test for Practice* that follows.

1. (a) Draw the graph of the relation defined by $3x - 2y = 12$, $-3 \leq x \leq 3$.
   (b) Write the range of the relation.
   (c) Explain why the relation is a function.

2. Each of the following pairs of lines has equal y-intercepts. Find the value of k.
   (a) $y = 2x + 4$
       $y = x + 4k$
   (b) $4x + y - 4 = 0$
       $2x + ky + 5 = 0$

3. (a) Draw the graph given by $g: x \longrightarrow \frac{2}{3}x - 1$.
   (b) What is the slope in (a)?

4. Find the slope, x-intercept and y-intercept for each of the following.
   (a) $2x - 3y - 12 = 0$ (b) $3x - 4y = 0$

5. The vertices of a rhombus ABCD are given by A(−7, 3) B(2, 8) C(7, −1) D(−2, −6).
   (a) Show that the opposite sides are parallel.
   (b) Show that the diagonals are perpendicular.

6. The vertices of a right triangle are shown. The point S(2, 0) is the same distance from the vertices. Show that S lies on PR.

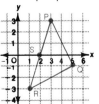

7. The lines $3x + 2y - 8 = 0$ and $2x + y = k$ have equal y-intercepts. Find the value of k.

8. (a) Show why the triangle with vertices given by (−6, −2), (−1, 0) and (1, −5) is right-angled.
   (b) At which vertex is the right angle?

9. The line segment with end points A(k, −5) and B(2, 3) has slope $-\frac{3}{4}$. Find k.

10. Find k if the line passing through (k, −5) and (4, −4) is parallel to the line joining (k, 4) and (−2, 3).

11. (a) Find the value of k if (−1, 1) is on the line $y = kx - 5$.
    (b) Determine the value of k if $\left(2, \frac{3}{4}\right)$ is on the line $ky = kx - 1$.

12. Find the value of t if the following lines have the same x-intercept.
    $3x - y = 1$ $3x - 2y - t = 0$

13. The segment AB given by A(2, 3), B(−10, −1) is divided into 4 equal parts. Find the points of division.

14. A region is given by $x \geq 0$, $y \geq 0$, $2x + 3y \leq 18$, $2x + y \leq 10$. Find the maximum value of P if P is given by $P = 2x + 5y$.

15. Steve and Sandra work for a sports firm which manufactures racquets. In 1 h Steve can put grips on 4 tennis racquets and 2 squash racquets while Sandra can put grips on 2 tennis racquets and 3 squash racquets. Each day grips must be put on a minimum of 24 tennis racquets and 20 squash racquets. Steve earns $6.00/h and Sandra earns $5.00/h.
    (a) How long should each person spend putting grips on each racquet to minimize the cost?
    (b) How many of each racquet will have grips put on them at this minimal labour cost?

test for practice

Try this test. Each *Test for Practice* will be based on the mathematics you have learned in the chapter. Try this test later in the year as a review. Keep a record of those questions that you were not successful with, find out how to answer them and review them periodically.

1. Write the domain and range for each relation.
   (a) (4, −1), (1, 1), (6, 0), (5, −4), (1, 3)
   (b) (2, 2), (−1, 5), (−4, 4), (1, −1), (3, 2)
   (c) Which relation is a function, (a) or (b)?

2. (a) Draw the graph of the relation $2x - 5y = 10$, $-1 \leq x \leq 4$.
   (b) Write the range of the relation.
   (c) Explain why the relation is a function.

3. If $f: x \longrightarrow x^2 - 1$, find each of the following.
   (a) $f(0)$ (b) $f(1)$ (c) $f(-2)$

4. (a) Find the intercepts for the line defined by $3x - 4y = 12$.
   (b) The lines given by $2x - y + 1 = 0$ and $2y - 4x + k = 0$ have equal y-intercepts. Find k.

5. Show that the line through A(−1, −1) and B(2, 3) is parallel to the line through C(−4, 0) and D(2, 8).

6. Use your skills with slope to determine which three points are collinear.
   (a) (−1, 3), (1, 8), (−3, −2)
   (b) (−2, −3), (4, 0), (10, 5)
   (c) (1, 12), (4, −3), (5, −8)

7. One vertex of a triangle is (4, 1). The slope of the side passing through this vertex is $-\frac{1}{2}$. Determine the value of b if $(b, 2)$ lies on the side.

8. Find the value of k if the line given by $2x - ky = -3$ has slope $\frac{1}{2}$.

9. The line passing through (5, k) and (−4, −2) is parallel to the line passing through (3, −k) and (1, 2). Find k.

10. Four lines, $x - 4y + 12 = 0$, $9x + y - 40 = 0$, $5x - y + 22 = 0$ and $x + 5y + 20 = 0$ intersect to form a quadrilateral. Determine whether or not the quadrilateral is a parallelogram.

11. Show that $y = m(x + 6) + \frac{1}{2}$ defines a family of lines passing through $\left(-6, \frac{1}{2}\right)$.

12. The point (−4, −9) is on the line $(k^2 - 3)y = \frac{1}{2}x - (3k + 1)$. Find the value of k.

13. Find the point on the y-axis which determines a line perpendicular to the line $x + 2y = 7$ when joined to (−2, 4).

14. Find the equation of the line passing through the intersection of the lines $3x = 5 - 2y$, $15x + 4y = 1$ and passing through (−3, 5).

15. Find the value of t if the following lines have the same x-intercept.
    $x - y + 13 = 0$ $tx - y = -6$

16. ABCD is a rectangle with vertices A(6, 3), B(−12, 9), C(−16, −3), and D(2, −9). Show that the quadrilateral formed by joining the midpoints of the sides is a parallelogram.

17. The midpoint of EF is (5, 1) and an endpoint is given by E(−1, 0). Find the co-ordinates of F.

18. A region is given by $x \geq 0$, $y \geq 0$, $x + y \leq 8$, $x + 2y \leq 10$. Find the maximum value of P if P is given by $P = 4x + 2y$.

math is plus applications: interpreting graphs

In the business and scientific world, people use their skills to interpret data that have been recorded on a graph. From the graph, they can make intelligent observations and derive useful conjectures. Often the graph is only sketched from the acquired data and the observations made about the graph may suggest a certain trend or conclusion. For example, the data collected for the profits of a company are shown on the following graph.

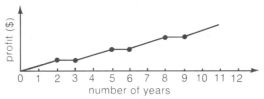

What observations can be made about the profit picture of the company in their first 10 a? (Remember: all we can do is make educated guesses based on the information we know.)

Each of the following graphs is a sketch obtained from the available data for distances travelled for a road trip. Describe what is happening during each trip.

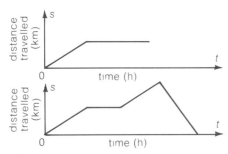

The following are the temperature graphs of various animals. Normal temperature is given by the broken line.

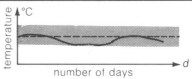

When the graph extends out of the shaded area, the animal is in critical condition.

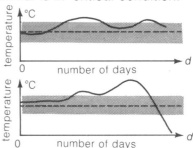

What information can be obtained from each graph?

The following is the graph of the temperature at a local swimming pool. How would you interpret the information given by the graph?

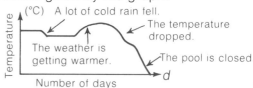

For each of the following situations, sketch a graph that describes the information. On your graph, record your interpretation.

A Water is taken from a tap and placed in a kettle and then heated.
B the change in length of a shadow throughout the day.
C the size of an apple growing on a tree.
D The temperature of a body of water through which icebergs, and other masses of ice, float periodically.
E The amount of sunshine received by a crop over a summer period.
F The number of people in attendance at a beach.
G The mass of the gasoline in the tank of a car during a road trip.

7 analytic geometry: distance, area, and applications

Distance, area, and applications; concept of locus, properties of circles, translations.

7.1 introduction

In studying mathematics or solving real-life problems, we often encounter problems that require a distance be calculated. To calculate the co-ordinates of a splashdown requires the knowledge of how to calculate distances very precisely.

In the telephone industry the charges for long distance calls are based on a formula that involves the assigning of co-ordinates to each place in a province and then finding the distance between the two places.

To calculate the area of a triangle, or any polygon, skill in calculating distances is required. From an aircraft the co-ordinates of the vertices of a triangular region may be determined and the area of the triangular region calculated directly. All that *just from knowing the vertices!* The work in this chapter makes use of algebra as a powerful partner with geometry to solve many problems using the methods and strategies of analytic geometry.

7.2 distance between two points

To solve many problems about analytic geometry we need to find the length of line segments.
- Finding the length of a line segment parallel to the x- or y-axis is straightforward.
- Finding the length of a line segment not parallel to either axis, requires the use of an important property of a right-angled triangle.

We may use the Pythagorean Theorem for a right-angled triangle to find the distance between two points. For any two points P and Q, on the plane, we may draw the associated right-angled triangle for the points as shown.

PC∥x-axis
QC∥y-axis
For △PCQ, $\angle C = 90°$.
$\Delta x = x_2 - x_1$
$\Delta y = y_2 - y_1$

We may use Δx and Δy to represent the directed distances in the above diagram.

Example 1
Find the length of the line segment joining P(−4, −2) and Q(2, 2).

Solution

From the diagram
length of PC
$= |2 - (-4)|$
$= |2 + 4| = 6$
Length of CQ
$= |2 - (-2)|$
$= |2 + 2| = 4$

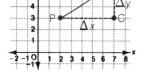

In △PQC, $\angle C$ is a right angle. Thus
$PQ^2 = PC^2 + CQ^2 = 36 + 16 = 52$
$PQ = \sqrt{52} = 2\sqrt{13}$ (length is positive)
Thus the length of the line segment joining P and Q is $2\sqrt{13}$ units.

finding a distance formula

We may find a general formula to show the distance between any two points.

$P_1(x_1, y_1)$ and $P_2(x_2, y_2)$ are any two points on the plane.
Locate the point $P_3(x_2, y_1)$ as shown in the diagram.

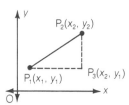

From the diagram $\angle P_2P_3P_1 = 90°$.
$P_1P_3 = |x_2 - x_1| \quad P_3P_2 = |y_2 - y_1|$
$\quad\quad |\Delta x| \quad\quad\quad\quad\quad |\Delta y|$

Since △$P_2P_3P_1$ is a right-angled triangle then
$P_1P_2^2 = P_1P_3^2 + P_3P_2^2$
$\quad\quad = (|x_2 - x_1|)^2 + (|y_2 - y_1|)^2$
$\quad\quad = (x_2 - x_1)^2 + (y_2 - y_1)^2$
$P_1P_2 = \sqrt{(x_2 - x_1)^2 + (y_2 - y_1)^2}$
or $P_1P_2 = \sqrt{|\Delta x|^2 + |\Delta y|^2}$ where $|\Delta x| = |x_2 - x_1|$
$\quad\quad\quad\quad\quad\quad\quad\quad\quad\quad\quad\quad\quad |\Delta y| = |y_2 - y_1|$

When solving a problem, it is useful to draw a diagram to help us visualize the problem as illustrated in Example 2.

Example 2
Find a point on the x-axis which is equidistant from A(−1, 5) and B(6, −2).

Solution Remember: Draw a diagram to help you interpret what the problem asks.

Let the co-ordinates of the point on the x-axis be P(x, 0).
Use PA = PB.

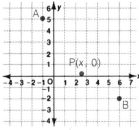

$PA = \sqrt{[x - (-1)]^2 + (0 - 5)^2}$
$\quad = \sqrt{x^2 + 2x + 1 + 25}$
$\quad = \sqrt{x^2 + 2x + 26}$

$PB = \sqrt{(x-6)^2 + [0-(-2)]^2}$
$= \sqrt{x^2 - 12x + 36 + 4}$
$= \sqrt{x^2 - 12x + 40}$

Use PA = PB.
$\sqrt{x^2 + 2x + 26} = \sqrt{x^2 - 12x + 40}$

Square both sides and collect terms.
$2x + 12x = 40 - 26$
$14x = 14$
$x = 1$

Thus the point P with co-ordinates (1, 0) is equidistant from A and B.

Remember: You should check the answer.

Throughout the following exercise, draw a diagram to help you clearly understand the problem, and answer the two questions:
- What information are we given?
- What information are we asked to find?

7.2 exercise

Throughout this exercise, unless indicated otherwise, you may leave your final answer in simplified radical form.

1 Find the length of each line segment.

(a) AH (b) CD
(c) OF (d) OA
(e) DJ (f) EI
(g) DF (h) AC

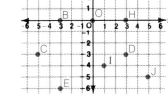

2 Find the distance from each point to the origin.

(a) (6, 8) (b) (4, 5) (c) (−1, 8)
(d) $\left(\frac{3}{4}, 1\right)$ (e) $(\sqrt{3}, 1)$ (f) $\left(\frac{-\sqrt{2}}{2}, \frac{\sqrt{2}}{2}\right)$

Find the length of the line segment joining each pair of points.

(a) (0, 0), (8, 6) (b) (0, 2), (3, 3)

(c) (−3, 0), (8, −5) (d) (−2, −2), (8, 0)
(e) (8, −8), (4, −1) (f) (6, 8), (−6, −8)

4 Find the distance from each point to (1, 4).
(a) (−1, 7) (b) (−2, 6) (c) (4, 6)
(d) What do you notice about your answers?

5 Which line segment is longer?
AB: A(−3, 1), B(3, 3)
CD: C(−1, −6), D(1, 1)

B

6 Find the perimeter of each triangle.
(a) △ABC A(1, 5) B(1, 2) C(5, 2)
(b) △PQR P(4, 2) Q(−5, −10) R(4, −10)
(c) △DEF D(8, 10) E(−7, −10) F(−7, 10)
(d) △STU S(2, 9) T(9, −15) U(9, 9)

7 A triangle has vertices P(−1, 2), Q(2, 6), R(−4, 4).
(a) Find the perimeter.
(b) Classify the triangle as scalene, isosceles, or equilateral.

8 Three vertices of rectangle ABCD are A(−8, 0), B(4, 4), and C(6, −2).
(a) Find the lengths of the sides.
(b) Find the length of the diagonal.

9 If A(−2, 2), B(7, 5), C(9, −3) and D(−4, −2) are the vertices of a quadrilateral, determine which diagonal is longer.

10 Calculate the area of the rectangle whose co-ordinates are P(−3, 2), Q(2, 4), R(4, −1), and S(−1, −3).

11 Decide whether the points P(−2, −1) and Q(5, −8) lie on the same circle with centre C(1, −5). Give reasons for your answer.

12. Find the co-ordinates of the point on the x-axis which is equidistant from A(−4, 6) and B(4, 10).

13. Find the length of the line segment with end points given by the points where the line $3x - 4y + 24 = 0$ cuts the x- and y-axis.

14. Given that P(2, −1), Q(−4, 7) and R(3, 6) lie on a circle, show that the point C(−1, 3) is the centre of that circle.

15. The vertices of an isosceles triangle are given by A(1, 7), B(−5, 1), and C(7, 1). Determine whether the triangle formed by joining the midpoints of the sides of △ABC is also isosceles.

C

16. If (x_1, y_1) and (x_2, y_2) are two points on the line $y = mx + b$, then show that the distance between the two points is given by the expression $|x_2 - x_1|\sqrt{1 + m^2}$.

problem/matics

Often, we are not able to prove something in mathematics because we have perhaps forgotten, or do not clearly understand, the meanings of the words given in the problem. For example, read the following.

> Prove that the sum of the squares of the lengths of the segments from any point P to two opposite vertices of a rectangle is equal to the sum of the squares of the lengths of the segments from P to the remaining vertices.

To solve the above problem you must understand clearly the meaning "sum of the squares", "opposite vertices" and "from any point P". Now solve the problem.

applications: solving problems about distance

17. The pilot of a jet calculates that there is just enough fuel to fly 37 km. The jet's location is at (12, 3) and the closest airport is at (17, −9). Does the jet have enough fuel to reach the airport or will it have to make an emergency landing (crash)?

18. The airport control tower locates two planes on its radar screen. Plane A is located at (−3, 7) and plane B at (8, 2). If the two planes are travelling at the same speed and the control tower is located at (1, 2), which plane will land first?

19. A small pleasure boat located at (1, 6) breaks its rudder and sends out a distress signal to the coast guard. Rescue boat A is at (4, −2) and rescue boat B is at (6, 3). Which rescue boat should the coast guard send out to aid the distressed boat?

20. A boat sails from its harbour at (6, 1) to an island at (12, −5). It then sails from the island to meet another sailboat anchored at (3, 4).
    (a) How far has the boat sailed in reaching the second sailboat?
    (b) How much farther did the boat sail than it would have if it had headed directly for the second boat?

21. A group of canoeists must find a campsite before it gets much darker. Their map indicates two possible campsites: camp A at (1, 7) and camp B at (8, −1). If the canoeists are at (−1, 1), which campsite is closer?

22. A co-ordinate system is superimposed on a billiard table. Jack has a red ball at co-ordinates (0, 0) and is going to "bank" it off the side rail at (4, 2) and into the pocket at (0, 4). How far will the red ball travel?

7.3 choosing strategies for solving problems

As we learn additional concepts and skills in analytic geometry, we learn different strategies for solving problems. One strategy may be more advantageous than another when we are confronted with a problem we have never seen before. Sometimes we need to combine strategies to solve the problem. However, to solve any problem we must read the problem carefully and be able to answer Questions A and B.

A: What information are we given?
B: What are we asked to find?

Then we can use our different strategies and the information in A to find B. To solve the next problem, we could use slopes from our earlier work. However, we may now use our work with distance.

Example 1
Show that $A(-6, -3)$, $B(-2, -1)$, and $C(2, 1)$ are collinear.

Solution

Draw a diagram to help organize your solution.

One way of proving A, B, and C are collinear, is to show $AB + BC = AC$.

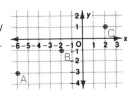

$AB = \sqrt{[-2 - (-6)]^2 + [-1 - (-3)]^2}$
$= \sqrt{16 + 4}$
$= \sqrt{20}$ or $2\sqrt{5}$

$BC = \sqrt{[2 - (-2)]^2 + [1 - (-1)]^2}$
$= \sqrt{16 + 4}$
$= \sqrt{20}$ or $2\sqrt{5}$

Thus, $AB + BC = 2\sqrt{5} + 2\sqrt{5} = 4\sqrt{5}$

$AC = \sqrt{[2 - (-6)]^2 + [1 - (-3)]^2}$
$= \sqrt{64 + 16}$
$= \sqrt{80}$
$= 4\sqrt{5}$

Since $AB + BC = AC$ then A, B, and C are collinear.

In this chapter, we will learn yet other concepts and skills to solve problems involving analytic geometry.
- The use of a diagram to translate the given information may save us needless steps and work.
- Before we tackle a problem we must plan our work, since a different strategy for solving the same problem may require a lot less work but accomplish the same result.

PLAN AHEAD

7.3 exercise

For each of the problems that follow
- choose a method of solving the problem.
- decide which other strategy could be used to solve the problem.

A

1 △ABC is given by the vertices
   A(9, 0)   B(5, 2)   C(7, −4)

   (a) Show that △ABC is right-angled by finding the lengths of sides.
   (b) Show that △ABC is right-angled by finding the slopes of sides.

2 The diagonals of a rhombus intersect at right angles.

   (a) Use the above fact to prove that the figure PQRS given by P(1, 2), Q(−2, −1), R(1, −4), and S(4, −1) is a rhombus.
   (b) Prove the result in (a) in another way.

3. (a) Use the distance formula to prove that the points D(−1, 0), E(2, 1), and F(−4, −1) are collinear.
   (b) What other strategy might be used to prove the result in (a)?

4. Show that P(−1, −5), Q(3, −3) and R(7, −1) are collinear.

5. Prove that P(2, 3) is the midpoint of the line segment DE given by D(9, 0), E(−5, 6).

6. The vertices for figure ABCD are given by A(−5, −9), B(−10, 3), C(2, 8), D(7, −4). Show that figure ABCD is a square.

7. Prove that the points P(−3, 1), Q(0, 2) and R(3, 3) are not the vertices of a triangle.

8. Prove that the midpoint of the line segment AB whose end points are A(−3, −2) and B(−1, 0) is (−2, −1).

9. Prove that △PQR is a right-angled triangle given the vertices P(4, 7), Q(−1, 2), R(2, −1).

10. (a) Prove that the vertices P(−5, −7), Q(−7, −1) R(2, 2) and S(4, −4) determine a rectangle.
    (b) What simple test would quickly prove that the figure in (a) is not a square?

11. Show that the points P(−8, 2) and Q(−2, −10) are the end points of the diameter of a circle with centre C(−5, −4).

12. Three consecutive vertices of a parallelogram are (−4, 1), (2, 3) and (8, 9). Find the co-ordinates of the fourth vertex.

13. (a) The lines $8x - 7y + 20 = 0$, $x + 5y = 21$, $7y = 8x - 27$ and $x = -5y - 26$, intersect to form a quadrilateral. Show that a parallelogram is formed.
    (b) How could you determine if the parallelogram is a rhombus?

14. The lines $x - 3y + 7 = 0$ and $x + 2y - 3 = 0$ intersect at A. the lines $4x - y = 28$ and $5x + 2y = 35$ intersect at B.
    (a) Find the midpoint of AB.
    (b) Determine whether the point in (a) lies on the line segment with end points P(−5, −1) and Q(7, 2).

15. • P is the intersection of $7x - 2y + 29 = 0$ and $4x + 5y + 35 = 0$.
    • Q is the intersection of $x + 4y = 19$ and $2x + 3y = 13$.
    (a) Find the midpoint M of PQ.
    (b) Show that △SMT is right-angled if the co-ordinates of S are (0, 5) and of T are (1, −2).

16. The lines defined by $4y = 5x - 11$, $4x + 5y - 17 = 0$, $x = \frac{4}{5}y - 6$ and $5y = -4x + 58$ intersect to form a quadrilateral. Determine whether the quadrilateral is a square.

17. The streets bordering a city block are represented by the lines $2y = x + 9$, $2x + y + 6 = 0$, $x = 2y + 6$ and $2x + y = -8$. Determine whether the city block is square.

18. (a) Find the vertices of the quadrilateral formed by the four lines $7y - 3x + 45 = 0$, $7x = 47 - 3y$, $7y - 3x = 13$, and $3y = 15 - 7x$.
    (b) Classify the quadrilateral.

19. A(−5, 3), B(−4, −2), C(7, −1) and D(6, 4) are the vertices of a parallelogram.
    (a) Find the midpoints P and Q of AB and CD respectively.
    (b) Show PQ is parallel to AD and BC.
    (c) If PQ intersects BD at X, does X lie on AC?

applications: distances in Canada

We may use a co-ordinate system to calculate distances between cities. The co-ordinate system is positioned with the origin on the boundary of Alberta and Saskatchewan as shown. Each place is then assigned co-ordinates with respect to the axes.

Edmonton E(−3, −2)   Winnipeg W(11, −6)

To calculate the straight line distance between Edmonton and Winnipeg, we may use the distance formula.

$$EW = \sqrt{[11 - (-3)]^2 + [-6 - (-2)]^2}$$
$$= \sqrt{196 + 16}$$
$$= \sqrt{212}$$
$$\doteq 14.6 \text{ (to 1 decimal place)}$$

From the map, the distance between Edmonton and Winnipeg is 14.6 units.

If each unit on the map represents approximately 80 km, what is the approximate distance between Edmonton and Winnipeg?
14.6 units corresponds to 1168 km.
This is approximately the distance given on the map.

20. Look at the map. What are the co-ordinates of each place? Express your co-ordinates to the nearest integer.
    (a) Vancouver   (b) Calgary   (c) Regina
    (d) Brandon   (e) Churchill (f) Dawson Creek
    (g) Saskatoon (h) Prince George

21. If 1 unit represents approximately 80 km, calculate the distance between each of the following to the nearest 10 km.
    (a) Manitoba: Churchill to Winnipeg
    (b) Saskatchewan: Regina to Uranium City
    (c) Alberta: Lethbridge to Peace River
    (d) British Columbia: Fort Nelson to Victoria

22. A private jet flies the following route.
    Victoria to Edmonton to Regina
    (a) Calculate the total distance of the trip.
    (b) How much shorter would the trip be if the plane flew directly from Victoria to Regina?

23. During the Hot Air Contest, a manned balloon drifted from York Factory, Manitoba, to Fort Nelson, British Columbia. Calculate the distance that the balloon drifted.

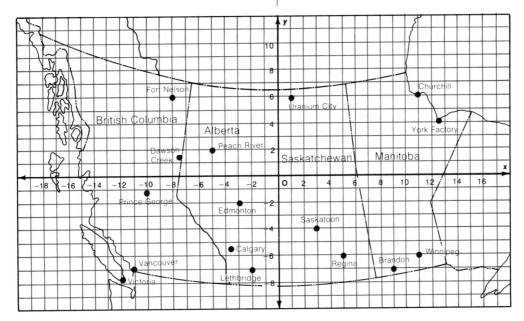

7.4 finding areas of triangles

To calculate the area, A, of a triangle, we may use the formula
$A = \frac{1}{2}bh$.

$A = \frac{1}{2}bh$

However, to calculate the area of a triangle whose vertices are as shown on a Cartesian plane, we must develop a method suitable for working with co-ordinates. To do so, we need to use the formula $A = \frac{1}{2}bh$ to develop such a method.

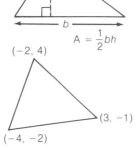

Example 1 uses a method involving numerical co-ordinates to suggest a general method of calculating the area of a triangle whose vertices are defined by co-ordinates.

Example 1
A triangular island is located on a grid with vertices P(2, 6), Q(4, 2), R(7, 5). Calculate its area.

Solution
Step 1
Draw the triangle on the Cartesian plane.

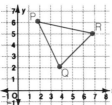

Step 2
To calculate the area of △PQR, construct the rectangle associated with the triangle. Then determine the co-ordinates of A, B, and C as shown.

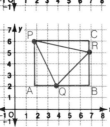

Step 3
From the diagram

Area of △PQR = Area of rect PABC − Area of △PAQ − Area of △RQB − Area of △PRC

$$= \underset{lw}{4 \times 5} \quad \underset{\frac{1}{2}bh}{= \frac{1}{2} \times 2 \times 4} \quad \underset{\frac{1}{2}bh}{= \frac{1}{2} \times 3 \times 3} \quad \underset{\frac{1}{2}bh}{= \frac{1}{2} \times 1 \times 5}$$

$= 20 \quad = 4 \quad = \frac{9}{2} \quad = \frac{5}{2}$

$= \left(20 - 4 - \frac{9}{2} - \frac{5}{2}\right)$ square units
$= 9$ square units

Using the above method to calculate the area of a triangle, given the co-ordinates of the vertices, may be time consuming. The purpose of mathematics is to examine a specific example carefully and try to develop a general method.

Thus the above method of calculating the area of a triangle, with given co-ordinates, provides the basis for developing a formula for the area of a triangle when the co-ordinates of the vertices are known. Compare the Steps 1, 2, 3 above to those that follow. Throughout our work we will use △ABC to represent *the area of* △ABC. You will know from the context whether the reference is to the "area" of △ABC.

formula: the area of a triangle

Step 1
△PQR has vertices with co-ordinates $P(x_1, y_1)$, $Q(x_2, y_2)$, $R(x_3, y_3)$ taken in counterclockwise direction.

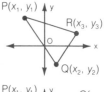

Step 2
Draw rectangle PABC and determine the co-ordinates of A, B, and C.

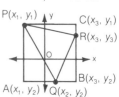

Step 3
From the diagram, in square units,
△PQR = rect PABC − △PAQ − △RQB − △PRC
Area of rect PABC
$$= (x_3 - x_1)(y_1 - y_2)$$
$$= x_3y_1 - x_1y_1 - x_3y_2 + x_1y_2$$

△PAQ $= \frac{1}{2}bh$
$$= \frac{1}{2}(x_2 - x_1)(y_1 - y_2)$$
$$= \frac{1}{2}x_2y_1 - \frac{1}{2}x_1y_1 - \frac{1}{2}x_2y_2 + \frac{1}{2}x_1y_2$$

△RQB $= \frac{1}{2}bh$
$$= \frac{1}{2}(x_3 - x_2)(y_3 - y_2)$$
$$= \frac{1}{2}x_3y_3 - \frac{1}{2}x_2y_3 - \frac{1}{2}x_3y_2 + \frac{1}{2}x_2y_2$$

△PRC $= \frac{1}{2}bh$
$$= \frac{1}{2}(x_3 - x_1)(y_1 - y_3)$$
$$= \frac{1}{2}x_3y_1 - \frac{1}{2}x_1y_1 - \frac{1}{2}x_3y_3 + \frac{1}{2}x_1y_3$$

△PQR $= x_3y_1 - x_1y_1 - x_3y_2 + x_1y_2$
$- \left(\frac{1}{2}x_2y_1 - \frac{1}{2}x_1y_1 - \frac{1}{2}x_2y_2 + \frac{1}{2}x_1y_2\right)$
$- \left(\frac{1}{2}x_3y_3 - \frac{1}{2}x_2y_3 - \frac{1}{2}x_3y_2 + \frac{1}{2}x_2y_2\right)$
$- \left(\frac{1}{2}x_3y_1 - \frac{1}{2}x_1y_1 - \frac{1}{2}x_3y_3 + \frac{1}{2}x_1y_3\right)$

$= x_3y_1 - x_1y_1 - x_3y_2 + x_1y_2$
$- \frac{1}{2}x_2y_1 + \frac{1}{2}x_1y_1 + \frac{1}{2}x_2y_2 - \frac{1}{2}x_1y_2$
$- \frac{1}{2}x_3y_3 + \frac{1}{2}x_2y_3 + \frac{1}{2}x_3y_2 - \frac{1}{2}x_2y_2$
$- \frac{1}{2}x_3y_1 + \frac{1}{2}x_1y_1 + \frac{1}{2}x_3y_3 - \frac{1}{2}x_1y_3$

$= \frac{1}{2}x_3y_1 - \frac{1}{2}x_3y_2 + \frac{1}{2}x_1y_2 - \frac{1}{2}x_2y_1 + \frac{1}{2}x_2y_3 - \frac{1}{2}x_1y_3$

Reorganize it into a more suitable form.

$= \frac{1}{2}x_1y_2 + \frac{1}{2}x_2y_3 + \frac{1}{2}x_3y_1 - \frac{1}{2}x_1y_3 - \frac{1}{2}x_3y_2 - \frac{1}{2}x_2y_1$

To obtain a positive number, use the absolute value

△PQR
$$= \frac{1}{2}|x_1y_2 + x_2y_3 + x_3y_1 - x_1y_3 - x_3y_2 - x_2y_1|$$

To calculate the area of a triangle using the above formula, we may use the following steps which simplifies the procedure.

Step 1 — Record the vertices in a counter-clockwise order

(x_1, y_1)
(x_2, y_2)
(x_3, y_3)
(x_1, y_1)

Step 2 — Calculate the down-products as shown.

$x_1 \searrow y_1$
$x_2 \searrow y_2$
$x_3 \searrow y_3$
$x_1 \searrow y_1$

Step 3 — Calculate the up-products as shown.

$x_1 \nearrow y_1$
$x_2 \nearrow y_2$
$x_3 \nearrow y_3$
$x_1 \nearrow y_1$

(Repeat the first pair.)

△PQR $= \frac{1}{2} \left| \begin{pmatrix}\text{Sum of Down}\\\text{Products}\end{pmatrix} - \begin{pmatrix}\text{Sum of Up-}\\\text{Products}\end{pmatrix} \right|$

$= \frac{1}{2}|x_1y_2 + x_2y_3 + x_3y_1 - (x_1y_3 + x_3y_2 + x_2y_1)|$

Example 2

Calculate the area of △PQR with vertices P(2, 4), Q(2, −3), R(−5, 2).

Solution

Draw a diagram to help plan the solutions.

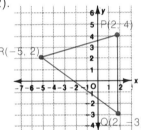

List the co-ordinates in a counterclockwise order.

Coordinates	Calculate down-products	Calculate up-products
(2, 4)	2 4	2 4
(−5, 2)	−5 2	−5 2
(2, −3)	2 −3	2 −3
(2, 4)	2 4	2 4

△PQR $= \frac{1}{2}|(4 + 15 + 8) - (-6 + 4 - 20)|$

$= \frac{1}{2}|27 + 22| = \frac{49}{2}$

The area of △PQR is $\frac{49}{2}$ square units.

Note:—In example 1, if the co-ordinates were not listed in a counterclockwise manner then

the calculation would be
$$\triangle PQR = \frac{1}{2}|-49| = \frac{49}{2}$$

A negative quantity is obtained between the absolute value bars.

7.4 exercise

A

1. △ABC has vertices A(−2, −5), B(3, −6), and C(7, 2).
   (a) Sketch the triangle.
   (b) Calculate the area of △ABC by listing the vertices in a clockwise order.
   (c) Calculate the area of △ABC by listing the vertices in a counterclockwise order.
   (d) How are your solutions in (b) and (c) similar? How are they different?

2. Calculate the area of the triangle given by each of the following sets of vertices.
   (a) (5, 2)   (−2, 0)   (1, −5)
   (b) (4, 2)   (−5, 1)   (9, −3)
   (c) (4, 3)   (7, 4)   (10, −2)
   (d) (−3, 5) (−6, −3,) (−1, −4)

3. Which has the greater area, △ABC or △DEF?
   △ABC: A(3, −3) B(1, 5) C(−1, 7)
   △DEF: D(3, 1) E(6, 5) F(5, −4)

4. △PQR has vertices P(5, 5), Q(−5, −5), and R(1, 1).
   (a) Calculate the area of △PQR.   (b) Plot △PQR.
   (c) Use your answer in (b) to provide a reason for your answer in (a).

B

5. Use the formula for calculating area to show that the points D(4, 9), E(−1, −1), F(−7, −13) are collinear.

6. The vertices of △STU are S(−1, −5), T(5, −2), and U(3, −3).
   (a) Calculate the area of △STU.
   (b) Based on your results in (a), what conclusion can you make about the points S, T, and U?

7. Determine which sets of points are collinear. Use your skill in finding the area of a triangle.
   (a) S(6, 0),   T(2, −3)   U(−2, −6)
   (b) P(−2, −1) Q(−6, −3) R(2, 1)
   (c) D(−1, 2)   E(3, 6)   F(−7, −3)
   (d) G(2, 4)   H(−4, 0)   K(−7, −2)

8. The vertices of △PQR are P(6, −5), Q(2, 5), and R(−8, −3).
   (a) Find the co-ordinates of S, the midpoint of RP.
   (b) Calculate the area of △QRS.
   (c) Calculate the area of △PQS.
   (d) Use your results in (a) to (c) to explain why "the median QS bisects the area of △PQR."

9. The equations of the sides of △PQR are $\ell_1: x + 3y − 14 = 0$, $\ell_2: 3x − y − 22 = 0$, and $\ell_3: 2x + y + 2 = 0$. Find the insersection point of
   (a) ℓ_1, ℓ_2 (b) ℓ_2, ℓ_3 (c) ℓ_3, ℓ_1
   (d) Use the intersection points in (a) to (c) to calculate the area of △PQR.

10. The sides of a triangular island are defined by $\ell_1: 2x − y + 1 = 0$, $\ell_2: x + 2y − 7 = 0$, and $\ell_3: 3x + y − 16 = 0$.
    (a) Draw a sketch of the island.
    (b) Calculate the area of the island.

11. A triangle is determined by the points A(−2, 1), B(−3, −3), and C(x, y). What equations will x and y satisfy if the area of the triangle is 17 square units?

12. The points A(3, 4), B(−5, −2), and C(x, y) form a triangle with area 25 square units. If C is equidistant from A and B, find the co-ordinates of C.

13 A triangle has vertices at P(−3, 2), Q(5, −4), and R(x, y). R is equidistant from P and Q. If the area of the triangle is 50 square units, find the values of x and y.

14 Find k such that $3x - ky = 12$, when intersected with the positive direction of the axes, makes a triangle whose area is 12 square units.

15 A triangle has vertices (4, 5), (4, −3) and (x, y). If the area of the triangle is 16 square units, determine the possible values of x.

16 The points P(1, y), Q(3, −5) and R(2, 4) form a triangle with area 2 square units. Find the possible values for y.

17 The points P(−6, −2), Q(6, 3), and R(−1, y) (where y > 1) form a triangle with area $\frac{59}{2}$ square units. Find y.

18 The points P(x, 2), Q(−2, 1), and R(2, −2), (where x > 0) form a triangle with area $\frac{13}{2}$ square units. Find x.

C

19 The vertices of △PQR are given by P(−4, 5), Q(−2, −5), R(4, 3).
   (a) Calculate the area of △PQR.
   (b) Find the midpoints A, B, C, of PQ, QR and RP respectively.
   (c) Show that $\triangle ABC = \frac{1}{4} \triangle PQR$.

20 The sides of △DEF are defined by
   DE: $x - 3y + 8 = 0$
   EF: $3x + y - 16 = 0$
   DF: $x + y - 12 = 0$
   If A, B, C are respectively the midpoints of DE, EF and DF, show that $4\triangle ABC = \triangle DEF$.

extending our work: area of polygon

To find the area of a polygon, we can use the same procedure that we used to find the area of a triangle. The following questions illustrate the method.

21 A rectangle PQRS is given by the vertices
   P(−5, 3) Q(4, 3) R(4, −3) S(−5, −3)
   (a) Draw a sketch of the rectangle.
   (b) Calculate the area of the rectangle, using the formula $A = b \times h$, base, b, height, h.
   (c) Calculate the area of the rectangle by using the same procedure for calculating the area of a triangle except include the 4th vertex as shown for the list of vertices.

Vertices of rect PQRS	
−5	3
4	3
4	−3
−5	−3
−5	3

22 Repeat the steps in Question 21 for each of the following polygons.
   (a) quadrilateral PQRS: P(2, 1), Q(1, 7), R(10, 3), S(14, −11).
   (b) pentagon PQRST: P(4, 2), Q(6, −4), R(5, −6), S(0, −8), T(2, −1).

23 The vertices of various polygons are given. Calculate the area.
   (a) A(0, 0), B(3, 2), C(1, 4), D(−2, 3)
   (b) S(−5, 2), T(−5, −3), U(0, −5), V(3, −2)
   (c) M(−3, 3), P(−3, −2), Q(0, −4), R(3, −2), V(3, 3)

   As a check on your results, calculate the area of each of the polygons above using a different method.

24 (a) Show that the four lines
      $3x + 2y + 7 = 0$ $3x + 2y + 20 = 0$
      $2x = 3y + 17$ $2x = 3y + 4$
      intersect to form a square.
   (b) Find the area of the square.

7.5 distance from a point to a line

A method of mathematics is as follows:

We utilize what we learn from specific examples → to develop a general method or formula.

There are many situations that require the calculation of the distance from a point to a line. By distance from a point to a line we mean the length of the perpendicular from the point to the line.

Example 1 finds the distance from a point to a line numerically and identifies the steps required to develop the general formula. Basic principles of analytic geometry are used in solving the following problem.

Example 1

Calculate the distance from the point $P(4,6)$ to the line defined by $\ell: 2x - y = 7$.

It is helpful to draw a sketch to include the given information so that you may plan your solution.

Solution

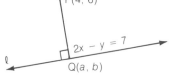

Step A
Let (a,b) represent the co-ordinates of Q at the foot of the perpendicular.
$Q(a,b)$ satisfies the line ℓ,
$$2a - b = 7 \quad ①$$

Step B
Since slope of ℓ is 2 and $PQ \perp \ell$, then
$$\text{slope } PQ = -\frac{1}{2}.$$
Equation of PQ is therefore
$$\frac{y-6}{x-4} = -\frac{1}{2} \quad \text{or} \quad x + 2y = 16$$

Since $Q(a, b)$ is on this line then
$$a + 2b = 16 \quad ②$$

Step C
Solve equations ① and ②.

$$\begin{array}{rl} 2a - b = 7 & ① \\ a + 2b = 16 & ② \end{array}$$

$2 \times ①$ $\quad 4a - 2b = 14 \quad ③$
$② + ③$ $\quad \overline{ 5a = 30}$
$ a = 6$

From ② $\quad 6 + 2b = 16$
$ b = 5$

The co-ordinates are given by $Q(6, 5)$.

Step D $\quad PQ = \sqrt{(4-6)^2 + (6-5)^2}$
$ = \sqrt{4 + 1}$
$ = \sqrt{5}$

The distance from $P(4, 6)$ to the line, ℓ, is $\sqrt{5}$ units. The closest that the point P is to the line ℓ is $\sqrt{5}$ units.

In Example 1 we identify what steps are needed to develop a general method of calculating the distance from a point to a line.

Example 2

Derive a formula for obtaining the distance from the point $P(x_1, y_1)$ to the line, ℓ, given by $Ax + By + C = 0$.

Solution

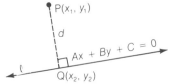

Step A
Let $Q(x_2, y_2)$ represent the point at the foot of the perpendicular. $Q(x_2, y_2)$ satisfies the line, ℓ.
$$Ax_2 + By_2 = -C \quad ①$$

Step B
Since the slope of ℓ is $-\frac{A}{B}$ and $PQ \perp \ell$
then $\text{slope}_{PQ} = \frac{B}{A}$.

Equation of PQ is
$$\frac{y - y_1}{x - x_1} = \frac{B}{A}$$
Since $Q(x_2, y_2)$ is on the above line then
$$\frac{y_2 - y_1}{x_2 - x_1} = \frac{B}{A}$$
or $\quad Bx_2 - Ay_2 = Bx_1 - Ay_1 \quad$ ②

Step C
Solve equations ① and ② for x_2 and y_2 to obtain.
$$x_2 = \frac{B^2x_1 - ABy_1 - AC}{A^2 + B^2} \quad y_2 = \frac{-ABx_1 + A^2y_1 - BC}{A^2 + B^2}$$
Use the co-ordinates for $Q(x_2, y_2)$ to calculate the length of PQ in Step D.

Step D
$$PQ = \sqrt{(x_1 - x_2)^2 + (y_1 - y_2)^2}$$
$$= \sqrt{\left[x_1 - \frac{B^2x_1 - ABy_1 - AC}{A^2 + B^2}\right]^2 + \left[y_1 - \frac{-ABx_1 + A^2y_1 - BC}{A^2 + B^2}\right]^2}$$
which simplifies to
$$PQ = \sqrt{\frac{(Ax_1 + By_1 + C)^2(A^2 + B^2)}{(A^2 + B^2)^2}}$$
$$= \sqrt{\frac{(Ax_1 + By_1 + C)^2}{(A^2 + B^2)}}$$
$$= \frac{|Ax_1 + By_1 + C|}{\sqrt{A^2 + B^2}}$$

Thus, to find the distance, d, from a point $P(x_1, y_1)$ to a line $Ax + By + C = 0$, we use the formula
$$d = \frac{|Ax_1 + By_1 + C|}{\sqrt{A^2 + B^2}}$$

In Example 2 note
- how the co-ordinates $P(x_1, y_1)$ are utilized.
- that the equation of the line is written in standard form.

Use the above formula to check the distance in Example 1.

Example 3
Find the altitude of △PQR from R to QP if the vertices are given by P(1, 2), Q(10, −1), and R(8, 3).

Solution
Draw a sketch.

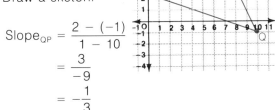

$$\text{Slope}_{QP} = \frac{2 - (-1)}{1 - 10}$$
$$= \frac{3}{-9}$$
$$= -\frac{1}{3}$$

Equation through P and Q
$$\frac{y + 1}{x - 10} = -\frac{1}{3} \quad \text{or} \quad x + 3y - 7 = 0$$

Distance, d, from R to PQ
$$d = \frac{|Ax_1 + By_1 + C|}{\sqrt{A^2 + B^2}} \quad \begin{array}{l} A = 1, B = 3, \\ C = -7 \\ x_1 = 8, y_1 = 3 \end{array}$$
$$d = \frac{|8 + 3(3) - 7|}{\sqrt{1^2 + 3^2}}$$
$$= \frac{10}{\sqrt{10}} = \sqrt{10}$$

Thus, the length of the altitude is $\sqrt{10}$ units.

In Section 7.7, we will see other ways of solving problems as well as developing other methods of deriving the formula for the distance from a point to a line.

7.5 exercise

Questions 1 to 10 examine skills needed to find the distance from a point to a line.

A

1 △PQR is given by the vertices
   P(1, 2)   Q(5, −4)   R(−1, −2).

   (a) Write the equation of the line through QR.

   (b) Use the formula to calculate the distance from P to QR.

   (c) Use the formula to calculate the altitude from Q to RP.

2. △STU is given by the vertices
S(8, 3)   T(4, −5)   U(−2, −3).
(a) Sketch the triangle.
(b) Calculate the length ST.
(c) Calculate the length of the altitude from U.
(d) Use your results in (b) and (c) to calculate the area of the triangle.
(e) Calculate the area of △STU by using the formula. Compare your results to that in (d).

3. (a) Calculate the distance from (3, 2) to $2x - y + 1 = 0$.
(b) Calculate the distance from (−1, 4) to $2x - y + 1 = 0$.
(c) Draw a sketch of your results in (a) and (b).

4. Calculate the distance from the origin to each of the following lines.
(a) $3x + 4y = 6$
(b) $6x - 3y = 2$
(c) $y = 4x - 5$
(d) $2(x - y) = 9$

5. Calculate the distance from (−3, 2) to each line.
(a) $3x - 2y = 8$
(b) $3x + 2y = 12$

6. Calculate the length of the perpendicular from (−2, 3) to each line.
(a) $5x - y - 8 = 0$
(b) $y = \dfrac{3x - 1}{2}$

7. (a) Calculate the distance from Q(3, −2) to the line $2x - 3y = 12$.
(b) Interpret your answer in (a).

8. Calculate the distance from the point P(3, 6) to the line through A(−5, 4) and B(2, −3).

9. For each point, a corresponding line is given. Calculate the distance from the point to the line.
(a) (4, 0) $2x = y + 3$
(b) (0, −3) $4x + 3y = 8$
(c) (−3, −4) $2(x - 2) = 3y$
(d) (0, 0) $3x + 11y + 64 = 0$
(e) (4, 3) $y = 8$ (f) (−3, 1) $x = 6$

10. Refer to the development of the formula
$$d = \frac{|Ax_1 + By_1 + C|}{\sqrt{A^2 + B^2}}$$
(a) Solve $Ax_2 + By_2 = -C$
$Bx_2 - Ay_2 = Bx_1 - Ay_1$
to obtain expressions for x_2 and y_2. Check your answers with the expressions in Example 2.
(b) Use the values of x_2 and y_2 from (a) to simplify the expression for PQ.
$PQ = \sqrt{(x_1 - x_2)^2 - (y_1 - y_2)^2}$
Check your answer with the expressions in Example 2.

B

11. (a) Find a point that satisfies the equation $3x - 4y = 12$.
(b) Use the point in (a) to find the distance between the two lines
$3x - 4y = 12$ $3x - 4y = 18$

12. (a) Why are the lines $2x - 3y + 8 = 0$ and $6x - 9y - 14 = 0$ parallel?
(b) Calculate the distance between the two lines.

13. Calculate the distance between each pair of lines.
(a) $3x - 2y = 6$; $3x - 2y = 12$
(b) $x = 6 - 2y$; $x + 2y + 12 = 0$
(c) $3x - y = -4$; $y = 3x + 8$
(d) $3(x + 2) = 4y$; $6x - 8y = 9$

14. The vertices of △MNP are
M(10, 3)   N(6, −5)   P(0, −3)
(a) Calculate the length of the altitude from M to NP.
(b) Calculate the lengths of the other two altitudes.

15. (a) Calculate the distance between Q(3, 7) and the intersection of ℓ_1: $x + 2y - 7 = 0$ and ℓ_2: $5x + y - 8 = 0$.

(b) What is the distance of Q to each line in (a)?

16 A triangle is defined by $x = 3$, $y = 0$ and $3x + y - 16 = 0$. Calculate the lengths of the altitudes.

17 $\triangle$STM is given by the vertices
S(4, 6)   T(−6, 0)   M(0, −2).

(a) Calculate the altitude from S to TM.
(b) Calculate the area by using 2 different methods.

18 (a) Calculate the distance from the point of intersection of $x + y = 0$ and $x - 3y + 8 = 0$ to the line given by $7x - y + 16 = 0$.
(b) Interpret your results in (a).

C

19 The distance from the point (−1, 4) to the line $x + By - 1 = 0$ is $\sqrt{10}$ units. Find B if $B > 0$.

20 (a) The perpendicular distance from the point A(−k, 2k) to the line $3x + 4y - 5 = 0$ is 4 units. Find k.
(b) Interpret your results in (a).

applications: problems about distance

To solve each problem follow *The Steps for Solving Problems*. As an aid in understanding the problem draw a diagram to help you plan the solution.
When solving problems given in kilometres, let each square of the co-ordinate plane represent 1 km². Use the units of distance given in the appropriate problem.

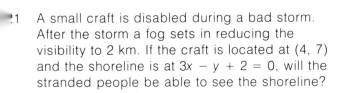

21 A small craft is disabled during a bad storm. After the storm a fog sets in reducing the visibility to 2 km. If the craft is located at (4, 7) and the shoreline is at $3x - y + 2 = 0$, will the stranded people be able to see the shoreline?

22 A group of canoeists decide it is time to find a campsite before it gets much darker. They spot shoreline A defined by $2x + 3y - 21 = 0$ and shoreline B defined by $2x - y - 9 = 0$. If their location is at (3, 2), which shoreline is closer?

23 A hovercraft on a path given by $x - 3y + 8 = 0$ is searching a swamp for a canoe. If the range of visibility is 2 km, will the hovercraft spot the canoe located at (3, 5)?

24 A ship travels an ocean route represented by the line $2x - 2y + 7 = 0$. A lighthouse is situated at the point (5, −4). If the lighthouse light can be seen anywhere within a radius of 10 km, will the ship see the light?

25 (a) A grid system is used to locate a wrecked oil freighter at sea. Coastguard Ship A travels along the path given by $2x - y - 9 = 0$. Coastguard Ship B travels along the path given by $x + 2y - 7 = 0$. If the visibility distance is 2.75 km, and the freighter is at (4, 5), which ship will spot the freighter?
(b) Draw a sketch of the system to locate the freighter.

26 A comet is travelling along the path represented by the line $8x - 6y - 3 = 0$. The sun is located at (2, 2). If one unit of distance is equivalent to 1 Astronomical Unit (1 A.U.), what is the point of closest approach, in A.U. s, between the comet and the sun?

math tip

Often a problem may have more than one answer. Sometimes it is possible to draw more than one diagram which satisfies the conditions of the problem. For the following problem, draw the different possible diagrams to predict how many answers might be obtained. Then solve the problem.
$\triangle$PQR has vertices P(−5, 1), Q(−1, 1) and R(x, y). If the area of $\triangle$PQR is 10 square units, calculate y.

7.6 satisfying conditions: locus of points

If you had to walk a path which was the same distance from two walls, which path would you follow? If you had to walk a path which was always a fixed distance from a pole, which path would you follow?

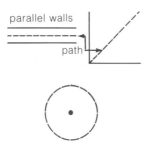

A **locus** is a set of points, and the only points, which satisfy certain given conditions.

The planets of our solar system follow paths that satisfy certain conditions. In astronomy, we refer to these paths as orbits. The scientist may think of a locus as the path traced by a point moving under certain conditions. We will use our skills with analytic geometry to solve problems about locus and find equations that define the locus (or the "path").

In Example 1 we will find points that satisfy certain conditions.

Example 1

Find a point(s) on the y-axis which is 4 units from the line given by $3x - 4y + 12 = 0$.

Solution

From the sketch, we see that there are 2 *possible* points that satisfy the given condition.

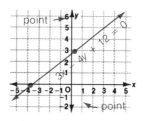

Let the co-ordinates of the point on the y-axis be $P(0, b)$. Distance, d, from P to the line, ℓ.

$$d = \frac{|Ax_1 + By_1 + C|}{\sqrt{A^2 + B^2}}$$

$$4 = \frac{|3(0) - 4(b) + 12|}{\sqrt{3^2 + (-4)^2}}$$

$$4 = \frac{|-4b + 12|}{5}$$

Thus $|-4b + 12| = 20$

$-4b + 12 = 20$ or $-4b + 12 = -20$
$-4b = 8$ $-4b = -32$
$b = -2$ $b = 8$

Thus the two required points are $(0, 8)$ and $(0, -2)$.

You may check that the distance from $(0, 8)$ or $(0, -2)$ to the line ℓ is 4 units.

In Example 2 a sketch is made from the given information to obtain information about the locus.

Example 2

Find the equation of the locus of points that are equidistant from the lines defined by $\ell_1: 2x - 3y - 6 = 0$ and $\ell_2: 3x + 2y - 6 = 0$.

Solution

Let $P(a, b)$ be any point on the locus. From the sketch PQ = PS.

Distance from P to ℓ_1
$$\frac{|2a - 3b - 6|}{\sqrt{4 + 9}}$$

Distance from P to ℓ_2
$$\frac{|3a + 2b - 6|}{\sqrt{9 + 4}}$$

Since PQ = PS, then
$$\frac{|2a - 3b - 6|}{\sqrt{13}} = \frac{|3a + 2b - 6|}{\sqrt{13}}$$
$$|2a - 3b - 6| = |3a + 2b - 6|$$

Thus
$2a - 3b - 6 = 3a + 2b - 6$
$-a - 5b = 0$
$a + 5b = 0$

Or $2a - 3b - 6 = -(3a + 2b - 6)$
$2a - 3b - 6 = -3a - 2b + 6$
$5a - b - 12 = 0$

Since $P(a, b)$ is any point on the locus, then the locus of all points equidistant from ℓ_1 and ℓ_2 is defined by the equations $x + 5y = 0$ and $5x - y - 12 = 0$.

Thus, from Example 2, we see that the locus of points equidistant from 2 lines is given by 2 straight lines that are the bisectors of the angles formed by the intersecting lines. We may now rephrase the above question.

Find the equation of the bisectors of angles formed when the lines defined by $2x - 3y - 6 = 0$ and $3x + 2y - 6 = 0$ intersect.

In the sections that follow we will study the graphs and equations of various loci (plural of locus).

7.6 exercise

A

1. Which of the following points are on the locus defined by $x - 5y + 3 = 0$?
   A(2,1)   B(15, 2)   C(7, 2)

2. A locus is defined by the equation $y = 2x^2 - 1$. Which of the following points are on the locus?
   A(−1, 1)   B(−2, −7)   C(3, 17)
   D(0,1)   E(2, 3)   F(1, −1)

3. The equation $x^2 + y^2 = 25$ defines a locus of points. Which of the following points is *not* on the locus?
   A(3, 4)   B(20, 5)

4. A locus is defined by the equations $x + y = 14$ and $x^2 + y^2 = 100$. Which of the following points are on the locus?
   A(8, 6)   B(10, 4)   C(12, 2)

5. A locus is defined by $x + 1 = y^2$. Decide which point is not on the locus.
   E(−2, 1)   F(8, 3)

6. A locus is defined by the equation $3kx - 2y = 1$. If P(−2, 3) is a point on the locus find the value of k.

7. A locus of points is defined by $9 - 3ky = (k - 1)x$. If P(−2, 3) is a point on the locus, find the value of k.

8. A locus of points is given by $y = 3x + b$.
   (a) What is the significance of b?
   (b) Find the value of b if the point (−1, 2) is on the locus.

9. A locus of points is defined by $y = mx + 6$.
   (a) What is the significance of m?
   (b) Find the equation of the locus if the point (−3, 0) is a point on the locus.

B

10. Find two points on the y-axis each of which is 4 units from the line $9x - 12y + 5 = 0$.

11. A point P satisfies the following conditions:
    - lies on the x-axis.
    - is 10 units from the line given by $7x + 24y - 30 = 0$.
    Find the co-ordinates of P.

12. What are the co-ordinates of points on the x-axis each of which is 1 unit from the line $x + y = \sqrt{2}$?

13. The distance from the line $7x - 24y - 5 = 0$ to the x-axis is 25 units at 2 points on the x-axis. Find the co-ordinates of the points.

14. Two lines given by ℓ_1: $x + 5y - 3 = 0$ and ℓ_2: $5x - y + 11 = 0$ intersect. Find the equation of the locus of points equidistant from each line.

15 Lines ℓ_1 and ℓ_2 intersect.
    ℓ_1: $3x - y - 5 = 0$ ℓ_2: $x - 3y + 4 = 0$
    (a) Find the equation of the locus of points that are equidistant from the lines.
    (b) What are the equations of the bisectors of the angles formed by the intersecting lines in (a)?

16 Find the equation of the bisectors of the angles formed when the two lines, p and q, intersect.
    p: $3x - 4y - 10 = 0$ q: $4x - 3y + 12 = 0$

17 Find the equation of the locus of points that are 2 units from each line.
    (a) $3x - 4y - 16 = 0$ (b) $x + 5y - 6 = 0$

18 Find the equation of the locus of points equidistant from the points A(3, 4) and B(−3, −4).

19 Find the equation of the locus of points equidistant from each pair of points.
    (a) (−5, −2) (1, 4)   (b) (6, 8) (−8, 6)

20 Find the equation of the locus of points equidistant from $(a, 0)$ and $(−a, 0)$?

C
21 Find the equation of the locus of points equidistant from the points $(−m, n)$ and $(m, −n)$.

22 Find the co-ordinates of the points on $3x + 5y = 15$ which are equidistant from the co-ordinate axes?

---

problem/matics

The word *locus* is a Latin word meaning *place* or *location*. Problems about locus involve finding the equation of points that satisfy certain conditions. Find the equation of the locus of a point P which satisfies $PA^2 - PB^2 = AB^2$ for A(1,3) and B(2,5).

---

applications: families and areas

The equation of a family of lines is $y = -\frac{3}{2}x + b$. One member of this family of lines meets the co-ordinate axes to form a triangle.

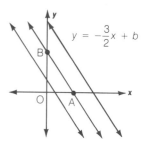

The following conditions are imposed on the member of the family.
- The area of △BOA is 3 square units.
- The x-intercept is positive.

What is the equation of the member of this family satisfying the above conditions?

From the diagram, the co-ordinates of A and B are obtained.

$y = 0$ $0 = -\frac{3}{2}x + b$

$\frac{3}{2}x = b$

$x = \frac{2}{3}b$ $x > 0$, thus $b > 0$.

$x = 0$ $y = b$.

Co-ordinates of A are $\left(\frac{2}{3}b, 0\right)$ and of B are (0, b).

In △BOA, base = $\frac{2}{3}b$ units, height = b units

Area of △BOA = $\frac{1}{2}$ (base × height)

$3 = \frac{1}{2}\left(\frac{2}{3}b\right)(b)$

$9 = b^2$

$b = \pm 3$

Since $b > 0$, then $b = 3$.

Thus the equation of the member of the family satisfying the conditions is

$y = -\frac{3}{2}x + 3$ or $3x + 2y - 6 = 0$

23. (a) Write the equation of the family of lines with slope −3.
    (b) From this family, find the member that forms a triangle of area 1 square unit with the co-ordinate axes and has a positive y-intercept.

24. (a) Write the equation of the family of lines with slope 3.
    (b) Select the member of the family which forms a triangle of area $\frac{3}{2}$ square units with the axes and passes through the 2nd quadrant.

25. If a family of lines has slope $\frac{1}{2}$, find the particular member that
    • forms a triangle of area 16 square units with the axes.
    • has a negative x-intercept.

26. A family of lines has slope 5. Find the member that forms a triangle of area $\frac{9}{10}$ square units with the axes and passes through the 2nd quadrant.

27. (a) Write the equation of the family of lines with slope $\frac{2}{3}$.
    (b) Select the member(s) of the family that form a triangle of area 75 square units with the axes.

28. A family of lines has slope −5. Find the members of this family that form a triangle of area 40 square units with the axes.

problem/matics

It is helpful to sketch a diagram to visualize some problems. Sketch a diagram and then solve this problem.

Find the equation of the line through the point $(3,-2)$ which meets the x-axis at the point $(a,0)$ and the y-axis at the point $(0,b)$, where $5ab = 1$.

7.7 solving problems: analytic geometry

In solving a problem, there are often different strategies available. Even when deriving a formula, such as the distance formula in Section 7.2, there are alternative methods of doing so. Whatever the method, it is helpful to sketch a diagram of the given information in order to develop a strategy for solving the problem. To solve any problem, we must clearly be able to answer the questions:

I: What information do we know (or can we obtain)?
II: What are we asked to find (or to prove)?

What is important is the *process* of utilizing the information in I and then of obtaining the answer to II.

For example, to derive a formula for d in the diagram, we examine Questions I and II above. From the diagram

$$\triangle PTS = \frac{1}{2}d \times TS$$

or $\quad d = \dfrac{2\triangle PTS}{TS}$

where T, S, are the points where the line defined by $Ax + By + C = 0$ crosses the axes. For the expression, $d = \dfrac{2\triangle PTS}{TS}$ we ask

• Can we obtain the area of $\triangle PTS$?
• Can we obtain the length of TS?

To find the area of $\triangle PTS$ and the length of TS, we need to find the co-ordinates of T and S.

$$y = 0 \quad Ax + By + C = 0$$
$$Ax + B(0) + C = 0$$
$$x = \frac{-C}{A}$$

Co-ordinates of S are $\left(-\dfrac{C}{A}, 0\right)$.

$x = 0 \quad Ax + By + C = 0$
$ A(0) + By + C = 0$
$ y = -\dfrac{C}{B}$

Co-ordinates of T are $\left(0, -\dfrac{C}{B}\right)$.

Since we know the co-ordinates of T and S, we can now work backwards and calculate
- the area of $\triangle$PTS.
- the length of TS.

Then we substitute in $d = \dfrac{2\triangle\text{PTS}}{\text{TS}}$ to obtain a formula.

Many important results in mathematics have been developed in similar ways by

In the exercise, Questions 1 to 3 develop the steps for deriving the formula. The exercise also provides practice in solving problems based on the skills we have acquired in analytic geometry. The problems are not organized in any special order, because one important step of problem-solving is to decide *which skills* are needed to solve the problem.

7.7 exercise

Questions 1 to 3 use the following diagram to develop an alternate proof to show that the distance, d, from $P(x_1, y_1)$ to $Ax + By + C = 0$ is given by
$$d = \dfrac{|Ax_1 + By_1 + C|}{\sqrt{A^2 + B^2}}$$
Refer to the diagram on the previous page.

A

1 Use the co-ordinates of T and S to show that the area of $\triangle$PTS is given by
$$\triangle\text{PTS} = \dfrac{1}{2}\left|\dfrac{C(Ax_1 + By_1 + C)}{AB}\right|$$

2 Use the co-ordinates of T and S to show that
$$TS = \left|\dfrac{C\sqrt{A^2 + B^2}}{AB}\right|.$$

3 Use the formula $d = \dfrac{2\triangle\text{PTS}}{\text{TS}}$ and your results in Questions 1 and 2 to show that
$$d = \dfrac{|Ax_1 + By_1 + C|}{\sqrt{A^2 + B^2}}$$

For Questions 4 to 25
- draw a sketch as needed to help you plan the solution.
- check your answer to see if it is reasonable.

4 If $(1, 2)$ is a point on the line defined by $2kx + 5y - 6 = 0$, find the value of k.

5 A line is defined by $(1 + m)x + my = m + 2$. If $P(-1, 1)$ is a point on the line, find the value of m.

6 Find the values of k if $P(3, 5)$ is on the line defined by $(1 + k^2)x + (1 - k^2)y = (k + 2)^2$.

7 Each pair of lines is parallel. Find each value of k.
   (a) $x + 3y = 8, \quad 2x + ky = 5$
   (b) $kx + (k + 1)y = 9, \quad 2x + 5y = 1$

8 Each pair of lines is perpendicular. Find each value of k.
   (a) $x + 3y = 8, \quad 2x + ky = 5$
   (b) $(1 - k)x + (1 + k)y = 3, \quad 2x - 3y = 1$

9 The lines given by $(2 - k)x + ky = 5$ and $3x + (2 + k)y = 7$ are parallel. Find two values of k.

10 The lines given by $(k + 1)x - 3y = 11$ and $(k - 1)x + (k + 1)y = 5$ are perpendicular. Find the value of k.

11 Show that the points A(4, −1), B(−2, 3) and C(−5, 5) are collinear by using
   (a) slopes of line segments.
   (b) the area formula. (c) the distance formula.

12 Given the points A(4, 1), B(1, 3), and C(2, 3). Find the equation of the line through C that is perpendicular to AB.

13 Find the point P on the y-axis which is equidistant from (2, 3) and (−2, 1).

B

14 The vertices of △MPN are
   M(0, 0)   P($2\sqrt{3}$, 0)   N($\sqrt{3}$, 3)
   (a) Find the lengths of all sides.
   (b) Classify △MPN.

15 What is the distance between the point of the intersection of $2x - y + 1 = 0$ and $3x + y - 11 = 0$ and the point of intersection of $2x - 3y - 14 = 0$ and $x + y - 2 = 0$?

16 The vertices of a quadrilateral are (−1, 1), (6, −5), (5, 0), and (−3, −7), but are not given in order.
   (a) Draw a sketch of the quadrilateral.
   (b) Calculate the area of the quadrilateral.

17 (a) The midpoint of AB is (2, 3). If the line $x - 2y + 4 = 0$ forms a right angle at the midpoint of AB, find the equation of the line AB.
   (b) If A is the point (−4, 0), find the co-ordinates of B.

18 Which of the following triangles has the greater perimeter?
   △ABC   A(6, 9)   B(−2, 1)   C(12, 3)
   △PQR   P(0, 1)   Q(9, 4)   R(13, −8)

19 A triangle is determined by the points A(2, 7), B(x, 5), and C(−3, 3). If the area of the triangle is 6 square units, find the possible values for x.

20 A boat leaves port A and travels 20 km due east to port B. From there it travels 40 km due north to port C and then heads east again to port D which is 30 km away.
   (a) How much shorter would it have been if the boat had sailed from port A to port D via port C only?
   (b) If the boat had gone directly from port A to port D via port B only, how much shorter would this route have been than the total route?

21 Given vertices P(−2k, 6), Q(3, 2) and R(k, −6). If the area of △PQR is 30 square units, find k.

22 A harbour police boat is located at map co-ordinates (6, −2) and is travelling towards the harbour whose map co-ordinates are (−1, 1). A distress call is received from a cruiser. The cruiser is travelling along a route whose equation is $x + 2y - 10 = 0$ and its engines fail at the point of intersection with the y-axis. What will be the equation of the route the harbour police will have to take in order to reach the cruiser?

23 (a) The vertices of △ABC are A(−4, 2), B(6, 8) and C(10, −4). Determine the equations of the 3 medians.
   (b) Show that the medians are concurrent.

24 A plane is flying into a small airport with a single runway. The runway starts at map co-ordinates (−3, 0) and ends at (0, −5). If the plane is currently at map co-ordinates (2, 4), what is
   (a) the equation of the path which the plane must follow in order to arrive at the beginning of the runway?
   (b) the equation of the runway?

C

25 Find the equation of the line which passes through the origin and bisects the part of the line $5x - 4y - 20 = 0$ contained between the axes.

7.8 writing proofs

In order to solve this problem using analytic geometry,

> The vertices of △QRP are P(−10, 8), Q(−8, 2) and R(4, 6). Prove that the midpoint of RP is equidistant from all 3 vertices.

we may organize the given information in answer to the two following important questions for problem solving

I What are we given or what do we know?
II What are we asked to find or what are we asked to prove?

under the following headings.

Given or *Hypothesis*
The information in the given problem in answer to Question I is reorganized and placed with this heading.

Required to Prove or *Conclusion*
The information in the given problem in answer to Question II is reorganized and placed with this heading.

Proof
The steps for using the information in I to show II is described under this heading.

When solving a word problem in algebra, we introduce variables and equations and use these to translate the given problem into a form that we can use to plan and obtain the solution.

When solving a word problem in geometry, we may also introduce a diagram, labels, or points, to translate the given problem into a form that we can use to plan the solution and in this case write the proof.

For example, to organize the proof of the previous problem involving analytic geometry, a diagram is made and the label M is used as an aid in translating the problem and writing the steps of the solution.

Given: △QRP with vertices P(−10, 8), Q(−8, 2), R(4, 6). M is the midpoint of RP.
Required to Prove:
MP = MQ = MR
Proof: Find the co-ordinates of M, the midpoint of RP.

$$M\left(\frac{-10+4}{2}, \frac{8+6}{2}\right) = M\left(\frac{-6}{2}, \frac{14}{2}\right)$$
$$= M(-3, 7)$$

$$MP = \sqrt{[-10-(-3)]^2 + (8-7)^2}$$
$$= \sqrt{(-7)^2 + (1)^2}$$
$$= \sqrt{50} = 5\sqrt{2}$$

$$MQ = \sqrt{[-8-(-3)]^2 + (2-7)^2}$$
$$= \sqrt{(-8+3)^2 + (-5)^2}$$
$$= \sqrt{50} = 5\sqrt{2}$$

$$MR = \sqrt{[4-(-3)]^2 + (6-7)^2}$$
$$= \sqrt{(4+3)^2 + (-1)^2}$$
$$= \sqrt{50} = 5\sqrt{2}$$

Thus MP = MQ = MR = $5\sqrt{2}$.

Thus, the midpoint of RP is equidistant from the vertices P, Q, and R.

Be sure to write a final statement.

The above steps provide a method of organizing our work and planning a solution to the given problem. The procedure of writing a proof as shown above is an important step in our subsequent work.

7.8 exercise

For each of the following problems, organize your work using the headings
Given   Required to Prove   Proof

A
1 Prove that the points A(6, 1), B(2, 0), C(−6, −2) are collinear.

2 Prove that △PQR is a right-angled triangle given by the vertices P(−7, 1), Q(−8, 4) and R(−1, 3).

3 Points A(2, 3), B(2, 0) and C(0, 4) are given. Prove that $AC = \frac{1}{2}BC$.

4 Prove that $\triangle PQR$ is isosceles where P(−2, 1), Q(1, 5) and R(5, 2).

5 Prove that $\triangle XYZ$, whose vertices are X($\sqrt{3}$, 1), Y(2$\sqrt{3}$, 2) and Z($\sqrt{3}$, 3), has sides of integral length.

B

6 A quadrilateral PQRS has co-ordinates P(5, −6), Q(3, 0), R(−1, 2) and S(−5, −4). Prove that the midpoints of the sides are the vertices of a parallelogram.

7 Prove that the points (9, −3), (8, 6) and (−1, 5) lie on a circle with centre at (4, 1).

8 The lines $2x - 3y = 0$ and $3x - 2y + 10 = 0$ intersect at A. The lines $6x - y + 1 = 0$ and $x + 3y + 16 = 0$ intersect at B. Prove that the midpoint of AB has co-ordinates $\left(-\frac{7}{2}, -\frac{9}{2}\right)$.

9 Prove that the four lines $2y = 5x + 15$, $4x + 5y - 21 = 0$, $5x - 2y - 18 = 0$, $3x + 2y + 2 = 0$ define the sides of a trapezoid.

10 Prove that the quadrilateral defined by the lines $4y = 3x - 6$, $4x + 3y = 33$, $4y = 3x + 19$, and $4x + 3y - 8 = 0$ is a square.

11 In $\triangle PQR$ the vertices are given by P(k − 5, −2), Q(1, 4) and R(3k + 4, 2). Prove that if the area of $\triangle PQR$ is 23 square units then k is equal to $-\frac{19}{4}$ or 1.

12 A triangle is given by the vertices P(−5, 4), Q(1, 8) and R(−1, −2). Prove that the perpendicular from P to RQ bisects RQ.

13 Prove that the quadrilateral given by the vertices (−4, 1), (5, 4), (7, −5) and (−2, −8) has opposite sides parallel.

14 The vertices of a quadrilateral are (−2, 6), (2, 2), (−2, −2) and (−6, 2). Prove that the diagonals of the quadrilateral are perpendicular bisectors of each other.

15 In $\triangle ABC$, the vertices are given by A(4, 7), B(8, 1) and C(−2, 3). Prove that the perpendicular bisector of CB passes through the vertex of A.

16 The vertices of $\triangle MPQ$ are given by M(−5, −3), P(3, 1) and Q(−1, −11). Prove that the median from M to PQ is co-incident with the perpendicular bisector of PQ.

17 (a) Prove that the intersections of the lines defined by $4y = 3x + 27$, $y - 6 = 0$, $3x - 4y + 12 = 0$, $y - 3 = 0$ are the vertices of a parallelogram.
    (b) Is the quadrilateral in (a) a rhombus? Why or why not?

C

18 A triangle has vertices S(−1, −3), T(5, −4) and U(k, −2). If the triangle has area 5 square units, prove that the values of k satisfy the equation $k^2 + 14k - 51 = 0$.

19 The line segment joining (−1, 5) and (2, 3) forms the base of a triangle. If the slope of the side containing (−1, 5) is $\frac{3}{5}$ and of the side containing (2, 3) is $\frac{5}{2}$, prove that the co-ordinates of the third vertex are (4, 8).

problem/matics

Use the following program in BASIC to find the distance from a point to a line.

```
10  INPUT X1, Y1, A, B, C
20  LET M = A*X1 + B*Y1 + C
30  LET N = ABS(M)
40  LET P = A↑2 + B↑2
50  LET Q = SQR(P)
60  LET D = N/Q
70  PRINT "THE DISTANCE FROM
    THE POINT TO THE LINE IS", D
80  END
```

applications: points of a triangle

The centre of the **inscribed circle** called the **incentre**, is the intersection of the bisectors of the angles of a triangle.

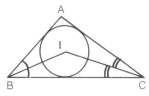

20 △MNP has vertices M(−5, 2), N(1, 4) and P(−1, 6). Prove that the bisectors of the angles are concurrent.

21 △PQR has vertices P(−3, −2), Q(3, 0), R(1, 2).
   (a) Find the co-ordinates of the incentre.
   (b) Find the radius of the inscribed circle.

For a triangle, the **orthocentre**, O, is the intersection of the altitudes of △ABC.

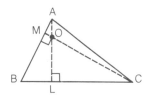

22 △ABC has vertices A(3, 4), B(4, −3), C(−4, −1). Prove that the altitudes intersect.

23 △ABC has vertices A(2, 8), B(6, 2), C(−3, 2).
   (a) Find the co-ordinates of the orthocentre.
   (b) What is the distance of the orthocentre from AB?

For a triangle, the intersection of the medians is the **centroid**, C.

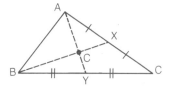

24 △DEF has vertices D(−2, 12), E(−10, 4) and F(2, 8). Prove that the medians are concurrent.

25 △PQR has vertices P(−12, 6), Q(4, 0) and R(−8, −6).
   (a) Find the co-ordinates of the centroid.
   (b) Draw a sketch of the triangle and show the medians and centroid.

26 If a triangular shape is suspended at its centroid it will be balanced horizontally. To locate the centroid of a triangular plate, it is placed on square paper. If the vertices are placed at (12, 18), (18, 6), and (3, 12), find the co-ordinates of the centroid.

27 A large triangular-shaped design, is placed on squared paper so that the co-ordinates of the vertices are at (8, 12), (12, 4), and (2, 8). Find the co-ordinates of the point at which we can suspend the triangular shape so that it balances.

problem/matics

To solve a problem, we must understand the meaning of each word. Otherwise we will not be able to solve the problem. Read the following problem. Why can it not be solved?

A line given by $2x + y = 4$ intersects the axes to form a triangle. How many lattice points are inside the triangle?

Lattice points on the co-ordinate plane are those that have integral co-ordinates. (3,4), (−2,0), and (−3,−4) are examples of lattice points. Now solve the problem!

It is important that you learn the vocabulary of mathematics when solving problems.

7.9 circle: analytic properties

What is the locus of all points, P, that are equidistant from a given fixed point C? The locus of all points satisfying the above condition is a circle.

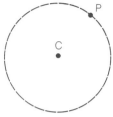

Using skills with analytic geometry, we may derive the equation of the circle and study its properties analytically.

To show that our equation represents a locus, we
I show that any point of the locus satisfies the equation.
II show that any point satisfying the equation is on the locus.

Example 1
Find the equation of the locus of all points that are 4 units from the origin O(0, 0).

Solution
I Let P(x, y) represent any point on the locus satisfying the given condition.
Then
$$OP = \sqrt{(x - 0)^2 + (y - 0)^2}$$
$$= \sqrt{x^2 + y^2}$$
But OP = 4.
Thus   or by squaring
$\sqrt{x^2 + y^2} = 4$ $x^2 + y^2 = 16$

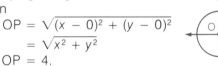

II Conversely, if P(x, y) is a point that satisfies $x^2 + y^2 = 16$ then P(x, y) also satisfies the condition
$$\sqrt{(x - 0)^2 + (y - 0)^2} = 4$$

Thus, the distance of P(x, y) to the origin is 4 units.
The equation of the circle with centre O(0, 0) and radius 4 units is $x^2 + y^2 = 16$.

For the above circle, given by $x^2 + y^2 = 16$
x-intercept
Let $y = 0$
$x^2 + y^2 = 16$
$x^2 + 0^2 = 16$
$x = \pm 4$

y-intercept
Let $x = 0$
$x^2 + y^2 = 16$
$0^2 + y^2 = 16$
$y = \pm 4$

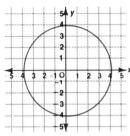

Domain
The domain is the set of all x values for $x^2 + y^2 = 16$ for all real values of y.
$x^2 + y^2 = 16$
$y^2 = 16 - x^2$
$y = \pm\sqrt{16 - x^2}$
Since $16 - x^2 \geq 0$ then $16 \geq x^2$.
Thus $|x| \leq 4$
The domain is $\{x \mid |x| \leq 4\}$ or $\{x \mid -4 \leq x \leq 4\}$.

Range
Similarly, the range is given by
$\{y \mid |y| \leq 4\}$ or $\{y \mid -4 \leq y \leq 4\}$.

In general, the equation of the circle with radius, r, and centre the origin is given by
$x^2 + y^2 = r^2, r > 0$.
This equation represents a family of circles with radius, r. In this case, r is a parameter. For the above circle

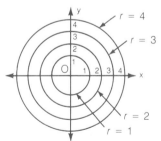

x-intercepts $\pm r$
y-intercepts $\pm r$
Domain $\{x \mid |x| \leq r\}$ or $\{x \mid -r \leq x \leq r\}$
Range $\{y \mid |y| \leq r\}$ or $\{y \mid -r \leq y \leq r\}$

We may use our skills with analytic geometry to learn about some properties of the circle as in Example 2.

Example 2

A circle with centre (0, 0) has diameter AB where A(−6, −8) and B(6, 8). If C(−8, −6) is another point on the circle, show that $\angle BCA$ is a right angle.

Solution

Draw a sketch to plan the solution.

To show $\angle BCA = 90°$, we show $(m_{CA})(m_{CB}) = -1$.

$$m_{CA} = \frac{-8-(-6)}{-6-(-8)} \qquad m_{CB} = \frac{8-(-6)}{6-(-8)}$$

$$= \frac{-2}{2} = -1 \qquad\qquad = \frac{14}{14} = 1$$

$(m_{CA})(m_{CB}) = (-1)(1) = -1$

Thus $\angle BCA = 90°$ as required.

In Example 2 we see that for a *particular* circle, $\angle BCA$ subtended by the diameter is a right angle. In later work, we will use the methods of plane and analytic geometry to prove that, in *general*, the result is true for *any circle*.

7.9 exercise

Questions 1 to 9 examine skills needed to study the analytic properties of circles.

A

1 The radius is given for a circle with centre (0, 0). Write its equation.
   (a) 3 (b) $\sqrt{3}$ (c) $2\sqrt{5}$ (d) $2r$

2 Two circles with centre (0, 0) are given by
   A: $x^2 + y^2 = 16$ B: $x^2 + y^2 = 50$
   For each circle write the
   (a) radius.   (b) x-intercepts.
   (c) y-intercepts.   (d) domain. (e) range.

3 The centre of each circle is at (0, 0). Find the equation of the circle passing through the point.
   (a) (6, 8)   (b) (−3, 2)   (c) (−4, 3)
   (d) ($\sqrt{2}$, 3)   (e) (−1, −1)   (f) (−$\sqrt{3}$, −$\sqrt{5}$)

4 The equation of a circle with centre the origin is given by $x^2 + y^2 = 100$. Find the values of the missing co-ordinates.
   (a) (−6, ?)   (b) (?, 8)   (c) (10, ?)
   (d) (0, ?)   (e) (?, 5)   (f) (?, $5\sqrt{2}$)

5 Which of the following points are *not* on the circle defined by $x^2 + y^2 = 25$?
   (a) (−5, 0)   (b) (0, 25)   (c) (0, 5)
   (d) (3, −4)   (e) $\left(12\frac{1}{2}, 12\frac{1}{2}\right)$ (f) (−$2\sqrt{2}$, 4)

6 Write an equation for the locus of all points that are 6 units from (0, 0).

7 Write an equation of the locus of all points equidistant from (0, 0) and passing through (−3, 2).

8 The equation of a locus is given by $4x^2 + 4y^2 = 1$.
   (a) Explain why this is the equation of a circle with centre the origin.
   (b) What is the radius of the circle?

B

9 Write an equation of the circle, centre (0, 0), and
   (a) with radius 8 units.
   (b) passing through the point (−4, 3).
   (c) passing through the point ($\sqrt{3}$, −2).
   (d) with x-intercept −5. (e) with y-intercept 4.

10 The locus of points is given by the equation $x^2 + y^2 = 36$.
   (a) Find m if the point $(m, 3)$ is on the locus.
   (b) Find k if the point $(-\sqrt{6}, k)$ is on the locus.
   (c) Show why the point with co-ordinates $(8, k)$ cannot be a point on the locus.

11 Two points on a circle have co-ordinates P(−3, −2) and Q(3, 2).
   (a) Calculate the length of the chord PQ.
   (b) If the centre of the circle is (0, 0), what other name is given to the chord PQ?

Questions 12 to 15 examine some properties of circles. In Chapter 9, we will prove these properties in general.

12. A locus is defined by the equation $x^2 + y^2 = 64$.
    (a) Show that the locus is a circle.
    (b) Show that $P(8, 0)$ and $Q(-4\sqrt{2}, 4\sqrt{2})$ are points on the circle.
    (c) Find the co-ordinates of the midpoint M of the chord PQ.
    (d) Write the equation of the line through M and perpendicular to PQ.
    (e) Show that the line in (d) passes through the centre of the circle.

13. On a circle, with centre (0, 0), three points are given by $P(2, -3)$, $Q(3, 2)$, $R(-3, -2)$.
    (a) Draw a sketch of the circle.
    (b) Find the equation of the perpendicular bisector of PQ.
    (c) Find the equation of the perpendicular bisector of RQ.
    (d) Find the intersection point given by the equations in (b) and (c).
    (e) Based on your results in (d) suggest a method of locating the centre of a circle.

14. A chord PQ is given in a circle with centre $O(0, 0)$ with co-ordinates $P(-6, 4)$ and $Q(4, -6)$.
    (a) Find the co-ordinates of the midpoint M of PQ.
    (b) Calculate the slopes of OM and PQ. What do you notice about your results?
    (c) Write a probable conclusion based on your work in (a) and (b).

15. Two chords are given in a circle with centre (0, 0).
    AB: $A(-7, -3)$ $B(3, -7)$
    CD: $C(3, 7)$ $D(-7, 3)$
    (a) Calculate the length of each chord.
    (b) Calculate the distance from the centre of the circle to each chord.
    (c) Based on your results in (a) and (b), write a probable conclusion.

16. A circle with centre (0, 0) has a diameter AB where the co-ordinates of B are $(-3, \sqrt{5})$.
    (a) Find the equation of the circle.
    (b) What are the co-ordinates of A?

17. The diameter PQ of a circle with centre (0, 0) is given by the co-ordinates $P(-4, -5)$ and $Q(4, 5)$. If $R(-4, 5)$ is a point on the circle show $\angle QRP$ is a right angle.

18. A circle is given by the equation $x^2 + y^2 = 20$. For the points $A(-4, -2)$ and $B(-2, -4)$, prove that the right bisector of AB passes through the centre of the circle.

A **tangent**, t, to the circle with centre O is a line which touches the circle in one point at P.

C
19. A circle with centre at the origin is tangent to the line $2x - y - 9 = 0$. Find the radius of the circle.

20. A circle with centre at the origin is tangent to the line passing through the points $(4, -2)$ and $(2, -6)$. Find the radius of the circle.

problem/matics

To solve some problems we may need to just exhaust all the numbers to find which satisfy the conditions. (A computer works somewhat like this.) Solve the problem.

The number 121 has the property that by crossing out the digit 2, the remaining number, 11, is the square root of the original number. Find all other numbers less than 1000 that have the same property.

extending our work

In the previous pages, the circles had centres at (0, 0). We may extend our work to examine circles with centres not at the origin. For example, what is the locus of a circle with radius 3 units and centre at C(3, −2)?

Let P(x, y) be any point on the locus. Then
$$CP = 3$$
$$CP = \sqrt{(x-3)^2 + (y+2)^2}$$
Thus $\sqrt{(x-3)^2 + (y+2)^2} = 3$
By squaring, we obtain the equation
$$x^2 - 6x + 9 + y^2 + 4y + 4 = 9$$
$$x^2 + y^2 - 6x + 4y + 4 = 0$$

21. Find the equation of each circle.
    (a) centre (3, 0), radius 4
    (b) centre (0, −4), radius 4
    (c) centre (−2, 3), radius 5

22. Write the equation of the locus of all points that are 6 units from the point (−3, 5).

23. Find the equation of the circle with
    (a) centre (−3, 5) and tangent to the x-axis.
    (b) centre (5, −4) and tangent to the y-axis.

24. A diameter of a circle has end points (−6, 0) and (6, 2). Find the equation of the circle.

25. The equation of a circle is given by
    $$x^2 + y^2 + kx - 2y + 6 = 0$$
    If the circle passes through the point (5, 1) find the value of k.

26. A diameter RS of a circle has co-ordinates R(−3, 8), S(−5, −4). If T(2, 3) is a point on the circle show that ∠RTS is a right angle.

27. Find the equation of the circle if the end points of the diameter AB are given by the co-ordinates A(−7, 1) and B(1, 3).

28. A circle passes through the points P(−4, 0), Q(−6, −6), R(−2, −6) and S(−8, −4).
    (a) Find the equation of the line through the midpoint of PQ and perpendicular to PQ.
    (b) Find the equation of the line through the midpoint of RS and perpendicular to RS.
    (c) Solve the equation in (a) and (b) to obtain the co-ordinates of the centre of the circle.

29. A circle with centre C(1, 4) is drawn through the point D(2, 7). For what value(s) of x does E(x, 5) lie on the circle?

The **circumcircle** of a triangle is the circle that passes through all the vertices of the triangle. The **circumcentre** is the centre of the circumcircle.

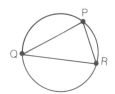

30. The vertices of △PQR are P(2, −2), Q(0, 10) and R(−6, 0). What is the radius of the circumcircle?

31. A triangle has vertices A(4, 6), B(8, 14) and C(16, 2).
    (a) Find the equation of the right bisector of AB.
    (b) Find the equation of the right bisector of BC.
    (c) Solve the equations in (a) and (b) to find the co-ordinates of the circumcentre.

32. Find the co-ordinates of the circumcentre of a triangle with vertices (−4, −6), (6, 14), (10, 2).

---

problem/matics

Use the following program in BASIC to calculate the area of a triangle given the vertices.
```
10 INPUT X1, Y1, X2, Y2, X3, Y3
20 LET D = X1 * Y2 + X2 * Y3 + X3 * Y1
30 LET U = X2 * Y1 + X3 * Y2 + X1 * Y3
40 LET A = ABS (D − U)/2
50 PRINT "AREA OF TRIANGLE IS =" A
60 END
```
Can you write a BASIC program to calculate the area of a quadrilateral?

7.10 regions and inequations: circles

A circle shown on the grid is defined by the equation
$$x^2 + y^2 = 16.$$

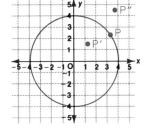

There are 3 regions associated with a circle.

I Any point P on the circle satisfies the condition OP = 4 and the equation $x^2 + y^2 = 16$.

II Any point P′ in the interior of the circle satisfies the condition
$$|OP'| < 4.$$
Thus, the point P′(x, y) satisfies the inequation
$$x^2 + y^2 < 16.$$
For example, the point R(−1, 0) is in the interior of the circle.
$$x^2 + y^2 = (-1)^2 + (0)^2$$
$$= 1$$
Since $x^2 + y^2 < 16$ then R is in the interior.

III Any point P″ in the exterior of the circle satisfies the condition
$$OP'' > 4.$$
Thus, the point P″(x, y) satisfies the inequation
$$x^2 + y^2 > 16.$$
For example, the point Q(6, −1) is in the exterior of the circle since
$$x^2 + y^2 = (6)^2 + (-1)^2$$
$$= 36 + 1$$
$$= 37$$
Since $x^2 + y^2 > 16$ then Q is in the exterior.

In general, for a circle with centre (0,0) and radius r
- the circle is defined by $x^2 + y^2 = r^2$.
- the interior of the region is defined by $x^2 + y^2 < r^2$.
- the exterior of the region is defined by $x^2 + y^2 > r^2$.

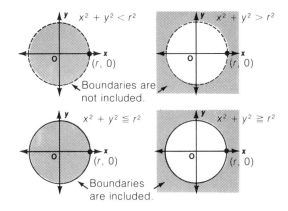

7.10 exercise

A

1. Write an inequation that defines the interior of a circle with centre (0, 0) and radius
   (a) 4   (b) $\sqrt{3}$ (c) $2\sqrt{2}$

2. Write an inequation that defines the exterior of a circle with centre (0, 0) and radius
   (a) 5   (b) $\sqrt{2}$ (c) $\frac{1}{2}$

3. A circle is defined by $x^2 + y^2 = 16$. Which points
   A are in the exterior? B are in the interior?
   C are on the circle?
   (a) (−3, −2)   (b) $2\sqrt{2}, -2\sqrt{2})$ (c) (2, −4)
   (d) (−4, 0)   (e) ($\sqrt{2}, -\sqrt{5}$)   (f) (3, −3)

4. A circle with centre (0, 0) passes through the point (4, 3). Write an inequation that defines
   (a) the interior of the circle.
   (b) the exterior of the circle.

B

5. P(x, y) are points that are within 5 units of the origin. Write an inequation to define the region.

6. Write an inequation that defines the region consisting of all points that are farther than 6 units from the origin.

7 Show that the region defined by $x^2 + y^2 < 36$ is the interior of a circle with centre (0, 0) and radius 6 units.

8 Write the inequation that defines each region.

(a)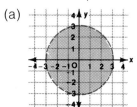

(b)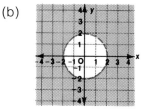

(c)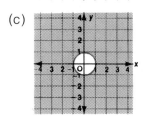

(d)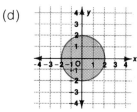

9 Draw a graph of the region defined by each of the following.
(a) $x^2 + y^2 > 9$
(b) $x^2 + y^2 < 16$
(c) $x^2 + y^2 \geq 25$
(d) $x^2 + y^2 \leq 49$

10 (a) Draw the region defined by $x^2 + y^2 \leq 25$.
(b) If $y \geq 0$, draw the region given by $x^2 + y^2 \leq 25$.

The shaded region is the set of all points $P(x, y)$ that satisfies $x^2 + y^2 \geq 1$ and $x^2 + y^2 \leq 9$.

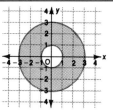

C
11 Draw the region that shows the set of all $P(x, y)$ that satisfies
(a) $x^2 + y^2 \leq 49$ and $x^2 + y^2 > 25$
(b) $x^2 + y^2 \leq 36$ and $x \geq 0$
(c) $x^2 + y^2 \geq 1$ and $y \geq 0$
(d) $x^2 + y^2 \geq 1$ and $x, y \geq 0$
(e) $x^2 + y^2 \leq 25$ and $y \leq x$

7.11 translations and co-ordinates

We can describe a common property of certain points by using analytic geometry to write the equation that is satisfied by the co-ordinates of all the points that have that property in common.

Linear Relation   Circular Relation

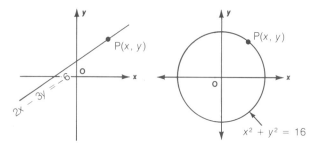

Another way of relating points on the plane is by studying transformations. A **transformation** of the plane is a one-to-one mapping of the plane onto itself. Subsequently we will be studying different transformations and their properties and developing new methods of solving problems.

On the plane, a point P is mapped into P'. We may use the following notation.

We can map a figure on the plane.
△ABC is mapped onto △A'B'C'.

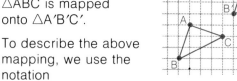

To describe the above mapping, we use the notation
[A, B, C] ⟶ [A', B', C']

A **translation** or **glide** is shown as
△ABC ⟶ △A'B'C'
if AA' = BB' = CC' and AA'∥BB'∥CC'.

Example 1

△ABC is given by the vertices
A(1, 4)   B(2, −1)   C(−3, 2).

(a) For the mapping given by
$(x, y) \longrightarrow (x + 1, y - 2)$
find the co-ordinates of the vertices of △A′B′C′.

(b) Show that the above mapping is a translation.

Solution

(a) $(x, y) \longrightarrow (x + 1, y - 2)$
A(1, 4) ⟶ A′(2, 2)
B(2, −1) ⟶ B′(3, −3)
C(−3, 2) ⟶ C′(−2, 0)

(b) $m_{AA'} = \dfrac{2 - 4}{2 - 1} = -2$

$m_{BB'} = \dfrac{-3 - (-1)}{3 - 2} = -2$

$m_{CC'} = \dfrac{0 - 2}{-2 - (-3)} = -2$

Thus AA′ ∥ BB′ ∥ CC′.

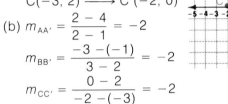

$AA' = \sqrt{(2 - 1)^2 + (2 - 4)^2}$
$= \sqrt{1^2 + (-2)^2} = \sqrt{5}$

$BB' = \sqrt{(3 - 2)^2 + [-3 - (-1)]^2}$
$= \sqrt{(1)^2 + (-2)^2} = \sqrt{5}$

$CC' = \sqrt{[-2 - (-3)]^2 + (0 - 2)^2}$
$= \sqrt{(1)^2 + (-2)^2} = \sqrt{5}$

Thus AA′ = BB′ = CC′.
Thus the mapping given by
$(x, y) \longrightarrow (x + 1, y - 2)$
is a translation.
For the above translation note that
$AB = \sqrt{(1 - 2)^2 + [4 - (-1)]^2}$
$= \sqrt{(-1)^2 + (5)^2} = \sqrt{26}$

$A'B' = \sqrt{(2 - 3)^2 + [2 - (-3)]^2}$
$= \sqrt{(-1)^2 + (5)^2} = \sqrt{26}$

Thus AB = A′B′.

Also $m_{AB} = \dfrac{4 - (-1)}{1 - 2}$ $m_{A'B'} = \dfrac{2 - (-3)}{2 - 3}$

$= \dfrac{5}{-1}$ $= \dfrac{5}{-1}$

$= -5$ $= -5$

Thus $m_{AB} = m_{A'B'}$.

In a similar way, we may show that
BC = B′C′ and AC = A′C′
$m_{BC} = m_{B'C'}$ and $m_{AC} = m_{A'C'}$

Based on Example 1, we see that for a translation the length and the slope of line segments are *preserved*. Since the calculation of area is based on lengths, then area is also preserved. You may check this for △ABC and △A′B′C′.

In general, a mapping on the plane, such as
$(x, y) \longrightarrow (x + a, y + b)$, $a, b \in R$
defines a translation. We shall see in our later work that in general for a translation
- length is invariant (preserved).
- slope is invariant.
- area is invariant, and so on.

We may use our work with translations to simplify our work in analytic geometry. For example, to calculate the area of a triangle, we may translate the triangle to the origin as shown in Example 2.

Example 2

Find the area of the triangle given by the co-ordinates (−2, 4), (−4, −2) and (4, 1).

Solution

Sketch the triangle and write the co-ordinates in a counterclockwise order.

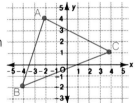

Since area is preserved for a translation, then translate the triangle to the origin. Use
$(x, y) \longrightarrow (x + 4, y + 2)$.

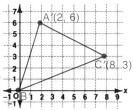

$\triangle A'B'C' = \frac{1}{2}|48 - 6|$
$= \frac{1}{2}|42|$
$= 21$

$\begin{array}{cc} 2 & 6 \\ \searrow & \searrow \\ 0 & 0 \\ \searrow & \searrow \\ 8 & 3 \\ \searrow & \searrow \\ 2 & 6 \end{array}$ $\begin{array}{cc} 2 & 6 \\ \nearrow & \nearrow \\ 0 & 0 \\ \nearrow & \nearrow \\ 8 & 3 \\ \nearrow & \nearrow \\ 2 & 6 \end{array}$

The area of $\triangle A'B'C'$ is 21 square units. Thus, the area of the given triangle is 21 square units.

Although Example 2 has been worked in detail we may, with practice, perform some of the above steps mentally and obtain the area of a given triangle with fewer calculations.

Skills and concepts in translations may be applied to the study of the properties of geometric figures. Also, the applications of translations will simplify our work in analytic geometry.

7.11 exercise

A

1 For the mapping given by
   $(x, y) \longrightarrow (x + 2, y - 3)$
   find the image of each of the following points.
   (a) (0, 3)   (b) (−4, 0)   (c) (−1, 3)
   (d) (−2, 3) (e) (4, 2)   (f) (−5, 6)

2 For the mapping given by
   $(x, y) \longrightarrow (x - 1, y + 2)$
   the co-ordinates of image points are given. Find the co-ordinates of the pre-image points.
   (a) (−3, 0) (b) (0, 5)   (c) (−1, 3)
   (d) (0, 0)   (e) (−3, 5)   (f) (4, −5)

B

3 A line segment has end points A(2, 3) and B(−1, 4).
   (a) Find the image of AB under the mapping
       $(x, y) \longrightarrow (x - 3, y + 4)$.
   (b) Show that the mapping in (a) is a translation.

4 A mapping $\triangle ABC \longrightarrow \triangle A'B'C'$ is given by
   $(x, y) \longrightarrow (x - 2, y + 3)$.
   If the co-ordinates of $\triangle ABC$ are given by A(−3, 4), B(0, −1), C(5, 1), then show that the mapping is a translation.

5 A square is given by the vertices
   P(−5, 1)   Q(−3, −7)   R(5, −5)   S(3, 3).
   Show that the mapping
   [P, Q, R, S] $\longrightarrow$ [P', Q', R', S'],
   defined by $(x, y) \longrightarrow (x - 2, y + 1)$, is a translation.

6 Three points are given.
   P(−7, 2)   Q(−3, −1)   R(1, −4)
   (a) Use the slopes to determine whether P, Q, and R are collinear.
   (b) Use the distance formula to determine whether P, Q, and R are collinear.
   (c) Apply the translation
       $(x, y) \longrightarrow (x - 3, y + 2)$
       to the points P, Q, and R to find the co-ordinates of the image points P', Q', R'.
   (d) Determine whether P', Q', and R', are collinear by using
       A: slopes   B: distance formula

7 (a) Find the midpoints of the sides of $\triangle STU$ whose vertices are
       S(−4, −6)   T(6, 14)   U(10, 2).
   (b) Find the midpoints of the sides of $\triangle S'T'U'$ whose vertices are given by the mapping,
       $(x, y) \longrightarrow (x + 2, y - 3)$.

8 △ABC has vertices A(2, 5), B(6, 4), C(−1, −7).
   (a) Determine whether ∠CAB is a right angle.
   (b) Apply the mapping
        (x, y) ⟶ (x − 4, y + 1)
        to △ABC to find △A'B'C'.
   (c) Determine whether ∠C'A'B' is a right angle.

9 △PQR has vertices P(−6, −2), Q(14, 10), R(6, −6).
   (a) Calculate the area of △PQR.
   (b) Apply the translation
        (x, y) ⟶ (x + 6, y + 2)
        to △PQR to find △P'Q'R'.
   (c) Which vertex of △P'Q'R' is at the origin?
   (d) Calculate the area of △P'Q'R'.

10 For each triangle, apply a translation to the triangle so that one vertex is at the origin. Then calculate the area of the triangle.
   (a) △ABC   A(1, 5)   B(2, −1)   C(−3, 1)
   (b) △PQR   P(3, 2)   Q(1, −3)   R(−4, −2)
   (c) △MNP   M(2, 7)   N(6, −2)   P(−4, 4)
   (d) △DEF   D(6, 2)   E(2, −6)   F(−4, −4)

11 (a) Calculate the distance from P(−3, 2) to
       ℓ: x − y − 4 = 0.
   (b) P' and ℓ' are found by applying the mapping
        (x, y) ⟶ (x − 2, y + 2)
        to the line ℓ and point P in (a).
   (c) Calculate the distance from P' to the line ℓ'. Compare your answer with that in (a).

problem/matics

Often a problem may not appear to have enough information to solve it. Does the following puzzle have enough information?

"George's age, at the time of his death, was $\frac{1}{29}$ of the year of his birth. He saw action in the First World War but died before the Great Stock Market Crash. When was George born?"

equations and mappings

In our work with analytic geometry we have found the equations of lines which satisfy given conditions. We shall now find the equation of a line translated under a mapping.

Example

A line given by $2x - 3y = 6$ is mapped under the translation
$(x, y) \longrightarrow (x - 4, y + 1)$. Find the resulting equation of the image line.

Solution 1

We may apply our principles of co-ordinate geometry and find the resulting equation by choosing 2 points on the original line.

Original line, ℓ   Translated line, ℓ'
(3, 0)   ⟶   (−1, 1)
(0, −2)   ⟶   (−4, −1)

Find the equation of ℓ'.

$\frac{y-1}{x+1} = \frac{2}{3}$ ⟵ $m_{\ell'} = \frac{-1-1}{-4-(-1)} = \frac{-2}{-3} = \frac{2}{3}$

or $2x - 3y = -5$

Note: that we might have used the fact that slopes of lines are preserved for a translation.

Solution 2

The following approach is a more general approach since this method can be extended to finding the resulting equations of curves *other than straight lines* which have been translated.

Let P'(X, Y) be the co-ordinates of a point on the image line corresponding to P(x, y).

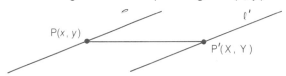

Thus $(x, y) \longrightarrow (X, Y)$.
But $(x, y) \longrightarrow (x - 4, y + 1)$.
Thus $X = x - 4$ $Y = y + 1$
      $x = X + 4$ $y = Y - 1$

But P(x, y) was a point on the line
$2x - 3y = 6$. Substitute for x and y in the pre-image line.
$$2x - 3y = 6$$
$$2(X + 4) - 3(Y - 1) = 6$$
$$2X + 8 - 3Y + 3 = 6$$
$$2X - 3Y = -5$$
But P'(X, Y) was any point on the image line. Thus, in general, the equation of the image line is given by $2x - 3y = -5$.

12 A line is given by $p: 3x + 5y = 15$.
(a) Find the slope of the line.
(b) Find the equation of the image line for the mapping given by $(x, y) \longrightarrow (x - 2, y + 3)$.
(c) Calculate the slope of the image line in (b).

13 A line is defined by the equation $3x + 2y = -6$. A translation $(x, y) \longrightarrow (x + 3, y - 5)$ is applied to the line. Find the image of each of the following points on the line.
(a) (0, −3)   (b) (−2, 0)   (c) (−4, 3)
(d) Use two of the image points of the points above to calculate the slope of the image line.
(c) Write the equation of the image line.

14 A line, given $3x - y = 8$, is translated under the mapping $(x, y) \longrightarrow (x + 4, y - 3)$. Find the equation of the image line.

15 For each line, find the equation of the image line for the given translation.
(a) $3x - y = 5$ $(x, y) \longrightarrow (x, y - 1)$
(b) $2(x - 5) = y$ $(x, y) \longrightarrow (x - 3, y)$
(c) $\frac{x - 5}{2} + y = 4$ $(x, y) \longrightarrow (x + 2, y + 1)$

16 Each of the following lines defined by the equations is translated under the mapping $(x, y) \longrightarrow (x - 4, y + 3)$. Find the equation of each corresponding image line.
(a) $4x - 3y = 12$ (b) $y = 2x + 8$
(c) $2(y - x) = 3$ (d) $2(y - 1) + 6 = x$

skills review: distance on the plane

1 Find the distance between each pair of points.
(a) A(−5, −3) B(5, 1)
(b) C(−2, 6)   D(6, 2)

2 △PQR is given by P(−2, 1), Q(10, −5), R(12, −1).
(a) Show that △PQR has a right angle.
(b) Find the co-ordinates of the midpoint of the hypotenuse.

3 Find the co-ordinates of a point on the y-axis which is equidistant from each pair of points.
(a) (3, 5)   (2, 4)   (b) (1, 4) (−1, −4)
(c) (−3, 5) (2, 4)   (d) (0, 6) (4, −2)

4 Find the co-ordinates of a point on the x-axis that is 4 units from $2x + 3y = 6$.

5 Find the distance between the lines given by $3x - 4y = 6$ and $3x - 4y = 12$.

6 Calculate the altitude from P to RQ for △PQR given by P(−3, 4), Q(−6, 3) and R(3, 1).

7 A circle has a diameter with end points A(−3, 2) and B(4, 3). Calculate the radius of the circle.

8 A circle with centre (0, 0) is tangent to $3x - 4y = 12$. Calculate the radius of the circle.

9 Two points (4, 6) and (−3, 2) determine a chord of a circle. Find the length of the chord.

10 Find the perpendicular distance from the point (5, 2) to the line $6x - 3y + 1 = 0$.

11 If the perpendicular distance from the point (−4, 1) to the line $Ax + 4y - 4 = 0$ is $\sqrt{2}$ units, then find A.

problems and practice: a chapter review

1. Show that the triangle with vertices $(0, \sqrt{6})$, $(\sqrt{2}, -\sqrt{6})$, and $(2\sqrt{2}, \sqrt{6})$ is isosceles.

2. The distance from the point $(2, 3)$ to the line $x + y + C = 0$ is $\sqrt{2}$ units. Find C.

3. A triangle has vertices $A(-2, 4)$, $B(4, 4)$ and $C(x, y)$. If the area of $\triangle ABC$ is 18 square units, find y.

4. A very famous triangle is known as the Bermuda Triangle.

   (a) For the given co-ordinates, calculate the area of $\triangle MBP$.

   (b) If a unit on the grid is 200 km, calculate the distance from Miami to Puerto Rico.

5. (a) Write the equation of the family of lines with slope $-\frac{1}{5}$.

   (b) Select the member(s) of this family that form a triangle of area 10 square units with the axes.

6. The line $3x - 4y + 10 = 0$ is perpendicular to the line $Ax + By - 40 = 0$. If the lines intersect at $(2, 4)$, find A and B.

7. A circle with centre at $(3, -1)$ is tangent to the line given by $x + 2y - 7 = 0$.

   (a) Find the radius of the circle.

   (b) Write the equation of the circle.

8. A circle is defined by $x^2 + y^2 = 12$ and meets the y-axis at R and S. $T(2\sqrt{2}, 2)$ is a point on the circle. Show that $\angle RTS$ is a right angle.

9. The path of a meteor is represented by the line $3x + 4y + 7 = 0$. If the earth is situated at $(4, -1)$ and each unit of distance is equivalent to 20 000 km, find the point of closest approach of the meteor to the earth.

10. An iceberg has floated into the Gulf of St. Lawrence and lodged itself at map co-ordinates $(4, 1)$. A passenger liner is at map co-ordinates $(-2, 6)$ and heading for $(5, -1)$.

    (a) If the liner does not alter course, will it crash into the iceberg?

    (b) If the iceberg is not in the liner's path, how close does the liner pass by the iceberg?

11. A tangent as shown, defined by the line $x - 3y - 2 = 0$, meets the circle with centre $C(4, 4)$ at P.

    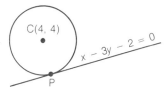

    (a) Find the distance from C to P.

    (b) What is the length of the radius of the circle?

12. (a) Show that the lines represented by $2x + 3y - 1 = 0$, $3x - 2y + 2 = 0$, $2x + 3y + 3 = 0$ and $3x - 2y + 6 = 0$, are the sides of a square.

    (b) Find the co-ordinates of the centre.

13. Calculate the area of $\triangle PQR$ whose sides are given by $PQ: x - 3y + 8 = 0$
    $QR: 2x - y - 9 = 0$
    $RP: 3x + y - 16 = 0$

14. A line defined by $3x - 4y = 12$ is translated by the mapping $(x, y) \longrightarrow (x - 3, y + 2)$. Find the equation of the image line.

test for practice

Try this test. Each *Test for Practice* will be based on the mathematics you have learned in the chapter. Try this test later in the year as a review. Keep a record of those questions that you were not successful with, find out how to answer them and review them periodically.

1. In △ABC the vertices are given as A(3, 5), B(6, −2) and C(−4, −4). If D is the midpoint of BC, determine the distance from D to each of the vertices.

2. Find the value of k such that the point $(k, -1)$ is equidistant from B(−2, −5) and C(5, −1).

3. Show that the triangle given by the vertices A(1, 1), B(2, 5), C(6, 6) is isosceles.

4. Determine the possible values of y if a triangle with area 15 square units has vertices at (−3, −1), (2, −1), and (x, y).

5. Find the distance from the point (−1, 1) to the line $x - 2y - 2 = 0$.

6. From a point P(4, k) the perpendicular distance to the line given by $3x - 4y - 1 = 0$ is 7 units. Find k.

7. (a) Write the equation of the family of lines with slope $\frac{2}{3}$.
   (b) Find the members of this family that form a triangle of area 108 square units with the axes.

8. There are two points on the x-axis, each of which is $\sqrt{13}$ units from the line $3x + 2y - 10 = 0$. Find these points.

9. Tom must propel his boat very slowly through the middle of two shallow areas in the lake. If his starting point is (2, 1) and the shallow points are at (1, 4) and (5, 2), determine the path he must steer his boat along in order to pass through the middle of these two shallow points.

10. (a) Show that the points P(3, 2), Q(3, −2), R(−3, −2), S(−3, 2) lie on the circle defined by $x^2 + y^2 = 13$.
    (b) Show that RP is a diameter of the circle.
    (c) Show that PQ ⊥ QR and PS∥QR.

11. A circle with centre C(−3, 4) is drawn through the point D(4, 2). For what value(s) of y does E(−5, y) lie on the circle?

12. (a) Draw the region shown by
    $$x^2 + y^2 < 25, \quad x, y \in R.$$
    (b) Which point is in the region, A(3, 2) or B(−3, −5)?
    (c) Draw the region in (a) if the restriction $y \geq 0$ is given.

13. A square is given by the vertices A(−1, −1), B(1, −8), C(9, −6), D(7, 2). Show that the mapping [A, B, C, D] ⟶ [A', B', C', D'] defined by $(x, y) \longrightarrow (x + 2, y - 1)$ is a translation.

14. A line defined by $2x - 3y = 6$ is translated under the mapping $(x, y) \longrightarrow (x - 4, y + 3)$. Find the equation of the image line.

math tip

We should have a plan when solving a problem. To devise a plan for some problems, it is often helpful to draw a diagram for the given conditions. Draw a diagram for the following problem.

A quadrilateral has vertices P(3, −8), Q(10, −5), R(7, 2), S(0, −1). What type of quadrilateral is it?

Now use the diagram to plan your work.

applications with calculators

If a computer is handy, we can use it to save our time on tedious calculations. For example: for the Remainder Theorem, we can write a straightforward computer program in BASIC to find the remainder when
$f(x) = ax^3 + bx^2 + cx + d$ is divided by $x - k$.

```
10  INPUT A, B, C, D, K
20  LET R = A * K ↑ 3 + B * K ↑ 2 + C * K + D
30  PRINT "THE REMAINDER IS", R
40  END  ← The remainder is f(k).
```

But if a computer is not handy, a calculator can still aid us. A calculator is an invaluable tool for doing the tedious computations in each of the following.

Heron's Formula

We already know various ways of calculating the area of a triangle. However, if we are just given the sides of a triangle, we can calculate the area using the following method.

Heron's Formula: For $\triangle ABC$, the semi-perimeter s is defined as
$$s = \frac{a + b + c}{2}.$$

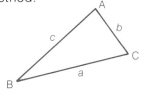

The area, A, of $\triangle ABC$ is given by
$$A = \sqrt{s(s - a)(s - b)(s - c)}$$

A Use the formula and calculate the area of each triangle with sides.
  (a) 6 m   8 m   10 m
  (b) 20 m 48 m   52 m

You can check the areas of these triangles by using another method. Why?

B Use the formula to calculate the area of each triangle to 1 decimal place.
  (a) 20 m 13 m 21 m   (b) 52 m 56 m 60 m
  (c) 40 m 42 m 45 m   (d) 27 m 40 m 42 m

To calculate the area, A, of a triangle inscribed in a circle of radius r, we may use the following formula.
$$A = \frac{abc}{4r}$$

C Draw a circle of radius 10 cm and draw a triangle in it.
  (a) Measure the sides to 1 decimal place.
  (b) Use the above formula to calculate its area, to 1 decimal place.
  (c) Use Heron's Formula to calculate the area again. Compare your answers in (b) and (c).

D The distance d, in metres, to the horizon from a building with height h, in metres, above sea level is given by the following formula.
$$d = \sqrt{2rh}$$ r is the radius of the earth.

How far is the horizon from a building, 100 m high?

Many bridges have circular arches. To calculate the radius, r of a circular arch of a bridge we can use the formula

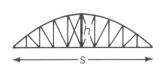

$$r = \frac{h}{2} + \frac{s^2}{8h}$$ where h, s are given in the diagram.

E Calculate the radius of a circular arch if the length of the bridge is 8.5 m long and the height of the arch is 2.3 m.

F Now for a challenge! The skills needed to derive the formula basically involve the manipulation of the Pythagorean Theorem. Can you do it?

8 statistics, probability and their applications

A study of the skills and concepts of statistics and probability; applications

8.1 introduction: statistics

We collect and study data continuously.

N.H.L. Canadian Teams Scoring Leaders

	GP	G	A	P	PiM
Gretzky, Edmonton	71	42	74	116	19
Lafleur, Montreal	65	47	66	113	8
Larouche, Montreal	68	48	37	85	14
Sittler, Toronto	63	36	48	84	49
MacDonald, Edmonton	72	41	42	83	4
Shutt, Montreal	68	39	41	80	26
Cloutier, Quebec	61	35	39	74	10
Smyl, Vancouver	68	29	40	69	164
Lukowich, Winnipeg	69	31	35	66	69

Statistics is a branch of mathematics which involves

A: collecting data
B: organizing and analyzing the data
C: interpreting and making inferences, predictions and decisions about the data.

The study of statistics is applicable to many fields; medicine, consumerism, advertising, insurance, pollution control, population, predictions and so on. We have seen many uses of statistics, such as the Gallup Poll.

GALLUP POLL
Majority of Canadians plan vegetable gardens

Today, 57 per cent of Canadians plan to grow vegetables this summer in their gardens.

Personal, in-home interviews were completed with 1 033 adult Canadians, 18 and over, during the first week of May. A sample of this

Each day we are bombarded by statistics on the radio, TV, newspaper. We are continually influenced by the process of statistics.
In this chapter we will learn skills and concepts about statistics and their applications. Many of the algebraic skills, as well as skills in graphing are used to solve problems about statistics.

8.2 statistics and sampling

If we were to ask our friends what statistics means, we would probably get a variety of meanings. Certainly some people think of statistics as tables of figures, such as sports standings of scoring leaders. Someone may think of statistics as information collected by the Government to show how well or how poorly the economy is progressing. We will think of statistics as a branch of mathematics that involves the process of

A: collecting data
B: organizing and analyzing data
C: making predictions, inferences and decisions about the data.

The following will illustrate this process.

A manufacturer of cassette tapes wishes to check the quality of the tapes before they are shipped to the retail outlets.

Step A Cassette tapes are mass produced by the millions. Each cassette tape cannot be checked individually for quality or defects. Thus a sample of the cassette tapes is chosen. In order that a proper decision will be reached
- the sample should be *representative* of the complete shipment.
- the tapes of the sample should be selected at random. (By **random sample**, we mean each member of the sample has an equal chance of being selected.)

All the tapes are referred to as the **population** and the tapes we select at random are called a **sample** from the population.

Step B Once we have collected the sample, we then carry out experiments on the sample and record the results. We may draw a chart or graph to aid us in analyzing the information we have obtained from the sample.

Step C Based on the information we have obtained from the sample we can then decide what percentage of the tapes in the sample meet the required standards. Then, a decision can be made about all the tapes (the population) based on the sample studied. The results we obtain are *probably* more reliable.

The above example illustrates the steps involved in studying statistics, namely: using the basic skills of statistics to make inferences about a population based on a random sample from that population.

8.2 exercise

To solve a problem involving statistics, we need to collect data. Questions 1 to 3 outline different methods of collecting data.

A

1. In order to determine the most popular comic strips, a company made a telephone survey.

    (a) What are some advantages of making a telephone survey to collect data?
    (b) What are some disadvantages of this method?
    (c) Suggest other ways of determining the most popular comic strip.

2. In order to determine the most popular songs a questionnaire was sent to a number of teenagers.

    (a) What are some advantages of using a questionnaire to collect data?
    (b) What are some disadvantages of this method?
    (c) Suggest other ways of determining the most popular songs for teenagers.

3 Experiments were conducted to collect information about the effectiveness of thermostats.

(a) What are some advantages of using experiments to collect data?

(b) What are some disadvantages of this method?

4 List two ways of collecting data, other than those mentioned in Questions 1 to 3.

B

5 Explain why each of the following would not produce a random sample.

(a) A survey on sports violence at a hockey game.

(b) A survey at an airport about the favourite type of candy.

(c) A survey of every second person buying a hamburg about their favourite movie star.

6 A large newspaper surveyed its readers about the projected results of an election. Explain why this sample is not a good one.

7 To test the quality of a type of television converter, a sample of 200 was chosen randomly from different lots. From the sample it was found that 2 were defective.

(a) How many converters, in a shipment of 1000 converters would you predict to be defective? Give reasons why.

(b) Do you think the above sample is representative? Why or why not?

8 A school board received a shipment of 10 000 compass sets to sell to students.

A: First, 10 sets were randomly selected and no compass sets were found to be defective.

B: Then, 100 sets were randomly selected and 1 set was found to be defective.

C: Then, 1000 sets were randomly selected and 9 sets were found to be defective.

(a) Based on A, would you be accurate in saying that none of the sets in the shipment was defective. Why or why not?

(b) Which of the following statements is more likely to be accurate? Give reasons for your answer.
S_1: 1% of the sets are defective.
S_2: 0.9% of the sets are defective.

(c) In the total shipment of 10 000 compass sets, how many sets would you estimate to be defective? Give reasons for your answer.

9 To predict the outcome of an election for mayor, which of the following methods would most likely provide data on which to make a prediction?

A: 100 persons interviewed in a specific neighbourhood.

B: 100 telephone calls to different parts of the city.

C: 100 completed questionnaires from a survey distributed across the city.

D: 100 phone calls to children living in the city.

10 Often it is not possible nor desirable to use a sample from a population to make a decision, but rather to use all of the population to make the decision. For each of the following, decide whether to use a sample or all of the population to make a decision.

(a) testing the air system in submarines

(b) determining the popularity of a particular magazine

(c) determining the quality of the picture tubes in a shipment of TV sets

(d) determining the quality of a number of parachutes

(e) deciding on the effectiveness of a new type of headache pill

(f) predicting the number of young people who will get married next year

(g) determining the chemical composition of a cooking oil for chicken
(h) checking the quality of the pistons in the engine of a car
(i) determining the number of potential buyers of a brand of TV set
(j) determining the attendance at a football game
(k) predicting the amount of oil in a new discovery

C
11 As an experiment, paper cups are tossed to determine the percentage of cups that fall in these positions.

 land up   land sideways

Three experiments A, B, and C are completed and the following data are recorded.

Experiment	Cups tossed	Cups land up	Cups land sideways
A	10	3	7
B	100	36	64
C	1000	351	649

(a) Based on the data above, which experiment would you use to predict how many cups will land up if 500 are tossed — A, B, or C?
(b) Calculate the percentage of cups that will land sideways for A, B, and C.
(c) The above results for A, B, and C are combined to obtain the figures for D.

Experiment	Cups tossed	Cups land up	Cups land sideways
D	1110	390	720

Explain why the data for D are more acceptable than for A, B, or C if all conditions for performing the experiment remain unchanged.

8.3 sampling: stratified and clustered

Random sampling is the most common type of sampling used to ascertain information about a population. Remember: a **random sample** is a relatively small group of items chosen randomly to represent a larger group called the **population**.

However, there are other types of specialized sampling that may be used to obtain data to solve a problem.

TV Audience research by Nielsen takes a sample (about 0.002%) of households—city, town, farm, etc. One source of their data is an *audimeter*. It records the channels, times, and lengths of time the sets are switched on. The data are "picked up" by a computer (by special telephone line). The results are then applied to the whole country.

Clustered Sampling

A manufacturer of farm equipment wants to decide in which magazine, A, B, or C, to spend its advertising budget. The manufacturer should not sample all the citizens to determine which magazine is preferred but rather survey farmers to see which one of A, B, or C is read. Thus, the manufacturer will take a *clustered sample* of only farmers. Thus, if a sample is taken from a particular part of a population, then it is said to be *clustered*. An example of a bad use of clustered sampling is the following:

> To determine whether a city official would be elected, voters in polling station A were

surveyed. Of these voters 75% said they would vote for the city official.

Would you accept the statement that the city official will be elected? Probably not, since the sample does not provide information about all the other polling stations and is therefore not a very useful conclusion.

Stratified Sampling
We have seen that to obtain the most probable results, the sample must be random. A statistician decides on the method of obtaining the sample. For example, to poll the popularity of 3 political leaders across Canada, how should the sample be constructed? If 1000 people are polled across Canada, and the sample is random it is possible that the same number of people are chosen from each province. This sample might give misleading results, since the number of voters in each province *differs*. Thus, to ensure that the sample is useful the sample should be stratified by choosing the number of people in each province *in proportion to* the population of each province.

Destructive Sampling
To test the quality of orange juice, cartons are chosen at random from the production line and tested. Once these cartons of orange juice are opened they are destroyed. This type of sampling is said to be *destructive* since the sample cannot be reincorporated into the population after the sampling has been completed. In destructive sampling, the entire population cannot be used.

A most important skill is the ability to design a sampling technique which will allow us to obtain the best possible sample. In our later work in statistics, we will learn additional types of sampling.

8.3 exercise

A

1 To test the quality of radial tires, they are chosen at random from a production line and then stretched. The rubber in the tire is then reused to construct new tires.
Explain why this type of sampling is referred to as destructive sampling.

2 A major fast-food company wishes to determine the suitability of a plaza for a fast-food outlet. Ten interviewers are hired to conduct a poll at the plaza. Why is this type of sampling referred to as clustered sampling?

B

3 To find an answer for each of the following, decide whether destructive (D) or non-destructive (ND) sampling is required.

   (a) testing the accuracy of calculators
   (b) testing how many words can be written with a ball point pen
   (c) checking the mass of cereals
   (d) testing the tensile strength of steel by stretching and breaking samples chosen from a batch
   (e) tasting the quality of wine
   (f) polling for an election
   (g) testing the crunch factor of a batch of crackers
   (h) checking the quality of gold charms for bracelets
   (i) testing firecrackers for brilliance

4 Each of the following samples is collected by using clustered sampling. Which of the following samples are good (G) and which samples are bad (B)?

   (a) going to a high school to determine the most popular song
   (b) asking senior students *only* about the format for graduation

(c) asking doctors about the value of a new medical procedure
(d) asking compact car owners about energy conservation policy
(e) asking hockey coaches about the quality of hockey sticks
(f) asking health food enthusiasts about their opinion of fast-food outlets

5. In each example, indicate whether a stratified sample should or should not be used.
   (a) At a camp, the director is to decide whether any funds should be spent on improving the swimming facilities.
   (b) A shipment of 10 000 ball point pens is to be checked to see how many are defective.
   (c) At a club, an opinion poll is to be conducted on the exercise facilities.
   (d) A nation is to hold a general plebiscite to decide an issue. A sample of 1000 is chosen to predict the outcome.
   (e) In an organization there are 750 women and 250 men. A sample of 20 is to be taken to determine the type of social night to be planned.
   (f) An opinion poll is to be conducted for a national election as to who is likely to be the new Prime Minister.

6. Choose one of these terms
   random, destructive, clustered, non-destructive, stratified
   as a description of the type of sample needed to obtain information for each of the following.
   (a) the amount of dues payable by plumbers of a trade union
   (b) testing the effervescence of soda pop
   (c) determining the preferable fast food
   (d) checking the radio installed in cars at a production line
   (e) obtaining information about the preferred brand of diapers
   (f) testing the sweetness of oranges
   (g) obtaining opinions about the best brand of tractor
   (h) establishing how many have seen a particular movie
   (i) determining the most popular hit parade song
   (j) obtaining an opinion of the results of a recent student council election
   (k) testing the durability of coffee cups

C
7. In a school the number of students in Grade 9 is 300, in Grade 10 is 200, in Grade 11 is 150, and in Grade 12 is 100. A sample of 100 students is used to determine which of 3 formats for the yearbook should be used. How many of the 100 opinions should be
   (a) Grade 9s?   (b) Grade 11s?

8. At a camp there are 600 girls, 250 boys, 50 adults. A poll is to be taken to decide on what improvements in the entertainment are needed. If 50 opinions are solicited, how many should be
   (a) girls?   (b) boys?   (c) adults?

applications: hypothesis testing

We often hear a rumour and wonder whether the rumour is true. To decide whether an "opinion", a rumour, etc., is probably true, we may design a sample and obtain information. The opinion or statement that we wish to verify is referred to as the **hypothesis**. For example, we may wish to check the following.

Hypothesis: 30% of the students eat pizza on Friday nights.

To test the hypothesis, we obtain a sample.

Sample A: 10 students were asked; 2 students eat pizza on Friday nights.

Sample B: 100 students were asked; 28 students eat pizza on Friday nights.

Based on Sample B, we might *accept* the hypothesis. Based on Sample A, we would *reject* the hypothesis.

In Questions 9 to 14 a hypothesis is given as well as a sample. Indicate whether you would accept (A) the hypothesis or reject (R) the hypothesis.

9 *Hypothesis:* Most Canadians exercise enough.
   Sample: 200 Grade 11 students were interviewed and most agreed with the statement.

10 *Hypothesis:* It was predicted that the next Prime Minister would not be married.
   Sample: 200 students in a primary grade classroom raised their hands.

11 *Hypothesis:* 98.9% of the light bulbs meet the required standards.
   Sample: 1000 light bulbs were tested daily and about 11 were found defective each time.

12 *Hypothesis:* 18% of the students at Erindale Secondary have blue eyes.
   Sample: 36 students in class 11GB completed a questionnaire and 18% of them had blue eyes.

13 *Hypothesis:* Fish and Tackle is read by most fishermen.
   Sample: 1000 people who fish were polled. Most of them indicated they read the magazine.

14 *Hypothesis:* Teen Teen, a magazine for teenagers, is read by very few teenagers.
   Sample: 200 people in downtown Winnipeg were interviewed. Only 3 people read the magazine, but 23 were aware of it.

8.4 frequency distributions and histograms

During a baseball season, the batting averages of 50 ball players were recorded. In its present form, it is difficult to tell how the averages were distributed.

0.282	0.277	0.260	0.225	0.218	0.265	0.271
0.284	0.251	0.248	0.245	0.302	0.272	0.292
0.274	0.271	0.225	0.274	0.301	0.279	0.265
0.236	0.251	0.267	0.261	0.292	0.252	0.282
0.252	0.265	0.221	0.265	0.233	0.251	0.259
0.208	0.274	0.241	0.248	0.241	0.275	0.281
0.245	0.265	0.231	0.262	0.265	0.202	0.272
0.265						

By examining the data we see that the smallest average is 0.202 and the greatest one is 0.302. The data could be rearranged from smallest to greatest in an array as shown.

0.202	0.208	0.218	0.221	0.225	0.225	0.231
0.233	0.236	0.241	0.241	0.245	0.245	0.248
0.248	0.251	0.251	0.251	0.252	0.252	0.259
0.260	0.261	0.262	0.265	0.265	0.265	0.265
0.265	0.265	0.265	0.267	0.271	0.271	0.272
0.272	0.274	0.274	0.274	0.275	0.277	0.279
0.281	0.282	0.282	0.284	0.292	0.292	0.301
0.302						

The array above helps us to make certain observations more easily than we could if we used the original recorded averages. For example, most of the averages are about 0.260 to 0.280.

In order to simplify our study of a large array of numbers, such as the array above, the statistician can compress the data by placing the data into non-overlapping **classes** (class intervals). For the above data we choose a convenient number of classes. The smallest and greatest numbers in each class are called the **class limits.** The number of observations in each class is called the **frequency** of that class.

Class	Class Limits	Tally	Frequency
1	0.201 to 0.210	II	2
2	0.211 to 0.220	I	1
3	0.221 to 0.230	III	3
4	0.231 to 0.240	III	3
5	0.241 to 0.250	⊮ I	6
6	0.251 to 0.260	⊮ II	7
7	0.261 to 0.270	⊮ ⊮	10
8	0.271 to 0.280	⊮ ⊮	10
9	0.281 to 0.290	IIII	4
10	0.291 to 0.300	II	2
11	0.301 to 0.310	II	2

We may denote the class frequency of class 1 as f_1, of class 2 as f_2, and so on. Thus
$$f_3 = 3 \qquad f_7 = 10$$
When the data are displayed in a table showing the class limits and the frequencies, such as the one above, we refer to the table as a **frequency distribution.**

For the above frequency distribution, the width of each class is 10. Once the data are organized in a frequency distribution, it is much easier to make observations about the data.
A Which class(es) has the greatest frequency?
B Which class(es) has the least frequency?

It is often desirable to provide a pictorial representation of data in a graphical form. One type of graph that is used to show the data contained in a frequency distribution is a **histogram.** A histogram such as the one drawn below for the previous data is a bar graph with no spaces between the bars.

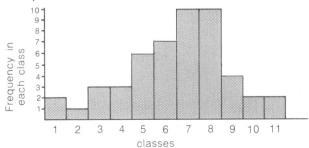

For this histogram, the height of each bar in the graph is determined by the class frequency which is marked along the vertical axis. The histogram is a visual way of organizing data to reveal any characteristics the data might have. The class boundaries separate one class from another.

A **frequency polygon** is made from the histogram as shown. The midpoints of the top of each bar are connected. How are the histogram and the frequency polygon alike? Different?

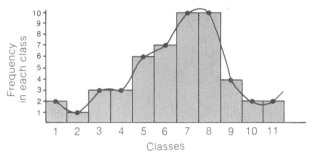

A smooth curve may be drawn to represent the frequency distribution and is called a **frequency distribution curve** or just a **frequency curve.**

8.4 exercise

Questions 1 to 5 are based on the following.
An agriculturalist was testing the effectiveness of a new chemical fertilizer by determining the yield, in units, of 40 sections of land. The test results are shown in the following histogram.

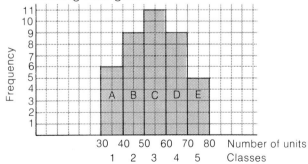

A
1 (a) What are the classes?
(b) What are the class boundaries?

2 (a) What is the value of f_2? f_5?
   (b) Calculate the ratios $\frac{\text{area D}}{\text{area E}}$ and $\frac{f_4}{f_5}$.
      What do you notice about your answers?

3 (a) In which class do most units occur?
   (b) In which two classes do the least number of units occur? Why?

4 Based on the results shown in the histogram, how would you describe these data?

5 What are the advantages of using a histogram to show numerical data?

Questions 6 to 11 are based on the following information.

A group of 40 people visited a doctor's office and had their blood pressure taken. The systolic pressure for the 40 people was recorded in the following table. (The *systolic pressure* is the maximum pressure obtained at each heart beat.)

```
121 132 125 118 102 132 122 126 124 141
109 123 120 126 132 119 131 127 119 124
128 120 135 133 126 118 125 149 129 129
126 127 122 116 141 125 142 132 127 121
```

B

6 (a) Why is it difficult to make observations about the above data in their present form?
   (b) Rearrange the data from smallest to greatest.

7 Use your answer in Question 6 to make a frequency distribution for the data.

8 Decide on class intervals for the data and construct a histogram. (The horizontal axis need not start at 0.)

9 Based on your histogram, answer these questions.
   (a) In which of your classes does the most common systolic pressure occur?
   (b) In which of your classes does the least common systolic pressure occur?

10 What percentage of the patients had a systolic reading over 120?

11 (a) If you record the systolic readings for 1000 patients, how many would you expect to have a systolic reading in the range 121-125? Give a reason for your answer.
   (b) What assumptions do you make in obtaining your answer?

For Questions 12 to 18 use the data in the following.
The following table records the number of shots on goal taken in 35 hockey games by the Weyburn Greyhounds.

```
23 21 33 48 43 28 51 30 32 24 33 32
19 42 40 31 19 25 15 20 34 17 42 28
16 30 31 49 15 53 24 28 26 41 33
```

12 Divide the data into 8 classes with interval widths of 5 and construct a histogram.

13 In which class do the fewest data occur?

14 In which class do the most data occur?

15 Divide the data into 4 classes with interval widths of 10 and construct a histogram.

16 In which way are the results of Questions 12 and 15 different? the same?

17 Draw the frequency polygon for each of the histograms.

18 How many times did the team take 31 shots?
   (a) using the histogram from Question 15?
   (b) using the histogram from Question 12?
   (c) using the table?

8.5 representatives of data

To make predictions and solve problems based on data, we group the data by various means: tables, charts, frequency distributions, graphs, and so on.

Each of the following frequency distributions has compressed the information by classifying the data so that we can more easily make predictions based on the data. Each of the following sets of data seems to "cluster" at some central point.

- How are these three frequency distributions alike?
- How are these three frequency distributions different?

Graph A: This frequency distribution curve is symmetrical. (CT seems to be an axis of symmetry.) Note how the data are evenly distributed. We often refer to this type of distribution as a **normal distribution** and the curve is often called a **bell curve**.

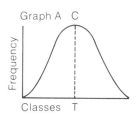

Graph B: Compared to Graph A, this frequency distribution curve is not symmetrical. We refer to the distribution as "skewed to the left". Most of the data are below or to the left of CT.

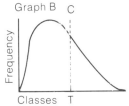

Graph C: Compared to Graph A, this frequency distribution curve is not symmetrical. We refer to this distribution as "skewed to the right". Most of the data are above or to the right of CT.

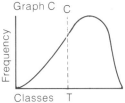

However, it would be useful to have some method of analyzing how the data cluster without having to draw a graph. It is useful, when studying statistics, to obtain a *representative* of the data in order to typify the data. *Three representatives* of data that are used to simply describe the "centre" of the data are

- the mean.
- the median.
- the mode.

Mean
The (arithmetic) mean includes *all* the data. It is what most people refer to as the "average".

Find the mean for the following sample of test results.
62, 75, 72, 68, 54, 82, 62, 62, 66

$$\bar{x} = \frac{62 + 75 + 72 + 68 + 54 + 82 + 62 + 62 + 66}{9} = 67$$

the mean ← the number of results ← the sum of the results

The mean is 67.

Note that there is no score which is 67. That is, the mean, 67, is not necessarily a member of the sample or data collected.

Median
The median for the previous sample represents the middle mark when they are arranged in order, from least to greatest.

54 62 62 62 66 68 72 75 82
smallest   middle number   greatest

The median is an easy representative of data to obtain. It is not influenced by the value of the greatest and least numbers. For example, if the top mark was 100, the median would remain *unchanged*. However, the mean would change (increase in this case).

If there is an even number of test results, the median is the average of the two middle numbers.

54 62 62 62 66 72 75 82 ← greatest
smallest   middle numbers

The median is 64. ← $\frac{62 + 66}{2} = 64$

Mode

The mode is the representative of the data which occurs most frequently. Like the median, the mode does not take into consideration the values of all the data.

For the previous sample, the most frequent mark is 62. Thus the mode is 62.

For the following results, there are 2 modes.

54 <u>60 60</u> 62 64 <u>65 65</u> 67

two modes 60 and 65

We refer to this data as **bimodal**.

The values of the mean, the median and the mode indicate where the centre of the data is in the frequency distribution. Thus, we refer to them as the **measures of central tendency**.

8.5 exercise

A

1 For each set of data find the mean, median, and mode. Express your answers, as needed, to 1 decimal place.

  (a) 8, 10, 6, 10, 12, 9, 14, 13, 12
  (b) 68, 52, 64, 69, 73, 62, 58, 60
  (c) 4, 0, 6, 7, 8, 5, 6, 12, 9, 3, 9

2 (a) The mean for a sample of data is 36. Illustrate with an example whether 36 is, or is not, one of the data.
  (b) The mode for a sample of data is 36. Illustrate with an example whether 36 is, or is not one of the data.
  (c) The median for a sample of data is 36. Illustrate with an example whether 36 is, or is not, one of the data.

3 (a) Find the mean and median for each set of data.
    M: 10, 12, 14, 16, 18, 98, 99, 100, 120.
    N: 10, 12, 14, 16, 94, 96, 98, 100, 120.
  (b) Are these numbers useful in describing the data?

  (c) Are the data in M and N shaped as A or B?
    A:  B:

4 (a) For the following data, find the mode, and the median for each set of data.
    P: 6, 20, 22, 30, 28, 0, 12, 18, 0.
    Q: 12, 18, 0, 30, 28, 6, 20, 2, 20.
  (b) How do the data in P and Q differ?
  (c) Which has a greater difference, the mode or median for each of P and Q?
  (d) Are the mode and median useful in describing the data for P or Q? Why or why not?
  (e) Is it reasonable to say that the mode or median are satisfactory measures of the location of the "centre" of the data? Why or why not?

B

5 Various persons at the airline ticket booth were asked one of A, B, or C.
    "Is the time you spend in line
    A: less than the average time?
    B: more than the average time?
    C: an average amount of time?"

  (a) If the number of replies for A, B, C are collected, explain why the mean cannot be found for this sample.
  (b) Why would the mode best represent the data?
  (c) How would you rephrase the question so that the *mean* could be calculated?

6 In July, the temperatures for 20 different cities of the world were determined at random.

22°C	28°C	28°C	27°C	25°C	34°C	30°C
26°C	27°C	20°C	24°C	32°C	32°C	22°C
22°C	33°C	20°C	28°C	28°C	28°C	

  (a) Find the mean, median, and mode for the data.
  (b) Which of the answers in (a) is useful in describing the distribution of the above data?
  (c) Would you say there was a "best" measure of central tendency? Why or why not?
  (d) Which quantity was the hardest to calculate?

8.6 predictions, inferences and estimations

The study of statistics enables us to make predictions, inferences, and estimations that are probably true. For example, by collecting data, weather forecasters are able to predict what the weather for tomorrow or the weekend will probably be.

One major application of statistics is in estimating. For example, out of a shipment of 10 000 light bulbs, a sample of 100 is randomly drawn. If the average (mean) life of a bulb is 55 h, what is the average (mean) life for the whole population? Clearly, the mean of the population should be somewhere around 55 h, but how close? We cannot answer that question precisely but based on certain properties of the sample, such as sample size, we can provide an interval in which the mean will probably occur. As another example of statistical inference, a certain street corner has more traffic flowing south than north based on the data collected at random times for a year.
Is it reasonable to use the data and determine decisions for traffic flow?
Suppose a coin is tossed 100 times on the floor and 55 heads and 45 tails turn up. Is it reasonable to say the coin is fair?
In the following exercise we will investigate estimation and predictions based on a knowledge of the mean, median, and mode.

8.6 exercise

Questions 1 to 5 are based on the following.

To make predictions about the spending habits of vacationers, 56 people were asked what was the average amount of money they spent each day while on vacation. The following data, in dollars, were collected.

```
98 72 52 91 48 51 64 83 69 61 61 55 56 87
82 86 77 58 65 64 73 92 92 59 96 77 88 67
86 77 67 73 66 85 71 61 42 91 68 69 56 63
81 57 74 76 73 66 84 77 76 73 58 62 61 68
```

1. Calculate the mean, the median, and the mode for the above data.

2. Use the data in (a) to predict what the histogram would look like for the above dollar values.

3. Construct the histogram. How does your constructed histogram compare to the one predicted in (b)?

4. In order to make decisions about the spending habits of vacationers for the next year predict how many people will spend more than
   A: $60/d.   B: $80/d.

5. If you were going on a vacation, based on the above results, how much money would you, on the average, spend in a week?

6. Based on collected data, a sports magazine issued the following statement. *The average football player uses a brand X helmet.* Which of the following measures of central tendency was probably used to make the above conclusion: the mean, the median, or the mode? Why?

7. A random sample of the masses, in kilograms, of 20 football players in a football league was recorded as follows.

```
70 74 68 70 68 73 75 75 77 74
72 76 72 72 69 72 73 72 73 73
```

Based on the sample, estimate the mean mass of the players in the entire league.

8 Two dice were rolled 144 times and the sums of the numbers obtained were recorded as follows.

Sum	2	3	4	5	6	7	8	9	10	11	12
Frequency	5	8	12	17	22	23	20	16	11	8	2

The sum of 4 turned up 12 times in the 144 tosses of the dice.

Based on the above sample, if the dice were rolled 1000 times, estimate how many times

(a) a sum of 5 would turn up.
(b) a sum of 7 would turn up.
(c) a sum of 11 would turn up.

9 Explain which of the following estimates or predictions are based on using the mean, the mode, or the median.

(a) The average salary of students is $32.50 per week.
(b) The middle mass of the football players is 70 kg.
(c) Most snowmobile accidents occur in winter.
(d) The average speed on our highways is 85 km/h.
(e) The favourite song is Sunshine.
(f) The middle mark was 64.
(g) The average score at hockey games is 3.2 goals per team.
(h) More people golf than play billiards.

applications: predictions and experimental statistics

It is quite clear that to improve on estimates or a prediction, it is preferable to obtain a large sample. It also seems that if we use more than one sample we will be in a position to improve our predictions. For example, in polls conducted by newspapers, the more samples that are taken the more confident the editors are in making a prediction.

10 (a) Fill a box with a different number of red, green, and blue chips, to total 100 in all.
   (b) Have someone take a random sample of 10 chips and have them predict the percentage of red, green and blue chips in the box. Return the chips to the box.
   (c) Repeat step (b) for another 10 chips and again predict the percentage of red, green, and blue chips.
   (d) Have the person repeat step (b) until an accurate prediction can be made about the number of red, green, and blue chips in the box.

11 (a) Toss 2 coins 10 times into a box and determine the number of times 2 heads occur.
   (b) Use this to estimate the number of times 2 heads occur in 1000 tosses.
   (c) Repeat (a) for another 10 tosses and answer (b) again.
   (d) Continue the steps in (c) until you can make a reasonable prediction as to how many times 2 heads will occur in 1000 tosses of 2 coins.

12 Four different shapes are shown.

   (a) Which of the above shapes do you think most people will choose?
   (b) Use the above shapes and ask 10 people to choose one of them.
   (c) Use the data in (b) to make a prediction. Which of the 4 shapes is
        M: most likely to be chosen?
        L: least likely to be choosen?
   (d) Repeat step (b) three more times.
   (e) Based on the samples, predict the percentage, of a group of 100 people, that will choose each shape.

3 (a) Write down any letter from A to L inclusive.
   (b) Predict how many students in your class will write the same letter as you.
   (c) Have the students record any letter from A to L. What are the 3 most frequently written letters?
   (d) Use your data in (c) to estimate which 3 letters would be chosen (and how many times) from all the students in your school. How could you check your prediction?

4 A thumbtack can land in one of 2 ways, ⋋ or ⊥. Devise an experiment to establish whether the following statement is more likely to be true or false, "When a thumbtack is dropped it will more likely land with its point down".

5 (a) Decide what you think are the 5 most popular TV shows.
   (b) Design a questionnaire to collect data on what other persons think are the 5 most popular TV shows. Use the data from the questionnaire to estimate or predict the 5 most popular TV shows.
   (c) Use another method, such as a telephone survey, to collect data on what other persons think are the 5 most popular TV shows. Use this data to estimate or predict the 5 most popular TV shows.
   (d) How do your answers in (a), (b), and (c) compare? Which of the above methods would you use to predict the 5 most popular TV shows across Canada?

6 Repeat the steps in Question 15 to establish predictions for the following.
   (a) the most popular hockey player.
   (b) the 3 most popular songs.
   (c) the 2 most popular movies of all time.
   (d) the 5 most popular male singers.

8.7 measures of dispersion

Two sets of data may have the same mean and same median, but can vary quite significantly in how the data is distributed.
For example, the following two samples have the same mean and median, but are dispersed (distributed) quite differently.

A: 2, 10, 12, 12, 12, 14, 22
   Mean 12
   Median 12

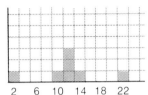

B: 6, 8, 8, 12, 16, 16, 18
   Mean 12
   Median 12

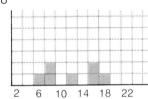

To indicate how widely the data of a sample are spread, other representatives of the data are needed. One representative of the data used is the **range**.

Range: The range of a set of data is the difference between the greatest and smallest data obtained.

For example, for the above samples A and B.

Sample A	*Sample B*
Greatest data = 22	Greatest data = 18
Smallest data = 2	Smallest data = 6
Range = 22 − 2	Range = 18 − 6
= 20	= 12

For a sample, the mean, median and mode give some useful information about where the data are *clustered* and although the range will give us some information as to how the data are *dispersed*, it is not usually a satisfactory statistic.

Example 1

For each sample, draw a histogram and calculate the range.

Sample M: 1, 4, 3, 4, 5, 2, 4, 4, 6, 3, 7, 3, 4, 5, 5
Sample N: 6, 5, 4, 1, 3, 1, 6, 2, 7, 3, 5, 1, 2, 6, 7

Solution

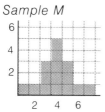

Sample M

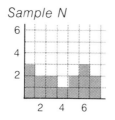

Sample N

Range of the data
7 − 1 = 6

Range of the data
7 − 1 = 6

Although the data of the samples M and N are dispersed quite differently, the range is the same and thus may provide misleading information when comparing the two samples. Thus, the range for the above samples is not a useful representative of the data. Like the median and mode, the range does not consider how the data may be clustered with respect to the rest of the data.

standard deviation

To describe the data for M and N above, we might say

M: the data are clustered, in this case, about its mean.

N: the data are dispersed

As a *measure* of the amount of dispersion, a representative of the data, called the **standard deviation** is calculated.

For the data in Sample M, x_1 and x_2 are the same "absolute distance" from the mean ($\bar{x} = 4$).

Sample M

For x_1
$x - \bar{x} = -3$

For x_2
$x_2 - \bar{x} = 3$

Note difference in sign.

To avoid negative results in calculating a measure of the dispersion, the deviations are squared, the mean of the squares is found and the square root of the last result is taken.

To calculate the standard deviation for a set of data, the following steps are completed.

Step A: Calculate the mean.
Step B: Calculate the deviation (difference) between each number and the mean.
Step C: Square the results in *Step B*.
Step D: Calculate the mean of the squares in *Step C*.
Step E: Calculate the square root of the mean in *Step D*.

The above steps are illustrated in Example 2. Question 13 of the exercise shows the significance of squaring the differences.

Example 2

A machine, packaging candy in 90-g packages, is thought to be faulty. A sample of 10 packages is randomly selected and the actual masses in grams are

86 91 89 88 92 90 93 90 90 91

The standard deviation must be less than 1.3 g for the data.

(a) Calculate the standard deviation.
(b) Is the packaging machine faulty?

Solution

(a) Use a chart to record the calculations in order.

Use x_i ← to represent each item of data

Thus $x_3 = 89$.

$\bar{x}$ ← to represent the mean

i	Masses x_i	Deviation from the mean $(x_i - \bar{x})$	Deviation squared $(x_i - \bar{x})^2$
1	86	−4	16
2	88	−2	4
3	89	−1	1
4	90	0	0
5	90	0	0
6	90	0	0
7	91	1	1
8	91	1	1
9	92	2	4
10	93	3	9

sum = 900   sum = 36

$$\bar{x} = \frac{900}{10} \qquad \text{mean of the squares} = \frac{36}{10}$$

Standard Deviation = $\sqrt{3.6}$
 = 1.9 (to 1 decimal place)

(b) Thus the standard deviation is 1.9 g and the machine is faulty.

The symbol, σ, is used to represent standard deviation. The standard deviation, σ, for a set of data $x_1, x_2, x_3, \ldots, x_n$, which has a mean $\bar{x}$ is given by

$$\sigma = \sqrt{\frac{(x_1 - \bar{x})^2 + (x_2 - \bar{x})^2 + \ldots + (x_n - \bar{x})^2}{n}}$$

We may use the symbol Σ, called sigma (the summation symbol), to write the above in a compact form.

$$\sum_{i=1}^{n} (x_i - \bar{x})^2 = (x_1 - \bar{x})^2 + (x_2 - \bar{x})^2 + \ldots + (x_n - \bar{x})^2$$

Thus

$$\sigma = \sqrt{\frac{\sum_{i=1}^{n} (x_i - \bar{x})^2}{n}}$$

8.7 exercise

A

1 If $x_1, x_2, \ldots, x_6$ represent 3, 6, 10, 10, 12, 13 and $\bar{x}$ is the mean of the data, then calculate

(a) $\sum_{i=1}^{6} (x_i - \bar{x})$ \qquad (b) $\sum_{i=1}^{6} (x_i - \bar{x})^2$

2 If $x_1, x_2, \ldots, x_8$ represent 16, 18, 19, 23, 23, 24, 26, 28 and $\bar{x}$ is the mean of the data, then calculate

(a) $\sum_{i=1}^{8} (x_i - \bar{x})$ \qquad (b) $\sum_{i=1}^{8} (x_i - \bar{x})^2$

Questions 3 to 7 are based on the following data.
  Sample P: 12, 6, 12, 18, 12, 11, 13

3 Calculate the mean $\bar{x}$.

4 Calculate $x_i - \bar{x}$ where $x_1, x_2, \ldots, x_7$ represent the data from least to greatest.

5 (a) Calculate the sum of the deviations, given by $S_1 = \sum_{i=1}^{7} (x_i - \bar{x})$.

   (b) What is the average of the deviations in (a)? (Namely, calculate $S_1 \div 7$.)

6 (a) Calculate the sum of the squares of the deviations, given by $S_2 = \sum_{i=1}^{7} (x_i - \bar{x})^2$.

   (b) Calculate the average of the sum of the squares of the deviations in (a). Namely, calculate $S_2 \div 7$.

7 Compare your answer in Questions 5 and 6.

Questions 8 to 11 are based on the following data.
  Sample Q: 8, 16, 18, 12, 6, 18, 6

8 Calculate the mean $\bar{x}$.

9 Calculate $x_i - \bar{x}$ where $x_1, x_2, \ldots, x_7$ represent the data from least to greatest.

10 (a) Calculate the sum of the deviations, given by $\sum_{i=1}^{7} (x_i - \bar{x})$.
    (b) What is the average of the deviations in (a)?

11 (a) Calculate the sum of the squares of the deviations, given by $\sum_{i=1}^{7} (x_i - \bar{x})^2$.
    (b) What is the average of the squares of the deviations in (a)?

12 Use your results in Questions 3 to 11.
    (a) Compare the ranges for the Samples P and Q.
    (b) Compare the values of the sum of the deviations for Samples P and Q.
    (c) Compare the values of the sum of the squares of the deviations for Samples P and Q.
    (d) What is the standard deviation for Sample P?
    (e) What is the standard deviation for Sample Q?

13 (a) Why is the range not a satisfactory statistic for Samples P and Q?
    (b) Why is the sum of the deviations for Samples P and Q not a satisfactory statistic?
    (c) Use your result in (b). Why is the sum of the squares of the deviations used in the calculation of the average deviation, rather than the sum of only the deviations?

Unless indicated otherwise, write answers to 1 decimal place.

14 A set of data $x_1, \ldots, x_6$ are recorded in the chart. The mean, for the data, is 64 (i.e. $\bar{x} = 64$).
    (a) What is the range for the data?
    (b) Copy and complete the table.
    (c) Use the results in (b) and calculate the standard deviation of the data.

x_i	$x_i - \bar{x}$	$(x_i - \bar{x})^2$
60		
62		
63		
65		
66		
68		

15 Another set of the data $x_1, \ldots, x_6$ is recorded in the chart. The mean for the data is $\bar{x} = 64$.
    (a) What is the range for the data?
    (b) Copy and complete the table.
    (c) Use the results in (b) and calculate the standard deviation of the data.

x_i	$x_i - \bar{x}$	$(x_i - \bar{x})^2$
54		
56		
57		
71		
72		
74		

16 For the data in Questions 14 and 15 the means are the same ($\bar{x} = 64$), but the standard deviations are different. Why is this so?

B

17 Given the set of masses
    90 g, 102 g, 100 g, 106 g, 112 g, 94 g, 82 g.
    (a) Calculate the mean.
    (b) Calculate the standard deviation.
    (c) Is the difference between the mean and the standard deviation within 5 g?

18 The numbers of hours worked by students on a part-time basis are listed below.

    | 14 | 8 | 16 | 8 | 14 | 18 | 8 | 5 |
    | 15 | 12 | 18 | 5 | 4 | 12 | 8 | |

    (a) Find the range. Is the range a useful representative of the above data?
    (b) Calculate the mean.
    (c) Find the standard deviation.
    (d) Is the difference between the mean and standard deviation within 1.5 h?

19 For 7 weeks, the amount of gold mined at Bruce Bay is given in kilograms.

    28   34   34   32   27   31   35

    (a) Find the range. (b) Calculate the mean.
    (c) Find the standard deviation.

0. For the next 7 weeks, the amount of gold mined at Bruce Bay, in kilograms, is given by

   30 28 31 34 26 30 31

   (a) Predict the mean.
   (b) Do you think the standard deviation will be large or small?
   (c) Calculate the mean.
   (d) Find the standard deviation.

1. As an experiment, the length of time a bulb burned continuously for a D-cell battery is given by the following data, in minutes, for 10 batteries.

   109 118 119 136 119 127 132 124 117 111

   (a) Calculate the mean, $\bar{x}$.
   (b) Calculate the standard deviation, σ (in minutes).
   (c) How many batteries burned continuously within 1 standard deviation of the mean (i.e. how many are in the interval $\bar{x} \pm \sigma$)?

2. The masses of 12 students on the wrestling team are

   55 kg   60 kg   61 kg   66 kg   57 kg   61 kg
   59 kg   61 kg   60 kg   61 kg   60 kg   63 kg

   (a) Calculate the standard deviation, σ, of the masses (to the nearest tenth of a kilogram).
   (b) How many of the wrestlers are within 1 standard deviation of the mean?

---

math tip

The following data are given in grouped form.

A	B	C
Class (ages)	Frequency	Total of each class
14	5	70
15	9	135
16	13	108
17	10	170
18	3	54
Total	40	637

To determine the mean, multiply the frequency in Column B by the corresponding number in Column A. The results in Column C are added and divided by 40, the sum of Column B. Thus $\bar{x} = 637 \div 40 = 15.9$.

8.8 decision making and standard deviation

The standard deviation is an important measure of dispersion if the data are distributed normally. In the previous section, 10 packages had the following masses.

86 g   91 g   88 g   89 g   92 g
90 g   93 g   90 g   90 g   91 g

The standard deviation was calculated to be 1.9 g (to 1 decimal place). What is the significance of this statistic for the above data?

If the data for the above packages are distributed normally, then the following normal distribution curve occurs, appearing in the shape of a bell.

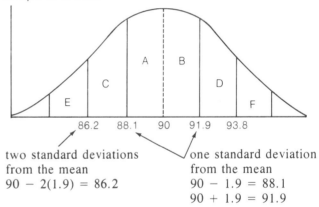

two standard deviations from the mean
$90 - 2(1.9) = 86.2$

one standard deviation from the mean
$90 - 1.9 = 88.1$
$90 + 1.9 = 91.9$

For the normal distribution

I the data represented by A and B ($\bar{x} \pm \sigma$) is 68% of the population.

II The data represented by A, B, C, D, ($\bar{x} \pm 2\sigma$) is 95% of the population.

III The data represented by A to F ($\bar{x} \pm 3\sigma$) is 99.7% of the population.

To interpret the results for the packages this means

I based on the sample, 68% of the packages have a mass between 88.1 g and 91.9 g.

II based on the sample, 95% of the packages have a mass between 86.2 g and 93.8 g.

Thus if 10 packages is a sample of 1000 packages which are produced each hour then the number of packages that have a mass within 1 standard deviation of the mean is given by
 68% of 1000 = 680.
Thus 680 packages have a mass between 88.1 g and 91.9 g.

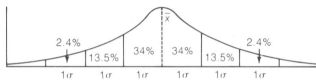

Thus, if the data are distributed normally and we use σ to represent 1 standard deviation, then the data is distributed approximately as shown in the following diagram.

Example 1
The mean life of a tire is given as 30 000 km. The standard deviation is 2500 km.

(a) What percentage of the tires will have a life that exceeds 27 500 km?
(b) A company with a fleet of cars purchases 1400 tires. How many of the tires do you expect will last for more than 25 000 km? How many will last less than 25 000 km?

Solution To understand a problem more fully a sketch of the normal distribution curve is useful to plan your solutions.

The mean, $\overline{x}$ = 30 000 km.
Standard deviation, σ = 2500 km.
The normal distribution curve is sketched.

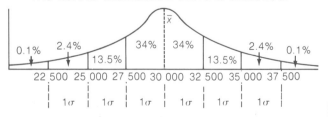

(a) Since 1σ = 2500 km then the percentage of tires with a life more than 27 500 km is given by
 34% + 34% + 13.5% + 2.4% + 0.1% or 84%.
(b) From the diagram 97.5% of the tires will last more than 25 000 km.
 97.5% of 1400 = 1365 tires

Thus it is expected that 1400 − 1365, or 35, tires will not last more than 25 000 km.

A knowledge of standard deviation is an important skill in solving problems in industry, medicine, engineering, traffic and so on.

8.8 exercise

Questions 1 to 6 are based on the following information.

A manufacturer of watches records information about how fast the watches are in a 1-month period.

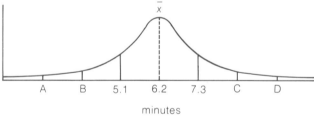

1. What is the average of the data?
2. What is the standard deviation?
3. If the data for 1000 watches were recorded, what percentage are within 1 standard deviation of the mean?
4. What are the values of A, B, C, and D?
5. If another batch of 6000 watches has the same results, how many watches do you expect will be running faster by 5.1 min in a month?
6. How many of the watches would you expect to be running faster by at least 5.1 min in a month?

Questions 7 to 9 are based on the following information.

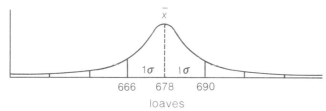

B

7 How many loaves have a mass within 1 standard deviation of the mean?

8 How many loaves exceed the required mass by at least 2 standard deviations?

9 Internal quality control regulations require that not more than 3% of the loaves be less than 650 g. Does the above batch meet the required specifications?

10 The girls on the speed skating team have heights as follows.

| 169 cm | 171 cm | 167 cm | 167 cm | 163 cm |
| 175 cm | 173 cm | 172 cm | 171 cm | 172 cm |

(a) Calculate the standard deviation.
(b) How many girls are within 1 standard deviation of the mean?

11 From a production line, the mass of a cookie is given as 40 g. The standard deviation of a sample is 2.5 g.
(a) If 10 000 cookies/d are produced, how many cookies are within 2.5 g of the required mass?
(b) How many cookies are rejected, if they are less than 35 g or greater than 45 g?

12 The average time required to serve a customer at a bank is 2 min 20 s. The standard deviation is 20 s. If 148 customers are served at the noon break, how many of them would expect to be served
(a) in less than 2 min?   (b) in more than 3 min?

13 Twenty random pollution counts were made of the air on 20 consecutive days and recorded as follows.

14, 10, 14, 11, 10, 15, 14, 9, 10, 16, 11
10, 11, 12, 8, 11, 14, 10, 12, 12

(a) Calculate the standard deviation for the data.
(b) For how many days was the pollution count within 1 standard deviation of the mean?
(c) If the pollution count exceeds 2 standard deviations of the mean, the electronic air cleaners must be activated. On how many days was this done?

C

14 The pulse rate (per minute) of 20 adults was recorded as

72 74 75 66 72 75 68 74 73 74
73 74 71 73 90 77 75 96 74 76

(a) Calculate the standard deviation for the above data.
(b) How many persons have a pulse rate within 1 standard deviation of the mean?
(c) What are the limits of the pulse rate for the persons in (b)?
(d) If the above sample is representative for a town with 6840 adults, how many of the adults will have a pulse rate that exceeds 75?

---

problem/matics

Use the following computer program written in BASIC to calculate the mean, M, of a set of data.

```
10  INPUT N
15  DIM X(100)
20  LET S = 0
30  FOR I = 1 TO N
40  INPUT X(I)
50  LET S = S + X(I)
60  NEXT I
70  LET M = S/N
80  PRINT "THE MEAN IS", M
90  END
```

A computer program in BASIC to calculate the standard deviation for a set of data can be found on page 291.

8.9 probability: skills and concepts

Many of the answers we obtain to questions often involve uncertainty. The answers are often only *probably* true.

- There is a 75% chance of rain tomorrow.
- Based on the advance polls, the plebiscite will probably take place.
- If you are older than 55 a, you will probably live another 22 a.

Poll shows Ryan ahead by 12 points

Using our skills with statistics, we collect data and organize the data to make inferences. The mathematics of chance or prediction is called **probability theory**. Probability is that branch of mathematics that predicts the outcome of events. The example of rolling a die will illustrate various terms associated with probability.

If a single fair die is rolled there are 6 *equally likely* outcomes. One of 1, 2, 3, 4, 5, or 6 will be rolled.  The 2 is called the **outcome** of rolling the die. We refer to the set of possible outcomes, namely {1, 2, 3, 4, 5, 6}, as the **sample space**. When the die is rolled, it is **equally likely** that any one of the 6 faces will turn up. When the elements of the sample space, S, are equally likely we refer to it as the **elementary** sample space. An **event** is any subset of a sample space, S, including the null set and S itself. That is, if the event, E, is rolling a 3, then we may write
$$E = \{3\}.$$

The notation used to indicate the probability of rolling a 3 is given by $P(3)$.

$$P(3) = \frac{1}{6}$$

The numerator shows the number of times the outcome can occur, (only 1 number on 1 roll).

We read this as "The probability of rolling a 3 is $\frac{1}{6}$."

The denominator shows the total number of possible outcomes — 6 possible numbers might turn up.

If the event is rolling either a 1 or a 2 then the event, E, is $E = \{1, 2\}$ and we may write
$$P(E) = \frac{2}{6} \leftarrow \text{roll one of 1, 2}$$
$$ \leftarrow \text{6 possible numbers}$$
$$= \frac{1}{3}$$

Examples such as the above suggest the following definition for probability.

$$\frac{\text{Probability of}}{\text{an event, } E} = \frac{\text{number of favourable outcomes, } F}{\text{total number of possible outcomes, } T}$$

We may write the above in a compact form.
$$P(E) = \frac{F}{T} \begin{array}{l} \leftarrow \text{number of favourable outcomes} \\ \leftarrow \text{total number of possible outcomes} \end{array}$$

Example 1
What is the probability of drawing a jack from a deck of 52 ordinary playing cards?
Solution
There are 4 jacks in a deck. $E = \{4 \text{ jacks}\}$.
There are 52 cards in a deck. $S = \{52 \text{ cards}\}$.
$$P(E) = \frac{4}{52} = \frac{1}{13}$$

Rolling a 7 on a die is impossible.
$$P(7) = \frac{0}{6} \begin{array}{l} \leftarrow \text{There are no faces with 7.} \\ \leftarrow \text{There are 6 possible numbers} \end{array}$$
$$= 0 \qquad \text{to be rolled.}$$

Thus the probability of an **impossible event** or outcome is 0.

Rolling a number less than 7 on a die is a certainty.
$$P(E) = \frac{6}{6} = 1 \begin{array}{l} \leftarrow \text{Any number appearing on the fac} \\ \text{of the die is a favourable outcom} \end{array}$$

The probability of a **certainty** is 1.

Theoretical Probability

When tossing a coin, we would obtain either a head, H, or a tail, T (assuming each event is equally likely). Thus, we can write the theoretical probability of obtaining a head as

$$P(H) = \frac{1}{2}. \begin{cases} \text{number of heads,} \\ \text{(favourable outcomes)} \end{cases}$$
← number of possible outcomes

Experimental Probability

Sometimes we cannot calculate the theoretical probability but must rely on the data we obtain from an experiment. Thus, we will need to use our skills in working with statistics to solve problems about probability. For example, we may conduct an experiment to find the probability of tossing a head on a coin.

Total number of tosses	100
Number of heads	48
Number of tails	52

The experimental probability of tossing a head based on the above data is

$$P(H) = \frac{48}{100} \leftarrow \text{number of heads} \atop \leftarrow \text{number of tosses}$$

$$= \frac{12}{25} \text{ or } 0.48 \text{ (to 2 decimal places)}$$

For some situations, we can calculate the theoretical probability and then compare it to the experimental probability. We may thus develop skills for probability by comparing these answers. Sometimes, however, it is only from experimental probability that we obtain the answers, as the following example illustrates.

Example 2

A thumbtack when tossed may land

up or down

What is the probability of tossing a thumbtack so that it lands up?

Solution

To answer the question, we must collect data.

Number of thumbtacks tossed	100
Number of thumbtacks face up (U)	61
Number of thumbtacks face down (D)	39

Based on the results of the data

$$P(U) = \frac{61}{100} = 0.61$$

Thus, the probability of tossing a thumbtack face up is 0.61. (Remember that the result is obtained by *experimental* probability.)

In Example 2
Sample space, S = {thumbtack up, thumbtack down}
Event space, U = {thumbtack up}
$P(U) = \frac{61}{100}$ based on the experimental data.

8.9 exercise

Throughout the exercise, we may, in the context of the problem, ascertain whether it refers to experimental or theoretical probability. Throughout the exercise, the term probability will be used.

A

1. The 26 letters of the alphabet are placed in a box and all have an equal chance of being drawn. What is the probability of drawing

   (a) t?   (b) a, b, c?   (c) the vowels?
   (d) letters that sound like "b" (bee)?

2. A single die is rolled. What is the probability of rolling

   (a) a 3?   (b) an even number?
   (c) a number greater than 1?
   (d) a number less than 4?
   (e) a number greater than 6?

3   In a parking lot there are 42 compact cars and 36 full size cars. All the keys are placed in a box. What is the probability of taking a key from the box for a compact car?

4   The probability of rolling a 4 on a die is $\frac{1}{6}$ and the probability of *not rolling* a 4 on a die is $1 - \frac{1}{6}$ or $\frac{5}{6}$.

   (a) What is the probability of rolling a number greater than 2 on a die?
   (b) Use your answer in (a) to calculate the probability of rolling a number not greater than 2.

5   A pair of football tickets was won by a class. There are 34 students and names are to be drawn to find the winner of the tickets.

   (a) If there are 14 girls, what is the probability of drawing a girl's name?
   (b) What is the probability of *not drawing* a girl's name?

B

6   From experimental results, the probability that an electronic calculator has a defect is 0.04.

   (a) If 1246 calculators were manufactured, how many would you expect to be defective?
   (b) What is the probability that you will choose a calculator that will work?

7   In a class experiment, the data for tossing coins, 100 times by each of 36 students are recorded.

Total number of tosses	3600
Number of heads (H)	1753
Number of tails (T)	1847

   36 × 100 tosses = 3600

   Based on the above data, what is the probability of tossing
   (a) a head, $P(H)$?   (b) a tail, $P(T)$?

8   Another class conducted the same experiment as in Question 7 and obtained the following results.

Total number of tosses	3200
Number of heads (H)	1523
Number of tails (T)	1677

   Based on the above data, what is the probability of tossing
   (a) a head, $P(H)$?   (b) a tail, $P(T)$?

9   (a) What is the theoretical probability of tossing a head? a tail?
   (b) Explain why your answers in parts (a) of Questions 7 and 8 differ.

10 Based on the number of games yet to play, the probability of the Eskimos finishing in first place is 0.46. What is the probability that they will not finish in first place?

11 Based on the data collected over the year the probability of the Canucks winning the game is 0.76 and of tying the game is 0.12. What is the probability that the Canucks will lose?

charts and sample spaces

Have you ever tossed a coin to decide who buys the coffee? The probability that you will have to buy the coffee is $\frac{1}{2}$. However, if three people toss a coin and the odd person out has to buy the coffee, what will be the probability that you will have to buy the coffee? A sample space is required in order to calculate the probability of a person buying coffee. We may draw a diagram to show the different possibilities, which is the sample space.

John	Penny	H	H	H	H	T	T	T	T
George	Nickel	H	H	T	T	H	H	T	T
Susan	Dime	H	T	H	T	H	T	H	T

John buys the coffee for these two situations.

Questions 12 to 14 refer to the above chart.

12 Use the above chart. What is the probability that
(a) John will buy the coffee?
(b) Susan will buy the coffee?
(c) George will not buy the coffee?

13 What is the probability of tossing
(a) 3 tails?   (b) 2 tails and a head?
(c) 1 tail and 2 heads?   (d) all the same?

14 What is the probability of
(a) tossing 2 heads? (b) not tossing 2 heads?
(c) not tossing 3 heads?
(d) not tossing a tail on the dime?
(e) not tossing a head on the nickel?

15 We may also draw a tree diagram to obtain the sample space for tossing the above 3 coins.

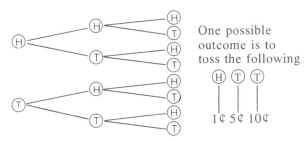

(a) Construct a sample space for tossing 4 coins.
(b) Construct a sample space for tossing 1 coin and rolling a die.

16 A pair of dice is rolled and a chart is used to record the sample space.

4 turns up and 3 turns up. The sum is 7.

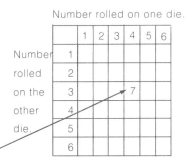

Copy and complete the chart.

17 Use the chart in Question 16. Copy and complete the following.

Sum of	2	3	4	5	6	7	8	9	10	11	12
Frequency											

18 From the data in Question 17, what is the probability of rolling a sum of
(a) 3?   (b) 6?   (c) 2?   (d) 11?
(e) Which of the sums has an equal probability of being obtained on a roll of the dice?

19 Which sums are more likely to be rolled?
(a) a 3 or a 7?   (b) a 5 or an 8?
(c) a 6 or a 9?   (d) a 2 or an 11?

20 What is the probability of not rolling
(a) a sum of 2?   (b) a sum of 7?
(c) a sum of 11?

21 What is the probability of not rolling
(a) an even sum?   (b) an odd sum?

22 What is the probability of rolling a sum
(a) greater than 8?   (b) less than 4?

23 An experiment consists of shuffling a deck of 52 cards and selecting a card. The elementary sample space consists of 52 equally likely outcomes, (namely, drawing each of 52 cards). The probability of selecting a black ace is given by

$$E = \{ \spadesuit, \clubsuit \} \qquad P(E) = \frac{2}{52} = \frac{1}{26}$$

Based on the above, what is the probability
(a) of picking the 3 of hearts?
(b) of not picking the 3 of hearts?
(c) of picking the ace?
(d) of not picking the ace?
(e) of picking a black jack?
(f) of not picking a black jack?

8.10 independent and dependent events

How many heads of a coin can we toss in a row? Suppose we toss a coin 5 times and obtain 5 heads, what is the probability that we will obtain another head on the next toss? Even though we have obtained 5 heads already, the probability of getting a head on the next toss is
$P(H) = \frac{1}{2}$.
Tossing a coin 5 times and getting 5 heads does not influence the outcome of the 6th toss. The 6th toss is *independent* of the previous results.

We say that experiment A is **independent** of experiment B if the results of one do not influence the results of the other.

Example 1
A coin is tossed and a die is rolled. What is the probability of tossing a head on the coin and rolling a six on the die?

Solution
To calculate the probability we need to list the sample space.
- For the coin, the sample is $S_1 = \{H, T\}$.
- For the die, the sample is $S_2 = \{1, 2, 3, 4, 5, 6\}$.

We may use a tree diagram to list the sample space for tossing a coin and rolling a die.

$S_3 = \begin{Bmatrix} (H,1), (H,2), (H,3), (H,4), (H,5), (H,6), \\ (T,1), (T,2), (T,3), (T,4), (T,5), (T,6) \end{Bmatrix}$

The probability of tossing a head and rolling a 5 is
$P(H,5) = \frac{1}{12}$.

Obtaining a head on a coin and tossing a 5 on a die are **independent events**.
$P(H) = \frac{1}{2} \qquad P(5) = \frac{1}{6}$
Note that $P(H)P(5) = \frac{1}{2} \times \frac{1}{6} = \frac{1}{12}$

For the independent events, E_1, and E_2, we calculate the probability of both events occurring, $P(E_1, E_2)$, by calculating the individual probabilities.

$P(E_1 \text{ and } E_2) = P(E_1) \times P(E_2)$

probability of event E_1 and E_2 occurring

product of the probability of each event

8.10 exercise

A

1. A coin is tossed and a die is rolled. Copy and complete the chart.

	A Probability of tossing	B Probability of rolling	C Probability of tossing and rolling
(a)	a tail	a 5	a tail and a 5
(b)	a head	a 2	a head and a 2
(c)	a tail	an odd number	a tail and an odd number

(d) How are the probabilities in Column A and Column B related to those in Column C?

2. In an experiment, a nickel and a dime are tossed and a die is rolled. Construct the sample space for the experiment.

3. Use the sample space in Question 2. What is the probability of tossing and rolling
   (a) two heads and a 6?
   (b) a tail, a head and a 4?
   (c) two tails and an odd number?

4. (a) Construct a sample space for tossing 4 coins.
   (b) What is the total number of possibilities of tossing 4 coins?

5. Use the sample space in Question 4. What is the probability of tosssing
   (a) all heads?   (b) all tails?
   (c) two heads, two tails?   (d) 3 heads, 1 tail?

6. A coin is tossed twice. We may draw a tree diagram to illustrate the sample space.

   Use a sample space to calculate the probability of tossing
   (a) 2 heads in a row   (b) 1 head and 1 tail
   (c) Calculate the above probabilities using products.

B

7. Three coins are tossed in a row. What is the probability of tossing
   (a) three heads?   (b) 2 heads and a tail?
   (c) 2 tails and a head?

8. (a) In addition to 3 coins tossed in Question 7, a die is rolled. What is the probability of
   (a) tossing 3 heads in a row and rolling a 4?
   (b) tossing 2 heads and a tail and rolling an even number?

9. Two coins are tossed and two die are rolled. Calculate the probability of
   (a) tossing 2 heads and rolling a sum of 6.
   (b) tossing a head and a tail and rolling a sum of 9.
   (c) tossing a head and tail and rolling a sum greater than 8.

10. A deck of 52 cards is shuffled. After one card is drawn from the deck it is returned to the pack before the next draw. Calculate the probability of
    (a) drawing the ace of diamonds and then a 4 of clubs.
    (b) drawing a 10 and then a 5.
    (c) drawing a queen and then an ace.
    (d) drawing a jack and not drawing another jack.
    (e) drawing a black card followed by another black card.

dependent events and conditional probability

Twenty people have qualified for a draw in which two people will be selected for two vacation-paid trips.

For the first draw, the probability of Jean, J, being selected is

$$P(J) = \frac{1}{20} \begin{array}{l} \leftarrow 1 \text{ successful choice} \\ \leftarrow 20 \text{ possible choices} \end{array}$$

Of course, you are not allowed to win a second trip so only 19 persons will qualify for the second draw.

For the second draw, the probability of Tina, T, being selected is

$$P(T) = \frac{1}{19} \begin{array}{l} \leftarrow 1 \text{ successful choice} \\ \leftarrow 19 \text{ choices left} \end{array}$$

If two experiments influence each other then we say that the corresponding events are **dependent**. Thus, the second draw is dependent on the first draw since the winner of the first draw does not participate in the second draw. We may show that the probability of Tina, T, winning is dependent on Jean, J, winning by using the notation

$$P(T/J) = \frac{1}{19}$$

The notation is read, "The probability that T occurs given that J occurs" and is referred to as **conditional** probability.

The probability of both Jean, J, and Tina, T, winning is given by

$$P(J \text{ and } T) = P(J) \times P(T|J)$$
$$= \frac{1}{20} \times \frac{1}{19}$$
$$= \frac{1}{380}$$

We may write $P(J \text{ and } T)$ as $P(J \cap T)$.

Throughout Questions 11 to 13, unless indicated otherwise, you may leave your answers in product form.

11 (a) Out of 40 applications, what is the probability of Terry's application being randomly selected?
   (b) What is the probability of Jennifer's application being selected after Terry's has been selected?
   (c) What is the probability that both Terry's and Jennifer's applications will be randomly selected?

12 50 balls are numbered 1, 2, 3, ..., 50.
   (a) What is the probability of randomly selecting ball number 3?
   (b) What is the probability that ball number 5 will then be selected?
   (c) What is the probability that the balls selected will be numbers 1, 2, and 3 in that order?

13 For a class lottery, 26 tickets were sold labelled A, B, ..., Z. Joan bought tickets labelled H and T.

   (a) What is the probability that a ticket drawn will be Joan's?
   (b) What is the probability that both tickets drawn in order will be Joan's?

8.11 applications: probability and statistics

We may apply our skills with probability and statistics to solve problems. In Example 1 statistical methods were used to collect the data, and skills in probability were then applied to answer the questions.

Example 1

A random sample was obtained from 1000 people for their responses to the question "Are television series becoming too violent?"
The results were: 686 *agreed* (A), 235 *disagreed* (D) and the remainder had *no comment*, (NC). What is the probability that another person selected at random will

(a) *agree?*   (b) *disagree?*
(c) have *no comment?*

Solution
(a) Probability of *agree* (A)
$$P(A) = \frac{686}{1000} \begin{matrix}\leftarrow\text{number that } agreed \\ \leftarrow\text{total number}\end{matrix}$$
$$= 0.686$$

(b Probability of *disagree* (D)
$$P(D) = \frac{235}{1000} = 0.235$$

(c) Probability of *no comment* (NC)
$$P(NC) = \frac{79}{1000} \begin{matrix}\text{number of people} \\ \text{with } no\ comment\end{matrix}$$
$$= 0.079$$

We may thus interpret the above results and say that the next person sampled randomly is *more likely* to *agree* than *disagree*.

8.11 exercise

Questions 1 to 4 are based on the following data. To determine the popularity of various cars the following data were randomly collected.

Car	A	B	C	D	E	F	G	H	I	J
Frequency of students responses	21	4	11	11	9	1	25	3	6	26

What is the probability that a person chosen at random from the above students has selected
(a) Car C? (b) Car G?

What is the probability that two persons in a row will choose car D?

(a) What is the mode for the above data?
(b) What is the probability that from a random selection, a person will choose the mode?

Questions 4 to 9 are based on the following sample.

On an experimental agriculture farm, 60 plants are selected at random. Their heights in centimetres after 3 weeks of growth are recorded as follows.

```
 7 15 19 13 21 12 20 10 14 14 17 12 14 22 12
17 18 14 10 13 12 15 16 11 11  9 16 13 16 11
11 19 16 15 18 14 22 19 13 15 16 10 10 12 15
 9 20 17  8 23 12 10 11 13 14 18 15 20 12 14
```

(a) Construct a frequency distribution of the data.
(b) Construct a histogram of the data. Use class intervals 4 – 5, 6 – 7, 8 – 9, 10 – 11, etc.

(a) For the data, calculate the mean.
(b) What percentage of the plants are within 2 cm of the mean?

(a) Calculate the mode.
(b) What percentage of the plants are within 2 cm of the mode?

7 (a) Calculate the median.
   (b) What percentage of the plants are within 2 cm of the median?

8 To calculate the standard deviation for data given in classes, we use the mean value of the class.
   (a) Calculate the standard deviation of the data.
   (b) What percentage of the plants are within 1 standard deviation of the mean?
   (c) If there are 20 000 plants on the farm, how many are within 1 standard deviation of the mean?

9 If you walk in a field and randomly select plants, calculate the probability that you will pick a plant which is
   (a) 12 – 13 cm high.
   (b) between 10 – 15 cm inclusive.
   (c) as long in length as the mode.
   (d) less in length than the mean.
   (e) longer in length than the median.
   (f) within 1 standard deviation of the mean.

problem/matics

Use the following BASIC program to calculate the standard deviation of a set data.

```
 10  INPUT N
 15  DIM X(100)
 20  LET S1 = 0
 30  FOR I = 1 TO N
 40  INPUT X(I)
 50  LET S1 = S1 + X(I)
 60  NEXT I
 70  LET M = S1/N
 80  LET S2 = 0
 90  FOR I = 1 TO N
100  FOR S2 = S2 + (X(I) – M) ↑ 2
110  NEXT I
120  LET S3 = S2/N
130  LET D = SQR(S3)
140  PRINT "THE STANDARD DEVIATION IS", D
150  END
```

Questions 10 to 13 are based on the following data.

To harvest a mushroom crop, measurements are made of the width of the cap. The data obtained are shown in the chart and are used to determine whether the crop is ready to market.

Width in Centimetres	Frequency
1.4 – 1.5	1
1.6 – 1.7	3
1.8 – 1.9	7
2.0 – 2.1	10
2.2 – 2.3	13
2.4 – 2.5	22
2.6 – 2.7	25
2.8 – 2.9	23
3.0 – 3.1	18
3.2 – 3.3	13
3.4 – 3.5	7
3.6 – 3.7	3
3.8 – 3.9	2
4.0 – 4.1	1

10. (a) Determine the modal width of the mushrooms.
    (b) What percentage of the mushrooms is within 0.4 cm of the modal width?

11. (a) Calculate the mean width of the caps.
    (b) Calculate the standard deviation.

12. The value of the crop is given by the following chart.

Standard Deviation	Profit per square metre
0.2 cm	$ 16.85
0.3 cm	$ 14.62
0.4 cm	$ 12.60
0.5 cm	$ 11.50

Calculate the approximate value of the crop if the mushroom farm has 360 m² of active growing surface.

13. If you pick a mushroom at random, what is the probability that the width of the cap is
    (a) between 2.0 cm – 2.1 cm?
    (b) between 2.4 cm – 2.7 cm?
    (c) less than 2.0 cm?
    (d) greater than 2.9 cm?

Questions 14 to 16 are based on the following information.

Another mushroom farmer was in a hurry to harvest the mushroom crop and chose 10 mushrooms at random and recorded the following measurements.

   1.8 cm   2.0 cm   1.8 cm   2.8 cm   3.8 cm
   2.6 cm   1.6 cm   1.4 cm   2.4 cm   2.5 cm

14. (a) Determine the modal width of the mushrooms.
    (b) What percentage of the mushrooms are within 0.4 cm of the modal width?
    (c) How does your answer in (b) compare with your answer in Question 10 for the same crop?

15. (a) Calculate the mean for the data.
    (b) Calculate the standard deviation.

16. (a) Use the data above and answer Question 12 again.
    (b) Compare the answer obtained for Question 12 using the above data and the earlier data. What do you notice?

problem/matics

As Jean walked down the escalator (at a uniform rate), she counted 50 steps but Simons counted 75 steps. Since she walked 3 times as fast (uniformly), down the escalator. If the escalator stopped, how many escalator steps would show?

skills review: statistical calculations

1. Calculate the
   - mean
   - median
   - mode

   for each set of data.

   (a) 8, 6, 10, 11, 13, 14, 18, 16, 6
   (b) 25, 30, 32, 35, 31, 21, 22, 34, 21, 22, 25, 22, 20

2. Two teams of 5 basketball players are compared for height (in centimetres).

   Wildcats 160 171 170 205 160
   Trojans   169 175 169 169 174

   (a) Find the mean and mode of the height for each team.
   (b) Which is more representative of the data for each team?

3. Calculate the
   - range
   - standard deviation

   for each set of statistical data.

   (a) 8, 8, 6, 2, 10, 14, 12, 12, 16, 12
   (b) 121, 109, 107, 101, 117, 111, 113, 115, 119, 105, 103, 123

problems and practice: a chapter review

1. A survey was made to determine the number of homes in a city that had central air conditioning. It was determined that 10 homes out of 100 had central air conditioning.

   (a) How many homes out of 1000 would you expect to have central air conditioning?
   (b) Do you think the above sample is representative of all homes? Why or why not?
   (c) What assumption(s) do you make in stating your answer in (a)?

2. The annual rainfall, in millimetres, for three regions is given below.

	J	F	M	A	M	J	J	A	S	O	N	D
Region A	800	730	630	350	250	200	50	50	220	420	750	780
Region B	130	130	200	300	400	550	460	380	380	320	220	160
Region C	450	400	370	370	390	390	390	400	400	420	410	420

   (a) Find the mean, median, and mode for each region.
   (b) Which measure best represents the data? Why?

3. The data for the life length in days, of a fibre, randomly sampled were

   222 234 248 264 254 238 272 238 236 218.

   (a) Calculate the mean, $\bar{x}$.
   (b) Calculate the standard deviation (in days).
   (c) How many fibres had a life length within 1 standard deviation of the mean?
   (d) How many fibres out of 10 000 would be within 1 standard deviation of the mean?

4. A box contains 36 red chips, 24 blue chips and 40 green chips.

   (a) Write a sample space for the experiment: "Taking a chip from the box."
   (b) Assign probabilities for each event.
   (c) What is the probability of selecting a red chip?
   (d) What is the probability of not selecting a red chip?
   (e) What is the probability of not selecting a green chip?
   (f) What is the probability of selecting a blue chip?

> ## math tip
> It is important to clearly understand the vocabulary of mathematics when solving problems. Make a list of all the new words you have met in this chapter. Provide a simple example to illustrate each word.

test for practice

Try this test. Each *Test for Practice* is based on the mathematics you have learned in the preceding chapter. Try this test later in the year as a review.

1. Use an example
   (a) to illustrate the three main steps when dealing with statistical information.
   (b) to show the meaning of *sample, population*.

2. List 2 methods of collecting data. Indicate the advantages and disadvantages of each method.

3. For each of the following, describe the method or type of sampling you might use to obtain a sample:
   (a) deciding how may people oppose the building of the hydroelectric dam.
   (b) testing the life of light bulbs.
   (c) obtaining an opinion about a new surgical procedure.
   (d) obtaining an opinion on a new brand of diapers.
   (e) deciding on the suitability of a new text-book for high school.

Questions 4 to 7 are based on the following data. The masses of 50 students are listed below.

```
60 64 55 71 50 71 54 65 67 70 51 71 67
57 61 62 45 57 44 59 62 44 67 59 45 52
72 67 60 46 43 66 61 63 64 58 48 59 48
64 50 47 44 69 48 61 53 52 68 69
```

4. Determine the class intervals for 7 classes of widths and construct a histogram.

5. Draw the frequency polygon for the data.

6. Which class contains the most students?

7. What percentage of the students has a mass in class f_4?

8. For the data, determine the mean, median, and mode.
   (a) 16, 18, 19, 20, 20
   (b) 30, 31, 34, 42, 31, 37, 33
   (c) 25, 21, 20, 22, 47, 21
   (d) 3.2, 4.2, 3.6, 4.1, 3.8, 4.5, 3.8, 4.0

9. The attendance each day for an exhibit is given.
   180   204   200   212   224   188   164
   (a) Calculate the mean.
   (b) Calculate the standard deviation.
   (c) Is the difference between the mean and standard deviation within 20?

10. (a) Draw a chart to show the different possibilities of tossing 3 coins.
    (b) What is the probability of tossing 2 heads and a tail?
    (c) What is the probability of *not* tossing 3 tails?

11. A deck of 52 cards are shuffled. Each time a card is taken from the deck, it is returned to the pack. Calculate the probability of
    (a) drawing 2 queens in a row.
    (b) drawing a king and not drawing a 3 of spades on the next draw.

---

problem/matics

Often the answer to a problem can differ significantly depending on what assumptions are made. Solve each problem.

A. What is the probability of choosing a 3 of hearts followed by a 4 of hearts from a deck? Assume that the first card is returned to the deck and reshuffled.

B. What is the probability of choosing a 3 of hearts followed by a 4 of hearts from a deck? Assume that the first card is not returned to the deck.

thinking inductively about circles

It is imperative to know the vocabulary when doing mathematics relating to the circle. We can use the index to help us find the meanings of words.

> radius, circumference, chord, arc, semicircle, segment, sector, inscribed angle, central angle, secant, tangent, interior of the circle, exterior of the circle

Often, mathematical experiments allow us to make conjectures.

Experiments, Investigations, Particular examples → the process of thinking inductively → Conjectures

Then as we develop and add to our store of algebraic and geometric skills we may use these skills to prove our conjectures.

To explore a relationship inductively, we need to design an experiment. After investigating a number of examples involving circles with different radii, we may make a conjecture. For example, in the following, angles subtended on the diameter are inscribed in a circle.

particular examples

Conjecture
The measure of the angle inscribed in a circle and subtended on the diameter is 90°

For each of the following,
- devise an experiment to investigate each.
- make a conjecture

A A line segment joining the centre of the circle and the midpoint of a chord *appears* to be perpendicular to the chord.

B a line drawn through the centre of a circle and perpendicular to the chord *appears* to bisect the chord.

C Inscribed angles drawn on the same side of the arc *appear* to be equal.

C The angle inscribed in a circle and subtended on a diameter *appears* to be a right angle.

E The inscribed angle drawn on an arc *appears* to be $\frac{1}{2}$ of the measure of its corresponding central angle (on the same arc).

We say that quadrilateral ABCD is inscribed in a circle if all the vertices of the quadrilateral are on the circumference of a circle. F and G are based on inscibed quadrilaterals.

F It appears that a quadrilateral in which the opposite angles are supplmentary can be inscribed in a circle.

G The opposite angles of a quadrilateral inscribed in a quadrilateral appear to be supplementary.

Tangents occur often in studying orbits. For example, if a ball at the end of a string is twirled and then let go, the ball leaves on a tangential path.

When a space probe leaves an orbit, it so often does on a tangential path.

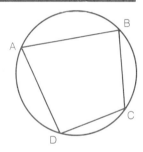

A tangent to a circle is a line drawn to a circle so that it touches the circle at only one point. Investigations H, and I are about tangents.

H A tangent drawn to a circle *appears* to be perpendicular to the radius.

I Tangents drawn to a circle from an exterior point *appear* to have the same length.

Once we have developed the appropriate mathematical tools we may then begin to prove each of the above conjectures.

Often the results we obtain can be used to develop useful constructions.

1. Use the results of D and H to draw a tangent from P to any circle.

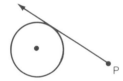

2. To place a hole exactly at the centre of the base of a can, we need to know where the centre is. Use one of A to I to devise a method for finding the centre of any circle.

Now that we have inductively investigated some properties of circles, we may now use our algebraic skills to find the missing values in each of the following diagrams. Once we have completed these numerical exercises, we can refer to Section 9.9, pages 320 and 321 where the conjectures we have introduced here are proved using the strategies and skills we have learned.

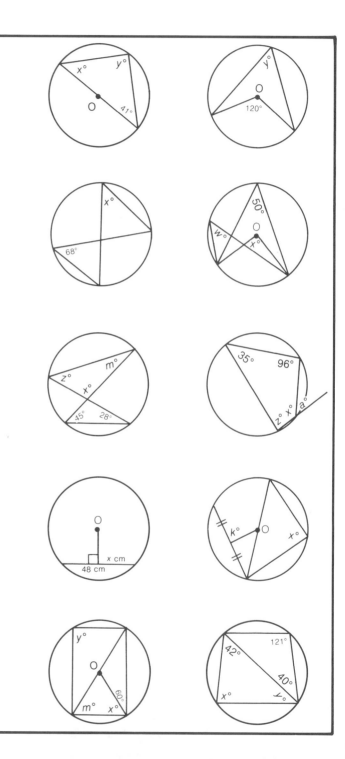

9 aspects of geometry: proof, properties, and solving deduction

Deductive thinking and analyzing deductions; theorems on congruence and parallelism, aspects of transformational geometry

9.1 language of a deductive system

One of the main reasons for studying mathematics is to develop thinking skills. We use many types of logical reasoning to arrive at a conclusion. For example, in the *Math Is Plus*, on pages 295 and 296, we dealt with inductive reasoning. In our work in geometry, we will explore the nature of deductive reasoning and utilize it to solve problems in the following forms of geometry.
- plane • co-ordinate • transformational

The achievements of the Ancients which are of a geometric nature are very visible and not difficult to recognize.

However, it is often difficult to recognize their intangible achievements. Euclid and his followers began to build much of the foundation of our work in geometry.

He organized his ideas logically into a deductive system. Euclid and his followers began with

298

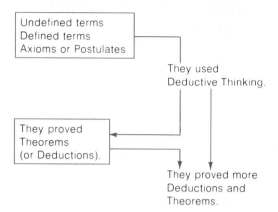

In our study of algebra, there are some words that we have accepted as undefined terms, such as *set*. In geometry, we also accept undefined terms.

undefined terms: Not all words can be defined, so we must accept some of them as undefined. Some of the undefined terms or words used in geometry are *point*, *line*, *plane*.

defined terms: We use these undefined terms to write definitions. For example, we define the word circle as the set of all points that are the same distance from a fixed point. We may then give the special name, **centre**, to the fixed point.

Once we have defined the terms, we may then use the terms to define new ideas. For example, concentric circles are circles with the same centre.

axioms or postulates: As well as undefined terms in our deductive system, we also make certain assumptions that we accept as true. We call these basic assumptions **axioms** or **postulates**.
- A line contains at least two points.
- For every two different points there is exactly one straight line that contains them.

In our study of plane geometry, we will learn other axioms and postulates. Statements to be proved are referred to as **deductions**.

Some deductions may be used to prove further statements. We refer to these proven statements as **theorems**.

Fundamental to our earlier study of geometry is a clear understanding of the vocabulary in geometry. We will meet again many of the following in our work in geometry. How many are you familiar with? Which ones are
- defined?
- undefined?

These terms are not in any special order.

point	line segment	inscribed angle
line	arms of angle	collinear points
plane	coplanar points	
ray	right bisector	congruent angles
angle	angle bisector	
vertex	equidistant	congruent figures
polygon	parallel lines	
triangle	transversal	alternate angles
pentagon	hypothesis	
hexagon	conclusion	interior angles
octagon	converse	
altitude	indirect proof	mean proportional
median	corollary	
theorem	hypotenuse	complementary angles
circle	straight angle	
chord	reflex angle	vertically opposite angles
radius	adjacent angles	
arc	quadrilateral	vertically opposite angles
segment	regular polygon	
sector	postulate	remote interior angles
diameter	circumference	
diagonal	subtended	congruent line segments
midpoint	exterior angle	
		corresponding angles

triangles: acute, equiangular, equilateral, right, obtuse, scalene, isosceles, similar

quadrilaterals: trapezoid, parallelogram, rectangle, rhombus, square

9.2 congruence in geometry

As we have seen, the components of a deductive system include the following.

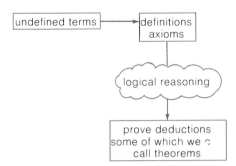

The concept of **congruence** is basic in our study of geometry.

Congruent line segments: If two line segments are congruent, then their lengths are equal. Thus $\overline{AB} \cong \overline{CD}$ if and only if $\overline{AB} = \overline{CD}$.

$\overline{AB}$ is congruent to $\overline{CD}$.   The lengths of AB and CD are equal.

Throughout our work, we will write $\overline{AB} = \overline{CD}$, since $\overline{AB} = \overline{CD} \Leftrightarrow \overline{AB} \cong \overline{CD}$.
  if and only if

Symbolism in mathematics is important, but often the symbolism may create reading difficulties. For this reason, we will use the simplified symbolism
AB = CD for $\overline{AB} \cong \overline{CD}$ (since we know that if AB = CD, then they are congruent).

Congruent angles: If two angles are congruent, then their measures are equal. Thus $\angle ABC \cong \angle DEF$ if and only if $\angle ABC = \angle DEF$.

$\angle ABC$ is congruent to $\angle DEF$.   The measures of $\angle ABC$ and $\angle DEF$ are equal.

In a similar way, the simplified symbolism $\angle ABC = \angle DEF$ will be used throughout.

Congruent triangles: Two triangles are congruent if and only if all corresponding angles and sides are congruent. Since △ABC and △DEF are congruent, we write

 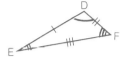

   The congruence relation shows which vertices are related.

The corresponding measures for △ABC and △DEF are equal.

Corresponding Sides	Corresponding Angles
AB = DE	$\angle A = \angle D$
BC = EF	$\angle B = \angle E$
CA = FD	$\angle C = \angle F$

congruence assumptions

The condition of congruence for triangles requires that all 6 parts be congruent. However, each of the following congruence assumptions may be made since, if certain parts are congruent, then they indicate that other parts are congruent.

side side side congruence assumption (SSS)
If three sides of one triangle are respectively congruent to three sides of another triangle, then the two triangles are congruent.

    △ABC ≅ △DEF

side angle side congruence assumption (SAS)
If two sides and the contained angle of one triangle are respectively congruent to two sides and the contained angle of another triangle, then the two triangles are congruent.
If 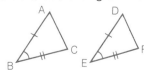 then △ABC ≅ △DEF.

angle side angle congruence assumption (ASA)
If two angles and the contained side of one triangle are respectively congruent to two angles and the contained side of another triangle then the two triangles are congruent.
If then △ABC ≅ △DEF.

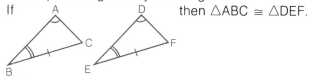

Another form of the above congruence assumption is given by the diagram.
If then △ABC ≅ △DEF.

right triangle congruence postulate (HS)
If the hypotenuse and a side of a right-angled triangle are congruent to the corresponding hypotenuse and side of another right-angled triangle, the triangles are congruent.

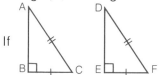

If then △ABC ≅ △DEF.

We also obtain congruent triangles if certain corresponding parts have equal measures. For △ABC and △DEF, if
 AB ≅ DE then AB = DE.
 BC ≅ EF then BC = EF.
 AC ≅ DF then AC = DF.
 Then △ABC ≅ △DEF.

writing proofs

A combination of the above congruence assumptions with other facts we know can be used to prove deductions in geometry. To solve a problem or prove a deduction requires that we clearly understand the answer to two important questions.
 I What do we know? (What are we given?)
 II What are we asked to find? (What are we asked to prove?)

In Example 1, we are given certain facts, which, along with logical reasoning, will allow us to reach a conclusion or prove the statement. In order that others may understand the reasoning or the process, we must organize our work in a clear and concise way.

 Given (or hypothesis): ← Information given in I.
 Required to prove: ← Information to be
 (or conclusion)   shown in II.
 Proof: The process, steps, procedure, by which we use the information in I to arrive at the conclusion in II.

Example 1
For quadrilateral XYZW, if WX = WZ and XY = ZY, prove that ∠WXY = ∠WZY.

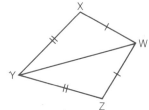

Solution
Given: Quadrilateral XYZW,   This information is
 WZ = WX, XY = ZY   given in the original
Required to prove: question. Read
 ∠WXY = ∠WZY   carefully.

Give reasons why you may deduce each line of the proof.

Proof: In △WXY and △WZY | Reason
 WX = WZ | given   S
 XY = ZY | given   S
 WY = WY | common side   S
 Thus △WXY ≅ △WZY | side-side-side
 | congruence
 | assumption
 | or SSS for short
 ∠WXY = ∠WZY | congruent triangles

The process of solving problems is similar in different branches in mathematics. Compare the similarities in the following.

In Algebra
• translate the problem
• construct the equation
• solve the problem
• make a final statement

In Geometry
• translate the deduction
• construct the diagram
• write the proof
• make a final statement

In the information in Example 2, it is important to translate the problem accurately and record it *on the diagram*. In this way, we are able to carefully plan the solution and apply the needed congruence assumptions to complete the proof.

Example 2
In quadrilateral PQRS, PQ = PS, and QR = SR. Prove that QS ⊥ RP.

Solution
Construct a diagram with the given information.

Given: Quadrilateral PQRS, PQ = PS, QR = SR

Required to prove: QS ⊥ RP

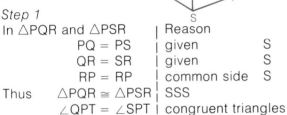

Proof:

Step 1

In △PQR and △PSR	Reason	
PQ = PS	given	S
QR = SR	given	S
RP = RP	common side	S
Thus △PQR ≅ △PSR	SSS	
∠QPT = ∠SPT	congruent triangles	

Step 2

In △PQT and △PST	Reason	
PQ = PS	given	S
∠QPT = ∠SPT	proven	A
PT = PT	common side	S
Thus △PQT ≅ △PST	SAS	
∠QTP = ∠STP	congruent triangles	
∠QTP + ∠STP = 180°	QTS straight line	
Thus ∠QTP = 90°		
and QS ⊥ RP.		

As seen in Example 2, we may need to recognize different overlapping triangles in a diagram, and use them to prove the deductions.

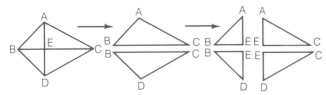

9.2 exercise

A

1. An important skill in proving deductions is recognizing pairs of congruent triangles. For the following triangles
   - write the congruence relation for pairs of congruent triangles.
   - give reasons for your answer.

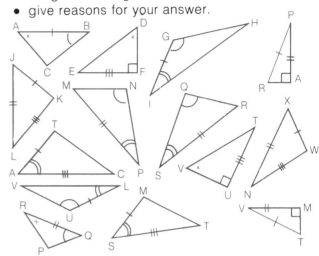

2. To prove triangles congruent, we may use SSS, SAS, ASA or HS. For each of the following, what additional information is needed in order to prove that the triangles in each pair are congruent? Give reasons for your answers.

There may be more than one answer.

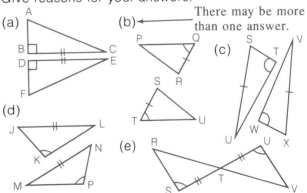

In Questions 3 to 7, the diagrams are given for each question. Use the given information to prove each deduction.

3 In the diagram AB bisects ∠CAD and AD = AC. Prove CB = DB.

4 Prove that in the diagram. ∠ARS = ∠BRS.

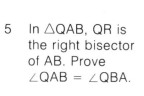

5 In △QAB, QR is the right bisector of AB. Prove ∠QAB = ∠QBA.

6 In quadrilateral TPAH ∠THP = ∠APH and ∠TPH = ∠AHP. Prove that AH = TP.

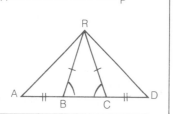

7 From the diagram, RC = RB and CD = BA. If ∠RBC = ∠RCB, prove that △RAD is an isosceles triangle.

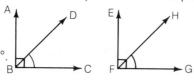

We are often asked to prove a general statement in geometry. Thus, we need not only construct a diagram, but also label the diagram to organize the steps of the solution.

For each of the following, use the given diagram to prove the theorem.

B

8 *Complementary Angle Theorem (CAT)*: If two angles are equal then their complements are equal.

Two angles are complementary if the sum of their measures is 90°.

If ∠DBC = ∠HFG then ∠ABD = ∠EFH.

9 *Supplementary Angle Theorem (SAT)*: If two angles are equal then their supplements are equal.

Two angles are supplementary if the sum of their measures is 180°

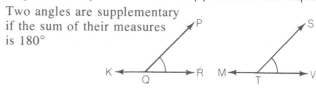

If ∠PQR = ∠STV then ∠PQK = ∠STM.

10 *Vertically Opposite Angle Theorem (VOAT)*: If two lines intersect, the vertically opposite angles are equal.

∠AOD and ∠COB, ∠AOC and ∠DOB are pairs of vertically opposite angles.

When proving deductions, we will use the symbols as shown to refer to these *theorems*.
- complementary angle theorem   CAT
- supplementary angle theorem   SAT
- vertically opposite angle theorem VOAT

11 If TP = SP and SQ = TR prove that ∠Q = ∠R.

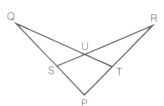

12 For the diagram, RB ⊥ AB TA ⊥ AB and BR = AT. Prove that ∠ATC = ∠BRC.

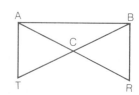

13 If A is the midpoint of CD and BT, and CT = BD, prove ∠C = ∠D.

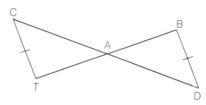

14. △ABC and △DBC are on the same side of the common base BC. If ∠ABC = ∠DCB and ∠DBC = ∠ACB, prove that AB = DC.

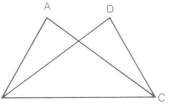

15. Line segments XZ and WY bisect each other at O. Prove XW = ZY.

16. Given △CAB. CX bisects ∠C and is perpendicular to AB. CA is produced to Y and AB is produced to Z. Prove ∠YAX = ∠CBZ.

17. Given △XYZ with AY being a bisector of ∠Y and A being a point of XZ. If XY = ZY, prove XA = ZA.

18. XYZA is a quadrilateral where YA and ZX intersect at C, YA = ZX, and ∠CYZ = ∠CZY. Prove that YX = ZA.

19. The four points BCDE are placed on a circle in the given order with BC = DE. F is the centre of the circle. Prove that
    (a) ∠BFC = ∠EFD
    (b) BE = DC

math tip

We usually use a variety of skills when solving a problem. We often combine our knowledge of geometry and our algebraic skills to solve the problem.
- Solve the following problem.
- List the skills in geometry you required.
- List the skills in algebra you required.

    The length of the perimeter of a right-angled triangle is 60 cm. The length of the altitude perpendicular to the hypotenuse is 12 cm. Find the sides.

In this chapter we will learn other methods of analyzing problems, as well as methods of proving them.

9.3 analyzing deductions

To solve a problem we may analyze it by asking the following questions in the reverse order.

I What are we asked to prove?
II What are we given?

For example, in the given diagram, we are asked to prove AB = DC.

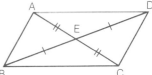

To do so, we may organize our thinking in reverse by beginning with what we are asked to prove and working backwards.

	I can	If I can
1	I can prove AB = CD	if I can prove △AEB ≅ △CED.
2	I can prove △AEB ≅ △CED	if I can use either ASA HS SSS SAS.

Analyze which congruence assumption to use.

ASA — There are no equal angles given. *Probably* I can not use ASA.

SSS — None of the sides we are asked to prove are given. I *cannot* use SSS.

SAS — Two sides are given, I will likely be able to use SAS if I can prove ∠AEB = ∠CED.

HS — There is no right angle. I *cannot* use HS.

3	I can prove ∠AEB ≅ ∠CED	But ∠AEB = ∠CED since the angles are vertically opposite angles.

Thus, we have analyzed the thought process in a reverse order, starting with

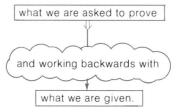

We may then proceed to write up a complete proof.

Given: Quadrilateral ABCD, AE = CE, BE = DE
Required to prove: AB = DC.

Proof: AC and BD intersect at E.

	Reasons
∠AEB = ∠CED	VOAT
In △AEB and △CED	
AE = CE	given S
∠AEB = ∠CED	proven A
BE = DE	given S
Thus △AEB ≅ △CED	SAS
AB = CD	congruent triangles
or AB = DC	

Often, in proving some deductions, an efficient analysis of the problem may be done mentally. Thus, once the problem has been analyzed, the solution can be recorded directly. However, if the deduction involves a number of steps, it is often useful to record a similar analysis using *I can — If I can* so that the required steps of the proof can be seen.

9.3 exercise

For each of the Questions 1 to 5, analyze the steps of the solution by using the headings,

| I can prove | | if I can prove. |

Once the deduction has been analyzed, write the complete proof.

A

1. In △PSR and △PQR
   ∠RPS = ∠RPQ
   ∠PRS = ∠PRQ.
   Prove that △PSQ is isosceles.

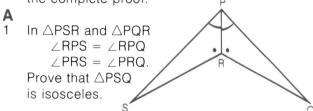

2. In the diagram,
   QS = RS
   ∠PSQ = ∠PSR.
   Prove that RP = QP.

3. In the diagram,
   AD = CD
   AB = CB.
   Prove that AE = CE.

4. For quadrilateral ABCD
   AD = CB
   DC = BA.
   Prove that ∠B = ∠D.

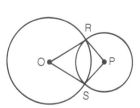

5. Two circles with centres O and P intersect at R and S. Prove that RS ⊥ OP.

For Questions 6 to 14
- analyze each problem and record the information under the headings

| I can prove | | if I can prove. |

- draw an accurate diagram to help plan your work.
- then write a complete proof.

B

6. In △PQR, RP = QP and S is the midpoint of RQ. Prove that ∠PSR = 90°.

7. In quadrilateral STUV, ST = SV and VU = TU. Prove that ∠STU = ∠SVU.

8. Line segments SU and TV intersect at M so that MU = MS and ∠TSM = ∠VUM. Prove that M is the midpoint of VT.

9. Prove that, in an isosceles triangle, the bisector of the vertical angle is the right bisector of the base.

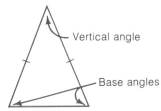

10. Given quadrilateral ABCD with AD = CB. BD is a diagonal. If ∠ADB = ∠CBD, prove AB = CD.

11. In quadrilateral RSTU, ∠SRT = ∠UTR and ∠URT = ∠STR. Prove that ∠S = ∠U.

C

12. A rhombus is a quadrilateral with all sides equal. Prove that the diagonals are perpendicular to each other.

13. A quadrilateral has both pairs of opposite sides equal. Prove that the opposite angles are equal.

14. Prove that the diagonals of a square are of equal length.

problem/matics

The strategies and skills of solving a problem recur each time we study a different branch of mathematics.

A: In solving a problem we often make the problem more difficult because we don't read carefully or we interpret the problem in only one way. Solve this problem.

> A circular field is to be divided into 4 equal parts using 3 fences. If the fences are of equal length, how will you do it?

B: To solve some problems, we often just need to use common sense and not advanced algebraic or geometric skills. Solve this problem.

> A hovercraft, travelling at 75 km/h, and a freighter, travelling at 15.5 km/h, head towards each other on open sea for a rendezvous. How far apart are they exactly 1 h before they meet?

9.4 isosceles triangle theorem

In drawing a number of isosceles triangles, we may conclude *inductively* that the angles at the base of each isosceles triangle seem to have equal measures.

The method of mathematics is to state a conjecture and then attempt to prove it *deductively*.

> conjecture: If a triangle is isosceles, then the angles opposite the equal sides are equal.

To prove the above conjecture a particular isosceles triangle is drawn and labelled. However on analyzing the diagram we find that to prove two angles equal we sometimes require two triangles. To obtain two triangles, we complete a useful construction as shown in the example.

Given: △ABC, AB = AC

Required to prove: ∠B = ∠C

Proof: Construct the bisector of ∠A to meet BC at D.

In △ABD and △ACD	Reasons	
AB = AC	given	S
∠BAD = ∠CAD	constructed	A
AD = AD	common side	S
Thus △ABD ≅ △ACD	SAS	
Then ∠B = ∠C.	congruent triangles	

Thus, in an isosceles triangle, the angles opposite the equal sides are equal. ← We make a final statement.

The above proven statement is an invaluable aid in proving many other deductions. Since it is so important we call it a **theorem**.

Isosceles Triangle Theorem (ITT)

Statements written in the form **if-then** are called **conditional statements**.

If a triangle is isosceles	then the angles opposite the equal sides are equal.
The hypothesis is the information given.	The conclusion is the information we are asked to prove.
I What do we know?	II What do we want to prove?

Conditional statements of the form *if p then q* (written symbolically as $p \Rightarrow q$) occur in algebra as well as in our everyday conversation.

A: If $\underbrace{x + 2 > 5}_{p}$, then $\underbrace{x > 3}_{q}$.

B: If two triangles are congruent, then their corresponding angles are equal.

C: If I walk in the rain, then I will get wet.

The **converse** of a statement is the statement obtained by reversing the hypothesis and conclusion as follows.

If the angles opposite two sides in a triangle are equal	then the sides opposite these angles are equal.

The converse of a statement may not be true, so we need to prove the converse.

Given: △PQS with
∠Q = ∠S.
Required to prove:
PQ = PS
Proof: Construct PT ⊥ QS as shown.

```
In △PQT and △PST    | Reasons
    ∠Q = ∠S          | given              A
  ∠PTQ = ∠PTS        | constructed        A
    PT = PT          | common side        S
Then △PQT ≅ △PST     | AAS
    PQ = PS          | congruent triangles
```

We may use the *Isosceles Triangle Theorem* (ITT) and its converse to write **biconditional statements** that use "if and only if". For example:
In an isosceles triangle, the angles opposite two sides are equal if and only if the sides are equal.

We use **iff** as a symbol. For example, $\left.\begin{array}{l}\text{if } p \text{ then } q \\ \text{if } q \text{ then } p\end{array}\right\} p$ **iff** q

Example 1
In quadrilateral EFGH, if EF = EH and GF = GH prove that ∠F = ∠H.

Solution
Given: Quadrilateral EFGH, EF = EH, GF = GH

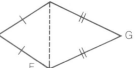

Required to prove: ∠F = ∠H

Since our work thus far depends on using triangles, we perform a construction to help us.

Proof:
```
Join FH                                        | Reasons
In △EFH,    EF = EH                            | given
Then       ∠EFH = ∠EHF       ①                | ITT
In △FGH,    GF = GH                            | given
Then       ∠GFH = ∠GHF       ②                | ITT
From ① and ②
    ∠EFH + ∠GFH = ∠EHF + ∠GHF                  | addition
    Then ∠F = ∠H as required.                  | of equal
                                               | measures
```

A 9.4 exercise

1 Write the converse of each statement. Which converse is true (T)? Which is false (F)?

(a) If $2y - 1 = 9$, then $y = 5$.
(b) If an equilateral triangle has each side 5 cm, then the perimeter is 15 cm.
(c) If $a, b, > 0$ then $ab > 0$.
(d) If $x > 9$, then $x > 5$.
(e) If $y = mx + b$ then $m = \frac{y - b}{x}$.
(f) If x and y are odd then $x + y$ is even.
(g) If two triangles are congruent, then the corresponding angles are congruent.
(h) If $a > 0, b < 0$ then $ab < 0$.
(i) If m and n are even, then mn is even.

Questions 2 to 6 provide practice in using the *Isosceles Triangle Theorem ITT* to prove a deduction. Redraw each diagram and record the appropriate information.

2 In the diagram, PS = RS and SQ is the bisector of ∠PSR.
   (a) Prove that △PQR is isosceles.
   (b) Prove that ∠QPR = ∠QRP.

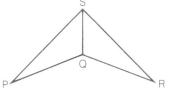

3 In the diagram, D is the midpoint of BC, and ∠ADB = ∠ADC. Prove that △ABC is isosceles.

4 From the given figure, SA = SB and AU = BT Prove that △STU is isosceles.

5 Two circles with centres O and P intersect at A and B. Prove that ∠OAP = ∠OBP.

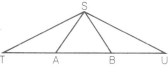

6 For the diagram, ∠TUV = ∠WXU and TU = TV = WX. Prove that ∠TVU = ∠WXV.

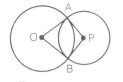

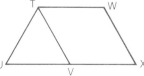

7 (a) Write the converse of the following: If the diagonals of a quadrilateral bisect each other then the quadrilateral is a rhombus.
   (b) Prove whether the converse in (a) is true or not.

8 (a) Write the converse of the following statement. If a point is on the right bisector of a line segment, then it is equidistant from the end points of the line segment. (We refer to this result as the *Right Bisector Theorem RBT*.)
   (b) Show whether the converse is true or not.

B

9 In quadrilateral RSTU, RS = RU and ∠UST = ∠TUS. Prove that ∠S = ∠U.

10 In △QNP, A is a point on NP so that AN = AP. If ∠NAQ = ∠PAQ, prove that QN = QP.

11 In △TSR, P is the midpoint of SR. If TP ⊥ SR prove that △TSR is isosceles.

12 MNP is an isosceles triangle with MN = MP. A is the midpoint of NP. Prove MA ⊥ NP.

13 △AXY is isosceles with ∠X = ∠Y. XY is produced to C and YX is produced to B so that XB = YC. Prove ∠B = ∠C.

14 △SMP is isosceles with SM = SP. R is an interior point of △SMP such that SR bisects ∠MSP. Prove that ∠RMP = ∠RPM.

15 In rhombus RSTU, prove that the opposite angles are equal, that is, ∠S = ∠U and ∠R = ∠T.

C

16 In a circle with centre O, XY and ST are equal chords. Prove that the perpendicular distance from the centre to each chord is equal.

17 Given △XYZ with point M in XY and point N in XZ so that MY = NZ and MZ = NY. Prove that △XZY is isosceles.

18 A quadrilateral MNPQ has the properties MN = MQ, and ∠N = ∠Q. Join MP and NQ to intersect at O. Prove ∠ONP = ∠OQP, and MP bisects NQ at right angles.

19 An isosceles triangle ABC is constructed with ∠B = ∠C. AB is produced to W and AC is produced to K so that BW = CK. CW and BK are joined. Prove ∠W = ∠K.

20 △XYZ is equilateral. Point A is outside △XYZ so that △AXY is equilateral. Similarly B and C are outside △XYZ so that △YBZ and △XZC are equilateral. Prove ZA = XB = YC.

exterior angle theorem

A triangle has 3 interior angles. We may draw the exterior angles 1, 2, 3, 4, 5 and 6 of a triangle as shown. From the diagram, the vertically opposite angles are equal.

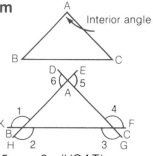

∠1 = ∠2, ∠3 = ∠4, ∠5 = ∠6   (VOAT)

Although from a diagram the following result seems true, it is necessary for us to prove deductively that in any triangle, the exterior angle is greater than either of the remote interior angles. (The theorem is given the symbol EAT.)

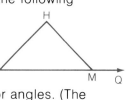

To prove this theorem, we perform a construction so that we end up with two triangles that we can work with.

Given: △HKM and exterior angle ∠HMQ

Required to prove:
(a) ∠HMQ > ∠H   (b) ∠HMQ > ∠K

Proof: Construct P, the midpoint of HM. Produce KP to R so that KP = PR. Join MR.

	Reasons
Since HM and KR intersect at P, then	
∠HPK = ∠MPR	VOAT
In △HPK and △MPR	
HP = MP	constructed S
∠HPK = ∠MPR	proven A
KP = RP	constructed S
Thus △HPK ≅ △MPR	SAS
∠H = ∠HMR	congruent triangles
But ∠HMQ = ∠HMR + ∠RMQ	
Thus ∠HMQ > ∠HMR	whole is greater than its parts
∠HMQ > ∠H	∠H = ∠HMR

Similarly, we may show that ∠HMQ > ∠K.

21 For △ABC, write the remote interior angles for each exterior angle.

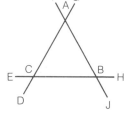

(a) ∠CAF   (b) ∠CBJ
(c) ∠ACE   (d) ∠HBA
(e) ∠GAB

22 Use the *Exterior Angle: Theorem* (EAT) to write an inequality involving each angle.

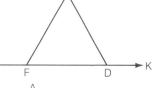

(a) ∠KDE > ?
(b) ∠DEF < ?
(c) ∠EFH > ?
(d) ∠EFD < ?
(e) ∠EDF < ?

23 If AB ⊥ CD prove ∠ABC > ∠ADB.

24 If AB = AC, prove that ∠ACD > ∠EBF.

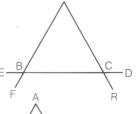

25 For the diagram, AB = AC. Prove that ∠ABE > ∠DFC.

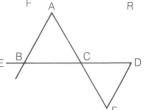

26 In △PQR, RQ is produced to T and QR is produced to S. Prove that ∠PRS + ∠PQT > 180°.

27 In an isosceles triangle prove that the base angles are acute.

9.5 parallel lines

The following is the structure of a deductive system.

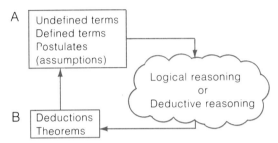

Parallel lines: Two lines are parallel if they lie in the same plane and do not intersect.

We must develop the following vocabulary.

A **transversal** is a line that intersects 2 or more lines in the plane. ST is a transversal for UK and VM.

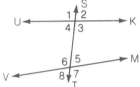

When a transversal intersects two or more lines certain angles are named as follows.

Corresponding Angles	Alternate (Interior) Angles	Interior Angles on the same side of the transversal
∠1, ∠6	∠4, ∠5	∠4, ∠6
∠4, ∠8 etc.	∠3, ∠6	∠3, ∠5

The above angles are characterized by their positions.

Corresponding Angles

Alternate (Interior) Angles   Interior Angles on the same side of the transversal.

Certain angles, in the above diagrams, appear to be equal.
If two lines are drawn parallel, and a transversal cuts them, certain pairs of angles appear to be equal.

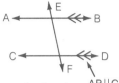

AB||CD

Corresponding Angles   Alternate Angles

Since it appears more acceptable that corresponding angles are equal, we accept the following postulate as true.

Parallel Postulate: If two parallel lines are cut by a transversal then the corresponding angles are equal.

We may now prove the following theorem using the *Parallel Postulate*.

Example 1

Prove that if two parallel lines are cut by a transversal, then the alternate angles are equal.

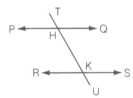

Solution

Given: PQ||RS. TU is a transversal that cuts PQ and RS.

Required to prove:
    (a) ∠QHK = ∠HKR   (b) ∠PHK = ∠HKS

Proof: Since PQ||RS, TU is |Reasons
        a transversal, then
          ∠THP = ∠HKR   ① |parallel postulate
        PQ, TU are intersecting
        lines at H. Thus
          ∠THP = ∠QHK   ②|(VOAT)
        From ① and ②
          ∠HKR = ∠QHK |Equality: If
                                                          |$a = b, c = b$,
                                                         |then $a = c$

Which is what we wanted to prove.

310

In a similar manner, we may prove
∠PHK = ∠HKS. We call the following theorem
the *Parallel Lines Theorem* (PLT).

If a transversal intersects two parallel lines, then
- the corresponding angles are equal.   (postulate)
- the alternate angles are equal.   (proved above)
- the interior angles on the same side of the transversal are supplementary.   (proved in the exercises)

We may apply the *Parallel Lines Theorem* (PLT) to solve additional deductions.

Example 2

In the diagram, PQ∥RS, and PQ = RS. Prove that M is the midpoint of PS and QR.

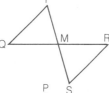

Solution

Given: PQ∥RS
PQ = RS

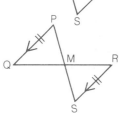

Required to prove:
PM = SM
QM = RM

Write the proof after you have done an analysis of the problem.

Proof:	Reasons
QR is a transversal for PQ∥RS.	
Then ∠PQR = ∠QRS.	PLT
PS is a transversal for PQ∥RS.	
Then ∠QPM = ∠MSR.	PLT
In △PQM and △SRM	
∠PQM = ∠SRM	proved A
PQ = SR	given S
∠QPM = ∠RSM	proved A
△PQM ≅ △SRM	ASA
PM = SM	congruent
QM = RM	triangles

9.5 exercise

A

1 For the diagram, name pairs of
   (a) corresponding angles.
   (b) alternate angles.
   (c) interior angles on the same side of the transversal.

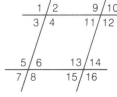

2 For the diagram in Question 1, name pairs of angles that are
   (a) supplementary
   (b) vertically opposite

3 In the diagram, PQ∥RS.

   (a) Prove that ∠PHK + ∠RKH = 180°.
   (b) Prove that ∠QHK + ∠SKH = 180°.
   (c) Why may the following be written as a theorem?

   If a transversal intersects two parallel lines, then the interior angles on the same side of the transversal are supplementary.

4 Which pairs of angles are equal for each diagram?

   (a) 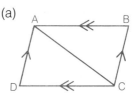   (b)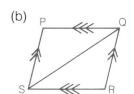

5 For each diagram, find the values of *m* and *n*. Give reasons for your results.

   (a) 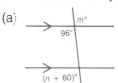   (b)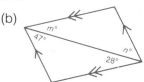

6 Calculate the values of m and n, if PQ = PR and QR||ST.

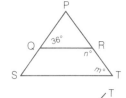

B

7 For the diagram, PQ||RS. Prove that ∠TUP = ∠WVS.

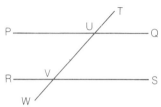

8 Prove that ∠AME = ∠BPF.

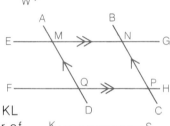

9 For the diagram, SV||KL and KV is the bisector of ∠SKL. Prove that KS = SV.

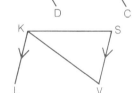

10 If TL||RS and M is the midpoint of LR, prove that M is also the midpoint of TS.

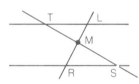

11 If AB||DC and DA||CB, prove that ∠A = ∠C.

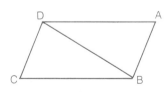

12 For the diagram, △PQR is isosceles with PQ = PR. If QR||ST prove that QS = RT.

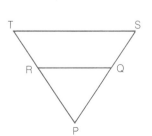

13 For the diagram, PQ||BC.

   (a) Why is ∠1 + ∠2 + ∠3 = 180°?
   (b) For △ABC, prove that ∠BAC + ∠B + ∠C = 180°.

This fact is an important theorem called the *Angle Sum Triangle Theorem* (ASTT). In a triangle, the sum of the angles of a triangle is 180°.

14 A rhombus is a quadrilateral with all sides equal. Prove that, in any rhombus, the opposite angles are equal.

15 If RS and PQ are both perpendicular to AB prove that RS||PQ.

16 In quadrilateral PQRS, PQ||SR. If the diagonal QS bisects ∠PSR, prove that PS = PQ.

17 In isosceles triangle PRT, RP = TP. If a line, drawn parallel to the base, meets PT and PR in Q and S respectively, prove that △PQS is also isosceles.

18 A parallelogram is a quadrilateral with opposite sides parallel. Prove the following properties of a parallelogram.

   (a) The opposite angles are equal.
   (b) The opposite sides are equal.
   (c) The diagonals bisect each other.

19 In parallelogram RSTV, the diagonal RT bisects ∠VRS. Prove that RSTV is a rhombus.

20 ABC is an isosceles triangle with AB = AC and BA is produced to D forming exterior ∠CAD. AP||BC. Prove ∠DAP = ∠PAC.

21 In square PQRS points M and N are chosen in PQ and QR respectively so that PM = NR. Prove
   (a) QS ⊥ MN.   (b) QS bisects MN.

sum of the angles of a triangle (ASTT)

In our earlier work, we showed inductively that the sum of the measures of the angles of a triangle is 180°.

We may use the *Parallel Lines Theorem* to prove deductively that the sum of the measures of the angles of a triangle is 180°.

Use this symbol.

Angle Sum of a Triangle Theorem (ASTT).
The sum of the measures of the angles of a triangle is 180°.

Given: △ABC

Required to Prove:
∠A + ∠B + ∠C = 180°

Proof: Draw PQ∥BC through A.   Reasons

Since PQ∥BC	
then ∠PAB = ∠B	PLT alternate
∠QAC = ∠C	angles
Thus	
∠A + ∠B + ∠C	
= ∠A + ∠PAB + ∠QAC	
= ∠PAQ = 180°	straight angle

Often a single fact may be directly deduced from a theorem. For example, based on ASTT we may show that if two angles of a triangle are congruent to two angles of another triangle then the third pair of corresponding angles is also congruent. We refer to the above result as a **corollary** of the theorem.

Corollaries of ASST

A   If two angles of a triangle are congruent to two angles of another triangle, then the remaining angles are congruent.

B   Each angle of an equilateral triangle is 60°.

22 Prove the corollaries shown above.

23 In △ABC, ∠ACF is an exterior angle of the triangle and ∠A and ∠B are the corresponding interior and opposite angles. Prove that ∠A + ∠B = ∠ACF.

24 For the diagram, show that ∠C is a right angle.

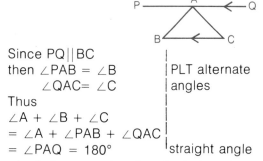

25 For each triangle, find the missing measures.

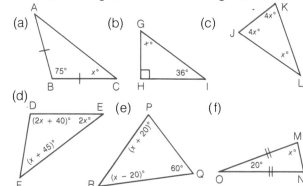

26 The angles of a triangle are 40° and 100°.
    (a) What is the measure of the third angle?
    (b) What type of triangle is it?

27 In △ABC ∠A is 20° more than ∠C and ∠B is 20° more than ∠A. Find the measures of the angles.

28 Find the values of x and y for each of the following diagrams.

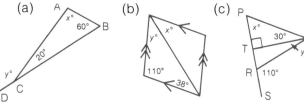

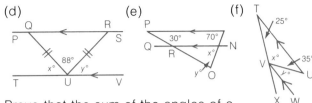

29. Prove that the sum of the angles of a quadrilateral is 360°.

30. If P is any point in △DEF such that PD = PE = PF, prove that ∠EPF = 2∠EDF.

31. Prove that in an equilateral triangle, the median (line drawn from vertex to midpoint of opposite side) to any side is also the altitude.

32. In △ABD, ∠B = 2∠D. If C is the right bisector of AD, prove that AC = AB.

33. In △ABC, V and W are points in AB and AC respectively. If VW∥BC and AB = AC prove that △AVW is isosceles.

34. If, in a triangle, the altitude from a vertex bisects the angle at the vertex, then prove that the triangle is isosceles.

35. In a parallelogram ABCD, AD = AB. Prove that the diagonals are right bisectors of each other.

36. In △ABC, the midpoint S is found for AC. If AS = BS, prove that ∠ABC is a right angle.

---

problem/matics

A golf hole has a diameter of 8 cm. What are the dimensions of the greatest square pole that can be placed in the hole?

9.6 parallel line theorem

In the previous section, we proved:

> If a transversal intersects a pair of parallel lines then the alternate angles are equal.

The converse of the above theorem, may or may not be true.

> If a transversal intersects two lines and the alternate angles are equal then the lines are parallel.

method of indirect proof

Proving theorems or deductions gives us practice in the process of deductive reasoning. There are other methods of reasoning that we may use to prove deductions. One such method is the method of **indirect proof** based on the following observation.

If there are two possibilities for a given statement, then
A: the statement is true   or
B: the statement is not true.

The basis of indirect proof is to choose one of the possibilities. If the chosen possibility leads to a contradiction of a fact or theorem, (a fact or theorem being a statement which is true) then the alternative must be true.

For example, if we assume A to be not true, and this assumption leads us to the contradiction of a fact, then our assumption is wrong and thus A must be true since A must be either true or false.

We use this method as well as other theorems to prove the converse of the *Parallel Lines Theorem* (PLT).

Example 1

Prove that if a transversal intersects two lines and the alternate angles are equal, then the lines are parallel.

Solution
Given: AB a transversal for PQ and RS.
∠PDE = ∠SED.

Required to prove:
PQ∥RS

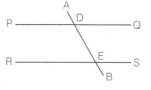

Proof: Either PQ∥RS or PQ ∦ RS. Make the assumption that PQ ∦ RS. Then PQ and RS meet at the point M, as shown.

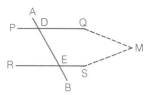

In △ MDE, ∠PDE is an exterior angle, ∠MED is a remote interior angle, and ∠MED = ∠PDE	Reasons
	given
This is impossible since for △MDE ∠MED < ∠PDE	EAT

Thus, our assumption is incorrect and PQ∥RS.

Once we have proved the above theorem, we may easily show the following corollaries. If

(a) the corresponding angles are equal or
(b) the interior angles on the same side of the transversal are supplementary

then the lines are parallel.

Example 2
In △ABC, CA = CB and BC is produced to D. If exterior ∠ACD is bisected by CE, prove that CE∥BA.

Translate the given information and construct a diagram to plan the solution.

Solution
Given: In
△ABC, CA = CB,
∠ACE = ∠ECD.
Required to prove:
BA∥CE

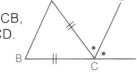

Proof:
In △ABC	Reasons
AC = BC	given
Then	
∠CAB = ∠CBA	ITT
∠ACD = ∠ACE + ∠ECD	
= 2 ∠ACE ①	construction
In △ABC	
∠ACD = ∠CAB + ∠CBA	ASTT corollary
= 2 ∠CAB ②	∠CAB = ∠ABC
From ① and ②	
2 ∠ACE = 2 ∠CAB	
∠ACE = ∠CAB	
Since AC is a transversal for AB and CE and	
∠ACE = ∠CAB,	proven
then BA∥CE.	PLT, alternate angles

Throughout the exercise, we use the new theorems to prove further deductions.

9.6 exercise

A

1 In the diagrams, which lines may be proven parallel?

(a) 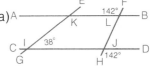 (b)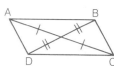

2 Prove that if AB∥CD and CD∥PQ, then AB∥PQ.

3 If ∠PQA = ∠SRD, prove AB∥CD.

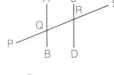

4 Use the diagram to prove that if two line segments PQ and RS bisect each other then PSQR is a parallelogram.

5 Use the indirect method of proof to prove that AB∥CD.

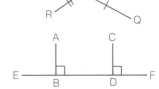

6   If M is the midpoint of both CB and AD, prove that AB∥CD.

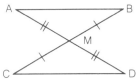

7   For △PQR, PQ = PR. If MQ = NR prove that MN∥QR.

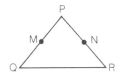

8   Prove the corollaries.
   (a) If a transversal intersects two lines and the corresponding angles are equal then the lines are parallel.
   (b) If a transversal intersects two lines and the interior angles on the same side of the transversal are supplementary, then the lines are parallel.

For Questions 9 to 21, try to use, where possible, the method of indirect proof to prove the deductions.

9   Prove that if ∠QBC = ∠RCB then PQ∥RS.

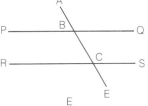

10 In the diagram, ∠BFG + ∠FGD ≠ 180°. Prove that AB ∦ CD.

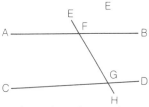

11 (a) Prove that the equal angles of an isosceles triangle are acute.
   (b) △PQR is a scalene triangle and PS ⊥ QR where S is in QR. Prove PS is not a median.

12 In △PQR, S is the midpoint of PR and T is the midpoint of QR. Prove PT and QS cannot bisect each other.

13 Prove that two lines perpendicular to the same line cannot intersect.

14 A quadrilateral PQRS is given so that ∠SRQ = ∠RSP. If RQ = SP prove RS∥QP.

15 In △URS, V and T are points in UR and US respectively. The perpendicular from T to VS bisects VS and VS bisects ∠S. Prove VT∥RS.

16 PQRS is a quadrilateral so that SP = PQ = QR and ∠PSQ = ∠RQS. Prove PQRS is a rhombus.

17 The quadrilateral SOGN has SN = OG and ∠G = ∠N. The midpoints of OS and GN are T and C respectively. Prove SO∥NG.

18 For △TAC, TA = TC. On TA there is a point Z and on TC point Q so that ZQ∥AC. Prove △TZQ is isosceles.

19 Prove that in any quadrilateral, if a pair of opposite sides are equal and parallel then the quadrilateral is a parallelogram.

20 (a) If the diagonals of a quadrilateral bisect each other, prove that the quadrilateral is a parallelogram.
   (b) If the diagonals of a quadrilateral are right bisectors of each other prove that the quadrilateral is a rhombus.
   (c) If the diagonals of a parallelogram bisect the angles, prove it is a rhombus.

21 △TUS is constructed so that TL = LV and SK = KW. If K and L are midpoints of UT and SU respectively, prove points V, U and W are collinear.

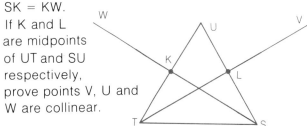

9.7 applications: properties of parallelograms

We dealt with some of the properties of parallelograms in the previous sections. Certain quadrilaterals have special properties which are useful in helping us to prove various deductions. Some definitions that we should be familiar with are summarized below:

- A quadrilateral is a polygon with 4 sides.
- A trapezoid (or trapezium) is a quadrilateral with a pair of opposite sides that are parallel and a pair that are not parallel.
- A parallelogram is a quadrilateral with both pairs of opposite sides parallel.
- A rhombus is a quadrilateral with all sides equal.
- A rectangle is a parallelogram with a right angle.
- A square is a rhombus with a right angle.

We could have defined the rectangle in terms of a quadrilateral but the definition would have been complex. Thus, the definition of the rectangle is based on the parallelogram as given above.

9.7 exercise

Questions 1 to 5 deal with the properties of parallelograms. Give a proof for each of the following. Collectively we refer to these properties as the *Properties of Parallelogram Theorem* PPT.

A

1. In a parallelogram prove that
   (a) the opposite sides are equal.
   (b) the opposite angles are equal.

2. In a parallelogram, prove that the diagonals bisect each other.

3. If both pairs of opposite sides of a quadrilateral are equal then the quadrilateral is a parallelogram.

4. If one pair of opposite sides of a quadrilateral is equal and parallel, then the quadrilateral is a parallelogram.

5. If the diagonals of a quadrilateral bisect each other then the quadrilateral is a parallelogram.

6. Use the definition of a rhombus to prove that any rhombus is a parallelogram.

7. Prove that the opposite angles of a rhombus are equal.

8. Prove that the diagonals of a rhombus are perpendicular bisectors of each other.

9. Prove that if the diagonals of a quadrilateral are perpendicular bisectors of each other, then the quadrilateral is a rhombus.

10. Prove that if a quadrilateral has a right angle and opposite sides are equal then the quadrilateral is a rectangle.

11. Prove that, for any rectangle, the diagonals are not perpendicular.

12. Prove that the diagonals of a rectangle are equal.

B

13. In the diagram, QT = TS and BT = TM. Prove that ∠XSM = ∠MQY.

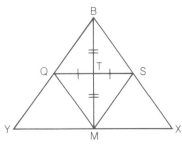

14. The line segment QX in the parallelogram CQTX bisects ∠CQT. Prove that CQTX is a rhombus.

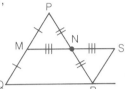

15. In the diagram RP = RQ. If RS bisects ∠PRT (RS≠PQ) then prove quadrilateral PQRS is a trapezoid.

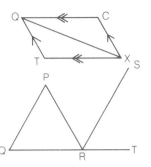

16. In quadrilateral ABCD, AC ⊥ DB. If the midpoints of BD and AC are coincident with the intersection point of AC and BD, prove that ABCD is a rhombus.

17. The midpoints of the sides of a square ABCD are P, Q, R, and S in turn. Prove that quadrilateral PQRS is a rhombus.

18. ECAF is a parallelogram in which B is the midpoint of FE and S is the midpoint of AC. Prove that BASE is a parallelogram.

a method of proof: using constructions

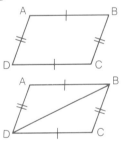

Earlier, we constructed a line segment as an aid in proving a deduction. For example, to use the information given in the diagram, we might want to obtain 2 triangles. Thus a diagonal DB could be constructed and we might then use one of the congruence assumptions for △ABD and △CDB.

Some theorems, such as the *Midpoint Parallel Triangle Theorem* (MPTT), involve constructions.

The line segment that joins the midpoints of two sides of a triangle is
- parallel to the third side.
- equal to $\frac{1}{2}$ of the third side.

The proof of the above theorem illustrates how the various strategies are organized to prove a deduction. Provide a reason, denoted by A, B, C, D, and so on, for each step.

Given: △PQR, PM = MQ, PN = RN.

Required to prove:
MN∥QR
MN = $\frac{1}{2}$QR

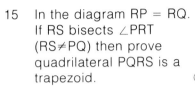

Proof: Complete the construction so that MN is produced to S and MN = SN.

	Reasons
Since PR and MS intersect at N then ∠PNM = ∠RNS	A
In △PNM and △RNS	
PN = RN	B
∠PNM = ∠RNS	C
MN = SN	D
Thus △PNM ≅ △RNS	E
Thus PM = RS	F
∠PMN = ∠RSN	G
MS is a transversal for PQ and SR.	
∠PMN = ∠RSN	H
Thus PQ∥SR or MQ∥SR	I
Since PM = RS	
PM = MQ	
then SR = MQ	J
In quadrilateral MSRQ	
MQ = SR, MQ∥SR.	K
Thus quadrilateral MSRQ is a parallelogram.	L
MS = QR	M
or MN = $\frac{1}{2}$QR and MN∥QR	

19 Use the indirect method of proof to prove that if a line bisects one side of a triangle and is parallel to another side, then the line bisects the remaining side.

20 For the diagram, AB∥DC, and S and T are the midpoints of AD and BC respectively. Prove

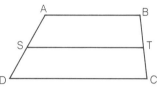

(a) ST∥AB∥DC   (b) ST = $\frac{1}{2}$(AB + DC)

21 SAID is a trapezoid with SD∥AI. T is the midpoint of AS and B is a point on ID so that TB∥AI. Prove that B is the midpoint of ID.

22 In quadrilateral PQRS, the midpoints A, B, C, D of each side are located. Prove that quadrilateral ABCD is a parallelogram.

23 For the parallelogram APSD where L is the midpoint of SP and J the midpoint of DA. DL intersects AS at W and JP intersects AS at Z.

(a) Show whether DL∥JP. Give reasons for your answer.

(b) Prove that WS = ZW = AZ.

problem/matics

Often the answer to a problem is surprising; one we do not expect. Solve the following problem.

An island is in the shape of an equilateral triangle. Where should a cottage be built so that the sum of the distances from the cottage to all shores is a minimum?

problem/matics

Use the method of indirect proof, dealt with in Section 9.6 to prove that $\sqrt{2}$ is an irrational number.

9.8 inequalities for triangles

Often, everyday problems may be solved using the properties of triangles. For example, if we have a garden in the shape of △ABC and ∠A < ∠B then we would need more fence along side AC than along side BC.

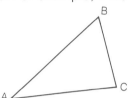

Triangle Inequality
The angle opposite the longest side of a triangle is the greatest angle; the angle opposite the shortest side of a triangle is the smallest angle.

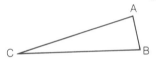

If AC > AB then ∠B > ∠C.

Given: △ABC, AC > AB
Required to prove:
  ∠ABC > ∠ACB
Proof: On AC locate Q such that AQ = AB.

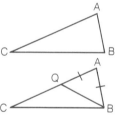

In △AQB	Reasons
AQ = AB	constructed
Thus ∠AQB = ∠ABQ	ITT
∠AQB > ∠ACB	exterior angle of triangle
∠ABQ > ∠ACB	since ∠ABQ = ∠AQB
∠ABC > ∠ABQ	
∠ABC > ∠ACB	

Which is what we wanted to prove.

We may ask whether the converse statement is true.

Converse of Triangle Inequality
The side opposite the larger angle of a triangle is greater in measure than the side opposite the smaller angle. To prove this theorem we may use the method of indirect proof.

9.8 exercise

A

1. For △PQR, ∠Q > ∠P. Prove that RP > QR.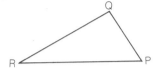

2. To prove the converse of the triangle inequality use the following diagram.

    Given: △ABC, ∠C > ∠B.
    Required to prove:
    AB > AC
    Proof: One of
    AB < AC, AB = AC,
    or AB > AC must be true.

    Use the method of indirect proof and show that

    (a) the assumption AB = AC gives a contradiction.
    (b) the assumption AB < AC gives a contradiction.

3. To prove that the sum of any two sides of a triangle is greater than the third, complete the following proof by writing the authorities (reasons) for each step.

    Given: △PQR where PM bisects ∠QPR.
    Required to prove:
    PQ + PR > QR
    Proof:

	Reasons
∠QMP > ∠MPR	A
∠MPR = ∠MPQ	B
∠QMP > ∠MPQ	C
PQ > QM	D
∠PMR > ∠QPM	E
∠QPM = ∠MPR	F
∠PMR > ∠MPR	G
PR > MR	H
PQ + PR > QM + MR	I
QM + MR = QR	J
PQ + PR > QR	K

4. △ABC is a scalene triangle. D is any point in AC. Prove
    AB + BC > AC.

B

5. In quadrilateral OAFL prove that
    LA + OF > OL + AF.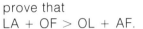

6. Prove that the sum of any 3 sides of a quadrilateral is greater than the fourth side.

7. Given △EFG with EF > EG. If the bisectors of ∠F and ∠G intersect in T, then prove FT > TG.

8. Given △XYZ with YV the bisector of ∠Y. Prove that if XY > XV, then YZ > VZ.

9. △XYZ is right-angled at Y. Prove that ∠X and ∠Z are acute.

C

10. ATEK is a quadrilateral in which ∠KTE = ∠E. Prove that
    TA + AK > KE.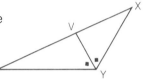

11. The diagonals of quadrilateral PQRS intersect on N. Prove that
    PQ + QR + RS + SP > QS + PR.

12. △PMN is isosceles with PM = PN. MN is produced to Q. Prove that PQ > PN.

13. Use the indirect method of proof to show that PS is the shortest segment to QR.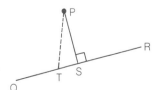

9.9 properties of circles

We have already explored the properties of circles using inductive methods. Now we will use our skills in deductive geometry to prove the same properties about circles.

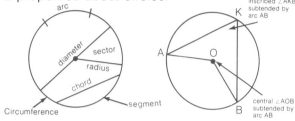

Example 1

Prove that the perpendicular bisector of a chord of the circle passes through the centre of the circle.

Solution
Given: AC = BC,
∠ACO = ∠BCO = 90°

Required to prove:
O is on EC.

Proof:	Reasons
OA = OB	radii
O is equidistant from A and B.	
Thus O is on the perpendicular bisector of AB.	RBT

In our inductive work we stated the hypothesis: The measure of the central angle is twice the measure of the inscribed angle subtended by the same arc.

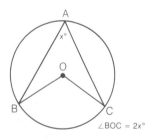

∠BOC = 2x°

We may deduce this result using the various theorems we have proved. In the following proof, we use 3 different diagrams to represent 3 different cases or situations that are possible.

Given: Inscribed ∠ABC and central ∠AOC both subtended by arc ARC.

Required to prove: ∠AOC = 2 ∠ABC

Proof:	Reasons
Case I	
OA = OB	radii
∠OAB = ∠OBA ①	ITT
∠AOK = ∠OAB + ∠OBA ②	EAT
∠AOK = 2 ∠OBA ③	from ① and ②
OB = OC	radii
∠OBC = ∠OCB ④	ITT
∠COK = ∠OBC + ∠OCB ⑤	EAT
∠COK = 2 ∠OBC ⑥	
Thus, ∠AOK + ∠COK = 2 ∠OBA + 2 ∠OBC	from ③ and ⑥
∠AOC = 2(∠OBA + ∠OBC)	factoring
∠AOC = 2 ∠ABC	∠OBA + ∠OBC = ∠ABC

Case II	
OB = OA	radii
∠OAB = ∠OBA ①	ITT
∠AOC = ∠OAB + ∠OBA ②	EAT
∠AOC = 2 ∠ABC	From ① and ②

Case III	
∠KOA = 2 ∠KBA	from Case II
∠KOC = 2 ∠KBC	from Case II
Thus ∠KOC − ∠KOA = 2 ∠KBC − 2 ∠KBA	
∠AOC = 2 ∠ABC	

We may write a corollary for the above theorem.
Corollary: The angle subtended by the diameter of a circle is 90°.

9.9 exercise

The results of Questions 1 to 5 have been suggested in our earlier work. Write a complete proof.

A

1. Prove that perpendicular bisectors of the chords of a circle pass through the centre of the circle.

2. Prove that the line containing the centre of the circle and the midpoint of a chord is perpendicular to the chord.

3. Prove that the line that contains the centre of the circle and is perpendicular to a chord, bisects the chord.

4. Prove that chords that are the same distance from the centre of the circle are equal in length.

5. Prove that chords that are equal in length are equidistant from the centre of the circle.

To obtain the numerical answers to Questions 6 and 7, we use our earlier skills in algebra and geometry.

6. Find the missing values in each diagram.

(a) 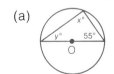 (b) (c)

(d) (e) 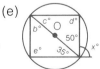 (f)

(g) (h)

7. Calculate the measures of the line segments. All measures are in centimetres

(a) 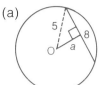 (b) 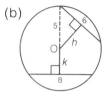 (c)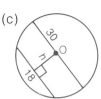

Provide a complete proof for each of the following deductions.

B

8. Given: 2 circles with centres O and P intersecting at A and C.
   Required to prove: OP is the perpendicular bisector of AC.

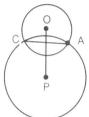

9. Given: Circle with centre O. AB∥CD, AP = BP
   Required to prove: CQ = DQ.

10. P is a point on the circumference of the circle with centre O. OP is perpendicular to chord BC. Prove that PB = PC.

11. ∠QPR has measure 80° and is subtended by arc QR in the circle with centre O. Calculate the measure of
    (a) ∠QOR (b) ∠OQR (c) ∠ORQ

12. MNPQ is an inscribed quadrilateral. Prove that the sum of the measures of opposite angles is 180°.

13. Prove that ∠A = ∠DCE.

C

14. Each point of quadrilateral PQRS is on the circle with centre O such that PS∥QR. Prove that PR = QS and ∠PQR = ∠SRQ.

geometry from different points of view

9.10 transformational geometry: a deductive system

A deductive system as we have seen may be illustrated by the following diagram.

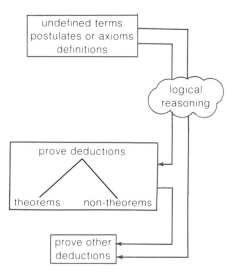

In the following sections, the definitions and properties of translations, reflections, rotations, and dilatations are dealt with.

To prove deductions from another point of view we will use our results in plane geometry concerning the properties of certain figures and the properties of transformations. The more strategies and methods we can use, the better we will be able to solve problems or deductions we have not seen before.

In our earlier work with reference to a co-ordinate system, we expressed a translation in terms of a mapping. Similarly, we will express other transformations in terms of mappings.

9.11 translations: properties and proofs

We may relate two points, A and B, on a plane in three ways.

- with respect to a co-ordinate system.

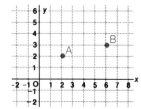

- with respect to a directed line segment. The directed line segment AB is shown by the symbol $\vec{AB}$.

$\vec{AB}$ has a magnitude and a direction. The magnitude of $\vec{AB}$ is given by the length of $\vec{AB}$. The positive direction of the arrow is from A to B.

We name the *directed line segment* a **vector**. Later we will learn that if two vectors are equal, then they have equal magnitudes and the same direction.

$\vec{AB} = \vec{CD} \longrightarrow AB = CD$ and $AB \| CD$

- by using transformations. For example, in this section, we will explore one of the transformations, namely, the translation. Earlier (in Section 7.11) the meaning of translation was introduced and then applied to the solution of problems, and to the finding of the equations of image lines. In this section we will explore the properties of translations and apply them to the proof of deductions.

defining translations

Points on a plane are related by a correspondence as shown.

A ⟶ A'   B ⟶ B'
C ⟶ C'   and so on

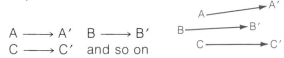

The correspondence A ⟶ A′, B ⟶ B′, C ⟶ C′, and so on is a translation if and only if AA′ = BB′ = CC′ and AA′ ∥ BB′ ∥ CC′

Using directed line segments or vectors we may write this compactly as $\overrightarrow{AA'} = \overrightarrow{BB'} = \overrightarrow{CC'}$

The translation is shown by the notation

[A, B, C, ...] ⟶ [A′, B′, C′, ...]

pre-image ↗ A is mapped onto A′. translation image

As in plane geometry, we can use the properties of translations to prove deductions. To show the properties of translations, we use the results we proved earlier in plane geometry.

Example 1

Show that if [A′, B′] is the translation image of [A, B] then AB = A′B′.

Once we prove the above property, we may use it as a theorem for our subsequent work.

Solution

Given: [A′, B′] is the translation image of [A, B].

Required to prove: AB = A′B′

Proof: Join AB and A′B′.

	Reasons
Since [A, B] ⟶ [A′, B′]	
Then AA′ = BB′	property of
AA′ ∥ BB′	translation
In quadrilateral AA′B′B	
AA′ = BB′	
AA′ ∥ BB′	
Then AA′B′B is a parallelogram.	property of a parallelogram
Thus AB = A′B′.	property of a parallelogram

From the above result, since a translation is a transformation that preserves length, we may refer to the translation as a **congruent transformation** or **isometry**.

An isometry is a transformation that preserves length.

Example 2

If △A′B′C′ is the translation image of △ABC, then prove that △ABC ≅ △A′B′C′.

Solution

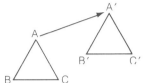

Given: △A′B′C′ is the translation image of △ABC.

Required to prove: △ABC ≅ △A′B′C′

Proof:

	Reasons
Since [A′, B′, C′] is the translation image of [A, B, C] then	
[A, B] ⟶ [A′, B′] and AB = A′B′	property of translation
[B, C] ⟶ [B′, C′] and BC = B′C′	property of translation
[A, C] ⟶ [A′, C′] and AC = A′C′	property of translation
In △ABC and △A′B′C′	
AB = A′B′	proved
BC = B′C′	proved
AC = A′C′	proved
Thus △ABC ≅ △A′B′C′	SSS

From Example 2, we observe the following properties of translations.

- Length is preserved. We may also say length is invariant for a translation.

- Area is preserved under a translation.

- The vertices of △ABC and △A′B′C′ have the same sense, namely counter-clockwise. Since the sense of the congruent figures is preserved we say that the translation is a **direct isometry**.

- Since ∠A′B′C′ is the translation image of ∠ABC then the measures of angles are preserved under a translation.

- In the diagram in Example 2, AA′B′B is a parallelogram. Thus slopes of lines are preserved under a translation.

We may extend the results of Example 2 and prove that if [A', B', C', ...] is the translation image of [A, B, C, ...] then all of the above properties are true. In our subsequent work, to prove deductions we may refer to the above properties as the *Translation Properties Theorem*. We may now use the *Translation Properties Theorem* for proving deductions.

9.11 exercise

A

1. For the diagram, PP'||RR'||QQ' and PP' = RR' = QQ'. Prove that △PQR ≅ △P'Q'R'.

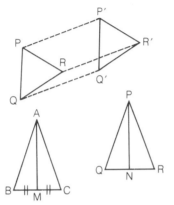

2. For the diagram, [P, Q, N, R] is the translation image of [A, B, M, C].

   (a) Prove that AM = PN and AM||PN.
   (b) Write the result in (a) as a general statement.

3. A translation is given by [P, Q, R] ⟶ [P', Q', R']. Prove that ∠PQR = ∠P'Q'R'.

4. The translation image of [S, T, U, V] is given by [S', T', U', V']. Prove that STUV ≅ S'T'U'V'.

5. Complete the proof of the following property. If A' is the translation image of A, then the translation of any line through A is parallel to the translation image of the corresponding line through A'.

   Given: [A', B'] is the translation image of [A, B]. (Line S' is the translation image of the line S.)

   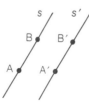

   Required to prove: AB||A'B' (i.e. S||S').

B 6. In the diagram, P' is the translation image of P. ℓ is a line parallel to PP' as shown.

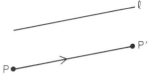

   (a) Prove that for the same translation the translation image of the line ℓ is itself.
   (b) Write a general statement for the above result in (a).

7. If [S, T, U] is the translation image of [A, B, C], prove that

   (a) the midpoint of AB is mapped onto the midpoint of ST.
   (b) the perpendicular from A to BC is mapped onto the perpendicular from S to TU.

8. Prove that the translation image of a square is congruent to the pre-image.

9. (a) Prove that the translation image of a triangle is a triangle.
   (b) Prove that the translation image of a straight line is a straight line.

10. The circle with centre P is translated. P ⟶ P'. Prove that the circle with centre P is congruent to the circle with centre P'.

11. Two circles with centres P and P' are congruent. If P' is the translation image of P, prove that the circle with centre P' is the translation image of the circle with centre P.

Translations may be used to relate points on a grid or on the Cartesian plane as shown in Questions 12 to 14.

12 On a grid, directed line segments represent translations. Which directed line segments represent the same translation?

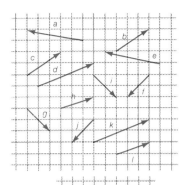

13 A directed line segment is shown. Find the translation image for each of the following.

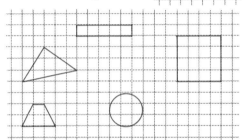

4 Earlier, we defined a translation on the Cartesian plane by a mapping of the form, $(x, y) \longrightarrow (x + a, y + b)$. For each mapping, find the translation image of the given figure.

figure	translation
(a) A(2, 3), B(−1, 2)	$(x, y) \longrightarrow (x − 3, y + 2)$
(b) P(2, 2), Q(−3, −2), R(3, −3)	$(x, y) \longrightarrow (x + 2, y − 4)$

proving deductions: translations

Previously, in plane geometry, we proved various theorems and applied them to the proof of deductions. Similarly, the properties of a translation have been established in Questions 1 to 11 and these properties may also be used to prove deductions. Our knowledge of transformations provides yet another strategy in proving deductions.

Example 3

Two congruent circles with centres P and Q intersect at E and F. A line through E, parallel to PQ, intersects the circle with centre P at R and the circle with centre Q at S. Prove that RE = ES.

Solution

Given: Congruent circles with centres P and Q. RS∥PQ

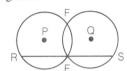

Record the given information on a diagram.

Required to prove: RE = ES

Proof: Since the circles with centres P and Q are congruent, then Q is the translation image of P for the translation given by $\overrightarrow{PQ}$. For the translation $\overrightarrow{PQ}$, then [R, E] $\longrightarrow$ [E, S].   ← property of translation
Since [E, S] is the translation image of [R, E] then RE = ES.

15 Use the diagram in Example 3.
   (a) Use the theorems established in plane geometry to prove the above result. How does your solution compare with the one given above?
   (b) Use the theorems of plane geometry to prove that RP∥EQ.
   (c) Use the properties of translations to prove that RP∥EQ. How does your method compare with the result in (b)?

In proving a deduction, often there are different strategies for proving the same result. As we have seen, the proof of the above deduction could have been obtained by our earlier methods. In algebra, there are also often different methods of obtaining the same result.

For the following deductions, use the properties of translations to prove each of the following.

16. Two parallelograms, STUV and STRQ, are shown, with a common base ST. If a translation is applied, write the translation image determined by the translation $\vec{ST}$.

    (a) [S, V]   (b) [Q, V, S]
    (c) Prove that △QVS ≅ △RUT.

17. In the diagram, the circles are congruent. RP = UQ. If PQ∥ST, prove that

    (a) RS = UT   (b) RP∥SQ

18. A parallelogram PQRS is drawn with midpoints in PS and QR respectively as M and N. Prove that for the translation defined by $\vec{SR}$, N is the translation image of M.

19. In the diagram, P and Q are the centres of congruent circles. If QR∥PT prove that SQ∥UP.

20. Two congruent rectangles PQRS and PQVT have a common base PQ, but are on opposite sides of PQ. The midpoints of SV, PV and SP are A, B, and C respectively. The midpoints of RT, TQ and QR are D, E, and F respectively. Prove that △ABC ≅ △DEF.

21. In the diagram, △PQR is translated as shown.
    [P, Q, R] ⟶ [P', R, R']
    Prove that

    (a) △RP'P ≅ △PQR   (b) △P'RR' ≅ △PQR
    (c) ∠P + ∠Q + ∠PRQ = 180°

9.12 reflections: properties and proofs

When we think of a reflection we might immediately think of a mirror. In mathematics, another transformation, called a reflection, is defined below for points on a plane and involves a mirror line.

For the points shown, the mapping
A ⟶ A'
B ⟶ B'
C ⟶ C'

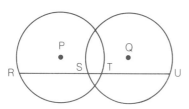

is a reflection in the line m if and only if the line m is the right bisector of AA', BB', and CC'.

We refer to the line m as the **mirror line** or **reflection line**. Thus from the definition

- if A and A' are corresponding points of a reflection in the mirror line m, then the mirror line m is the right bisector of AA'.

AN = NA'
AA' ⊥ mirror line m

- if A and A' are two points such that a line m is the right bisector of AA' then A and A' are corresponding points of a reflection in the mirror line m.

In mathematics, it is essential that we use symbols and language consistently. Thus, the terminology for a reflection is similar to the terminology for a translation.

A reflection of △ABC is shown on the grid. △ABC is mapped onto △A'B'C'.

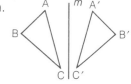

We write
[A, B, C] ⟶ [A', B', C']   ← Called the reflection image of [A, B, C].

We developed the properties of a translation by using our skills in plane geometry. We will now develop the properties of a reflection in a similar way.

Example 1

[A′, B′] is the reflection image of [A, B] in the mirror line m. Prove that AB = A′B′.

Solution

Given: [A, B] and its reflection image [A′, B′].

Required to prove: AB = A′B′

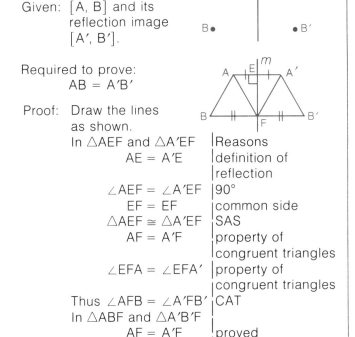

Proof: Draw the lines as shown.

In △AEF and △A′EF | Reasons
AE = A′E | definition of reflection
∠AEF = ∠A′EF | 90°
EF = EF | common side
△AEF ≅ △A′EF | SAS
AF = A′F | property of congruent triangles
∠EFA = ∠EFA′ | property of congruent triangles
Thus ∠AFB = ∠A′FB′ | CAT
In △ABF and △A′B′F |
AF = A′F | proved
∠AFB = ∠A′FB′ | proved
BF = B′F | definition of reflection
△ABF ≅ △A′B′F | SAS
Thus AB = A′B′ ← Which is what we wanted to prove.

Thus in Example 1, we have proved that, for a reflection, *length is preserved*. That is, length is invariant for a reflection. Thus:

a reflection is an isometry.

By using our theorems of plane geometry, we may prove other properties of reflections and then use these properties as another strategy in proving deductions. For example, we may use the result:

If [A, B] ⟶ [A′, B′] then AB = A′B′.

to deduce the result:

If [A, B, C] ⟶ [A′, B′, C′] then △ABC ≅ △A′B′C′.

Other properties of reflections may also be deduced. (See Exercise 9.12.)

properties of reflections

For a reflection in a mirror line m

- if [A′, B′] is the reflection image of [A, B] the mirror line m is the perpendicular bisector of AA′ and BB′.
- length is invariant.
- measures of angles are invariant.
- sense of figures is not preserved.
- if $\ell \| p$ then $\ell' \| p'$.
- if ℓ' is the reflection image of ℓ, then the mirror line bisects the angle between ℓ and ℓ', $\ell \not\| \ell'$.
- any point, P, on the mirror line is its own reflection image P′. We say that the point P is an **invariant point** of the reflection.

9.12 exercise

Questions 1 to 14 deal with proving the properties of reflections. These properties may then be used to prove deductions.

A

1. Prove that the reflection image of a straight line is a straight line.

2. In the diagram, OF is the bisector of ∠GOH, the angle between lines ℓ and k. Use the diagram to prove that if two lines intersect they are the reflection images of each other in the bisector of the angle between them.

3. In the diagram, [A′, C′] is the reflection image of [A, C]. B is any point on AC. Use the diagram to prove that the reflection image of a straight line is a straight line.

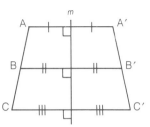

4. (a) Use the definition of reflection to prove that if ∠A′B′C′ is the reflection image of ∠ABC then ∠A′B′C′ = ∠ABC.
   (b) Write a general statement for your result in (a).

5. (a) Prove that if [A′, B′, C′] is the reflection image of [A, B, C] then △ABC ≅ △A′B′C′.
   (b) Write a general statement for your result in (a).

B

6. For a reflection, prove that
   (a) length is invariant.
   (b) the measure of an angle is invariant.

7. If two lines ℓ and p are parallel and their reflection images are ℓ′ and p′, prove that ℓ′∥p′.

8. If ℓ′ is the reflection image of ℓ, then prove that the mirror line bisects the angle between ℓ and ℓ′. (Note ℓ ⊁ ℓ′.)

9. Prove that a straight line, perpendicular to the mirror line of a reflection, is its own image.

10. Prove that if two circles are reflection images of each other then
    (a) they are congruent.
    (b) their centres are reflection images of each other.

11. Prove that two congruent circles are the reflection images of each other.

12. (a) For the diagram, the two circles are reflection images of each other. Prove that the mirror line is the right bisector of the segment PP′.
    (b) Write a general statement for your result in (a).

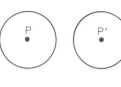

13. (a) Two congruent circles intersect in the common chord RS. Prove that the two circles are reflection images of each other in the line through RS.
    (b) Write a general statement for your result in (a).

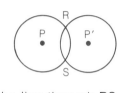

14. Prove that a circle is its own reflection image with the diameter as the mirror line.

15. Use squared paper to find the reflection image of each shape in the mirror line m.

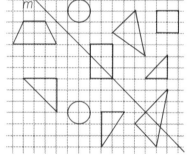

16. Find the image of each point with reflection in the x-axis.
    (a) (3, 2)   (b) (−1, 3)   (c) (−2, −5)

17. Find the image of each point with reflection in the y-axis.
    (a) (1, 5)   (b) (3, −2)   (c) (−4, −3)

18. Show that the mapping (x, y) ⟶ (x, −y) is a reflection in the x-axis.

19. Show that the mapping (x, y) ⟶ (−x, y) is a reflection in the y-axis.

proving deductions: reflections

Throughout the exercise, we may use the properties of reflections and translations to prove the deductions.

Example
Two congruent circles intersect in a common chord with end points P and Q. A line perpendicular to PQ and passing through Q meets the circles at S and T. Prove that ∠SPQ = ∠TPQ.

Solution
Given: Congruent circles, PQ ⊥ ST
Required to prove: ∠SPQ = ∠TPQ

Proof:

	Reasons
Since the circles are congruent, PQ is a mirror line. P and Q are their own reflection images,	Congruent circles are reflection images in a common chord.
P ⟶ P ① Q ⟶ Q ②	Points on the mirror line are invariant points of the reflection.
Since PQ ⊥ SQT, then T is the reflection image of S ③ S ⟶ T	PQ is the mirror line of the congruent circles.
Thus, [P, Q, T] is the reflection image of [P, Q, S]. △PQS ≅ △PQT	From previous correspondence ① ② ③. A reflection is an isometry congruent transformation.
∠SPQ = ∠TPQ	Properties of congruent triangles.

20 For the diagram, the centre of the circle is O, CH is a chord and OP ⊥ CH. Use your work with reflections to prove that CP = HP.

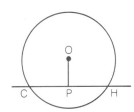

21 Use the diagram. If CP = PH, use your work with reflections to prove that OP ⊥ CH.

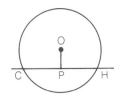

22 Use the diagram. PQ||SR and ∠S = ∠R. Prove that ∠P = ∠Q.

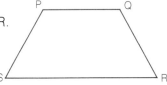

23 Use the diagram. The two given circles are congruent. If PS ⊥ AB and TV ⊥ AB prove that △PQT ≅ △SRV.

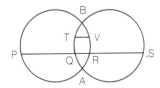

24 For quadrilateral PQRS the diagonals intersect at T so that PT = QT and TS = TR. Prove that △PSR ≅ △QRS.

25 A reflection is given by [P, Q, R] ⟶ [P', Q', R']. If the midpoints of △PQR are A, B, and C respectively and the midpoints of △P'Q'R' are D, E, and F respectively, prove that △ABC ≅ △DEF.

26 Circles with centres P and Q intersect at R and S. Prove that PQ is the right bisector of RS.

9.13 working with rotations

A rotation is another correspondence between points on a plane. To specify a rotation we need to know

- the rotation or turn centre. For △ABC ⟶ △A'B'C', the rotation centre is O.
- the rotation angle. ∠BOB' is the rotation angle.

For a rotation, △ABC ⟶ △A'B'C'. Then, in particular,
OB = OB', OA = OA', OC = OC',
∠BOB = ∠AOA' = ∠COC'.

In particular, a half-turn is defined by
∠BOB' = ∠AOA' = ∠COC' = 180°.

In the exercise, we will explore some properties of half-turns. The properties of a rotation are as follows.

- A rotation is an isometry (length is invariant).
- The measures of angles are invariant.
- The sense is preserved.
- The centre of rotation lies on the perpendicular bisector of the segment joining a point and its image.

9.13 exercise

Questions 1 to 9 develop properties of rotations in general, and properties of half-turn rotations in particular.

A

1. [A', B'] is the rotation image of [A, B]. Prove that AB = A'B'.

2. [A', B', C'] is the rotation image of [A, B, C]. Show that
   (a) △ABC ≅ △A'B'C'.
   (b) the sense of △ABC is preserved.

3. For a reflection with turn centre O, A ⟶ A'. Prove that O lies on the right bisector of AA'.

4. [A', B'] is the rotation image of [A, B] for a half-turn. Prove that AB∥A'B'.

5. Prove that the rotation image of a straight line is a straight line.

B

6. The line ℓ passes through P. If P is the rotation centre for a half-turn, prove that the reflection image of ℓ is itself.

7. Two parallel lines p and q are shown. T is a point half way between the lines. Prove that, for a half-turn, p and q are rotation images of each other.

8. (a) The centre of a circle is O. RS is a diameter. Use the diagram to show that, for a half-turn, R and S are images of each other about O.
   (b) Write a general statement about your result above.

9. (a) The circles with centres P and P' are congruent. For a half-turn, prove that the circles are rotation images of each other about the midpoint of PP'.
   (b) Write a general statement about your result in (a).

10. Use squared paper to find the rotation image of each of the following for $\frac{1}{4}$-turn, $\frac{1}{2}$-turn about the turn centre P.

11. For a rotation of 90° counter-clockwise about the origin, find the rotation image of
    (a) (−3, 4)   (b) (5, −3)   (c) (−4, −5)

12. For a rotation of 180° clockwise about the origin, find the rotation image of
    (a) (3, 5)   (b) (−2, 3)   (c) (−4, −3)

proving deductions

Use the properties of rotations to prove the following deductions.

3. △PQR is rotated through a half-turn about the midpoint S of PQ so that R ⟶ R′. Prove that
   (a) R, S, R′ are collinear.   (b) RS = R′S.
   (c) Use the above results to show that the diagonals of a parallelogram bisect each other.

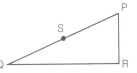

4. For the figure, PQ∥RS and QO = RO. Prove that PQ = SR.

5. In a circle with centre O, two chords PQ and RS are drawn so that QOR is a diameter and RP∥SQ. Prove that P, O, S, are collinear.

6. PQRS is a parallelogram and M is the midpoint of QS. Prove that M is the midpoint of the diagonal RP.

7. For the diagram, P and Q are centres of the congruent circles. If SV∥YW prove that SY∥VW.

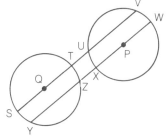

9.14 deductions and analytic geometry

In proving deductions in geometry, we have used strategies involving theorems in *plane geometry*, and theorems in *transformational geometry*. Another strategy for proving deductions is to use the *methods of analytic geometry*. We have already proved deductions involving numerical co-ordinates. In this section, we will develop a general strategy for using our skills in analytic geometry.

Example 1

We proved this theorem in our earlier work.

In a triangle, the line segment joining the midpoints of two sides of a triangle is parallel to the third side and equal in length to one half of the third side.

When using analytic geometry we position a general triangle on the plane. We may place the triangle in a convenient position, such as, one of the vertices at the origin and one side along a co-ordinate axis.

Solution

Given: △ABC, M, N are midpoints of AB and AC respectively.

Required to prove:
(a) $MN = \frac{1}{2}BC$
(b) $MN \parallel BC$

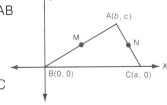

Proof: Co-ordinates of M are $\left(\frac{b}{2}, \frac{c}{2}\right)$.

Co-ordinates of N are $\left(\frac{a+b}{2}, \frac{c}{2}\right)$.

(a) $MN = \sqrt{\left(\frac{b}{2} - \frac{a+b}{2}\right)^2 + \left(\frac{c}{2} - \frac{c}{2}\right)^2} = \frac{a}{2}$

$BC = \sqrt{(a-0)^2 + (0-0)^2} = a$

Thus $MN = \frac{1}{2}BC$.

(b) Slope of MN = $\dfrac{\frac{c}{2} - \frac{c}{2}}{\frac{a+b}{2} - \frac{b}{2}} = 0$

Slope of BC = $\dfrac{0 - 0}{a - 0} = 0$

Thus MN∥BC.

In the above example, △ABC was placed in a position on the plane that simplified the algebra. As we have seen when we translate or rotate a figure, its properties remain invariant. Thus the general result is true no matter how we transform the figures. Thus, in proving the following deductions, place the figures so that one vertex is at the origin and a side is co-incident with an axis.

9.14 exercise

A 1 △ABC is a right-angled triangle. M is the midpoint of AB. Use the diagram to prove that MA = MB = MC.

2 Use the diagram to prove that the diagonals of a square are equal in length.

3 Use the diagram to prove that the sum of the measures of the 3 medians of a triangle is less than the perimeter of the triangle.

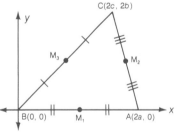

B
4 Prove that the diagonals of a parallelogram bisect each other.

5 Prove that the diagonals of a square are perpendicular to each other.

6 Prove that in an isosceles triangle, the median to the base is perpendicular to the base.

7 Prove that, in a triangle, if a line bisects one side and is parallel to a second side, then the line is a bisector of the remaining side.

8 Prove that, in a parallelogram, the opposite sides are equal.

9 In a quadrilateral, a pair of opposite sides are equal and parallel. Prove that the quadrilateral is a parallelogram.

10 △PQR is isosceles with QP = RP. If S and T are the midpoints of QP and RP respectively, prove that TQ = SR.

11 In a right-angled triangle, prove that the midpoint of the hypotenuse is the centre of the circle passing through the vertices of the triangle

12 For any quadrilateral, prove that the midpoints of the sides are the vertices of a parallelogram.

13 △PQR is isosceles so that QP = RP. If M, N, and P are the midpoints of QP, RP, and QR respectively prove that △MNP is isosceles.

14 Prove that the three medians of a triangle are concurrent (intersect at the same point).

C
15 Prove that the median from a vertex to the opposite side of a scalene triangle is not perpendicular to that side.

math tip

It is important to clearly understand the vocabulary of mathematics when solving problems. In this chapter we have learned these words; deductive system, transformation, proof, and so on.
- Make a list of all the words you have met in this chapter.
- Provide a simple example to illustrate each word.

skills review: making conjectures

For each of the following,
A write a conjecture about the diagram, based on the information given.
B prove your conjecture in A.

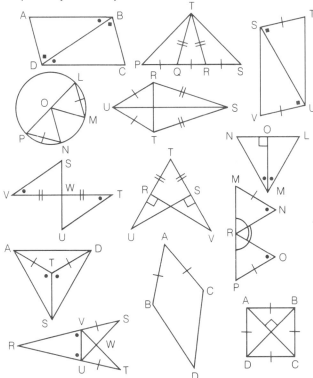

problems and practice: a chapter review

You have learned different strategies for proving a deduction. For each of the following, use the method most appropriate to prove the deduction.

1. Given: Circle with centre O. AB = CD.
   Required to prove: ADBC is a rectangle.

2. In △PQR, PQ = PR. On PQ and PR points S and T respectively are chosen so that PS = PT. Prove that ∠QSR = ∠RTQ.

3. In △PQR, QR is produced to T and RQ is produced to S so that ∠PRT = ∠PQS. Prove that PQ = PR.

4. If AB||CD and transversal TQ intersects AB and CD at S and R respectively, prove that the bisectors of ∠RSA and ∠SRD are parallel.

5. Prove that the bisectors of the opposite angles of a parallelogram are parallel.

6. In △ABC if ∠C > ∠B, prove that AB > AC.

7. △ABC is isosceles so that AB = AC. PS is drawn parallel to BC meeting AB and AC at Q and T respectively. Prove that BQ = CT.

8. PQRS is a quadrilateral. If the diagonals intersect at T prove that
   (a) PR + QS > PR + RS
   (b) PR + QS > PS + QR

9. Prove that the translation image of a triangle is congruent to the pre-image.

10. O is the centre of a circle and P is a point outside the circle. PQ intersects the circle at S and T and PR intersects the circle at U and V. PO bisects ∠QPR. Prove that △TUP ≅ △VSP.

11. Use co-ordinate geometry to prove that in any quadrilateral, the line segments joining the midpoints of the opposite sides of the quadrilateral bisect each other.

test for practice

1. Write 6 facts that are true if △PQS ≅ △ABC.

2. For each diagram, what fact is required so that the pairs of triangles are congruent?

   (a) 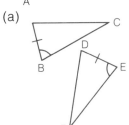   (b)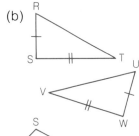

3. In the diagram, ∠S = ∠Q and SP = QP. Prove that QR = SR.

4. Use the diagram to prove that PT = QT.

   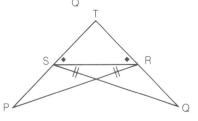

5. Use the diagram. If PQRS is a parallelogram, prove that △WSR ≅ △VQP.

   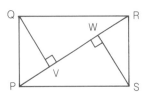

6. In the diagrams, which lines may be proven parallel?

   (a) 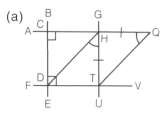   (b)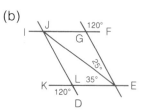

7. △PQR is given by vertices P(2, 4), Q(2, 1), and R(5, 4). On the same set of axes, show a rotation of △PQR of 90° clockwise about the turn centre, T(3, 5).

8. Quadrilateral ABCD is inscribed in a circle with centre O such that BD is a diameter, ∠ABD = 60° and ∠BDC = 25°. Calculate the measure of

   (a) ∠BAC   (b) ∠DCA   (c) ∠BCA   (d) ∠DBC

   Prove the following deductions.

9. △ABC is isosceles with ∠B = ∠C. BC is produced to N and CB is produced to M, so that MB = NC. Prove that △AMN is isosceles.

10. Circles with centres A and B intersect at P and Q. Prove that the line joining the centres A and B is perpendicular to PQ.

11. In △ABC, D is a point on AC such that DC = DA = DB. Prove that ∠ABC is a right angle.

12. A line passes through the points A and B. If A′ and B′ are the translation images of A and B, prove that AB∥A′B′.

13. Use transformations to prove that PQ = SR. O is the centre of the circle.

14. Prove that a reflection is an isometry.

15. Prove, using co-ordinate geometry, that in any rectangle, the midpoints of the sides are the vertices of a rhombus.

looking back: a cumulative review
Chapters 5 to 8

1. What are the number(s) such that when they are squared and then added to each term of the ratio 4:21, the resulting ratio is equivalent to 4:5?

2. Solve for m and n.
$$\frac{2m}{2n+1} = \frac{3-2m}{4n+2} = \frac{m+1}{2n+3}$$

3. A direct variation is given by $A = kW$. k is constant. Find A_1 when $W_1 = 42$, $A_2 = 320$, $W_2 = 8$.

4. Find the missing value for each inverse variation.

	s_1	t_1	s_2	t_2
(a)	25	16	?	100
(b)	8	?	4	16
(c)	5	100	4	?

5. The surface area of a metal ball varies directly with the square of its radius. A ball with radius 5 cm has surface area 314 cm². What is the surface area of a ball with radius 7 cm?

6. The slopes of pairs of lines are given. Find the value of k.
   (a) $\ell_1 \parallel \ell_2$ $m_1 = \frac{4-k}{8}$, $m_2 = \frac{3}{5}$
   (b) $\ell_1 \perp \ell_2$ $m_1 = \frac{k-5}{6}$, $m_2 = -\frac{9}{4}$

7. The sides of a quadrilateral are given by the equations $y = 4x - 20$, $x + 4y - 22 = 0$, $4x - y + 14 = 0$, and $x + 4y = 5$. Which polygon best describes the quadrilateral: square, rhombus, parallelogram or rectangle?

9. Calculate the distance from the point of intersection of the lines $2x + 3y - 31 = 0$ and $3x - y - 8 = 0$ to
   (a) the x-axis   (b) the y-axis   (c) the origin

10. For each line, find the equation of the image line for the given translation.
    (a) $2x + y = 8$ $(x, y) \longrightarrow (x - 2, y - 4)$
    (b) $3(x - 5) = y$ $(x, y) \longrightarrow (x - 4, y + 5)$

11. The number of hours of practice by various teams are listed below.

    | Barons | 14 | Rams | 9.1 |
    | Raiders | 9.5 | Warriors | 8.5 |
    | Braves | 9.1 | Hawks | 7.8 |
    | Spartans | 8.3 | Bulls | 8.3 |

    (a) Find the mode. Why is the mode not a good measure of central tendency for the above data?
    (b) Which team is not represented well by the mean?
    (c) What is the median?
    (d) What is more representative of the centre of the data: the mean, the median, or the mode?

12. For each of the following, describe the method or type of sampling you might use when obtaining a sample.
    (a) testing the quality of grapefruit
    (b) deciding how many people would purchase from a discount store
    (c) deciding how many people would be interested in installing a swimming pool
    (d) obtaining an opinion about recent car repair prices
    (e) determining the most popular TV star
    (f) checking the quality of tires for a car

10 geometry: more concepts and skills

A study of area, similar figures, dilatations and vector methods of proof.

10.1 the nature of mathematics

To solve a problem in mathematics, we may often require a combination of skills we have learned. As we learn different strategies and skills to solve problems, we improve our ability to solve a problem we have never met before. However, basic to solving any problem is understanding clearly the answers to these two questions.

A: What information are we given in the problem?
B: What information are we asked to find?

In this chapter we integrate the theorems and properties we learned about figures and apply them to learn about the areas of figures.

In mathematics, we may prove a deduction in more than one way, depending on which strategy is more easily applied. To prove a deduction, we have used the methods and theorems of
- plane geometry
- transformational geometry
- analytic geometry

In this chapter, we learn another strategy for proving deductions, based on vector methods.

The concepts and skills we learn in one topic of mathematics often parallel the concepts and skills we have learned in another topic of mathematics.

For example,
- In the previous chapter we explored the properties of *congruent triangles* and then investigated and used the properties of *congruent transformations* to prove deductions.
- In this chapter, we explore the properties of *similar triangles* and then use these properties to investigate *similarity transformations*, called dilatations.

10.2 area: parallelograms and triangles

Previously we used formulas to calculate the areas of various figures. In this section, we will learn skills to develop formulas, and to prove deductions involving area.

If two triangles are congruent, then their areas are equal. We write
$$\triangle ABC \cong \triangle DEF \longrightarrow \triangle ABC = \triangle DEF$$

We will use this symbol to mean that $\triangle ABC$ and $\triangle DEF$ have equal areas.

But the converse is not true. It is possible to have triangles of various shapes that are equal in area but are not congruent. As an aid in proving theorems involving area we use the following assumption.

The area of a rectangle with base, b, and height, h, is given by
$$A = bh$$

To develop a formula for the area of a parallelogram, we use the formula for the area of a rectangle and the theorems of plane geometry. Be sure there is a reason for each step of the proof in Example 1.

Example 1
Prove that the area, A, of a parallelogram with base, b, and height, h, is given by $A = bh$.

Solution
Given: ||gm PQRS, base, b, height, h
Required to prove: ||gm PQRS = bh

We use this symbol to indicate "the area of ||gm PQRS is bh."

Proof: Construct the diagram as shown.

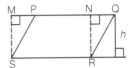

	Reasons
SM ⊥ QP produced	
RN ⊥ PQ	
Since ∠SMP = 90°	Why?
∠RNP = 90°	Why?
∠SMP + ∠RNP = 180°	Why?
and MS∥NR	Why?
Thus MNRS is a rectangle.	Why?
In △MPS and △NQR	
MS = NR	
∠SMP = ∠RNQ	Why?
SP = RQ	Why?
△MPS ≅ △NQR	Why?
△MPS = △NQR	Why?
In the diagram	
||gm PQRS	
= quad SPNR + △NQR	Why?
= quad SPNR + △MPS	Why?
= rect SMNR = bh	
Then ||gm PQRS = bh.	

From the above theorem, we can obtain a number of important corollaries. For example:

Corollary: The area, A, of a triangle with base, b, and height, h, is given by $A = \frac{1}{2}bh$.

For any triangle, we may construct a corresponding parallelogram.

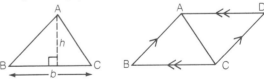

In ||gm ABCD we have proved that when a diagonal AC is drawn △ABC ≅ △CDA. Thus △ABC = △CDA, since
△ABC + △CDA = ||gm ABCD. Thus
$$\triangle ABC = \frac{1}{2} \|gm\ ABCD = \frac{1}{2}bh$$

The importance of the above result is that any polygon can be shown to be constructed of triangles only. Thus, we may use the above result to calculate the area of any polygon.

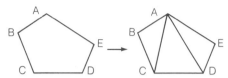

In the following exercise we will prove other important results involving the areas of various figures.

10.2 exercise

Throughout the exercise, round answers to 1 decimal place, as needed.

A
1 Find the area of each of the following.

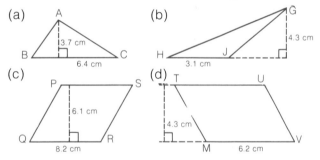

2 Calculate the missing information.

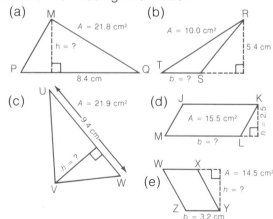

B
3 Find the area of each figure.

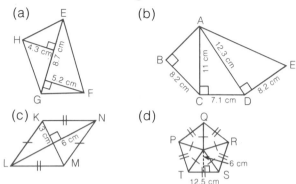

4 Find the area of the shaded region for each of the following (measures in centimetres).

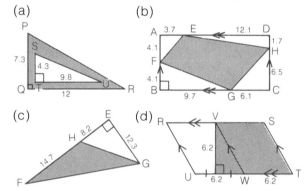

5 In △ABC, AD is the altitude to BC and BE is the altitude to AC. If AD = 5 cm, BC = 8 cm, and BE = 6 cm, find the measure of AC.

6 In △PQR, PS is the altitude to QR. If PS = 3.4 cm, QR = 7.2 cm, and PR = 8.4 cm, find the length of QT, the altitude to PR.

7 PQRS is a rhombus, and QT ⊥ SR where T is a point in SR. If the area of the rhombus is 98 cm² and QT = 7 cm, then find the length of PS.

8 PQRS is a parallelogram, and A and B are points in SR and RQ respectively so that PB ⊥ RQ and SR ⊥ PA. If PA = 8 and SR = 12, find RQ if PB = 4.

9. Use the following facts.
   - The opposite sides of a parallelogram are equal.
   - The distance between parallel lines is constant.

   Prove that $\triangle ABC = \frac{1}{2}bh$.

   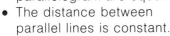

10. A trapezoid is a quadrilateral with a pair of parallel sides, a and b, and height, h. Prove that the area, A, of a trapezoid is given by $A = \frac{1}{2}(a + b)h$.

    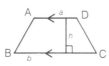

11. Use the formula for the area of a trapezoid to calculate the area of each polygon.

    (a) (b) (c) (d)

    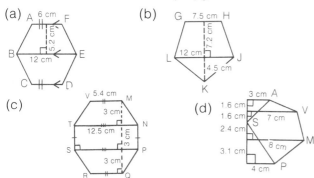

12. Express the area of $\triangle ABC$ in terms of a.

    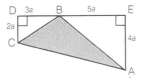

problem/matics

A problem contains important clue-words which imply information required to solve the problem. What is the important clue word that occurs in this problem?

$\triangle ABC$ has vertices $A(-4,4)$, $B(0,-4)$ and $C(x,y)$. The area of $\triangle ABC$ is 20 square units. If C is equidistant from the other vertices, what are the co-ordinates of C?

Now solve the problem.

10.3 theorems about area

As we prove more theorems about figures on the plane, we continue to build our deductive system and our understanding of the principles of plane geometry.

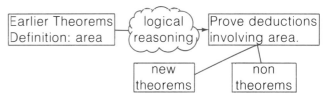

Example 1

If two parallelograms are on the same base and between the same parallel lines, prove that their areas are equal.

Solution

Given: ||gm ABCD and ||gm EBCF, AF||BC.

Required to prove: ||gm ABCD = ||gm EBCF

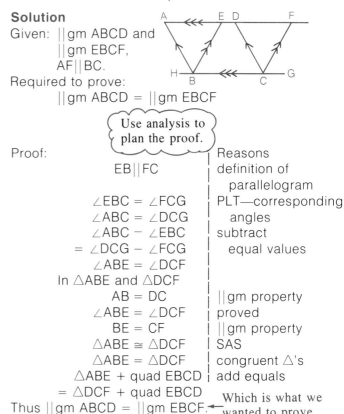

Use analysis to plan the proof.

Proof: | Reasons
EB||FC | definition of parallelogram
$\angle EBC = \angle FCG$ | PLT—corresponding
$\angle ABC = \angle DCG$ | angles
$\angle ABC - \angle EBC$ | subtract
$= \angle DCG - \angle FCG$ | equal values
$\angle ABE = \angle DCF$ |
In $\triangle ABE$ and $\triangle DCF$
AB = DC | ||gm property
$\angle ABE = \angle DCF$ | proved
BE = CF | ||gm property
$\triangle ABE \cong \triangle DCF$ | SAS
$\triangle ABE = \triangle DCF$ | congruent $\triangle$'s
$\triangle ABE$ + quad EBCD | add equals
$= \triangle DCF$ + quad EBCD
Thus ||gm ABCD = ||gm EBCF. Which is what we wanted to prove.

We may use a different strategy, based on our knowledge of transformations, to prove the above theorem. For the above diagram, we can show that a transformation gives the correspondence [A, B, E] ⟶ [D, C, F]. Thus △ABE ≅ △DCF since a translation is an isometry.

We refer to the result in Example 1 as the *Area of a Parallelogram Theorem (APT)*. If two parallelograms are on the same base and between the same parallel lines then the parallelograms are equal in area.

An important *corollary* that is included with the above theorem is

Corollary: If two parallelograms are on bases of equal length and are between the same parallel lines then the parallelograms are equal in area.

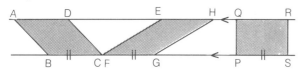

||gm ABCD = ||gm EFGH = ||gm PQRS

Example 2

Prove the following theorem:

Area of a Triangle Theorem (ATT): If two triangles are on the same base and between the same parallel lines, then the triangles are equal in area.

Give reasons for each step of the proof.

Solution
Given: △VBA and △SBA, on same base, BA.
VS||BA
Required to prove:
△VBA = △SBA

Proof:
Construct TB||SA,
QA||VB

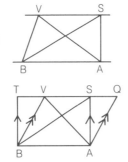

	Reasons
In ||gm VBAQ	
△VBA = $\frac{1}{2}$||gm VBAQ ①	Why
In ||gm TBAS	
△SBA = $\frac{1}{2}$||gm TBAS ②	Why?
But ||gm TBAS and ||gm VBAQ are on the same base and between same parallels. Thus	Why?
||gm VBAQ = ||gm TBAS	Why?
Thus	
$\frac{1}{2}$||gm VBAQ = $\frac{1}{2}$||gm TBAS	Why?
△VBA = △SBA	

For the theorem in Example 2, we may prove the following corollaries.

1) If two triangles are on different bases that are equal and between the same parallel lines, the areas of the triangles are equal.

2) A median bisects the area of a triangle.

3) If a triangle and parallelogram are on different bases of equal length and lie between the same parallel lines then the area of the triangle is one-half of the area of the parallelogram.

The theorem in Example 2 and its corollaries are then used in Example 3 to prove a deduction.

Example 3
In ||gm PQRS,
PM = MS.
Prove △PQT = △MRT.

Solution
Given: ||gm PQRS, PM = MS
Required to prove: △PQT = △MRT

Proof: In △PTS, TM is the median.	Reasons
△TPM = △TMS ①	Area of a Triangle Theorem (ATT)
△PMQ and △MSR are on different bases of equal length,	

PM and MS, and between
the same parallel lines.
△PMQ = △MSR   ② Corollary of ATT
Add ① and ②.
△PMQ + △TPM   equal sum
= △MSR + △TMS
△PQT = △MRT ← Which is what we wanted
                to prove.

10.3 exercise

For the following questions, use the theorems and corollaries dealt with in this section to prove deductions.

Use analysis to plan the proof.

Which figures have equal areas? Give reasons for your answer.

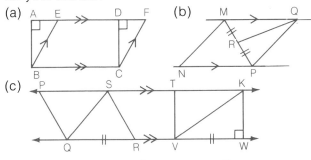

For each of the following, give reasons for the answer.

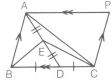

(a) △ABD = △ADC   (b) △ABE = △BDE
(c) △ACE = △ECD   (d) △ABE = $\frac{1}{2}$ △ABD
(e) △ABC = △APC   (f) △ACE = $\frac{1}{4}$ △ABC
(g) △ABC = $\frac{1}{2}$ ||gm APCB (h) △ABE = △DEC
(i) △ABE + △EDC = $\frac{1}{2}$ △ABC

B

3 Use the diagram to prove the theorem:
   If two parallelograms are on different bases of equal length, and are between the same parallel lines, then the areas of the parallelograms are equal.

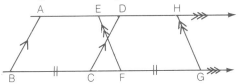

4 Use the diagram to prove the theorem:
   If two triangles are on different bases of equal length, and are between the same parallel lines, then the areas of the triangles are equal.

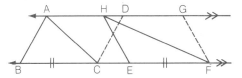

5 Use the diagram to prove that the median of a triangle bisects the area of a triangle.

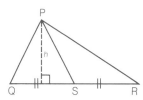

6 Prove that the diagonal of any parallelogram bisects its area.

7 In △PQR, PS is a median to QR. Prove that △PQT = △PRT.

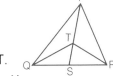

8 If MB = BA and NA = RA prove that △BMR = $\frac{1}{4}$ △MNR.

9 A median PS to QR is produced to T so that PS = ST. Prove that △PQT = △PQR.

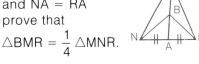

10 If PS ∥ TR
    prove that
    △QST = △PQR.

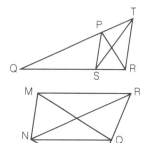

11 For the diagram,
    MR ∥ NQ.
    Prove that
    △RNP = quad MNPQ.

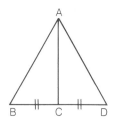

12 Use the result in Question 11 to construct a triangle equal in area to a square.

13 Draw any quadrilateral. Construct a triangle equal in area to the quadrilateral.

14 S and T are midpoints of MN and MR respectively in △MNR. Prove that
    (a) △SNR = △TNR   (b) △SNT = △SRT

15 S and T are midpoints of PQ and PR in △PQR. QT and RS intersect at V. Prove that
    (a) △SVQ = △TVR   (b) quad PSVT = △VQR

16 The diagonals AC and BD of ∥gm ABCD intersect at E. Prove that
    △ABE + △BEC = △CED + △DEA.

17 The midpoints of the sides of ∥gm PQRS are T, U, V, and W in order.
    Prove that ∥gm TUVW = $\frac{1}{2}$ ∥gm PQRS.

problem/matics

To solve some problems about geometric figures, we need to use our algebraic skills. Solve this problem.

The bisectors of the exterior angles at B and C of △ABC meet at D. Prove that
$$\angle BDC = 90° - \frac{A}{2}$$

ratios of triangular areas

For the results of earlier theorems we may write ratios of their areas. Why are the following true?

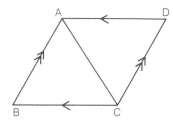

$\dfrac{\triangle ABC}{\triangle ACD} = 1$ $\dfrac{\triangle ABC}{\parallel\text{gm ABCD}} = \dfrac{1}{2}$

We may use the theorems about area to deduce other theorems about the ratios of areas. For the following theorem, give reasons for each step of the proof.

Theorem: If two triangles have equal altitudes, then the areas of the triangles are proportional to their bases.

Given:
△ABC and △DEF,
AG = DH = h,
BC = a, EF = b.

Required to prove:
$\dfrac{\triangle ABC}{\triangle DEF} = \dfrac{a}{b}$

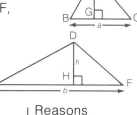

Proof: | Reasons

△ABC = $\frac{1}{2}ah$ ① | Why?

△DEF = $\frac{1}{2}bh$ ② | Why?

From ① and ②

$\dfrac{\triangle ABC}{\triangle DEF} = \dfrac{\frac{1}{2}ah}{\frac{1}{2}bh}$ | Why?

$\dfrac{\triangle ABC}{\triangle DEF} = \dfrac{a}{b}$ ← Which is what we wanted to prove.

8 Prove the following corollary.
   If the bases of two triangles are equal, prove that the ratio of their areas is equal to the ratio of their corresponding altitudes.

Use the results of the theorem, which we refer to as the *Proportional Areas of Triangles Theorem* (PATT) in the following work.

9 Express as a ratio
   (a) △PQS:△PSR.
   (b) △PQS:△PQR.
   (c) △PSR:△PQR.

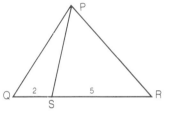

10 Find the ratios for each of the following.
   (a) △PQS:△QSR
   (b) △PQT:△PTS
   (c) △PQS:||gm PQRS
   (d) △PST:||gm PQRS
   (e) △PQT:||gm PQRS

   What are the ratios for
   (a) △SON:△QST?
   (b) quad SNOM:△STQ?

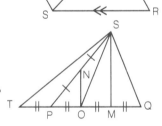

Prove that △PST:||gm PQRS = 1:7.

Prove that, if PQ = SQ, then △ZSQ:△OZS = 1:4.

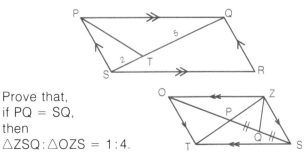

24 Prove that $\dfrac{\text{quad DENG}}{||\text{gm DEFG}} = \dfrac{4}{7}$.

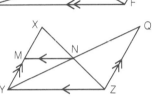

25 Prove that $\dfrac{XY}{MX} = \dfrac{QY}{NY}$.

26 In quad XYZW, △XYW = △YZW. Show that YW bisects XZ.

27 The diagonals PR and QS of quadrilateral PQRS intersect at T. Prove △PTS:△PTQ = △STR:△QTR.

28 If in any quad XYZW, XZ and YW intersect at U show that $\dfrac{\triangle XUW}{\triangle XYW} = \dfrac{\triangle ZUW}{\triangle ZYW}$.

29 A trapezoid ABCD is given with AD||BC. If AC and BD intersect at X show that △AXB = △CXD.

C

30 For △ABC, P is the intersection of the medians. Prove that $\dfrac{\triangle ABP}{\triangle ABC} = \dfrac{1}{3}$.

31 In △PQR, S divides QR internally in the ratio QS:SR = 1:3 and M divides PS internally in the ratio PT:TS = 2:1. Prove that $\dfrac{\triangle QTS}{\triangle PQR} = \dfrac{1}{12}$.

problem/matics

Sometimes, at first glance, it appears that a problem does not have enough information to solve it. Often this is not so. Solve the following problem.

On a circular ride, the seats shown are 2 m apart. If the person in Seat A goes twice as fast as the person in Seat B, how much further does the person in Seat A go in one loop of the ride?

10.4 parallel proportion triangle theorem

Earlier we proved:

If RS ∥ BC
and AR = RB
then AS = SC.

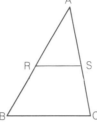

Thus the ratio $\frac{AR}{RB} = \frac{AS}{SC}$ suggests that a more general result might be obtained if R is not the midpoint.

Example 1
If a line intersects two sides of a triangle and is parallel to the third side, prove that the line divides the two sides in the same ratio.

Solution
Given:
△PQR with ST ∥ QR.
Required to prove:
$\frac{PS}{SQ} = \frac{PT}{TR}$

Proof:	Reasons
Join TQ and SR. Since △SQT and △TRS are on the same base ST and between the same parallel lines, then △SQT = △TRS.	Why?
Thus $\frac{\triangle SQT}{\triangle PST} = \frac{\triangle TRS}{\triangle PST}$ ①	Why?
Since △SQT and △PST have the same altitude, then their areas are proportional to their bases. $\frac{\triangle SQT}{\triangle PST} = \frac{SQ}{PS}$ ②	Why?
Similarly, $\frac{\triangle TRS}{\triangle PST} = \frac{TR}{PT}$. ③	Why?
Thus from ② and ③ $\frac{SQ}{PS} = \frac{TR}{PT}$ or $\frac{PS}{SQ} = \frac{PT}{TR}$	Which is what we wanted to prove.

We refer to the above theorem as the *Parallel Proportion Triangle Theorem* (PPTT).

A method of mathematics is, after having proved a theorem, to investigate whether the converse is true, as shown in Example 2. We use the method of indirect proof.

Example 2
If a line divides two sides of a triangle into the same ratio, then the line is parallel to the third side.

Solution
Given:
△PQR and MN such that $\frac{PM}{MQ} = \frac{PN}{NR}$.
Required to prove:
MN ∥ QR

Proof:	Reasons
Either MN ∥ QR or MN ∦ QR. Assume MN ∦ QR. Then construct MS ∥ QR.	
Then $\frac{PM}{MQ} = \frac{PS}{SR}$.	Why?
But $\frac{PM}{MQ} = \frac{PN}{NR}$.	Why?
Then $\frac{PS}{SR} = \frac{PN}{NR}$.	
This means that two different points divide PR in the same ratio. This is a contradiction. Thus MN ∥ QR.	

We may combine the results of Examples 1 and 2.

Parallel Proportion Triangle Theorem (PPTT)
A line divides two sides of a triangle in the same ratio if and only if the line is parallel to the third side.

Example 3
In trapezoid PQRS, a line drawn parallel to SR cuts PS and QR at M and N respectively. Prove that PM:MS = QN:NR.

Solution   Translate the given information on a diagram.

Given: Trapezoid PQRS, PQ||SR, MN||SR
Required to prove:
  PM:MS = QN:NR
Proof: | Reasons
Join QS to cut MN at T.
In △QSR, TN||SR.
Then $\frac{QN}{NR} = \frac{QT}{TS}$.   ① | PPTT
In △PQS, MT||PQ.
Then $\frac{PM}{MS} = \frac{QT}{TS}$.   ② | PPTT
From ① and ②
  $\frac{QN}{NR} = \frac{PM}{MS}$ ← Which is what we wanted to prove.

10.4 exercise

A

1 In algebra if $\frac{a}{b} = \frac{c}{d}$ then $\frac{a+b}{b} = \frac{c+d}{d}$.
Use this result to prove that in △PQR
(a) $\frac{PQ}{SQ} = \frac{PR}{TR}$ (b) $\frac{PQ}{PS} = \frac{PR}{PT}$

2 Use the diagram to prove that AD:AB = CE:CB.

3 Find k for each of the following.
(a) 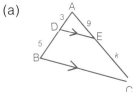 (b)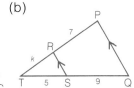

(c) 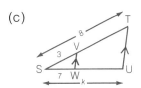 (d)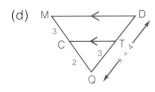

4 Which of the following lines are parallel?
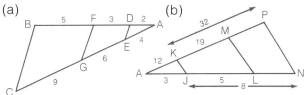

B proving deductions

5 For the diagram, prove that ZA:AX = YB:BX.

6 Prove that DA||YZ.

7 In △PQR, ST||UV||QR. Prove $\frac{PS}{UQ} = \frac{PT}{VR}$.

8 Prove that, if PQ||RS, then PT:ST = QT:RT.

9 In △MNP, AB||NC, AN||BC. Prove that MA:AN = NC:CP.

10 If MP||BA and NP||CA prove that BC||MN.

11 Trapezoid MNPQ has MN||QP. AB is drawn parallel to MN where A is the midpoint of MQ and B is a point on NP. Prove NB = BP.

12 Prove that the line joining the midpoints of the non-parallel sides of a trapezoid is parallel to the parallel sides.

13 D is the midpoint of PS and C is the midpoint of QR in ||gm PQRS. If DQ and SC intersect PR at E and F respectively, prove
    (a) DQ||SC   (b) PE = EF = FR

14 In △XYZ, A and B are in XY and XZ respectively so that AB||YZ. XY is produced to T so that BY||ZT. Prove that XA:XY = XY:XT.

15 N is any point in QR of △PQR. NM||QP with M on PR, NS||RP with S on QP. Prove
    $$\frac{\triangle SQN}{\triangle SPM} = \frac{\triangle SPM}{\triangle MNR}.$$

16 P, Q, R, and S are the midpoints of AB, BC, CD and AD respectively in quadrilateral ABCD. Prove quadrilateral PQRS is a parallelogram.

17 In △PQR, ∠Q = 90° and S is the midpoint of RP. Prove that SQ = PS = RS.

C

18 In trapezoid PQRS, PQ||SR, N is in PS and M is in QR such that $\frac{PN}{NS} = \frac{QM}{MR}$. Use the indirect method of proof to show NM||PQ.

19 P is any point inside quadrilateral VWXY. Join P to each vertex of the quadrilateral. N is any point on PY. M is on PX such that NM||YX. R is on PW such that MR||WX. S is on PV such that RS||WV. Prove SN||VY.

20 In quadrilateral PQRS, M is any point in PS. E is in PM and G is in MS such that PE:EM = SG:GM. N is in MQ such that EN||PQ. T is in MR such that GT||SR. Prove NT||QR.

10.5 internal and external division

When a line is drawn parallel to a side of a triangle, we see that $\frac{PR}{RQ} = \frac{PT}{TS}$.

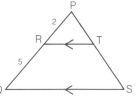

We say that R divides PQ internally, such that PR:RQ = 2:5.

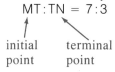

In the following, MN is divided externally if the point of division, T, is not between M and N but on the line containing MN as shown. MN is divided externally at T in the ratio
MT:TN = 7:3

initial point   terminal point

T is external to MN. T is the point of division.

The ratio determines on what side of the segment the point of division, T, is placed.

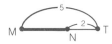 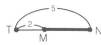

T divides MN externally in the ratio MT:TN = 5:2 (MN is produced to T).

T divides MN externally in the ratio MT:TN = 2:5 (NM is produced to T).

We can apply our skills with ratios in the following example.

Example 1

The point P divides ST internally in the ratio 5:2. If the length of ST is 35 m, find the position of P.

Solution Draw a diagram to translate the problem.

Let $\dfrac{SP}{PT} = \dfrac{m}{n}$.

Then $m = 5k$, $n = 2k$,
for k constant.
$5k + 2k = 35$
$7k = 35$
$k = 5$
Thus $m = 25$, $n = 10$.
Thus P is 25 m from S (or P is 10 m from T).

To divide a given line into a certain ratio we may use the *Parallel Proportion Triangle Theorem* proved in the previous section and applied in the next example.

Example 2
Divide the given line SQ internally at B in the ratio 3:2.

Solution
1 Draw SP for an acute angle.

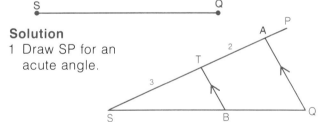

2 On SP mark off two segments ST and TA so that ST = 3 cm, TA = 2 cm (i.e. ST:TA = 3:2). Join AQ.

3 Draw TB∥AQ as shown. Then
ST:TA = SB:BQ. ← By the parallel proportion triangle theorem (PPTT)
But ST:TA = 3:2.
Thus SB:BQ = 3:2.
The required point B divides SQ internally in the ratio 3:2.

You may use a similar procedure to find the point of division when a line segment is divided externally in a ratio.

10.5 exercise

A

1 Indicate whether the following ratios describe an internal or external division of PQ.

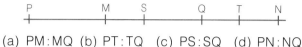

(a) PM:MQ (b) PT:TQ (c) PS:SQ (d) PN:NQ

2 Write the ratio for each of the following.

(a) MN:NP (b) MN:NQ (c) MP:PN (d) MQ:QP
(e) QP:PN (f) PQ:QN (g) NM:MQ (h) NP:PM
(i) NQ:QP (j) MQ:QN (k) QP:PM (l) MP:PQ

3 PQ is divided externally at R in the ratio of 5:3. Find the ratios for

(a) PR:RQ (b) PQ:QR (c) RQ:QP (d) RP:PQ

4 If RT is 25 cm long and S is a point on RT, find the length of RS and ST where RS:ST = 7:3.

5 If PQ is 28 cm long and N is a point on PQ, find the length of PN and NQ where PN:NQ = 9:2.

B

6 Point D divides the line EF internally in the ratio of 8:3. The length of EF is 46 cm. What are the lengths of ED and DF?

7 If a line AB, 28 m in length, is divided internally at point C in a ratio of 6:4, what are the lengths of AC and CB?

8 The length of a line DF is 16 cm. If E divides DF externally in the ratio 5:3, what are the lengths of DE and EF?

9 Point Z divides the line OP externally in the ratio 6:2. If OP is 44 cm in length, what are the lengths of OZ and ZP?

10 (a) PQ is 12 cm long. Divide PQ internally at R in the ratio of 5:4.
    (b) AB is 15 cm long. Divide AB internally at C in the ratio 2:5.
    (c) MN is 12 cm long. Divide MN externally at P in the ratio 1:3.
    (d) ST is 15 cm long. Divide ST externally at V in the ratio 7:3.

11 Draw any line segment PQ. Find a point that divides the line segment
    (a) internally in the ratio 3:2.
    (b) externally in the ratio 5:2.
    (c) externally in the ratio 2:5.

12 Draw any line segment. Find its points of trisection.

13 Base QR of △PQR is divided internally at N in the ratio 4:1 and externally at M in the ratio 6:1. Write the ratio for
    (a) $\dfrac{\triangle PQN}{\triangle PNM}$ (b) $\dfrac{\triangle PQR}{\triangle PQM}$ (c) $\dfrac{\triangle PRN}{\triangle PRM}$ (d) $\dfrac{\triangle PQM}{\triangle PQN}$

14 △PQR has perimeter 20 cm. If the ratio of 3 sides is PQ:QR:RP = 5:3:7, construct the triangle.

15 In △PQR, base QR is divided internally at S in the ratio 5:2. Write the ratio for
    (a) $\dfrac{\triangle PQS}{\triangle PSR}$ (b) $\dfrac{\triangle PQS}{\triangle PQR}$

16 In △MNP, NP is divided internally at Q in the ratio 3:7. Write the ratio for
    (a) $\dfrac{\triangle MQN}{\triangle MNP}$ (b) $\dfrac{\triangle MQN}{\triangle MPQ}$ (c) $\dfrac{\triangle MPQ}{\triangle MNP}$

17 YZ of △XYZ is divided externally at T in the ratio YT:TZ = 7:3. Write the ratios for
    (a) $\dfrac{\triangle XYZ}{\triangle XYT}$ (b) $\dfrac{\triangle XYZ}{\triangle XZT}$ (c) $\dfrac{\triangle XZT}{\triangle XYT}$

C
18 Use the method of indirect proof to prove that there is one and only one point that divides a line segment internally in a given ratio.

19 In △DEF, EF is divided internally at G in the ratio 4:3. DG is divided internally at H in the ratio 3:1. Write the ratio for
    (a) △DEG:△DEF   (b) △DEG:△DEH
    (c) △DHF:△HGF   (d) △HEG:△DHF

working with co-ordinates

20 The line segment PQ given by P(−4, −3) and Q(8, 5) is divided internally in the ratio 2:1 by M. Find the co-ordinates of M.

21 PQ is divided externally by R in the ratio 3:2. If the co-ordinates are P(5, 4), Q(−7, 0), find the co-ordinates of R.

22 If RS given by R(−12, −2), S(9, 10) is divided internally in the ratio 3:2 by the point T, what are the co-ordinates of T?

23 ST is divided externally by R in the ratio 2:3. If the co-ordinates are S(3, −1), T(−1, 2), find the co-ordinates of R.

24 Into what ratio does the point (3, 5) divide AB internally, given the co-ordinates A(−6, −7) and B(9, 13)?

25 The vertices of a parallelogram are given by A(−1, −3), B(1, 5), C(5, 6) and D(3, −2).
    (a) If M is the midpoint of CD, find its co-ordinates.
    (b) P divides BD internally in the ratio 2:1. Find the co-ordinates of P.
    (c) Prove that M is on the line AP produced.

10.6 similar triangles AAA~

In Chapter 9 we used the congruence of triangles to develop theorems for the deductive system. In this section the mathematical meaning of similar figures is used to again develop theorems involving triangles.

Two triangles are similar if corresponding angles are equal and corresponding sides are proportional.

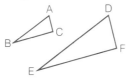

If $\angle A = \angle D$, $\angle B = \angle E$, $\angle C = \angle F$, and $\dfrac{AB}{DE} = \dfrac{AC}{DF} = \dfrac{BC}{EF}$, then $\triangle ABC \sim \triangle DEF$. (is similar to)

Example 1

In the figure, $\triangle ABC \sim \triangle ADE$. Find the value of x.

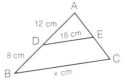

Solution

Since $\triangle ABC \sim \triangle ADE$ then $\dfrac{AB}{AD} = \dfrac{BC}{DE} = \dfrac{AC}{AE}$.

$\dfrac{20}{12} = \dfrac{x}{18}$ or $x = \dfrac{18 \times 20}{12} = 30$

In studying congruence we found that certain minimum conditions are sufficient to show that two triangles are congruent. A comparable result is proved for similar triangles in Example 2.

Example 2

If the 3 angles of a triangle are equal to 3 corresponding angles of another triangle, then the triangles are similar.

Solution
Given:
$\triangle PQR$ and $\triangle ABC$
$\angle P = \angle A$, $\angle Q = \angle B$
$\angle R = \angle C$

Required to prove:
$\triangle PQR \sim \triangle ABC$

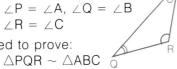

Use analysis to plan the proof.

Proof:

Construct $\triangle PMN$ as shown so that $PM = AB$, $PN = AC$.

	Reasons
$\angle MPN = \angle BAC$	given
Thus $\triangle PMN \cong \triangle ABC$	SAS
and $\angle PMN = \angle ABC$ ①	congruent $\triangle$'s
But $\angle PQR = \angle ABC$ ②	given
Thus $\angle PQR = \angle PMN$	from ① and ②

Since $\angle PMN$ and $\angle PQR$ are corresponding angles for MN and QR with transversal PQ, then $MN \parallel QR$. | proved
| PLT ← Parallel lines theorem

In $\triangle PQR$, M and N are points on sides PQ and PR respectively such that $MN \parallel QR$. Thus
$\dfrac{PM}{PQ} = \dfrac{PN}{PR}$ | proved
| PPTT ← Parallel proportion triangle theorem

Since $PM = AB$, $PN = AC$ then $\dfrac{AB}{PQ} = \dfrac{AC}{PR}$. ③

In a similar way, we may use the construction shown to prove that
$\dfrac{AC}{PR} = \dfrac{BC}{QR}$ ④

From ③ and ④
$\dfrac{AB}{PQ} = \dfrac{BC}{QR} = \dfrac{AC}{PR}$

Thus $\triangle PQR \sim \triangle ABC$.

Since the sum of the angles of a triangle is 180°, then only two corresponding angles of the two triangles need be equal to apply the above result. We may use the following.

Corollary: If two angles of a triangle are equal to two corresponding angles of another triangle, then the triangles are similar.

The above results are combined as the *Angle-angle-angle similarity theorem* (AAA~)

Example 3

In △PQR, PQ ⊥ RQ and U is a point in RP and V is a point in RQ so that UV ⊥ RP. Prove that UR:QR = UV:QP.

Solution

Given:
  △PQR, UV ⊥ RP,
  PQ ⊥ QR
Required to prove:
  UR:QR = UV:QP

Proof:

	Reasons
∠Q = 90°	PQ ⊥ QR
∠RUV = 90°	UV ⊥ RP
Then ∠RUV = ∠Q	equal angles
In △PQR and △VUR	
∠Q = ∠RUV	proved
∠R = ∠R	common angle
∠P = ∠UVR	ASTT
△PQR ~ △VUR	AAA~
Thus $\frac{UR}{QR} = \frac{UV}{QP}$	Property of similar triangles

Later we will prove other theorems that will further allow us to prove that triangles are similar.

10.6 exercise

A

1. Why are the following pairs of triangles similar? Write the corresponding ratios for the sides.

   (a)

   (b)

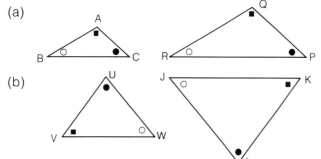

2. (a) Prove that △ABE ~ △DCE.
   (b) Prove that AB:DC = AE:DE.

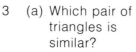

3. (a) Which pair of triangles is similar?
   (b) Write corresponding equal ratios.

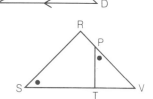

4. Find the value of the variables.
   (a) (b) (c)
   (d) (e) (f)

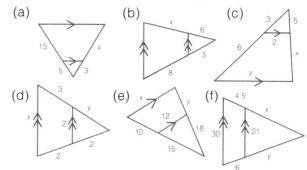

5. If DE∥BC, AE = 7, EC = 4, DE = 4, find BC.

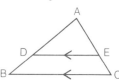

6. If ST∥VU, SW = 6, WV = 2.5, UW = 4, find TW.

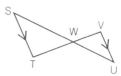

7. If PR = 8, SQ = 5, QR = 3.2, ∠PRQ = 90°, ∠PSR = 90°, find SR.

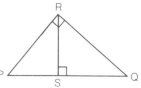

B

8. We are given △XYZ with ∠X = 90°. Point A is on XZ and point B is on YZ so that AB ⊥ YZ. Prove that XY × ZB = AB × XZ.

9. In △PQR, F, G, H are the midpoints of PQ, QR, and RP respectively. Prove that △PQR ~ △GHF.

10. In △PQR, QX ⊥ PR, PY ⊥ QR where Y and X are points on QR and PR respectively. If QX and PY intersect at Z, prove that △PZX ~ △PRY.

11. △MOT is right-angled at T. If Q is on TM and R on OM so that QR ⊥ OM prove TO:QR = TM:MR.

12. In △XYZ the points M and N are on XY and XZ respectively such that MN∥YZ. Prove
    (a) XM × ZN = YM × XN
    (b) XM × YZ = XY × MN

13. For two similar triangles, prove that the corresponding altitudes are proportional to corresponding sides.

14. A clubhouse has a roof whose slant height is 13 m and whose base is 24 m. Two extra supports 2 m and 3 m from the peak need to be constructed. Find the length of each support.

15. Gene is 1.7 m above the ground and stands 1.3 m from a puddle. The puddle is 20 m from a pole in which the light is reflected. The reflection property of a puddle gives ∠PDM = ∠BDF. With this information find PM, the height of the pole.

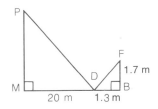

16. In ∥gm XYZW, A is a point on XW so that AZ bisects ∠YZW. YX and ZA are produced to meet at B. Prove $\frac{WA}{AZ} = \frac{YB}{BZ}$.

applications: similar triangles

There are many devices that have been constructed that are based on the properties of similar triangles. In the following, the bases of the similar triangles are angles. For example, the supports AD and CB open up so that the triangles formed are similar.

17. From the picture, prove that AB∥DC.

18. The instrument shown below is used to make an exact copy on a larger or smaller scale. The four rods are connected at A, B, C, D to form a parallelogram so that OD = DC and CB = BP. A tracing point is at P and a pencil at C. To obtain a larger copy, the tracing point is at P and a pencil at C and reversed for a smaller copy.

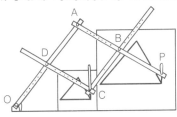

    (a) Prove that △DOC ~ △CBP.
    (b) What ratio gives the magnification of the drawing?

19. For light hitting a mirror, the angle of incidence equals the angle of reflection. The top of a street light pole is reflected into a puddle 25.0 m away. A person 5 m away on the other side of the puddle sees the reflection. If the person's eyes are 2.1 m above street level, how high is the light?

20 In a lumberjack contest two poles are secured for a pole-climbing event. The poles are of different heights. The wires from the top of the poles to the ground are parallel as shown in the diagram. If the smaller pole is 8 m with its wire secured 9.2 m from the base, and the guide wire to the taller pole is 16 m, find the height of the taller pole.

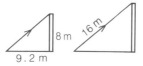

21 A plane climbs 30 m for every 100 m the plane travels forward. If the plane is travelling at 225 km/h, how high will the plane be after 12 s?

22 Two people attached to kites jump from two different heights. Their descent paths are parallel. If Ian's cliff was 28 m high and he lands 52.8 m from his cliff, find the height of the cliff Shelly jumped from if she landed 120 m from her cliff.

23 A ski tow rises 38 m for a horizontal distance of 100 m. If a skier is towed at the rate of 6 km/h, how far will the skier have risen after 9 s?

24 In 600 B.C. the Greek mathematician, Thales, used the following method to find the height of the Great Pyramid. The shadow of the top of the pyramid of Cheops, Son of Seneferu, was 280 m (to the centre of the pyramid). At the tip of the shadow a stick 2.2 m was placed having a shadow 4.2 m long. Calculate the height of the Great Pyramid to the nearest metre.

25 A radio antenna has two parallel wires attached to it at different points and secured to the ground. The first wire is attached at a height of 14.0 m from the ground and is anchored at 15 m from the base of the antenna. The second wire is anchored 25.7 m from the antenna. How high above the base of the antennae is the second wire attached?

applications: proving theorems

As we go along, each additional skill we learn in mathematics, can be used to derive more important facts. For example, we may use the AAA~ theorem to prove the Pythagorean Theorem from a different point of view. The following is an outline of the proof. Give reasons for each step.

$\triangle ABC$ is a right-angled triangle with $\angle C = 90°$. To prove that $a^2 + b^2 = c^2$, construct $CD \perp AB$ and use $AD = x$, $DB = y$.

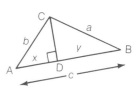

In $\triangle ACD$ and $\triangle ABC$
$\angle DAC = \angle A$ (?) $\angle DCA = \angle B$ (?)
Thus $\triangle ACD \sim \triangle ABC$ Similarly, $\triangle CBD \sim \triangle ABC$
$\dfrac{b}{x} = \dfrac{c}{b}$ or $b^2 = cx$ $\dfrac{a}{y} = \dfrac{c}{a}$ or $a^2 = cy$ (?)
Thus $a^2 + b^2 = cx + cy = c(x + y)$ (?)
$= c \times c = c^2$

Now prove each of the following important facts that we have used in our earlier work.

A *Theorem:* If two lines ℓ_1 and ℓ_2, have slopes m_1 and m_2 respectively and are parallel, then $m_1 = m_2$.
Converse: If two lines, ℓ_1 and ℓ_2 have slopes m_1 and m_2 respectively and $m_1 = m_2$, then $\ell_1 \| \ell_2$.

B: *Theorem:* If two lines ℓ_1 and ℓ_2, have slopes m_1 and m_2 respectively and are perpendicular, then $m_1 = -\dfrac{1}{m_2}$.
Converse: If two lines, ℓ_1 and ℓ_2 have slopes m_1 and m_2 such that $m_1 \times m_2 = -1$, then $\ell_1 \perp \ell_2$.

C *Theorem:* If $P_1(x_1, y_1)$ and $P_2(x_2, y_2)$ are any two points in the plane, prove that

(a) the co-ordinates of the midpoint of $\overline{P_1P_2}$
$\left(\dfrac{x_1 + x_2}{2}, \dfrac{y_1 + y_2}{2}\right)$.

(b) the co-ordinates of the point dividing $\overline{P_1P_2}$ in the ratio $a:b$ is
$\left(\dfrac{ax_2 + bx_1}{a + b}, \dfrac{ay_2 + by_1}{a + b}\right)$.

10.7 dilatations: similarity transformations

When a movie is shown on a screen, we obtain an enlarged shape that is similar to the original image.

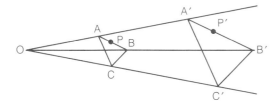

A dilatation is a transformation which changes the size of a figure but not its shape and is defined as follows: A **dilatation**, with dilatation factor k, is a transformation which maps any point P onto a point P', so that $OP' = |k|(OP)$ and O, P, and P' are collinear.

We call O the centre of dilatation and often refer to the constant k as the scale factor. The above correspondence [A, B, C] ⟶ [A', B', C'] is a dilatation with dilatation factor k. Thus
$$OA' = k(OA) \quad OB' = k(OB) \quad OC' = k(OC)$$
If $k = 2$ then $OA' = 2(OA)$, and so on.

We may use our skills with similar figures to prove properties of a dilatation.

Example 1
For a dilatation with factor k, $k > 0$, prove that $AC \| A'C'$, $AB \| A'B'$, $BC \| B'C'$.

Use above diagram.

Solution
Given: [A, B, C] ⟶ [A', B', C'], dilatation with factor k.
Required to prove: $AC \| A'C'$, $AB \| A'B'$, $BC \| B'C'$.

Proof: | Reasons
In the diagram | definition of
$OA' = k(OA) \quad OC' = k(OC)$ | dilatation
$\dfrac{OA'}{OA} = k \quad \dfrac{OC'}{OC} = k$ | equality
Thus $\dfrac{OA'}{OA} = \dfrac{OC'}{OC}$. |
In $\triangle OA'C'$, $\dfrac{OA'}{OA} = \dfrac{OC'}{OC}$. | proved
Thus $AC \| A'C'$. | PPTT
Similarly by using $\triangle OB'C'$ and $\triangle OA'B'$, we may prove $BC \| B'C'$ and $AB \| A'B'$.

Thus, for a dilatation, the image of any line is a parallel line. Other properties we may prove about dilatations are as follows:

- The image of a figure is a similar figure. In particular, the image of a triangle is a similar triangle.
- Ratios are preserved for a dilatation.
- Measures of angles are invariant.
- The image of a line segment AB is $A'B' = k(AB)$ for a dilatation with factor k.

Since the images for a dilatation with factor k are similar, we often refer to dilatations as **similarity transformations**. In the exercise that follows we will explore the properties of similarity transformations using our skills with mappings and analytic geometry.
  Note: If $k < 0$, then P and P' are on opposite sides of the dilatation centre O.

For $k < 0$, the corresponding lengths are related by a factor $|k|$.

10.7 exercise

A

1. For the diagram at the beginning of the section, prove that
   (a) △ABC ~ △A'B'C'   (b) BC||B'C'
   (c) ∠A'B'C' = ∠ABC   (d) A'B' = k(AB)

2. Copy each of the following onto a grid. Use the origin as the dilatation centre and find the image for each figure for the dilatation factor shown.
   (a) 2   (b) $\frac{1}{2}$ (c) −2   (d) $-\frac{1}{2}$

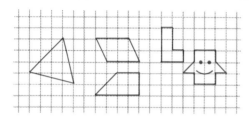

3. A dilatation, f, with factor k, k > 0 is given by the mapping f: P(x, y) ⟶ P'(x', y'). Prove that the co-ordinates of P' are given by (kx, ky).

4. A mapping, f, is defined by f:(x, y) ⟶ (3x, 3y). Prove that f is a similarity transformation with magnification factor 3.

5. For the mapping, g, defined by g:(x, y) ⟶ (−2x, −2y), prove that g is a similarity transformation and corresponding lengths are related by OP' = −2(OP).

B

6. A triangle has vertices A(2, 3), B(1, 5), C(3, 4). Find △A'B'C' for the dilatation given by
   (a) (x, y) ⟶ (3x, 3y).
   (b) (x, y) ⟶ (2x, 2y)   (c) (x, y) ⟶ (−x, −y)
   (d) Calculate the ratios of corresponding sides for △ABC and △A'B'C' for each dilatation above. What do you notice?

7. For the dilatation given by (x, y) ⟶ (4x, 4y), prove that for A(4, 2), B(−3, 1) and where A' and B' are the image points of A and B respectively,
   (a) AB||A'B'   (b) A'B' = 4(AB)

8. In △PQR, M and N are the midpoints of PQ and PR respectively. If QN and RM intersect at S, prove that the correspondence [M, N] ⟶ [R, Q] is a dilatation with centre S and dilatation factor −2.

9. A triangle has vertices A(2, 0), B(2, 10) and C(6, 5).
   (a) Calculate the area of △ABC.
   (b) Calculate the area of the image of △ABC under the dilatation with centre at the origin and scale factor 2.

10. A square has vertices A(2, 1), B(5, 1), C(5, 4) and D(2, 4).
    (a) Calculate the area of square ABCD.
    (b) Calculate the area of the image of this square under the dilatation with centre at the origin and scale factor $-\frac{1}{3}$.

11. Based on your results in Questions 9 and 10 write a conclusion about the ratio of the areas of a figure and its image under a dilatation with centre at the origin and scale factor k.

12. Find the equation of the image line given by 5x − 4y = 60 under the dilatation (x, y) ⟶ $\left(\frac{3}{2}x, \frac{3}{2}y\right)$.

C

13. The equation of the image of the line given by y = 3x − 8 under a dilatation with centre at the origin and scale factor k is 3x − y + 4 = 0. Find k.

10.8 theorems for similar triangles: SAS~, SSS~

In mathematics, we use what we know or have proved to develop more knowledge and prove other theorems.

The theorem proved about similarity of triangles in the previous section may be combined with our earlier theorems about triangles to prove yet other useful theorems about the similarity of triangles.

For similar figures or triangles, we know
I corresponding angles are equal.
II corresponding sides are in the same ratio.

In the previous section, AAA~ shows that *if I is true then II is true*. In the example that follows we prove the converse of the above, namely: *if II is true then I is true*, and thus the triangles are similar.

Example 1
Prove the following theorem:
If the sides of two triangles are in the same ratio, then the triangles are similar, (SSS~ Theorem).

Solution
Given: △ABC and △PQR such that
$\frac{AB}{PQ} = \frac{AC}{PR} = \frac{BC}{QR}$.
Required to prove:
△ABC ~ △PQR

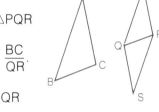

Proof:	Reasons
Construct △SRQ as shown, so that ∠SRQ = ∠C, ∠SQR = ∠B.	①
Thus ∠A = ∠S.	② ASTT
In △ABC and △SQR, corresponding angles are equal. Thus △ABC ~ △SQR	AAA~

	Reasons
$\frac{AB}{SQ} = \frac{BC}{QR}$	③ property of similar triangles
But $\frac{AB}{PQ} = \frac{BC}{QR}$	④ given
From ③ and ④	
$\frac{AB}{SQ} = \frac{AB}{PQ}$	equality
SQ = PQ	⑤ equal ratios
Similarly, we can prove	
PR = SR	⑥
In △PQR and △SQR	
PQ = SQ	proved ⑤
PR = SR	proved ⑥
QR = QR	common side
Thus △PQR ≅ △SQR	SSS
Thus ∠P = ∠S, ∠PQR = ∠SQR, ∠PRQ = ∠SRQ	⑦ property of congruent triangles
From ①, ②, and ⑦	
Thus ∠A = ∠P, ∠B = ∠PQR, ∠C = ∠PRQ	
△ABC ~ △PQR	AAA~

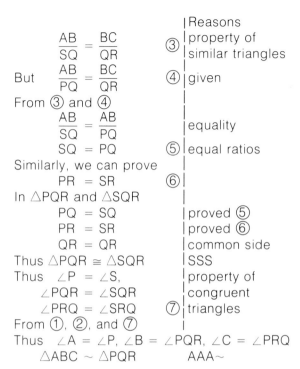

In mathematics, the theorems we prove may suggest other theorems for similar figures. For example, for congruent triangles we proved SAS. Can we use our earlier theorems to prove a theorem for similarity which compares with SAS, namely SAS~?

Example 2
Prove the SAS~ theorem:
If two triangles have a pair of equal angles and the sides containing the equal angles are in the same ratio, then the triangles are similar.

Solution
Given: △ABC and △PQR,
∠A = ∠P, $\frac{AB}{PQ} = \frac{AC}{PR}$.
Required to prove:
△ABC ~ △PQR

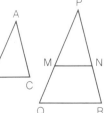

Proof: Construct △PMN as shown
such that PM = AB,
PN = AC. | Reasons

$$\frac{AB}{PQ} = \frac{AC}{PR}$$ | given

Thus $\frac{PM}{PQ} = \frac{PN}{PR}$ ①

In △PQR, $\frac{PM}{PQ} = \frac{PN}{PR}$ | from ①

Then   MN ∥ QR. | PPTT
Thus   ∠PMN = ∠Q,   ② | PLT,
       ∠PNM = ∠R   ③ | corresponding angles

In △ABC and △PMN
       AB = PM | constructed
       ∠A = ∠P | given
       AC = PN | constructed
       △ABC ≅ △PMN | SAS
Thus   ∠B = ∠PMN   ④ | property of
       ∠C = ∠PNM   ⑤ | congruent triangles

Thus   ∠B = ∠Q | from ② and ④
       ∠C = ∠R | from ③ and ⑤

In △ABC and △PQR
       ∠B = ∠Q | proved
       ∠C = ∠R | proved
       ∠A = ∠P | given
Thus   △ABC ~ △PQR | AAA~

A corollary to the above theorem that is a useful result is
   If BC ∥ DE then
   △ABC ~ △ADE.

To prove deductions, we may use the theorems for similarity.

AAA~   angle-angle-angle similar triangle theorem
SSS~   side-side-side similar triangle theorem
SAS~   side-angle-side similar triangle theorem

Note the comparisons to the theorems for congruency of triangles.

Use a diagram to translate the problem.

Example 3
In △PQR, PS ⊥ QR where S is a point in QR. If QS:PS = PS:SR, prove that ∠QPR is a right angle.

Solution
Given: △PQR with
       PS ⊥ QR,
       QS:PS = PS:SR.
Required to prove:
       ∠QPR = 90°

You may draw the triangles to correspond as shown to aid your solution.

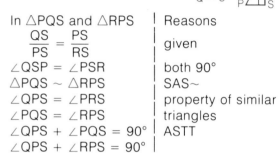

Proof:
   In △PQS and △RPS | Reasons
   $\frac{QS}{PS} = \frac{PS}{RS}$ | given
   ∠QSP = ∠PSR | both 90°
   △PQS ~ △RPS | SAS~
   ∠QPS = ∠PRS | property of similar triangles
   ∠PQS = ∠RPS
   ∠QPS + ∠PQS = 90° | ASTT
   ∠QPS + ∠RPS = 90°

The above result combined with its converse is called the *Mean Proportional Theorem* (MPT).
  In a right-angled triangle, the altitude to the hypotenuse is the mean proportional between the segments of the hypotenuse.

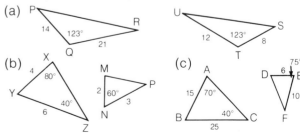

$$\frac{QS}{PS} = \frac{PS}{SR}$$

10.8 exercise

A
1 Which pairs of triangles are similar?

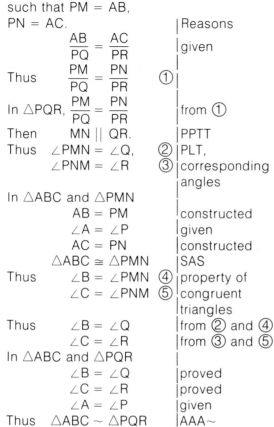

2 The lengths of sides of triangles are shown in each column. Which corresponding triangles are similar?

   (a) 18, 12, 9   30, 20, 15
   (b) 12, 8, 6   18, 12, 9
   (c) 15, 20, 25   21, 28, 35
   (d) 12, 24, 30   14, 28, 35
   (e) 15, 20, 25   18, 24, 28

3 Find the values of a, b, and c.

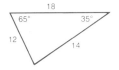

4 Find the values of x and y. Give reasons for your answer.

   (a)   (b)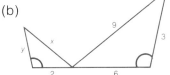

   (c)   (d)   (e)

5 △ABC is similar to △DEF and the sides of △ABC are 5 cm, 6 cm, 7 cm. The perimeter of △DEF is 360 cm. Find the lengths of the sides of △DEF.

6 △ABC has a perimeter of 45 cm. △TEC has sides 20 cm, 21 cm, and 25 cm. If △ABC ~ △TEC, what are the lengths of the sides of △ABC (to 1 decimal place)?

7 Find the value of x.

   (a)   (b)

8 For the diagram, show that YZ : AB = 5 : 1.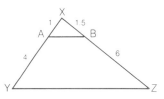

9 In △XYZ, A and B are points on XZ and YZ respectively such that AB ⊥ XZ. If ∠Y = 90°, AZ = 12, YZ = 20 and XY = 14, find AB.

proving deductions

B

10 For the diagram, if NY : AN = MN : XN, then prove NY : AN = MY : AX.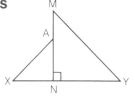

11 In the diagram, SQ bisects ∠PQR and T is a point in SQ so that $\frac{PQ}{SQ} = \frac{PT}{SR}$. Prove that ∠RSQ = ∠TPQ.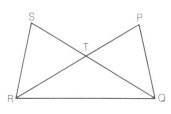

12 In △PST, Q and R are points in PS and PT respectively so that PR : PQ = PT : PS. Prove that QR ∥ ST.

13 In △XYZ, M and N are on XZ and XY respectively so that $\frac{XM}{XZ} = \frac{XN}{XY}$. Prove $\frac{XM}{XZ} = \frac{MN}{YZ}$.

14 In △PQR, ∠P = 90°, PS ⊥ QR and S is a point in QR. Prove that

   (a) △SPQ ~ △SRP.   (b) $\frac{QS}{SP} = \frac{SP}{SR}$.

   The above result and the converse in Example 3 are called the *Mean Proportional Theorem*.

15 Prove that the corresponding medians of similar triangles are proportional to corresponding sides.

16. Prove that the lengths of corresponding altitudes of similar triangles are proportional to the lengths of the corresponding sides.

17. In △ABC, P is a point in AB so that $AC^2 = AB \times AP$. Prove that ∠ACP = ∠ABC.

18. In △PQR, A and B are points in PQ and PR respectively such that $\frac{PB}{PA} = \frac{PQ}{PR}$. Prove that ∠PBA = ∠PQR.

19. Prove that for any △ABC the triangle formed by vertices that are the midpoints of the sides is similar to △ABC.

20. In quadrilateral PQRS, the diagonals intersect at O. If $\frac{PO}{OQ} = \frac{RO}{OS}$ and $\frac{PO}{OS} = \frac{RO}{OQ}$, prove that the quadrilateral is a parallelogram.

C

21. Prove that if the corresponding sides of two quadrilaterals are in the same ratio and a pair of corresponding angles are equal, then the quadrilaterals are similar.

22. Given a quadrilateral XYZW with XZ and YW intersecting at A so that XA:ZA = YA:WA. Prove the quadrilateral is a trapezoid.

23. In quadrilateral XYZW with ∠Y = ∠W and $\frac{XW}{ZW} = \frac{YZ}{XY}$, prove that XW = YZ.

10.9 areas of similar triangles

The need to know the relationship among areas of similar figures on drawings is important in making decisions when designing buildings. Particular examples such as the following suggest a generalization about the areas of triangles.

$$\frac{\triangle DEF}{\triangle ABC} = \frac{6}{24} = \frac{1}{4}$$

$$\frac{EF^2}{BC^2} = \frac{4^2}{8^2} = \frac{1}{4}$$

Example 1

Prove that if two triangles are similar, then the ratio of their areas is equal to the ratios of the squares of corresponding sides.

Solution

Given:
$$\triangle ABC \sim \triangle PQR$$

Required to prove:
$$\frac{\triangle ABC}{\triangle PQR} = \frac{AB^2}{PQ^2} = \frac{AC^2}{PR^2} = \frac{BC^2}{QR^2}$$

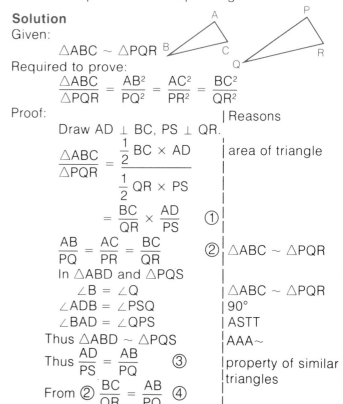

Proof: | Reasons
Draw AD ⊥ BC, PS ⊥ QR. |
$\frac{\triangle ABC}{\triangle PQR} = \frac{\frac{1}{2} BC \times AD}{\frac{1}{2} QR \times PS}$ | area of triangle
$= \frac{BC}{QR} \times \frac{AD}{PS}$ ① |
$\frac{AB}{PQ} = \frac{AC}{PR} = \frac{BC}{QR}$ ② | △ABC ~ △PQR
In △ABD and △PQS |
∠B = ∠Q | △ABC ~ △PQR
∠ADB = ∠PSQ | 90°
∠BAD = ∠QPS | ASTT
Thus △ABD ~ △PQS | AAA~
Thus $\frac{AD}{PS} = \frac{AB}{PQ}$ ③ | property of similar triangles
From ② $\frac{BC}{QR} = \frac{AB}{PQ}$ ④ |

Use ③ and ④ in ①.

$$\frac{\triangle ABC}{\triangle PQR} = \frac{BC}{QR} \times \frac{AD}{PS}$$
$$= \frac{AB}{PQ} \times \frac{AB}{PQ} \quad \leftarrow \text{from ③}$$
$$= \frac{AB^2}{PQ^2} \quad \leftarrow \text{Which is what we wanted to prove.}$$

Similarly, we may prove that
$\frac{\triangle ABC}{\triangle PQR} = \frac{AC^2}{PR^2}$ and $\frac{\triangle ABC}{\triangle PQR} = \frac{BC^2}{QR^2}$.

The results for triangles may be extended to the ratios of areas of any pair of similar figures. For example,

If two rectangles ABCD and PQRS are similar then
$$\frac{\text{area ABCD}}{\text{area PQRS}} = \frac{AD^2}{PS^2}$$
and so on.

10.9 exercise

In the exercise, △ABC is used to show *the area of △ABC*.

A

1 The ratios of corresponding sides of similar triangles are given. What is the ratio of their areas?
   (a) 3:4   (b) 5:2   (c) $3:\frac{1}{2}$

2 In △ABC, XY ∥ BC.
   If △AXY = 4 cm², find
   (a) △ABC   (b) quad XYCB

3 For the diagram,
   △AXY = 4 cm² and
   trapezoid XYCB = 12 cm².
   If AX = 5 cm, find XB.

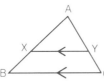

4 If △XYZ ~ △MNQ, XY = 6, MN = 5, find △XYZ:△MNQ.

5 Two triangles have the property that △OXB ~ △TAF, OB = 16, and TF = 13. Find △OXB:△TAF.

6 If △ABC ~ △RST and $\frac{\triangle ABC}{\triangle RST} = \frac{16}{25}$, find AB if RS = 10.

B

7 Prove that if two triangles are similar then the ratio of their areas is equal to the ratio of the squares of the corresponding medians.

8 In △QRS, ∠R > ∠S. If X is a point on QS so that ∠QRX = ∠S, prove that $\frac{QR}{QS} = \frac{RX}{RS}$.

9 Show that the ratio of the area of a triangle to the area of the triangle formed by joining the midpoints of the sides of the original triangle is in the ratio 4:1.

10 In △PQR, two medians PA and RB are drawn to intersect at C. What are the following ratios?
   (a) △ABC:△CPR   (b) △ABC:△PQR

11 In △XYZ, M is the midpoint of XY and N is the midpoint of XZ. If MZ and NY intersect at P, then find
   (a) $\frac{\triangle MPN}{\triangle YPZ}$ (b) $\frac{\triangle YPX}{\triangle XYZ}$ (c) $\frac{\triangle MPN}{\triangle XYZ}$

C

12 In a rectangle PQRS, PS > PQ. A and B are points in QS so that PB ⊥ QS and RA ⊥ QS, making QB = BA = AS. Prove that QR² = 2PQ².

13 If rectangles PQRS and ABCD are similar, prove that rect PQRS:rect ABCD = PQ²:AB².

14 For two similar quadrilaterals, prove that the ratio of their areas is equal to the ratio of the squares of the corresponding sides.

10.10 vectors on the plane: another point of view

When sailing a boat we must take into consideration the *direction* of the current and the *speed* of the current. There are other examples of quantities that are represented by *direction* as well as *magnitude*. For example:

- pushing a broom
- sky diving
- throwing a football
- shooting an arrow

A **vector quantity** is a quantity that has both magnitude and direction. We may represent it by a **directed line segment** and refer to it as a vector.

The arrow indicates the direction. The magnitude is shown by the length of the directed line segment.

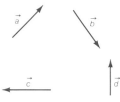

Quantities specified only by magnitude are then called **scalar quantities** or **scalars**. For example,

- the amount of gas in a tank,
- the number of people in a car.

We may apply our skills with plane geometry, as well as co-ordinate geometry, to the study of vectors.

A vector AB may be shown on a grid, and indicated algebraically by the symbols: A, the initial point, and B, the terminal point.

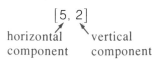

horizontal component   vertical component

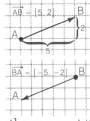

Two vectors given by [a, b], [c, d] are equal if a = c, b = d. On the Cartesian plane a vector may be specified by co-ordinates.

Vector $\overrightarrow{AB}$, where A(−3, 1), B(2, 3).

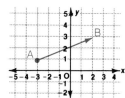

Magnitude
Use the distance formula to calculate the magnitude of the vector. $|\overrightarrow{AB}|$ is used to show the magnitude of the vector. Since $|\overrightarrow{AB}|$ represented the length of AB then we will use AB to show the length or magnitude of the vector.

$$AB = \sqrt{[2-(-3)]^2 + (3-1)^2}$$
$$= \sqrt{25 + 4} = \sqrt{29}$$

We may represent the above vector by $\vec{a} = [5, 2]$. Other vectors on the plane, equal to $\vec{a}$, are shown, since they have the same magnitude and direction.

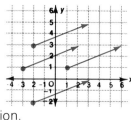

$-\vec{a}$ is called the additive inverse of $\vec{a}$ since $\vec{a} + (-\vec{a}) = \vec{0}$ where $\vec{0} = [0, 0]$.
$\vec{a}$ and $-\vec{a}$ are vectors of equal magnitude but opposite direction. Thus we may define subtraction of vectors as
$\vec{a} - \vec{b} = \vec{a} + (-\vec{b})$.
We may add parallel vectors.
$\vec{a} + \vec{b} = \vec{c}$
$\vec{c}$ is called the resultant vector.

To add vectors that are not parallel, we may think of $\vec{a} + \vec{b}$ as the vector $\vec{a}$ followed by the vector $\vec{b}$.
$\vec{a} + \vec{b} = \vec{c}$

Triangle Law of Vectors
For any two vectors $\overrightarrow{AB}$ and $\overrightarrow{BC}$
$\overrightarrow{AB} + \overrightarrow{BC} = \overrightarrow{AC}$.

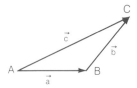

Parallelogram Law of Vectors
For any two vectors
$\vec{a}$ and $\vec{b}$
$\vec{c} = \vec{a} + \vec{b}$.

On a grid, if two vectors are given by $[a, b]$ and $[c, d]$, then $[a, b] + [c, d] = [a + c, b + d]$. In the following exercise, we will explore the properties of vectors.

10.10 exercise

A

1. If $\vec{a} = [5, 7]$ and $\vec{b} = [-1, 3]$ draw a diagram to show each vector.
   (a) $\vec{a}$ (b) $\vec{b}$ (c) $-\vec{a}$ (d) $-\vec{b}$
   (e) $\vec{a} - \vec{b}$ (f) $\vec{b} - \vec{a}$ (g) $-\vec{a} - \vec{b}$

2. If $\vec{a} = [2, -3]$, draw a diagram to show each vector.
   (a) $\vec{a}$ (b) $-\vec{a}$ (c) $3\vec{a}$ (d) $-2\vec{a}$
   (e) Calculate the magnitudes of the vectors in (a) to (d).

3. A vector $\overrightarrow{PQ}$ is given by each of the following set of co-ordinates. Express each vector in the form $[m, n]$ and calculate the magnitude of each vector.
   (a) P(-2, 3), Q(3, 5)   (b) P(3, 2), Q(-5, 2)
   (c) P(4, -5), Q(-3, -2) (d) P(-2, 3), Q(-5, 1)

4. Points on the plane have co-ordinates P(-4, -2), Q(2, 2), R(3, -2), and S(-5, 3). Express each vector in the form $[m, n]$ and calculate its magnitude.
   (a) $\overrightarrow{PQ}$ (b) $\overrightarrow{QP}$ (c) $\overrightarrow{RS}$ (d) $\overrightarrow{QS}$

5. (a) If $\vec{a} = [2, 3]$, $\vec{b} = [-1, 3]$, use the triangle method to find the sum $\vec{a} + \vec{b}$.
   (b) Write $\vec{a} + \vec{b}$ in the form $[m, n]$.
   (c) Calculate the magnitude of $\vec{a} + \vec{b}$.

6. (a) If $a = [-1, 5]$, $b = [4, 5]$, use the parallelogram method to find the sum $\vec{a} + \vec{b}$.
   (b) Write $\vec{a} + \vec{b}$ in the form $[m, n]$ and calculate its magnitude.
   (c) Find $\vec{a} - \vec{b}$ and calculate its magnitude.

7. Write each of the following as a single vector $[m, n]$.
   (a) $3[2, -1] + 5[3, -1]$
   (b) $-2[3, -1] - 3[5, -2]$
   (c) $-4[5, 1] + 2[3, 1] - 5[2, 1]$

8. Find m, n for each of the following.
   (a) $[-2, 3] + [3, 5] = [m, n]$
   (b) $[2, -3] + [m, n] = 3[3, 5]$
   (c) $2[4, 5] - [m, 4] = [6, n]$

9. If $\vec{a} = [a_1, a_2]$, $\vec{b} = [b_1, b_2]$ and $\vec{c} = [c_1, c_2]$ prove that
   (a) $\vec{a} + \vec{b} = \vec{b} + \vec{a}$
   (b) $\vec{a} + (\vec{b} + \vec{c}) = (\vec{a} + \vec{b}) + \vec{c}$

geometric vectors and polygons

We may use geometric vectors to represent the sides of a polygon. For △ABC, we may write
$\overrightarrow{BC} + \overrightarrow{CA} = \overrightarrow{BA}$

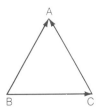

For rectangle ABCD, we may write
$$\vec{AB} + \vec{BC} = \vec{AC}$$
$$\vec{AD} + \vec{DC} = \vec{AC}$$

In the diagram
$\vec{AC} = -\vec{CA}$, $\vec{DA} = -\vec{AD}$ and so on.

The skills with geometric vectors may be used in the next section to develop a strategy for proving deductions using another approach.

B

10 For the rectangle, which of the following are true (T). Which are false (F)?

(a) $\vec{DB} + \vec{BC} = \vec{DC}$ (b) $\vec{CB} + \vec{DB} = \vec{BC}$
(c) $\vec{AB} = -\vec{BA}$ (d) $\vec{AB} - \vec{AC} = -\vec{BC}$
(e) $\vec{BC} - \vec{DB} = \vec{CD}$ (f) $\vec{DB} - \vec{DB} = \vec{AD} - \vec{AD}$
(g) $\vec{CB} + \vec{BA} = -\vec{AC}$ (h) $\vec{DB} = \vec{AD} + \vec{DB}$

11 For the parallelogram, give reasons why,

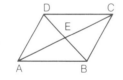

(a) $\vec{AD} = \vec{BC}$ (b) $\vec{AB} = \vec{DC}$
(c) $\vec{AC} \neq \vec{DB}$ (d) $\vec{DE} = \vec{EB}$
(e) $\vec{AE} = \vec{EC}$ (f) $\vec{AD} + \vec{DE} = \vec{AE}$
(g) $\vec{AB} + \vec{BC} = \vec{AC}$ (h) $\vec{AD} + \vec{AB} = \vec{AC}$

12 $\vec{a}$ and $\vec{b}$ are vectors shown by the diagram. Express each of the following in terms of $\vec{a}$ and $\vec{b}$.

(a) $\vec{DC}$ (b) $-\vec{CD}$ (c) $\vec{AD}$
(d) $\vec{AC}$ (e) $\vec{DB}$ (f) $-\vec{BD}$

13 For the diagram, write a single vector to show each of the following.

(a) $\vec{AB} + \vec{BC}$ (b) $\vec{DB} + \vec{BC}$
(c) $\vec{DB} + \vec{BA}$ (d) $\vec{DC} + \vec{CB} + \vec{BA}$

Multiples of vector $\vec{a}$ are shown.

$$\vec{AC} = \vec{a} + \vec{a} + \vec{a}$$
$$= 3\vec{a}$$

The vector AC is written as the product of the scalar 3 and the vector $\vec{a}$.

For the vector $r\vec{a}$, where $r \in R$, then

• $r\vec{a}$ is vector with the same direction as $\vec{a}$, and magnitude $r|\vec{a}|$.
• if $r > 0$, $r\vec{a}$ has the same direction as $\vec{a}$.
• if $r < 0$, $r\vec{a}$ has the opposite direction to $\vec{a}$.
• if, for any two vectors $\vec{a}$ and $\vec{b}$, $\vec{a} = r\vec{b}$, then $\vec{a} \| \vec{b}$ (i.e. $\vec{a}$ and $\vec{b}$ are parallel vectors).

14 For the diagram, M and N are the midpoints of PQ and PR respectively. Use $\vec{MP} = \vec{a}$ and $\vec{PN} = \vec{b}$. Express each of the following in terms of $\vec{a}$ and $\vec{b}$.

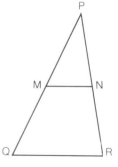

(a) $\vec{NR}$ (b) $\vec{PR}$ (c) $\vec{QR}$

15 In $\triangle PQR$, S is a point on QR so that RS:SQ = 2:1. Express each of the vectors in terms of $\vec{PQ}$ and $\vec{PR}$.

(a) $\vec{QR}$ (b) $\vec{QS}$ (c) $\vec{RS}$ (d) $\vec{PS}$

applications: vectors and internal division

Using co-ordinate geometry, we have developed a method of finding the co-ordinates of a point of internal division.

Using vectors, we may now divide a line segment internally using a different strategy.

In the diagram, O is the origin, and P divides AB internally in the ratio
 AP:PB = 4:3.

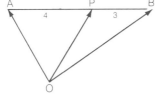

Thus
$$\overrightarrow{AP} = \frac{4}{7}\overrightarrow{AB}, \text{ since } \overrightarrow{AP} \| \overrightarrow{AB}.$$

Use $\overrightarrow{OP} = \overrightarrow{OA} + \overrightarrow{AP}$

$$= \overrightarrow{OA} + \frac{4}{7}\overrightarrow{AB} \longrightarrow \overrightarrow{AB} = \overrightarrow{AO} + \overrightarrow{OB}$$
$$\phantom{= \overrightarrow{OA} + \frac{4}{7}\overrightarrow{AB}} = -\overrightarrow{OA} + \overrightarrow{OB}$$
$$= \overrightarrow{OA} + \frac{4}{7}(-\overrightarrow{OA} + \overrightarrow{OB})$$
$$= \overrightarrow{OA} - \frac{4}{7}\overrightarrow{OA} + \frac{4}{7}\overrightarrow{OB}$$
$$= \frac{3}{7}\overrightarrow{OA} + \frac{4}{7}\overrightarrow{OB}$$

Note, how the coefficients are related to how AB is divided internally.

If the co-ordinates are A(4, 5) and B(3, 8), then $\overrightarrow{OA} = [4, 5]$, $\overrightarrow{OB} = [3, 8]$.

Then $\overrightarrow{OP} = \frac{3}{7}[4, 5] + \frac{4}{7}[3, 8]$

$$= \left[\frac{12}{7}, \frac{15}{7}\right] + \left[\frac{12}{7}, \frac{32}{7}\right]$$
$$= \left[\frac{24}{7}, \frac{47}{7}\right]$$

Thus the co-ordinates of P are $\left(\frac{24}{7}, \frac{47}{7}\right)$.

Use the vector method to find the co-ordinates of the point of internal division for each of the following.

16 P divides AB internally in the given ratio. Express $\overrightarrow{OP}$ in terms $\overrightarrow{OA}$ and $\overrightarrow{OB}$ where O is the origin.

 (a) 3:2 (b) 1:4 (c) 5:4

17 P is a point of division of AB so that AP:PB = 3:4. If A(−4, 4), B(3, 7), find the co-ordinates of P.

18 Find the co-ordinates of P, if P divides AB, where A(6, 12) and B(−4, 3), into each of the following ratios.

 (a) 2:1 (b) 1:2 (c) 2:3

---

problem/matics

Often in mathematics a numerical pattern suggested by a problem may often lead to making a conjecture that applies to all situations. For example, the table is completed.

polygon	number of sides	sum of the interior angles
triangle	3	180°
quadrilateral	4	360° = 2 × 180°
pentagon	5	540° = 3 × 180°
hexagon	6	720° = 4 × 180°

Based on the numerical pattern in the above results we may make the conjecture:

 For a polygon of n sides, the sum of the interior angles is (n − 2) × 180°.

Try the conjecture for an octagon. Then use another method to check your result.

However, unless we can prove the result, it will remain a conjecture. Is the above conjecture true? (You can answer this question by proving or disproving the result.)

10.11 proving deductions by vector methods

Previously we used different strategies to prove deductions based on the following.

- plane geometry
- analytic geometry
- transformational geometry

There is another strategy, based on vectors, for proving deductions. We may derive those properties of geometric figures which we derived earlier by using a different approach. For example, in △ABC, M, N are the midpoints of AB and AC respectively. By using vector geometry, we may prove that MN||BC and MN = $\frac{1}{2}$BC.

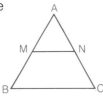

Since M and N are midpoints
then $\vec{BM} = \vec{MA}$ or $\vec{BA} = 2\vec{MA}$ ①
and $\vec{AN} = \vec{NC}$ or $\vec{AC} = 2\vec{AN}$ ②

In △ABC, $\vec{BC} = \vec{BA} + \vec{AC}$ triangle law of vectors
       = $2\vec{MA} + 2\vec{AN}$ from ②
       = $2(\vec{MA} + \vec{AN})$ from ①

From the diagram, we notice that $\vec{MA} + \vec{AN}$ is related to △AMN.

In △AMN, $\vec{MN} = \vec{MA} + \vec{AN}$ ③
Thus $\vec{BC} = 2(\vec{MA} + \vec{AN})$
     $\vec{BC} = 2(\vec{MN})$

which means BC||MN and MN = $\frac{1}{2}$BC.

When proving deductions we may apply the following skills with vectors.

- If $\vec{AB} = \vec{CD}$ then $\vec{AB} = -\vec{DC}$.
- In △ABC
   $\vec{BC} = \vec{BA} + \vec{AC}$ triangle law of vectors

Note the position of the points A, B, C.
If $\vec{AB} = \vec{CD}$ then AB = CD and AB||CD.

The converse of this is also true.
- We may replace a vector by an equal vector. For example, in square ABCD, AB = DC, AB||DC. Then $\vec{AB} = \vec{DC}$.
Thus $\vec{AB} + \vec{BC} = \vec{AC}$ or $\vec{DC} + \vec{BC} = \vec{AC}$.

10.11 exercise

To prove the following deductions, use the methods of vector geometry.

A

1. In quadrilateral ABCD, AB||DC and AB = DC. Explain why
   (a) $\vec{BA} = \vec{CD}$
   (b) $\vec{BC} = \vec{BA} + \vec{AC}$
   (c) $\vec{AD} = \vec{AC} + \vec{CD}$
   (d) Use the above facts to prove that $\vec{AD} = \vec{BC}$ and thus AD = BC and AD||BC.

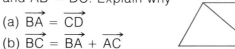

2. Use the diagram. AM = MB and MN||BC. Explain why
   (a) $\vec{BA} = 2\vec{MA}$
   (b) $\vec{MA} + \vec{AN} = \vec{MN}$
   (c) $\vec{BA} + \vec{AC} = \vec{BC}$ (d) Prove that AN = NC.

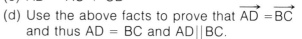

3. In rectangle PQRS, A and B are midpoints of PS and QR respectively. Prove that $\vec{AB} = \vec{PQ}$

B

4. PQRS is a parallelogram with A and B midpoints of PS and QR. Prove that AQ = SB.

5. In a quadrilateral PQRS, PS = QR and PS||QR. Prove that
   (a) PQ = SR   (b) PQ||SR

6. In quadrilateral ABCD, E is the midpoint of AC. If AB = DC, prove that D, E, and B are collinear.

7 Two line segments AB and CD bisect each other at E. Prove that ACBD is a parallelogram.

8 In △PQR, the median PS is drawn to QR. Prove that $\vec{PS} = \frac{1}{2}(\vec{PQ} + \vec{PR})$.

9 Prove that if two sides of a quadrilateral are equal and parallel then the quadrilateral is a parallelogram.

10 Prove that in any quadrilateral, the midpoints of the sides are the vertices of a parallelogram.

11 In trapezoid PQRS, A and B are the midpoints of PQ and RS respectively, and QR∥PS. Prove that AB∥QR.

12 Prove that if the diagonals of a quadrilateral bisect each other then the quadrilateral is a parallelogram.

13 In △PQR, A, B, and C are midpoints PQ, QR, and PR respectively. If S is any point, then prove SA + SB + SC = SP + SQ + SR.

14 Prove that if two triangles are similar then the ratio of their areas is equal to the ratio of the squares of the corresponding altitudes.

15 If $\vec{a} = [8, 9]$, $\vec{b} = [2, 5]$, draw a diagram to show the vector
(a) $\vec{a} + \vec{b}$. (b) $\vec{a} - \vec{b}$.
(c) Find the magnitude of the vectors in (a) and (b).

C
16 For the trapezoid PQRS, PQ∥SR. If M and N are the midpoints of PS and QR respectively, prove that $\vec{MN} = \frac{1}{2}(\vec{PQ} + \vec{SR})$.

17 Use a vector method to prove the following deduction: The quadrilateral formed by joining the midpoints of the sides of a quadrilateral, in succession, is a parallelogram.

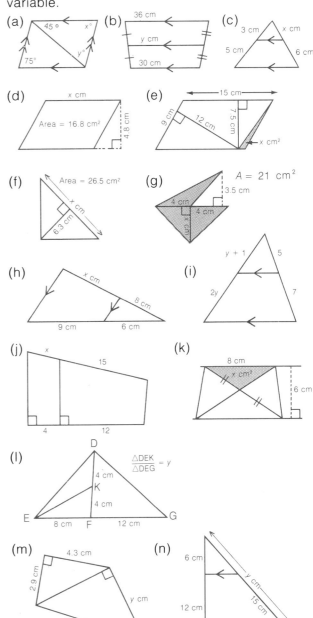

skills review: solving problems

For each diagram, find the values of the variable.

problems and practice: a chapter review

1. Find the area of the shaded area.

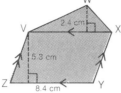

2. Prove that the area of a rhombus is the product of the length, b, of one side and the length, h, of the altitude to that side.

3. If the area of $\triangle OSM = 5$, find $\triangle MIT$.

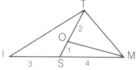

4. Find x.

   (a)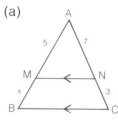

   (b)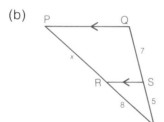

5. If HG is 18 cm long, find the length of FG and HF where HF:FG = 11:3.

6. Find the co-ordinates of C if AB, given by $A(-1, -2)$ and $B(7, 10)$, is divided externally in the ratio 5:4.

7. GF is divided internally by X in the ratio 3:8. If GF is 62 cm, what are the lengths of GX and XF?

8. $\triangle ABC$ is an isosceles triangle with a right angle at B. $\triangle XYZ$ is an isosceles triangle with a right angle at Y. Prove $\triangle ABC \sim \triangle XYZ$.

9. A and B are midpoints of sides XW and YZ respectively of ||gm XYZW. XB and AZ intersect YW at M and N respectively. Prove that YM = MN = NW.

10. Trapezoid ABCD in which AB||DC has G as the midpoint of DC. If E and F are midpoints of AG and BG respectively, prove $\triangle ADE = \triangle BFC$.

11. $\triangle PQR$ and $\triangle SQR$ are as shown. N is any point on QR. NM||QP with M on PR. NG||QS with G on SR. Prove PS||MG.

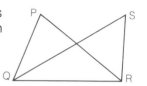

12. If $\dfrac{AX}{XB} = \dfrac{XM}{XY}$ prove $\angle A = \angle B$.

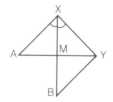

13. S is a point of division of PQ so that PS:SQ = 3:4. If the co-ordinates are given by $P(-1, 6)$, $Q(6, 9)$, find the co-ordinates of S.

14. Prove that for any vector $\vec{a}, \vec{b}, \vec{c}$
    (a) $\vec{a} + \vec{b} = \vec{b} + \vec{a}$
    (b) $(\vec{a} + \vec{b}) + \vec{c} = \vec{a} + (\vec{b} + \vec{c})$

15. Use a vector method to prove the following deduction. In quadrilateral ABCD, BA = CD and BA||CD. If the diagonals intersect at E and EA = EC prove that EB = ED.

problem/matics

What method for solving the following problem is suggested by the facts given in the problem?

In $\triangle XYZ$, A, B and C are the midpoints of XY, YZ and ZX respectively. O is any other point. Prove
$\vec{OA} + \vec{OB} + \vec{OC} = \vec{OX} + \vec{OY} + \vec{OZ}$.

test for practice

1. Calculate the shaded area.
   (a)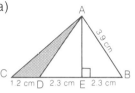
   (b)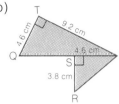

2. △MPR is right-angled at P. If MP = 10 cm, PR = 24 cm and MR = 26 cm, find the length of PN, the altitude to MR.

3. Find
   (a)
   (b)
   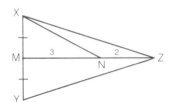

4. Prove that the area of a trapezoid is one half the product of the sum of the lengths of the parallel sides and the perpendicular distance between these sides.

5. Prove that △MNR : ||gm MNPQ = 3 : 14.

6. T is the midpoint of PQ in ||gm PQRS. Prove △PST = $\frac{1}{4}$ ||gm PQRS.

7. Find x.
   (a)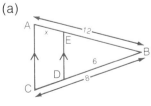
   (b)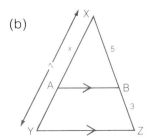

8. A is any point in MN of △MNR. AP||NR. S is on MN produced such that RS||PN. Prove $\frac{MA}{AN} = \frac{MN}{NS}$.

9. NP is divided externally at Q in the ratio of 3:2. Find the ratio for
   (a) NP:PQ (b) NQ:QP (c) PQ:QN
   (d) PN:NQ (e) QN:NP (f) QP:PN

10. If DF is 35 cm long, find the lengths of DE and EF where DE:EF = 5:2.

11. AD is 16 cm long. What are the lengths of BD and AB if B divides AD internally in the ratio 1:3?

12. The line BK is 11 cm and C is a point that divides it externally in the ratio 9:4. What is the length of BC?

13. In the diagram, AG||BF||CE. AG = 10, DG = 50, BF = 5, CE = 2. Find EF.

problem/matics

When solving problems, often the first important step is realizing that more than one strategy can be used to solve the problem. Once we realize this then we have to decide which strategy is the most straightforward. What general methods do you know of solving the following problem? Are there any facts given that suggest one method over another? Solve this problem.

Prove that any isosceles triangle whose sides have slopes m_1, m_2, m_3 such that $m_1 m_2 + m_2 m_3 + m_3 m_1 = -3$ is equilateral.

math is plus — matrices

The following is a quick glimpse and overview of a topic of mathematics that is relatively young compared to many of the other branches of mathematics. In fact, matrix algebra was organized as recently as 1857, by Arthur Cayley, an Englishman. Data may be written in a concise way using a matrix.

$$\begin{bmatrix} 3 & 5 \\ 8 & 9 \end{bmatrix} \quad \begin{bmatrix} 4 & 5 \\ 6 & 7 \\ 8 & 9 \end{bmatrix}$$

Column of the matrix — Row — 2×2 matrix — 3×2 matrix — order of the matrix

We may operate with matrices.

$$\begin{bmatrix} 2 & -3 \\ 4 & 5 \end{bmatrix} + \begin{bmatrix} 4 & 3 \\ -8 & -2 \end{bmatrix} = \begin{bmatrix} 2+4 & -3+3 \\ 4-8 & 5-2 \end{bmatrix} = \begin{bmatrix} 6 & 0 \\ -4 & 3 \end{bmatrix}$$

$$\begin{bmatrix} 3 & 4 \\ 5 & 6 \\ 4 & 5 \end{bmatrix} - \begin{bmatrix} 2 & 1 \\ 3 & -2 \\ -2 & 3 \end{bmatrix} = \begin{bmatrix} 3-2 & 4-1 \\ 5-3 & 6-(-2) \\ 4-(-2) & 5-3 \end{bmatrix} = \begin{bmatrix} 1 & 3 \\ 2 & 8 \\ 6 & 2 \end{bmatrix}$$

$$2 \begin{bmatrix} 3 & 1 & -2 \\ -4 & 1 & 6 \end{bmatrix} = \begin{bmatrix} 6 & 2 & -4 \\ -8 & 2 & 12 \end{bmatrix} \quad 2(6) = 12$$

We may also multiply matrices; the rule is shown for 2 examples.

$$[a \; b \; c] \begin{bmatrix} e \\ f \\ g \end{bmatrix} = [ae + bf + cg] \quad \begin{bmatrix} a & b \\ c & d \end{bmatrix} \begin{bmatrix} e \\ f \end{bmatrix} = \begin{bmatrix} ae + bf \\ ce + df \end{bmatrix}$$

We may use a matrix as an operator on a point A in the plane. For example a point A(3, 2) on the plane is shown in matrix form $A \begin{bmatrix} 3 \\ 2 \end{bmatrix}$. If we multiply by the matrix $\begin{bmatrix} -1 & 0 \\ 0 & 1 \end{bmatrix}$ an interesting result is obtained.

$$\begin{bmatrix} -1 & 0 \\ 0 & 1 \end{bmatrix} \begin{bmatrix} 3 \\ 2 \end{bmatrix} = \begin{bmatrix} -3 \\ 2 \end{bmatrix}$$

It appears that $\begin{bmatrix} -1 & 0 \\ 0 & 1 \end{bmatrix}$ has reflected the point (3, 2) in the y-axis to obtain the reflection image (−3, 2). Thus the matrix $\begin{bmatrix} -1 & 0 \\ 0 & 1 \end{bmatrix}$ as an operator maps $\begin{bmatrix} 3 \\ 2 \end{bmatrix} \longrightarrow \begin{bmatrix} -3 \\ 2 \end{bmatrix}$.

These matrices are given as operators.

$$A = \begin{bmatrix} 0 & 1 \\ -1 & 0 \end{bmatrix} \quad B = \begin{bmatrix} 1 & 0 \\ 0 & -1 \end{bmatrix} \quad C = \begin{bmatrix} 0 & -1 \\ 1 & 0 \end{bmatrix}$$

$$D = \begin{bmatrix} -1 & 0 \\ 0 & 1 \end{bmatrix} \quad E = \begin{bmatrix} -1 & 0 \\ 0 & -1 \end{bmatrix}$$

A number of transformations are listed.
  I rotation of 180° counter-clockwise about the origin.
 II reflection in the y-axis
III rotation of 90° clockwise about the origin
 IV reflection in the x-axis
  V rotation of 90° counter-clockwise about the origin

Match the operators A to E with the transformations I to V.

In mathematics, a new skill often can be applied to solve an earlier problem from a different point of view. For example, we may write a system of equations in matrix form. Check it.

Equations   Matrix form
$2x - 3y = 7$
$3x + 2y = 4$ $\begin{bmatrix} 2 & -3 \\ 3 & 2 \end{bmatrix} \begin{bmatrix} x \\ y \end{bmatrix} = \begin{bmatrix} 7 \\ 4 \end{bmatrix} \quad A \begin{bmatrix} x \\ y \end{bmatrix} = \begin{bmatrix} 7 \\ 4 \end{bmatrix}$

To solve this equation, we need only multiply by the inverse of A, A^{-1}, and obtain the following.

$$A^{-1} A \begin{bmatrix} x \\ y \end{bmatrix} = A^{-1} \begin{bmatrix} 7 \\ 4 \end{bmatrix} \quad \text{Thus} \quad \begin{bmatrix} x \\ y \end{bmatrix} = A^{-1} \begin{bmatrix} 7 \\ 4 \end{bmatrix}$$

If $A = \begin{bmatrix} a & b \\ c & d \end{bmatrix}$ then the inverse of A, namely A^{-1} is given by the rule. $A^{-1} = \dfrac{1}{ad - bc} \begin{bmatrix} d & -b \\ -c & a \end{bmatrix}$

Solve these two equations by using the matrix form.
(a) $3x - 2y = 10$ (b) $2x + y = 6$
      $x + 5y = -3$ $3x - 2y = 9$

Check your solutions by using previous methods you have learned.

11 elements of trigonometry; concepts, skills, and applications

Basic concepts and skills of trigonometry, drawing graphs, sine and cosine laws, solving problems, applications and making decisions.

11.1 trigonometry: today and tomorrow

The work we have done in Euclidean geometry and analytic geometry is now applied to the study of the branch of mathematics called **trigonometry**.

The early Greeks used the study of angles and sides to solve problems involving triangles. Trigonometry is derived from the Greek words

　　tri—three
　　gonia—angle　　　　*trigonometry*
　　metria—measurement

A main application of trigonometry is the solution of problems involving distances. The early astronomers used the concepts in trigonometry to aid their work. Trigonometry is used today in these and other ways:

In Surveying

In Navigation

In Medicine

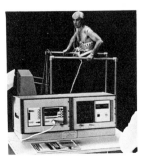

In Traffic Control

11.2 trigonometry: concepts and skills

The Cartesian plane is fundamental in the study of trigonometry. Angles are related to the co-ordinate axes and associated with rotation.

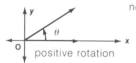

The vertex of the angle is placed at the origin and the initial arm placed along the x-axis. ∠AOC is said to be in **standard position**.

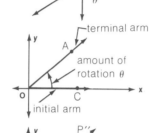

A point P(x, y) is marked on the terminal arm. From the diagram these ratios are equal.

$$\frac{PQ}{OP} = \frac{P'Q'}{OP'} = \frac{P''Q''}{OP''}$$

Thus △OPQ ~ △OP'Q' ~ △OP''Q''.
This observation results in the following definitions which are essential to the study of mathematics.

The trigonometric ratios of angle θ are defined. The trigonometric values are *independent* of the choice of P on the terminal arm, since △OPQ ~ △OP'Q' ~ △OP''Q''.

Primary Trigonometric Ratios

$$\text{sine } \theta = \frac{y}{r} \quad \text{cosine } \theta = \frac{x}{r} \quad \text{tangent } \theta = \frac{y}{x}$$

$$\sin \theta = \frac{y}{r} \quad \cos \theta = \frac{x}{r} \quad \tan \theta = \frac{y}{x}$$

By writing the reciprocals of the above ratios, other trigonometric ratios are defined.

$$\text{cosecant } \theta = \frac{r}{y} \quad \text{secant } \theta = \frac{r}{x} \quad \text{cotangent } \theta = \frac{x}{y}$$

$$\csc \theta = \frac{r}{y} \quad \sec \theta = \frac{r}{x} \quad \cot \theta = \frac{x}{y}$$

The co-ordinate axes divide the plane into four quadrants. An angle is shown in each quadrant. Clockwise angles are, by convention, negative and counter clockwise angles are positive.

To calculate the trigonometric ratios we need find only a point on the terminal arm. We may use any point on the terminal arm.

Example 1
The point (3, −4) is on the terminal arm of angle θ as shown. Calculate the trigonometric ratios.

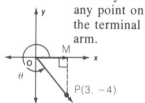

Solution
From the diagram $r = \sqrt{x^2 + y^2}$
$= \sqrt{(3)^2 + (-4)^2} = 5$

Thus $x = 3, y = -4, r = 5$.

$$\sin \theta = \frac{y}{r} = -\frac{4}{5} \quad \bigg| \quad \cos \theta = \frac{x}{r} = \frac{3}{5} \quad \bigg| \quad \tan \theta = \frac{y}{x} = -\frac{4}{3}$$

$$\csc \theta = \frac{r}{y} = -\frac{5}{4} \quad \bigg| \quad \sec \theta = \frac{r}{x} = \frac{5}{3} \quad \bigg| \quad \cot \theta = \frac{x}{y} = -\frac{3}{4}$$

If we know the value of one of the trigonometric ratios and the quadrant that the angle is in, we may calculate the remaining trigonometric ratios.

Example 2
A positive angle θ is in the third quadrant and $\cos \theta = -\frac{8}{17}$. Calculate the primary trigonometric values.

Solution

From the diagram $r = 17$, $x = -8$.

Since $r = \sqrt{x^2 + y^2}$
then $17 = \sqrt{(-8)^2 + y^2}$
$289 = 64 + y^2$
$225 = y^2$
or $y = \pm 15$

Remember: To solve the problem, draw a sketch of the given information.

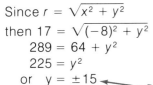

Since P is a point in the third quadrant, $y = -15$.

Use $r = 17$, $x = -8$, $y = -15$.

$\sin\theta = \dfrac{y}{r}$ $\cos\theta = \dfrac{x}{r}$ $\tan\theta = \dfrac{y}{x}$

$= \dfrac{-15}{17}$ $= \dfrac{-8}{17}$ $= \dfrac{15}{8}$

Example 3

Find $\cos\theta$ if θ is positive and $\sin\theta = -\dfrac{3}{5}$.

Solution

If θ is a positive angle and $\sin\theta = -\dfrac{3}{5}$, then θ may be angle in the third or fourth quadrant.

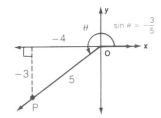   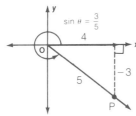

Third Quadrant   Fourth Quadrant
$x = -4, y = -3, r = 5$ $x = 4, y = -3, r = 5$
$\cos\theta = -\dfrac{4}{5}$ $\cos\theta = \dfrac{4}{5}$

11.2 exercise

Throughout the exercise you may leave answers in fractional or radical form.

A

1. Draw a sketch of each angle in standard position.
   (a) 120° (b) −60° (c) 135°
   (d) −225° (e) 210° (f) −135°

2. For each angle θ, a point on the terminal arm is given. Calculate each value of θ.

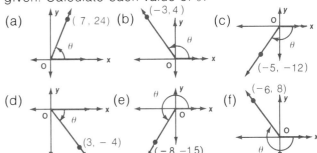

3. For each angle, a point on the terminal arm is shown. Calculate the trigonometric ratios.

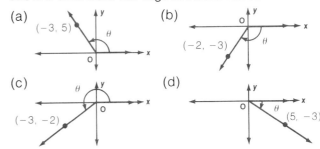

4. The point P(−3, −4) lies on the terminal arm of θ. For θ, write its
   (a) sine (b) cosine (c) tangent

5. The point Q(−8, 15) lies on the terminal arm of α. For α, calculate its
   (a) cosecant (b) secant (c) cotangent

6. An angle θ is in standard position in the second quadrant. If $\cos\theta = -\dfrac{3}{4}$, draw the terminal arm of θ.

7. For each θ, construct the terminal arm.
   (a) $\tan\theta = -\dfrac{3}{4}$, θ in fourth quadrant
   (b) $\sin\theta = \dfrac{7}{25}$, θ in second quadrant
   (c) $\csc\theta = \dfrac{17}{-8}$, θ in third quadrant

The angles referred to in Questions 8 to 22 are in standard position.

B

8. $P(x, y)$ is a point on the terminal arm of α. If $OP = r$, then find the primary trigonometric values.
   (a) $P(3, k)$, $r = 5$
   (b) $P(k, -8)$, $r = 10$
   (c) $P(3, k)$, $r = \sqrt{13}$
   (d) $P(-\sqrt{3}, k)$, $r = 2$

9. α is an angle in the third quadrant and $\cos \alpha = \dfrac{-\sqrt{3}}{2}$.
   (a) Write the co-ordinates of a point on the terminal arm.
   (b) Find $\sin \alpha$ and $\tan \alpha$.

10. θ is an angle in the second quadrant and $\csc \theta = \dfrac{17}{15}$.
    (a) Write the co-ordinates of a point on the terminal arm.
    (b) Find $\cos \theta$, $\sec \theta$, and $\cot \theta$.

11. β is an angle with its terminal arm in the second quadrant. If $\sin \beta = \dfrac{4}{5}$, find the values of $\cos \beta$ and $\tan \beta$.

12. If $\cos \theta = -\dfrac{5}{13}$, and θ is a second quadrant angle, then calculate
    (a) $\sin \theta \cos \theta$
    (b) $2 \cot \theta$
    (c) $\sin^2 \theta$
    (d) $\cot \theta \tan \theta$
    $\sin^2 \theta$ means $(\sin \theta)^2$

13. θ and β are angles in standard position. θ has its terminal arm in the first quadrant; β has its terminal arm in the second quadrant. If $\cos \theta = \dfrac{3}{5}$ and $\cos \beta = -\dfrac{3}{5}$, what is the value of $\cos \beta + \cos \theta + \sin \beta + 2 \sin \theta + \tan \beta$?

14. β is an angle and has its terminal arm in the first quadrant. If $\sec \beta = \dfrac{5}{3}$, find the value of $\sec \beta + \sin \beta + \cos \beta - \tan \beta$.

15. Given that $\cos \theta = -\dfrac{7}{25}$.
    (a) In which possible quadrants can the terminal arm be placed?
    (b) Draw a diagram to show each case in (a).
    (c) Calculate the trigonometric values of $\sin \theta$.

16. (a) If $\sin \theta = \dfrac{-8}{17}$, find two values of $\cos \theta$.
    (b) Given that $\cot \alpha = -\dfrac{12}{5}$. Find two values of $\sin \alpha$.
    (c) For $\sec \beta = -\dfrac{25}{7}$, find $\tan \beta$.
    (d) θ is in standard position. If $\cos \theta = \dfrac{-\sqrt{3}}{2}$, find $\cot \theta$.

signs of $\sin \theta$, $\cos \theta$, $\tan \theta$

17. In which quadrant does the terminal arm of θ lie if
    (a) $\sin \theta$ and $\cos \theta$ are both positive?
    (b) $\tan \theta$ is positive and $\sin \theta$ is negative?
    (c) $\cos \theta$ is negative and $\sin \theta$ is positive?
    (d) $\csc \theta$ and $\tan \theta$ are both negative?
    (e) $\sec \theta$ and $\tan \theta$ are both negative?
    (f) $\cot \theta$ and $\sin \theta$ are both positive?

18. In which possible quadrant(s) does the terminal arm of θ lie if
    (a) $\sin \theta$ is positive?
    (b) $\cos \theta$ is negative?
    (c) $\tan \theta$ is positive?
    (d) $\csc \theta$ is negative?
    (e) $\cot \theta$ is positive?
    (f) $\sec \theta$ is positive?

19. For each quadrant, what conclusion can be made about the sign (positive or negative) for each of the following?

(a) $\sin \theta$ (b) $\cos \theta$ (c) $\tan \theta$
(d) $\csc \theta$ (e) $\sec \theta$ (f) $\cot \theta$
(g) Write a rule for the signs of $\sin \theta$, $\cos \theta$, and $\tan \theta$ based on your results in Questions 17 to 19.

20. For any angle θ, show why
(a) $\sin \theta = \dfrac{1}{\csc \theta}$ (b) $\cos \theta = \dfrac{1}{\sec \theta}$
(c) $\tan \theta = \dfrac{1}{\cot \theta}$ (d) $\csc \theta = \dfrac{1}{\sin \theta}$
(e) $\sec \theta = \dfrac{1}{\cos \theta}$ (f) $\cot \theta = \dfrac{1}{\tan \theta}$

21. If $\sin \theta = -\dfrac{3}{5}$, prove that $\sin^2 \theta + \cos^2 \theta = 1$.

22. (a) If $\cot \theta = -\dfrac{15}{8}$, show that $\dfrac{\sin \theta}{\cos \theta} = \tan \theta$.
(b) If $\cos \theta = -\dfrac{\sqrt{3}}{2}$, show that $\dfrac{\sin \theta}{\cos \theta} = \tan \theta$.
(c) What probable conclusion seems true based on your results in (a) and (b)?

problem/matics

In the past, students of plane geometry did not have as many methods of algebra to aid them as we do today because the methods of algebra were not as well developed as they are today. Some of the algebraic stategies and skills have taken years to develop. Use algebra to solve this geometric type problem.
The distance from Acton to Beeton through Eaton is the same as the distance from Acton to Beeton through Croton. How far is Acton from Eaton?

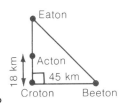

11.3 coterminal and special angles

Three positive angles in standard position are shown, each having a terminal arm that passes through $(-3, 4)$.

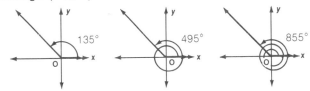

We say that the angles are coterminal since they have the same terminal arm. Associated with each angle is the principal angle, 135°, which is referred to as the smallest positive angle. For any angle, the related principal angle θ is $0° \leq \theta \leq 360°$.

Three coterminal angles are shown (clockwise).

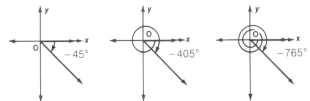

The smallest positive angle, related to the above negative angles is shown in the diagram.
$0° \leq \theta \leq 360°$

Thus, the principal angle for the coterminal angles $-45°$, $-405°$ and $-765°$ is 315°.

Example 1
For the angle 765°
(a) draw a sketch in standard position.
(b) write the principal angle.

Solution

The curved arrow is used to indicate the number of rotations the terminal arm moves through.

Since $765° = 2(360°) + 45°$ then the principal angle is 45°.
($0° \leq 45° \leq 360°$)

Based on Example 1, any angle, coterminal with 45°, may be written in general as
  45° + k(360°) where k is an integer.
Thus, coterminal angles are given by
  k = 1   k = −1   k = 2   k = −2
  405°   −315°   765°   −675°

Thus, cos 405° = cos(−315°) and so on since 405° and −315° are coterminal angles.

Associated with each rotation angle is a reference triangle as shown in the diagrams.

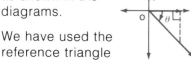

We have used the reference triangle to calculate the trigonometric values of angles.

For acute angles in the first quadrant, we introduce a vocabulary to refer to the sides of triangles

Acute Angle   Triangle

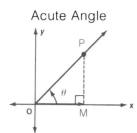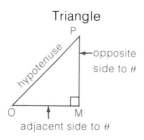

To calculate the trigonometric value of 45°, we may construct the following acute triangle. △ABC is isosceles.
Thus ∠B = ∠A = 45°.
In △ABC
  $AB^2 = AC^2 + BC^2$
  $AB^2 = 1^2 + 1^2$
  $AB = \sqrt{2}$
Thus, from the diagram,
  $\sin B = \dfrac{1}{\sqrt{2}}$ $\cos B = \dfrac{1}{\sqrt{2}}$ $\tan B = 1$
  $\csc B = \sqrt{2}$ $\sec B = \sqrt{2}$ $\cot B = 1$

To calculate the trigonometric values of 30° and 60° an equilateral triangle of sides 2 units is drawn as shown.
In △ABC
  AD ⊥ BC, BD = CD
  ∠BAD = ∠DAC = 30°
  Thus $AB^2 = AD^2 + BD^2$
  $(2)^2 = AD^2 + 1$
  $AD = \sqrt{3}$

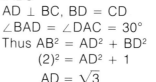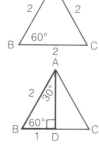

From △ABD we may calculate the primary trigonometric values.

$\sin 60° = \dfrac{\sqrt{3}}{2}$ $\sin 30° = \dfrac{1}{2}$

$\cos 60° = \dfrac{1}{2}$ $\cos 30° = \dfrac{\sqrt{3}}{2}$

$\tan 60° = \dfrac{\sqrt{3}}{1} = \sqrt{3}$ $\tan 30° = \dfrac{1}{\sqrt{3}}$

In studying mathematics, we frequently refer to the trigonometric values of 45°, 30° and 60°. Thus, it is useful to memorize these diagrams so that we can mentally obtain the trigonometric values of 45°, 30° and 60°.

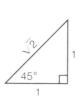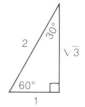

Example 2
Calculate the value of
sin 45° cos 45° + cos 30° tan 60°.

Solution
The trigonometric values are derived from the above triangles.
sin 45° cos 45° + cos 30° tan 60°
$= \left(\dfrac{1}{\sqrt{2}}\right)\left(\dfrac{1}{\sqrt{2}}\right) + \left(\dfrac{\sqrt{3}}{2}\right)(\sqrt{3}) = \dfrac{1}{2} + \dfrac{3}{2} = 2$

We may also use these triangles to calculate trigonometric values. For example, to show 225° in standard position, we use the reference triangle to locate the position of the terminal arm.

Example 3
(a) Find the value of sin 225°.
(b) Sketch 225° in standard position.

Solution To calculate trigonometric values we may choose any convenient point on the terminal arm.

The reference triangle shown in the diagram has sides marked. For the diagram
$x = -1, y = -1, r = \sqrt{2}$.
$\sin \theta = \dfrac{y}{r} \qquad \sin 225° = \dfrac{-1}{\sqrt{2}}$

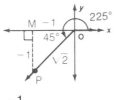

11.3 exercise

What is the value of θ in each diagram? The reference triangle is marked.
(a) ∠POM = 60°   (b) ∠POM = 45°   (c) ∠POM = 30°

(d) ∠POM = 60°   (e) ∠POM = 45°   (f) ∠POM = 30°

Write 2 negative angles that are coterminal with each angle.
(a) 45°   (b) 120°   (c) 380°

3 Write 2 positive angles that are coterminal with each angle.
(a) −45°   (b) −120°   (c) −420°

4 For each angle, write the value of the principal angle, θ, $0° \leq \theta \leq 360°$.
(a) 390°   (b) −315°   (c) 415°   (d) −105°
(e) −120° (f) 800°   (g) 780°   (h) 1090°
(i) −340° (j) 940°   (k) −215° (l) −460°

5 Draw a sketch of each angle in standard position.
(a) 120°   (b) −225°   (c) −390°
(d) −420° (e) 780°   (f) −840°

6 A point is shown on the terminal arm of each angle. Write the primary trigonometric values.

(a) 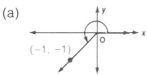   (b)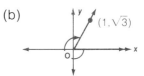

(c) 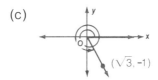   (d)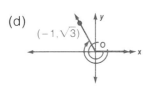

B
7 Calculate each of the following.
(a) sin 45°   (b) cos 60°   (c) cos 30°
(d) sin 60°   (e) sec 45°   (f) cot 30°
(g) tan 30°   (h) csc 60°   (i) cot 45°
(j) sec 30°   (k) csc 45°   (l) cot 60°

8 (a) Draw a sketch of 300° in standard position.
   (b) Calculate the primary trigonometric values for 300°.

9 (a) Draw a sketch of −225° in standard position.
   (b) Calculate the reciprocal trigonometric values of −225°.

10 Calculate each of the following.
   (a) sin (−60°)   (b) sin 300°
   (c) What do you notice about your answers in (a) and (b)? Give a reason for your answer.

11 Calculate each of the following.
   (a) cos 225°   (b) sin 120°   (c) tan (−150°)
   (d) tan 135°   (e) sec (−30°) (f) cot 330°
   (g) sec (−45°) (h) tan (−225°) (i) csc 240°

12 Calculate each of the following.
   (a) tan 390°   (b) cos (−480°) (c) tan 510°
   (d) csc (−495°)(e) tan 585°   (f) cos 780°

13 Calculate each of the following.
   (a) $\sin 45° \cos 45° + \tan^2 225°$
   (b) $\sin 30° \cos 60° + \sin 60° \cos 30°$
   (c) $2 \sin 60° \cos 60°$
   (d) $\sin^2 60° + \cos^2 60°$

14 Calculate each of the following.
   (a) $\cos 45° \sin 225° + \cos 330°$
   (b) $\csc 315° \sin (−120°) \cot 225°$
   (c) $\tan^2 225° − \sin 60° \cos 30°$

15 θ is an angle in the second quadrant and $\cos \theta = -\frac{1}{2}$.
   (a) Find tan θ.
   (b) What is the value of θ, $0° \leq \theta \leq 360°$?

16 α is an angle in the fourth quadrant and $\tan \alpha = -\sqrt{3}$.
   (a) Find cos α.
   (b) What is the value of α, $0° \leq \alpha \leq 360°$?

17 (a) If $\cos \theta = -\frac{1}{2}$ and $0° \leq \theta \leq 360°$, find sin θ.
   (b) What are the possible values of θ?

18 For each of the following $0° \leq \theta \leq 360°$. Find possible values of θ.
   (a) $\cos \theta = \frac{\sqrt{3}}{2}$ (b) $\sin \theta = \frac{1}{2}$
   (c) $\cot \theta = 1$ (d) $\csc \theta = -\sqrt{2}$
   (e) $\sec \theta = -2$ (f) $\sin \theta = -\frac{1}{\sqrt{2}}$

19 If β is an angle in the third quadrant and $\cos \beta = -\frac{1}{\sqrt{2}}$, then calculate $\sin \beta \cos \beta + \tan^2 \beta$.

20 The terminal arm of α is in the fourth quadrant. If $\cot \alpha = -\sqrt{3}$, then calculate $\sin \alpha \cot \alpha − \cos^2 \alpha$.

C
21 Prove the identities.
   (a) $\cos 60° \sec 30° = \tan 30°$
   (b) $\cos 45° \sin 45° + \sin 45° \cos 45° = 1$

22 Prove that
   (a) $\sin 60° + \sin 60° \neq \sin 120°$
   (b) $\cos 45° + \cos 60° \neq \cos 105°$

---

math tip

Computers are an invaluable tool as an aid in solving problems, and reducing the tedious calculations involved in many advanced problems. For example, the trigonometric values given in the tables are derived from formula such as the following.

$$\sin x = x - \frac{x^3}{3!} + \frac{x^5}{5!} - \frac{x^7}{7!} + \ldots$$

This means $5! = 5 \times 4 \times 3 \times 2 \times 1$. A computer program is then designed to tabulate the values of sin x and cos x. For example, the following BASIC program uses 3 terms of the above expression where x is expressed in radians.

10 INPUT X
20 LET S = X − X ↑ 3/6 + X ↑ 5/120
30 PRINT " THE VALUE OF SIN X IS", S
40 END

Try the above computer program.

11.4 trigonometric values: use of tables

For certain angles, we may compute the trigonometric values from a triangle, such as

$\sin 60° = \dfrac{\sqrt{3}}{2}$ $\cos 45° = \dfrac{1}{\sqrt{2}}$

$\tan 60° = \sqrt{3}$ $\tan 45° = 1$

To calculate sin 48° there is no special triangle except one that is constructed by measurement.

$\sin 48° = \dfrac{PM}{OP}$

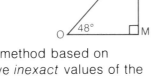

However, the above method based on measurement will give *inexact* values of the trigonometric values. Because accurate tables of trigonometric values are required for scientific work, the trigonometric values are generated by using a computer. A part of the table, found on page 456, is shown as follows:

degrees	sin	cos	tan	csc	sec	cot
45	0.7071	0.7071	1.0000	1.4142	1.4142	1.0000
46	0.7193	0.6947	1.0355	1.3902	1.4396	0.9657
47	0.7314	0.6820	1.0724	1.3673	1.4663	0.9325
48	0.7431	0.6691	1.1106	1.3456	1.4945	0.9004
49	0.7547	0.6561	1.1504	1.3250	1.5243	0.8693
50	0.7660	0.6428	1.1918	1.3054	1.5557	0.8391
51	0.7771	0.6293	1.2349	1.2868	1.5890	0.8098
52	0.7880	0.6157	1.2799	1.2690	1.6243	0.7813
53	0.7986	0.6018	1.3270	1.2521	1.6616	0.7536
54	0.8090	0.5878	1.3764	1.2361	1.7013	0.7265

The table gives the trigonometric values for each degree $0° \leq \theta \leq 90°$. The values are an approximation to 4 decimal places (except for a few exact values such as tan 45° = 1.0000, etc.). For example,
when $\theta = 48°$, sin 48° = 0.7431.
when $\theta = 52°$, tan 52° = 1.2799.

We may also use the tables to find the measures of angles when the trigonometric values are given. For example,
if cos θ = 0.6157 then θ = 52° from the table.
if sec θ = 1.4396 then θ = 46° from the table.

However if we wish to find the values for θ for the trigonometric values which do not occur in the table, we may still use the table to find the value of θ.

Example 1
(a) If sin θ = 0.7859, find θ.
(b) If cos α = 0.6311, find α.

Solution
From the above tables,
(a) sin 51° = 0.7771 ⎫ difference 0.0088
   sin θ = 0.7859 ⎬
   sin 52° = 0.7880 ⎭ difference 0.0021
   Thus $\theta \doteq 52°$. since sin θ is nearer sin 52° than sin 51°

(b) cos 50° = 0.6428 ⎫ difference 0.0117
    cos α = 0.6311 ⎬
    cos 51° = 0.6293 ⎭ difference 0.0018
    Thus $\alpha \doteq 51°$.

If the trigonometric value is given in fractional form, we must convert the fraction to a decimal in order to use the tables to obtain the value of the angle.

From the table, we note that $0° \leq \theta \leq 90°$. As θ increases, the value of sin θ increases. From a diagram, we may note the reason

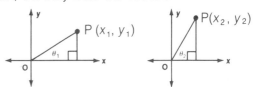

Since $\theta_2 > \theta_1$, then $y_2 > y_1$.
Thus $\dfrac{y_2}{r} > \dfrac{y_1}{r}$ and sin θ_2 > sin θ_1.

11.4 exercise

For Questions 1 to 4, use the trigonometric tables shown on the previous page.

A

1. What is the value of each of the following?
   - (a) sin 47°
   - (b) tan 48°
   - (c) csc 45°
   - (d) cos 50°
   - (e) sec 47°
   - (f) cot 45°
   - (g) csc 54°
   - (h) sin 48°
   - (i) cos 52°
   - (j) sin 53°
   - (k) cot 52°
   - (l) csc 51°

2. As θ increases in value from 45° to 54° does the value of each of the following increase (I) or decrease (D)?
   - (a) sin θ
   - (b) cos θ
   - (c) tan θ
   - (d) csc θ
   - (e) sec θ
   - (f) cot θ

3. Find θ for each of the following.
   - (a) sin θ = 0.7547
   - (b) tan θ = 1.1504
   - (c) cot θ = 0.8391
   - (d) cos θ = 0.6691
   - (e) sec θ = 1.6243
   - (f) sin θ = 0.8090

4. Find θ for each of the following.
   - (a) sin θ = 0.7443
   - (b) sec θ = 1.4718
   - (c) cot θ = 0.9391
   - (d) tan θ = 1.2662
   - (e) csc θ = 1.3478
   - (f) cos θ = 0.6320

For each of the following, use the tables for trigonometric values on page 456.

5. Find the value of each of the following.
   - (a) cos 23°
   - (b) sin 46°
   - (c) sec 79°
   - (d) csc 63°
   - (e) cot 72°
   - (f) tan 81°
   - (g) cos 42°
   - (h) sec 11°
   - (i) sin 55°

6. Find the value θ for each of the following.
   - (a) sin θ = 0.1736
   - (b) sec θ = 1.0515
   - (c) csc θ = 1.0223
   - (d) cot θ = 0.4663
   - (e) sin θ = 0.9925
   - (f) cos θ = 0.9063
   - (g) tan θ = 0.7265
   - (h) cot θ = 2.1445

7. Find the value of θ for each of the following.

   (a) 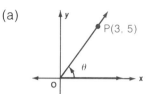 (b)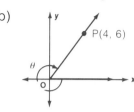

B

8. Find the value of θ.
   - (a) sin θ = 0.3439
   - (b) sec θ = 1.1143
   - (c) tan θ = 3.7219
   - (d) csc θ = 2.7821
   - (e) cos θ = 0.2856
   - (f) cot θ = 0.4563
   - (g) tan θ = 0.1853
   - (h) sin θ = 0.9860

9. $P(x, y)$ is a point on the terminal arm of θ. Find θ to the nearest degree.
   - (a) P(2, 1)
   - (b) P(3, 2)
   - (c) P(4, 3)
   - (d) P(6, 9)
   - (e) P(1, 2)
   - (f) P(2, 7)

10. θ is a first quadrant angle. Find each value of θ.
    - (a) sin θ = $\frac{1}{2}$
    - (b) cos θ = $\frac{2}{3}$
    - (c) tan θ = $\frac{3}{2}$
    - (d) cot θ = $\frac{4}{3}$
    - (e) csc θ = $\frac{5}{2}$
    - (f) sec θ = 3

11. θ is a first quadrant angle. As θ increases describe the change in values of
    - (a) sin θ
    - (b) cos θ
    - (c) tan θ
    - (d) csc θ
    - (e) sec θ
    - (f) cot θ

12. If θ is a first quadrant angle, why is
    - (a) $0 \leq \sin \theta \leq 1$?
    - (b) $0 \leq \cos \theta \leq 1$?

13. Without referring to the tables of trigonometric values, which of the following are false, for $0° \leq \theta \leq 90°$?
    - (a) cos θ = 3.2151
    - (b) tan θ = 3.2151
    - (b) sec θ = 3.2151
    - (d) sin θ = 3.2151
    - (e) cot θ = 3.2151
    - (f) csc θ = 3.2151

14 Show that
   (a) $2 \sin 32° \neq \sin 64°$
   (b) $\sin 20° + \sin 40° \neq \sin 60°$
   (c) $\dfrac{\tan 75°}{3} \neq \tan 25°$

C

15 If $0° < \theta < 90°$, then find θ if
   (a) $\sin 2\theta = 0.3420$
   (b) $\cos (\theta + 10°) = 0.1045$
   (c) $\tan (90° - 2\theta) = 1.7321$

16 Solve for θ and α if $0° < \theta + \alpha < 90°$.
   (a) $\sin (\theta + \alpha) = 0.9397$
       $\cos (\theta - \alpha) = 0.9848$
   (b) $\tan (\theta + \alpha) = 3.7321$
       $\tan (\theta - \alpha) = 0.2679$

trigonometric values of all angles

The tables on page 456 give the trigonometric values for θ, $0° \leq \theta \leq 90°$. However, we can use the tables to find the trigonometric values for angles greater than $90°$. For example, if θ is in the second quadrant, draw the angle in standard position. From the diagram the values of $\cos \angle POM$ and $\cos \theta$ differ only in sign.

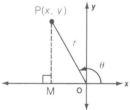

$$\cos \angle POM = \dfrac{MO}{OP} \qquad \cos \theta = \dfrac{x}{r}$$

As a particular example, if $\theta = 123°$, then
   $\cos \angle POM = \cos 57°$
   $\qquad\qquad\quad = 0.5446$
In the second quadrant, $\cos 123°$ is negative.
Thus $\cos 123° = -0.5446$.

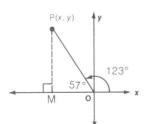

Thus, to calculate all the trigonometric values, we use the following observation about the signs of the trigonometric values.

As a memory aid $0° \leq \theta \leq 360°$, the trigonometric values of θ as shown are positive.

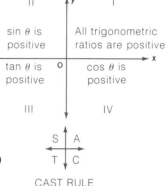

17 Use the information in each diagram to calculate the primary trigonometric values of θ.

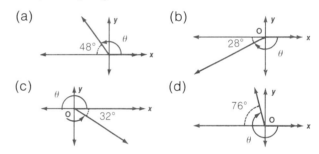

18 Use the information in each diagram to calculate the reciprocal trigonometric values of θ.

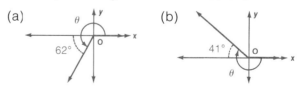

19 For each diagram, find the trigonometric value of θ shown.

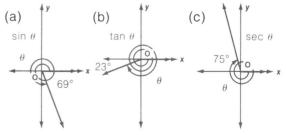

20 Find the value of
   (a) sin 127°   (b) cos 272°   (c) cos (−45°)
   (d) cos 135°   (e) sec (−57°)   (f) csc 160°
   (g) tan 229°   (h) cot (−172°)   (i) tan 159°

21 For each θ, the quadrant is given. Find the value of θ, $0° \leq \theta \leq 360°$.
   (a) cos θ = −0.9063 II   (b) tan θ = 0.5543 III
   (c) csc θ = −1.1547 III (d) sin θ = −0.9962 IV
   (e) sec θ = 1.7434 I   (f) cos θ = −0.8572 III

22 For each α, the quadrant is given. Find the value of α, $0 \leq \alpha \leq 360°$.
   (a) tan α = −3.7161 II   (b) cos α = −0.9636 III
   (c) sin α = 0.1923 II   (d) sec α = 1.0954 IV
   (e) csc α = −1.0752 III (f) cos α = −0.4461 II

If sin θ = −0.9063, then we can draw two diagrams to show θ in standard position, $0° \leq \theta \leq 360°$. Since sin θ is negative then θ is either a third or fourth quadrant angle.

23 Draw the diagrams and find the 2 values of θ.

24 (a) If cos θ = −0.3420, find θ.
   (b) Given that tan α = −0.4663, find α.

25 Find two possible values of θ for each of the following $0° \leq \theta \leq 360°$.
   (a) sin θ = −0.8910   (b) sec θ = 1.0785
   (c) tan θ = −4.3315   (d) cos θ = −0.4384
   (e) csc θ = 2.4586

26 Find θ to the nearest degree for each of the following, $0° \leq \theta \leq 360°$.
   (a) sin θ = −0.9685   (b) tan θ = 0.2761
   (c) csc θ = 1.1829   (d) cos θ = 0.3361
   (e) cot θ = −1.6340   (f) sec θ = −3.3196

11.5 working with radian measure

In our previous work, we expressed the measures of the angles in degrees. This is referred to as **sexagesimal** measure. To construct a degree, we define a complete rotation to be 360°. Thus $\frac{1}{360}$ of a rotation has a measure of 1°. Another useful measure of angles, called the **radian**, is also related to the circle. We know that for a circle, the angle subtended at the centre of the circle is proportional to the arc AB.

A radian is the measure of the angle subtended at the centre of the circle by an arc equal in length to the radius of the circle.

The measure of θ is defined to be 1 radian.

measure of an angle in radians = $\frac{\text{length of arc subtending the angle}}{\text{length of radius}}$

Measure of θ = $\frac{2r}{r}$ ← number of radii in arc AB
= 2

Measure of θ is 2 rad.

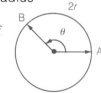

For a circle with radius r, the circumference has the measure $2\pi r$. The radian measure of each angle θ is calculated.

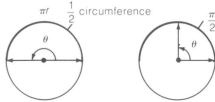

Measure of θ is given by $\frac{\text{length of arc}}{\text{length of radius}}$ $= \frac{\pi r}{r} = \pi$	Measure of θ is given by $\frac{\text{length of arc}}{\text{length of radius}}$ $= \frac{\frac{\pi}{2}r}{r} = \frac{\pi}{2}$

The relationship between the two angle units of measure, the radian and the degree, is determined from the following diagram.

 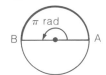

From the diagram
∠AOB = 180°   ∠AOB = π rad
Thus π rad = 180° (π rad = 180°).

Throughout our work, the measure of an angle, in radians, is written as follows:

2 rad or 2
π rad or π ⟵ radians understood

Example 1
Express each of the following in radian measure. Use $\pi \doteq 3.14$.

(a) 60°   (b) 28°

Solution
π rad = 180 degrees
or 180 degrees = π rad
1 degree = $\frac{\pi}{180}$ rad

(a) $60° = 60\left(\frac{\pi}{180}\right)$ rad
$= \frac{\pi}{3}$ rad $\left(\text{or } \frac{\pi}{3}\right)$

(b) $28° = 28\left(\frac{\pi}{180}\right)$ rad
$\doteq 28\left(\frac{3.14}{180}\right)$ rad
$\doteq 0.488$ rad
$= 0.49$ rad (to 2 decimal places)

We may write radian measures as degree measures.

Example 2
Convert each of the following radian measures into degree measures to the nearest degree. Use $\pi \doteq 3.14$.

(a) $\frac{4\pi}{3}$ (b) 2.3

Solution
(a) 1 rad = $\frac{180}{\pi}$ degrees

$\frac{4\pi}{3}$ rad $= \frac{4\pi}{3}\left(\frac{180}{\pi}\right)$ degrees
$= 240$ degrees or 240°

(b) 1 rad = $\frac{180}{\pi}$ degrees

2.3 rad $= 2.3\left(\frac{180}{\pi}\right)$ degrees
$\doteq 2.3\left(\frac{180}{3.14}\right)$ degrees
$\doteq 131.8$ degrees

2.3 rad $= 132°$ to the nearest degree

Throughout our work, we will write π rad as π.

As we shall see, the radian is a more useful measure in working with some applications of trigonometry.

11.5 exercise

Throughout the exercise, use $\pi \doteq 3.14$.

A

1 Write each radian measure as a degree measure.

(a) $\frac{\pi}{3}$ (b) $\frac{\pi}{2}$ (c) $\frac{\pi}{4}$ (d) $\frac{\pi}{6}$
(e) $\frac{3}{4}\pi$ (f) $-\frac{\pi}{2}$ (g) $-\frac{2}{3}\pi$ (h) $\frac{5\pi}{6}$
(i) $\frac{4}{3}\pi$ (j) $-\frac{3}{4}\pi$ (k) $-\frac{5}{3}\pi$ (l) $\frac{3\pi}{2}$
(m) 2π (n) -4π (o) 3π (p) $-\frac{3}{2}\pi$

2 Write each degree measure as a radian measure.
(a) 180° (b) 360° (c) 90° (d) 45°
(e) −60° (f) −150° (g) 30° (h) 240°
(i) −330° (j) 270° (k) −90° (l) 120°

3 Write the value of θ in radian measure.

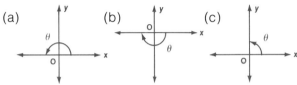

4 Sketch each angle, given in radian measure, in standard position.
(a) π (b) $\frac{1}{2}\pi$ (c) $-\frac{1}{4}\pi$ (d) $\frac{5}{4}\pi$
(e) $\frac{2}{3}\pi$ (f) $-\frac{5}{6}\pi$ (g) $\frac{9}{4}\pi$ (h) $-\frac{3}{4}\pi$

5 In which quadrant does the terminal arm of each angle lie?
(a) $\frac{3}{4}\pi$ (b) $-\frac{2}{3}\pi$ (c) $\frac{7}{6}\pi$ (d) $-\frac{7}{4}\pi$
(e) $-\frac{\pi}{3}$ (f) $\frac{7}{4}\pi$ (g) $-\frac{5}{4}\pi$ (h) $-\frac{7}{6}\pi$

6 Sector angles are drawn in a unit circle. Find the measure of the arc of the circle that subtends an angle measuring
(a) 90° (b) 180° (c) 30°
(d) 1 rad (e) $\frac{\pi}{2}$ rad (f) 2.6 rad

7 Each radian measure is given to 1 decimal place. Express to the nearest degree.
(a) 1.1 (b) 0.9 (c) −0.3 (d) 1.5
(e) −0.8 (f) 0.4 (g) 1.9 (h) 2.3

8 Convert each of the following to radian measure.
(a) $\frac{1}{2}$ revolution (b) $\frac{2}{3}$ revolution
(c) 2 revolutions (d) 10 revolutions

B

9 Express the measure of angle θ in radians.

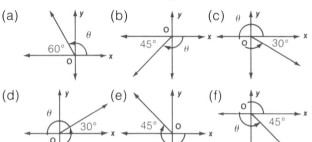

10 A point is given on the terminal arm of each angle. Calculate the measure of the principal angle in radians to 1 decimal place.
(a) (2, 3) (b) (−3, 1) (c) (−2, −5)
(d) (3, −5) (e) (−2, −3) (f) (−4, −2)

11 The radian measures of angles are shown. Write the measure of the coterminal angle θ, $-2\pi \leq \theta \leq 2\pi$.
(a) $\frac{\pi}{3}$ (b) $-\frac{\pi}{4}$ (c) $\frac{\pi}{6}$ (d) $-\pi$
(e) $\frac{3}{4}\pi$ (f) $-\frac{3}{2}\pi$ (g) $-\frac{5}{4}\pi$ (h) $\frac{2}{3}\pi$

12 Write each degree measure as a radian measure expressed to 4 decimal places.
(a) 1° (b) 8° (c) 25° (d) 65°
(e) 70° (f) 170° (g) −235° (h) −390°

13 Two angles α and $-\frac{3}{4}\pi$ are in standard position and have a terminal arm in common. Find the measure of α if $-2\pi \leq \alpha \leq 2\pi$.

14 (a) Write $\frac{\pi}{3}$ rad in sexagesimal measure.
(b) Calculate $\cos \frac{\pi}{3}$.

15 (a) Write 2 rad in sexagesimal measure.
    (b) Calculate sin 2.

16 Calculate each of the following.
    (a) cos 2   (b) tan 1   (c) sec 3   (d) cot (−1)
    (e) sin (−2)   (f) csc $\frac{1}{2}$ (g) sin 1.1   (h) cos 2.1

trigonometric values of $\frac{\pi}{3}, \frac{\pi}{4}, \frac{\pi}{6}$.

We may use the triangles as memory aids to find the trigonometric values of

$\frac{\pi}{4} = 45°$ $\frac{\pi}{3} = 60°$ $\frac{\pi}{6} = 30°$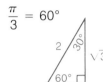

17 Use the above diagram. Calculate.
    (a) cos $\frac{\pi}{4}$ (b) sin $\frac{\pi}{6}$ (c) cos $\frac{\pi}{3}$
    (d) tan $\frac{\pi}{4}$ (e) cot $\frac{\pi}{6}$ (f) sec $\frac{\pi}{3}$

18 Find the value of each of the following.
    (a) sin $\frac{3}{4}\pi$ (b) cos $(-\frac{\pi}{4})$ (c) tan $\frac{2}{3}\pi$
    (d) csc$(-\frac{4}{3}\pi)$ (e) sec $\frac{5}{6}\pi$ (f) cot $(-\frac{\pi}{6})$
    (g) cos $(-\frac{3}{4}\pi)$ (h) sec $\frac{5}{4}\pi$ (i) tan $\frac{5}{3}\pi$
    (j) cot $(-\frac{4}{3}\pi)$ (k) csc $(-\frac{7}{6}\pi)$ (l) sin $(-\frac{5}{6}\pi)$

19 If $0 \leq \theta \leq 2\pi$, find two values of θ for each of the following.
    (a) sin $\theta = \frac{1}{\sqrt{2}}$ (b) cos $\theta = \frac{\sqrt{3}}{2}$ (c) sec $\theta = 2$

20 Find values of α, $0 \leq \alpha \leq 2\pi$ for each of the following.
    (a) cos $\alpha = -\frac{\sqrt{3}}{2}$ (b) tan $\alpha = -1$ (c) csc $\alpha = -\frac{2}{\sqrt{3}}$

11.6 identities

For particular examples of angles we make the following observations.

$\theta = 45°$ $\sin^2 \theta + \cos^2 \theta = \sin^2 45° + \cos^2 45°$
$= \left(\frac{1}{\sqrt{2}}\right)^2 + \left(\frac{1}{\sqrt{2}}\right)^2$
$= \frac{1}{2} + \frac{1}{2} = 1$

$\theta = 120°$ $\sin^2 \theta + \cos^2 \theta = \sin^2 120° + \cos^2 120°$
$= \left(\frac{\sqrt{3}}{2}\right)^2 + \left(-\frac{1}{2}\right)^2$
$= \frac{3}{4} + \frac{1}{4} = 1$

The results of particular examples suggest a hypothesis.

Is it true that $\sin^2 \theta + \cos^2 \theta = 1$ for all θ?

In this section we will prove the above result, as well as other identities in trigonometry. Remember, an **identity** is true for all values of the variable. An angle, θ, is shown in standard position. $P(x, y)$ is any point on the terminal arm. From the basic definition of the trigonometric ratios, we may derive the fundamental trigonometric identities for all θ.

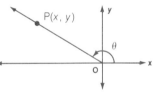

$\sin \theta = \frac{y}{r}$ $\csc \theta = \frac{r}{y}$
$= \frac{1}{\frac{r}{y}}$ $= \frac{1}{\frac{y}{r}}$
$= \frac{1}{\csc \theta}$ $= \frac{1}{\sin \theta}$

Similarly we may show that

$\cos \theta = \frac{1}{\sec \theta}$ $\sec \theta = \frac{1}{\cos \theta}$
$\tan \theta = \frac{1}{\cot \theta}$ $\cot \theta = \frac{1}{\tan \theta}$

For any angle, we may use algebra to show other identities in trigonometry. For all θ,

$$\frac{\sin \theta}{\cos \theta} = \frac{\frac{y}{r}}{\frac{x}{r}} = \frac{y}{x} = \tan \theta.$$

Thus
$$\frac{\sin \theta}{\cos \theta} = \tan \theta.$$

From the diagram we may write the identity

$$y^2 + x^2 = r^2 \quad \text{(Pythagorean Theorem)}$$

Since $r \neq 0$, then we may write the identity in different forms.

$$y^2 + x^2 = r^2$$
$$\frac{y^2}{r^2} + \frac{x^2}{r^2} = \frac{r^2}{r^2} \quad \text{or} \quad \left(\frac{y}{r}\right)^2 + \left(\frac{x}{r}\right)^2 = 1$$

From the diagram $(\sin \theta)^2 + (\cos \theta)^2 = 1$
or $\sin^2 \theta + \cos^2 \theta = 1$

Similarly, we may write

$$y^2 + x^2 = r^2 \qquad\qquad y^2 + x^2 = r^2$$
$$\frac{y^2}{x^2} + \frac{x^2}{x^2} = \frac{r^2}{x^2} \qquad \frac{y^2}{y^2} + \frac{x^2}{y^2} = \frac{r^2}{y^2}$$
$$\tan^2 \theta + 1 = \sec^2 \theta \qquad 1 + \cot^2 \theta = \csc^2 \theta$$

We refer to these identities as the **Pythagorean Identities**. For all θ.

$$\sin^2 \theta + \cos^2 \theta = 1$$
$$\tan^2 \theta + 1 = \sec^2 \theta$$
$$1 + \cot^2 \theta = \csc^2 \theta$$

Once we have proved the above fundamental trigonometric identities we may use them to deduce other results and prove identities.
To show that the following identity is true, we must simplify the expression and show that LS = RS.

Example 1

For all α, prove that $\dfrac{1}{\cot \alpha} = \sin \alpha \sec \alpha$.

Solution

Since the RS has $\sin \alpha$, we attempt to express LS and RS so that they involve $\sin \alpha$.

$$\text{LS} = \frac{1}{\cot \alpha} \qquad\qquad \text{RS} = \sin \alpha \sec \alpha$$
$$= \tan \alpha \qquad\qquad\quad = \sin \alpha \frac{1}{\cos \alpha}$$
$$= \frac{\sin \alpha}{\cos \alpha} \qquad\qquad\quad = \frac{\sin \alpha}{\cos \alpha}$$

Thus LS = RS.
For all α, $\dfrac{1}{\cot \alpha} = \sin \alpha \sec \alpha$. ← Make a final concluding statement.

To prove some identities we may observe that one side has only $\sin \theta$ and $\cos \theta$. Thus, we may begin with the other side and attempt to express it in terms of $\sin \theta$ and $\cos \theta$.

Example 2

For all θ, prove that
$$\tan \theta + \frac{1}{\tan \theta} = \frac{1}{\sin \theta \cos \theta}.$$

Solution

$$\text{LS} = \tan \theta + \frac{1}{\tan \theta}$$
$$= \frac{\sin \theta}{\cos \theta} + \frac{1}{\frac{\sin \theta}{\cos \theta}} \qquad \tan \theta = \frac{\sin \theta}{\cos \theta}$$
$$= \frac{\sin \theta}{\cos \theta} + \frac{\cos \theta}{\sin \theta} \quad \left\{\begin{array}{l}\text{We use our earlier skills} \\ \text{in algebra to find a} \\ \text{common denominator.}\end{array}\right.$$
$$= \frac{(\sin \theta)(\sin \theta) + (\cos \theta)(\cos \theta)}{\cos \theta \sin \theta}$$
$$= \frac{\sin^2 \theta + \cos^2 \theta}{\cos \theta \sin \theta}$$
$$= \frac{1}{\cos \theta \sin \theta} \qquad \text{But } \sin^2 \theta + \cos^2 \theta = 1.$$

$$\text{RS} = \frac{1}{\sin \theta \cos \theta} = \frac{1}{\cos \theta \sin \theta}$$

Thus LS = RS

For all θ, $\tan \theta + \dfrac{1}{\tan \theta} = \dfrac{1}{\sin \theta \cos \theta}$.

11.6 exercise

Questions 1 to 9 examine skills needed to prove identities in trigonometry.

A

1. Use the ratios $\sin\theta = \dfrac{y}{r}$, $\cos\theta = \dfrac{x}{r}$ and so on, to prove
   (a) $\sin\theta = \dfrac{1}{\csc\theta}$
   (b) $\cot\theta = \dfrac{1}{\tan\theta}$

2. For all θ, prove that
   (a) $\cos\theta = \dfrac{1}{\sec\theta}$
   (b) $\sec\theta = \dfrac{1}{\cos\theta}$
   (c) $\tan\theta = \dfrac{1}{\cot\theta}$
   (d) $\csc\theta = \dfrac{1}{\sin\theta}$

3. Use the Pythagorean identity
   $\sin^2\theta + \cos^2\theta = 1$
   to show that
   (a) $\tan^2\theta + 1 = \sec^2\theta$ (b) $1 + \cot^2\theta = \csc^2\theta$

4. (a) Prove that $1 + \cot^2\alpha = \csc^2\alpha$ for all α.
   (b) Use $\alpha = 120°$ to illustrate the identity in (a).

5. Use the ratios $\sin\theta = \dfrac{y}{r}$, $\cos\theta = \dfrac{x}{r}$, and so on, to prove
   (a) $\tan\theta \cos\theta = \sin\theta$
   (b) $\cot\theta \sec\theta = \csc\theta$
   (c) $\dfrac{1 + \cot^2\theta}{\csc^2\theta} = 1$

6. Factor each of the following.
   (a) $1 - \cos^2\theta$
   (b) $1 - \sin^2\theta$
   (c) $\sin^2\theta - \cos^2\theta$
   (d) $\sin\alpha - \sin^2\alpha$
   (e) $\tan^2\alpha - \cot^2\alpha$
   (f) $\sec^2\theta - 1$

7. Express each of the following in terms of $\sin\theta$ or $\cos\theta$ or both.
   (a) $\dfrac{1}{\sec\theta}$
   (b) $\sin^2\theta + \dfrac{1}{\sec^2\theta}$
   (c) $\cos\theta \dfrac{1}{\sec\theta}$
   (d) $\tan\theta \cos\theta$
   (e) $1 - \csc^2\theta$
   (f) $\dfrac{1 + \cot^2\theta}{\cot^2\theta}$

8. (a) To show that
   $\cos\alpha \cot\alpha = \dfrac{1}{\sin\alpha} - \sin\alpha$
   express the LS in terms of $\sin\alpha$.
   (b) Prove the identity.

9. (a) To show that
   $\dfrac{\cos^2\theta}{1 - \sin\theta} = 1 + \sin\theta$
   write $\cos^2\theta$ in terms of $\sin\theta$.
   (b) What are the factors of $1 - \sin^2\theta$?
   (c) Use (a) and (b) to prove the identity.

B

10. (a) Prove that for all angles θ
    $\dfrac{\tan^2\theta}{1 + \tan^2\theta} = \sin\theta \cos\theta$.
    (b) Check the identity in (a) using $\theta = 45°$.

11. (a) Prove that for all α,
    $\dfrac{\tan^2\alpha}{\sin^2\alpha} = 1 + \tan^2\alpha$.
    (b) Check the identity in (a) using $\alpha = 120°$.

12. Prove each identity.
    (a) $\tan\theta \cos\theta = \sin\theta$
    (b) $1 - \cos^2\theta = \dfrac{\cos\theta \sin\theta}{\cot\theta}$
    (c) $\dfrac{\sec\theta}{\cot\theta} = \dfrac{\sin\theta}{\cos^2\theta}$

13. Prove each identity.
    (a) $\cot\theta \cot\theta \times \tan\theta \tan\theta = 1$
    (b) $\tan\theta \sin\theta + \cos\theta = \sec\theta$
    (c) $\sec\theta \cos\theta + \sec\theta \sin\theta = 1 + \tan\theta$

14 Prove each of the following identities.

(a) $\dfrac{\sin^2 \theta}{1 - \cos \theta} = 1 + \cos \theta$

(b) $\dfrac{1 - \tan^2 \theta}{\tan \theta - \tan^2 \theta} = 1 + \dfrac{1}{\tan \theta}$

(c) $\sin^4 \alpha - \cos^4 \alpha = \sin^2 \alpha - \cos^2 \alpha$

15 Prove each of the following identities. (Be aware of any restrictions that might be involved in completing your proof.)

(a) $\cos^2 \theta = (1 - \sin \theta)(1 + \sin \theta)$

(b) $\dfrac{1}{\sec \theta} + \dfrac{\sin \theta}{\cot \theta} = \dfrac{1}{\cos \theta}$

(c) $(1 - \sec \theta)(1 + \sec \theta) = -\tan^2 \theta$

(d) $\csc \theta (1 + \sin \theta) = 1 + \csc \theta$

(e) $\dfrac{1 - \tan \theta}{\tan \theta} = \dfrac{\cos \theta - \sin \theta}{\sin \theta}$

(f) $\sec \theta + \dfrac{1}{\cot \theta} = \dfrac{1 + \sin \theta}{\cos \theta}$

(g) $\dfrac{\cos \theta}{\csc \theta} - \dfrac{\sin \theta}{\tan \theta} = \dfrac{\sin \theta - 1}{\sec \theta}$

(h) $\dfrac{\sec^2 \theta}{\sin^2 \theta} = \dfrac{1}{\sin^2 \theta} + \dfrac{1}{\cos^2 \theta}$

(i) $\sin^2 \theta \cot^2 \theta + \dfrac{\cos^2 \theta}{\cot^2 \theta} = 1$

(j) $\dfrac{1}{\cot \theta} + \sin \theta = \sin \theta \left(\dfrac{\cos \theta + 1}{\cos \theta} \right)$

(k) $\dfrac{\cos^2 \theta - \sin^2 \theta}{\cos^2 \theta + \sin \theta \cos \theta} = \dfrac{\cot \theta - 1}{\cot \theta}$

(l) $\sin \theta \tan \theta + \cos \theta - \sec \theta + 1 = \sec^2 \theta \cos^2 \theta$

C

16 Prove for all θ, $\sin^4 \theta - \cos^4 \theta = \dfrac{2 - \sec^2 \theta}{-\sec^2 \theta}$.

17 For all α, prove
$$\tan \alpha = \dfrac{\sin \alpha + \sin^2 \alpha}{\cos \alpha (1 + \sin \alpha)}.$$

18 Prove that for all θ
$$\dfrac{\csc \theta}{1 + \csc \theta} + \dfrac{\csc \theta}{1 - \csc \theta} = -\dfrac{2 \sin \theta}{\cos^2 \theta}.$$

11.7 drawing graphs: trigonometric functions

There are many phenomena that are periodic in nature. For example
- the swing of the pendulum of a clock.
- the bobbing up and down of a ship.

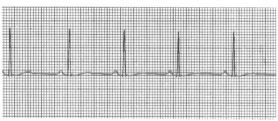

Heartbeats are periodic also, as shown in this cardiogram.

A study of the trigonometric functions will enable us to understand better the nature of periodic phenomena.

The skills and strategies we have learned earlier for $y = mx + b$ extend to the study of the graph of trigonometric functions given by

$\sin: \theta \longrightarrow \sin \theta$ $\cos: \theta \longrightarrow \cos \theta$ $\tan: \theta \longrightarrow \tan \theta$
$y = \sin \theta$ $y = \cos \theta$ $y = \tan \theta$

To draw the graph, we again construct tables of values. In particular, for the sine function, we may use the table of values for
$0° \leq \theta \leq 360°$ (or, in radians, $0 \leq \theta \leq 2\pi$).

To calculate the trigonometric values of 0°, 90°, and so on, we choose any point on the terminal arm as shown.

For 0° or 0 rad

$\sin \theta = \dfrac{y}{r}$ $y = 0, x = 1$
           $r = \sqrt{x^2 + y^2}$
           $r = 1$

P(1, 0)

$\sin 0° = \dfrac{0}{1}$ $\cos 0° = \dfrac{1}{1}$ $\tan 0° = \dfrac{0}{1}$

$\sin 0° = 0$ $\cos 0° = 1$ $\tan 0° = 0$

$\sin 0 = 0$ $\cos 0 = 1$ $\tan 0 = 0$

For 90° or $\frac{\pi}{2}$ rad

$\sin \theta = \frac{y}{r}$ $y = 1, x = 0$
$r = \sqrt{x^2 + y^2}$
$r = 1$

$\sin 90° = \frac{1}{1}$ $\cos 90° = \frac{0}{1}$ $\tan 90° = \frac{1}{0}$
$\quad = 1$ $\quad = 0$ undefined

$\sin \frac{\pi}{2} = 1$ $\cos \frac{\pi}{2} = 0$ $\tan \frac{\pi}{2}$ is undefined.

From the table of values, the values of tan θ increase without limit as θ increases in value. Similarly we can calculate the trigonometric values of 180°, 270°, or π, $\frac{3}{2}\pi$ rad, and so on.

Label the axes when drawing the graph of the function
$\sin: \theta \longrightarrow \sin \theta$ or
$y = \sin \theta$.

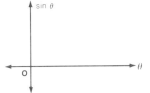

Suitable values of θ are chosen, and the corresponding value of sin θ is recorded to 2 decimal places in the chart.

θ	0°	30°	60°	90°	120°	150°	180°
sin θ	0	0.5	0.87	1	0.87	0.5	0
θ	210°	240°	270°	300°	330°	360°	
sin θ	−0.5	−0.87	−1	−0.87	−0.5	0	

The corresponding ordered pairs, (θ, sin θ), are plotted.

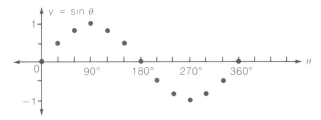

We may join the points using a smooth curve.
(For every θ there is a corresponding value sin θ.)

For radian measure the axes would be labelled as shown.

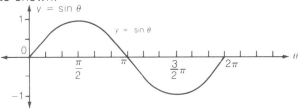

For the above portion of the graph,
  domain $\{\theta \mid 0 \leqq \theta \leqq 2\pi\}$ (as radian)
  $\{\theta \mid 0° \leqq \theta \leqq 360°\}$ (as degrees)
  range $\{y \mid -1 \leqq y \leqq 1\}$
    where $y = \sin \theta$
  intercepts: y-intercept = 0
    x-intercepts 0°, 180°, 360° or 0, π, 2π

To draw the graph for all θ, the ordered pairs (θ, sin θ) are plotted to obtain the graph in the following form.

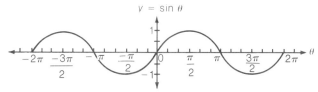

From the graph, we may make these observations.
- The graph extends indefinitely in either direction, as (θ, sin θ) are graphed for all θ.
- The graph recurs in cycles. The graph for
  $0° \leqq \theta \leqq 360°$ or $0 \leqq \theta \leqq 2\pi$
  is one complete cycle. We say the function, sin, is *periodic*. The period is 2π.
- The y-intercept is 0.
- The x-intercepts are integral multiples of π.
  In degrees ..., −360°, −180°, 0°, 180°, 360°, ...
  In radians ..., -2π, $-\pi$, 0, π, 2π, ...
  In general, the x-intercepts are given by $\pi + n\pi, n \in I$
- Every value of θ determines a unique value of sin θ. Thus, we may say that $\sin: \theta \longrightarrow \sin \theta$ is a function.

In the following exercise, we will investigate some properties of the trigonometric functions.

11.7 exercise

Throughout the questions, use the approximation $\pi \doteq 3.14$ as needed. Express answers to 1 decimal place or to the nearest degree.

A

1. Calculate the trigonometric values of each of the following.

   (a) $\cos 180°$ (b) $\sec 90°$ (c) $\cos 90°$
   (d) $\sec 270°$ (e) $\csc 90°$ (f) $\cot (-180°)$
   (g) $\tan 270°$ (h) $\cos 360°$ (i) $\sec (-360°)$

2. Calculate each of the following.

   (a) $\cos \pi$ (b) $\sin \left(-\dfrac{\pi}{2}\right)$ (c) $\tan \dfrac{3}{2}\pi$
   (d) $\sec 2\pi$ (e) $\csc (-\pi)$ (f) $\cot \dfrac{\pi}{2}$
   (g) $\sec \left(-\dfrac{3}{2}\pi\right)$ (h) $\sin \dfrac{3}{2}\pi$ (i) $\cos (-\pi)$

Questions 3 to 7 develop properties of the trigonometric functions.

3. Copy and complete each table of values for $(\theta, \sin \theta)$.

   (a)
degrees	0°	30°	60°	90°	120°	150°	180°
radians							
sin θ							

   (b)
degrees							
radians	π	$\dfrac{7}{6}\pi$	$\dfrac{4}{3}\pi$	$\dfrac{3}{2}\pi$	$\dfrac{5}{3}\pi$	$\dfrac{11}{6}\pi$	2π
sin θ							

   (c)
degrees	0°	-30°	-60°	-90°	-120°	-150°	-180°
radians							
sin θ							

   (d)
degrees							
radians	$-\pi$	$-\dfrac{7}{6}\pi$	$-\dfrac{4}{3}\pi$	$-\dfrac{3}{2}\pi$	$-\dfrac{5}{3}\pi$	$-\dfrac{11}{6}\pi$	-2π
sin θ							

B

4. (a) Use the table of values in Question 3(a) and (b) to construct a graph of the function $\sin: \theta \longrightarrow \sin \theta$, $0 \leq \theta \leq 2\pi$.
   (b) Use the table of values in Question 3(c) and (d) to construct a graph of the function $\sin: \theta \longrightarrow \sin \theta$, $-2\pi \leq \theta \leq 0$.

5. For the graph drawn of the function $\sin: \theta \longrightarrow \sin \theta$, $-2\pi \leq \theta \leq 2\pi$, what is
   (a) the domain?   (b) the range?
   (c) the value(s) of the y-intercept?
   (d) the value(s) of the x-intercept?
   (e) the maximum value of $\sin \theta$ for $-2\pi \leq \theta \leq 2\pi$?
   (f) the minimum value of $\sin \theta$ for $-2\pi \leq \theta \leq 2\pi$?
   (g) the value(s) of θ when $\sin \theta$ is a maximum value?
   (h) the value(s) of θ when $\sin \theta$ is a minimum value?

6. (a) Study the graph in Question 5 and use it to sketch the graph of the trigonometric function $y = \sin \theta$, $-4\pi \leq \theta \leq 4\pi$.
   (b) Extend the graph in (a) to include the domain $-6\pi \leq \theta \leq 6\pi$.
   (c) Why may we refer to the graph given by $y = \sin \theta$ as periodic?

7. (a) What is meant by the statement "The graph of the trigonometric function, $\sin: \theta \longrightarrow \sin \theta$ is periodic."?
   (b) What is the period of the sine function?

8. Refer to Question 3. Complete a table similar to those in parts (a) to (d) for values of $\cos \theta$.

9. (a) Use the tables of values for $\cos \theta$ in Question 8 to construct a graph of the function
   $$\cos: \theta \longrightarrow \cos \theta, \ 0 \leq \theta \leq 2\pi.$$

(b) Use the table of values for cos θ to construct the graph of the cosine function for $-2\pi \leq \theta \leq 0$.

10 For the graph of cos: $\theta \longrightarrow \cos \theta$, $-2\pi \leq \theta \leq 2\pi$, drawn in Question 9, what is

(a) the domain? (b) the range?
(c) the value(s) of the y-intercept?
(d) the value(s) of the x-intercept?
(e) the maximum value of cos θ for $-2\pi \leq \theta \leq 2\pi$?
(f) the minimum value of cos θ for $-2\pi \leq \theta \leq 2\pi$?
(g) the value(s) of θ at which the maximum values of cos θ occur?
(h) the value(s) of θ at which the minimum values of cos θ occur?

11 Use your work in Questions 8 to 10.

(a) Extend the graph of cos: $\theta \longrightarrow \cos \theta$ for the values $-4\pi \leq \theta \leq 4\pi$.
(b) Why may we refer to the graph given by $y = \cos \theta$ as periodic?
(c) Why may we say that cos: $\theta \longrightarrow \cos \theta$ is a function?

12 (a) What is meant by the statement "The graph of the trigonometric function cos: $\theta \longrightarrow \cos \theta$ is periodic."?
(b) What is the period of the cosine function?

13 Compare the graphs of the two functions. sin: $\theta \longrightarrow \sin \theta$, cos: $\theta \longrightarrow \cos \theta$

(a) How are the graphs alike?
(b) How do the graphs differ?

Questions 14 to 16 develop the properties for the trigonometric function, tan: $\theta \longrightarrow \tan \theta$.

14 (a) Construct a table of values for $(\theta, \tan \theta)$, $0 \leq \theta \leq 2\pi$.

(b) Use the results in (a) to draw the graph given by tan: $\theta \longrightarrow \tan \theta$.

15 Refer to the graph in Question 14 for tan: $\theta \longrightarrow \tan \theta$, $0 \leq \theta \leq 2\pi$.

(a) What is the domain?
(b) Give the value(s) of the x-intercept.
(c) Give the value(s) of the y-intercept.
(d) Why may we say that tan: $\theta \longrightarrow \tan \theta$ represents a function?

16 Use your graph in Question 14 to justify each of the following statements for the function given by tan: $\theta \longrightarrow \tan \theta$

(a) The tangent function has no maximum or minimum value.
(b) The function is periodic. The period is π.
(c) There is no y-intercept.

trigonometric function values

We may use different symbols to write the same function
$$f: \theta \longrightarrow \sin \theta$$
$$\sin: \theta \longrightarrow \sin \theta$$
$$s: \theta \longrightarrow \sin \theta$$
We may write $f(\theta) = \sin \theta$ or $s(\theta) = \sin \theta$
the value of / the value of /
the function f, at θ the function s, at θ

If $f: \theta \longrightarrow \sin \theta$ then we say that $f(\theta)$ is the *value of the trigonometric function*, f, at the value θ. In short we will refer to $f(\theta)$ or sin θ as the *trigonometric value*.

17 If $s: \theta \longrightarrow \sin \theta$, then $s(\theta) = \sin \theta$. Calculate each of the following.

(a) $s(\frac{\pi}{3})$ (b) $s(\frac{\pi}{4})$ (c) $s(\frac{\pi}{2})$
(d) $s(-\frac{\pi}{2})$ (e) $s(\frac{3}{2}\pi)$ (f) $s(\pi)$
(g) $s(\frac{7}{6}\pi)$ (h) $s(\frac{3}{4}\pi)$ (i) $s(-\frac{2}{3}\pi)$

18 If $f(\theta) = \sin \theta$, then calculate
  (a) $f(30°)$ (b) $f(60°)$ (c) $f(90°)$
  (d) $f(-90°)$ (e) $f(225°)$ (f) $f(120°)$

19 For $c: \theta \longrightarrow \cos \theta$, then $c(\theta) = \cos \theta$. Calculate each of the following.
  (a) $c(\frac{\pi}{6})$ (b) $c(\frac{\pi}{4})$ (c) $c(\frac{\pi}{2})$
  (d) $c(-\frac{\pi}{6})$ (e) $c(-\frac{\pi}{4})$ (f) $c(-\frac{3}{2}\pi)$
  (g) $c(2\pi)$ (h) $c(\pi)$ (i) $c(-\frac{5\pi}{4})$
  (j) $c(30°)$ (k) $c(60°)$ (l) $c(-180°)$
  (m) $c(225°)$ (n) $c(330°)$ (o) $c(-45°)$

20 For $t: \theta \longrightarrow \tan \theta$, then $t(\theta) = \tan \theta$, calculate each of the following.
  (a) $t(\frac{\pi}{3})$ (b) $t(\frac{2}{3}\pi)$ (c) $t(-\frac{\pi}{2})$
  (d) $t(210°)$ (e) $t(-135°)$ (f) $t(-\frac{3}{2}\pi)$

applications: tides

The effect of the moon on the tides of the earth is periodic. The cycle of the tides repeats every 24 h.

The Fundy tides rise more than 8m around the "Flower Pot Rocks" at Hopewell Cape, New Brunswick

A graph of the tidal motion is shown for a cycle of any 24-h period.

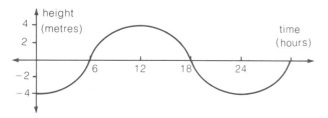

Questions 21 to 24 are based on the above graph.

21 What will be the height of the tide at each of the following times?
  (a) 6 h   (b) 12 h   (c) 18 h
  (d) 3 h   (e) 9 h   (f) 21 h

22 At what times on the graph will each of the following heights of the tide occur?
  (a) 4 m   (b) -4 m   (c) 0 m
  (d) -2 m (e) 3.5 m (f) 2 m

23 (a) Extend the above graph to cover a 48-h period.
  (b) At what times on the graph will maximum tide occur?
  (c) At what times on the graph will minimum tide occur?

24 At a certain town, the tide has the same profile as that shown above. By measurement, it is noted that maximum tide occurs at 11:00.
  (a) Draw a graph to show the heights of the tide for a 24-h period.
  (b) Use the graph in (a). At what times does low tide occur?

problem/matics

Once we learn new skills in one branch of mathematics we may use them to derive additional mathematics. For example, we may calculate the area of a segment of a circle with radius r, given the following information.

$$A = \frac{\pi r^2 \theta}{360} - \frac{r^2 \sin \theta}{2}$$ where θ is the sector angle measured in radians.

Find the area of each segment.

(a)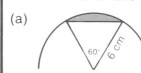

(b)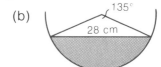

elements of applied trigonometry

11.8 solving right-angled triangles

In mathematics

I				II
We use what we know	=	Trigonometry	⇒	to show what we want to know.

The Ancients used the skills and concepts of trigonometry to solve problems involving distance. In this section and subsequent ones we will study the elements of applied trigonometry to solve problems based on triangles.

We refer to the sides of a right-angled triangle as shown, with respect to $\angle A$.

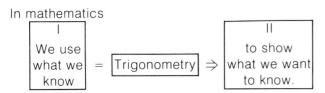

Primary Ratios of Trigonometry

$\sin A = \dfrac{\text{opposite}}{\text{hypotenuse}}$

$\cos A = \dfrac{\text{adjacent}}{\text{hypotenuse}}$

$\tan A = \dfrac{\text{opposite}}{\text{adjacent}}$

Reciprocal Ratios of Trigonometry

$\csc A = \dfrac{\text{hypotenuse}}{\text{opposite}}$

$\sec A = \dfrac{\text{hypotenuse}}{\text{adjacent}}$

$\cot A = \dfrac{\text{adjacent}}{\text{opposite}}$

What are the corresponding trigonometric ratios for $\angle B$?

We may use the tables of trigonometric values to calculate the missing parts of triangles. To solve these problems, we use our skills with algebra to solve equations derived by using our concepts in trigonometry.

In order to simplify the calculation we should choose trigonometric ratios that require the variable to be placed in the numerator. For example, we would use
$\dfrac{b}{36} = \sec 43°$ rather than $\dfrac{36}{b} = \cos 43°$.

Example 1

In $\triangle ABC$, $\angle C = 90°$, $\angle B = 42°$ and $c = 36$. Find b to 1 decimal place.

Solution Draw a diagram to record the given information and plan your solution.

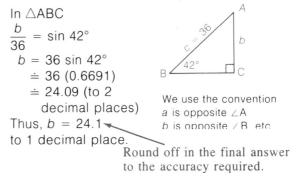

In $\triangle ABC$

$\dfrac{b}{36} = \sin 42°$

$b = 36 \sin 42°$

$\doteq 36 \,(0.6691)$

$\doteq 24.09$ (to 2 decimal places)

Thus, $b = 24.1$ to 1 decimal place.

We use the convention a is opposite $\angle A$, b is opposite $\angle B$, etc.

Round off in the final answer to the accuracy required.

We can use the table of trigonometric values to find the measure of the angle of a triangle.

Example 2

In $\triangle PQR$, $PQ \perp RQ$, $PQ = 28$, $RP = 39$. Find $\angle R$.

Translate the given information to a diagram.

Solution

In $\triangle PQR$, $\angle Q = 90°$.

$\sin R = \dfrac{PQ}{RP}$

$= \dfrac{28}{39}$

$\doteq 0.7179$ (to 4 decimal places)

From the tables

$\sin 45° = 0.7071$ ⎫ difference 0.0108
$\sin R = 0.7179$ ⎬
$\sin 46° = 0.7193$ ⎭ difference 0.0014

Thus $\angle R = 46°$.

392

To solve △ABC means to find the measures of all parts of △ABC.

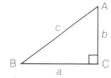

Example 3
In △ABC, ∠C = 90°, b = 8.0, ∠B = 36°. Solve △ABC. Express answers to 1 decimal place.

Solution

Translate the given information to a diagram.

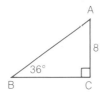

In △ABC, the sum of the angles is 180°.
∠A = 180° − (90° + 36°)
    = 54°

$\frac{c}{8}$ = csc 36° $\frac{a}{8}$ = cot 36°

c = 8 csc 36°   a = 8 cot 36°
  ≐ 8(1.7013)   ≐ 8(1.3764)
  ≐ 13.61 (2 decimal   ≐ 11.01 (2 decimal
       places)   places)

For the final answer round to the accuracy desired.

Thus, for the accuracy required △ABC is given by
a = 11.0   b = 8.0   c = 13.6
∠A = 54°   ∠B = 36°   ∠C = 90°

You may also summarize the final information on the triangle to be sure all parts have been determined.

In Example 3, to avoid possible errors, the values of c and a were calculated using the given information (side b) rather than using values that had been calculated. Be sure, where possible, to derive results from the given data.

The skills for solving triangles are now applied in the next section to solving problems based on our skills in trigonometry.

11.8 exercise

Unless indicated otherwise, express the measures of sides to 1 decimal and angles to the nearest degree.

A

1. For △ABC, which trigonometric ratio of ∠B is equal to

   (a) $\frac{AB}{AC}$? (b) $\frac{AC}{CB}$?
   (c) $\frac{CB}{AB}$? (d) $\frac{CB}{AC}$? (e) $\frac{AB}{CB}$? (f) $\frac{AC}{AB}$?

2. Write the values of the primary trigonometric ratios (in fractional form) for

   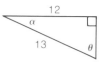

   (a) sin A   (b) sin B   (c) cos A   (d) tan B

3. Write the values of the reciprocal trigonometric ratios (in fractional form) for

   (a) sec θ   (b) cot α   (c) csc θ   (d) csc α

4. For △DEF, which trigonometric ratio of ∠E has each value?

   (a) $\frac{7}{24}$ (b) $\frac{25}{7}$
   (c) $\frac{24}{7}$ (d) $\frac{24}{25}$ (e) $\frac{7}{25}$ (f) $\frac{25}{24}$

5. For each triangle, calculate the measure of the missing side.

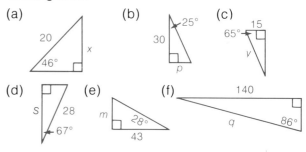

6 For each triangle, find the measure of the missing angle.

(a)   (b) 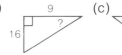 (c)

(d)   (e)   (f)

7 For each triangle, find $\sin \theta$.

(a)   (b)   (c) 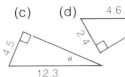 (d)

B

8 For △PQR, find the missing side.
(a) $\angle Q = 90°$, $\angle P = 61°$ $q = 14$.
(b) $\angle P = 90°$, $\angle Q = 38°$ $q = 2.6$.
(c) $\angle R = 90°$, $\angle Q = 62°$ $r = 160$.

9 In △ABC, find the missing angles
(a) $\angle A = 90°$, $a = 16$ $c = 13$
(b) $\angle C = 90°$, $c = 9.2$, $a = 5.3$
(c) $\angle B = 90°$, $c = 82$, $b = 120$

10 (a) In △ABC, $b = 12$, $a = 16$, $\angle C = 90°$. Find $\angle B$.
(b) In △PQR, $\angle R = 35°$, $\angle Q = 90°$, $RQ = 7.6$. Find RP.
(c) In △STU, $t = 20$, $\angle T = 90°$, $\angle S = 28°$. Find s.
(d) In △PQR, $p = 21.1$, $\angle R = 19°$, $\angle Q = 90°$. Find r.

11 In △PQR, $q = 9$, $p = 6$, $\angle Q = 90°$.
(a) Find r. (b) Find $\angle P$.
(c) Write the 6 trigonometric ratios for $\angle R$.

12 Solve △PQR.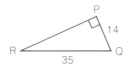

13 Solve △PQR.

14 Solve △PMR.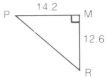

15 Solve each triangle.
(a) △PQR, $\angle Q = 90°$, $r = 8$, $p = 6$.
(b) △STU, $\angle T = 90°$, $\angle U = 28°$, $s = 14$.
(c) △ABC, $\angle B = 90°$, $\angle C = 30°$, AC = 4.8.
(d) △DEF, $\angle E = 90°$, DE = 14, DF = 25.
(e) △RMP, $\angle R = 90°$, $p = 120$, $\angle P = 38°$.

16 (a) In △ABC, AB = 12.8, AC = 14.6 and $\angle B = 90°$. Find the measures of the missing sides and angles.
(b) In △DEF, $\angle E = 90°$, $e = 28.1$ and $\angle F = 39°$. Solve △DEF.
(c) In △GHI, HI = 143, $\angle I = 90°$ and $\angle G = 43°$. Solve △GHI.

17 In △STU, $\angle U = 90°$, $t = 15$, $u = 17$, show that
(a) $\sin^2 T + \cos^2 T = 1$. (b) $\sec^2 S = 1 + \tan^2 S$.

18 In △PRT, an altitude PS is drawn so that PS ⊥ RT and S is in RT. If PT = 14, $\angle R = 72°$, and $\angle T = 61°$, then find the length of RT.

C

19 A quadrilateral RSTU is drawn so that ST ⊥ RT and $\angle U = 90°$. If $\angle RTU = 32°$ and $\angle SRT = 42°$, find RS when UT = 25.

problem/matics

We may use trigonometry to develop new formulae for the area of a triangle. Prove that the area A of △ABC is given by

$A = \dfrac{1}{2}ab \sin C$

$A = \dfrac{1}{2}ac \sin B$

$A = \dfrac{1}{2}bc \sin A$

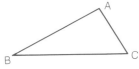

Use the formula to find the areas of triangles in the above exercise.

11.9 solving problems based on right angles

We can use our skills in trigonometry to calculate the lengths of inaccessible distances. For example, to measure the width of a river which is inaccessible the following measurements may be made.

The points A, B, and C are located on the banks of the river. The measureable distance, AC, is found and a line of sight, AB, determines the measure of ∠BAC = 79°.

Example 1
Use the above diagram to calculate the width of the river to 1 decimal place.

Solution
Let the width, d, in metres, represent the width of the river. From the diagram

$$\frac{d}{26.5} = \tan 79°$$
$$d = 26.5 \tan 79°$$
$$d \doteq 26.5(5.1446)$$
$$\doteq 136.33 \text{ (2 decimal places)}$$

Thus, the width of the river is 136.3 m to 1 decimal place.

Angles determined by lines of sight are measured with surveying equipment. For example,

From a point A an observer can measure an angle of elevation and the distance d, to calculate the height of a hydro tower.

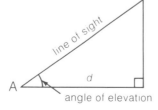

From a tower B, an observer can measure the angle of depression and use the height, d, of the tower to calculate distances.

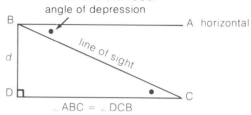

∴ ∠ABC = ∠DCB

In trigonometry the given information in word problems is *translated* with the aid of a diagram as can be seen in Example 2.

Example 2
An airplane involved in an air rescue mission determines the angle of depression of a disabled freighter to be 12°. If the airplane is at an altitude of 1200 m, how far away is the freighter? Express your answer to the nearest 10 m.

Solution
Let d, in metres, represent the horizontal distance of the freighter from the plane.

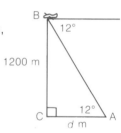

In △ABC
$$\frac{d}{1200} = \cot 12°$$
$$d = 1200 \cot 12°$$
$$\doteq 1200(4.7046)$$
$$\doteq 5645.5 \text{ (to 1 decimal place)}$$

Thus, the freighter is 5650 m away.

Although trigonometry is used today to solve problems involving distance, it is also used to solve problems in economics, navigation, medicine and so on.

11.9 exercise

A

1. To measure the width of a swamp the following measurements are noted on the diagram. What is the width of the swamp?

2. Measurements are made for a triangular lot as shown.

Calculate the amount of frontage on Portage Avenue.

3. Calculate the area of each triangle.

(a) 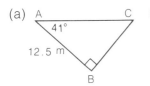 (b)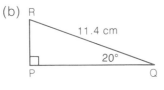

4. The Rowden Dam stands 700 m high. If the holding surface is angled back 10° from the vertical, calculate

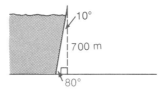

   (a) the length of the holding surface.
   (b) the area of the holding surface if the river is 150 m wide.

5. The slope of a ski hill is 0.84. Find the angle that the hill makes with the horizontal.

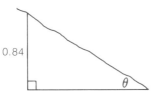

6. The diagram shows the supporting lines for the mast of a sailboat. If the mast is 10 m high, calculate the length of each supporting line.

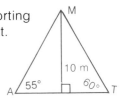

B

7. From a lifeguard station at a beach, the deck is 18 m above the sea level. A shark is seen at an angle of depression of 18°. How far is the shark from the lifeguard station?

8. A search plane at an altitude of 2000 m determines the angle of depression of a disabled power launch to be 16°. What is the horizontal distance of the plane from the power launch?

9. The angle of elevation of a favourite tree at Victoria Park in Sydney is 66° when a distance of 14 m is marked off from the base of the tree. Calculate the height of the tree.

10. A mountain road is inclined at 17° to the horizontal. If you travel 3 km up the road, by how many metres has your altitude increased?

11. From a traffic helicopter, 620 m in the air, an accident is located at an angle of depression of 28°. How far along the highway is the accident?

12. A ramp to a deck for wheelchairs is 12 m long. If the ramp is inclined at an angle of 12° to the horizontal, how high is the deck?

13. The largest advertising sign ever erected was that of the name CITROEN on the Eiffel Tower from 1925 to 1936. The diagonal member of the letter "N" was 39 m long and angled at 35° to the horizontal. What was the height of the letter?

14. From a hot air balloon, the angle of depression of a town is 7°. If the observation deck of the balloon is 250 m high, how far away, horizontally, is the town?

15. From an office tower overlooking the Assiniboine River, the angle of depression of a row boat upstream is 6°. If the observation was made 28 m above the river, how far from the base of the building is the row boat?

16. For speed skating, the minimum angle is shown by the skater in the photograph.

Calculate the measure of θ.

17. For highway construction, a 2% gradient means that the road rises 2 units for each 100 units of horizontal travel.
    (a) Calculate the angle of inclination of the road with the horizontal.
    (b) What is the angle of inclination for a 5% gradient?

18. From a tower, 25 m in height, near Barrington Passage, the angle of depression of 2 people is 11°. How far are the 2 people from the tower?

19. The world's tallest fountain is at Fountain Hills, Arizona. The column of water reaches a height of 172 m. If Jim is 1.5 m tall and stands 100 m away from the base of the column, what is Jim's angle of elevation up to the top of the column?

20. A lighthouse casts a beam which meets the surface of the water at angle of 3° and 250 m from the shore. How high is the top of the lighthouse above the surface of the water?

11.10 the Law of Sines

To solve a problem involving a right triangle, we may solve the triangle using the trigonometric ratios. To solve oblique triangles, the **Law of Sines** for triangles is developed. An oblique triangle does not have a right angle and may either be

acute   obtuse

Given: $\triangle ABC$ is an oblique triangle.

Required to prove: $\dfrac{a}{\sin A} = \dfrac{b}{\sin B} = \dfrac{c}{\sin C}$.

Referred to as the Law of Sines.

Proof:
Case 1 $\triangle ABC$ is acute.
Draw $AD \perp BC$.
In $\triangle ABD$ $\dfrac{AD}{c} = \sin B$
$AD = c \sin B$
In $\triangle ADC$ $\dfrac{AD}{b} = \sin C$
$AD = b \sin C$

Thus $c \sin B = b \sin C$ or $\dfrac{c}{\sin C} = \dfrac{b}{\sin B}$.

Case 2 $\triangle ABC$ is obtuse.
Draw $AD \perp BC$ produced.
In $\triangle ABD$ $\dfrac{AD}{c} = \sin B$
$AD = c \sin B$
In $\triangle ACD$ $\dfrac{AD}{b} = \sin(180° - C)$
$= \sin C$
$AD = b \sin C$

But $\sin(180° - C) = \sin C$

Thus $c \sin B = b \sin C$ or $\dfrac{c}{\sin C} = \dfrac{b}{\sin B}$.

Thus for any $\triangle ABC$ $\dfrac{c}{\sin C} = \dfrac{b}{\sin B}$.

Similarly, by drawing perpendiculars to the other sides, it may be proved that
$$\frac{a}{\sin A} = \frac{b}{\sin B} = \frac{c}{\sin C}.$$

We may use the Law of Sines to find the measures of the other parts of the triangle if we are given one of the following.
- two angles and a side opposite one of the given angles.

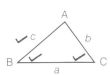

$$\frac{b}{\sin B} = \frac{c}{\sin C}$$
or $b = \dfrac{c \sin B}{\sin C}$

- two angles and the contained side

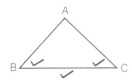

Since $\angle B$ and $\angle C$ are known, then
$\angle A = 180° - (\angle B + \angle C)$
$$\frac{b}{\sin B} = \frac{a}{\sin A}$$
or $b = \dfrac{a \sin B}{\sin A}$

In Example 1, the given parts are used to find a side of the triangle.

Example 1
In $\triangle ABC$, $BC = 20$, $\angle A = 31°$, $\angle C = 104°$. Find AC to 1 decimal place.

Solution

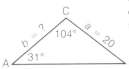

Record the given information on a rough diagram to aid your understanding of the problem.

To determine AC, $\angle B$ is needed.
$\angle B = 180° - (31° + 104°)$
$\quad\; = 45°$

Use $\dfrac{b}{\sin B} = \dfrac{a}{\sin A}$.

$\dfrac{b}{\sin 45°} = \dfrac{20}{\sin 31°}$

Multiply both sides by $\sin 45°$ to isolate b.

$b = \dfrac{20 \sin 45°}{\sin 31°}$
$\quad \doteq \dfrac{20(0.7071)}{(0.5150)}$
$\quad \doteq 27.46$ (to 2 decimal places)

Thus, AC is 27.5 (to 1 decimal place).

In Example 2, the given parts of the triangle are used to find an angle of the triangle.

Example 2
In $\triangle ABC$, $\angle A = 105°$, $b = 15.1$, $a = 20.3$. Find $\angle B$ to the nearest degree.

Solution
Use the Law of Sines.
$\dfrac{\sin B}{b} = \dfrac{\sin A}{a}$

$\sin B = \dfrac{b \sin A}{a}$

$\quad\quad = \dfrac{15.1 \sin 105°}{20.3} \doteq \dfrac{15.1(0.9659)}{20.3}$

$\quad\quad \doteq 0.7185$ from the sin tables

From the tables $\angle B \doteq 46°$.

To *solve an oblique triangle* means to use the measures that are given for the sides or angles to determine the measures of the remaining sides and angles.

11.10 exercise

Throughout the exercise, unless indicated otherwise, express the measures of sides to 1 decimal place and angles to the nearest degree.

A

1. The Law of Sines is written in the compact form $\dfrac{a}{\sin A} = \dfrac{b}{\sin B} = \dfrac{c}{\sin C}$. Write 3 different equations represented by the above compact form.

2. In △ABC, use the Law of Sines to complete each of the following.
   (a) $\dfrac{a}{\sin A} = ?$ (b) $\dfrac{\sin B}{b} = ?$ (c) $\dfrac{a}{b} = ?$
   (d) $\dfrac{c}{b} = ?$ (e) $a \sin B = ?$ (f) $b \sin C = ?$

3. In △PQR, use the Law of Sines to complete each of the following.
   (a) $\dfrac{\sin Q}{q} = ?$ (b) $\dfrac{p}{q} = ?$ (c) $\dfrac{r}{p} = ?$
   (d) $p \sin Q = ?$ (e) $r \sin P = ?$ (f) $q \sin P = ?$

4. In △ABC sin A = 0.3, sin B = 0.5, and b = 20. Find the value of a.

5. In △PQR, p = 10, sin P = 0.30, and sin R = 0.24. Find the value of r.

6. In △DEF, e = 3, f = 4 and ∠F = 30°. Find sin E.

7. For each triangle, find the missing side indicated.
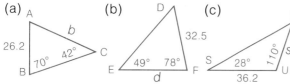

8. For each triangle, find the missing angle indicated.
   (a) 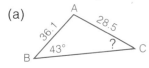 (b)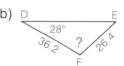
   (c) 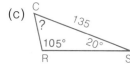 (d)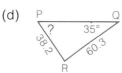

9. For each triangle, find the value of each variable.
   (a) 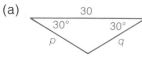 (b)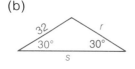

10. Use the Law of Sines to find AB.

11. If △PQR is obtuse, prove that $\dfrac{p}{\sin P} = \dfrac{q}{\sin Q}$.

12. If △DEF is acute, prove that $\dfrac{f}{\sin F} = \dfrac{e}{\sin E}$.

B

13. Solve each triangle
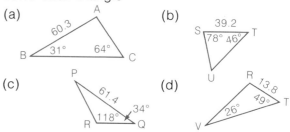

14. (a) △ABC, ∠A = 31°, ∠C = 81°, c = 96.3. Find a.
    (b) △DEF, e = 3, ∠F = 64°, ∠E = 46°. Find ∠D.
    (c) △PQR, ∠P = 46°, ∠Q = 26°, r = 123. Find p and q.

15. In △PQR, find the value of q if ∠R = 83°, ∠Q = 40° and r = 25.

16. In △PQR, p = 24, ∠Q = 49°, ∠R = 45°. Find q, r, and ∠P.

17. If 0° ≤ A ≤ 180°, find two values for ∠A for which
    (a) sin A = 0.7431   (b) sin A = 0.3907
    (c) sin A = 0.7169   (d) sin A = 0.8686

18. In △ABC, a = 11.4, c = 7.1 and ∠C = 36°. Find the two values for ∠A.

19. Solve △DEF if f = 10.1, e = 6.9, ∠E = 40°. (You should obtain 2 different triangles.)

20. In △PQR, r = 12, q = 9.1, ∠Q = 76°. Find ∠P.

21 In △ABC, AB = 7.1, BC = 7.7 and ∠C = 42°. Find ∠A.

22 Solve each triangle.
(a) △PQR, ∠P = 46°, ∠Q = 61°, p = 98.
(b) △ABC, ∠A = 26°, ∠C = 79°, a = 23.1.
(c) △PQR, ∠P = 59°, ∠Q = 46°, r = 14.1.
(d) △DEF, ∠E = 64°, ∠F = 51°, e = 12.1.
(e) △MNP, ∠M = 109°, m = 20.1, ∠N = 15°.
(f) △ABC, ∠A = 42°, ∠C = 13°, a = 32.

23 In △DEF, $\frac{d}{e} = \frac{\sqrt{2}}{2}$ and ∠D = 30°. Find ∠E.

24 (a) If ∠A = 45° and ∠B = 30°, find a value for $\frac{a}{b}$ in △ABC. Express your answer in radical form.
(b) In △PQR, ∠Q = 60°, and ∠R = 45°. Find a radical expression for the value of $\frac{q}{r}$.

25 In △PQR, PQ = 45, ∠P = 70°, ∠R = 49°. Find the length of the altitude from R to PQ.

26 In a triangle, two angles measure 24° and 69°, and the longest side is 55 m. Find the length of the shortest side, correct to the nearest tenth of a metre.

27 Use the Law of Sines to prove that if the measures of two angles of a triangle are equal, then the lengths of the sides opposite these angles are equal.

28 Use the Law of Sines to prove that in any triangle a line drawn parallel to a side of a triangle divides the other sides proportionately.

problem/matics

If $\tan A = \frac{2xy}{x^2 - y^2}$ where $x > y > 0$ and $0 < A < \frac{\pi}{2}$, find an expression for $\sin A$.

11.11 solving problems: Law of Sines

To solve a problem involving the Law of Sines, the problem must be interpreted correctly. We must understand two important questions.
A: What information we are given?
B: What information we are asked to find?

Example 1

From an observation tower near Lake Winnipeg, the angle of elevation of a weather balloon is 65°. In the same plane, 35 km away, the balloon is sighted from another location with an angle of elevation of 47°. Calculate the distance from the weather balloon to the observation tower to the nearest tenth of a kilometre.

Solution

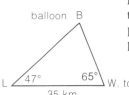

Draw a diagram to record the information and help plan the solution of the problem.

To calculate BW, use the Law of Sines.

$$\frac{BW}{\sin L} = \frac{LW}{\sin B} \qquad \angle B = 180° - (68° + 47°) = 65°$$

$$BW = \frac{LW \sin L}{\sin B}$$
$$= \frac{35 \sin 47°}{\sin 65°}$$
$$\doteq \frac{35(0.7314)}{0.9063}$$
$$\doteq 28.2 \text{ (to 1 decimal place)}$$

Thus, the distance from the weather balloon to the observation tower is 28.2 km.

To solve the following problem, we also need to use our earlier skills with solving problems that involve right-angled triangles.

Example 2

To calculate the width of a river, a surveyor marks a base line AB, 250 m in length, along the river bank. From each end of the base line, an object, C, is sighted on the other bank of the river, making angles of 60° and 74° with the base line, as shown. Find the width of the river to the nearest metre.

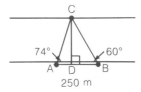

Solution

From the diagram, use the Law of Sines to obtain an expression for CB in △ABC.

$$\frac{CB}{\sin 74°} = \frac{250}{\sin 46°}$$

$$CB = \frac{250 \sin 74°}{\sin 46°}$$

In △BCD

$$\frac{CD}{CB} = \sin 60°$$

$$CD = CB \sin 60°$$
$$= \frac{250 \sin 74°}{\sin 46°} \times \sin 60°$$
$$= \frac{250(0.9613)(0.8660)}{0.7193}$$
$$\doteq 289.3 \text{ (to 1 decimal place)}$$

Thus, the width of the river is 289 m (to the nearest metre).

Throughout the exercise, unless indicated otherwise record lengths accurately to 1 decimal place and angles to the nearest degree.

11.11 exercise

A

1. Abel Island and Bolton Island are 3 km apart. How far is Carter Island from Abel Island and Bolton Island on the given map?

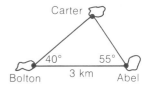

2. A post is supported by two wires, as shown, in opposite directions forming an angle of 80° at the top of the post. The ends of the wire at the ground are 12 m apart with one wire forming an angle at 40° with the ground. Find the lengths of the wires.

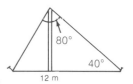

3. Along one bank of a river with parallel banks, a surveyor places a base line measuring 200 m as shown. From each end of the base line, a rock is sighted on the other bank of the river. The lines of sight of the rock make angles of 46° and 69° with the base line. Find the width of the river.

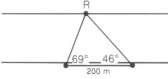

4. A rack for drying fish nets has five posts each 1.7 m long. How many metres of the net can be exposed in one wrap around the rack?   OB = 1.7 m

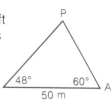

5. Jack's golf ball lands 50 m left of Arnies' as shown. If Jack is approaching the pin from a 48° angle and Arnie from a 60° angle, how far is each from the pin at P?

6. Two sightings at A and B are made by a steamer as it passes an iceberg at I. How close does the ship come to the iceberg?

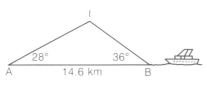

For each of the following problems, draw a diagram to record the information accurately. Then solve the problem.

B

7. Two scuba divers are swimmng 6 m below the surface. When they are 20 m apart they see a shark directly below them. If the angle of depression from the first diver to the shark is 47° and the angle of depression from the second diver to the shark is 40°, how far is each diver from the shark?

8. A pulley is suspended from the ceiling by two chains. One chain 6.2 m in length forms an angle of 55° with the ceiling. Find the length of the other chain which forms an angle of 30° with the ceiling.

9. Two tracking stations 500 km apart simultaneously notice a U.F.O. above them. The two stations and the U.F.O. are in the same vertical plane. The angles of elevation from stations A and B to the U.F.O. are 57° and 40° respectively. How far is each station from the U.F.O?

10. An isosceles triangle ABC has a base of length 18 cm. The vertical angle A of the triangle measures 38°.
    (a) Find the measure of ∠B and ∠C.
    (b) Find the lengths of the equal sides of the triangle.

11. To avoid heavy snow loads on the roof, a ski chalet plan calls for a 36° vertical angle. Assuming the width of the base of the chalet to be 14 m, determine the slant height of the roof on each side. (The chalet is in the shape of an isosceles triangle.)

12. In the solar system, Earth, Mars and the Sun form a triangle. The angle from Earth to Mars to the Sun in 40°. The angle from the Sun to Earth to Mars is 79°. If the distance from Earth to the Sun is defined as 1 Astronomical Unit (1 A.U.), how many astronomical units is Mars from the Sun (to 2 decimal places)?

13. In order to calculate the height of the chimney stack on top of a generating station, a group of environmentalists measured the angle of elevation of the top of the stack to be 15° from point A. They then advanced over level ground to a point B from which the angle of elevation of the top of the chimney was 20°. If the distance from A to B is 40 m, find the height of the chimney stack.

C

14. A conical funnel having a vertical angle of 52°, rests inside a glass of height 7 cm and diameter 3 cm, internal measurements. Find the height of the apex of the funnel above the base of the glass.

15. In order to fit a business building to the available land it is necessary to build it in the form of a quadrilateral with inside corner angles of 100°, 50°, 80°, and 130° respectively, in clockwise order. The sides are 98 m, x m, p m, and q m in the same order. The side 98 m long is between the 100° and 50° angles. Find x and p if $p = q$.

applications: golf and the Law of Sines

There are many problems in golf that can be solved using the Law of Sines. The Law of Sines may improve your golf score!

16. A golf hole has a "dog leg" around a pond as shown in the diagram.

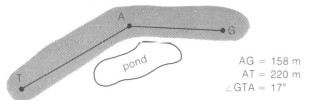

AG = 158 m
AT = 220 m
∠GTA = 17°

A golfer would like to drive from the tee, T, to the green, G, without using the fairway. How far would the golfer have to drive the ball?

17. A golf hole is designed to have a "dog leg" as shown in the diagram.

With ∠CTA = 28°, ∠ACT = 39° and TC = 422 m, find

(a) AT (b) AC

18. At Meadowbank Golf Course, one of the courses is designed as shown in the diagram with the tee at T, and the hole at H. The trees obscure the hole, H, from T so the course is ideally completed in two shots: T to A and A to H.

If AH is 102 m, ∠ATH = 22°, and ∠AHT = 41° find the first distance TA that a golfer must hit the ball.

11.12 Law of Cosines

The parts of triangles that establish the congruence of triangles also fix the shape and size of a triangle. Thus, we say that these parts determine the triangle. In the previous section, we used the Law of Sines to solve a triangle when given

- two angles and the contained side (ASA).
- two angles and a side (AAS).

To solve a triangle when given

- two sides and the contained angle (SAS) or
- three sides (SSS)

we may derive another property of triangles known as the **Law of Cosines** which states

For any $\triangle ABC$,
$a^2 = b^2 + c^2 - 2bc \cos A$
$b^2 = c^2 + a^2 - 2ac \cos B$
$c^2 = a^2 + b^2 - 2ab \cos C$

To prove the Law of Cosines, we need to express one side in terms of the other sides and the contained angle.
Thus in $\triangle ABC$, if both $\angle B$ and $\angle C$ are acute, construct $AP \perp BC$.
In $\triangle APB$
$c^2 = h^2 + m^2$
$= h^2 + (a - n)^2$ ← (Since $m = a - n$)
$= h^2 + a^2 - 2an + n^2$
$= \underline{h^2 + n^2} + a^2 - 2an$ ← In $\triangle APC$ $\frac{n}{b} = \cos C$
In $\triangle APC$ $b^2 = h^2 + n^2$ $n = b \cos C$
Thus, $c^2 = b^2 + a^2 - 2ab \cos C$.
In a similar manner, we may prove
$a^2 = b^2 + c^2 - 2bc \cos A$
$b^2 = a^2 + c^2 - 2ac \cos B$

If $\angle C$ is obtuse, we draw $AP \perp BC$ produced as shown. We develop the proof using the fact that
$\cos \angle ACP = \cos (180° - C) = -\cos C$.

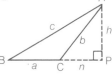

In △APB
$c^2 = h^2 + (a + n)^2$
$= h^2 + a^2 + 2an + n^2$
$= h^2 + n^2 + a^2 + 2an$

In △APC
$\dfrac{n}{b} = \cos(180° - C)$
$n = -b \cos C$

In △APC $b^2 = h^2 + n^2$

Thus $c^2 = b^2 + a^2 - 2ab \cos C$.

The Law of Cosines is given in the useful forms

Given 2 sides, contained angle	Given 3 sides
$a^2 = b^2 + c^2 - 2bc \cos A$	$\cos A = \dfrac{b^2 + c^2 - a^2}{2bc}$
$b^2 = a^2 + c^2 - 2ac \cos B$	$\cos B = \dfrac{a^2 + c^2 - b^2}{2ac}$
$c^2 = a^2 + b^2 - 2ab \cos C$	$\cos C = \dfrac{a^2 + b^2 - c^2}{2ab}$

Example 1
In △ABC, ∠C = 63°, b = 4.2, a = 5.3.
Find c to 1 decimal place.

Solution   Record the given information on a sketch to help plan the solution.

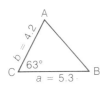

In △ABC
$c^2 = a^2 + b^2 - 2ab \cos C$
$= (5.3)^2 + (4.2)^2 - 2(5.3)(4.2) \cos 63°$
$\doteq 28.09 + 17.64 - 44.52 (0.4540)$ ← $\cos 63° \doteq 0.4540$
$\doteq 28.09 + 17.64 - 20.21$
$\doteq 25.52$
$c \doteq 5.05$ (to 2 decimal places)
Thus, $c \doteq 5.1$ (to 1 decimal place).

We may find the measures of the angles of a triangle if we are given 3 sides as shown in Example 2.

Example 2
Find ∠B if a = 12, b = 18, and c = 15 in △ABC.

Solution   Draw a diagram to record the given information

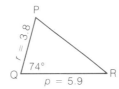

Use the Law of Cosines.
$\cos B = \dfrac{a^2 + c^2 - b^2}{2ac}$
$= \dfrac{12^2 + 15^2 - 17^2}{2(12)(15)}$
$= \dfrac{144 + 225 - 289}{360}$
$= \dfrac{80}{360} \doteq 0.2222$ (to 4 decimal places)

From the cosine tables
∠B $\doteq$ 77° (to the nearest degree)

To solve a triangle, we may need to use the Law of Cosines and the Law of Sines as shown in Example 3

Example 3
Solve △PQR if ∠Q = 74°, p = 5.9 and r = 3.8.

Solution

Record the given information on the diagram.

Use the Law of Cosines.
$q^2 = p^2 + r^2 - 2pr \cos Q$
$= (5.9)^2 + (3.8)^2 - 2(5.9)(3.8) \cos 74°$
$\doteq 34.81 + 14.44 - 44.84 (0.2756)$
$\doteq 34.81 + 14.44 - 12.36$
$\doteq 36.89$
$q \doteq 6.07 \doteq 6.1$ (to 1 decimal place)
Use the Law of Sines.
$\dfrac{\sin P}{p} = \dfrac{\sin Q}{q}$

$$\sin P = \frac{p \sin Q}{q}$$
$$\doteq \frac{5.9 \sin 74°}{6.1}$$
$$\doteq \frac{5.9 (0.9613)}{6.1}$$
$$\doteq 0.9297$$

From the sine tables
$$\angle P \doteq 68° \text{ (to the nearest degree)}$$
$$\angle R = 180° - (68° + 74°)$$
$$= 38°$$
Thus $p = 5.9 \qquad q = 6.1 \qquad r = 3.8$
$\qquad \angle P = 68° \qquad \angle Q = 74° \qquad \angle R = 38°$

↑ List all the parts of the triangle so that you know the triangle has been solved.

11.12 exercise

Throughout the exercise, unless indicated otherwise, express the measures of sides to 1 decimal place and angles to the nearest degree.

A

1 Prove the Law of Cosines can be given as
   (a) $a^2 = b^2 + c^2 - 2bc \cos A$.
   (b) $b^2 = a^2 + c^2 - 2ac \cos B$.
   (c) $\cos C = \frac{a^2 + b^2 - c^2}{2ab}$

2 In $\triangle PQR$ if $\angle Q$ is obtuse, prove that
   $$\cos Q = \frac{p^2 + r^2 - q^2}{2pr}.$$

3 In $\triangle DEF$, if $\angle F$ is obtuse prove that
   $$d^2 = e^2 + f^2 - 2ef \cos D.$$

4 In each triangle, find the missing side.
   (a)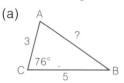
   (b)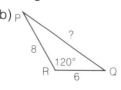

5 In each triangle, find the missing angle indicated.
   (a)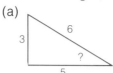
   (b)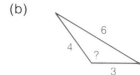

6 For each triangle, find the value of the variables.
   (a)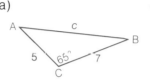
   (b)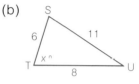

7 (a) Draw a sketch of $\triangle DEF$, marking the formation $d = 5$, $e = 8$, $f = 11$ on the diagram.
   (b) Calculate $\angle E$.

8 $\triangle ABC$ has the property that $\angle A = 60°$, $b = 15$ and $c = 24$. Find a.

9 In $\triangle PQR$, $p = 10$, $q = 12$, $\cos R = \frac{1}{5}$. Find r.

10 In $\triangle TUV$, $t = 10$, $u = 14$, $v = 12$. Find $\cos U$.

11 In $\triangle STU$, $t = 20$, $u = 24$, $\cos S = -\frac{1}{3}$. Find s.

12 In $\triangle DEF$, $d = 3$, $e = 4$, and $\angle F = 120°$. Find f. Express your answers in radical form.

As an important skill for problem-solving, record the given information on a sketch. Then plan the solution.

B

13 For each triangle, find the missing part.
   (a) $\triangle MPN$, $m = 8$, $n = 8$, $\angle P = 48°$, $p = ?$.
   (b) $\triangle PQR$, $p = 4$, $q = 3$, $r = 4$, $\angle P = ?$.
   (c) $\triangle ABC$, $a = 5$, $b = 7$, $c = 8$, $\angle C = ?$.

14. For each triangle, calculate the side opposite the given angle.
    (a) $\triangle PQR$, $p = 4.1$, $r = 6.3$, $\angle Q = 53°$
    (b) $\triangle DEF$, $d = 9.6$, $f = 8.3$, $\angle E = 126°$

15. (a) In $\triangle PQR$, $p = 16$, $q = 12$, $r = 10$. Find $\cos Q$.
    (b) In $\triangle ABC$, $a = 10$, $b = 16$, and $c = 14$. Find $\angle C$.

16. In $\triangle PQR$, $p = 2\sqrt{3}$, $q = 2$, and $r = 4$. Find $\angle P$.

17. Solve $\triangle ABC$ given in the diagram.

18. Solve $\triangle PQR$ given in the diagram.

19. Solve each triangle.
    (a) $\triangle TUV$, $t = 7$, $v = 6$, $\angle U = 43°$.
    (b) $\triangle JKF$, $j = 4.5$, $k = 6.3$, $f = 5.8$.
    (c) $\triangle PHW$, $p = 4.9$, $w = 6.3$, $\angle H = 136°$.

20. If in $\triangle DEF$, $d = 6$, $e = 10$, and $f = 12$, find the cosine of the largest angle.

21. In $\triangle PQR$, $p = 11.9$, $\angle Q = 53°$, $r = 15.4$. Find
    (a) q (b) $\angle R$

22. In a triangle, two sides measure 43 cm and 28 cm and the contained angle is 112°. Find the measure of the smallest angle.

23. If two sides of a triangle measure 30 cm and 48 cm and the contained angle measures 60°, find the third side.

math tip
It is important to clearly understand the vocabulary of mathematics when solving problems.
- Make a list of all the words you have met in this chapter.
- Provide a sample example to illustrate each word.

24. For a parallelogram, two adjacent sides measure 12 cm and 10 cm. If the contained angle measures 120°, then find the length of the longer diagonal.

C
25. In $\triangle ABC$, $c^2 = a^2 + b^2$ where a, b, and c are the sides of the triangle. Use the Cosine Law to prove that $\triangle ABC$ is a right-angled triangle.

Note: This is the converse of Pythagoras Theorem.

26. For the diagram, prove that
$c^2 = a^2 + b^2 - 2ab \cos C$.
(Hint: You may use the distance formula.)

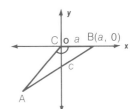

applications: solving problems

To solve any problem, we must accurately interpret the problem and translate the word problem into a mathematical model. Then we can apply the skills we have already developed to solve the problem.

To solve problems in trigonometry, we must draw a diagram that shows accurately
A: the information we are given.
B: the information we are asked to find.

Example 1
From T, a golfer aims a ball towards the hole at H which is 100 m away. But the ball is actually sliced in a direction 30° off course and lands at M, 60 m away, as shown in the diagram. If the next shot is hit 50 m towards the hole, will the ball go in the hole?

Solution

The information is recorded in the diagram to help plan the solution.

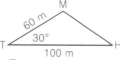

From the diagram, we use the Law of Cosines to find t.

$t^2 = m^2 + h^2 - 2mh \cos T$
$t^2 = 100^2 + 60^2 - 2(100)(60) \cos 30°$
$\doteq 10\,000 + 3600 - 12\,000(0.8660)$
$\doteq 13\,600 - 10\,392$ ← to the nearest whole number
$= 3208$
$t = \sqrt{3208},\ t > 0$
$\doteq 57$ (to the nearest whole number)

Thus, the golf ball will not go in the hole but will be about 7 m short of the hole.

27 Find the perimeter of the triangular plot of land as shown.

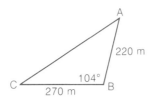

28 Find the length of Lake Temagog if the following measurements are made at Gerald's Inn.

29 A golf hole at H has a "dog leg" as shown in the diagram. How far is the tee, at T, from the hole at H?
TA = 210 m,
AH = 72 m, ∠TAH = 118°.

30 The Law of Cosines may be used to solve astronomical distances. Refer to the diagram. The earth is 1 A.U. (astronomical unit), from the sun and the asteroid Ceres is 2.8 A.U. from the sun. If the angle from earth to sun to Ceres is 38°, find the distance from earth to Ceres.

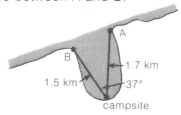

31 A campsite as shown in the diagram at the end of a point in Lake Tabor. From the campsite to A is 1.7 km and to B 1.5 km. If the angle between the sightings from the campsite is 37°, find the distance between A and B.

32 From a point on a plain the distances from Jean's eyes to the peaks of two mountains at the same height are 5 km and 14 km. If the angle between her lines of sight is 102°, find to the nearest kilometre, the distance between the peaks.

33 The distance between the bases in a baseball diamond is 30 m. If the third baseman picks up a fair ground ball 3 m from third base, and on the line from second to third base, how far will be the throw to first base?

34 To find the distance between two points P and Q separated by a marsh, a station at R was established. The distances RP and RQ were found to be 70 m and 76 m and ∠PRQ was measured to be 62°. Find to the nearest metre the distance P to Q, (the width of the marsh).

35 A ship passing an island establishes by sonar a distance of 3.5 km from the ship to one end of the island and 5.1 km to the other end. The angle contained between the tip of the island is 115°. Find the length of the island to one tenth of a kilometre.

36 Find the smallest angle of a triangle having sides 3 cm, 5 cm, and 7 cm.

37 The diagonals of a parallelogram are 24 cm and 28 cm and intersect at an angle of 67°. Calculate the shorter side of the parallelogram to the nearest centimetre.

38 Two jets were sighted from Vancouver Airport in the same vertical plane as the airport. The sighting distance from the planes to the airport are 15 km and 21 km with an angle of 130° between the sightings. Find the distance between the jets.

39 The big hand on a kitchen clock is 10 cm long and the little hand is 7 cm. long. How far is the tip of the big hand from the tip of the little hand at 08:00?

40 Sightings are made for a triangular island. Two coasts of the island measure 9 km and 12 km and the angle contained between the two coasts is 77°.
   (a) Find the total length of the coast to one tenth of a kilometre.
   (b) Calculate the area of the triangle to the nearest square kilometre.

41 Two sides of a triangular plot of land measure 140 m and 250 m. If the angle contained between these two sides is 75°, find its area.

42 A golfer hits a tee shot on a 350 m long golf hole. The ball is sliced 18° to the right. If the ball travelled 200 m, how far is the ball from the golf hole, (to the nearest metre)?

applications: the Cosine Law and direction

We may use the Law of Cosines to solve problems that involve finding direction or bearing. For the two ships located in the diagram at S_1 and S_2, we may state their bearing with respect to each other.

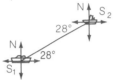

- The bearing of ship S_2 from S_1 is either E28°N or N62°E.
- The bearing of ship S_1 from ship S_2 is W28°S or S62°W.

43 In an airport control tower A, 2 planes at B and C are located at the same altitude on a radar screen. The range finder determines one plane to bear N60°E at 100 km while the other bears S50°E at 160 km. How far apart are the planes from each other?

44 Two planes left Regina at the same time. One travelled due east a distance of 225 km while the second travelled 120° north of the eastern flight a distance of 130 km. How far apart are the airplanes?

45 A ship headed due east from Halifax a distance of 35 km. At the same time a second ship travelled in a direction 40° north of east from Halifax a distance of 23 km. How far apart are the ships?

46 Jean is a cross-country skier and skis 10 km in a direction N40°E of the ski lodge. At this point she turns and skis S10°E for 4 km and arrives at a chalet. How far is Jean from the lodge?

47 Two ships take separate bearings on the same island. From Ship A, the island is N49°E and from ship B it is N44°W. If ship A and ship B are respectively 70 km and 75 km from the island, find the distance between the two ships.

11.13 making decisions to solve problems

Although the context of a problem may vary, the steps for solving it usually do not. Review these steps that we have used many times.

> Steps for Solving Problems

Step A Read the problem carefully.
Ask yourself these two questions:
 I What information do we know?
 II What information are we asked to find?
Step B Translate from words to mathematical symbols.
A diagram may be drawn. Equations may need to be recorded.
Step C Solve the equations. Solve the triangle.
Step D Check the answers in the original problem. Are the answers reasonable?
Step E Write a final statement as the answer to the problem.

To solve problems based on solving triangles, we must accurately translate the given information and record it on a diagram. Then we must make a decision as to which property of triangles we need to apply.
• Cosine Law • Sine Law • Pythagorean Theorem
The exercise that follows will provide practice in making the right decision for solving problems.

11.13 exercise

Unless indicated otherwise, express lengths to 1 decimal place and angles to the nearest degree.

A

1. If $d = 10$, $e = 18$, and $\cos F = \frac{7}{15}$, in $\triangle DEF$, find f.

2. In $\triangle PQR$, $p = 130$, $\angle P = 39°$, $\angle Q = 65°$. Find q.

3. In $\triangle PQR$, $\angle Q = 90°$, $\angle R = 23°$, $p = 4.1$. Find q.

4. In $\triangle PQR$, $\angle P = 30°$, $\angle Q = 45°$, and $p = 10$. Find the value of q. Express your answer in radical form.

5. In $\triangle PQR$, $p = 36$, $r = 18$, and $\sin P = \frac{2}{5}$. Find $\sin R$.

6. Find the cosine of the largest angle in a triangle with sides that measure 8.1 cm, 10.3 cm and 12.5 cm.

B

7. Jim wanted to swim from his dock across the lake to Julie's dock. He estimated the distance by first standing at a point 200 m down the beach from his dock, and from this point he found that his dock was N70°E and Julie's dock was N20°E. He then went back to his dock and found that Julie's dock was N10°W of it. What is the distance between the docks?

8. Scuba divers, 40 m below the water surface, see their boat at an angle of elevation of 30°. How far are they from the boat?

9. Two ships are 50 km apart. Ship A sights a distress flare at S5°E whereas ship B sights the same flare at S13°W. If ship A is N45°W of ship B, find the distance each ship is from the distressed ship.

10. Cathy brings her glider in for landing at an angle of 16° to the horizontal. If she is 500 m above the ground when she begins her descent, calculate the length of her glide-in descent.

11. To meet the solar heating tolerances, a house roof line must be constructed to exact specifications. If the house is 15 m wide and the roof rafters must make 24° and 62° with the joists, find the length of each rafter.

C

12. The perimeter of a triangle is 150 cm and the angles are in the ratio of 1:4:7. Find the length of each side.

11.14 solving 3 dimensional problems

We may use the Sine Law and the Cosine Law to solve problems that involve 3 dimensional diagrams. The most important step is to interpret the problem correctly and draw the appropriate diagram.

Example 1
An observer in a lighthouse 100 m high observes one edge of an oil slick to have an angle of depression of 33° and the other edge with an angle of depression of 46°. The angle subtended by the two edges of the oil slick at the eye of the observer is found to be 56°. Use the given information to find the width of the oil slick to the nearest metre.

Solution

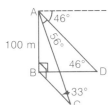

The given information is recorded on the diagram so that the steps of the solution may be planned

From the diagram
$\dfrac{AC}{100} = \csc 33°$

$AC = 100(1.836) \doteq 183.6$ (to 1 decimal place)

$\dfrac{AD}{100} = \csc 46°$

$AD \doteq 100(1.390) \doteq 139.0$ (to 1 decimal place)

From the diagram and using the Law of Cosines
$DC^2 = AC^2 + AD^2 - 2\,AC \times AD \times \cos 56°$
$\doteq (183.6)^2 + (139.0)^2 - 2(183.6)(139.0)(0.5592)$
$\doteq 33\,709 + 19\,321 - 28\,542$ (rounded to the nearest whole number)
$\doteq 24\,488$
$DC \doteq 156.5$ (to 1 decimal place)

Thus, the width of the oil slick is 157 m to the nearest metre.

11.14 exercise

Express distances to 1 decimal place and angles to the nearest degree unless you are told otherwise.

A

1. An engineer wants to find the height of an inaccessible cliff and takes measurements as shown in the diagram. △ACB is a horizontal plane.

   Find the height of the cliff, DB.

   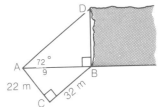

B

2. Adam Dimitrich, a Swiss mountaineer, paused on a ledge of Mount Alpein to estimate his altitude. He determined the angle of depression to the town of Glostein to be 25°. Glostein is S56°W of the mountain and Dromadin is S66°E of the mountain and 4 km distant. If Dromadin and Glostein are 5 km apart, calculate Adam's altitude.

3. The captain of an ocean liner is positioned 42 m above sea level. He sights the angle of depression of the water level at one end of an iceberg to be 22° and the angle of depression of the water level at the other end to be 16°. If the angle subtended at the observers' eyes by the iceberg is 75°, how long is the iceberg? Express your answer to the nearest metre.

4. To estimate the usable lumber in a Redwood tree in California, the company officials must first estimate the usable height of the tree. A certain tree has angles of elevation of 41° and 52° respectively determined from points that are 50 m apart. If the angle formed at the base of the tree by the positions of the two sightings is a right angle, find the height of the tree.

skills review

1. Calculate each of the following.
   (a) cos 30°   (b) csc 60°   (c) sec 45°
   (d) sec 60°   (e) cot 60°   (f) tan 30°
   (g) sin 45° + cos 45° + cos 45° sin 45°

2. Calculate.
   (a) cos 225°   (b) sec (−30°)   (c) cot 330°
   (d) cos (−300°) (e) csc 240°   (f) tan (−225°)

3. Calculate.
   (a) sec $\frac{\pi}{6}$ (b) sin $\frac{\pi}{4}$ (c) cot $\frac{\pi}{3}$
   (d) cot $\frac{\pi}{4}$ (e) tan $\frac{\pi}{6}$ (f) cos $\frac{\pi}{3}$

4. Calculate.
   (a) csc $\frac{3}{4}\pi$ (b) cot $\frac{2}{3}\pi$ (c) cos $\frac{5}{6}\pi$
   (d) tan $(-\frac{\pi}{6})$ (e) cos $(-\frac{3}{4}\pi)$ (f) sec $(-\frac{4}{3}\pi)$

5. Calculate.
   (a) cos 90°   (b) csc 90°   (c) cos 360°
   (d) sin (−90°) (e) sec 270°   (f) tan (−180°)

6. Calculate.
   (a) cos $\frac{\pi}{2}$ (b) tan $\frac{3}{2}\pi$ (c) csc (−π)
   (d) sec $(-\frac{3}{2}\pi)$ (e) csc $\frac{3}{2}\pi$ (f) tan $(-\frac{3}{2}\pi)$

Solve each of the following triangles for the given information.

7. In △PQR ∠R = 90°, PQ = 14.2, ∠PQR = 29°.
8. In △ABC, AB = 15.1, CB = 34.9, ∠A = 90°.
9. In △ABC, ∠A = 29°, ∠C = 79°, and c = 96.3.
10. In △PQR, ∠P = 48°. ∠Q = 27°, r = 125.
11. In △ABC, b = 3, a = 5, ∠C = 77°.
12. In △PQR, r = 6, q = 3, p = 5.
13. In △JKF, j = 4.6, k = 6.2, f = 5.6.

problems and practice: a chapter review

1. An angle θ is in standard position in the second quadrant. If cos θ = $-\frac{3}{4}$, draw the terminal arm of θ.

2. Calculate the secant of each of the following angles.
   (a) 135°   (b) −240°   (c) 210°

3. β is an angle such that cot β = −1 and is in the fourth quadrant.
   (a) Find sec β.
   (b) What is the value of β, 0° ≤ β ≤ 360°?

4. If sin θ = $-\frac{1}{\sqrt{2}}$ and 0° ≤ θ ≤ 360°, find the values of θ.

5. Find the values of
   (a) cot 245°   (b) csc 273°   (c) sin (−128°)

6. Find 2 values of θ for cot θ = 0.2679, 0°≤ θ ≤ 360°.

7. Prove that
   (a) $\frac{1}{\cot \theta} + \cot \theta = \frac{1}{\sin \theta \cos \theta}$
   (b) $\left(\tan \theta - \frac{1}{\cos \theta}\right)\left(\tan \theta + \frac{1}{\cos \theta}\right) = -1$

8. For purposes of safety, the suggested angle of elevation a ladder makes with the ground is 75°.
   (a) What should be the length of the ladder to reach 10 m up a wall?
   (b) If the base of a ladder is 1.5 m from the wall, how far up the wall will the ladder reach?

9. In △ABC, ∠A = 80°, a = 8.3, b = 5.1. Find ∠B.

test for practice

Try this test. Each *Test for Practice* is based on the mathematics you have learned in the preceding chapter. Try this test later in the year as a review. Keep a record of those questions that you were not successful with and review them periodically.

1. If $\cot \theta = -\dfrac{13}{12}$ and θ is in the second quadrant then construct the terminal arm of θ.

2. Calculate.
   (a) $\cos 60°$ (b) $\sec 45°$ (c) $\cot 30°$

3. Calculate the sine of each of the following angles.
   (a) $225°$ (b) $-120°$ (c) $150°$

4. Find 2 values of θ for $\csc \theta = 2.4586$, $0° \leq \theta \leq 360°$.

5. Express each of the following in terms of $\sin \theta$ and $\cos \theta$.
   (a) $\dfrac{1 + \tan \theta}{\sec \theta}$ (b) $\sec^2 \theta - \tan^2 \theta$

6. Prove that
   (a) $\sec \theta \tan \theta = \dfrac{\sin \theta}{\cos^2 \theta}$
   (b) $\dfrac{\tan x - 1}{\tan x} = \dfrac{\sin x - \cos x}{\sin x}$

7. Calculate the value of
   (a) $\csc \pi$ (b) $\cos(-\pi)$ (c) $\tan 2\pi$

8. If $f(\theta) = \sin \theta$, calculate
   (a) $f\left(\dfrac{3}{4}\pi\right)$ (b) $f(120°)$ (c) $f(90°)$

9. Solve $\triangle PQR$ if $\angle Q = 90°$, $\angle R = 30°$, and $RP = 4.8$.

10. For each triangle, calculate the measure of the missing side as shown.
    (a) 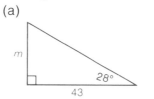 (b)

11. A cable, 200 m in length, is secured to the top of an FM transmitting antenna. If the cable makes an angle of 37° with the ground, how high is the antenna?

12. A pole 4.6 m tall leans and forms an angle of 98° with the ground. A wire is attached to prevent it from falling, forming an angle of 42° with the ground. How long is the wire?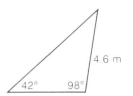

13. (a) Draw a sketch of $\triangle PQR$ given the following information. $\angle Q = 63°$, $p = 6$, $r = 5$.
    (b) In $\triangle PQR$, find q.

14. The distance from Las Vegas to Brampton is 3100 km, from Las Vegas to Chicago is 2450 km, and from Chicago to Brampton is 700 km. With the cities as vertices of a triangle, what is the angle at Brampton?

15. The angle of elevation of the top of a building from a point A, 51 m from the building, is 58°. A flagpole is on top of the building. The angle of elevation of the top of the flagpole is 62°. What is the length of the flagpole?

16. At a recent construction site in downtown Ottawa, two tunnels were excavated starting at the same point. One tunnel was 400 m long and the other tunnel was 250 m long. Find the distance between the ends of the tunnels if the angle contained between them is 84°.

12 the quadratic function and its applications

A study of the quadratic function and quadratic equation, applications, and problems involving maximum and minimum

12.1 introduction

Earlier we studied the linear function
$$y = mx + b \quad \text{←degree 1}$$
and its properties.

The path of a motor cycle leap

or the shape of the arches of many bridges

are examples of parabolas.

The general form of the equation of a parabola is $y = ax^2 + bx + c$, $a \neq 0$. ←second degree

Equations involving the variable in the second degree occur in many branches of science.

area of circle	energy
$A = \pi r^2$	$E = mc^2$
free fall in space	orbits of space probes
$s = ut + \frac{1}{2}gt^2$	$y = ax^2$

To study the quadratic function we may now apply our earlier mathematical skills, such as analytic geometry and transformations.

The concepts and skills relating to the quadratic function occur frequently in our study of mathematics.

12.2 quadratic functions and their graphs

When an object is dropped from a building or when a parachutist jumps from a plane, the distance fallen, in metres, and the length of time, in seconds, are related mathematically. It is believed that Galileo made some of his important discoveries by dropping objects from the Leaning Tower of Pisa.

If measurements are made and the results plotted

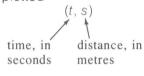

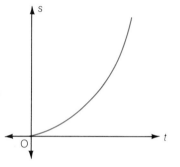

a graph, showing the relationship, is drawn. We will study this graph in detail in this and subsequent sections. The relationship shown in the graph is **non-linear**, and is given in general by

$$f:x \longrightarrow ax^2 + bx + c, \quad a \neq 0, \quad a, b, c \in R.$$

The above function is called a **quadratic function** (*quadratus* is Latin for "square").

We may use a table of values to draw the graph of a quadratic function, as shown in Example 1. In this example, $a = 1$, $b = 0$ and $c = 0$.

> To study this function you may wish to refer to the **Function Plus, techniques for graphing**, pages 447 to 450.

Example 1
A function, f, is given by
$$f:x \longrightarrow x^2, \quad x \in R.$$
Draw the graph of f for $-3 \leq x \leq 3$.

Solution
Choose selected points.

x	$f(x) = x^2$
-3	9
-2	4
-1	1
0	0
1	1
2	4
3	9

From the table of values, the points are plotted. A smooth curve is drawn through the points.

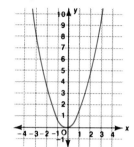

From the above graph of f, we note the following:

domain
  The domain is $-3 \leq x \leq 3$.

range
  The range is $0 \leq y \leq 9$. If the domain for f were all $x \in R$, then the range would be $0 \leq y, y \in R$.

axis of symmetry
  For every point on the curve given by $A(a, b)$ there is a corresponding point $A'(-a, b)$. The equation $y = (-x)^2$ defines the same curve as $y = x^2$. Thus the graph of f is a reflection of itself in the y-axis. The y-axis is the **axis of symmetry** of the graph of f. Thus the mapping $(x, y) \longrightarrow (-x, y)$ maps the parabola onto itself.

vertex
  For the graph of f, the point $(0, 0)$ shows a **minimum point**. We call the point $O(0, 0)$ for f, a **vertex**. We say that the function, f, has a minimum value since the minimum value of all the ordinates, y, is 0. In general, the vertex of a parabola is at the intersection of the graph and its axis of symmetry.

The graph of the function, g, in Example 2 illustrates a function with a maximum point. For this quadratic function $a = -1$, $b = 6$, $c = -5$.

Example 2

(a) Draw a graph of the function, g, defined by $y = -x^2 + 6x - 5$, $-1 \leq x \leq 7$.

(b) Write the domain, range, co-ordinates of the vertex and equation of the axis of symmetry.

Solution

Choose selected points.

x	$g(x)$
-1	-12
0	-5
1	0
2	3
3	4
4	3
5	0
6	-5
7	-12

From the table of values, the points are plotted. A smooth curve is drawn through the points.

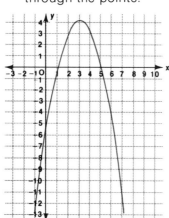

From the graph
domain: $\{x \mid -1 \leq x \leq 7\}$
range $\{y \mid -12 \leq y \leq 4\}$

From the graph, if the domain of g is $x \in R$, then the range of g would be $\{y \mid y \leq 4, y \in R\}$.

co-ordinates of vertex: (3, 4)
equation of axis of symmetry: $x = 3$

From the above graph, we notice that for all x the maximum value of the ordinate y is 4. Since $y \leq 4$ for all y we say that the graph of g has a **maximum point**. The *maximum value* of g is 4.

The above graph of g also intersects the x-axis at the points (1, 0) and (5, 0). Thus the x-intercepts of the graph are the values 1 and 5. We may also calculate the intercepts algebraically by solving the quadratic equation related to the function g. For example,

Find x for $y = 0$ or $g(x) = 0$.
$-x^2 + 6x - 5 = 0$
$x^2 - 6x + 5 = 0$
$(x - 1)(x - 5) = 0$
$x = 1$ or $x = 5$

Thus $g(1) = 0$ and $g(5) = 0$. Since the function values for g are zero when $x = 1$ or $x = 5$, we refer to 1 and 5 as **zeroes of the function g.**

The study of the quadratic function and its properties is important in our study of mathematics.

12.2 exercise

Throughout the exercise, the domain of each quadratic function is R unless indicated otherwise.

A

1. Which of the following define a quadratic function? Give reasons for your answer.

   (a) $f : x \longrightarrow 3x^2$
   (b) $g : x \longrightarrow 2x^3 - 5$
   (c) $h : x \longrightarrow 2x^2 - x + 1$
   (d) $F : x \longrightarrow 3x - \dfrac{1}{x}$

2. Express each of the following quadratic functions in the form $y = ax^2 + bx + c$.

   (a) $y = 3(x - 1)^2$
   (b) $y = -2(x + 1)^2$
   (c) $y = -2(x - 3)^2 - \dfrac{4}{3}$
   (d) $y = \dfrac{1}{2}(x - 4)^2 + 5$

3. A graph of a quadratic function is shown.

   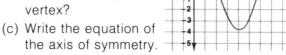

   (a) What is the domain? range?
   (b) What are the co-ordinates of the vertex?
   (c) Write the equation of the axis of symmetry.
   (d) What are the zeroes of the function, g?
   (e) What are the x-intercepts of the graph?
   (f) Does the curve have a maximum or minimum point?

We may describe the graph of a parabola as

concave upward   concave downward

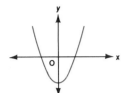

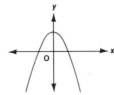

The parabola has   The parabola has
a minimum point.   a maximum point.

B

4 (a) Draw a sketch of each parabola.
   A: $y = x^2$ B: $y = -x^2$
   (b) Which parabola has a minimum point? maximum point?
   (c) Which is concave upward?

5 For each function, determine which is concave upward (U), or concave downward (D).
   (a) $f: x \longrightarrow 2x^2 + \frac{1}{2}$ (b) $g: x \longrightarrow -3x^2 + 1$
   (c) $h: x \longrightarrow 2(x - 1)^2 - 1$

6 (a) Draw a sketch of the parabola given by the equation $y = \frac{1}{2}x^2 - 3$.
   (b) What are the co-ordinates of the intersection of the axis of symmetry and the curve?
   (c) Does the graph have a minimum or maximum point? Write its co-ordinates.

7 (a) Draw the graph of the function defined by $f: x \longrightarrow 3x^2 + 2$.
   (b) What are the domain, range, co-ordinates of the vertex and the equation of the axis of symmetry?

8 (a) Draw the graph of the function given by $g(x) = -2x^2 + 4$.
   (b) What are the domain, range, co-ordinates of the vertex and the equation of the axis of symmetry?

   (c) Does the graph have a minimum or maximum point?

9 (a) Sketch the parabola $y = x^2 - 3$.
   (b) Use the graph in (a) to obtain a maximum or minimum value of the expression $x^2 - 3$. What are the co-ordinates of the vertex?

10 (a) Sketch the parabola $y = -2(x - 1)^2 + 3$.
   (b) Use the graph in (a) to obtain a maximum or minimum value of the expression $-2(x - 1)^2 + 3$. What are the co-ordinates of the vertex?

11 (a) Which of the following points are on the parabola given by $y = 2x^2 - 1$?
   A(1, 1)   B(2, 8)   C(-2, 7).
   (b) Write the equation of the axis of symmetry of the graph in (a).
   (c) Write the co-ordinates of the mirror image of the points in (a) which are on the parabola.

12 (a) Draw the graph of $g: x \longrightarrow x^2 - 9$.
   (b) Write the co-ordinates of the vertex and the equation of the axis of symmetry.
   (c) Write the co-ordinates of the intersection of the graph of g with the x-axis.
   (d) What are the zeroes of the function g?

13 From an experiment, it is found that the ordered pairs $(x, f(x))$ satisfy the functional relationship
   $f(x) = 2.1x^2 + 3.5x + 1.2$.
   Draw a graph of f, and estimate the values of x to 1 decimal place for which $f(x) = 0$.

14 Find the zeroes of the function given for each of the following to 1 decimal place.
   (a) $g(x) = 0.9x^2 + 8.1x + 9.7$
   (b) $F(x) = 1.7x^2 - 0.4x - 12.6$

property of a locus

An important property of a locus is used to solve problems: If a point lies on a locus then the co-ordinates of the point satisfy the defining equation of the locus.

15 A quadratic function is given by
$$f : x \longrightarrow ax^2 + b.$$
If the two points A(2, 2) and B(0, −8) are on the graph of the function, then find a and b.

16 The equation of a locus is given by $y = ax^2 + b$. If the points M(0, 1) and N(−2, −3) are points on the locus, find the defining equation.

17 If $f(x) = ax^2 + b$ and $f(5) = -20$ and $f(0) = -5$, calculate a and b.

18 A quadratic function, f, is defined by
$$f : x \longrightarrow ax^2 + bx + c.$$
If the three points, A(3, 10), B(−1, 0) and C(0, 1) are on the graph of the function, find a, b, and c.

19 The equation of a parabolic locus is given by $y = ax^2 + bx + c$. If the three points P(2, 0), Q(−1, −6) and C(1, 4) are points on the locus, then find the defining equation.

20 If $f(x) = ax^2 + bx + c$ and $f(2) = 3$, $f(-2) = 15$ and $f(1) = -6$, then calculate a, b, and c.

investigating graphs of quadratic functions

21 Use the same co-ordinate axes and draw the graph of each of the following.
(a) $y = x^2$ (b) $y = 2x^2$ (c) $y = 3x^2$
(d) $y = 4x^2$ (e) $y = \frac{1}{2}x^2$ (f) $y = \frac{1}{3}x^2$
(g) How are the graphs in (a) to (f) the same? How are they different?

22 Use the same co-ordinate axes and draw the graph of each of the following.
(a) $y = -x^2$ (b) $y = -2x^2$ (c) $y = -3x^2$
(d) $y = -4x^2$ (e) $y = -\frac{1}{2}x^2$ (f) $y = -\frac{1}{3}x^2$
(g) How are the graphs in (a) to (f) the same? How are they different?

23 Use your results in Questions 21 and 22 and the graph given by $y = ax^2$, $a \neq 0$.
(a) What is the direction of opening if $a > 0$?
(b) What is the direction of opening if $a < 0$?
(c) If $a > 0$, does $y = ax^2$ have a maximum or a minimum point?
(d) If $a < 0$, does $y = ax^2$ have a maximum or a minimum point?

24 Use the same co-ordinate axes and draw the graphs of each of the following.
(a) $y = x^2$ (b) $y = x^2 + 3$ (c) $y = x^2 - 3$
(d) Write the co-ordinates of each vertex and the equation of each axis of symmetry for the graphs in (a) to (c).
(e) How are the graphs given in (a) to (c) the same? How are they different?

25 Repeat the instructions in Question 24 for the graphs of each of the following.
(a) $y = -x^2$ (b) $y = -x^2 + 3$ (c) $y = -x^2 - 3$

26 (a) Draw the graph of each of the following on the same set of axes.
A: $y = x^2$ B: $y = (x - 1)^2$
(b) How are the graphs of A and B alike? How do they differ?

27 (a) Draw the graph of each of the following on the same set of axes.
A: $y = 2x^2 - 1$ B: $y = 2(x - 3)^2 - 1$
(b) How are the graphs of A and B alike? How do they differ?

28 (a) Draw the graphs of each of the following on the same set of axes.
    A: $y = -2(x - 1)^2$ B: $y = -2(x - 1)^2 + 4$
    (b) How are the graphs of A and B alike? How do they differ?

29 Based on your results in Questions 26 to 28 how might the graphs of A and B be related in each of the following?
    (a) A: $y = 2x^2$ B: $y = 2(x - 3)^2 + 4$
    (b) A: $y = -3x^2$ B: $y = -3(x - 1)^2 - 2$
    Check your answers by drawing the graphs of A and B on the same set of axes in each case.

30 Use your results answer the questions about the graph defined by $y = ax^2 + c$.
    (a) If $a > 0$ and $c > 0$, describe the position of the graph.
    (b) If $a > 0$ and c increases in value, describe the graphs given by $y = ax^2 + c$.
    (c) If $a < 0$ and $c > 0$, describe the position of the graph.
    (d) If $a < 0$ and c decreases in value, describe the graphs given by $y = ax^2 + c$.

31 (a) What mapping $(x, y) \longrightarrow (?, ?)$ maps the graph of $y = x^2$ onto the graph of $y = x^2 - 1$?
    (b) What mapping maps the graph of $y = 3x^2$ onto the graph of $y = 3x^2 + 4$?

applications: sports and quadratics

The path of a ball thrown forward is a familiar sight.
As the ball travels forward, gravity acts on it and makes the ball return to earth. At any time, t, in seconds, the height, h, in metres, of a ball thrown at a certain speed is given by
  $h = 8 + 6t - 2t^2$
If the ball is thrown at different angles or speeds the equation will vary.

Questions 32 to 35 are based on the above equation.

32 (a) Draw a graph of the ordered pairs (t, h) when $t \geq 0$.
    (b) From the graph estimate the maximum height reached by the ball.

33 (a) Find the height, h, when $t = 0$.
    (b) How high was the ball above the earth before it was thrown?

34 Find the height of the ball at each time.
    (a) 1 s   (b) 2 s   (c) 1.5 s

35 When $h = 0$, the ball has returned to earth. Use the above equation to determine how long the ball is in the air.

Questions 36 to 41 are based on the following equation.

If an object is thrown vertically upwards with a starting speed of v m/s from an altitude of H, in metres, then its height, h, in metres, at any time, t, in seconds is given by
  $h = vt - 4.9t^2 + H$
The height a basketball player jumps in the air obeys the above mathematical law. If the player jumps at a speed of 6 m/s then $v = 6$ (m/s). Since the player began on the floor then $H = 0$. Thus the equation describing the height, h, of the player above the floor at time, t, is given by
  $h = 6t - 4.9t^2$ ($v = 6$, $H = 0$)

36 Find the height of the player above the floor after
    (a) 0.5 s   (b) 1 s

37 Use the equation to determine how long it will take the player to return to the floor. (Use $h = 0$.)

38 A ball is thrown upward with a velocity of 14.7 m/s by a person 1.4 m tall. How long is the ball in the air before it is caught again?

39 A missile is launched vertically with a velocity of 2450 m/s from a base 500 m above sea level. How long will it be before the missile descends to an altitude of 500 m above sea level?

40 A diver jumps from a tower 30 m above the water with a velocity of 4.9 m/s. How long does it take for the diver to reach a point 0.6 m above the water?

41 Wilbur was crop dusting in his single engine plane at an altitude of 50 m when the propeller fell off. The height of a falling object is given by
$$h = A - 4.9t^2$$
where A is the initial height of the object and t is the number of seconds elapsed.

(a) How far above the ground is the propeller after 3 s?

(b) Will the propeller have hit the ground after the 4th second?

---

problem/matics

To solve some problems, we may obtain an insight into the solution if we draw a sketch of the given information. Solve the following problem.

Points $P(1, 2)$ and $Q(-F, -2)$ are given and R is a point on the line $x = -1$. Find the co-ordinates of point R if $PR + RQ$ is a minimum.

We should remember this strategy throughout the study of the mathematics of this chapter.

12.3 the method of completing the square

In mathematics it is important to study the characteristics of a general quadratic function, f, given by
$$f: x \longrightarrow ax^2 + bx + c, \quad a \neq 0.$$
Once the properties are identified for the *general* quadratic function, we may then apply them to specific quadratic functions in other branches of mathematics or disciplines, such as science, economics, sports, and so on.

We have already seen, by plotting various functions, f, given by $f(x) = ax^2 + bx + c$, that

if $a > 0$ then f has a minimum.
if $a < 0$ then f has a maximum.

Our skills in algebra are often applied to our study of geometry. From algebra, the method of completing the square provides a technique for obtaining useful information *without* drawing the graph of f.

For the function, $f: x \longrightarrow ax^2 + bx + c, \; x \in R$, we may complete the square to express the equation of the function in the form
$$y = a(x - p)^2 + q.$$

Example 1

Express $y = x^2 - 4x + 3$ as $y = a(x - p)^2 + q$.

Solution

$y = x^2 - 4x + 3$ ⟵ Add a value that completes the square
$ = x^2 - 4x + 4 - 4 + 3$
$ = x^2 - 4x + 4 - 1$
Thus $y = (x - 2)^2 - 1$.

$x^2 - 4x + 4 - 4$
$= (x - 2)^2 - 4$

In Example 1 in order to decide what to add to $x^2 - 4x$ to make a perfect square we must note that $x^2 - 4x = x^2 - 4x + 4 - 4$

Compute $\frac{1}{2}$ of the coefficient of x and find its square. $\left\{ \begin{array}{l} \frac{1}{2}(-4) = -2 \\ (-2)^2 = 4 \end{array} \right.$

We refer to the above procedure as **completing**

the square. To write the quadratic equation in the form $y = a(x - p)^2 + q$ if $a \neq 1$, we may apply the same procedure.

Example 2

Use the method of completing the square to write $y = 2x^2 - 12x + 13$ in the form $y = a(x - p)^2 + q$.

Solution

$y = 2x^2 - 12x + 13$
$= 2(x^2 - 6x) + 13$
$= 2(x^2 - 6x + 9 - 9) + 13$
$= 2[(x - 3)^2 - 9] + 13$
$= 2(x - 3)^2 - 18 + 13$
$= 2(x - 3)^2 - 5$

Add a value to complete the square
$x^2 - 6x$
$= x^2 - 6x + 9 - 9$
$= (x - 3)^2 - 9$

When the graphs of functions are given by equations in the form $y = a(x - p)^2 + q$, we may obtain useful information as follows. Each of the following graphs are plotted from a table of values.

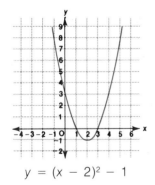
$y = (x - 2)^2 - 1$

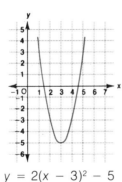
$y = 2(x - 3)^2 - 5$

From the graph we may obtain by observation,

axis of symmetry	$x = 2$	How are these facts related to the equation?	$x = 3$
minimum value, $a > 0$	-1		-5
co-ordinates of vertex	$(2, -1)$		$(3, -5)$

To obtain the x-intercepts for each of the above, use $y = 0$ and solve the equation.

For the quadratic function, f, given by
$f: x \longrightarrow a(x - p)^2 + q$ or
$y = a(x - p)^2 + q$
we may obtain the equation of the axis of symmetry and the co-ordinates of the vertex for the equation by inspection.

Axis of symmetry: $x - p = 0$ or $x = p$.
Co-ordinates of vertex: (p, q).

From our earlier work,
if $a > 0$, then the minimum value is q.
if $a < 0$, then the maximum value is q.

Once the equation for a function is expressed in the form $y = a(x - p)^2 + q$, we can draw a sketch of the curve *without constructing* a table of values.

This form of the defining equation of a quadratic function

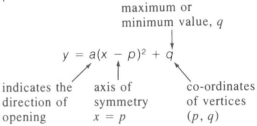

is a more useful form than $y = ax^2 + bx + c$.

From our earlier work with translations, we will show that if $y = ax^2$ is translated for the mapping $(x, y) \longrightarrow (x + p, y + q)$, the equation of the image is $y = a(x - p)^2 + q$. Thus we may use our skills with translations to sketch parabolas.

12.3 exercise

A

1 Find the value of c that will make each of the following a perfect square.

(a) $x^2 + 4x + c$ (b) $x^2 - 4x + c$
(c) $x^2 - 6x + c$ (d) $x^2 + 8x + c$
(e) $x^2 + 5x + c$ (f) $x^2 - 3x + c$

2. Find the values of p and q that will make each of the following true.

   (a) $x^2 + 4x = (x - p)^2 + q$
   (b) $x^2 - 6x = (x - p)^2 + q$
   (c) $x^2 - 9x = (x - p)^2 + q$
   (d) $x^2 + 5x = (x - p)^2 + q$

3. Write each of the following expressions in the form $a(x - p)^2 + q$.

   (a) $x^2 - 2x$
   (b) $x^2 + 4x$
   (c) $x^2 - 2x - 2$
   (d) $x^2 + 6x + 8$
   (e) $2x^2 - 4x + 1$
   (f) $-3x^2 + 6x - 4$
   (g) $4x^2 - 8x + 3$
   (h) $-6x^2 + 12x - 2$

4. Express each of the following in the form $a(x - p)^2 + q$.

   (a) $x^2 - 4x + 6$
   (b) $-2x^2 + 4x - 6$
   (c) $13 + 6x - 2x^2$
   (d) $6 + 4x - 3x^2$

5. For each of the following, what is
   - the equation of the axis of symmetry?
   - the co-ordinates of the vertex?
   - the maximum or minimum value?

   (a) $y = 3(x - 1)^2 - 2$
   (b) $y = -2(x + 1)^2 + 5$
   (c) $y = 2(x - 1)^2 - \frac{3}{2}$
   (d) $y = -\frac{2}{3}(x + 2)^2 + 6$
   (e) $y = -3(x - 3)^2$
   (f) $y = \frac{3}{4}(x + 5)^2 - \frac{3}{2}$

6. For each curve,
   - describe whether the curve is concave upwards (U) or downwards (D).
   - write the co-ordinates of the maximum or minimum point.

   (a) $y = 2(x - 1)^2 + 5$
   (b) $y = -3(x + 5)^2 - 2$
   (c) $y = -\frac{2}{3}\left(x + \frac{1}{2}\right)^2 - 3$
   (d) $y = \frac{4}{3}\left(x - \frac{3}{2}\right)^2 + \frac{5}{2}$

7. (a) Draw the graphs of A and B on the same set of axes.
   A: $y = x^2$ B: $y = (x - 3)^2 + 2$
   (b) How are the graphs in (a) alike? How do they differ?

8. (a) Draw the graphs of C and D on the same set of axes.
   C: $y = -2x^2$ D: $y = -2(x - 1)^2 - 3$
   (b) How are the graphs in (a) alike? How do they differ?

B

9. (a) A function, f, is given by $y = x^2 - 2x - 3$. Write an equation in the form
   $$y = a(x - p)^2 + q.$$
   (b) Write the equation of the axis of symmetry.
   (c) Write the co-ordinates of the vertex.
   (d) What is the minimum or maximum value of f?
   (e) Find the x- and y-intercepts of f.
   (f) Draw a sketch of the graph of f.

10. Use each of the following to complete Question 9 again.

    (a) $y = 3x^2 - 6x + 1$
    (b) $y = -2x^2 + 4x + 1$

11. For each parabola find the equation of the axis of symmetry and the co-ordinates of the vertex.

    (a) $y = x^2 + 4x + 3$
    (b) $y = x^2 - 6x + 5$
    (c) $y = 3x^2 - 6x - 1$
    (d) $y = -2x^2 + 4x - 3$

12. For each of the following find the equation of the axis of symmetry and the co-ordinates of the vertex.

    (a) $y = -x^2 - 4x - 2$
    (b) $y = 2x^2 - 4x - 1$
    (c) $y = 5x^2 - 10x + 4$
    (d) $y = -3x^2 + 6x - 2$

13. Draw a sketch of each of the following parabolas.

    (a) $y = 3x^2 - 6x$
    (b) $y = -2x^2 - 4x$
    (c) $y = x^2 + 6x + 8$
    (d) $y = -x^2 + 4x + 5$

14 Use the method of completing the square to write the minimum or maximum value of each of the following functions.
   (a) $y = x^2 - 6x + 8$ (b) $y = -x^2 + 4x - 6$
   (c) $y = 4x^2 - 8x + 3$ (d) $y = -6x^2 + 12x - 10$

15 For each of the following, write
   • the equation of the axis of symmetry.
   • the co-ordinates of the vertex.
   (a) $y = x^2 + 2x - 1$ (b) $y = 3x^2 - 6x + 4$
   (c) $y = -3x^2 + 6x + 1$ (d) $y = 4x^2 - 8x + 6$

16 Write each of the following in the form
   $y = a(x - p)^2 + q$.
   (a) $y = 3x^2 + 12x + 12$ (b) $y = 6 + 5x - 2x^2$
   (c) $y = 2x^2 - 5x + 3$ (d) $y = 15 - 5x - 3x^2$

   For each of the above what are the co-ordinates of the vertex?

17 Draw a sketch of the curve defined by each of the equations in Question 16.

C

18 A function, f, is given by the equation
   $y = a(x - p)^2 + q$.
   What conditions for a, p, and q are required for the parabola to have these characteristics?
   (a) concave downwards, touching the x-axis
   (b) concave upwards, intersecting the x-axis in two points
   (c) concave downwards, passing through the origin
   (d) concave upwards, passing through the origin
   (e) concave downwards, not intersecting the x-axis

> **math tip**
> The study of quadratic equations has evolved over a long period of time. As far back as 2000 B.C. the Babylonians had already been working with quadratic equations by completing the square and even substituting in formulas.'

formulas: for quadratic functions

Each time we have been given a function, f, in the form $y = ax^2 + bx + c$, we have used the method of completing the square to write it in the form
$$y = a(x - p)^2 + q.$$
We have found
• the axis of symmetry, $x = p$.
• the co-ordinates of the vertex, (p, q).
• if $a > 0$, q is a minimum value.
• if $a < 0$, q is a maximum value.

We may develop a formula to find directly the equation of the axis of symmetry and co-ordinates of the vertex.

General case
$$y = ax^2 + bx + c$$
$$= a\left(x^2 + \frac{b}{a}x\right) + c$$
$$= a\left(x^2 + \frac{b}{a}x + \frac{b^2}{4a^2} - \frac{b^2}{4a^2}\right) + c$$
$$= a\left[\left(x + \frac{b}{2a}\right)^2 - \frac{b^2}{4a^2}\right] + c$$
$$= a\left(x + \frac{b}{2a}\right)^2 - \frac{b^2}{4a} + c$$
$$= a\left(x + \frac{b}{2a}\right)^2 + \frac{4ac - b^2}{4a}$$

Numerical case
$$y = 2x^2 + 3x + 4$$
$$y = 2\left(x^2 + \frac{3}{2}x\right) + 4$$
$$y = 2\left(x^2 + \frac{3}{2}x + \frac{9}{16} - \frac{9}{16}\right) + 4$$
$$= 2\left[\left(x + \frac{3}{4}\right)^2 - \frac{9}{16}\right] + 4$$
$$= 2\left(x + \frac{3}{4}\right)^2 - \frac{9}{8} + 4$$
$$= 2\left(x + \frac{3}{4}\right)^2 + \frac{23}{8}$$

axis of symmetry
$$x = -\frac{3}{4}$$

minimum value
$$\frac{23}{8}$$

Thus, we have developed the following formulas.

equation of axis of symmetry $x = -\dfrac{b}{2a}$

maximum or minimum value $\dfrac{4ac - b^2}{4a}$

co-ordinates of the vertex $\left(-\dfrac{b}{2a}, \dfrac{4ac - b^2}{4a}\right)$

Use the formulas in the numerical example above as a check.
$y = 2x^2 + 3x + 4$, $a = 2$, $b = 3$, $c = 4$
axis of symmetry $x = -\dfrac{b}{2a}$ $x = -\dfrac{3}{4}$ ✓ checks
$a > 0$, minimum value
$\dfrac{4ac - b^2}{4a} = \dfrac{4(2)(4) - (3)^2}{4(2)} = \dfrac{32 - 9}{8} = \dfrac{23}{8}$
checks ✓

If we had a computer we would use the above formula to calculate the minimum or maximum value of the functions. However, it is important for us to learn the method of completing the square in case we cannot remember the formulas.

19 For each of the following, use the formula to determine a minimum or maximum value.
(a) $y = x^2 - 3x + 1$ (b) $y = 3 - 2x + x^2$
(c) $y = 5x^2 + 10x - 1$ (d) $y = -3x^2 + 5x - 1$

20 Use the formula.
• What is the equation of the axis of symmetry?
• What are the co-ordinates of the vertex?
(a) $y = x^2 - 3x - 3$ (b) $y = 2 - 3x + 2x^2$
(c) $y = 3x^2 - 2x + 1$ (d) $y = 5 - x - 3x^2$

21 Use the formula. Draw a sketch of each of the following parabolas.
(a) $y = x^2 + 6x + 5$ (b) $y = 3 + 2x - x^2$
(c) $y = 2x^2 - x - 15$ (d) $y = 2 + 3x - 2x^2$

problem/matics

A parabola is given by the equation $y = ax^2 + bx + c$.
What relationship must be true among the co-efficients of a, b, and c for the graph of the parabola
• to touch the x-axis?
• to intersect the x-axis in two points?

translations and congruent parabolas

In our earlier work we saw that a mapping given by $(x, y) \longrightarrow (x + a, y + b)$
defined a translation. By using our skills in completing a square, we can develop some useful properties about the graphs of quadratic functions or parabolas. For the parabola defined by $y = x^2$, we apply the translation given by the mapping
$(x, y) \longrightarrow (x + 1, y)$.
What is the resulting equation of the graph with respect to the x- and y-axis?

Let (X, Y) represent any point on the image, with respect to the x- and y-axis. From the mapping,
$x + 1 = X$ or $x = X - 1$ and $y = Y$.
Since (x, y) is a point on $y = x^2$ then substitute for x and y in $y = x^2$.
Thus $Y = (X - 1)^2$.

Remember: any point on the locus satisfies the equation of the locus.
$(x, y) = (X - 1, Y)$ is on the locus.

Since (X, Y) is any point on the image, then the equation of the image curve is $y = (x - 1)^2$ (with respect to the x- and y-axis). Since a translation is an isometry, then the parabolas given by $y = x^2$ and $y = (x - 1)^2$ are congruent.

In a similar way, the translation applied to $y = x^2$ given by
$(x, y) \longrightarrow (x + 1, y + 3)$
results in the equation $y = (x - 1)^2 + 3$. The parabolas defined by these equations are congruent. Why?

In general, if a translation given by
$(x, y) \longrightarrow (x + p, y + q)$
is applied to $y = ax^2$, the resulting equation of the image parabola is
$y = a(x - p)^2 + q$.

22. For the translation given by
    $(x, y) \longrightarrow (x - 3, y + 5)$
    find the equation of the resulting image for each of the following.
    (a) $y = x^2$ (b) $y = 2x^2$ (c) $y = -3x^2$
    (d) $y = (x - 2)^2$ (e) $y = (x + 4)^2 - 3$

23. The translation
    $(x, y) \longrightarrow (x - 3, y + 2)$
    is applied to $y = (x - 1)^2 + 3$.
    (a) Find the resulting equation of the image.
    (b) Compare the equations of the axes of symmetry. What do you notice?
    (c) Compare the minimum values. What do you notice?
    (d) Compare the domains and ranges. What do you notice?

24. (a) Find the equation of the image of $y = 5x^2$ under the translation
    $(x, y) \longrightarrow (x - 3, y + 4)$.
    (b) Why are the corresponding graphs congruent?
    (c) What is the domain for each graph? How do they compare?
    (d) What is the range for each graph? How do they compare?
    (e) How are the minimums related for each graph? How do they compare?

25. Find the values of a and b for the mapping
    $(x, y) \longrightarrow (x + a, y + b)$
    that maps parabola A into parabola B.

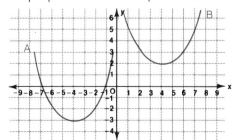

26. Use your results in Questions 23 to 25. Sketch the resulting curve when the translation
    $(x, y) \longrightarrow (x - 3, y + 2)$
    is applied to each of the following graphs.
    (a) 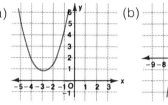 (b)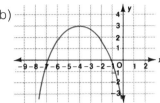

27. (a) Why are the graphs $y = 3(x - 1)^2 - 4$ and $y = 3(x + 2)^2 + 5$ congruent?
    (b) Find the mapping of the translation for the congruent curves in (a).

28. Find the translation given for each of the following.

Equation of original quadratic	Image curve
(a) $y = x^2$	$y = (x - 3)^2 + 5$
(b) $y = 2x^2$	$y = 2(x + 5)^2 - 3$
(c) $y = -3x^2$	$y = -3(x - 1)^2 + \frac{4}{3}$
(d) $y = -\frac{2}{3}x^2$	$y = -\frac{2}{3}(x + 6) - \frac{2}{3}$

29. Which of the following quadratic curves are congruent?
    (a) $y = (x - 3)^2 - 6$ (b) $y = 2(x - 3)^2 - 6$
    (c) $y = -3(x - 1)^2 + 4$ (d) $y = 2(x + 5)^2 - 3$
    (e) $y = -3(x + 1)^2 - 3$ (f) $y = (x + 4)^2 + 7$

    For each pair of congruent curves, write the corresponding translation mapping.

30. Which of the following parabolas are congruent to $y = 2x^2$?
    (a) $y = 2(x - 1)^2$ (b) $y = (x - 2)^2$
    (c) $y = -(x - 3)^2 + 2$ (d) $y = 2(x + 5)^2 - 2$

    Write the appropriate mapping to show related congruent parabolas.

12.4 applications: maximum and minimum

For a function given by
$$h = 1 + 3t - 2t^2$$
we can find the maximum value of h by the method of completing the square.
$$h = -2t^2 + 3t + 1$$
$$= -2\left(t^2 - \frac{3}{2}t\right) + 1$$
$$= -2\left(t^2 - \frac{3}{2}t + \frac{9}{16} - \frac{9}{16}\right) + 1$$
$$= -2\left[\left(t - \frac{3}{4}\right)^2 - \frac{9}{16}\right] + 1$$
$$= -2\left(t - \frac{3}{4}\right)^2 + \frac{9}{8} + 1$$
$$= -2\left(t - \frac{3}{4}\right)^2 + \frac{17}{8}$$

From the above expression
$$\left(t - \frac{3}{4}\right)^2 \geq 0 \text{ for all } t, \quad -2\left(t - \frac{3}{4}\right)^2 \leq 0 \text{ for all } t.$$
Thus, the maximum value of the expression occurs for $t = \frac{3}{4}$, namely, when $-2\left(t - \frac{3}{4}\right)^2 = 0$.
Thus, the maximum value of h is $\frac{17}{8}$. We may use our skills to solve problems involving maximum and minimum.

Example 1
The height, h, in metres, after the launching of a projectile, on the earth's surface at any time, t, in seconds is defined by
$$h = \frac{81}{4} + 9t - 3t^2.$$
(a) Find the maximum height reached by the projectile. At what time, t, does this occur after the launching?
(b) Find the values of t when $h = 0$. What is the significance of the result?

Solution
(a) $h = -3t^2 + 9t + \frac{81}{4}$
$$= -3(t^2 - 3t) + \frac{81}{4}$$
$$= -3\left(t^2 - 3t + \frac{9}{4} - \frac{9}{4}\right) + \frac{81}{4}$$
$$= -3\left[\left(t - \frac{3}{2}\right)^2 - \frac{9}{4}\right] + \frac{81}{4}$$
$$= -3\left(t - \frac{3}{2}\right)^2 + \frac{27}{4} + \frac{81}{4}$$
$$= -3\left(t - \frac{3}{2}\right)^2 + 27$$

Since $-3\left(t - \frac{3}{2}\right)^2 \leq 0$ for all t, then h has a maximum value, $h = 27$ when $t = \frac{3}{2}$.

Thus, the maximum height reached is 27 m at 1.5 s.

(b) For $h = 0$, then
$$-3\left(t - \frac{3}{2}\right)^2 + 27 = 0$$
$$-3\left(t - \frac{3}{2}\right)^2 = -27$$
$$\left(t - \frac{3}{2}\right)^2 = 9$$
$$t - \frac{3}{2} = \pm 3$$
$$t - \frac{3}{2} = 3 \quad \text{or} \quad t - \frac{3}{2} = -3$$
$$t = \frac{9}{2} \quad \quad t = -\frac{3}{2}$$
$$\quad \quad \quad \quad \quad \quad \text{inadmissible}$$

Thus $\frac{9}{2}$ s or 4.5 s after launching, $h = 0$.
Since h represents the height of the projectile after t seconds then, 4.5 s after launching, the projectile strikes the earth's surface.

To solve some problems, we may draw a diagram to help us plan the solution.

Example 2

In a conservation park, a lifeguard has used 620 m of marker buoys to rope off a safe swimming area. If one side of the park is adjacent to the beach, calculate the dimension of the swimming area so that it is a maximum.

Solution

Let w, in metres, represent the width of the swimming area. Then the length, in metres, is given by $620 - 2w$.

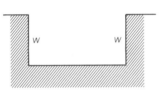

The area, A, is given by
$$A = w(620 - 2w)$$
$$= -2(w^2 - 310w)$$
$$= -2(w^2 - 310w + 24\,025 - 24\,025)$$
$$= -2[(w - 155)^2 - 24\,025]$$
$$= -2(w - 155)^2 + 48\,050$$

Since $-2(w - 155)^2 \leq 0$ for all w, then A is maximum for $w = 155$. The value of A is 48 050. Thus, the maximum swimming area is obtained if the dimensions are 310 m by 155 m.

12.4 exercise

A

1. What is the minimum or maximum value for each of the following? Give reasons for your choice.
   (a) $y = 3(x - 2)^2 - \frac{3}{2}$ (b) $y = -2(x - 5)^2 + \frac{12}{5}$
   (c) $y = -\frac{3}{2}(x - 1)^2 - 25$ (d) $y = \frac{4}{3}(x - 3)^2 + 16$

2. (a) Find the maximum value of V if $V = 4m - 3m^2 + 12$.
   (b) For what value of m is the maximum obtained?

3. (a) Find the minimum value of P if $P = 2k^2 - 3k + 5$.
   (b) For what value of k is the minimum obtained?

4. For what value of t is H a minimum?
   (a) $H = t^2 - 2t + 5$ (b) $H = 2t^2 - 3t + 16$

5. What value of t is H a maximum?
   (a) $H = 3t - 2t^2$ (b) $H = -3t^2 + 6t + 9$

6. What is the maximum value of A in each of the following?
   (a) $A = 24x - x^2$ (b) $A = 5 - 12x - x^2$
   (c) $A = 16 - 3x - 2x^2$ (d) $A = 25 - 5x - 3x^2$

7. Two numbers n and $n + 8$ are given. What are the numbers if their product is to be a minimum?

8. Find two numbers which have a difference of 7 and a product that is a minimum.

9. Find two numbers whose difference is 13 and whose squares when added together yield a minimum.

10. Determine the values of two positive numbers whose sum is 80 and whose product yields a maximum.

11. Two numbers are related in the manner $3x + y = 42$. Find the values of x and y so that their product is a maximum.

12. A fence is to be built around an area such as the one shown in the diagram.

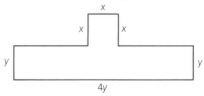

What must be the values of x and y if the area is to be a minimum and the perimeter is to be 300 m?

13. A frisbee is thrown straight up in the air from a position 2 m above ground level. Because of the wind pattern, the height, h, in metres, after time, t, in seconds, is given by the formula $h = 2 + 6t - 2t^2$.
    (a) What is the maximum height the frisbee will reach?
    (b) If it is caught 2 m above the ground, how long will it have been in the air?

14. On a sloping street, Lori gives Michael, who is on a skateboard, an abrupt push uphill. After time, t, in seconds, the distance, s, in metres, that Michael travels from the starting point is determined by the equation $s = 16t - 2t^2$.
    (a) Find the farthest distance Michael travels uphill before he starts to roll back down.
    (b) After how many seconds will he start to roll back down the street?

15. The Northern Resources Department wants to mark off an area as a conservation park. One side of the rectangular-shaped area will be a large lake. Not including this side, the lengths of the remaining 3 sides must not total more than 36 km. What must be the dimensions of the conservation park in order to obtain a maximum size?

16. A uniform patio is to be constructed around a rectangular pool such that the areas on each side of the pool are equal. The pool measures 30 m by 20 m.
    (a) Find the width of the patio.
    (b) Find the minimum size of a lot required for both pool and patio.

17. A handball court is to be surrounded by plastic stripping 30 m long. One side of the court does not need the stripping because it is a brick wall. What must be the dimensions of the court in order to obtain a maximum?

18. Twins Lisa and Lesley got $55 for Xmas. How did they divide the money so that the product of their portions was a maximum?

19. A landscaper wishes to enclose a rectangular rest area with trees planted 1 m apart along the boundary. The perimeter of the area is to be 120 m. What will be the maximum area of the rest area?

20. An arrow shot with a crossbow with an upward velocity of v m/s achieves a height, h, in metres, given by the formula.
    $$h = vt - \frac{1}{2}gt^2$$
    after time, t, in seconds ($g = 9.8$).
    (a) How long will the arrow take to return to earth if it is shot with an upward velocity of 14.7 m/s?
    (b) What is the maximum height reached?

21. Two consecutive odd numbers and a third number total 42. What must the numbers be so that the sum of their squares is a minimum?

applications: problems for Galileo

Because of Galileo we are able to determine a projectile's displacement h, in metres, at any particular time, t, in seconds, by using the formula $h = H + v_i t - 4.9t^2$.

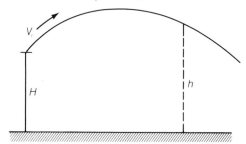

v_i is the initial velocity, in metres per second, and H is the initial height, in metres, from which the projectile is launched.

22 At Ralph's Rifle Range the clay pigeons are fired upward from ground level at an initial velocity of 14.7 m/s. What is the maximum height achieved by the clay pigeons?

23 Babe Ruth, the all-time Yankee slugger, was renowned for hitting high fly balls. One day he made contact with a ball 2 m above the ground and hit the ball at a speed of 29.4 m/s. Determine the maximum height of the fly ball.

24 A human cannonball is fired from a cannon at an initial speed of 9.8 m/s from a height of 6 m above the ground. If the cannonball is to pass through a hoop at the top of the flight, how high must the hoop be above the ground?

25 Jay used a slingshot to propel a rock vertically upward at 245 m/s from the edge of a cliff 2940 m above sea level. After how many seconds will the rock hit the water below?

problem/matics

Often when solving a problem we have never seen before, we must be able to recall quickly many skills and concepts for solving problems. It may be necessary to combine these skills and concepts in order to obtain a strategy for solving the problem.

For example, we have studied
- the solution of equations.
- factoring expressions.
- trigonometry

To solve the following equation, we must combine many of our earlier skills, as well as the skills we will learn in the next section.

Solve $\sin^2\theta - \sin\theta = 0$, $0 \leq \theta \leq 2\pi$.

Now solve the following trigonometric equations, $-2\pi \leq \theta \leq 2\pi$.
A $(\sin\theta - 1)(\cos\theta - 1) = 0$ B $\cos^2\theta - 1 = 0$
C $\sin^2\theta - 1 = 0$ D $\sin^2\theta + \sin\theta = 0$
E $\sin^2\theta - \sin\theta - 2 = 0$ F $2\cos^2\theta = 1 + \sin\theta$

12.5 solving quadratic equations

There are many situations that may be defined by quadratic functions.
- the rate of fall of the water in a hydro-electric dam
- the drop of a bomb
- the path of a sky-diver

and so on

To sketch the graph of a quadratic function, we have used
- the co-ordinates of the vertex
- the x-intercepts
- the equation of the axis of symmetry
- the y-intercept

To find the x-intercepts of the graph defined by
$f(x) = x^2 - 4x - 1$
we find the roots of the *corresponding quadratic equation* $x^2 - 4x - 1 = 0$.
This equation cannot be factored. Some equations such as $x^2 - 6 = 0$, which cannot be factored, can be solved in the following way.
$$x^2 - 6 = 0$$
$$x^2 = 6$$
$$x = \pm\sqrt{6}$$
The above solution suggests a method of solving $x^2 - 4x - 1 = 0$. We use the method of **completing the square**.

Example 1
Solve $x^2 - 4x - 1 = 0$ by completing the square.

Solution
$$x^2 - 4x - 1 = 0$$
$$x^2 - 4x = 1$$
$$x^2 - 4x + 4 = 1 + 4$$
$$(x - 2)^2 = 5$$
$$x - 2 = \pm\sqrt{5}$$
$$x = 2 \pm\sqrt{5}$$
Thus $x = 2 + \sqrt{5}$ or $x = 2 - \sqrt{5}$.

To complete the square use $\frac{1}{2}(-4) = -2$. Then $(-2)^2 = 4$.

Thus, for the function, f, defined by $y = x^2 - 4x - 1$, its x-intercepts are $2 + \sqrt{5}$ and $2 - \sqrt{5}$. Similarly the zeroes of the function, f, are also $2 + \sqrt{5}$ and $2 - \sqrt{5}$.

If we use $\sqrt{5} \doteq 2.24$, then we may obtain the roots of $x^2 - 4x - 1 = 0$ accurate to 2 decimal places as shown.

$x = 2 + \sqrt{5}$ | $x = 2 - \sqrt{5}$
$\doteq 2 + 2.24 \doteq 4.24$ | $\doteq 2 - 2.24 \doteq -0.24$

Thus the graph of f passes through the points (4.24, 0) and (−0.24, 0) on the x-axis. We may use these points to sketch the graph.
The method of completing the square is based on obtaining a perfect square that contains the variable as shown in Example 2.

Example 2
Use the method of completing the square to solve $3x^2 - 6x - 1 = 0$.

Solution
$$3x^2 - 6x - 1 = 0$$
$$3x^2 - 6x = 1$$
$$x^2 - 2x = \frac{1}{3}$$
$$x^2 - 2x + 1 = 1 + \frac{1}{3}$$
$$(x - 1)^2 = \frac{4}{3}$$
$$x - 1 = \pm \frac{2}{\sqrt{3}}$$
$$x = 1 \pm \frac{2}{\sqrt{3}}$$

We may write $1 \pm \frac{2\sqrt{3}}{3}$ or $\frac{3 \pm 2\sqrt{3}}{3}$.

When solving a quadratic equation, be sure to check first to see whether the equation is factorable. In Example 3, we do not need to use the method of completing the square.

Example 3
A quadratic function, f, is defined by $f(x) = x^2 - 3$.
Find m if $f(2m - 1) = f(m)$.

Solution
$f(x) = x^2 - 3 \qquad f(m) = m^2 - 3$
$f(2m - 1) = (2m - 1)^2 - 3$
$\qquad\qquad = 4m^2 - 4m + 1 - 3$
$\qquad\qquad = 4m^2 - 4m - 2$

Since $f(2m - 1) = f(m)$ then
$$4m^2 - 4m - 2 = m^2 - 3$$
$$3m^2 - 4m + 1 = 0$$
$$(3m - 1)(m - 1) = 0$$
$$3m - 1 = 0 \quad \text{or} \quad m - 1 = 0$$
$$m = \frac{1}{3} \qquad\qquad m = 1$$

Using the method of completing the square to solve a quadratic equation will enable us to develop a formula for the roots of a quadratic equation as shown on page 430.

12.5 exercise

Throughout the exercise, unless indicated otherwise, express the roots in radical form.

A

1 Solve each equation by factoring.
   (a) $x^2 - x - 6 = 0$ (b) $x^2 - 2x - 15 = 0$
   (c) $x^2 - 10x + 21 = 0$ (d) $2x^2 + x - 1 = 0$
   (e) $3x^2 - 7x + 4 = 0$ (f) $2x^2 - 11x + 15 = 0$

2 The method of completing the square has been used to rewrite each equation. Find the values of x.
   (a) $(x - 1)^2 = 3$ (b) $(x - 2)^2 = 5$
   (c) $2(x + 1)^2 = 8$ (d) $\left(x - \frac{2}{3}\right)^2 = \frac{5}{9}$
   (e) $2\left(x - \frac{1}{2}\right)^2 = \frac{3}{4}$ (f) $\frac{1}{2}\left(x - \frac{1}{3}\right)^2 = \frac{5}{16}$

3 Show that $-4 - \sqrt{13}$ is a root of $x^2 + 8x + 3 = 0$.

4 Show that one root of $x^2 + 5x - 3 = 0$ is $\frac{1}{2}(-5 + \sqrt{37})$.

5 Express each of the following in simplest form.
   (a) $\pm \dfrac{\sqrt{24}}{3}$
   (b) $\dfrac{3 \pm \sqrt{32}}{2}$
   (c) $\dfrac{4 \pm \sqrt{32}}{2}$
   (d) $\dfrac{-4 \pm \sqrt{96}}{4}$

6 (a) Use the method of completing the square to solve the equation $x^2 - 10x + 4 = 0$.
   (b) Verify your roots in (a).

B

7 Use the method of completing the square to solve each equation.
   (a) $x^2 - 3x + 1 = 0$
   (b) $x^2 - 5x + 2 = 0$
   (c) $2x^2 - 5x + 1 = 0$
   (d) $2c^2 - 3c - 2 = 0$

8 Solve each of the following.
   (a) $3x(x - 1) = 1$
   (b) $y^2 = 5 - 5y$

   When solving a quadratic equation, first check to see if the equation is factorable. If not, then use the method of completing the square.

9 Solve each of the following.
   (a) $x^2 + 5x - 3 = 0$
   (b) $x^2 + 2x - 15 = 0$
   (c) $3x^2 + 8x - 3 = 0$
   (d) $2x^2 + 8x + 1 = 0$
   (e) $3x^2 - 5x - 2 = 0$
   (f) $3x^2 - 17x + 10 = 0$
   (g) $6x^2 + 5x - 1 = 0$
   (h) $5x^2 - 6x - 1 = 0$

10 (a) Find the roots of $x^2 - 3x + 1 = 0$ by completing the square.
   (b) What is the relationship of your answer in (a) with the function defined by $y = x^2 - 3x + 1$?

11 A function is defined by $y = x^2 + 8x + 3$. Find the x-intercepts.

12 The function, g, is given by $g: x \longrightarrow 5x^2 + 3x - 2$. Find the co-ordinates of the points where the parabolic curve meets the x-axis.

13 Find the zeroes of the function, f, defined by $f(x) = 2x^2 - 3x - 5$.

14 Show that one zero of the function, f, defined by $f(x) = x^2 - x - 5$ is $\tfrac{1}{2}(1 + \sqrt{21})$.

15 Find the roots of each of the following equations to 1 decimal place.
   (a) $x^2 + 3x - 1 = 0$
   (b) $2x^2 - 5x + 1 = 0$
   (c) $3x^2 - 2x - 1 = 0$
   (d) $2(x^2 - x) - 3 = 0$

16 Find the x-intercept for each of the following to 1 decimal place.
   (a) $y = x^2 - 2x - 6$
   (b) $y = 3 - 2x - x^2$
   (c) $y = 2x^2 - 5x - 1$
   (d) $y = 3 - x - 3x^2$

17 For the functions given in Question 16, find the co-ordinates of the vertex to 1 decimal place and draw a sketch of the graph.

18 For the function, f, defined by $f(x) = x^2 - 14x + 48$ find k so that
   (a) $f(k) = 0$
   (b) $f(k) = 8$
   (c) $f(k) = 3$

19 The formula $S_n = \dfrac{n^2 + n}{2}$ represents the sum of the numbers $1, 2, 3, 4, 5, \ldots, n$. How many numbers are needed so that
   (a) $S_n = 36$?
   (b) $S_n = 120$?
   (c) $S_n = 465$?

problem/matics

The following computer program written in the language of BASIC, calculates the values of the function $f(g(x))$, $x \in R$, where $f(x) = a_1x^2 + b_1x + c_1$ and $g(x) = a_2x^2 + b_2x + c_2$

```
10 INPUT, A1, B1, C1, A2, B2, C2, K
20 LET G = A2 * K ↑ 2 + B2 * K + C2
30 LET F = A1 * G ↑ 2 + B1 * G + C1
40 PRINT "THE VALUE IS", F
50 END
```

20 A function, g, is defined by $g(x) = x^2 - 3x - 15$. Find the values of k if $g(k) = k$.

21 (a) For $f(x) = x^2 - 2x$, find k if $f(k) = 5$.
    (b) If $g(x) = x^2 - 3x$, find k if $g(k) = k + 1$.

22 If $f(x) = x^2 - 3x + 1$, then find m if $f(m - 1) = f(2m)$.

23 If $h(x) = x^2 - 2x + 14$, then find the values of m if $h(3m - 2) = 3h(m) - 2$.

C

24 Verify that $1 + \sqrt{1 - c}$ is a root of $x^2 - 2x + c = 0$.

25 Show that one root of $x^2 + bx - 4 = 0$ is $\dfrac{-b + \sqrt{b^2 + 16}}{2}$.

26 Find the roots of each of the following equations.
    (a) $x^2 + 2x + c = 0$ (b) $x^2 + bx - 3 = 0$
    (c) $ax^2 + 4x + 1 = 0$ (d) $x^2 + bx + c = 0$

the quadratic formula: roots of a quadratic equation

In many scientific experiments, quadratic equations with decimal coefficients are obtained.
$$2.1x^2 - 3.2x - 5.2 = 0$$
To find the roots of quadratic equations such as the one above we can use the method of completing the square to develop a formula for calculating the roots of the general quadratic equation given by
$$ax^2 + bx + c = 0, \ a \neq 0.$$
As is frequently the case, the procedure we use with equations with numerical coefficients also applies to those equations with literal coefficients. Compare the following.

$$2x^2 + 5x + 1 = 0$$
$$2x^2 + 5x = -1$$
$$x^2 + \frac{5x}{2} = -\frac{1}{2}$$
$$x^2 + \frac{5}{2}x + \frac{25}{16} = \frac{25}{16} - \frac{1}{2}$$
$$\left(x + \frac{5}{4}\right)^2 = \frac{17}{16}$$
$$x + \frac{5}{4} = \frac{\pm\sqrt{17}}{4}$$
$$x = -\frac{5}{4} \pm \frac{\sqrt{17}}{4}$$
$$x = \frac{-5 + \sqrt{17}}{4} \quad \text{or} \quad x = \frac{-5 - \sqrt{17}}{4}$$

$$ax^2 + bx + c = 0$$
$$ax^2 + bx = -c$$
$$x^2 + \frac{b}{a}x = -\frac{c}{a}$$
$$x^2 + \frac{b}{a}x + \frac{b^2}{4a^2} = \frac{b^2}{4a^2} - \frac{c}{a}$$
$$\left(x + \frac{b}{2a}\right)^2 = \frac{b^2 - 4ac}{4a^2}$$
$$x + \frac{b}{2a} = \frac{\pm\sqrt{b^2 - 4ac}}{2a}$$
$$x = -\frac{b}{2a} \pm \frac{\sqrt{b^2 - 4ac}}{2a}$$
$$x = \frac{-b + \sqrt{b^2 - 4ac}}{2a} \quad \text{or} \quad x = \frac{-b - \sqrt{b^2 - 4ac}}{2a}$$

The roots of the quadratic equation $ax^2 + bx + c = 0$, $a \neq 0$ are given by the formula
$$x = \frac{-b \pm \sqrt{b^2 - 4ac}}{2a}.$$
For $2x^2 + 5x + 1 = 0$, $a = 2$, $b = 5$, $c = 1$.
Then $x = \dfrac{-5 \pm \sqrt{25 - 4(2)(1)}}{2(2)} = \dfrac{-5 \pm \sqrt{17}}{4}$

These are the same roots as we obtained above.

27 For each quadratic equation, what are the values of a, b, and c?
   (a) $x^2 - 2x - 5 = 0$
   (b) $x^2 - 3x + 1 = 0$
   (c) $2x^2 - 5x - 1 = 0$
   (d) $3x^2 - 2x + 5 = 0$
   (e) $5x^2 - 3x = 8$
   (f) $2(x^2 - 2x) - 1 = 0$
   (g) $3(x^2 - 1) = 2x$
   (h) $3x - 2(x^2 - 1)$

28 Use the formula to solve each of the following.
   (a) $x^2 + 2x - 35 = 0$
   (b) $x^2 + 2x - 15 = 0$
   (c) $3x^2 - 10x + 3 = 0$
   (d) $2x^2 + 5x - 3 = 0$
   Check each of the above answers by factoring.

29 (a) Use the formula to find the roots of $2x^2 - 3x - 1 = 0$.
   (b) Check your answers in (a) by completing the square.

30 Use the formula to find the roots of the quadratic equation.
   (a) $2x^2 - 4x - 1 = 0$
   (b) $3x^2 - 6x - 5 = 0$

31 Find the x-intercepts of the graph given by
   (a) $y = 2x^2 - 5x - 1$
   (b) $y = 3x^2 - 3x - 4$

32 Find the zeroes of the function, f, defined by $f: x \longrightarrow 5x^2 - x - 3$.

33 Find the roots of each quadratic equation to 1 decimal place.
   (a) $3x^2 - 5x - 1 = 0$
   (b) $4x^2 - 6x - 3 = 0$

34 Use the formula to find the roots of $2.1x^2 - 3.2x - 5.2 = 0$ to 2 decimal places. (Remember this equation given earlier in the introduction?)

35 Find the x-intercepts of each of the following graphs to 2 decimal places.
   (a) $y = 1.8x^2 - 9.8x + 12.2$
   (b) $y = -1.2x^2 - 1.3x + 1.4$

12.6 solving problems: quadratic equations

Once we learn new concepts and skills we may apply them to solving problems. A quadratic equation may be used to solve the problem in Example 1.

We may use our earlier *Steps for Solving Problems* as a guide.

Steps for Solving Problems

Step A Read the problem carefully.
  I What information do we know?
  II What information are we asked to find?
  Then introduce any needed variables.
Step B Translate the problem into mathematics by writing an appropriate equation.
Step C Solve the equation.
Step D Check your answers in the original problem. Check for inadmissible roots.
Step E Write a final statement as the answer to the problem.

The strategies we have developed in other topics also may be applied to the solution of a problem. For example
• drawing diagrams
• using a chart to organize information
and so on.

On solving a quadratic equation we usually find 2 roots of the equation. It may be possible that one of the roots is not *admissible*, if it does not satisfy the conditions of the problem. For example, negative value for time or for the length of the diagonal of a quadrilateral is inadmissible, and thus the negative value does not satisfy the conditions of the original problem. As a result, it is important to check all the roots of a quadratic equation to determine whether they are admissible or not.

In Example 1, the given facts are recorded in a chart to organize the information.

Example 1

Each trip, a fishing trawler sails 120 km from shore to the Grand Banks. With a full cargo, the ship returns at a speed 10 km/h slower. If the return trip takes 2 h longer, find the total travelling time for each trip to the Grand Banks.

Solution

Let t, in hours, represent the time taken, to travel to the Grand Banks. Then $t + 2$, in hours, is the return time.

	distance (km)	rate (km/h)	time (h)
out	120	$\frac{120}{t}$	t
in	120	$\frac{120}{t+2}$	$t + 2$

Use the chart to organize the information.

From the chart, the difference in the rate is given by

$$\frac{120}{t} - \frac{120}{t + 2} = 10$$
$$120(t + 2) - 120t = 10t(t + 2)$$
$$120t + 240 - 120t = 10t^2 + 20t$$
$$10t^2 + 20t - 240 = 0$$
$$t^2 + 2t - 24 = 0$$
$$(t - 4)(t + 6) = 0$$
$$t = 4 \quad \text{or} \quad t = -6$$
$$\qquad\qquad\qquad\text{inadmissible}$$

Thus, the time out is 4 h and the time in is 6 h. The total travelling time is 10 h.

In Example 1, the root -6 is inadmissible since time cannot be negative.

Example 2

For Curran Park, a landscaper wishes to plant a boundary of tulips within a rectangular garden with dimensions 18 m by 12 m. To obtain a pleasing look the area of the tulip border should be half of the area of the garden. How wide should the border be, to 1 decimal place?

Solution

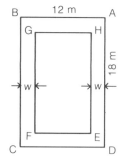

Let w, in metres, represent the width of the border.
Area of ABCD is given by 12 m × 18 m or 216 m².
Area of HEFG given by $(18 - 2w)(12 - 2w)$ m².

Use area of HEFG $= \frac{1}{2}$ (Area ABCD)

$$(18 - 2w)(12 - 2w) = \frac{1}{2}(216)$$
$$216 - 60w + 4w^2 = 108$$
$$4w^2 - 60w + 108 = 0$$
$$w^2 - 15w + 27 = 0$$

Use $w = \dfrac{-b \pm \sqrt{b^2 - 4ac}}{2a}$

$a = 1, b = -15, c = 27.$

$w = \dfrac{15 + \sqrt{225 - 108}}{2}$ or $w = \dfrac{15 - \sqrt{225 - 108}}{2}$

$= \dfrac{15 + \sqrt{117}}{2}$ $\qquad\sqrt{117} \doteq 10.82$ $\qquad= \dfrac{15 - \sqrt{117}}{2}$

$= \dfrac{15 + 10.82}{2}$ $\qquad\qquad\qquad\qquad\doteq \dfrac{15 - 10.82}{2}$

$\doteq 12.91$ $\qquad\qquad\qquad\qquad\qquad\doteq 2.09$

inadmissible root $\qquad\qquad$ (to 2 decimal places)

Thus, the width of the boundary is 2.1 m to 1 decimal place.

---

math tip

It is important to **understand** clearly the vocabulary of mathematics when solving problems involving the quadratic function. Make a list of all the new words you have met in this chapter. Provide a sample example to illustrate each word.

12.6 exercise

Express your answers to 1 decimal place, as needed.

A

1. Two positive consecutive integers, n and $n + 1$ are squared. If the sum of the squares is 145, what are the numbers?

2. A package designer wants to protect an expensive square book with a 5-cm wide cardboard rim. The area of the rim of the front cover is equal to the area of the front cover of the book. What is the length of the book?

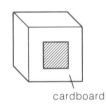
cardboard

3. The perimeter of a right-angled triangle is 60 cm. The lengths of 2 sides of a right-angled triangle, in centimetres, are shown. Find the lengths of all sides of the triangle.

4. In travelling from Arcola to Beamsville we can use either the super highway or connect through Pasqua as shown. On the super highway we can travel 20 km/h faster, and take 2 h less time. What is the average speed for each route?

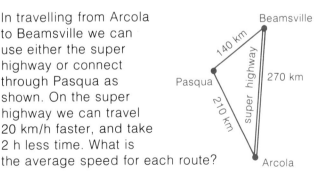

5. Two numbers differ by 4 and the sum of their squares is 208. What are the numbers?

B

6. When two consecutive integers are squared and the squares added their sum is 421. What are the possible numbers?

7. Find 3 consecutive positive odd integers such that the sum of the squares of the first two is 15 less than the square of the third.

8. A right-angled triangle has a height 8 cm more than twice the length of the base. If the area of the triangle is 96 cm², find the dimensions of the triangle.

9. For a certain model of a jet plane, aerodynamics requires that the tail have the profile of a right-angled triangle with area 7.5 m². If the height must be 2 m less than the base length for attachment purposes, calculate the base length of the tail.

10. A rectangular solar heat collecting panel has a length 3 m more than its width. If the area of the solar panel is 28 m², how long is the panel?

11. A right-angled triangle has a perimeter of 120 cm. If the hypotenuse is 50 cm, find the lengths of the other two sides.

12. A matte is to be placed around a painting in such a way that the area of the matted surface is twice the area of the picture. If the outside dimensions of the matte are 40 cm and 60 cm, find the width of the matte.

13. Each side of a square house is 6 m longer than each side of the square garage. If the combined area of house and garage is 180 m², find the dimensions of the house and the garage.

14. The "Fashion Up" magazine prints square photographs in two sizes. The larger photographs have sides 2 cm longer than the smaller photographs. The combined area of 3 of the smaller photographs is 12 cm² more than one larger photograph. Find the dimensions of each size of photograph.

15 A square swimming pool with a side measuring 16 m is to be surrounded by a rubberized floor covering. If the area of the floor covering equals the area of the pool, find the width of the rubberized covering.

16 Janet swims lengths at a local pool every morning. Last week she swam a total of 12 000 m. This represented 20 more lengths than the length of the pool. What is the length of the pool?

17 Davidson is on a cross Canada motorcycle trip and has just arrived at the foothills of the Rockies. He plans to take 1 h longer on the 245-km trip up the east side than on the 225-km trip down the west side. To do this he will have to average 20 km/h faster on the downhill side. How long will the trip through the Rockies take?

18 If a guitar maker increases the production rate by 3 guitars per month he will finish an order for 65 guitars in 2 months less time. What is the original output of the guitar maker (to the nearest whole guitar)?

19 John and Petra leave the cottage by two separate roads, each 3 km in length, to go to town. Petra walks 0.5 km/h faster than John. Along the road John meets a friend and stops to talk for 5 min so that Petra has to wait in town 10 min for John to arrive.

(a) How fast do John and Petra walk?
(b) How long does it take John and Petra to travel the distance?

20 The Caldwells start out on a 1520-km car trip to Wascana Park. On the first day they cover 960 km. On the second day they complete the trip, but a heavy rain storm causes them to reduce their average speed by 10 km/h. If the 2-d trip took a total of 20 h, what was the average speed on each day?

21 In the annual 60-km charity Walk-A-Thon, Mark and Tina leave at the same time. Tina walks 0.8 km/h faster than Mark, but stops to have her feet taped, thus losing 0.5 h. Even with this delay Tina finishes the race 2 h before Mark.

(a) How fast was each person walking?
(b) How long did it take each to walk the course?

C

22 A television screen is 40 cm high and 60 cm long. Using the vertical and horizontal control buttons, the picture is compressed to 62.5% of its original area, leaving a uniform dark strip around the outside. Find the dimensions of the smaller picture.

23 Amy loves pizza crust and Bill likes the cheese that gathers in the middle. They want to split a pizza with radius 30 cm so that Amy gets the outside ring and Bill gets the centre, yet they each get the same amount of pizza. How wide should Amy's ring be?

problem/matics

The following computer program, written in the language of BASIC, calculates the roots of the quadratic equation $ax^2 + bx + c = 0$ using the formula $x = \dfrac{-b \pm \sqrt{b^2 - 4ac}}{2a}$.

```
10 INPUT A, B, C
20 LET D = B ↑ 2 - 4 * A * C
30 IF D < 0 then 80
40 LET X1 = (-B + SQR(D))/(2 * A)
50 LET X2 = (-B - SQR(D))/(2 * A)
60 PRINT "THE ROOTS ARE", X1, X2
70 GO TO 90
80 PRINT "THE ROOTS ARE NOT REAL"
90 END
```

Use the above program to check the roots for the quadratic equations we obtained on pages 430 and 431. Use the program also to obtain the answers to the problems in this section.

12.7 the nature of the roots of quadratic equations

For the general quadratic equation
$$ax^2 + bx + c = 0$$
we have obtained a formula for the roots m and n.
$$m = \frac{-b + \sqrt{b^2 - 4ac}}{2a} \quad n = \frac{-b - \sqrt{b^2 - 4ac}}{2a}$$

From the above expressions, if the roots are real numbers, then we must have $b^2 - 4ac \geq 0$ since $\sqrt{b^2 - 4ac}$ will not be a real number if $b^2 - 4ac < 0$.

Much of our study of the nature of the roots of quadratic equations, as well as the theory of quadratic equations, is based on the values of the following expression called the **discriminant**, D.

$$D = b^2 - 4ac$$

For example:
If $b^2 - 4ac = 0$, then $m = n$. Thus the roots of the corresponding quadratic equation are equal. Graphically, we have seen that in this case the corresponding graph touches the x-axis.

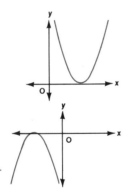

If $b^2 - 4ac > 0$ then there are 2 roots since the graph cuts the x-axis in two points.

If $b^2 - 4ac < 0$ then $\sqrt{b^2 - 4ac}$ is not a real number, and the graph does not meet the x-axis.

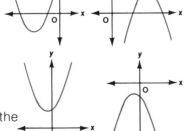

The roots of a quadratic equation may be characterized as follows:
  I real or not real
 II equal or unequal
III rational or irrational

By examining the values of $D = b^2 - 4ac$ we may learn about the characteristics of the roots *without requiring to find the roots*, as we shall see in the following exercise.

12.7 exercise

A

1. For each equation,
   A find the roots. B calculate $D = b^2 - 4ac$.
   (a) $x^2 - 5x - 6 = 0$ (b) $x^2 - 7x + 12 = 0$
   (c) $2x^2 + x - 1 = 0$ (d) $3x^2 + 8x - 3 = 0$
   How do your answers in A and B compare?

2. For each equation,
   A find the roots. B calculate $D = b^2 - 4ac$.
   (a) $x^2 - 4x - 1 = 0$ (b) $2x^2 - 3x - 1 = 0$
   (c) $3x^2 - 5x - 1 = 0$ (d) $2x^2 - 5x + 1 = 0$
   How do your answers in A and B compare?

3. For each equation,
   A find the roots. B calculate $D = b^2 - 4ac$.
   (a) $x^2 + 4x + 4 = 0$ (b) $x^2 - 8x + 16 = 0$
   (c) $9x^2 - 12x + 4 = 0$ (d) $4x^2 - 4x + 1 = 0$
   How do your answers in A and B compare?

4. For each equation,
   A find the roots. B calculate $D = b^2 - 4ac$.
   (a) $x^2 - x + 3 = 0$ (b) $x^2 - 2x + 5 = 0$
   (c) $2x^2 + x + 3 = 0$ (d) $3x^2 + 3x + 1 = 0$
   How do your answers in A and B compare?

5. Use your results in Questions 1 to 4. What is the value of the discriminant if the roots of the

equation are described as follows?

(a) real (b) not real
(c) equal, real (d) rational, not equal
(e) irrational, not equal

characteristics of roots

To describe the roots of the quadratic equation we must calculate the value of the discriminant. In the previous exercise we found the following.

I If $b^2 - 4ac > 0$ then $\sqrt{b^2 - 4ac}$ is a real number. Thus the roots are real numbers.
If $b^2 - 4ac < 0$, then $\sqrt{b^2 - 4ac}$ is not a real number. Thus the roots are *not* real.

II If $b^2 - 4ac = 0$, then the roots are equal.
Thus $m = -\dfrac{b}{2a}$ and $n = -\dfrac{b}{2a}$.

III If $b^2 - 4ac$ is a perfect square then $\sqrt{b^2 - 4ac}$ is a rational number. Thus the roots are rational.
If $b^2 - 4ac$ is not a perfect square, then the roots are irrational.

For Questions 6 to 18 use the above characteristics of roots to answer the questions.

6 (a) Calculate the discriminant of the equation $x^2 - 5x + 1 = 0$, and describe the characteristics of the roots as I, II, and III.
 (b) Check your answers in (a) by finding the roots in (a).

7 (a) Calculate the discriminant of the equation $16x^2 - 24x + 9 = 0$, and describe the characteristics of the roots.
 (b) Check your answer in (a) by finding the roots in (a).

8 For each equation,
• calculate the discriminant.
• describe the nature of the roots.

(a) $x^2 - 8x + 16 = 0$ (b) $x^2 - 7x + 6 = 0$
(c) $5x^2 + 7x + 2 = 0$ (d) $2x^2 - 5x + 6 = 0$
(e) $2x^2 - 3x - 7 = 0$ (f) $x^2 - 3x + 5 = 0$

B 9 For each function defined by the equation indicate whether the graph
A: touches the x-axis.
B: intersects the x-axis in 2 points.
C: does not meet the x-axis.

(a) $y = x^2 - 6x + 5$ (b) $y = 16 + 8x + x^2$
(c) $y = 2x^2 - 3x - 2$ (d) $y = 1 - 2x - 7x^2$
(e) $y = 4x^2 + 7x + 2$ (f) $y = 3x^2 + x + 1$
(g) $y = 17 + 11x - x^2$ (h) $y = 9 - x^2$

10 The equation $y = mx^2 - 4x + m$ does not cut the x-axis. What value or values must m have?

11 If the equation $4x^2 - 4x + k = 0$ has equal roots, what must be the value of k?

12 If $y = 3x^2 + 4x + c$ has 2 x-intercepts, what value or values must c have?

13 If the graph, defined by the equation $y = x^2 + 9x + c$, cuts the x-axis in 2 distinct points, what value or values must c have?

14 If the graph of $y = x^2 - mx + m + 24$ touches the x-axis, what must be the values of m?

15 Each equation has two real and distinct roots. What value or values must k have?
(a) $kx^2 - 4x + 5 = 0$ (b) $x^2 + kx + 5 = 0$

16 Each equation has non-real roots. What value or values must k have?
(a) $kx^2 - 10x + 12 = 0$ (b) $6x^2 + 6x + k = 0$

C
17 For the quadratic equation given by
$mx^2 + (2m + 1)x + m = 0$
show that $4m + 1$ must be zero for the equation to have equal roots.

18 For the quadratic equation given by
$(k + 1)x^2 + 2kx + (k - 1) = 0$
show that there are no real numbers of k for which the equation has equal roots.

12.8 solving linear quadratic systems

Some of the equations we have worked with define the orbits of the planets, satellites, comets, and space probes, etc.

Shape of orbit	Type of equation
Circular	$x^2 + y^2 = r^2$
Parabolic	$y = ax^2$

One area of interest to space explorers is the calculation of the intersection of the paths of various satellites. Although the actual equations are complex and are solved by using computers, the principles of solving a system of equations remain unchanged. The graph of a linear function may intersect a quadratic function as shown in 3 situations.

2 points of intersection

1 point of intersection

no points of intersection

We can find the intersection of a linear and quadratic function, graphically.

Example 1
Find the intersection of the graphs defined by $y = x^2 + 2x$ and $4x + y = -8$.

Solution
Draw the graphs defined by each equation. Use
$$y = x^2 + 2x$$
$$= x^2 + 2x + 1 - 1$$
$$= (x + 1)^2 - 1$$

axis of symmetry, $x = -1$; vertex $(-1, -1)$; x-intercepts $-2, 0$

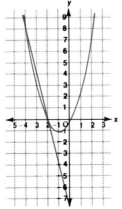

From the graph, the intersection points are $(-2, 0)$, and $(-4, 8)$.

However, as we have already noted in our work with linear functions, finding the co-ordinates of intersection points graphically is an inaccurate method. To solve the system algebraically, we use the *substitution method* (see page 128). The method of mathematics is to extend our work to apply to new situations.

$$x - 2y = -13 \quad ①$$
$$y = x^2 - 1 \quad ②$$

From ①. $x = 2y - 13$
Substitute for x in ②.
$$y = (2y - 13)^2 - 1$$
$$y = 4y^2 - 52y + 169 - 1$$
$$4y^2 - 53y + 168 = 0$$
$$(y - 8)(4y - 21) = 0$$
$$y - 8 = 0 \quad \text{or} \quad 4y - 21 = 0$$
$$y = 8 \quad \bigg| \quad y = \frac{21}{4}$$

Substitute for each value of y in ①.

$$x - 16 = -13 \quad \bigg| \quad x - \frac{21}{2} = -13$$
$$x = 3 \quad \bigg| \quad x = -\frac{5}{2}$$

Thus the intersecting points of the curves are given by $(3, 8)$ and $\left(-\frac{5}{2}, \frac{21}{4}\right)$.

Thus we have developed a method of solving a **linear-quadratic system**.

12.8 exercise

A

1. A linear-quadratic system is given by the equations
$$2x - y = 12 \qquad y = x^2 + 9x$$

(a) From the linear equation, express y in terms of x.
(b) Use the expression in (a) for y and substitute in the quadratic equation.
(c) Solve the system.

2 A linear-quadratic system is given by
   $y = x^2 - 6x$ $2x + y = -3$
   (a) Sketch a graph of the system. Find the co-ordinates of the intersection points.
   (b) Verify your answers in (a) by solving the system algebraically.

3 (a) Solve the linear-quadratic system given by
   $y = x^2 + 3x$ $2x - y = -12$
   (b) Verify your answers in (a).

B

4 A circle is defined by $x^2 + y^2 = 25$. The line given by $y = 2x + 5$ intersects the circle. Find the points of intersection.

5 Find the points of intersection of the quadratic function defined by $y = x^2 + x$ and the linear function given by $7x + y = -7$.

6 A parabola given by $y = x^2 + 5x$ is intersected by the straight line $6x + y = -28$. Find the co-ordinates of the points of intersection.

7 A circle defined by $x^2 + y^2 = 10$ is intersected by the line defined by $2x - y = 5$. Find the co-ordinates of the intersection points.

8 Solve each linear-quadratic system.
   (a) $x - y = -5$ (b) $y = x^2 + 6x$
       $y = -x^2 + 5$ $2x - y = -32$
   (c) $y = x^2 + 2x$ (d) $x + y = 5$
       $4x + y = 27$ $y = x^2 + 3$
   (e) $x^2 + y^2 = 18$ (f) $x^2 + y^2 = 41$
       $x = y$ $x + y = 9$

problem/matics

We may apply the same principles as in Section 12.8 to finding the intersection points of the graphs defined by 2 quadratic equations. Can you find the intersection points of these curves?

A $y = x^2 + 2$ B $x^2 + y^2 = 5$
   $y = 2x^2 + 1$ $x^2 - y^2 = 3$

C $x^2 + 5y^2 = 25$ D $2x^2 + 3y^2 = 17$
   $x^2 + y^2 = 9$ $x^2 + y^2 = 7$

For Questions 9 to 11, a grid is superimposed on the solar system. The equations are given with reference to the co-ordinate system.

9 A satellite travels along a path given by $y = x^2 - 6x$. A rocket travels along a path given by $4x - y = 21$. Find the intersection points.

10 A weather satellite travels a circular orbit given by $x^2 + y^2 = 25$. A space probe, travelling in a straight line defined by $x - y = 2$, is on a collision course with the satellite. Find the co-ordinates of the points where a collision might take place to 2 decimal places.

11 The orbit of a space probe is defined by $y = x^2 - x - 6$. A cluster of meteors from an explosion are travelling along a straight line path defined by $x + 2y + 6 = 0$. Find the co-ordinates of the possible intersection points.

quadratic regions

The graph of $y = x^2$ divides the plane into 3 regions.

A: on the graph of $y = x^2$

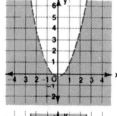

B: exterior region of the parabola defined by $y < x^2$

C: interior region of the parabola defined by $y > x^2$

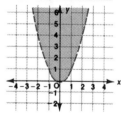

12 (a) Draw a graph of $y = x^2 - 3x$.
   (b) Draw a graph of the region given by $y > x^2 - 3x$.
   (c) Draw a graph of the region given by $y < x^2 - 3x$.

13 (a) Draw a graph of the region defined by $y > x^2 - 2$.

(b) Which of the following points are in the region?
A(2, 2)   B(0, 1)   C(3, 7)

14 (a) Draw a graph of the region given by
   A: $y \geq -2x^2 + 5$ B: $y < -x^2 + 5$
   C: $y > -x^2 + 5$
   (b) To which region does each point belong?
   P(−1, 6)   Q(3, −4)   R(−2, 0)

15 Sketch and shade a region defined by each of the following.
   (a) $y < (x - 1)^2 + 4$ (b) $y > 2x^2 - 3$
   (c) $y \leq (x - 1)^2 + 4$ (d) $y \geq 2x^2 - 3$
   (e) $y < -3(x - 1)^2$ (f) $y \geq -2(x + 1)^2 - 1$

16 Draw the region defined by
   (a) $x + y < 6$ and $y < x^2$.
   (b) $2x - y > 2$ and $y > 2x^2$.
   (c) $x - 2y \leq 3$ and $y \geq -x^2 + 5$.

17 Write the inequation that describes each region.
   (a)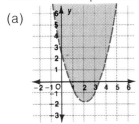
   (b)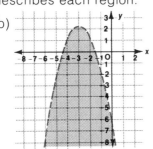

problem/matics

The general quadratic equation
$ax^2 + bx + c = 0$, $a \neq 0$
has roots m and n.
A: Show that the sum of the roots, $m + n$, is given by $-\dfrac{b}{a}$.
B: Show that the product of the roots, mn, is given by $\dfrac{c}{a}$.

skills review: quadratic functions and equations

1 Use the method of completing the square to find x.
   (a) $x^2 - 6x + 2 = 0$ (b) $2x^2 - 5x + 1 = 0$
   (c) $2x^2 + 8x + 1 = 0$ (d) $3x^2 - 2x - 1 = 0$

2 Use the quadratic formula to determine the roots of
   (a) $3x^2 - 2x + 5 = 0$ (b) $3x^2 - 5x - 1 = 0$

3 Find the x-intercepts of $y = 1.9x^2 - 9.7x + 12.1$.

4 Find the roots of the following equations.
   (a) $2x^2 - 5x + 2 = 0$ (b) $3x^2 - 7x + 4 = 0$
   (c) $2x^2 - 5x - 2 = 0$ (d) $5x^2 + 6x - 1 = 0$
   (e) $2x^2 - 13x + 10 = 0$ (f) $6x^2 - 7x + 2 = 0$

5 Find the equation of symmetry and the co-ordinates of the vertex of each of the following.
   (a) $y = 1 + 4x - 2x^2$ (b) $y = 3x^2 - 6x + 2$

6 Use the method of completing the square to find the maximum or minimum for each of the following.
   (a) $y = 5 - 6x + 3x^2$ (b) $y = x^2 - 6x + 9$
   (c) $y = -3x^2 - 6x + 5$

7 Find the discriminant for each equation.
   (a) $x^2 - 9x + 6 = 0$ (b) $6x^2 - 5x + 1 = 0$
   (c) $2x^2 - 3x - 1 = 0$ (d) $5x^2 - 3x - 2 = 0$
   (e) Use your answers above to describe the roots of each equation.

problems and practice: a chapter review

You deserve a rest!

test for practice

Try this test. Each *Test for Practice* will be based on the mathematics you have learned in the chapter. Try this test later in the year as a review. Keep a record of those questions that you were not successful with. Find out how to answer them and review them periodically.

1. A function, g, is defined by $g:x \longrightarrow 2(x - 1)^2 - 5$.
   (a) Draw the graph of the function, g.
   (b) Does the graph have a minimum or maximum point?

2. Draw a sketch of $y = 2x^2 + 8x - 6$. Write the equation of the axis of symmetry and the co-ordinates of the vertex.

3. If $A(0, 1)$ and $B(-2, -3)$ are points on the graph defined by $y = ax^2 + b$, find the values of a and b.

4. For each of the following functions, find its zeroes, co-ordinates of the vertex, maximum or minimum value, and the equation of the axis of symmetry.
   (a) $y = x^2 - 4x + 3$ (b) $y = -2x^2 + 9x - 4$

5. For the translation given by $(x, y) \longrightarrow (x + 1, y - 2)$ find the equation of the image of $y = 2x^2$.

6. Draw a graph of each curve.
   (a) $y = 2 + x - x^2$ (b) $y = 2x^2 - 5x + 2$

7. What is the minimum or maximum value of each of the following?
   (a) $y = -\frac{3}{4}(x - 2)^2 - 23$ (b) $y = 3(x + 1)^2 + 25$

8. Use the method of completing the square to find x for $2x^2 - 5x - 1 = 0$.

9. Find the roots of each of the following.
   (a) $2x^2 - 5x + 3 = 0$ (b) $3x^2 - 4x - 1 = 0$

10. A parabola is defined by $y = x^2 - 5x - 2$. Find the co-ordinates of the points where the parabola intersects the x-axis.

11. Find the zeroes of the function f, defined by $f:x \longrightarrow 3x^2 - 10x + 3$.

12. Calculate the discriminant for each equation.
    (a) $6x^2 - 13x + 6 = 0$ (b) $4x^2 - x - 1 = 0$
    (c) Describe the nature of the roots in (a) and (b).

13. Solve each system.
    (a) $y = x^2 + 6x$ (b) $x^2 - 2y = 11$
        $x - y = 4$ $x - y = 4$

14. In a newspaper contest the last two numbers of a combination to open a safe add up to 44. As an additional clue, the product of the two numbers is a maximum. What are the last two numbers?

15. The perimeter of a right-angled triangle is 36 cm. If the hypotenuse is 15 cm, find the length of the other two sides.

16. The hypotenuse of a right-angled triangle is 1 m more than the length of the second largest side. If the perimeter is 56 m, find the length of the other sides.

17. On a spending spree contest John spends $330 in 2 h less than his competitor Susan. Susan spends money at a rate of $15/h less than John. Find the rate of spending of each person, in dollars per hour to 1 decimal place

18. Dave and Jim entered the annual spring canoe race last year. The number of times they paddled the course during practice was one less than the number of kilometres in the course. If they paddled a total of 72 km during practice, how long was the course?

function plus

An overview of concepts and skills, investigations, transformations, and special functions.

f.1 the language of functions

The study of the function is fundamental to the study of mathematics. We have seen a variety of functions throughout our work.

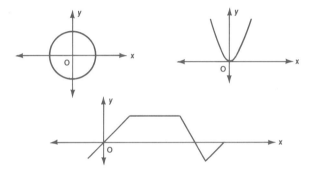

The method of mathematics is to look for relationships and patterns and explore the related mathematics. The study of the function relates some topics which may, on the surface, seem unrelated. To study functions we begin by defining a relation.

A relation is a set of ordered pairs. We can construct a set of ordered pairs called the **Cartesian product**.
   $A \times B = \{(a, b) | a \in A, b \in B\}$
For example, if $A = \{1, 2\}$ and $B = \{3, 4, 5\}$ then
$A \times B = \{(1, 3), (1, 4), (1, 5), (2, 3), (2, 4), (2, 5)\}$
The graph of $A \times B$ is shown. $A \times B$ is a relation.

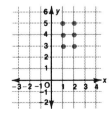

As we have seen, the domain of the relation is the set of all first components of the ordered pairs of the relation.
   domain = $\{1, 2\}$
The range of the relation is the set of all second components of the ordered pairs of the relation.
   range = $\{3, 4, 5\}$

We may also draw a mapping diagram for the above relation.

If D is the domain and R is the range then we may express the above mapping as $f: D \longrightarrow R$. If a relation has the property that for each first component there is a unique second component, then this special relation is called a **function**.

not a function   a function

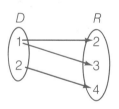 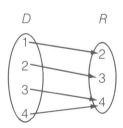

For a function, only one arrow leaves each member of the domain, although more than one arrow may point to a member in the range. In terms of ordered pairs, we may define a function as follows. *A relation is a function, f, such that, if $(a, b) \in f$ and $(a, c) \in f$, then $b = c$.* In other words, a function is a relation such that for each member of the domain there is exactly one corresponding member of the range.

On the Cartesian plane, we may use the vertical line test to test for a function.

not a function   a function

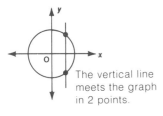
The vertical line meets the graph in 2 points.

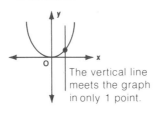
The vertical line meets the graph in only 1 point.

Vertical Line Test: If a vertical line passes through points of the domain and intersects the graph in exactly one point, then the relation is a function as shown by the above examples.

Since the function concept is used in many branches of mathematics, appropriate notation has been invented for the different ways of studying the function. For example, the linear function may be shown in different ways.

as a mapping
$f : x \longrightarrow x + 2, x \in R$

using function notation
$f(x) = x + 2, x \in R$

as a set of ordered pairs
$f = \{(x, y) | y = x + 2, x \in R\}$ or just
$y = x + 2, x \in R$

the defining equation of the function

by a graph

building vocabulary

One-to-one: A function, f, is said to be one-to-one if the function has the property that

- for each member of the range there is a corresponding member of the domain.
- for each member of the domain there is only one corresponding member of the range.

f, one-to-one function

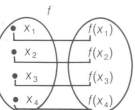

The above mapping diagram shows how the notation $f(x)$ is related to the mapping given by f.

Symmetry: We can learn some of the properties of a relation or function from its graph. For example, the graph of $P : x \longrightarrow x^2$ is shown. We see that the graph of P is *symmetric with*

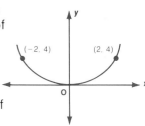

respect to the y-axis. For the point (2, 4) there is a corresponding point (−2, 4) on the graph P. In general, we may define the graph of a relation P to be *symmetric with respect to the y-axis* if for any (m, n) on the graph there is a $(-m, n)$ on the graph. Thus the y-axis acts as a mirror. The mapping $(m, n) \longrightarrow (-m, n)$ represents a reflection in the y-axis.

The relation at right is *symmetric with respect to the x-axis.* In general, the graph of a relation is symmetric with respect

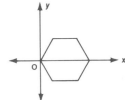

to the x-axis if for any (m, n) on the graph there is a $(m, -n)$ on the graph. Thus the x-axis acts as a mirror. The mapping $(m, n) \longrightarrow (m, -n)$ represents a reflection with respect to the x-axis.

Continuous: We may describe a function as continuous if its graph can be drawn continuously (without lifting a pen) as shown.

continuous   non-continuous

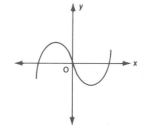
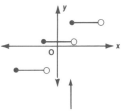

Why does this graph show a function?

Example

For the graph defined by $x^2 + y^2 = 16$

(a) what are the intercepts?

(b) is the curve symmetric with respect to the x-axis? y-axis?

(c) is the graph continuous or not?

Solution

(a) To find the x-intercept let $y = 0$.
$x^2 + y^2 = 16$
$x^2 + 0 = 16$
$x = \pm 4$
The x-intercepts are $+4$ and -4.

To find the y-intercept let $x = 0$.
$x^2 + y^2 = 16$
$0 + y^2 = 16$
$y = \pm 4$
The y-intercepts are $+4$ and -4.

(b) (x, y) and $(x, -y)$ satisfy $x^2 + y^2 = 16$. Thus the graph is symmetric with respect to the x-axis.

(x, y) and $(-x, y)$ satisfy $x^2 + y^2 = 16$. Thus the graph is symmetric with respect to the y-axis.

(c) The graph is continuous (as shown below).

Earlier we saw that $x^2 + y^2 = 16$ is the defining equation of a circle with centre $(0, 0)$ and radius 4. The circle is also *symmetric with respect to the origin* since for any (x, y) on the graph, $(-x, -y)$ is also on the graph.

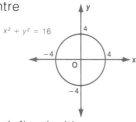

The term *symmetric* may be defined with respect to a line (axis), or with respect to a point (origin). Each of the following figures has symmetry. The triangle may be reflected in the line shown and mapped onto itself. The other figure may be rotated about the point O and mapped onto itself.

- A figure has **reflectional symmetry** in a line *m* if the figure may be mapped onto itself by reflection in the line, *m*, (called its axis or line of symmetry).

  $\triangle ABC$ has reflectional symmetry in the line *m*.

- A figure has **rotational symmetry** about a point, O, if the figure may be mapped onto itself by a rotation about the point O.

  Figure T has rotational symmetry about O. Since there are 3 rotations that map the figure onto itself, we say the figure has *rotational symmetry* of order 3.

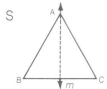

f.1 exercise

1. For each set of ordered pairs, which is a function and which is not? Give reasons for your answer.

   (a) $(4, -1), (5, 0), (-3, -1), (6, 1), (5, 1), (4, 0)$
   (b) $(-4, 2), (-1, -1), (-3, 1), (0, -1), (-2, 0)$

2. Solve for the components.

   (a) $(4, y) = (-x, 3)$ (b) $(x + 1, 4) = (-2, y - 1)$
   (c) $(x - 3, y + 1) = (2x - 4, 3y - 7)$

3. The graph of a function is given. Write the value of

   (a) $g(3)$ (b) $g(-3)$
   (c) $g(0)$ (d) $g(-1)$
   (e) Is the graph continuous? Why or why not?

   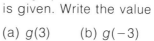
   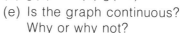
   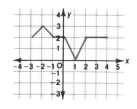

4. The graph of a function is shown. Write the value of

   (a) $h(-2)$ (b) $h(3)$ (c) $h(0)$
   (d) Is the graph continuous? Why or why not?

   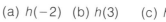

   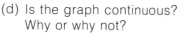
   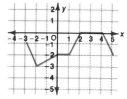

5. A graph of a relation is shown.

   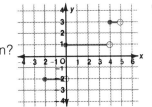

   (a) Is the relation a function? Why or why not?
   (b) Write the value of $f(-1)$, $f(3)$, $f(-2)$, $f(4)$.
   (c) Is the graph continuous? Why?

6. If $A = \{a, b\}$ and $B = \{1, 2\}$ list the members of
   (a) $A \times B$ (b) $B \times A$
   (c) For what conditions would $A \times B = B \times A$?

7. For each relation shown by a mapping diagram, write the set of ordered pairs.

   (a) 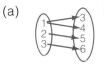 (b)   (c)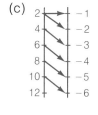

   (d)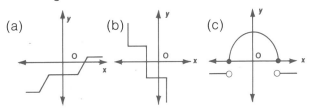

   (e) Write the domain and range for each relation.
   (f) Which of the above relations are functions?
   (g) Which functions are one-to-one?

8. Write each function in three other forms.
   (a) $f(x) = 3x - 2$ (b) $g: x \longrightarrow 2x^2 - 5$

9. All graphs on the plane are subsets of $R \times R$ where R is the set of real numbers. Which of the following are functions? Which are not?

   (a) (b) (c) (d) (e) (f) (g) (h) (i)

   Which of the above are continuous? Which are not?

10. If $f(x) = 2x$, $g(x) = 2x^2 - 3$, find the values of
    (a) $f(-1)$ (b) $g(-1)$ (c) $g(0)$ (d) $f^2(-2)$
    (e) $g(-3)$ (f) $f[g(2)]$ (g) $g[f(-3)]$ (h) $f[g(-3)]$
    — Find $g(2)$ first.

11. For the function defined by each of the following, find the image of the given point.
    (a) $f(x, y) \longrightarrow (x - 1, 3 + y)$ $(-2, 1)$
    (b) $g: (x, y) \longrightarrow (2x, y + 2)$ $(-5, -4)$
    (c) $H: (x, y) \longrightarrow (x - 1, -3y)$ $(6, -3)$

12. For the following functions, which are one-to-one? Give reasons why.
    (a) $g: x \longrightarrow 2x - 5$ (b) $y = 5 - 8x$
    (c) $g(x) = 2x^2 - 3$ (d) $f: x \longrightarrow 2x^2$
    (e) $y = 2x^2 - 1$ (f) $f(x) = 3x - 2$

13. The defining equation of various relations are given. Which are symmetric with respect to
    • the x-axis? • the y-axis? • the origin?

    (a) $y = 2x - 3$ (b) $f: x \longrightarrow \dfrac{1}{x^2}$
    (c) $x^2 + y^2 = 1$ (d) $y = (x - 1)^2 + 5$
    (e) $f(x) = 3x^2 - 5$ (f) $\dfrac{x^2}{4} + \dfrac{y^2}{9} = 1$

    (g) Draw a graph of each of the above.
    (h) Which are functions?

14 If $f(x) = (x - 1)(x + 2)$, find an expression for each of the following.
(a) $f(2a)$ (b) $f(a + 2)$ (c) $f(3 - a)$
(d) $f(-a)$ (e) $f(-2a)$ (f) $f(a^2 - 1)$

15 In rolling 2 dice, we may write ordered pairs to show the outcome. How many ordered pairs are there if $A = \{1, 2, 3, 4, 5, 6\}$, $B = \{1, 2, 3, 4, 5, 6\}$?

(2, 5) — number rolled on first cube, number rolled on second cube

16 Some Canadian teams of the National Hockey League (NHL) are
V Vancouver   W Winnipeg   M Montreal
E Edmonton   T Toronto
We use ordered pairs to list the combinations of games the teams will play.

(W, M) — Winnipeg plays Montreal in Winnipeg. First component is the home team.

(a) What is meant by (V, E)?
(b) How many different games are possible?
(c) Is the following true or false? Give reasons for your answer. "If $G = \{V, E, W, T, M\}$ then all games played by the teams are given by $G \times G$."

17 (a) Use a Cartesian product to show the different teams possible for tennis doubles if the following people are at the tennis court.
Barbara B, Michael M, Lori L, Peter P, Gerard G, Elaine E
(b) How many pairs are all girls?
(c) How many pairs have a boy and a girl?

18 Which of the following figures has reflectional symmetry? Rotational symmetry? Give the number of lines of symmetry or the order of rotational symmetry.

(a)   (b)   (c)   (d) (e) (f)

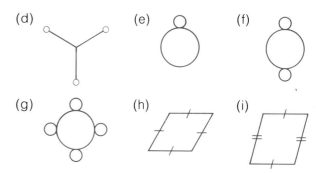

(g) (h) (i)

inverses of relations and functions

In mathematics, the development of vocabulary enlarges on the meanings of words used earlier. For example, in our earlier work in computation, we used the terms *inverse operations*, *add*, *subtract*, *multiply*, *divide* or other terms such as *additive* or *multiplicative inverse*. We now base the definition of **inverse functions** on our earlier understanding of the word *inverse*. For the domain D and range E, a relation or function f is shown.

If there is a relation that maps $f(x)$ onto x then we call that relation the inverse of f, denoted by f^{-1}, or simply called f inverse.

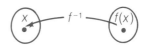

Note that f^{-1} does not mean $\dfrac{1}{f}$.

If $(x, y) \in f$ then $(y, x) \in f^{-1}$. Thus to find the inverse of f, we interchange the components of each ordered pair (x, y) of f.

Example 1

A function f is given by
$f = \{(-2, 2), (-1, 3), (0, 3)\}$.

(a) Find the graph of f and f^{-1}.
(b) Is f^{-1} a function?

Solution

(a) $f^{-1} = \{(2, -2), (3, -1), (3, 0)\}$
From the graph, we see that the graph of f^{-1} is the reflection of the graph of f in the line $y = x$.

(b) f^{-1} is not a function since $(3, 0)$ and $(3, -1) \in f^{-1}$.

From the above example, and also from the following exercise we will find that

- the graph of f^{-1} is the reflection of the graph of f in the line $y = x$.
- if f is a function, f^{-1} may or may not be a function.

19 For each relation, f, defined by an equation, draw its graph, as well as the graph of its inverse on the same axes.

(a) $y = 2x - 3$ (b) $y = x^2$
(c) $x = y^2$ (d) $x^2 + y^2 = 16$
(e) Which inverse relations above are functions?

20 The function f is shown.

(a) Draw a graph of f^{-1}.
(b) Find the values of $f^{-1}(-1), f^{-1}(0), f^{-1}(1)$.
(c) Is f^{-1} a function?

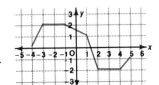

21 A function is given by $f: x \longrightarrow x^2$. The domain of f is $D = \{-2, -1, 0, 1, 2\}$.

(a) List the ordered pairs of f.
(b) Write the ordered pairs of f^{-1}.
(c) Draw the graph of f and f^{-1}.
(d) Is f^{-1} a function?

22 A function is defined by $g(x) = x^2 - 2$.

(a) Draw a mapping diagram to show f and f^{-1}.
(b) Draw the graphs of f and f^{-1}.
(c) Is f^{-1} a function?

23 A relation is given by the vertices of a quadrilateral $A(-2, 1), B(1, 2), C(4, 0), D(1, -1)$. Draw a graph of the inverse relation.

24 A function is given as ordered pairs. Write the domain and range of the inverse of f.

(a) $(-2, 3), (0, 2), (2, 1), (4, 0)$
(b) $(-3, 3), (-2, 1), (0, 0), (2, 1), (3, 3)$
(c) For which of the above is f^{-1} a function?

25 (a) If $f(x) = 2x + 1$, then complete the ordered pairs.
$(-1, ?)$ $(0, ?)$ $(?, 0)$

(b) Write the corresponding ordered pairs of f^{-1} for those in (a). Use these ordered pairs to draw the graph of f^{-1}.

(c) Write the equation of the graph of f^{-1}.

26 Two functions are given by $f(x) = x - 1$, $g(x) = 2x + 1$. Calculate.

(a) $f(1)$ (b) $f^{-1}(1)$ (c) $f[f^{-1}(1)]$
(d) $g(3)$ (e) $g^{-1}(3)$ (f) $g[g^{-1}(3)]$
(g) $f^{-1}[g^{-1}(-2)]$ (h) $g^{-1}[f^{-1}(-2)]$

27 A function, f, is given by the equation $f(x) = 2x - 3$.

(a) Draw the graph of f and its inverse f^{-1}.
(b) Find the defining equation of f^{-1}.

28 A function, h, is given by $y = x^2$.

(a) Draw the graph of h and its inverse h^{-1}.
(b) Find the defining equation of h^{-1}.

29 Find the defining equation of the inverse of each of the following.

(a) $3x - 2y = 6$ (b) $y = x^2 - 6$

30 Prove the following statement. The inverse of a linear function is an inverse function.

f.2 functions and their graphs: techniques for sketching

In studying a specific example of a function, such as the quadratic function, we made some observations that helped us draw its graph. We translated the curve and, using our skills with transformations, obtained the defining equations. From our work with transformations we say that

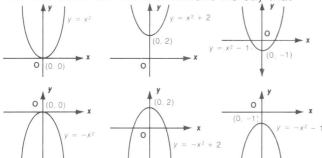

the mapping f given by $f: x \longrightarrow (x - p)^2 + q$ is a translation defined by $(x, y) \longrightarrow (x + p, y + q)$.

The method of mathematics is to interpret the *specific* results for a quadratic function and investigate the *general case* for any function, f, as given in the following exercise.

If a function, f, is defined by $y = f(x)$, we will investigate the effect on the function for each of the following in the exercises.

I $y = f(x) + a$	II $y = f(x + a)$
III $y = af(x)$	IV $y = f(ax)$
V $y = -f(x)$	VI $y = f(-x)$

The following skill is helpful to complete the exercises.

Example
(a) The diagram of the graph of $y = f(x)$ is given. What is the value of $f(2)$, $f(-3)$, $f(0)$?

(b) Draw the graph of $y = f(x + 1)$.

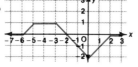

Solution
(a) From the graph, if $x = 2$, then $f(2) = 2$.
      if $x = -3$, then $f(-3) = 1$.
      if $x = 0$, then $f(0) = -2$.

(b) To draw the graph of $y = f(x + 1)$, construct a table of values.

x	−7	−6	−5	−4	−3	−2	−1	0	1	2
x + 1	−6	−5	−4	−3	−2	−1	0	1	2	3
f(x + 1)	0	1	1	1	0	−1	−2	−1	0	0

If $x = -6$, $x + 1 = -5$,
$f(x + 1) = f(-5)$ ← Use the value for the given graph.
$= 1$

Use the above table to draw the graph of $y = f(x + 1)$.

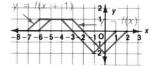

f.2 exercise

sketching $y = f(x) + q$, $y = f(x + p)$

Questions 1 to 5 investigate the relationship between $y = f(x)$ and $y = f(x) + q$, $q \in R$.

1. The graph defined by $y = f(x)$ is shown. On the same set of axes, draw the graphs of
   (a) $y = f(x) + 3$ (b) $y = f(x) + 2$
   (c) $y = f(x) - 1$ (d) $y = f(x) - 2$

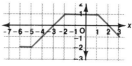

2. Based on your observations in Question 1.
   (a) what is the relationship between the graph of $y = f(x)$ and $y = f(x) + q$, if $q > 0$? If $q < 0$?
   (b) describe the effect on the graph of $y = f(x) + q$ if q increases. Decreases.
   (c) write a mapping to relate the graph of $y = f(x)$ and $y = f(x) + q$.
   (d) why may the mapping in (c) be described as a translation?
   (e) are there any invariant points of the mapping?

3. Use the graph $y = f(x)$ defined by $f(x) = 2x$. Sketch the graphs of $y = f(x) + 3$, $y = f(x) - 1$.

4. Use the graph of $y = f(x)$ defined by $f(x) = \frac{1}{2}x^2$ to sketch the graphs of $y = f(x) + 4$, $y = f(x) - 2$.

5. Use the graph of $y = f(x)$ defined by $f(x) = \sin x$ to sketch the graphs of $y = f(x) + 2$, $y = f(x) - 3$.

Questions 6 to 10 investigate the relationship of $y = f(x)$ and $y = f(x + p)$.

6. The graph defined by $y = f(x)$ is shown. On the same set of axes draw the graphs of

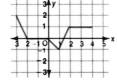

   (a) $y = f(x + 3)$ (b) $y = f(x + 2)$
   (c) $y = f(x - 1)$ (d) $y = f(x - 2)$

7. Based on your observations in Question 6,
   (a) what is the relationship between the graph of $y = f(x)$ and $y = f(x + p)$ if $p > 0$? If $p > 0$?
   (b) describe the effect on the graph of $y = f(x + p)$ if p decreases.
   (c) write a mapping to relate the graph of $y = f(x)$ and $y = f(x + p)$.
   (d) why may the mapping in (c) be described as a translation?
   (e) are there any invariant points of the mapping?

8. Use the graph of $y = f(x)$ defined by $f(x) = -3x$ to sketch the graphs of $y = f(x + 3)$, $y = f(x - 1)$.

9. Use the graph of $y = f(x)$ defined by $f(x) = -\frac{1}{2}x^2$ to sketch the graphs of $y = f(x + 4)$, $y = f(x - 2)$.

10. Use the graph of $y = f(x)$ defined by $f(x) = \sin x$ to sketch the graph of $y = f\left(x + \frac{\pi}{2}\right)$, $y = f\left(x - \frac{\pi}{2}\right)$.

11. Use your results in Questions 1 to 10 to answer the following questions.
    (a) How is the graph of $y = f(x)$ related to the following graphs?
        A $y = f(x) + q$ B $y = f(x + p)$
        C $y = f(x + p) + q$
    (b) Use a mapping to relate the graph of $y = f(x)$ to each of A, B, and C above.
    (c) Use the above results to sketch the graph of
        A $y = x^2 + 4$ B $y = (x - 2)^2$
        C $y = (x - 2)^2 + 4$

12. Draw a sketch of each graph.
    (a) $y = (x - 1)^2 + 3$ (b) $y = (x + 1)^2 - 2$
    (c) $y = \sin\left(x + \frac{\pi}{2}\right) - 3$ (d) $y = \cos\left(x - \frac{\pi}{2}\right) + 2$

sketching $y = af(x)$, $y = f(bx)$

Questions 13 to 16 investigate the relationship between $y = f(x)$ and $y = af(x)$.

13. Use the graph defined by $y = f(x)$ as shown. On the same set of axes, draw the following graphs.

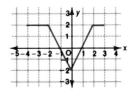

    (a) $y = 2f(x)$ (b) $y = 3f(x)$
    (c) $y = -2f(x)$ (d) $y = -3f(x)$

14. Based on your observations in Question 13,
    (a) what is the relationship between $y = f(x)$ and $y = af(x)$ if $a > 0$? If $a < 0$?
    (b) describe the effect of the graph of $y = af(x)$ if
        A: $a > 0$ and a increases
        B: $a < 0$ and a decreases.
    (c) write a mapping to relate the graph of $y = f(x)$ and $y = af(x)$.
    (d) are there any invariant points of the mapping?

15. Use the graph of $y = f(x)$ defined by $f(x) = x^2$ to sketch the graphs of $y = 2f(x)$, $y = -3f(x)$.

16. Use the graph of $y = f(x)$ defined by $f(x) = \sin x$ to sketch the graph of $y = 2f(x)$, $y = -2f(x)$.

Questions 17 to 20 investigate the relationship between $y = f(x)$ and $y = f(bx)$, $b > 0$.

17. Use the graph defined by $y = f(x)$ as shown. On the same set of axes, draw the graphs of

    (a) $y = f(2x)$ (b) $y = f(3x)$

    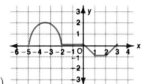

18. Based on your observations in Question 17
    (a) what is the relationship between $y = f(x)$ and $y = f(bx)$ if $b > 0$?
    (b) describe the effect on the graph of $y = f(bx)$ if $b > 0$ and b increases; b decreases.
    (c) write a mapping to relate the graph of $y = f(x)$ and $y = f(bx)$.
    (d) are there any invariant points of the mapping?

19. Use the graph of $y = f(x)$ defined by $f(x) = x^2$ to sketch the graphs of $y = f(2x)$, $y = f\left(\frac{1}{2}x\right)$.

20. Use the graph of $y = f(x)$ defined by $f(x) = \sin x$ to sketch the graph of $y = f(2x)$, $y = f\left(\frac{1}{2}x\right)$.

21. Use your results in Questions 13 to 20 to answer the following questions.
    (a) How is the graph of $y = f(x)$ related to the following graph, $b > 0$, $a \in R$?
    A $y = af(x)$ B $y = f(bx)$ C $y = af(bx)$
    (b) Use a mapping to relate the graph of $y = f(x)$ to each of A, B, and C above.

(c) Use the above results to sketch the graph of each of the following if $f(x) = x^2$.
A $y = 2f(x)$ B $y = f\left(\frac{1}{2}x\right)$ C $y = 2f\left(\frac{1}{2}x\right)$

22. Draw a sketch of each graph.
    (a) $y = 3 \sin (2x)$ (b) $y = 2 \sin \left(\frac{1}{2}x\right)$
    (c) $y = -2 \sin\left(\frac{1}{2}x\right)$ (d) $y = 2 \cos (2x)$
    (e) $y = -3 \cos \left(\frac{1}{2}x\right)$ (f) $y = -2 \cos (2x)$

sketching $y = f(-x)$, $y = -f(x)$

Questions 23 to 27 investigate the relationship of $y = f(x)$ to $y = f(-x)$ and $y = -f(x)$.

23. The graph defined by $y = f(x)$ is shown. On the same set of axes draw the graphs of

    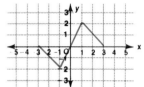

    (a) $y = -f(x)$ (b) $y = f(-x)$

24. Based on your observations in Question 23
    (a) what is the relationship of $y = f(x)$ to $y = -f(x)$ and to $y = f(-x)$?
    (b) write a mapping to relate the graph of $y = f(x)$ and $y = -f(x)$.
    (c) write a mapping to relate the graph of $y = f(x)$ and $y = f(-x)$.
    (d) are there any invariant points of the mapping?

25. Use the graph of $y = f(x)$ defined by $f(x) = x^2$ to sketch the graphs of $y = -f(x)$ and $y = f(-x)$.

26. Use the graph of $y = f(x)$ defined by $f(x) = \sin x$ to sketch the graph of $y = -f(x)$ and $y = f(-x)$.

27. Sketch the graph of each of the following.
    (a) $y = -\cos x$ (b) $y = \sin (-x)$
    (c) $y = \cos (-x)$ (d) $y = -\sin (-x)$

applying skills: sketching graphs

The results of previous investigations may be used to sketch the graph of a curve provided the fundamental curve $y = f(x)$ is known. For example, if $f(x) = x^2$ then we may draw a sketch of $y = -\frac{1}{2}(x - 3)^2 + 4$ using the steps on page 447.

- Sketch the graph of the fundamental or basic curve, $y = x^2$.
- Then apply Steps III or IV where applicable. Sketch the curve given by $y = \frac{1}{2}x^2$.
- Then apply Steps V or VI where applicable. Sketch the curve given by $y = -\frac{1}{2}x^2$.

- Then apply Steps I and II where applicable. Sketch the curve given by $y = -\frac{1}{2}(x - 3)^2 + 4$.

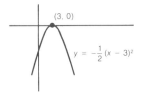 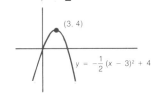

28 Sketch the graphs of each of the following on the same set of axes.
(a) $y = x^2$ (b) $y = 3x^2$ (c) $y = -3x^2$
(d) $y = 3(x - 1)^2$ (e) $y = 3(x + 2)^2$
(f) $y = -3(x - 1)^2$ (g) $y = 3(x + 2)^2 - 4$

29 Sketch the graph of each of the following.
(a) $y = 2(x - 1)^2 + 1$ (b) $y = -3(x + 1)^2 - \frac{1}{2}$
(c) $y = 2 \sin(-x) + 3$ (d) $y = -\frac{1}{2}\cos(-x) - 1$
(e) $y = -3\sin\left(x + \frac{\pi}{2}\right)$ (f) $y = 2\cos\left(x - \frac{\pi}{2}\right)$

f.3 the exponential function

Exponents occur frequently in the study of science, economics, populations, and so on.

Bacteria in a culture grow at a rate which is exponential. In a certain period of time the number of bacteria will double at a constant rate. Radioactive materials lose their radioactivity at an exponential rate.

Earlier we explored the properties of integral exponents.

Vocabulary

2^4 means $\underbrace{2 \times 2 \times 2 \times 2}_{\text{4 factors}}$ power → 2^4 ← base 2, exponent 4

$a^m a^n = a^{m+n}$ $a^m \div a^n = a^{m-n}$ $a^0 = 1$ $a^{-1} = \frac{1}{a}$

An exponential function, f, is a function such that the exponent contains the variable. For example,
$f: x \longrightarrow 2^x$ In general for any $a \in R$ $f: x \longrightarrow a^x$.

To sketch the graph of $f: x \longrightarrow 2^x$ construct a table of values.

x	-3	-2	-1	0	1	2	3
f(x)	1/8	1/4	1/2	1	2	4	8

As x decreases in value, 2^x decreases in but never becomes 0.

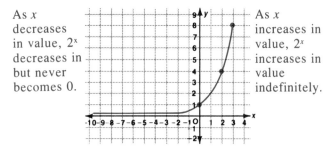

As x increases in value, 2^x increases in value indefinitely.

From the above graph we see that the function is one-to-one.

We may use function notation to write the above exponential function, f, as $f(x) = 2^x$. To calculate $f(2.5)$ we may use the graph. an estimated value
$f(2.5) = 2^{2.5} \doteq 5.5$ ← from the graph

f.3 exercise

To review your skills with the laws of exponents refer to your earlier work on pages 10 to 16.

1. From the graph of $f:x \longrightarrow 2^x$, calculate each of the following.
   (a) $f(0.5)$ (b) $f(-0.5)$ (c) $f(3.5)$
   (d) Check the values of each of the above using a calculator.

2. Calculate the value of each of the following.
   (a) $2^{1.5}$ (b) $2^{-1.5}$ (c) $2^{4.5}$
   (d) Check the values of each of the above using a calculator.

3. (a) Construct a table of integral values for $g:x \longrightarrow 3^x$, $-3 \leq x \leq 3$.
   (b) Draw the graph of $y = 3^x$.
   (c) From the graph calculate $3^{0.5}$, $3^{2.5}$.
   (d) What is the y-intercept for $y = 3^x$?

4. On the same set of axes sketch the graphs of
   (a) $y = 2^x$ (b) $y = 3^x$ (c) $y = 4^x$
   (d) How are the graphs alike? How do they differ?

5. On the same set of axes draw the graphs of
   (a) $y = 2^x$ (b) $y = 2^{-x}$
   (c) How are the graphs alike? How do they differ?

6. On the same set of axes draw the graphs of
   (a) $y = 2x$ (b) $y = \left(\dfrac{1}{2}\right)^x$
   (c) How are the graphs alike? How do they differ?

7. (a) For the exponential graph $y = a^x$, $1 < a$, describe the change in the graph as a increases.
   (b) For the graph $y = a^x$, $0 < a < 1$, describe the change in the graph as a decreases.

8. (a) Draw the graph of $y = f(x)$ given by $f(x) = 2^x$.
   (b) Use the graph in (a) to draw the graph of $y = f(x) + 1$.
   (c) What is the equation of the graph in (b)?

9. Draw a graph of each of the following.
   (a) $y = 2^x + 3$ (b) $y = 2^x - 2$
   (c) $y = 2^{x-1}$ (d) $y = 2^{x+1}$
   (e) $y = 2^{x-1} - 1$ (f) $y = 2^{x+1} + 2$

10. An exponential function is given by $y = a^x$. Use your earlier skills from Section f.2 to describe the graph of each of the following.
    (a) $y = a^x + q$, $q > 0$ (b) $y = a^x + q$, $q < 0$
    (c) $y = a^{x+p}$, $p > 0$ (d) $y = a^{x+p}$, $p < 0$
    (e) $y = a^{x+p} + q$ (f) $y = a^{x-p} - q$

11. For a certain culture in a laboratory, the number of bacteria doubles every 2.5 h.
    (a) If 100 bacteria are isolated, copy and complete the following table.

Time (h)	0	2.5	
Number of bacteria	100		

    (b) Draw a graph of the above data. Join the points with a smooth curve.
    (c) How many bacteria occur after 4 h, 10 h? Explain why the growth is called exponential.

12. A certain population of rabbits doubles every 14.2 d.
    (a) Draw a graph to show the growth of the population.
    (b) If 100 rabbits are counted at the start, how long will it take to have a population of 10 000 rabbits?
    (c) About how many rabbits will there be in 60 d?

f.4 investigating functions

The following investigations explore the nature and properties of various functions.

Investigation A

Refer to the definition of absolute value.
$|x| = x$ if $x > 0$ $|x| = 0$ if $x = 0$ $x \in R$
$|x| = -x$ if $x < 0$

1. (a) Construct a table of values to sketch the graph of the relation $y = |x|$, $x \in R$. Draw the graph.
   (b) Describe the properties of the graph in (a).
   (c) Is the relation $y = |x|$ a function? Why or why not?

2. Use the graph of $y = |x|$. Sketch each of the following graphs using your skills with transformations. (Refer to section f.2, pages 447 to 450.)
   (a) $y = |x| + 2$ (b) $y = |x| - 2$ (c) $y = |x + 2|$
   (d) $y = |x - 2|$ (e) $y = 2|x|$ (f) $y = \frac{1}{2}|x|$
   (g) $y = -|x|$ (h) $y = |-x|$

Investigation B

3. (a) Construct a table of values to sketch the graph of the relation $y = \sqrt{x}$, $x \in R$.
   (b) Describe the properties of the graph in (a).
   (c) Is the relation $y = \sqrt{x}$ a function? Why or why not?

4. Use the graph of $y = \sqrt{x}$. Sketch each of the following graphs using your skills with transformations.
   (a) $y = \sqrt{x} + 2$ (b) $y = \sqrt{x} - 2$
   (c) $y = \sqrt{x + 2}$ (d) $y = \sqrt{x - 2}$
   (e) $y = 2\sqrt{x}$ (f) $y = \frac{1}{2}\sqrt{x}$
   (g) $y = -\sqrt{x}$ (h) $y = \sqrt{-x}$

Investigation C

The following relation given by $y = [x]$ is a step relation called the **greatest integer function** and is defined as: $[x]$ is the greatest integer less than or equal to x. i.e. $[3.2] = 3$, $[3] = 3$, $[0.6] = 0$, $[-3.2] = -4$

5. (a) Construct a table of values to sketch the graph of the relation $y = [x]$, $x \in R$. Draw the graph. Describe the properties of the graph.
   (b) Is $y = [x]$ a function? Why or why not?

6. (a) Construct a definition of the **least integer function** based on your investigation above.
   (b) Draw a graph of the least integer function.

Investigation D

A square is drawn with its centre at the origin. A point P is on the perimeter of the square.

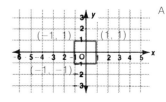

From S, P moves a directed distance, d, to a point P'. The co-ordinates of P' are given by $(c(d), s(d))$.
   If P moves counterclockwise then $d > 0$.
   If P moves clockwise then $d < 0$.
For example,
- P moves to (1, 1). Then the directed distance on the perimeter is $d = 1$. Thus, $c(1) = 1$, $s(1) = 1$.
- If P moves to (1, −1), then the directed distance on the perimeter is $d = -1$. Thus $c(-1) = 1$, $s(-1) = -1$.

As P travels to different points, the values of $c(d)$ and $s(d)$ are defined.

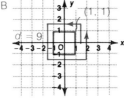 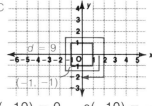

$c(9) = 1$ $s(9) = 1$ $c(-10) = 0$ $s(-10) = -1$

The above definition of c(d) and s(d) results in the two relations.

$$s: d \longrightarrow s(d) \qquad c: d \longrightarrow c(d)$$

7. (a) Construct a table of values given by $(d, s(d))$. Draw the graph.
   (b) Describe the properties of the graph in (a).
   (c) Is the relation given by $(d, s(d))$ a function? Why or why not?

8. (a) Use the steps in Question 7 and construct the graph given by $(d, c(d))$.
   (b) How are the graphs drawn in Questions 7 and 8 similar? How do they differ?

Investigation E

In the previous investigation, we might think of the directed distance, d, as a string winding around a square. In the following investigation we might think of winding string around a unit circle. In this way, we may relate points on the circumference of the circle in a similar way to that used in Investigation D.

Choose the same convention for d.
$d > 0$ if P moves counterclockwise.
$d < 0$ if P moves clockwise.

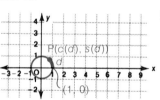

Circumference of the circle is 2π units.

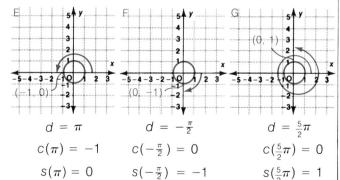

$d = \pi$
$c(\pi) = -1$
$s(\pi) = 0$

$d = -\frac{\pi}{2}$
$c(-\frac{\pi}{2}) = 0$
$s(-\frac{\pi}{2}) = -1$

$d = \frac{5}{2}\pi$
$c(\frac{5}{2}\pi) = 0$
$s(\frac{5}{2}\pi) = 1$

The following relations are used in this investigation.

$$s: d \longrightarrow s(d) \qquad c: d \longrightarrow c(d)$$

9. (a) Construct a table of values given by $(d, s(d))$. Draw the graph.
   (b) Describe the properties of the graph in (a).
   (c) Is the relation $(d, s(d))$ a function? Why or why not?

10. (a) Use the steps in Question 9 and construct the graph given by $(d, c(d))$.
    (b) How are the graphs drawn in Questions 9 and 10 similar? How do they differ?

The above functions, s and c, are often referred to as **circular** or **winding** functions. How are they related to $\sin d$ and $\cos d$?

Investigation F

In section f.3 we drew the graph of an exponential function. What does the graph of its inverse look like?

11. (a) Draw the graph of $y = 2^x$, $-4 \leq x \leq 4$.
    (b) Draw the graph of the inverse of $y = 2^x$. (Reflect the graph of $y = 2^x$ in the line $y = x$.) Is the inverse a function?
    (c) Write the defining equation for the inverse of $y = 2^x$.
    (d) Draw the graph of $y = x^2$ on the same set of axes. Compare the graphs of $y = x^2$ and $y = 2^x$.

The word *logarithm* is used as the name for the inverse of the exponential function.

exponential	logarithm
$y = 2^x$	$x = 2^y$

To write the equation of the logarithm so that y is expressed in terms of x, we write $y = \log_2 x$. Later in mathematics we will learn much of the uses and properties of the logarithmic function given by $y = \log_a x$.

TABLE OF TRIGONOMETRIC FUNCTIONS

ANGLE deg	sin	cos	tan	csc	sec	cot
0	0.0000	1.0000	0.0000	undefined	1.0000	undefined
1	0.0175	0.9998	0.0175	57.299	1.0002	57.2900
2	0.0349	0.9994	0.0349	28.654	1.0006	28.6363
3	0.0523	0.9986	0.0524	19.107	1.0014	19.0811
4	0.0698	0.9976	0.0699	14.336	1.0024	14.3007
5	0.0872	0.9962	0.0875	11.474	1.0038	11.4301
6	0.1045	0.9945	0.1051	9.5668	1.0055	9.5144
7	0.1219	0.9925	0.1228	8.2055	1.0075	8.1443
8	0.1392	0.9903	0.1405	7.1853	1.0098	7.1154
9	0.1564	0.9877	0.1584	6.3925	1.0125	6.3138
10	0.1736	0.9848	0.1763	5.7588	1.0154	5.6713
11	0.1908	0.9816	0.1944	5.2408	1.0187	5.1446
12	0.2079	0.9781	0.2126	4.8097	1.0223	4.7046
13	0.2250	0.9744	0.2309	4.4454	1.0263	4.3315
14	0.2419	0.9703	0.2493	4.1336	1.0306	4.0108
15	0.2588	0.9659	0.2679	3.8637	1.0353	3.7321
16	0.2756	0.9613	0.2867	3.6280	1.0403	3.4874
17	0.2924	0.9563	0.3057	3.4208	1.0457	3.2709
18	0.3090	0.9511	0.3249	3.2361	1.0515	3.0777
19	0.3256	0.9455	0.3443	3.0716	1.0576	2.9042
20	0.3420	0.9397	0.3640	2.9238	1.0642	2.7475
21	0.3584	0.9336	0.3839	2.7904	1.0711	2.6051
22	0.3746	0.9272	0.4040	2.6695	1.0785	2.4751
23	0.3907	0.9205	0.4245	2.5593	1.0864	2.3559
24	0.4067	0.9135	0.4452	2.4586	1.0946	2.2460
25	0.4226	0.9063	0.4663	2.3662	1.1034	2.1445
26	0.4384	0.8988	0.4877	2.2812	1.1126	2.0503
27	0.4540	0.8910	0.5095	2.2027	1.1223	1.9626
28	0.4695	0.8829	0.5317	2.1301	1.1326	1.8807
29	0.4848	0.8746	0.5543	2.0627	1.1434	1.8040
30	0.5000	0.8660	0.5774	2.0000	1.1547	1.7321
31	0.5150	0.8572	0.6009	1.9416	1.1667	1.6643
32	0.5299	0.8480	0.6249	1.8871	1.1792	1.6003
33	0.5446	0.8387	0.6494	1.8361	1.1924	1.5399
34	0.5592	0.8290	0.6745	1.7883	1.2062	1.4826
35	0.5736	0.8192	0.7002	1.7435	1.2208	1.4281
36	0.5878	0.8090	0.7265	1.7013	1.2361	1.3764
37	0.6018	0.7986	0.7536	1.6616	1.2521	1.3270
38	0.6157	0.7880	0.7813	1.6243	1.2690	1.2799
39	0.6293	0.7771	0.8098	1.5890	1.2868	1.2349
40	0.6428	0.7660	0.8391	1.5557	1.3054	1.1918
41	0.6561	0.7547	0.8693	1.5243	1.3250	1.1504
42	0.6691	0.7431	0.9004	1.4945	1.3456	1.1106
43	0.6820	0.7314	0.9325	1.4663	1.3673	1.0724
44	0.6947	0.7193	0.9657	1.4396	1.3902	1.0355
45	0.7071	0.7071	1.0000	1.4142	1.4142	1.0000

ANGLE deg	sin	cos	tan	csc	sec	cot
45	0.7071	0.7071	1.0000	1.4142	1.4142	1.0000
46	0.7193	0.6947	1.0355	1.3901	1.4396	0.9657
47	0.7314	0.6820	1.0724	1.3673	1.4663	0.9325
48	0.7431	0.6691	1.1106	1.3456	1.4945	0.9004
49	0.7547	0.6561	1.1504	1.3250	1.5243	0.8693
50	0.7660	0.6428	1.1918	1.3054	1.5557	0.8391
51	0.7771	0.6293	1.2349	1.2868	1.5890	0.8098
52	0.7880	0.6157	1.2799	1.2690	1.6243	0.7813
53	0.7986	0.6018	1.3270	1.2521	1.6616	0.7536
54	0.8090	0.5878	1.3764	1.2361	1.7013	0.7265
55	0.8192	0.5736	1.4281	1.2208	1.7435	0.7002
56	0.8290	0.5592	1.4826	1.2062	1.7883	0.6745
57	0.8387	0.5446	1.5399	1.1924	1.8361	0.6494
58	0.8480	0.5299	1.6003	1.1792	1.8871	0.6249
59	0.8572	0.5150	1.6643	1.1666	1.9416	0.6009
60	0.8660	0.5000	1.7321	1.1547	2.0000	0.5774
61	0.8746	0.4848	1.8040	1.1434	2.0627	0.5543
62	0.8829	0.4695	1.8807	1.1326	2.1301	0.5317
63	0.8910	0.4540	1.9626	1.1223	2.2027	0.5095
64	0.8988	0.4384	2.0503	1.1126	2.2812	0.4877
65	0.9063	0.4226	2.1445	1.1034	2.3662	0.4663
66	0.9135	0.4067	2.2460	1.0946	2.4586	0.4452
67	0.9205	0.3907	2.3559	1.0864	2.5593	0.4245
68	0.9272	0.3746	2.4751	1.0785	2.6695	0.4040
69	0.9336	0.3584	2.6051	1.0712	2.7904	0.3839
70	0.9397	0.3420	2.7475	1.0642	2.9238	0.3640
71	0.9455	0.3256	2.9042	1.0576	3.0716	0.3443
72	0.9511	0.3090	3.0777	1.0515	3.2361	0.3249
73	0.9563	0.2924	3.2709	1.0457	3.4203	0.3057
74	0.9613	0.2756	3.4874	1.0403	3.6280	0.2867
75	0.9659	0.2588	3.7321	1.0353	3.8637	0.2679
76	0.9703	0.2419	4.0108	1.0306	4.1336	0.2493
77	0.9744	0.2250	4.3315	1.0263	4.4454	0.2309
78	0.9781	0.2079	4.7046	1.0223	4.8097	0.2126
79	0.9816	0.1908	5.1446	1.0187	5.2408	0.1944
80	0.9848	0.1736	5.6713	1.0154	5.7588	0.1763
81	0.9877	0.1564	6.3138	1.0125	6.3925	0.1584
82	0.9903	0.1392	7.1154	1.0098	7.1853	0.1405
83	0.9925	0.1219	8.1443	1.0075	8.2055	0.1228
84	0.9945	0.1045	9.5144	1.0055	9.5668	0.1051
85	0.9962	0.0872	11.4301	1.0038	11.474	0.0875
86	0.9976	0.0698	14.3007	1.0024	14.336	0.0699
87	0.9986	0.0523	19.0811	1.0014	19.107	0.0524
88	0.9994	0.0349	28.6363	1.0006	28.654	0.0349
89	0.9998	0.0175	57.2900	1.0002	57.299	0.0175
90	1.0000	0.0000	undefined	1.0000	undefined	0.0000

answers

Answers for skills reviews, chapter reviews, and chapter tests: See math is/5 teacher's edition

CHAPTER 1

1.2 exercise/page 4

1a) $3, x$ **b)** $-4, x^2$ **c)** $2, y^2$ **d)** $-16, xy$ **e)** $-25, mn$ **f)** $3, m$ **g)** $\frac{1}{3} \cdot mn$ **h)** $25, p^2$ **i)** $\frac{2}{3} \cdot mn$ **2a)** $4m$ **b)** bx **c)** $6y$ **d)** $-6y$ **e)** $12p$ **f)** $-6q$ **g)** $3p$ **h)** $3r$ **3)** Monomial **4a)** $5x + 5y$ **b)** $4x + 4y$ **c)** $8x + 3y$ **d)** $4m + 2p$ **e)** $2p$ **5a)** $2x + 5y$, binomial **b)** q, monomial **c)** $2m - 5n$, binomial **d)** $5u - w$, binomial **e)** $-p - q + 7r$, binomial **f)** $-x$, monomial **g)** $-2x^2$, monomial **6a)** 2 **b)** -2 **c)** -1 **d)** 7 **e)** -3 **f)** 1 **g)** 2 **h)** -1 **7a)** $\{3, 5, 7, 9\}$ **b)** $\{14, 11, 8, 5\}$ **c)** $\{2, -1, 2\}$ **d)** $\{-4, -3, -2\}$ **e)** $\{0, -1, 0\}$ **8a)** $2m^2 - 3m$ **b)** 5 **9a)** 6 **b)** 4 **c)** 9 **d)** 34 **10a)** -19 **b)** -14 **c)** -11 **d)** -8 **11a)** $0, 3, 8, 15$ **b)** $18, 6, 0, 0$, **c)** $5, 1, -1, -1$ **12a)** $7\frac{1}{2}$ **b)** 7 **c)** True **13a)** $3\frac{1}{2}$ **b)** $3\frac{1}{2}$ **c)** False **14a)** T **b)** F **c)** F **d)** T **e)** T **f)** T **g)** F **h)** F.

1.3 exercise/page 7

3) $1, 4, 9, 6, 5, 6, 9, 4, 1, 0, 1, 4, 9, 6, 5, 6, 9, 4, 1, 0$, etc **4)** 55 **5a)** 140, **b)** 385 **6)** Inductive **7)** Deductive **8)** Inductive **9)** Deductive **10)** Inductive **11)** Deductive **12a)** 20 **b)** 132 patterned **13)** Deductive **14)** Inductive **15)** Deductive **16)** $22, 35$, patterned **17a)** Hexagonal numbers **b)** $28, 45$, patterned **18a)** 25 **b)** 625 Patterned **19)** Inductive **20)** Deductive **21a)** None **b)** 6 **c)** 12 **d)** 8 Patterned **22a)** 96 **b)** 48 **c)** 8 **23a)** $6, 4, 10$ **b)** 36 **24)** 105

1.4 exercise/page 11

1a) m^5 **b)** p^7 **c)** p^9 **d)** x^5 **e)** w^8 **f)** 2^4 **g)** 10^5 **h)** -1 **i)** -1 **j)** -32 **k)** x^{10} **l)** y^4 **2a)** x^3 **b)** a^7 **c)** y^4 **d)** 9 **e)** 256 **f)** 10 **g)** x^{4p} **h)** y^x **i)** m^{5x} **j)** 1 **k)** 1 **l)** -1 **m)** $-m^5$ **n)** a^2 **o)** a^8 **3a)** 729 **b)** 64 **c)** a^6 **d)** y^8 **e)** 1 **f)** a^6 **g)** m^2y^2 **h)** m^4y^2 **i)** m^3y^6 **j)** $\frac{x^2}{y^2}$ **k)** $\frac{a^6}{b^3}$ **l)** $\frac{b^2}{a^4}$ **m)** y^4 **n)** x^{10} **o)** m^{10} **p)** y^{12} **4a)** 128 **b)** 81 **c)** $100\,000$ **d)** 2187 **e)** 32 **f)** $\frac{8}{27}$ **g)** y^5 **h)** m^3 **i)** y^6 **j)** m^9 **k)** x^4y^7 **l)** $\frac{x^5}{y^5}$ **m)** p^{15} **n)** a^7 **o)** $\frac{a^3}{8}$ **p)** $8k^3$ **q)** k^8 **r)** m^5n^5 **5a)** $3^2, 1$ **b)** $3^4, 17$ **c)** $2^5, 7$ **d)** Equal **e)** $3^5, 118$ **f)** $3^6, 513$ **6a)** 2^5 **b)** 2^3 **c)** 2^6 **d)** 2^4 **e)** 2^9 **f)** 2^3 **g)** 2^{3m} **h)** 2^{5p} **i)** 2^{11} **j)** 2^{3p+3} **k)** 2^{3m} **l)** 2^m **m)** 2^{4k+2} **n)** 2^{8p-4} **o)** 2^{12p-2} **7a)** -2 **b)** 4 **c)** 4 **d)** 4 **e)** -8 **f)** -8 **8a)** -32 **b)** $-\frac{1}{2}$ **c)** -8 **d)** -8 **e)** -6 **f)** $-\frac{2}{3}$ **9a)** $4\frac{1}{2} \cdot 4\frac{1}{2}$ **d)** $\frac{16}{9}$ **10a)** $1, -1, 1, -1, 1, -1$ **b)** -1 **c)** 1 **11a)** T **b)** T **c)** T **d)** F **e)** F **f)** T **g)** F **h)** T **i)** T **j)** T **k)** F **l)** F **m)** T **n)** F **o)** F **p)** T **12a)** 3^{11} **b)** 5^3 **c)** 2^4 **d)** 10 **e)** 2^{2x+1} **f)** 3^{2y+1} **g)** 2^8 **h)** 3^{2y-1} **13a)** p^{2x+y} **b)** y^{2x-2p} **c)** p^{5x} **d)** p^{x-1} **e)** m^{2x} **f)** a^{2m-n} **g)** y^{n-m} **h)** k^{2m+3n} **i)** p^m **j)** a^{-m} **k)** x^{2a+2b} **l)** y^{a-b-2c} **14a)** -8 **b)** 12 **c)** -128 **d)** $-32\,768$ **e)** $-131\,072$ **f)** -2048 **g)** 1024 **h)** 1 **i)** 256 **15a)** a^5 **b)** m^3 **c)** a^{5k} **d)** x^p **e)** m^{3k+5} **f)** p^{6k} **16a)** 64 **b)** $\frac{1}{4}$ **c)** $-\frac{1}{2}$ **17a)** 1 **b)** $\frac{np}{m^2}$

18a) 72.1 kg **b)** 167.6 cm **19a)** 177 cm **b)** Very accurate **20a)** 143.3 kg **21a)** $4, 2, 1$ **b)** decreases rapidly **22)** 10 **23a)** 0.016 units **b)** 9 **24)** A **25a)** b^{5-p} **b)** P^{2b-3} **c)** $64, -1$, **26a)** 18 m **b)** increases 4 times **27a)** 24 m **b)** 8 m or by a third **28a)** 21.375 m **b)** $16, 625$ m **29a)** No **b)** Yes.

1.5 exercise/page 15

9a) -1 **b)** -1 **c)** -1 **d)** 1 **e)** 1 **f)** -1 **10a)** 10^{10} **b)** -10^2 **c)** -10^5 **d)** 10 **e)** 10^5 **f)** -10^0 **g)** -10^2 **h)** 10^2

1.6 exercise/page 17

1a) a^3b^3 **b)** $6a^4b^5$ **c)** x^4y^2 **d)** $-16x^3y^5$ **2a)** a^2 **b)** $4a^2$ **c)** $4a^3$ **d)** x^2y **e)** $-2xy$ **f)** $-4x^2y$ **3a)** $6x^3y^4$ **b)** -96 **4a)** $-3xy$ **b)** 6 **5a)** $16x^2y^2$ **b)** 576 **6a)** -18 **b)** -36 **c)** 36 **d)** 6 **e)** 12 **f)** 4 **7a)** 36 **b)** -72 **c)** 144 **d)** $\frac{8}{3}$ **e)** $\frac{16}{3}$ **f)** $-\frac{8}{3}$ **8a)** $12x^3$ **b)** $54x^8$ **c)** $2x^3y^3$ **d)** $3x^5y^7$ **9a)** $2x^2$ **b)** $-4x^2y^2$ **10b)** $4m + 3n$ **11b)** $3x$ **12b)** $9a^4b^4$ **13a)** $-12a^3b^2$ **b)** $-15m^5$ **c)** $-12m^3n^3$ **d)** $-18p^4q^4$ **e)** $-m^5n^4$ **f)** $-4p^5q^4$ **14a)** $-6xy$ **b)** $-12x^3$ **c)** $8x^2y^2$ **d)** $-6a^2b^2$ **e)** $6x^3y^3$ **f)** $-4p^4q^3$ **g)** $9x^3y^3$ **h)** $2a^3b^2$ **i)** $3m^4n^2$ **j)** $-3p^4q^2$ **15a)** $36a^2$ **b)** $9a^4$ **c)** $36x^6$ **d)** $9m^2n^2$ **e)** $36a^4b^8$ **f)** $16x^4y^2$ **16a)** $-4x$ **b)** $3mn$ **c)** $-8x^2y$ **d)** $-6pg$ **17a)** -648 **b)** -9 **c)** -27 **d)** -864 **e)** 9 **f)** -18 **18a)** $16m^2$ **b)** $36x^6y^6$ **c)** $16a^4b^2$ **d)** $729a^8b^6$ **19a)** $-3c$ **b)** $5n$ **c)** $-2a$ **d)** $-4m$ **e)** $-6x^2y^2$ **20a)** $a - b + b^2$ **b)** $-2y - 4$ **c)** $-7mn + 2$ **d)** $-3b + 4$ **e)** $-4ab + a^3b^3$ **f)** $-p + p^2 + 3q$ **g)** $-3mn + 2n + 6m$ **h)** $-b + 2a - 3a^2$ **21a)** 3 **b)** 12 **c)** -43 **22a)** $-15a^3b^2$ **b)** $8b$ **c)** $8mn$ **d)** $-2b$ **e)** $-3a^8b$ **f)** $-3x^5y^4$ **g)** $6a - 1$ **h)** $-2a^3b$ **i)** $-\frac{3}{4}a^2b$ **j)** a^4b^3 **k)** $-3x^2 + 6xy + 2y$ **l)** $-4x^2y^2 + xy$ **23a)** $-18a^5$ **b)** 576 **24a)** $2x^2$ **b)** $9m^2n^2$ **25a)** $12b^3$ **b)** $3a^2$ **26)** A **27)** 36 units^2

1.7 exercise/page 20

1a) $7x - 3y$ **b)** $9a - 2b$ **c)** $8x^2 - 2x$ **d)** $5xy - y^2$ **2a)** $3x - 15$ **b)** $-4y + 2$ **c)** $-9a + 15$ **d)** $12 - 4y$ **e)** $-3x + 6y$ **f)** $4x + 10y$ **g)** $6x^2 - 9x + 15$ **h)** $-6a^2 + 15a + 18$ **3a)** $2x^2 - 2x$ **b)** $-3a^2 + 6a$ **c)** $2m^3 - 4m^2$ **d)** $-3y^3 - 12y^2$ **e)** $-2t^3 + 4t^2 + 2t$ **f)** $-3m^2 + 6m^3 + 15m$ **g)** $6a^3b + 12a^2b^2$ **h)** $-3x^3y - 3xy^3$ **4a)** $-a - 4b$ **b)** $-a - 4b$ **5a)** $-2x - 3$ **b)** $y + 10$ **c)** $-3a - 6b$ **d)** $-6x + 5y$ **e)** $-4m - 2n + 9$ **6a)** $-x - 8y$ **b)** $-x - 8y$ **8a)** $-x + 6$ **b)** $5x - 12$ **9b)** $-15x + 12y$ **10a)** $\times$ **b)** $+$ **c)** $\times$ **d)** $\times$ **e)** T **f)** $\times$ **11a)** $2x - y$ **b)** $-x^2 - x + 1$ **c)** $-1 - 4ab + 7a$ **d)** $-2x^2 - 4xy - 3y^2$ **e)** $-4mp + np$ **12a)** $13 + 2x - x^2$ **b)** $-11 + 6x - 4x^2$ **c)** $24 - 5x + 4x^2$ **d)** $-2xy^2 - 4x^2y$ **13a)** $6a - 4c$ **b)** $-x$ **c)** $3a + 3b$ **d)** 0 **e)** $m - 8n$ **14a)** $10x - 2y$ **b)** $16x$ **c)** $12x + 9y$ **15a)** x **b)** $-x^2 + 5x + 28$ **c)** $-a - 5b$ **d)** $x^2 - 8x$ **16a)** $14x + 4$ **b)** $4a - 2b$ **c)** $3x$ **d)** $-8a + 18b - 38$ **e** $-5x + 7y - 12$ **f)** $-3x^2 + 7x + 6$ **g)** $7a^2 + 4a - 2$ **17a)** $-3x^2 + 30x - 15$ **b)** $-x^2 + 13x$

c) $-3x^2 - 5x - 10$ **18a)** $38 + 4y - 5y^2$ **b)** $18 - 21y - y^2$
c) $-3 + 16y - 2y^2$ **19a)** $-10a - 34b$ **b)** $-17y + 16$ **c)** $x^2 + 4x$
20a) $-2a$ **b)** $a + 11b$ **c)** $4a - 9b$ **d)** $a + 11b$; (b) and (d) are equivalent. **21a)** $8x - 25$ **b)** $-15y - 30$ **c)** $15b + 28$ **22)** 90 **23)** 3
24) 217 **25)** A **26)** A **27)** C **28)** $a + b - c$ **29a)** $6 - 4a + 2b$, $3 - 5b$ **b)** $-3a + 3m + 9$, $-4m - 5a - 8$ **30a)** 3877.5 m
b) 3510 m **c)** ground level **31)** 20 s **32a)** 437.5 m **b)** 26th and 27th
33a) 2160 m **b)** 397 m **c)** 421.4 m **34a)** 10.45 m **b)** 24.5 m **c)** No.

1.8 exercise/page 24

1) $14y - 31$ **2)** $6x^2 - 8x + 1$ **3)** $8x + 7$ **4)** $5y + 30$
5) $3xy^2 + 6xy + 3y^2$ **6)** 2 **7)** $9x + 4$ **8)** $9x^2 + xy - 6y$ **9)** $-11x + 19$
10) $11a - 31$ **11)** $-5y^2 + 30y$ **12)** $-x^2 - 5x - 4$ **13)** $2x + 10$
14) $10x - 10$ **15)** $2y - 7$ **16)** $-20a + 12$ **17)** $a + 14$ **18)** $m^2n^2 + 4mn$
19) $16y^2 + y + 26$ **20)** $-x + 3y - 2z$ **21)** $-6x^2 - 10x + 8$
22) $-x^2 - 6$ **23)** $-4y - 23$

1.9 exercise/page 26

1a) $\frac{1}{2}$ **b)** 3 **c)** 1 **d)** 9 **2a)** 3 **b)** -2 **c)** 1 **d)** 2 **3a)** 3 **4b)** -1 **5a)** $-\frac{19}{5}$
b) 1 **c)** -1 **d)** 1 **6a)** 2 **b)** 11 **c)** -2 **d)** 44 **e)** 1 **f)** 6 **g)** 41 **7a)** -2 **b)** 5
c) 0 **d)** $\frac{2}{3}$ **e)** -10 **8)** (c) **9a)** 3 **b)** $\frac{15}{7}$ **c)** 5 **10a)** 6 **b)** -5 **11)** (c)
12) -13 **13)** $-\frac{5}{2}$ **14)** -2 **15)** 2 **16)** 2 **17)** 2 **18)** 17 **19)** $\frac{2}{3}$ **20)** 9
21) -1 **22)** $\frac{5}{2}$ **23)** -1

1.10 exercise/page 29

9a) $s = \frac{t - 3p}{3}$ **b)** $w = \frac{P}{2} - l$
c) $t = \frac{V^2 - u^2}{2v}$ **d)** $a = \frac{2A}{t} - b$ **e)** $u = \frac{2s}{t} - v$ **f)** $v = u + at$
g) $M = -\frac{RU}{Gm}$ **h)** $n = \frac{PV}{RT}$ **i)** $D_o = \frac{fD_i}{D_i - f}$ **j)** $v = \frac{2d - at^2}{2t}$
10a) $y = 1 - 2x$ **b)** $y = \frac{1 + 2x}{3}$ **c)** $x = 3 - 2y$ **d)** $y = \frac{6 + 5y}{2}$
e) $y = 2x - 3$ **f)** $y = \frac{7x + 2}{2}$ **g)** $x = \frac{12 - 3y}{4}$ **h)** $y = \frac{4x + 12}{3}$
11a) 160 **b)** 150 **12a)** $v = \frac{2s}{t} - u$ **b)** 8, 56, 14.2 **13a)** $P = \frac{I}{rt}$ **b)** 500, 516, 420, 350 **14a)** $x = \frac{c - 3b}{3}$ **b)** $x = a(2 + 2b)$ **c)** $x = \frac{b}{a}$
d) $x = \frac{3b + a}{3m}$ **15a)** $x = \frac{4b - 3a}{13}$ **b)** $x = \frac{3p + 6}{13}$ **c)** $x = \frac{3b}{4}$
d) $x = \frac{7b - c}{6c}$ **16)** 13.44% **17)** $2508.96 **18)** 30 **19)** 172 r/min
20) 27.5 m/min **21)** 3.7 cm **22)** $\left(64 - \frac{\pi a^2}{16}\right)$ cm² **23)** $\left(a^2 b - \frac{\pi d^3}{2}\right)$ cm³
24) $\left(s^3 - \frac{\pi s^3}{6}\right)$ cm³ **25)** $(5k - 6\pi d^2)$ cm² **26)** 12 **27)** $\frac{\pi bk}{4s^2}$ blocks
28) $\frac{\pi b^2 h}{6}$ **29)** $\frac{2c^3}{3k^2 b}$ glasses **30)** $\left(\frac{\pi c^2}{4} - t^2\right)$ cm²
31) $\left(\frac{8ab - \pi b^2}{16}\right)$ cm²

1.11 exercise/page 33

7) Nicole 250, Paul 200, Sam 410 **8)** 28.8 m² **9)** 1000 m² **10)** 50, 150, 100 **11)** Bone $240, Knee $480, Foot $280 **12)** 3 h, 5 h, 7 h, 9 h **13)** 6$1, 7$2, 3$5 **14)** 2 h, 3 h, 4 h **15)** 36 compacts, 18 sports, 10 vans **16)** First quake $50 015 000, second quake $49 865 000, third quake $424 120 000 **17)** 420 m² **18)** Italy 10 d, France 5 d, Spain 6 d **19)** $500 **20)** Gas $79, Hydro $52, water $18
21) music 30 min, comedy 60 min, sports 50 min
22) $6000 @ 10%, $2000 @ 12% **23)** 21 **24)** 450 pennies, 18 dimes, 74 nickels **25)** 180 km

1.12 exercise/page 36

1a) $10x + 14$ **b)** $31 - 24m$ **c)** $8 - 12y$ **d)** $31 - 24k$ **e)** $-11 - 4k$
2a) $x - 12$ **b)** $x - 12$ **c)** $x - 12$ **d)** $-x + 3$; (a), (b), (c) are equivalent **3)** $-a - b$ **4)** $a + b$ **5a)** $\frac{a - 3b + 6}{6a}$ **b)** 1
c) $a - b$ **6)** -1 **7)** 2 **15)** 5 **16)** 6 **17a)** 5 **b)** 3 **c)** 11
18a) 14 **b)** 50 **c)** 5 **19a)** $\frac{1}{2}$ **b)** $-\frac{1}{2}$ **c)** $\frac{1}{2}$ **20a)** 0 **b)** 42 **c)** 12 **d)** 30
e) 54 **f)** 30 **21a)** 30 **b)** 225 **c)** 225 **22a)** 5 **b)** 2 **c)** $\frac{8}{3}$
23a) -1 **b)** 3 **c)** 12 **24)** -5 **25)** -1

CHAPTER 2

2.2 exercise/page 44

10a) $6y - 2y^2$
b) $5x^2 + 14x - 96$ **c)** $4x^2 + 5xy - 3y^2$ **d)** $-5a^2 + 4ab - 98b^2$
e) $-2m^2 + 6mn - 19n^2$ **f)** $7c^2 + 12cd + 13d^2$
g) $3x^3 - 8x^2y - 5xy^2 - 8y^3$ **11a)** $-x^2 - 12$ **b)** $-y^2 - 18y + 27$
c) $m^2 - 21m + 36$ **d)** $-x^2 + 6xy - 5y^2$ **e)** $9a^2 - 2ab - b^2$
12) $2x^2 - 24x + 3$ **13)** $6x^2 - 8x + 17$ **14)** $5x^2 - 6xy + 5y^2$
15a) $3x^2 - 18x + 24$ **b)** 24 units² **c)** $x = 1$ would make $x - 1 = 0$
16a) $14x^2 + 12x - 3$ **b)** $12x^2 + 4x - 1$ **17a)** 9 **b)** 43 **c)** 97 **18a)** 14
b) 25 **c)** 34 **d)** 41 **e)** 46, 49, 50 **19a)** -11 **b)** 146 **c)** 151 **d)** -34
(c) is the maximum **21b)** No. **22b)** All values of y **23a)** 4
b) -1 **c)** 0 **d)** $-\frac{1}{2}$ **e)** 3 **f)** 0 **25a)** -3 **b)** 7 **c)** 2 **d)** identity **e)** $\frac{1}{2}$

2.3 exercise/page 47

1a) $2b$ **b)** $2x$ **c)** $2y^2$ **d)** $6ac$ **e)** $2a^2b$ **2a)** a **b)** $8ab$ **c)** y **d)** $2x^3$ **e)** $3a^2$
f) 3 **3a)** $2xy(x - 3y)(x - y)$ **4a)** $2m$ **b)** a **c)** $3x$ **d)** $-2mn$ **g)** $2\pi m$

5a) $-13a$ **b)** $5xy$ **c)** $4m$ **d)** 3 **e)** x **f)** $2xy$ **g)** $2ab^2$ **h)** $3m^2n^2$
6b) $(a - b)(x + 2y)$ **7a)** $(a + b)(2x + y)$ **b)** $(x - y)(3m - k)$
c) $(m - n)(3y - 2)$ **d)** $(y + 3)(2m + n)$ **e)** $(3m - 2n)(3x + 2y)$
f) $(4a - ab)(4m - 3n)$ **8a)** $(x + y)(a + b)$ **b)** $(a + b)(x + y)$
9a) $(x + y)(a + b)$ **b)** $(2x - y)(a - b)$ **c)** $(a + c)(a - b)$
d) $(b - 3a)(3n - 2m)$ **e)** $(ac + b)(ac + d)$ **f)** $(y^3 + 2)(y + 1)$
10a) $4x(a - 2b)$ **b)** $9y^3(2y - 3)$ **c)** $7xy(7 - 2xy)$ **d)** $5ab(5 - 2b)$
e) $y(3y^2 - y + 1)$ **f)** $-3m(2m + m^2 + 3)$ **g)** $2a^3(1 + a + 4a^2)$
h) $a(a^2 - ab + b^2)$ **i)** $2xy^2(xy - 3x + y)$ **j)** $3x^2(3x^2 - 2xy + 4y^2)$
k) $(x - 1)(x - y)$ **l)** $(a - c)(a + b)$ **m)** $3x(m - n)(y - 2)$
11a) $P = 2(l + w)$ **b)** $S = n(n + 1)$ **c)** $S = n(4n + 3)$
d) $E = l(r + R)$ **e)** $S = 180(n - 2)$ **f)** $S = \frac{n}{2}(n + 1)$
g) $A = 2\pi r(h + r)$ **h)** $S_n = \frac{n}{2}(3n - 1)$ **12a)** 18 **b)** $A = \frac{1}{2}h(a + b)$
13a) 420 **b)** $S_n = n(n + 1), 420$ **14a)** 1657.9 cm² **b)** 383.1 cm²
15a) 441 **b)** 441 **16a)** 784 **b)** 1296 **c)** 3025 **17a)** $14\,400$ **b)** 3025
18a) 6669 **b)** $53\,900$ **c)** 3016

2.4 exercise/page 50

7a) $(x + 6)(x + 7)$ **b)** $(3m + 1)(m + 2)$
c) $(x - 2)(x + 1)$ **d)** $(y + 3)(y - 1)$ **e)** $3(m + 2)(m + 1)$
f) $(5m + 1)(m + 2)$ **g)** $(2y - 3)(2y + 1)$ **h)** $2(a + 2)^2$ **9a)** y^2 **b)** $4a^2$
c) $9x^2$ **d)** 1 **e)** $36x$ **f)** $-6\sqrt{10}a$ **10a)** $(m + 3)^2$ **b)** $(a - 4)^2$
c) $(2x + 1)^2$ **d)** $(3y - 2)^2$ **e)** $(5a - 4b)^2$ **f)** $(2 - 5m)^2$ **11a)** 9 **b)** 22
c) 36 **d)** 1 **e)** 12 **f)** 16 **12a)** $(a + 2)(a + 1)$ **b)** $(a - 2)(a + 6)$
c) $(x - 7)(x + 8)$ **d)** $(6x + 7)(x - 1)$ **e)** $2(5 - x)^2$
f) $(4x + 1)(5x - 6)$ **g)** $(4m - n)^2$ **h)** $(x - 12y)^2$ **i)** $x(x - 3)^2$
j) $(3x + 1)(x - 1)$ **k)** $(5 + 4x)(2 + 7x)$ **l)** $3(a + 3)^2$
m) $3(x + 3)(a + 4)$ **n)** $(5x + 1)(x - 6)$ **13a)** $(x + 3)(x + 5)$
b) $(a - 9)(a + 8)$ **c)** $(b + 4)(b + 2)$ **d)** $(2x - 7)(2x - 1)$
e) $(2x - 3)(2x + 1)$ **f)** $(a + 4)^2$ **g)** $(5t - 1)^2$ **h)** $2(x + 5)(x - 1)$
i) $(2x - 3)(3x - 7)$ **j)** $(3x + 2)(x - 3)$ **k)** $(10 + x)(1 + 2x)$
l) $(3x - 5)(x - 2)$ **14a)** 7 **b)** 8 **c)** $2, 14$ **d)** $5, 7$ **e)** $4, 6$ **f)** $7, 12, 15,$
16 **15a)** $x + 10, x \neq -5$ **b)** $a - 5, a \neq -7$ **c)** $y + 4, y \neq 5$
d) $m + 6, m \neq 15$ **e)** $x + 4y, x \neq 8y$ **f)** $t + 3, t \neq -4$
g) $a - 8b, a \neq -3b$ **h)** $x - 4y, x \neq -2y$ **i)** $2x + 1, x \neq 3$
j) $5k - 8, k \neq -6$ **k)** $2m + 5n, m \neq \frac{7}{3}n$ **l)** $2x - 35, x \neq \frac{1}{3}$
m) $x + 4y, x \neq \frac{3}{2}y$ **n)** $6x + 5y, x \neq \frac{1}{3}y$

2.5 exercise/page 53

3b) $(2x + y)^2 - c^2$ **c)** $(x + 3)^2 - 4y^2$ **d)** $4x^2 - (a + b)^2$
e) $9y^2 - (2a - b)^2$ **f)** $16x^2 - (m + 3n)^2$ **g)** $25y^2 - (1 - 3x)^2$
h) $(a - b)^2 - (x + y)^2$ **4b)** $(k - x)(k + x), (a + b - x)(a + b + x)$
5b) $(2k - x)(2k + x), (2a + 2b + x)(2a + 2b - x)$
6b) $(p - 3k)(p + 3k), (p - 3q - 3r)(p + 3q + 3r)$
7a) $(3a - b - x)(3a - b + x)$ **b)** $(1 - y + 2x)(1 - y - 2x)$
c) $(3a + 1 - 3b)(3a + 1 + 3b)$ **d)** $(10x - m - 1)(10x + m + 1)$
e) $(7x + 2m + 1)(7x - 2m - 1)$ **f)** $(5y + 1 - 3x)(5y - 1 + 3x)$
g) $(x + y + a - b)(x + y - a + b)$ **h)** $(7x - y - 1)(-x + y - 1)$
8a) $(m^2 + 1)(m + 1)(m - 1)$ **9a)** $(x - 2)(x + 2)(x - 1)(x + 1)$
11a) $(y - 4)(y + 4)$ **b)** $(m - 10)(m + 10)$ **c)** $(6 - x)(6 + x)$
d) $2(3y - 1)(3y + 1)$ **e)** $(3 - 2k)(3 + 2k)$ **f)** $(6x - y)(6x + y)$
g) $(5m - 4a)(5m + 4a)$ **h)** $(1 - 5y)(1 + 5y)$ **i)** $(ab - 1)(ab + 1)$
j) $m(1 - 4m)(1 + 4m)$ **k)** $(3m - 11)(3m + 11)$ **l)** $2a(7 - 6a)(7 + 6a)$
m) $(2a - 3b^2)(2a + 3b^2)$ **n)** $(4m^2 + 3n)(4m^2 - 3n)$
o) $(xy - 2)(xy + 2)$ **12a)** $(n - 6)(n + 6)$ **b)** $(3h - 7m)(3h + 7m)$
c) $(15m - 1)(15m + 1)$ **d)** $(xyz - 3)(xyz + 3)$ **e)** $(6m - 1)(6m + 1)$
f) $(x + y + 3)(x + y - 3)$ **g)** $(x + y + m)(x + y - m)$
h) $(y^2 + 9)(y + 3)(y - 3)$ **i)** $2(2x - 5y)(2x + 5y)$
j) $(y + a + h)(y - a - h)$ **k)** $(x - 3y + 2a)(x - 3y - 2a)$
l) $(3m + 2x - 2y)(3m - 2x + 2y)$ **m)** $\left(\frac{1}{2}x - \frac{1}{3}y\right)\left(\frac{1}{2}x + \frac{1}{3}y\right)$
n) $(a^2 + b^2)(a + b)(a - b)$ **o)** $(y + 0.4x)(y - 0.4x)$
p) $2\left(x - 2y + \frac{1}{2}m\right)\left(x - 2y - \frac{1}{2}m\right)$
q) $(x - y + 2p - 2q)(x - y - 2p + 2q)$
r) $(x + y - 3z + 2k)(x + y - 3z - 2k)$
s) $(x + y + 3a + 3b - 3c)(x + y - 3a - 3b + 3c)$
13a) $(a - b + m)(a - b - m)$ **b)** $(k + m - n)(k - m + n)$
c) $(a - y + 2x)(a - y - 2x)$ **d)** $(a - b + c^2)(a - b - c^2)$
e) $(x + y + z)(x - y - z)$ **f)** $(m^2 + x - y)(m^2 - x + y)$
g) $(x + y + a + b)(x + y - a - b)$ **h)** $(x + y - a + b)(x + y + a - b)$
i) $(x + w - y - z)(x + w + y + z)$ **14a)** $(4x - y)(4x + y)$
b) $(x - 7y)(x + 7y)$ **c)** $(2x - 1)(2x + 1)$ **d)** $(2x - 5)(2x + 5)$
e) $(x + 8)(x - 2)$ **f)** $(x + y + 3)(x - y + 3)$ **g)** $2(x - 3)(x + 3)$
h) $3(x + 5)(x - 5)$ **i)** $(a + b - 2c)(a - b + 2c)$ **j)** $(5 + x)(9 - x)$
k) $6(x - 3y)(x + 3y)$ **l)** $3(3x - y)(3x + y)$
m) $(3a - 1 + 3m)(3a - 1 - 3m)$ **n)** Impossible
o) $(m - 4)(m + 4)(m + 1)(m - 1)$ **p)** $(x + y - 3)(x - y + 7)$
q) Impossible **r)** $\left(\frac{x}{4} - \frac{y}{6}\right)\left(\frac{x}{4} + \frac{y}{6}\right)$ **s)** $(2a - 1)(2a + 1)(a - 3)(a + 3)$
t) $(3m^2n + y^2)(3m^2 - n - y^2)(9m^4n^2 + y^4)$
u) $(a - b + 3x + 3y)(a - b - 3x - 3y)$
v) $(m - n + p + q)(m - n - p - q)$
15a) 24 **b)** 6 **c)** 24 **d)** 10 **e)** 40 **f)** 36
16a) $(x^2 + 1)^2 - x^2 = (x^2 + x + 1)(x^2 - x + 1)$
b) $(a^2 + 4)^2 - a^2 = (a^2 + a + 4)(a^2 - a + 4)$
c) $(a^2 + 2)^2 - 4a^2 = (a^2 - 2a + 2)(a^2 + 2a + 2)$
d) $(m^2 - n^2)^2 - 9m^2n^2 = (m^2 + 3mn - n^2)(m^2 - 3mn - n^2)$
e) $(2y^2 - 1)^2 - 9y^2 = (2y^2 + 3y - 1)(2y^2 - 3y - 1)$
f) $(a^2 + 3b^2)^2 - 4a^2b^2 = (a^2 + 3b^2 + 2ab)(a^2 + 3b^2 - 2ab)$
17a) $(y^2 + 3y + 5)(y^2 - 3y + 5)$ **b)** $(x^2 + 2x + 3)(x^2 - 2x + 3)$
c) $(a^2 + a + 2)(a^2 - a + 2)$ **d)** $(x^2 + 5xy + y^2)(x^2 - 5xy + y^2)$
e) $(4x^2 + 2x - 1)(4x^2 - 2x - 1)$ **f)** $(3m^2 + 2m + 1)(3m^2 - 2m + 1)$
g) $(3a - 4b)(a + 2b)(3a + 4b)(a - 2b)$
h) $(2m^2 + 2m - 3)(2m^2 - 2m - 3)$

2.6 exercise/page 57

3a) $-3, -5$ **b)** $-10, 3$ **c)** $5, -5$ **d)** $-\frac{2}{5}$ **e)** $7, -3$ **f)** $\frac{1}{4}, -\frac{1}{4}$
g) $-\frac{1}{2}, -3$ **h)** $-\frac{1}{2}, -4$ **i)** $-\frac{1}{3}, -3$ **j)** $-\frac{3}{5}, 2$ **4b)** $3, 4$ **5a)** $8, -6$
b) $\frac{5}{3}, -6$ **c)** $-\frac{1}{3}, \frac{3}{2}$ **d)** $\frac{3}{2}, 2$ **e)** $-\frac{5}{2}, 3$ **f)** $3, -10$ **6b)** $4, -3$
7a) $-2, -4$ **b)** $-8, 9$ **c)** $\frac{3}{2}, \frac{7}{3}$ **d)** $-2, -3$ **e)** $-\frac{5}{2}, 3$ **f)** $\frac{5}{3}, 2$
8b) $-\frac{11}{2}, 3$ **9a)** $-\frac{2}{3}, 1$ **10a)** $-3, 7$ **11a)** $\frac{3}{2}, -5$ **12a)** $3, -2$ **b)** $-3, 5$
c) $-4, -5$ **d)** $6, 7$ **e)** $0, 8$ **f)** $\frac{-5}{2}, -7$ **g)** $4, -4$ **h)** $-2, 7$ **i)** $-\frac{1}{3}, 5$
j) $3, -7$ **k)** $\frac{7}{5}, -1$ **l)** $\frac{5}{4}, -\frac{5}{4}$ **13a)** $7, -3$ **b)** $2, -9$ **c)** $\frac{5}{3}, -3$ **d)** $\frac{7}{2}, 1$
e) $-6, 7$ **f)** $\frac{6}{5}, 4$ **g)** $0, 9$ **h)** $6, -6$ **14a)** $1, 2, 3$ **b)** $0, 1, -3$
c) $0, -2, -3$ **d)** $0, \frac{3}{2}, -5$ **e)** $0, \frac{5}{3}, -3$ **f)** $0, -3, 7$ **15a)** $2, -2, 4$
b) $3, -3, -1$ **c)** $3, -3, 4$ **16)** $1, -1, 3, -3$ **17)** $2, -2, 3, -3$
18a) $-\frac{1}{5}, 4$ **b)** $\frac{5}{3}, -3$ **c)** $4, -4$ **d)** -1 **e)** $-\frac{5}{2}, \frac{3}{4}$ **f)** $\frac{1}{5}, -\frac{1}{5}$ **g)** -5
h) $\frac{1}{6}, 1$ **i)** $\frac{10}{3}, -3$ **j)** -2 **k)** $\frac{3}{2}, -1$ **l)** $0, 4, -4$ **m)** $-\frac{5}{3}, \frac{3}{4}$ **n)** $\frac{1}{3}$
o) $-\frac{5}{2}, \frac{2}{3}$ **p)** $\frac{2}{3}, -\frac{1}{2}$ **q)** $\frac{4}{5}, -\frac{4}{5}$ **r)** $0, -2, 6$ **s)** $0, 3, -3$ **t)** -5
u) $16, 3, -3$ **v)** $7, 2, -2$ **19a)** $(2x-5)(2x+1); \frac{5}{2}, -\frac{1}{2}$
b) $2x^2 + x - 21 = 0; 3, -\frac{7}{2}$ **c)** $3x^2 + 8x - 3 = 0; (3x-1)(x+3) = 0$
d) $(3x-1)(x-3) = 0; \frac{1}{3}, 3$ **20a)** $x^2 - x - 6 = 0$ **b)** $x^2 + 10x + 24 = 0$
c) $2x^2 - 7x + 3 = 0$ **d)** $3x^2 + 7x - 6 = 0$ **e)** $x^2 - 4x = 0$
f) $6x^2 + x - 1 = 0$ **21a)** $x^3 - x = 0$ **b)** $3x^3 - 7x^2 - 6x = 0$
c) $x^3 + x^2 - 9x - 9 = 0$ **22a)** -12 **b)** $\frac{2}{3}$ **23a)** $-2, \frac{1}{3}$
b) $p = -5; -\frac{1}{2}$ **c)** $p = 25; \frac{1}{4}$ **d)** $p = 5; -5$ **24b)** $3, 2$ **25a)** $-2, 4$
b) $\frac{3}{2}, -\frac{5}{2}$ **c)** $1, -4$ **d)** $-5, -8$ **26)** 9 km **27)** $5, 9$
28) 4 cm $\times$ 4 cm, 7 cm $\times$ 7 cm **29)** 5 kg **30)** $12, 13$
31) 2 mm **32)** 13 **33)** 16 cm $\times$ 12 cm

2.7 exercise/page 60

1) (a) (c), (d) **2a)** 1 **b)** -1 **c)** 1 **d)** 1 **e)** 1 **f)** -1 **3a)** $-a-b$ **b)** $2x + y$
c) $(1-x)$ **d)** $\frac{5-a}{5+a}$ **4a)** $\frac{1}{4}$ **b)** $\frac{b}{5x}$ **c)** $\frac{-y}{t}$ **d)** $\frac{45x^3}{128y^2}$ **e)** $\frac{-2m^3}{9}$ **f)** 2
g) $\frac{-x^{11}}{4}$ **h)** $2x^5y^2$ **5a)** -2 **b)** $-\frac{1}{2}$ **c)** $\frac{1}{3(x+1)}$ **d)** $\frac{y}{(y-1)^2}$

6a) $\frac{x+3}{4}, x \ne 5$ **b)** $\frac{y+5}{y+2}, y \ne -2$ **c)** $\frac{-1}{y+1}, y \ne -1$
d) $\frac{y-5}{y+2}, y \ne -2$ **7a)** $\frac{x(x+2)}{(x-1)(x+7)}$ **b)** $\frac{x+3}{x-1}$ **c)** $\frac{(x-1)(x-3)}{2(x+3)}$
d) $\frac{3(a+b)}{2}$ **e)** 1 **f)** $\frac{2x-1}{2(x+4)}$ **8a)** 3 **b)** $\frac{-2}{5}$ **c)** 0 **d)** 1 **9a)** $\frac{(m+4)(m+5)}{2(m-2)}$
b) $\frac{x-2}{(2x+1)}$ **c)** $\frac{a-5}{a-3}$ **d)** -1 **e)** $\frac{1}{(a-1)(a+3)}$ **f)** $x(x-2)(x+1)$
10) Expression decreases from $-\frac{4}{3}$ to $-\frac{20}{3}$
11) Expression decreases from $-\frac{7}{8}$ to -2 **12a)** -3 **b)** 0 **c)** -1
d) $-4\frac{4}{5}$ **e)** -14

2.8 exercise/page 62

1a) $\frac{x+y}{x-y}$ **b)** $\frac{4+b}{4-b}$ **c)** $\frac{a^2}{-3(a+1)}$ **d)** $\frac{k+1}{k-1}$ **2a)** $\frac{-1}{3+x}$
b) $\frac{1}{x-3}$ **c)** $4-y$ **d)** $\frac{-(2a+1)}{a+1}$ **e)** x **f)** $x-y$ **g)** $\frac{-m}{5+m}$
h) $\frac{2a+5}{2a-5}$ **3a)** $x^2 + x$ **b)** $3a + 6$ **c)** $-(2-m)^2$ **d)** $2y^2 - 3y + 1$
e) $2pq + 5q^2$ **f)** $-3m^2 + 13m - 12$ **4a)** $(2x-5)^2$
b) $(y^2-1)(y+2)$ **c)** $(2m-1)(m+3)(3m-4)$ **d)** $a^2 - 25$
e) $6p(p+2)(4p-7)(2p+1)$ **5b)** $\frac{1}{x+1}$ **6)** $\frac{2y}{y-3}$ **7a)** $xy(x-y)$
b) $a^2 - 9$ **c)** $1 - 4y^2$ **d)** $(x+1)(x+2)(x+3)$ **8a)** $\frac{4-y}{20}$ **b)** $\frac{3a}{4}$
c) $\frac{2}{3x}$ **d)** $\frac{2-3y}{2y^2}$ **e)** $\frac{3a-1}{a(a+1)}$ **f)** $\frac{3p+1}{p(p+1)}$ **g)** $\frac{-x+11}{(x+1)(x-2)}$
h) $\frac{x-11}{12}$ **i)** $\frac{-3(a^2+b^2)}{ab}$ **j)** $\frac{3(a^2+b^2)}{(2a-3b)(3a+2b)}$ **k)** 2
9a) $\frac{x^2-4xy+y^2}{x^2-y^2}$ **b)** $\frac{-2(a+8)}{a^2-9}$ **c)** $\frac{-4ab}{a^2-b^2}$ **d)** $\frac{-3}{xy}$
e) $\frac{xy-3y^2}{(x-y)(x-2y)}$ **f)** $\frac{x^2-8x+3}{4(x-4)}$ **10a)** $\frac{4(y+1)}{1-y}$ **b)** 0
11a) $\frac{3x+2}{(x-6)(x+3)(x-2)}$ **b)** 0 **12a)** 1 **b)** $\frac{-4}{3}$ **13a)** $\frac{x(9x^2+x+10)}{15(x^2-1)}$
b) $\frac{9x^2+15x-10}{9x^2-4}$ **c)** $\frac{2a^2-a+12}{(a+1)(a+5)(a-4)}$ **d)** $\frac{6m-19}{(2m-5)^2}$
e) $\frac{3(5y-9)}{4}$ **f)** $\frac{x^2}{x^4-y^4}$ **g)** $\frac{-2(m^2+4m-3)}{(m+1)^2(m-1)^2}$ **h)** $6(x^2-6x+30)$
14) $g(x)$ decreases from $\frac{5}{2}$ to $\frac{5}{3}$ **15)** $f(y)$ decreases from $\frac{5}{2}$ to 2

2.9 exercise/page 66

1) 0 **b)** $x-1$ **2a)** 0 **b)** $x+2$ **c)** $x-3, x-4$ **3a)** 0 **b)** $x-2$
5a) $1, 2, 17, 34$ **c)** $(x-1)(x+2)(x-17)$ **6a)** $(x-1)(x-2)(x-3)$
b) $(y-4)(y^2+3y+3)$ **c)** $(3m-2)(9m^2+9m+5)$ **8a)** $x+3$
b) $(x-a)$ **c)** 0 **d)** x **9a)** -2 **b)** -14 **c)** 0 **10a)** $y+3$ **b)** $2x+1$
c) y^2-2y+1 **d)** $9m^2+9m+5$ **e)** a^2+3a+3 **f)** a^2+3a+3
11a) y^2+4y+3 **b)** $3m^2-2m+9$ **c)** y^2+y-6 **d)** m^2+2m+1

e) $y^2 - 9$ **f)** $2k^2 - 3k + 1$ **g)** $x^2 + 2x + 4$ **h)** $y^2 - 3y + 9$
12) $(x + 2)(x - 1)(x - 17)$ **13)** $(x - 2)^2(x + 1)$ **14a)** $f(-1) \neq 0$
b) $(x - 1), (x - 2), (x - 3)$ **15)** Yes **16)** $x + 3$ **18a)** Yes
b) $x - 1, x + 1, x + 6$ **19)** $x^3 - 4x^2 - 11x + 30$ **20)** $y^3 - 2y^2 - y + 2$
21) $x^3 + x^2 - 9x - 9$ **22)** a^2 **23a)** $x + 3$ **b)** $2x^3 - 9x^2 + 28x + 1$
24a) $(x - 4)(x^2 + 3x + 3)$ **b)** $(m + 1)(2m + 3)(m - 4)$
c) $(x + 1)^2(x + 3)$ **d)** $(x - 3)(3x^2 + 2x + 2)$ **e)** $(2y - 1)(3y^2 - 2y - 8)$
25a) -2 **b)** 1 **c)** 16 **26a)** 7 **b)** 0 **c)** 2 **27a)** 3 **b)** $3x - 5, x + 1$
28a) 3 **b)** $(x - 1)(3x^2 + 6x + 11)$ **29a)** 0 **b)** $(x + y), x^2 - xy + y^2$
30a) $(y + 1)(y^2 - y + 1)$ **b)** $(x + 2)(x^2 - 2x + 4)$
c) $(y - 1)(y^2 + y + 1)$ **d)** $(x - 2)(x^2 + 2x + 4)$
e) $(2x + 1)(4x^2 + 2x + 1)$ **f)** $(2y - 1)(4y^2 + 2y + 1)$
g) $(x + 4)(x^2 - 4x + 16)$ **h)** $(x - a)(x^2 + ax + a^2)$
i) $(x + y)(x^2 - xy + y^2)$ **j)** $(3x + 3y)(9x^2 + 3xy + y^2)$
k) $(1 - 3y)(1 + 3y + 9y^2)$ **l)** $(y + 2)\left(\frac{y^2}{8} - \frac{y}{4} + \frac{1}{2}\right)$
31a) $(a + b)(a^2 - 4b + 16b^2)$ **b)** $2(x - 1)(x^2 + x + 1)$
c) $2(1 - 3x)(1 + 3x + 9x^2)$ **d)** $y(y + 1)(y^2 - y + 1)$
e) $(x - y)(x + y)(x^4 - x^2y^2 + y^4)$ **f)** $(3a - 5)(9a^2 + 15a + 25)$
g) $27(3m - 1)(9m^2 + 9m + 1)$ **h)** $(10m - y)(100m^2 + 10my + y^2)$
32a) 1 **b)** 7 **33)** $-2, \frac{1}{2}$ **34a)** -4 or 3 **b)** -1 or -3 **c)** $-1, 3,$ or -3
d) 2 **35)** 5 **36a)** $(x - 3), (x + 3), (x + 2)$ **b)** $3, -3, -2$ **37b)** $-2, 1$
38) 1 **39a)** $1, 3, -1$ **b)** $-1, -3$ **c)** $4, 3, -3$ **d)** $1, -3, -\frac{1}{2}$
40a) Remove brackets **b)** $1, -\frac{4}{3}, -\frac{1}{2}$ **41a)** $1, -1, 2$ **b)** $-2, -1, 1$
c) $2, 3, -2$ **d)** 1 **e)** $-1, -\frac{1}{2}, 2$

2.10 exercise/page 70

1a) $x^2 - 1$ **b)** $3y(3y + 5)$ **c)** $m^2 - 36$ **d)** $(m - 3)^2$ **e)** $k^2 - k$
f) $(x^2 - 1)(x - 2)$ **g)** $(y - 2)(y^2 - 9)$ **2b)** $\frac{1}{3}$ **3a)** 10 **b)** $\frac{-10}{3}$ **4a)** 0
b) 3 **5a)** $\frac{9}{5}, 4$ **b)** $\frac{9}{5}, 4$ **6a)** $y(y - 3); \frac{3}{2}$ **b)** $5a; \frac{3}{2}$ **c)** $m + 2; 11$
d) $5(s - 1); \frac{28}{3}$ **e)** $(y + 1)(y + 2); \frac{-4}{5}$ **f)** $3(m - 4); \frac{60}{13}$ **7a)** $y \neq 0;$
$y = -1$ **b)** $m \neq 0, 1; m = -\frac{3}{2}$ **c)** $s \neq -\frac{1}{2}; s = \frac{-7}{11}$ **d)** $k \neq 0, -3;$
$k = 3$ **e)** $m \neq \pm 1; m = \frac{2}{3}, -2$ **f)** $y \neq 0, 1; y = 1, -3$ **8a)** $-2, 12$
9) $5, 10$ **10a)** 1 **b)** -2 **c)** $\frac{-4}{5}$ **d)** -1 **e)** 4 **f)** 5 **11a)** $-1, 3$ **b)** $\frac{-3}{2}$
c) $\frac{-21}{2}, 1$ **d)** 11 **e)** -1 **f)** 4 **12a)** $\frac{7}{2}, -2$ **b)** 3 **c)** $1, 4$ **d)** $\frac{-25}{36}$ **e)** $\frac{1}{2}$
f) -3 **g)** $\frac{-21}{2}, 1$ **h)** 12 **i)** $\frac{5}{3}$ **j)** $1, \frac{-21}{2}$ **k)** 6 **l)** $\frac{5}{3}$ **m)** $\frac{-13}{10}$ **13)** $\frac{29}{5}$
14a) $y \neq 5, 0$ **15a)** $x \neq 0, 3; 1, -2$ **b)** $x \neq 2, -2; 0, 1$
c) $x \neq 0, -1; 1$ **d)** $x \neq 2, -2; 0, 1$ **16b)** 6 **17)** $\frac{1}{2}, 3$ **18)** -5

2.11 exercise/page 73

2) 17, 12 or $-17, -12$ **3)** 22, 15 or $-22, -15$
4) $2, -6$ or $\frac{2}{3}, 8\frac{2}{3}$ **5b)** 100 km/h **6b)** 0.5 m/s **7b)** 10 km/h
8) 90 km/h **9a)** 6 km/h **b)** 8 km/h **10)** 5 h **11)** 14 h by bus,
11.2 h by train **12a)** Arnold: 55 km/h; Sheila: 60 km/h
b) Arnold: 6 h; Sheila: 5.5 h **13)** 250 km/h, 300 km/h
14a) 20 m/s, 25 m/s **b)** 26 s **15)** 18 km/h **16)** 3 h

Chapter 3

3.2 exercise/page 81

1a) 3 **b)** 26 **c)** 25 **d)** 36 **e)** 8 **f)** 48 **g)** 16 **h)** 52 **i)** 72 **j)** 30 **k)** 9 **l)** 8
2a) 16.1 **b)** 36.1 **c)** 26.5 **d)** 38.6 **e)** 21.4 **f)** 35.9 **g)** 24.4 **h)** 18.8
3a) 40 **b)** 21 **c)** 60 **d)** 25.5 **e)** 28.6 **4a)** 16.1 cm **b)** 17 cm
5a) 31.3 cm **b)** 28.5 cm **c)** 33.4 cm **d)** 18.2 cm **6)** 11.3 cm
7) 11.2 cm **8)** 14.8 cm **9)** 8.5 cm **10)** 10.4 cm **11)** 2.6 m
12a) 74.2 cm **b)** 19.4 cm **13)** 100 diagonals **14)** 15 m **15)** 130 m
16) Yes **17)** 135.8 km **18)** 4.9 m **19)** 4.8 m **20a)** 13.9 cm **b)** 6.9 cm
21) 8.7 cm **22)** 17.3 cm **23)** 10.8 cm **25a)** 1001 m **b)** 1000 m
c) 44.7 m **26)** 44.7 **27)** 31.6 m **28a)** 19.8 cm × 26.4 cm
b) 25.2 cm × 33.6 cm **c)** 29.4 cm × 39.2 cm
29a) 30.6 cm × 40.8 cm **b)** 1248 cm² **c)** 912 cm² **30)** 213.6 cm²
31) 13.6 cm **32)** 3.3

3.3 exercise/page 86

12a) $\frac{1}{4}$ **b)** $\frac{7}{9}$ **c)** $\frac{-25}{99}$ **d)** $\frac{13}{99}$
e) $2\frac{2}{15}$ **f)** $\frac{13}{999}$ **g)** $-3\frac{19}{30}$ **h)** $2\frac{424}{495}$ **i)** $\frac{1}{2}$ **13a)** $\frac{9}{50}$ **b)** $\frac{2}{11}$ **c)** $\frac{17}{90}$ **d)** 1
e) 0.2 **f)** 0.12 **15a)** $\frac{1}{2}$ **b)** $\frac{1}{2}$ **c)** equal **d)** $3\frac{3}{4}$ **16a)** 0.25, 0.24$\overline{9}$
b) 0.375, 0.374$\overline{9}$ **c)** $-1.125, -1.124\overline{9}$ **d)** 0.15, 0.14$\overline{9}$
e) 0.625, 0.624$\overline{9}$ **f)** $-0.24, -0.23\overline{9}$ **17a)** $\frac{2}{9}$ **b)** $\frac{1}{15}$ **c)** $\frac{4}{9}$ **d)** $\frac{25}{99}$
25a) F **b)** T **c)** F **d)** F **26)** (a), (b), (d), (e), (f) **29)** (a), (c),

3.4 exercise/page 90

1a) $y < 28$ **b)** $m < 1$ **c)** $m \geq 6$ **d)** $k \leq 35$ **e)** $y < 5$ **f)** $m > 4$
g) $p \leq 4$ **h)** $p \geq 3$ **2a)** 4 **b)** 8 **3a)** $x < -4$ **5a)** $\{y \mid y \leq 2\}$
b) $\{y \mid y > -2\}$ **6a)** $\{y \mid y \leq -5\}$ **b)** $\{y \mid y > 4\}$ **7a)** $y > -4, y \leq 2$
b) $\{y \mid -4 < y \leq 2\}$ **c)** $-4\,-3\,-2\,-1\,-0\quad 1\quad 2\quad 3$ **8a)** $\{x \mid x \leq 5\}$
b) $\{y \mid y > 41\}$ **c)** $\{m \mid m > 1\}$ **9a)** $-2\,-1\,\,0\,\,1\,\,2\,\,3\,\,4\,\,5\,\,6$
b) $-7\,-6\,-5\frac{11}{2}\,-5\,-4\,-3\,-2\,-1$ **c)** $-7\,-6\,-5\frac{1}{3}\,-5\,-4\,-3$ **10)** (a)
11a) $\{y \mid y \geq 10\}$ **b)** $\{m \mid m < 78\}$ **12a)** $\{x \mid x \geq -3\}$ **b)** $\{x \mid x \leq 2\}$
c) $-5\,-4\,-3\,-2\,-1\,0\,1\,2\,3$ **13)** $-2\,-1\,0\,1\,2\,3$ **14a)** $\{y \mid y > 0\}$
b) $\{y \mid y \geq -\frac{1}{2}\}$ **c)** $\{y \mid y \leq 8\}$ **15a)** $\{y \mid -\frac{11}{2} \leq y < 9\}$

b)$\{p\mid \frac{1}{2} < p < 3\}$ c)$\{k\mid k < \frac{1}{3}$ or $k > 1\}$ **16a)**$\{y\mid 0 < y \leq 3\}$
b)$\{m\mid -2 \leq m < 5\}$ c)(a)-2 -1 0 1 2 3 4 5
(b)-3 -2 -1 0 1 2 3 4 5 6 **17)**$m < \frac{7}{3}$ **18)**$x \leq -4$
19)$m \geq -2$ **20)**$b < -\frac{8}{7}$ **21)**$x < -\frac{5}{3}$ **22)**$k \geq \frac{25}{6}$ **23)**$y < \frac{1}{3}$

3.5 exercise/page 93

1a)1 **b)**2 **c)**$\frac{1}{2}$ **d)**3 **e)**$\frac{1}{3}$ **f)**1 **g)**-1 **h)**-10 **i)**$-\frac{1}{10}$ **j)**1 **k)**5 **l)**0.2
2a)5 **b)**5 **c)**$\frac{1}{5}$ **d)**16 **e)**$\frac{1}{2}$ **f)**$\frac{1}{16}$ **g)**2 **h)**2 **i)**-2 **j)**3 **k)**$\frac{1}{3}$ **l)**-3
3a)2 **b)**4 **c)**8 **d)**2 **e)**4 **f)**8 **4a)**5 **b)**-2 **c)**10 **d)**2 **e)**-3 **f)**4 **g)**4
h)16 **i)**-8 **j)**-2 **k)**25 **l)**-9 **5a)**$\sqrt[3]{27}$ **b)**$\sqrt[5]{243}$ **c)**$\sqrt[3]{-1}$ **d)**$\sqrt[4]{16}$
e)$\sqrt[5]{-32}$ **f)**$\sqrt[3]{64}$ **g)**$\sqrt[4]{625}$ **h)**$\sqrt[4]{2401}$ **6)**(a), (c), (d) **7a)**m^4 **b)**x^9
c)p^5 **d)**2 **e)**10^2 **f)**1 **g)**3^{-6} **h)**2^{-6} **i)**$-a^{-6}$ **j)**b^2a^4 **k)**x^5y^5 **l)**$64a^{-6}$
8a)2 **b)**-5 **c)**$\frac{1}{2}$ **d)**2 **e)**$\frac{5}{6}$ **f)**$-\frac{3}{5}$ **g)**$\frac{1}{32}$ **h)**$\frac{1}{2}$ **i)**-10 **j)**-1.2 **k)**$\frac{1}{9}$
l)-25 **m)**4 **n)**$\frac{-1}{4}$ **o)**$\frac{4}{5}$ **p)**0.04 **q)**$\frac{1}{0.04}$ **r)**32 **s)**$\frac{1}{2}$ **t)**8 **u)**-3
9a)2^{-1} **b)**-8 **c)**8^{-1} **d)**2^{-1} **e)**125^{-1} **f)**-2 **g)**4^{-1} **h)**0.04 **i)**2
j)-3^{-1} **k)**$6(7)^{-1}$ **l)**144 **m)**$5(7)^{-1}$ **n)**1 **o)**-8 **10a)**$3\frac{1}{2}$ **b)**$3\frac{2}{3}$
c)$\frac{-17}{20}$ **d)**$-7\frac{1}{2}$ **e)**$-4\frac{17}{40}$ **11a)**$\frac{2}{3}$ **b)**$2\frac{2}{3}$ **c)**$\frac{5}{6}$ **d)**1 **e)**5 **f)**4 **g)**$\frac{-9}{4}$
h)$7\frac{1}{2}$ **12a)**$a^{-1}b^9$ **b)**m^{-1} **c)**$(-1)^k x^{-2k} a^{-3k}$ **d)**x^{-p} **e)**$a^4 b^{-4}$ **f)**a^2
g)p^{-7k} **h)**1 **13a)**4 **b)**2 **c)**4 **d)**$\frac{1}{4}$ **e)**4 **f)**$\frac{1}{16}$ **g)**16 **h)**16 **i)**16
14a)$\frac{a^3}{b^2}$ **b)**$\frac{3bc}{a}$ **c)**$\frac{3am^2}{4b^2}$ **d)**$\frac{a^2b^2c^2}{3}$ **e)**$\frac{bc^2}{5a}$ **f)**$\frac{m}{a^3b^2}$ **15a)**$x \geq -\frac{1}{3}$
b)$y \geq \frac{3}{2}$ **c)**$|m| \geq 5$ **d)**$k \geq -1$ **e)**No restriction **f)**$y \geq -1$ **16a)**0
b)$\frac{1}{3}$ **c)**0 **d)**3 **e)**5 **f)**1 **17)**A

3.6 exercise/page 95

1a)$4\sqrt{2}$ **b)**$4\sqrt{3}$ **c)**$-3\sqrt{3}$ **d)**$-12\sqrt{2}$ **e)**$10\sqrt{2}$ **f)**$2\sqrt{2}$ **g)**$-3\sqrt{6}$
h)$-30\sqrt{6}$ **i)**$-21\sqrt{2}$ **j)**$-\sqrt{5}$ **k)**$-12\sqrt{3}$ **l)**$10\sqrt{3}$ **m)**$2\sqrt[3]{2}$
n)$-4\sqrt[3]{3}$ **o)**$2\sqrt[3]{3}$ **p)**$-3\sqrt{5}$ **q)**$2\sqrt[3]{3}$ **r)**$-6\sqrt{2}$ **2a)**$\sqrt{18}$ **b)**$-\sqrt{48}$
c)$\sqrt{20}$ **d)**$\sqrt{675}$ **e)**$\sqrt{288}$ **f)**$-\sqrt{108}$ **g)**$\sqrt[3]{24}$ **h)**$-\sqrt[3]{54}$ **i)**$\sqrt[3]{432}$
3)A,B,D **4)**B,C **5a)**$\sqrt{50}$ **b)**$\sqrt{48}$ **6a)**$A\sqrt{72}$, $B\sqrt{75}$ **b)**$B > A$
7a)$12\sqrt{2} + 6\sqrt{2} - 6\sqrt{2}$ **b)**$12\sqrt{2}$
8a)$24\sqrt{3} + 5\sqrt{3} - 4\sqrt{7} + 9\sqrt{7}$ **b)**$29\sqrt{3} + 5\sqrt{7}$ **9a)**$\sqrt{2}$
b)$2\sqrt{3} - \sqrt{2}$ **c)**$2\sqrt{3} - \sqrt{7} - \sqrt{3}$
10)A: $3\sqrt{2} - 5\sqrt{3}$, B: $3\sqrt{2} - \sqrt{3}$, C: $-5\sqrt{3} + 3\sqrt{2}$, A,C
equivalent **11a)**$6 - 6\sqrt{3}$ **b)**$9\sqrt{2} - 15$ **c)**$-4\sqrt{3} + 8\sqrt{2}$
d)$15\sqrt{5} + 25\sqrt{6}$ **e)**$-\sqrt{3} - 2\sqrt[3]{2}$ **f)**$6\sqrt[3]{2}$ **12a)**$\sqrt{3}$
b)$18\sqrt{2}$ **c)**$-23\sqrt{5}$ **d)**$5\sqrt[3]{2}$ **e)**$-37\sqrt{2}$ **f)**$31\sqrt{2}$ **13a)**$3\sqrt{2} - 4\sqrt{3}$
b)$-7\sqrt{2} - 2\sqrt{5}$ **c)**$6\sqrt{7} - 54\sqrt{2} - 2\sqrt{102}$ **d)**$-5\sqrt{3} - 18\sqrt{5}$

e)$3\sqrt{3} - 3\sqrt{5}$ **14a)**$-5\sqrt{3}$ **b)**$18\sqrt{2}$ **c)**$-16\sqrt{5}$
d)$-4\sqrt{5} - 48\sqrt{7}$ **e)**$16\sqrt{7}$ **f)**$15\sqrt[3]{5}$ **g)**$-22\sqrt{10}$ **h)**$4\sqrt[3]{2} - 12\sqrt[3]{3}$
15)A **16)**B **17)**A,C **18a)**$12\sqrt{2} - 45\sqrt{5}$ **b)**$12\sqrt{2} - 45\sqrt{5}$
19a)$20 - 25\sqrt{2}$ **b)**$\sqrt{2} - 5\sqrt{3}$ **c)**$7\sqrt{3} - 54\sqrt{2}$ **d)**$17\sqrt[3]{5}$
20a)$\sqrt{2} - 2\sqrt{3}$ **b)**$4\sqrt{2} - 6\sqrt{3}$ **c)**$14\sqrt{3} - 12\sqrt{2}$ **d)**$5\sqrt{2} - 8\sqrt{3}$
e)$10\sqrt{2} - 16\sqrt{3}$ **f)**$6\sqrt{2} - 7\sqrt{3}$ **21a)**$-4\sqrt{2} + 4\sqrt{3}$
b)$-2\sqrt{2} - 8\sqrt{3}$ **c)**$12\sqrt{3} - 2\sqrt{2}$ **d)**$-2\sqrt{2} + 12\sqrt{3}$
e)$12\sqrt{3} - 4\sqrt{2}$ **f)**$-6\sqrt{2} - 12\sqrt{3}$ **22a)**$3\sqrt[3]{2} + 5\sqrt[3]{3}$
b)$10\sqrt[3]{3} - 3\sqrt[3]{2}$ **c)**$3\sqrt[3]{2} + 30\sqrt[3]{3}$ **d)**$4\sqrt[3]{2} - 10\sqrt[3]{3}$
e)$3\sqrt[3]{2} + 10\sqrt[3]{3}$ **f)**$\frac{4}{3}\sqrt[3]{2} - 10\sqrt[3]{3}$ **23a)**$33\sqrt{2} - 4\sqrt{3}$ **b)**$-10\sqrt{3}$
c)$-\sqrt{5} - 18$ **d)**$46 - 2\sqrt{11} - 20\sqrt{5}$ **e)**$\sqrt[3]{3} - \sqrt{5}$
24a)$\sqrt{2}(7b - 6a)$ **b)**$\sqrt{3}(-13m - 6n)$ **c)**$\sqrt{5}(8y - 3x)$
d)$m(-29\sqrt{3} + 14\sqrt{2})$ **25)**17m **26)**$26x - 36y$ **27)**$-9a - 24b$
28)$9x$ **29a)**$\sqrt{2}(33a - 194b)$ **b)**$-487\sqrt{2}$ **30a)**$36, 36, -36$
31a)$2, 2, 2$ **32a)**$2\sqrt{2}\,|x|$ **b)**$7\sqrt{2x}\,|x|$ **c)**$3\sqrt{2}x^2$ **d)**$6\sqrt{b}\,|ab|$
e)$3a^2\,|b|\sqrt{3b}$ **f)**$6\sqrt{2x}\,x^2\,|y|$ **g)**$4m^2\,|n|\sqrt{5n}$ **h)**$8\,|x|z^2\sqrt{y}$
i)$4\,|a|b^2\sqrt{7a}$ **j)**$2\sqrt[3]{2}\,ab$ **k)**$-3xb\sqrt[3]{2x}$ **l)**$4ay\sqrt[3]{2a^2}$ **m)**$3\,|x|\sqrt{y}$
n)$2pq\sqrt[3]{3q^2}$ **o)**$4\,|st|\sqrt[4]{t}$ **33a)**$3\,|x| - 2\,|x|\sqrt{x} - 5x^2$
b)$2\,|b|\sqrt{b} - 3b^2\sqrt{b}$ **c)**$3\,|x|\sqrt{y} - 2x^2\,|y|\sqrt{y}$
d)$-3\,|a|b^2\sqrt{ab} - 2a^2\,|b|\sqrt{ab} + 5\,|a^3b|\sqrt{ab}$
e)$3xy^2\sqrt[3]{xy} - 6x^2y\sqrt[3]{xy} + 2x^2y^3\sqrt[3]{xy}$
34)$9\sqrt{2} + 10\sqrt{3}$ **35)**$9 - 6\sqrt{3}$
36a)$(4 + 3\,|b|)\sqrt{b} - (9 + 2\,|a|)\sqrt{a}$ **b)**$(4 + 3b)\sqrt{b} - (9 + 2a)\sqrt{a}$
37a)$(a - 3)\sqrt{b} - 2b\sqrt{a}$ **b)**$(1 - b - 3ab - a)\sqrt{ab}$
c)$(9 - 10a + 3ab)\sqrt{b}$ **d)**$(6a - 3)\sqrt{b} - (2b - 8)\sqrt{a}$
38a)$(7\,|b| + 2\,|a| - 2)\sqrt{b}$ **b)**$(56 - 21\,|b|)\sqrt{b} + (2 + |a|)\sqrt{a}$
c)$(7\,|a| + 26\,|ab| + 4)\sqrt{ab}$

3.7 exercise/page 101

1a)A: $-6\sqrt{6}$, B: $6\sqrt{18}$ **b)**B: $54\sqrt{2}$ **2a)**$-72\sqrt{6}$ **b)**$-72\sqrt{6}$ **3b)**360
5a)$6\sqrt{6}$ **b)**$-8\sqrt{30}$ **c)**$15\sqrt{15}$ **d)**$-\sqrt{15}$ **e)**$15\sqrt{15}$ **f)**$4\sqrt{5}$
g)$15\sqrt{30}$ **h)**-6 **i)**-72 **j)**$-6\sqrt{6}$ **6a)**$-108\sqrt{14}$ **b)**$-96\sqrt{10}$
c)$-96\sqrt{3}$ **d)**$-180\sqrt{3}$ **e)**$-240\sqrt{2}$ **f)**$-240\sqrt{30}$
7a)$6\sqrt{10} - 9\sqrt{15}$ **b)**$-27\sqrt{2} + 9\sqrt{6}$ **c)**$-24\sqrt{2} + 8\sqrt{15}$
d)$6\sqrt{6} - 12\sqrt{3}$ **e)**$-9\sqrt{10} + 12\sqrt{5}$ **f)**$10\sqrt{3} - 18$ **g)**$-36 + 18\sqrt{6}$
h)$6\sqrt{15} - 10\sqrt{10}$ **8b)**$48 - 36\sqrt{6}$ **9a)**0 **b)**$18\sqrt{6}$ **c)**-30 **d)**-252
e)$12 - 6\sqrt{6}$ **f)**$18 - 18\sqrt{6}$ **g)**$50 - 12\sqrt{15}$ **h)**$80 - 42\sqrt{10}$
10a)$3a^2 + 5ab - 2b^2$ **b)**$5\sqrt{6}$ **11a)**$29 + 6\sqrt{6}$ **b)**$66 - 36\sqrt{2}$
12a)$-1 - \sqrt{6}$ **b)**$21 - 9\sqrt{6}$ **c)**$42 - 12\sqrt{10}$ **d)**$-77 + 21\sqrt{15}$
e)$-57 + 15\sqrt{15}$ **13a)**$100 - 40\sqrt{6}$ **b)**$160 - 40\sqrt{15}$
c)$180 - 72\sqrt{6}$ **d)**$-12 + 12\sqrt{15}$ **14a)**$6\sqrt{10} - 40\sqrt{2}$
b)$10 + 3\sqrt{6}$ **c)**$12\sqrt{6} - 18$ **d)**$12\sqrt{6} - 30$ **15a)**$8 + 2\sqrt{15}$
b)$20 + 8\sqrt{6}$ **c)**$38 - 12\sqrt{10}$ **d)**8 **e)**$29 - 6\sqrt{6}$ **f)**$265 - 30\sqrt{70}$
g)$18 - 12\sqrt{2}$ **h)**$100 + 50\sqrt{3}$ **i)**$109 - 48\sqrt{5}$ **j)**$36 - 24\sqrt{2}$
k)$215 - 150\sqrt{2}$ **l)**$42 - 24\sqrt{3}$ **16a)**$8 + 2\sqrt{15}$ **b)**$8 - 2\sqrt{15}$
c)$67 - 16\sqrt{3}$ **d)**$67 + 16\sqrt{3}$ **e)**$14 - 4\sqrt{6}$ **f)**$80 + 10\sqrt{15}$
g)$18 - 12\sqrt{2}$ **h)**$28 - 12\sqrt{5}$ **17a)**$14 - 4\sqrt{6}$ **b)**2 **c)**$248 - 80\sqrt{6}$
d)$77 - 24\sqrt{10}$ **18a)**$-53 + 15\sqrt{15}$ **b)**$42 - 81\sqrt{2}$ **c)**$93 - 37\sqrt{6}$

d) $10 - 22\sqrt{15}$ e) $-1016 + 224\sqrt{3}$ 19b) -66 20a) -8 b) 75
c) $-45 + 18\sqrt{6}$ d) $375 - 96\sqrt{15}$ 21a) -2 b) 10 c) 39 d) -20
22a) 3 b) 37 c) -2 d) 135 e) -52 f) 61 g) 5 h) 46 i) -9 j) -270
k) -20 l) 1163 23a) $3\sqrt{3} - \sqrt{2}$ b) $6\sqrt{5} - 2\sqrt{2}$ c) $46 + 16\sqrt{3}$
d) $403 - 36\sqrt{3}$ e) $75\sqrt{6} - 30\sqrt{3}$ f) $-237 + 72\sqrt{6}$
24a) $594\sqrt{6} - 1768$ b) $438\sqrt{15} - 2524$ c) $-68\sqrt{6}$ 25a) $-6\sqrt{6}$
b) $18\sqrt{6}$ c) $-3\sqrt{6}$ d) 30 e) 6 f) -6 g) $18 - 12\sqrt{6}$ h) $12 + 12\sqrt{6}$
26a) $40 + 22\sqrt{5}$ b) $-60 - 33\sqrt{5}$ c) $4 - 32\sqrt{5}$ 27) $36 + 16\sqrt{6}$
34b) Yes 35b) Yes

3.8 exercise/page 105
1) (b), (c), (e),(f), (h), (i), (j), 2a) A: $\sqrt{30}$, B: $2\sqrt{15}$ 3) (c), (g)
4a) $6\sqrt{2}$ b) $\frac{2\sqrt{15}}{5}$ c) $\frac{\sqrt{6}}{3}$ d) $4\sqrt{3}$ e) $-3\sqrt{2}$ f) $\frac{\sqrt{3}}{3}$ g) $-\frac{\sqrt{10}}{3}$
h) $-\frac{\sqrt{2}}{2}$ i) $6\sqrt{5}$ j) $-2\sqrt{3}$ k) $\frac{-2\sqrt{15}}{15}$ l) $\frac{-\sqrt{5}}{10}$ 5a) $4\sqrt[3]{4}$ b) $12\sqrt[3]{3}$
c) $\frac{\sqrt[3]{28}}{2}$ d) $20\sqrt{20}$ e) $\sqrt[3]{8}$ f) $\frac{1}{2}$ 6a) $\frac{\sqrt{6}}{12}$ b) $16\sqrt{2}$ c) $-2\sqrt{7}$ d) $2\sqrt{5}$
e) 27 f) $\frac{3\sqrt{2}}{2}$ g) $\frac{\sqrt{15}}{3}$ h) $4\sqrt{3}$ i) $\frac{-1}{6}$ j) $\frac{-4\sqrt{5}}{3}$ k) 10 l) $\frac{\sqrt[3]{5}}{2}$ 7a) 1
b) 6 c) 13 d) 13 e) 34 f) 0 g) -43 h) -27 8a) $-3\sqrt{2}$ b) $\frac{-3}{2}$
c) $\frac{-5\sqrt{2}}{2}$ d) $-\frac{\sqrt{2}}{2}$ e) $\frac{-2\sqrt{30}}{9}$ f) $\frac{1}{4}$ g) $\frac{\sqrt{6}}{6}$ h) $-10\sqrt{2}$ i) $\frac{\sqrt[3]{20}}{5}$
j) -30 9a) $\frac{\sqrt{6} + \sqrt{10}}{2}$ b) $\sqrt{6} - 3$ c) $\frac{4 + \sqrt{6}}{2}$ d) $\frac{3\sqrt{10} - 2}{4}$
10a) $-4\sqrt{2} - 2\sqrt{6}$ b) $\frac{2 - 3\sqrt{2}}{4}$ c) $6\sqrt{6} - 1$ d) $\frac{6 - 3\sqrt{6}}{4}$
e) $\frac{-\sqrt{10} + 3}{3}$ f) $\frac{5\sqrt{10} - 4}{16}$ 11a) $\sqrt{5} + \sqrt{2}$ b) $10 - 3\sqrt{10}$
c) $5 - 2\sqrt{6}$ d) $4 - 2\sqrt{5}$ e) $\frac{11\sqrt{6} - 6}{47}$ f) $\frac{35 - 12\sqrt{6}}{19}$
12a) $8\sqrt{10} + 24$ b) $8\sqrt{10} + 24$ 13a) $\frac{18\sqrt{3} + 3\sqrt{2}}{106}$ b) $\sqrt{6} + 2$
c) $\frac{-8\sqrt{3} - 12\sqrt{2}}{15}$ d) $\frac{9\sqrt{2} + 2\sqrt{3}}{25}$ e) $2\sqrt{2} + \sqrt{6}$ f) $\frac{3\sqrt{5}}{25}$
g) $\frac{5\sqrt{6} + 12}{2}$ h) $\frac{-12\sqrt{15} - 15\sqrt{10}}{3}$ i) $\frac{8 - 3\sqrt{6}}{2}$ j) $\frac{-9 - 4\sqrt{6}}{15}$
14a) 2.12 b) -1.16 c) 2.45 d) -1.09 e) -2.50 f) 2.45 g) 0.78
h) 3.37 15a) 2.5 b) -13.3 c) 1.1 d) 8.2 e) 3.7 f) 0.5 16a) $4\sqrt{3}$

3.9 exercise/page 110
1a) $2x - 1$ b) $25 + 10\sqrt{x} + x$ c) $9x - 6\sqrt{x} + 1$
d) $64 - 32\sqrt{m} + 4m$ e) $-1 + 4\sqrt{x - 5} + x$ f) $y + 10\sqrt{y - 3} + 22$
g) $9m - 14 - 12\sqrt{m - 2}$ h) $55 - 48\sqrt{y - 1} + 9y$ 2a) 6 b) 2
c) -2 d) 2 e) 3 f) 7 g) 4 h) 3 3a) 17 b) 3 c) 3 d) 5 e) 5 f) 3 4a) N
b) Y c) Y d) N e) Y f) Y 5a) 16 b) None c) $2, 3$ d) $2, 42$ e) None
6a) 36, b) 9 c) ϕ d) 63 e) 10 f) 8 7a) 25 b) 17 c) 5 d) 8 e) 32
f) $\frac{49}{9}$ g) 36 h) 49 8) 6 9a) 2 b) 4 c) 1 d) 4 e) $\frac{-1}{2}, \frac{-3}{2}$ 10a) $0, 3$

b) 4 c) No solution d) 4 e) $2, 38$ 11a) $-1, 3$ b) 144 c) ϕ 12a) $0, 8$
b) 12 c) $\frac{9}{11}$ 13a) 13 b) -21 c) $\pm 2\sqrt{17}$ d) ± 4

3.10 exercise/page 112
1a) 8 units b) 6 units 2a) 162 units² b) 65.4 units² 3) AB: 4 units,
DF: 1 unit or AB: 8 units, DF: 5 units. 4) ST: 5 units, WY: 3 units
5) △MRN: $8\sqrt{3}$ units², △STU: $2\sqrt{205}$ units² 6) 4 7) 12 8) 6 9) 5

3.11 exercise/page 113
1a) -7 b) -7 2a) 1 b)   3a) $\frac{225}{16}$ b) $\frac{1}{2}$ 4a) 1 b) 3 c) -1
5a) -1 b) 5 6a) 1 b) -1 7a) 9 b) 2 c) 3 d) $\frac{11}{7}$ 9) (b) 10) (a) 11) (a)

Chapter 4

4.2 exercise/page 123
1) (a), (d) satisfy; (b), (c), (e), (f) do not satisfy 2) (a), (c), (d), (f)
3a) A,B b) D,F c) H,I 4a) -3 b) -6 5a) 3 b) 7 c) 4 d) 10 e) -11
f) 5 6b) A,C 7d) Same y-intercept 3 8d) equal slopes 9) Arnold,
Hazel 10) Cherokee, Starcraft 11) Golden, Banff, Calgary
12) Calgary, Red Deer 13a) $(3,2)$ b) $(-3,4)$ c) $(4,-3)$ d) $(-6,-2)$
14a) $(-1,1)$ b) $(-1,4)$ 15a) $(3,2)$ b) Tug c) Cruiser 16) $(-1,6)$
17) I: (b), (e); II: (a), (f); III: (d)

4.3 exercise/page 126
1b) $(3,-2)$. Point satisfies both equations. 3a) P,S b) P,S
c) $0(-2 -3)$ R$(3,2)$ 6c) They co-incide. d) B = 2xA 7a) B = 2A
b) B = $-$A c) B = 3A d) B = $\frac{1}{2}$A 8d) They all intersect at $(2,1)$
10a) $2x + y = -5$ or $-3y = 9$ b) $3x - y = -5$ or $-x - 3y = 5$
c) $3x - y = 16$ or $x + 3y = 18$ d) $5x + y = -4$ or $x + 3y = -12$
11a) Add b) Subtract c) Add

4.4 exercise/page 129
4a) $(1,1)$ 5a) $(3,-1)$ b) $(-2,5)$ c) $(-2,0)$ d) $(-4,-3)$ e) $(1,3)$
f) $(-1,-2)$ 6a) $\left(\frac{1}{2}, \frac{3}{2}\right)$ 7a) $(1,0)$ b) $(-1, -1)$ c) $(3,-2)$ d) $(-1,2)$
e) $(3,0)$ f) $(1,0)$ g) $(1,-1)$ h) $(-1,1)$ 8a) $(1,-3)$ 9a) $(5,-11)$
10a) $(3,-1)$ b) $(-1,3)$ c) $\left(\frac{2}{3}, 7\right)$ d) $\left(\frac{-4}{3}, \frac{1}{3}\right)$ e) $(-1,-3)$ f) $(3,1)$
g) $(-2,3)$ h) $(1,-1)$ 11a) $(0,1)$ 12a) $(-3,1)$ b) $(2,1)$ c) $(1,0)$ d) $(4,0)$
13a) $(1,0)$ b) $(-1,1)$ c) $(0,1)$ d) $(1,-1)$ e) $(1,4)$ f) $(-3,2)$ 14b) $(o,-2)$
15a) $(-1,3)$ b) $(6,2)$ c) $(0,-1)$ d) $(-3,0)$ e) $(1,2)$ 16b) $(4,-4)$
17a) $(6,1)$ b) $(3,4)$ c) $(3,6)$ d) $(3,-2)$ 18a) $x = -9, y = -10$ b) $(6,5)$
19) $(2,3)$ 20) $(3,4)$ 21) $(-1,1)$ 22a) $(2,3)$ b) $(-4,-3)$

4.5 exercise/page 133

5a) $(-1, 1)$ **b)** $(2, 0)$ **c)** $(-3, -2)$ **d)** $(4, -2)$ **e)** $(-3, -3)$ **f)** $(5, -1)$
6a) $(-1, 3)$ **b)** $(4, 2)$ **c)** $(-3, 0)$ **d)** $(4, 1)$ **7b)** $(-2, 0)$ **8a)** $(2, 1)$
b) $(3, 0)$ **c)** $(1, 1)$ **d)** $(1, 3)$ **e)** $\left(\frac{-3}{2}, 2\right)$ **f)** $(0, 0)$ **g)** $(1, 2)$ **h)** $(9, 1)$
i) $\left(2, \frac{1}{2}\right)$ **j)** $(2, 6)$ **9b)** $(4, -3)$ **10b)** $(2, -3)$ **11a)** $(4, -6)$ **b)** $(1, 8)$
c) $(3, 1)$ **d)** $(2, -1)$ **e)** $(4, -6)$ **f)** $\left(1, \frac{1}{3}\right)$ **g)** $(8, 1)$ **h)** $(2, -1)$
12) $(-8, 3)$ **14)** $P(-5, 3)$ $Q(3, 6), R(5, -2)$

4.6 exercise/page 136

1a) $B = 2A$ **b)** $B = 3A$ **c)** Removing brackets and collecting terms
d) Removing brackets **e)** $B = 2A$ **2b)** $(5, 3)$ **3b)** $(-2, 3)$ **4a)** $(-6, 1)$
b) $(2, -7)$ **c)** $(-2, -3)$ **d)** $\left(\frac{7}{12}, -\frac{1}{3}\right)$ **e)** $(1, 3)$ **5a)** $(-3, 0)$
6a) $\left(0, -\frac{1}{3}\right)$ **b)** $(-7, -3)$ **7b)** $(1, 2)$ **8a)** $(-1, 4)$ **b)** $\left(\frac{-2}{11}, \frac{-1}{2}\right)$
c) $(2, -5)$ **d)** $\left(\frac{7}{4}, \frac{1}{11}\right)$ **9a)** $\left(\frac{1}{8}, \frac{-2}{3}\right)$ **10a)** $\left(\frac{2}{5}, \frac{3}{5}\right)$ **b)** $\left(\frac{1}{4}, \frac{-1}{5}\right)$
c) $\left(\frac{1}{8}, \frac{-1}{2}\right)$ **d)** $\left(\frac{1}{3}, \frac{-2}{3}\right)$ **12b)** $y = \frac{cd - af}{bd - ae}$, $bd - ae \neq 0$ **14a)** D
b) C **c)** B **d)** A **15a)** $(a, 2a)$ **16a)** $\left(\frac{6}{a + b}, \frac{2b - 4a}{a + b}\right)$ **17a)** $(2a, -3a)$
b) $\left(\frac{1}{2}a, \frac{1}{4}a\right)$ **c)** $(-2a, b)$ **d)** $(3a, 4b)$ **e)** $\left(\frac{1}{2a + b}, \frac{-6a - 2b}{2a + b}\right)$
f) $\left(\frac{8 - 6a}{6 - a^2}, \frac{18 - 4a}{6 - a^2}\right)$ **18)** $\left(\frac{de - bf}{ad - bc}, \frac{af - ce}{ad - bc}\right)$ **19a)** $(-2, 1)$
b) $(1, -1)$ **c)** $(3, 1)$ **d)** $(-1, 1)$

4.7 exercise/page 140

6) 236 girls, 184 boys **7)** 14,39 **8)** Sam: $1300, Alex: $700
9) 950 cards **10)** 18 $1-bills, 6 $2-bills **11)** 16 pucks, 20 sticks
12) 15, 73 **13)** 13,20 **14)** 8 wins, 5 losses **15)** 30,10 **16)** 48,24
17) $1100 **18)** 3 small, 12 medium **19)** 50 children, 150 adults
20) 168 cm **21)** $110

4.8 exercise/page 142

1) Tony: 566.7 m, Ned: 933.2 m **2)** Henry: 53 a, Janice: 47 a
3) $360 @ 10%, $300 @ 12% **4)** 6000 L , 4000 L
5) 100 kg 90%, 50 kg 60% **6)** $1400 @ 9%, $400 @ 6%
7) $4000 @ 10%, $6000 @ 11% **8)** 36 L **9)** 5,25 **10)** 3 kg of 10%,
7 kg of 30% **11)** $1600 @ 10%, $2000 @ 8% **12)** 400 L premium,
600 L regular **13)** 4 L **14)** 6 pieces bacon, 3 eggs **15)** 225 mL
16) 1 L **17)** 6.24 mL **18)** $2900 @ 4%, $1300 @ 5%
19) 50 L of 20%, 200 L 50% **20)** 20 L dandelion, 30 L crabgrass
21) 0.8 kg of 14 k, 2.4 kg of 6 k **22)** 3.5 kg **23)** 7.5 g of 18 k,
10 g of 11 k, **24)** 312 kg **25a)** 21 g **b)** 10.5 g

4.9 exercise/page 146

Problem 1: 70 L of 20%, 30 L of 10% **Problem 2:** Sue 88 km/h,
Peter 68 km/h **1)** 240 L of 50%, 60 L of 30% **2)** 20 kg of 20%,
80 kg of 40% **3)** 4 L **4)** 83.$\dot{3}$ mL **5)** Grant 5 m/s, John 4.8 m/s
6) 137 km/h, 113 km/h **7)** plane 200 km/h, wind 40 km/h
8) 110 km/h, 120 km/h **9)** 9 m/s, 11 m/s **10)** 400 km/h, 450 km/h
11) $7000 @ 10% $3000 @ 9% **12)** 2.5 h **13)** 23 m/s, 27 m/s
14) 20 km/h, 21 km/h **15)** 3.$\dot{3}$ mL **16)** 60 kg @ $2.20,
40 kg @ $2.40 **17)** 11 L **18a)** TZ600 **b)** 11:25 **c)** 50 km/h
19) 17.5 km/h **20)** Helicopter: 140 km/h, wind: 20 km/h **21)** wind:
35 km/h, plane: 365 km/h **22a)** 585 km/h **b)** 45 km/h **23)** canoeist:
13.5 km/h, current: 3.5 km/h **24)** river: 5 km/h, tanker: 17 km/h
25) kayak: 23 km/h, current: 17 km/h

4.10 exercise/page 148

1) 70 kg yogurt, 30 kg peaches **2)** 12 full pages, 8 half pages
3) 300 mL **4)** 3.125 mL **5)** $216 @ 8%, $144 @ 6% **6)** wind:
36.4 km/h, plane: 254.5 km/h **7)** $700 from bank, $600 from trust
8) plane: 37.5 m/s, wind: 12.5 m/s **9)** 5 trout, 2 pickerel
10) 5 pages **11)** 65 sacks clover, 35 sacks lucerne **12)** 20 people
13) $140 000 **14)** 20 games **15)** 150 cm height, 100 cm width
16) $8000 in bonus, $2000 in chequing **17)** $400 annual, $6/h,
18) new: 75¢ each, used: 35¢ each. **19)** 125 g of 50%,
375 g of 70% **20a)** after 4 h **b)** 3000 km

4.11 exercise/page 151

1) 40 trees, 16 shrubs **2)** $1000 **3)** 651 biologists, 483 sociologists
4) 383 cm × 167 cm **5)** 15,10 **6)** 150 L of 40%, 50 L of 60%
7) 40 nickels, 30 dimes **8)** overhaul: $225, price: $300 **9)** 45 goals,
30 assists **10)** 3 converted, 4 goals

4.12 exercise/page 153

1b) A **c)** 5 off-hours, 9 peak hours **2b)** C **c)** $1250 @ $9\frac{1}{2}$ %,
$250 @ 5%. **3)** C; 105 adults, 200 students **4)** A;
250 km after hours, 900 km regular hours **5)** B; 330 km in-city,
450 km long distance **6)** B; 60 L of 30%, 40 L of 40% **7)** A; John:
7 a, Tom: 12 a **8)** C; sailboat 42 km/h, fishing boat 24 km/h
9) B; 70 kg **10)** O; 11 adults **11)** F; 2.5 m **12)** A; 12 badminton,
16 tennis **13)** L; 15 h **14)** K; $300 @ 2%, $200 @ $1\frac{1}{2}$% **15)** D: 350 men
16) P: 3 bottles **17)** N: 35 kg sunflower, 15 kg raisin

CHAPTER 5

5.1 exercise/page 161

1a) 6:1 **b)** 8:1 **c)** 15:2 **d)** 4:1 **e)** 8:1 **f)** 4:1 **g)** 20:1 **h)** 35:12
2a) 1:2 **b)** $2a$:1 **c)** a:$2b$ **d)** $-x$:2 **e)** 3:1 **f)** 1:2 **g)** 18:pq **h)** xy:27
i) $(x + 2y)$:x **j)** $3a$:$(a + 2b)$ **k)** a:b **l)** $(p - q)$:1 **m)** $(a + b)$:1
n) $(a^2 + b^2)$:$(a + b)$ **3a)** $\frac{n}{m}$ **b)** nx **c)** $\frac{y}{n}$ **4a)** $-\frac{1}{2}$ **b)** $-\frac{1}{3}$ **c)** $-\frac{1}{5}$ **d)** -5

5a) $\frac{3}{2}$ b) $\frac{3}{2}$ c) $\frac{9}{5}$ d) $\frac{7}{4}$ e) $\frac{4b+5}{5b-5}$ 6a) $\frac{35}{2}$ b) 10 c) $\frac{28}{15}$ d) $\frac{27}{4}$ e) $\frac{5}{2}$
f) $\frac{12}{7}$ 7a) 30 b) $\frac{27}{4}$ c) 2 d) $\frac{36}{5}$ e) $\frac{6}{7}$ f) 4 8) 3:4 9) −2:5 10a) 3:2
b) 5:4 c) 8:9 d) 3:4 e) 3:1 f) 4:3 g) 5:1 h) −3:1 11) $m = 30$, $n = 16$ 12a) $x = 6, y = 15$ b) $m = 15, n = \frac{4}{5}$ c) $p = 24, q = 20$
d) $r = 10, s = 5$ 14a) 5 b) $\frac{15}{2}$ c) $-\frac{1}{5}$ d) −2 15a) −5 b) −13 c) $-\frac{2}{5}$
16a) 0 b) 1 c) −23 17a) 0 b) −7 c) −21 d) 5 18a) 6 b) 2 c) −2
d) −1 e) 13 f) $-\frac{7}{3}$ g) 0 h) 5 19a) 2:1 b) 3:1 c) 3:8 d) −3:2
2a) 1:1 or 3:1 b) −5:1 or 2:1 c) 1:2 or 3:1 d) 2:5 or 1:2 21a) ±6,
b) ±9 c) ±mn d) ±$6m^2$ e) ±$4ab$ f) ±$8p^3$ g) ±3 h) ±$2a\sqrt{a}$ 22a) 16
b) 20 c) 32 d) 1 23a) $\frac{4}{3}$ b) 3 c) 4 d) $\frac{3}{2}$ e) 7 f) $\frac{3}{5}$ g) $\frac{1}{3}$ 24a) $\frac{9}{4}$ b) $\frac{5}{6}$
c) $-\frac{3}{8}$ d) $\frac{1}{5}$ 27) $\frac{12}{49}$ 28) −8:9

5.2 exercise/page 163

1) 35 cm 2) 75 3) $21 270 4) 8.8 mL 5) 204 cm 6) 205 goals
7) 56, 84 8) Tony 12, Colin 42, André 18. 9) Jane $240, Karen $180, Rose $120 10) 8, 12 11) 6, 10 12) 25, 20 or −25, −20
13) 14, 4 or −14, −4 14) 26, 8 or −26, −8 15) 19 16) 9
17) Ann 12 h, Bill 20 h, Chris 18 h 18) Bill 120 shares, Brad 80 shares, Bob 100 shares 21) 40 g 22) 8 g 23) 48 g
24) 21 g 25) 71 g 26) 68 g 27) 45 g 28a) 48 g b) 66 g

5.3 exercise/page 166

1a) 2:3:4 b) 3:−1:2 c) $x^2:xy:y^2$ d) $a:2b:3$ e) $(a+b):1:(a-b)$
f) $(x+y):1:(x-y)$ 2a) $a = 2, b = 24$ b) $a = 1, b = 2$ c) $a = 2, b = 10$ d) $a = 6, b = 5$ e) $a = 8, b = 15$ f) $a = \frac{3}{4}, b = 32$
g) $a = 10, b = 12$ h) $a = 6, b = 15$ 3a) 3:1:4 b) 3:−1:−4
4) $a = 2, b = 5$ 5) $a = 2, b = 4$ b) $a = 0, b = 9$ 7) $a = −4, b = −5$ 8a) $m = \frac{5}{2}, n = 0$ b) $m = 2, n = 3$ c) $m = \frac{2}{3}, n = \frac{3}{4}$
9) 8 wins, 4 losses, 6 ties 10) 40°, 40°, 100° 11) 28, 84, 140
12) 194.4 g protein, 277.8 g carbohydrates, 27.8 g water,
13) Jenny $20 000, Lori $30 000, Alex $50 000 14) 46, 69, 115
15) 12, 18, 30 16) 9, 12, 15, 17) 13.3 18) 22.6 kg 19) 4, 10, 14
20) 6, 8, 12 21) 64.3 g 22) 5.1 g carbon, 0.7 g hydrogen, 6.9 g oxygen 23) 35 g nitrogen, 120 g oxygen 24) 21 g carbon, 24 g oxygen 25a) 3 b) $-\frac{3}{13}$ 26a) 2 b) $\frac{19}{10}$ 27a) $-\frac{7}{3}$ b) $\frac{81}{25}$ c) $\frac{98}{39}$
28) 21 kg phosphoric acid, 18 kg potash 29) 8 kg nitrogen, 2 kg phosphoric acid 30) 1 kg potash 31) 2.7 kg potash, 1.3 kg nitrogen

5.4 exercise/page 169

1a) $A = kW$ b) 900 2a) $H = kW$ b) 58.9 3a) $\frac{1}{6}$ b) 12 c) 15 d) 10
e) 2 4a) (3, 12) b) (3, 18) c) (10, 150) d) (5, 60) and (10, 120)

5a) $k = 200, V = 7.5$ b) $k = 50, d = 600$ c) $k = 36, h = 720$.
d) $k = 15, x = 20$ e) $k = 12.5, t^2 = 6.2$ 6a) 675 b) 64 c) 6 d) 12
e) 53.2 f) 81.6 7a) $P = kl$ b) $d = st$ c) $p = ka$ d) $g = km$
e) $c = kr$ f) $m = kd^3$ 8a) $m = kl$ b) 250 kg 9a) 21 g b) 3.$\dot{3}$ cm
10a) 175 situps b) 12 s 11) 333.3 km 12) 880 km/h 13) $1.20
14) $32.50 15) 15 L

5.5 exercise/page 172

1) $\frac{1}{2}$ 2a) $s = kp^2$ b) 25 3) 30 4a) $k = 16$ b) 20 s 5a) $m = kd^2$
b) 100 g 6) 25 units 7) 576π cm² 8) 375 m 9a) 6 carats b)
10) 80 min 11) 5544 cm³ 12) 8.5 cm 13a) $k = 5$ b) 9 km
14a) $d = kt^2$ 487.5 m 15a) 43.875 m b) 78 m c) 34.125 m
16a) 34.125 m b) 43.875 m c) 9.75 m 17) 18 m 18) 81 m
19) 50.6 m 20) Yes 21a) 15 m b) 1.2 km/h, 0.6 km/h 22) 2.78 s
23) No

5.6 exercise/page 176

1a) 50 b) 64 c) 96 d) 288 2a) A,C b) A,B c) A,B,C d) B,C 3a) 324
b) 18 c) 6 4a) $H = \frac{k}{s}$ b) 21 5a) 25 b) 40 c) 8 6) 5 7) 5 8) 8 9) 54
10) 12.5 11) 12 12) 18 13) 64 14a) $va = k$ b) $V = kT$ c) $fw = k$
d) $c = kd$ e) $VP = k$ 15) 2.5 h 16) 3 units 17) 45 units 18) 2.25 s
19) 800 km/h 20) 160 units 21) $44.10 22) 1 h 23) 20 m

5.7 exercise/page 179

1) 384 2) 4 3a) 5400 b) 216 c) 6 4) 150 5a) 10 b) 100 6) 3
7) 33.75 8) 2.56 9) 2.5 10) 75 11a) $ld^2 = k$ b) $A = kd^2$ c) $f\lambda = k$
d) $gd^2 = k$ e) $M = kV$ 12) 21°C 13) 5.4 km/L 14) 5 cm 15) 49 mg
16) 24 cm 17) 833.$\dot{3}$ g 18) 16 s 19) 12 cm 20) $7.03 21) 84 cm³

5.8 exercise/page 181

1a) $V = kTQ$ b) $R = \frac{kT}{P}$ c) $U = \frac{kT^2}{S}$ d) $E = kMV^2$ e) $Q = k\frac{M}{p^2}$
2a) $J = \frac{kT}{p}$ b) $P = \frac{kQ}{R^2}$ c) $A = kBC^2$ d) $R = \frac{kG}{S^2}$ e) $V = \frac{kR}{T^2}$ 3a) 20
b) 125 4b) 1.75 5) (a), (d) 6) 6 7) 15 8) $Q_2 = 2520$ 9) 180
10) 121 cm² 11) 27 units 12) 3 m/s 13) 60 m/s² 14) 20 000 units

5.9 exercise/page 183

1) $m = 3, p = 13$ 2) $k = 40, p = 20$ 3) 1.5 4) $405 5) 301 players
6) $144.80 7) 9 persons 8) $1850 9) $21/h 10) 4.5 h 11) $10.20

5.10 exercise/page 184

1) $7500 2) 10 cm 3) $259.20 4) $\frac{1}{4}$ 5) 74 L 6) 50 625 km 7) $8.38
8) 80 cm² 9) 15 min 10) 5 s 11) 200 revolutions 12) 2.8 m 13) 1250
14) 7.8 L

CHAPTER 6

6.2 exercise/page 192

6a) $\{-1, 0, 1, 2\}$; $\{-2, -1, 0\}$ **b)** No **7b)** $\{0, 1, 2, 3\}$; $\{1, 2, 3, 4\}$ **c)** Yes. **8)** Functions: (b), (d); Non-functions: (a), (c) **9a)** $\{(0, -1), (1, 2), (2, 5), (3, 8), (4, 11)\}$. **b)** $\{-1, 2, 5, 8, 11\}$ **c)** Yes **10c)** R **b)** No **11)** $f(x) = 2x + 5$, $\{(x, y) | y = 2x + 5, x \in R\}$, $y = 2x + 5, x \in R$ **12a)** $\{-5, -2, 1, 10\}$ **b)** $\{5, 3, 1, -6\}$ **c)** $\{4, 1, 0, 9\}$ **d)** $\{1, -2, -3, 6\}$ **e)** $\{-\frac{1}{2}, -1, \text{no value}, \frac{1}{3}\}$ **f)** $\{-\frac{1}{7}, -\frac{1}{4}, -1, -\frac{1}{8}\}$ **13b)** Yes **14b)** Domain: $\{x \in R, x \neq 0\}$, Range: $\{y \in R, y \neq 0\}$ **c)** Yes **15b)** Domain: $\{x \in R, x \geq -3\}$, Range: R **c)** No **17b)** $\{y \in R | -5 \leq y \leq 7\}$ **18a)** 0 **b)** 3 **c)** 8 **d)** 0 **e)** 15 **f)** $\frac{5}{4}$ **19a)** 8, -17 **b)** 1, 9 **c)** 7, 7 **20a)** -1 **b)** -3 **c)** 2 **d)** $-\frac{1}{2}$ **e)** $\frac{-11}{2}$ **f)** 25 **21a)** -5 **b)** 17 **c)** 12 **d)** 3 **e)** 1 **f)** 4 **g)** 21 **h)** $\frac{7}{2}$ **22a)** 3 **b)** -1 **23a)** 5 **b)** 7 **c)** 20 **d)** 49 **e)** -10 **f)** 17 **g)** 50 **h)** 199

6.3 exercise/page 195

1) (a), (c), (d), (e) **2a)** $\{y \in R | -2 \leq y \leq 2\}$ **b)** $\{y \in R | -9 \leq y \leq -3\}$ **c)** $\{y \in R | -11 \leq y \leq 21\}$ **d)** $\{y \in R | -11 \leq y \leq -5\}$ **3a)** $x = -3$ **b)** $x = 5$ **c)** $x = 2$ **4a)** $y = 1$ **b)** $y = -4$ **c)** $y = -3$ **5)** They have the point $(0, 5)$ in common **6)** They have slope 5 i.e. they are parallel. **7a)** $2x - y + 1 = 0$ **b)** $x - 3y - 15 = 0$ **c)** $2x - 3y - 1 = 0$ **d)** $x - 2y - 15 = 0$ **e)** $3x - 2y - 36 = 0$ **f)** $2x - 3y - 6 = 0$ **g)** $3x - 2y + 41 = 0$ **h)** $2x - 3y - 9 = 0$ **8a)** $\frac{2}{3}, -2$ **b)** $5, -2$ **c)** $-\frac{1}{6}, 1$ **d)** $-6, 12$ **e)** $\frac{1}{2}, -\frac{1}{2}$ **f)** 5, no y-intercept **g)** no x-intercept, -6 **h)** $\frac{1}{3}, \frac{1}{4}$ **i)** $\frac{1}{3}, -1$ **j)** $\frac{1}{2}, -\frac{1}{3}$ **9a)** $\{y \in R | -15 \leq y \leq -6\}$ **b)** $\{y \in R | -5 \leq y \leq -3\}$ **c)** $\{y \in R | \frac{1}{3} \leq y \leq \frac{4}{3}\}$ **d)** $\{y \in R | \frac{-29}{2} \leq y \leq \frac{-11}{2}\}$ **10a)** $(3, 0), (0, 9)$, **b)** $(6, 0), (0, -4)$ **c)** $(5, 0), (0, -3)$ **d)** $(9, 0), (0, -3)$ **e)** $(4, 0), (0, -4)$ **f)** $(\frac{1}{3}, 0), (0, -\frac{1}{2})$ **g)** $(\frac{3}{2}, 0), (0, -3)$ **h)** $(\frac{3}{2}, 0), (0, -1)$ **11a)** 1 **b)** -2 **c)** -17 **d)** any real value **e)** 1 **12)** $-\frac{1}{2}$ **13)** 4 **14)** 6 **15)** 2 **17a)** 3, $\frac{3}{2}$ **b)** $-2, 4$ **c)** $2, -6$ **d)** 4, 8 **e)** $-2, 6$ **f)** $\frac{5}{3}, -5$

6.4 exercise/page 198

4a) 1 **b)** 1 **c)** $\frac{1}{5}$ **d)** $\frac{1}{5}$ **5a)** $\frac{-3}{2}$ **b)** 0 **c)** $\frac{2}{5}$ **d)** $\frac{-3}{7}$ **e)** $-\frac{1}{4}$ **f)** undefined **6a)** $m_{AB} = -\frac{3}{5}$, $m_{BC} = \frac{4}{7}$, $m_{AC} = \frac{7}{2}$ **b)** $m_{PQ} = \frac{1}{3}$, $m_{QR} = -5$, $m_{SR} = \frac{1}{3}$, $m_{PS} = -5$ **7)** (a), (d) **8a)** -1 **b)** 1 **c)** -2 **d)** $\frac{1}{2}$ **e)** $\frac{-3}{2}$ **f)** 2 **g)** $\frac{2}{3}$ **h)** $\frac{1}{3}$ **i)** -2 **j)** 0 **k)** undefined **l)** 0 **10a)** 1.06 **b)** $\frac{4\sqrt{2}}{\sqrt{3}}$ **c)** $-\frac{b}{2a}$ **d)** $\frac{y_2 - y_1}{x_2 - x_1}$ **e)** $\frac{-b}{a}$ **f)** $\frac{1}{(y - x)}$ **11)** $m_{\text{Œ}} = \frac{8}{9}$, m_{EF} = undefined, $m_{DF} = 0$ **12a)** $m_{PQ} = \frac{5}{8}$, $m_{QR} = \frac{7}{5}$, $m_{SR} = \frac{5}{8}$, $m_{PS} = \frac{7}{5}$ **b)** Opposite sides have equal slopes **c)** parallelogram **13a)** $m_{PQ} = \frac{4}{3}$, $m_{QR} = \frac{1}{5}$, $m_{SR} = \frac{4}{3}$, $m_{PS} = \frac{1}{5}$ **b)** Opposite sides have equal slopes **c)** ∥ **14)** $(-3, 0)$ **15)** $(4, -8), (6, -11)$ **16)** $A(-2, 3), B(1, 4)$ **17a)** $M(3, 7)$ **b)** $Q(2, 4)$ **c)** $(-3, -1)$ **d)** $A(-2, 3), B(1, 4)$ **18)** 2 **19)** 2 **20a)** -4 **b)** 0 **c)** $\frac{5}{7}$ **21)** 5 **22)** $-5, 2$ **23)** $\frac{-6}{7}$ **25)** 0.93 **b)** Good **26a)** 0.74 **b)** Poor

6.5 exercise/page 202

7a) 4, -3 **b)** $-2, 9$ **c)** $\frac{3}{2}, 3$ **d)** 2, 0 **e)** 1, -12 **f)** $\frac{1}{2}, -9$ **g)** $\frac{3}{8}, \frac{-9}{4}$ **h)** 0, 9 **i)** 1, -3 **j)** 1, $-\frac{1}{2}$ **8a)** $y = 2x - 4$; $y = 2x + \frac{5}{2}$, $y = 2x - 6$ **b)** equal slopes **9b)** same y-intercept 5 **10b)** equal slopes **c)** lines with equal slopes are parallel. **12a)** $(4, 0), (0, -3)$ **b)** $(\frac{1}{6}, 0), (0, -\frac{1}{2})$ **c)** $(16, 0), (0, -24)$ **d)** $(-2, 0), (0, \frac{4}{3})$ **e)** $(-2, 0), (0, 6)$ **f)** $(\frac{1}{3}, 0), (0, -\frac{1}{3})$ **13a)** $(3, 0), (0, 6)$ **b)** $(\frac{2}{3}, 0), (0, -2)$ **c)** $(16, 0), (0, -8)$ **d)** $(\frac{3}{4}, 0), (0, -\frac{3}{2})$ **e)** $(\frac{-1}{3}, 0), (0, 1)$ **f)** $(12, 0), (0, -18)$ **14)** $A \parallel G, B \parallel F$ **15)** 1 **16a)** -18 **b)** -23 **c)** 0 **d)** -1 **17)** 1 **18)** -2 **19)** -2 **20)** 0 **21)** $\frac{1}{2}, -5$ **22)** -2 **23)** 3 **24)** Yes. Opposite sides have equal slopes. **25)** No. Only 1 pair of opposite sides is parallel. **26a)** $y = -\frac{B}{A}x - \frac{C}{B}$ **b)** Slope $= -\frac{A}{B}$, y-intercept $= -\frac{C}{B}$ **27a)** $-2, 3$ **b)** 3, 5 **c)** $\frac{4}{3}, 0$ **d)** $\frac{3}{2}, -3$ **e)** 3, -2 **f)** $\frac{3}{2}, \frac{9}{2}$ **g)** $-1, 6$ **h)** $\frac{2}{3}, -\frac{4}{3}$ **28)** 3 **29)** 1

6.6 exercise/page 205

1) parallel: (a), (e); perpendicular: (b), (d), (h), (i) **2a)** 3, $\frac{-1}{3}$ **b)** $\frac{2}{3}, \frac{-3}{2}$ **c)** $-\frac{1}{2}, 2$ **d)** $-1, 1$ **e)** $\frac{5}{4}, \frac{-4}{5}$ **f)** 0, undefined **3a)** $\frac{-1}{3}$ **b)** $\frac{3}{2}$ **c)** $-\frac{1}{2}$ **d)** 2 **e)** undefined **f)** 0 **4a)** ∥ **b)** ⊥ **c)** N **d)** ⊥ **e)** ∥ **f)** N **5a)** $y = 2x - 5$ **b)** $y = -\frac{1}{3}x - 3$ **c)** $y = \frac{3}{2}x + 8$ **d)** $y = \frac{2}{3}x - 5$

e) $y = \frac{1}{2}x - \frac{11}{4}$ **f)** $y = 2x + \frac{11}{3}$ **g)** $y = 3x + 2$
h) $y = \frac{-2}{3}x - 3$ **i)** $y = \frac{3}{2}x - 8$ **j)** $y = \frac{1}{2}x + \frac{5}{2}$; (a)||(f), (c)||(i), (e)||(j); (b) ⊥ (g); (c)⊥ (h); (h) ⊥ (i) **7)** (a), (c), (d), (e) **8a)** $\frac{1}{8}$ **b)** 2
9a) $\frac{3}{2}$ **b)** −5 **10)** 9 **11)** 31 **12)** 5 **13)** $\frac{5}{6}$ **14)** 5, −3 **15)** −2, −9.

6.7 exercise/page 207
1) (a), (d) **2)** (b), (c) **8)** Yes ∠C = 90° **9)** Yes **11)** No

6.8 exercise/page 210
1a) y-intercept 6 **2a)** Slope $-\frac{1}{2}$ **3a)** $y = 2x + b$ **b)** $y = -3x + b$
c) $y = \frac{3}{2}x + b$ **d)** $y = b$ **e)** $y = \frac{-4}{3}x + b$ **f)** $x = a$ **4a)** $y = mx + 3$
b) $y = mx - 4$ **c)** $y = mx + \frac{1}{3}$ **d)** $y = mx - \frac{3}{2}$ **e)** $y = mx + \frac{4}{3}$
f) $y = mx$ **5a)** $y = m(x - 3) + 2$ **b)** $y = m(x + 2) + 5$
c) $y = m(x + 1) - 3$ **d)** $y = mx + 8$ **e)** $y = m(x - 4) - 5$
f) $y = m(x + 6)$ **g)** $y = m(x - a) + b$ **h)** $y = mx + b$
i) $y = m(x - a)$ **6a)** y-intercept 6 **b)** passing through (6, 4)
c) slope $-\frac{1}{2}$ **d)** x-intercept −2 **7a)** 0, undefined **b)** $y = b, x = a$
8a) $y = mx - 6$ **b)** $y = m(x - 4) - 3$ **c)** $y = -8x + b$ **d)** $x = a$
e) $y = -\frac{2}{3}x + b$ **f)** $y = mx + \frac{2}{3}$ **g)** $y = m(x - 3) - 4$ **h)** $y = b$
i) $y = m(x + 6)$ **9a)** $y = -2x + 3$ **b)** $y = -2x - 2$ **c)** $y = -2x - 8$
10a) $y = \frac{2}{3}x + 6$ **b)** $y = -\frac{11}{3}x + 6$ **c)** $y = -2x + 6$
11a) $y = m(x + 3)$ **b)** $y = \frac{1}{3}x + 1$ **12a)** $y = m(x - 3) - 4$
b) $y = -\frac{5}{6}x - \frac{3}{2}$ **13a)** $x = a$ **b)** $x = 3$ **14a)** $y = b$ **b)** $y = 8$
15a) $y = mx$ **b)** $y = -\frac{5}{2}x$ **16a)** $2x - 3y + 21 = 0$
b) $2x + 5y - 19 = 0$ **c)** $5x - y + 20 = 0$ **d)** $11x + 3y + 18 = 0$
e) $y = 5$ **f)** $x = -3$ **g)** $3x - 2y + 19 = 0$ **h)** $y = 5$ **17a)** −8 **b)** 6
18) −2 **19)** 3 **20)** 1, $\frac{-11}{14}$

6.9 exercise/page 212
1a) $y = -2x + b$ **b)** $y = \frac{1}{2}x + b$ **c)** $y = -\frac{3}{2}x + b$ **d)** $x = a$
2a) $y = \frac{1}{3}x + b$ **b)** $y = -\frac{2}{3}x + b$ **c)** $y = -\frac{2}{3}x + b$ **d)** $y = \frac{1}{3}x + b$
3a) $y = mx + 2$ **b)** $y = m(x + 3) + 8$ **c)** $y = \frac{4}{3}x + b$
d) $y = m(x + 3)$ **e)** $y = m(x - 2) - 6$ **f)** $x = a$ **4a)** $y = 2x + b$
b) $y = 2x + 7$ **c)** $y = 2x + 4$ **5a)** $y - 3 = 2x$ **b)** $y = 2x + 3$

6a) $y = m(x + 3) + 2$ **b)** A: $y = 2x + 8$; B: $y = -\frac{5}{3}x - 3$;
C: $= -x -1$, D: $x = -3$ **7a)** $2x - 3y - 6 = 0$ **b)** $y = -\frac{4}{3}x + 2$
c) $3x + y = 0$ **d)** $5y = 20x - 8$ **e)** $y = -4$ **f)** $x + 2y + 3 = 0$
g) $y = 3x + 2$ **h)** $y = -x - 2$ **i)** $2x + y + 8 = 0$ **j)** $y = \frac{1}{2}x - 2$
k) $y = 1$ **l)** $x + 3y + 8 = 0$ **m)** $y = -2x + 7$ **8)** $2x + y + 1 = 0$
9) $3x - 2y - 6 = 0$ **10)** $2x + 3y + 4 = 0$ **11)** $y = -3$
12) PQ: $x - 4y + 13 = 0$, SR: $x - 4y - 2 = 0$, PS: $4x - y + 7 = 0$,
QR: $4x - y - 8 = 0$ **13a)** $x - 3y + 3 = 0$ **b)** (−6, −1), (3, 2)
14a) $x + 3y - 10 = 0$ **b)** $x - 4y - 2 = 0$ **15)** From G:
$2x + 7y - 12 = 0$, From F: $3x - 4y - 15 = 0$, From E:
$5x + 3y - 27 = 0$ **16)** $\left(\frac{8}{5}, 0\right)$ **17)** $2x - y + 6 = 0$ **18)** $\left(0, \frac{9}{2}\right)$
19) $11x + y + 8 = 0$ **20)** $\left(0, \frac{8}{3}\right)$ **21)** $x + 4y = 0$ or $x + y + 6 = 0$
22) $4x - 3y - 8 = 0$ **23)** P(0,−2) **24)** (3,0)

6.10 exercise/page 215
1a) (−4, 4) **b)** (3, 2) **c)** $\left(\frac{3}{2}, 4\right)$ **d)** (3, 2) **e)** $\left(\frac{13}{2}, -\frac{11}{2}\right)$ **f)** (6, −2)
g) $\left(\frac{1}{2}, 4\right)$ **h)** $\left(\frac{-3}{2}, \frac{-3}{2}\right)$ **2a)** (−6, −2) **b)** (1, 5) **c)** $m_{MN} = 1$,
$m_{QR} = 1$ **d)** The same **3a)** $-\frac{5}{4}$ **b)** −1 **c)** −4 **d)** $-\frac{4}{5}$ **5)** AB||CD,
BC||AD **6a)** $x + 4y + 2 = 0$ **b)** $x - 2y + 4 = 0$
c) $2x - 3y - 4 = 0$ **d)** $x + 2y - 4 = 0$ **7a)** $y = -2$ **b)** $x = 4$
8) $t \in R$ **9)** $\frac{13}{6}$ **10)** −3 **11)** (0, 10) **12)** $\left(0, \frac{10}{3}\right)$ **13)** (−9, −7)
14) (6, −3) **15)** $(5\sqrt{2}, -3\sqrt{2})$ **16)** $3x + 2y + 7 = 0$
17) $5x + 3y + 14 = 0$ **18)** $3x + 2y - 5 = 0$ **19)** $3x - 5y + 10 = 0$
20) $3x - y = 0$ **21)** $p = -3, k = 53$ **22)** −1 **23)** (−1, 6)
24a) $3x - 2y + 17 = 0$ **b)** $7x - 2y - 11 = 0$, $x - 6y - 13 = 0$
c) $x - y + 2 = 0$ **25)** $x - 3y - 10 = 0$, $3x + y = 0$
26) $x + y - 4 - \sqrt{17} = 0$, or $x + y - 4 + \sqrt{17} = 0$
27) $3x + 2y - 11 = 0$ **28a)** $5x - 4y + 11 = 0$, $7x + y + 11 = 0$,
$2x - 5y + 20 = 0$ **b)** $\left(-\frac{5}{3}, \frac{2}{3}\right)$ **29a)** $x - 2y - 4 = 0$,
$2x + 3y + 1 = 0$, $4x - y - 7 = 0$ **b)** $\left(\frac{10}{7}, -\frac{9}{7}\right)$
30a) $x - y - 2 = 0$, $x + 3y - 2 = 0$, $3x + y - 6 = 0$ **b)** (2, 0)
32a) (1, −2) **b)** $x - 2y - 5 = 0$ **c)** (7, 1) **33a)** $\left(5, \frac{3}{2}\right)$ **b)** $\left(5, \frac{3}{2}\right)$

6.11 exercise/page 219
1) c **2)** (a), (c) **3)** A,B,C **4a)** $2x + 3y \geq 6$ **b)** $x - y > 3$
5a) $2x - 5y \leq 10$ **b)** $y < 3$ **c)** $2x + 5y > -10$ **d)** $4x + 3y \leq 10$
7a) $x - 3y + 8 \geq 0$, $x + y \leq 0$ **b)** $x - 3y + 8 < 0$, $5x + y - 8 \leq 0$
c) $2x - y + 1 > 0$, $x + 2y - 7 > 0$ **d)** $2x - y + 1 > 0$,

$5x + y - 8 \leq 0$ **9a)** $x + 3y - 4 \geq 0, x - y \leq 0$ **b)** $x - y + 2 \leq 0$, $x + y - 4 \leq 0$ **c)** $x - 4y + 6 > 0, 2x + 3y - 10 < 0$ **d)** $x - y + 5 \geq 0, x + 3y - 3 > 0$

6.12 exercise/page 221

1b) $(0, 0)$, $\left(10\frac{1}{3}, 0\right)$ $(5, 8)$ $\left(0, \frac{19}{3}\right)$ **2b)** $(0, 8)$ $(4, 3)$ $(10, 0)$ **3a)** 42 **b)** 29 **4a)** 18 **b)** 17 **5)** 6 **6)** 240 **7)** 33 **8)** 22 **9)** 60 **10)** 740 **11)** 16 pairs, 24 pairs **12a)** 30 kg of Husky, 20 kg of Vibrant **b)** $1390 **13)** 120 2-bulb lamps, 60 4-bulb lamps, profit $4500 **14)** 1 self-wind, 3 automatic, profit = $79 **15)** 5h, 5h **16a)** 3L, 5.5L **b)** $30.50

CHAPTER 7

7.2 exercise/page 229

1a) 4 **b)** 8 **c)** 5 **d)** 5 **e)** 252 **f)** $2\sqrt{5}$ **g)** $\sqrt{85}$ **h)** $\sqrt{113}$ **2a)** 10 **b)** $\sqrt{41}$ **c)** $\sqrt{65}$ **d)** $\frac{5}{4}$ **e)** 2 **f)** 1 **3a)** 10 **b)** $\sqrt{10}$ **c)** $\sqrt{146}$ **d)** $2\sqrt{26}$ **e)** $\sqrt{65}$ **f)** $4\sqrt{20}$ **4a)** $\sqrt{13}$ **b)** $\sqrt{13}$ **c)** $\sqrt{13}$ **5)** CD **6a)** 12 **b)** 36 **c)** 60 **d)** 56 **7a)** $5 + 2\sqrt{10} + \sqrt{13}$ **b)** scalene **8a)** $AB = 4\sqrt{10}, BC = 2\sqrt{10}$ **b)** $AC = 10\sqrt{2}$ **9)** BD **10)** 29 units² **11)** P, Q, lie on circle **12)** $(4, 0)$ **13)** 10 units **15)** Yes **17)** Jet has enough fuel. **18)** Plane A **19)** Boat B **20a)** $15\sqrt{2}$ units **b)** $12\sqrt{2}$ **21)** A **22)** $4\sqrt{5}$ units

7.3 exercise/page 231

12) $(2, 7)$ **13b)** Not a rhombus: diagonals are not perpendicular. **14a)** $(3, 1)$ **b)** Yes **15a)** $(-3, 1)$ **16)** Yes **17)** No **18a)** $(8, -3), (5, 4)$ $\left(\frac{33}{29}, \frac{68}{29}\right), \left(\frac{120}{29}, -\frac{135}{29}\right)$ **b)** Parallelogram **19a)** $P\left(\frac{-9}{2}, \frac{1}{2}\right)$, $Q\left(\frac{13}{2}, \frac{3}{2}\right)$ **c)** X (1, 1), yes

7.4 exercise/page 236

1b) 22 units² **c)** 22 units² **2a)** 20.5 units² **b)** 25 units² **c)** 10.5 units² **d)** 21.5 units² **3)** △DEF **4a)** 0 **c)** Points are collinear **6a)** 0 **b)** Collinear **7)** (a), (b), (d) **8a)** $S(-1, -4)$ **b)** 33 units² **c)** 33 units² **9a)** $(8, 2)$ **b)** $(4, -10)$ **c)** $(-4, 6)$ **d)** 80 units² **10b)** 10 units² **11)** $4x - y - 25 = 0, 4x - y + 43 = 0$ **12)** $(2, -3)$ or $(-4, 5)$ **13)** $(7, 7)$ or $(-5, -9)$ **14)** -2 **15)** 0 or 8 **16)** 9, or 17 **17)** 5 **18)** 1 **19a)** 38 units² **b)** $A(-3, 0), B(1, -1), C(0, 4)$ **21b)** 54 units² **22a)** 85 units² **b)** 30 units² **23a)** 10.5 units² **b)** 30.5 units² **c)** 30 units² **24b)** 13 units²

7.5 exercise/page 239

1a) $x + 3y + 7 = 0$ **b)** $\frac{7\sqrt{10}}{5}$ **c)** $\frac{14\sqrt{5}}{5}$ **2b)** $4\sqrt{5}$ **c)** $\frac{14\sqrt{5}}{5}$ **d)** 28 units² **e)** 28 units² **3a)** $\sqrt{5}$ **b)** $\sqrt{5}$ **4a)** $\frac{6}{5}$ **b)** $\frac{2\sqrt{5}}{15}$ **c)** $\frac{5\sqrt{17}}{17}$ **d)** $\frac{9\sqrt{2}}{4}$ **5a)** $\frac{2\sqrt{13}}{13}$ **b)** $\frac{17\sqrt{13}}{13}$ **6a)** $\frac{21\sqrt{26}}{26}$ **b)** $\sqrt{13}$ **7a)** 0 **b)** Q is a point on the line. **8)** $5\sqrt{2}$ **9a)** $\sqrt{5}$ **b)** $\frac{17}{5}$ **c)** $\frac{2\sqrt{13}}{13}$ **d)** $\frac{32\sqrt{130}}{65}$ **e)** 5 **f)** 9 **10a)** $x_2 = \frac{B^2 x_1 - ABy_1 - AC}{A^2 + B^2}$, $y_2 = \frac{A^2 y_1 - ABx_1 - BC}{A^2 + B^2}$ **11a)** $(4, 0)$ **b)** 1.2 units **12a)** equal slopes $\frac{2}{3}$ **b)** $\frac{38\sqrt{13}}{39}$ **13a)** $\frac{6\sqrt{13}}{13}$ **b)** $\frac{18\sqrt{5}}{5}$ **c)** $\frac{2\sqrt{10}}{5}$ **d)** 2.1 units **14a)** $\frac{14\sqrt{10}}{5}$ units **b)** from P to MN = $\frac{14\sqrt{5}}{5}$ units, from N to MP = $\frac{14\sqrt{34}}{17}$ units **15a)** $2\sqrt{5}$ units **b)** to $\ell_1 = 2\sqrt{5}$ units, to $\ell_2 = \frac{7\sqrt{26}}{13}$ units **16)** $\frac{7}{3}$ units, 7 units, $\frac{7\sqrt{10}}{3}$ units **17a)** $\frac{14\sqrt{10}}{5}$ three **b)** 28 units² **18a)** 0 units **b)** All lines are concurrent. **19)** 3 **20a)** 5, -3 **b)** There are two possible points. **21)** No **22)** B **23)** Yes **24)** Yes **25a)** Ship A **26)** 0.1 A.U.

7.6 exercise/page 243

1) A, C **2)** A, C **3)** B **4)** A **5)** E **6)** $\frac{-7}{6}$ **7)** 1 **8a)** y-intercept **b)** 5 **9a)** slope **b)** $y = 2x + 6$ **10)** $\left(10, -4\frac{7}{12}\right), \left(0, 5\frac{5}{12}\right)$ **11)** $(40, 0)$ or $\left(-31\frac{3}{7}, 0\right)$ **12)** $(2\sqrt{2}, 0)$ or $(0, 0)$ **13)** $(90, 0)$ or $\left(-88\frac{4}{7}, 0\right)$ **14)** $4x - 6y + 9 = 0$ and $6x + 4y + 3 = 0$ **15a)** $2x + 2y - 9 = 0$ and $4x - 4y - 1 = 0$ **b)** Same as in (a) **16)** $x + y + 22 = 0$ or $7x - 7y + 2 = 0$ **17)** $3x - 4y - 26 = 0$ or $3x - 4y - 6 = 0$ **b)** $x + 5y - 6 - 2\sqrt{26} = 0$ or $x + 5y - 6 + 2\sqrt{26} = 0$ **18)** $3x + 4y = 0$ **19a)** $x + y + 1 = 0$ **b)** $7x + y = 0$ **20)** $ax = 0$ **21)** $mx = ny$ **22)** $\left(\frac{15}{8}, \frac{15}{8}\right), \left(\frac{-15}{2}, \frac{15}{2}\right)$ **23a)** $y = -3x + b$ **b)** $y = -3x + \sqrt{6}$ **24a)** $y = 3x + b$ **b)** $y = 3x + 3$ **25)** $2y - x - 8 = 0$ **26)** $y = 5x + 3$ **27a)** $y = \frac{2}{3}x + b$ **b)** $2x - 3y + 30 = 0$ or $2x - 3y - 30 = 0$ **28)** $5x + y - 20 = 0$ or $5x + y + 20 = 0$

7.7 exercise/page 246

4) -2 **5)** -3 **6)** $\frac{2}{3}, -2$ **7a)** 6 **b)** $\frac{2}{3}$ **8a)** $-\frac{2}{3}$ **b)** $-\frac{1}{5}$ **9)** 1, -4 **10)** 4 **12)** $3x - 2y = 0$ **13)** $(0, 2)$ **14a)** MP: $2\sqrt{3}$, PN: $2\sqrt{3}$, MN: $2\sqrt{3}$ **b)** Equilateral **15)** $\sqrt{53}$ units **16b)** 48.5 units² **17a)** $2x + y - 7 = 0$ **b)** $(8, 6)$ **18)** △PQR **19)** $\frac{5}{2}, -\frac{7}{2}$ **20a)** 15.3 km **b)** 20 km **21)** 2, -8 **22)** $7x + 6y - 30 = 0$ **23a)** From A: $y = 2$, From B: $3x - y - 10 = 0$, From C: $x + y - 6 = 0$

24a) $4x - 5y + 12 = 0$ **b)** $5x + 3y + 15 = 0$ **25)** $5x + 4y = 0$

exercise 7.8/page 248

21a) $(-3, 0)$ **b)** $\frac{4\sqrt{10}}{15}$ **23a)** $(2, 8)$ **b** $\frac{16\sqrt{13}}{39}$ **25a)** $\left(\frac{-16}{3}, 0\right)$
26) $(11, 12)$ **27)** $\left(\frac{22}{3}, 8\right)$

7.9 exercise/page 252

1a) $x^2 + y^2 = 9$ **b)** $x^2 + y^2 = 3$ **c)** $x^2 + y^2 = 20$ **d)** $x^2 + y^2 = 4r^2$
2a) A:4, B:$5\sqrt{2}$ **b)** A 4, -4; B $5\sqrt{2}$, $-5\sqrt{2}$ **c)** A 4, -4
d) A:$-4 \leq x \leq 4$, B:$-5\sqrt{2} \leq x \leq 5\sqrt{2}$ **e)** A:$-4 \leq y \leq 4$,
B:$-5\sqrt{2} \leq y \leq 5\sqrt{2}$ **3a)** $x^2 + y^2 = 100$ **b)** $x^2 + y^2 = 13$
c) $x^2 + y^2 = 25$ **d)** $x^2 + y^2 = 11$ **e)** $x^2 + y^2 = 2$ **f)** $x^2 + y^2 = 8$
4a) 8 or -8 **b)** 6 or -6 **c)** 0 **d)** 10 or -10 **e)** $5\sqrt{3}$ or $-5\sqrt{3}$
f) $5\sqrt{2}$ or $-5\sqrt{2}$ **5)** (b), (e), (f) **6)** $x^2 + y^2 = 36$ **7)** $x^2 + y^2 = 13$
8b) $\frac{1}{2}$ **9a)** $x^2 + y^2 = 64$ **b)** $x^2 + y^2 = 25$ **c)** $x^2 + y^2 = 7$
d) $x^2 + y^2 = 25$ **e)** $x^2 + y^2 = 16$ **10a)** $3\sqrt{3}, -3\sqrt{3}$ **b)** $\sqrt{30}, -\sqrt{30}$
11a) $2\sqrt{13}$ **b)** Diameter **12c)** M$(4-2\sqrt{2}, 2\sqrt{2})$
d) $(1 - 2\sqrt{2})x - y = 0$ **13b)** $x + 5y = 0$ **c)** $3x + 2y = 0$ **d)** $(0, 0)$
14a) $(-1, -1)$ **b)** $m_{OM} = 1$, $m_{PQ} = -1$: They are negative reciprocals.
c) OM $\perp$ PQ **15a)** AB: $2\sqrt{29}$, CD: $2\sqrt{29}$ **b)** $\sqrt{29}$ **16a)** $x^2 + y^2 = 14$
b) $(3, -\sqrt{5})$ **19)** $\frac{9\sqrt{5}}{5}$ units **20)** $2\sqrt{5}$ units
21a) $x^2 + y^2 - 6x - 7 = 0$ **b)** $x^2 + y^2 + 8y = 0$
c) $x^2 + y^2 + 4x - 6y - 12 = 0$ **22)** $x^2 + y^2 + 6x - 10y - 2 = 0$
23a) $x^2 + y^2 + 6x - 10y + 9 = 0$ **b)** $x^2 + y^2 - 10x + 8y + 16 = 0$
24) $x^2 + y^2 - 2y - 36 = 0$ **25)** -6 **27)** $x^2 + y^2 + 6x - 4y - 4 = 0$
28a) $x + 3y + 14 = 0$ **b)** $3x - y + 10 = 0$ **c)** $\left(\frac{-22}{5}, \frac{-16}{5}\right)$
29) $4, -2$ **30)** 6.4 units **31a)** $x + 2y - 26 = 0$ **b)** $2x - 3y = 0$
c) $\left(\frac{78}{7}, \frac{52}{7}\right)$ **32)** $(-1, 5)$

7.10 exercise/page 255

1a) $x^2 + y^2 < 16$ **b)** $x^2 + y^2 < 3$ **c)** $x^2 + y^2 < 8$ **2a)** $x^2 + y^2 > 25$
b) $x^2 + y^2 > 2$ **c)** $4x^2 + 4y^2 > 1$ **3)** A:(c), (f); B:(a), (e); C:(b), (d)
4a) $x^2 + y^2 < 25$ **b)** $x^2 + y^2 > 25$ **5)** $x^2 + y^2 < 25$ **6)** $x^2 + y^2 > 36$
8a) $x^2 + y^2 < 9$ **b)** $x^2 + y^2 > 4$ **c)** $x^2 + y^2 > 1$ **d)** $x^2 + y^2 \leq 4$

7.11 exercise/page 258

1a) $(2, 0)$ **b)** $(-2, -3)$ **c)** $(1, 0)$ **d)** $(0, 0)$ **e)** $(6, -1)$ **f)** $(-3, 3)$
2a) $(-2, -2)$ **b)** $(-1, 3)$ **c)** $(0, 1)$ **d)** $(1, -2)$ **e)** $(-2, 3)$ **f)** $(5, -7)$
3a) A'$(-1, 7)$ B'$(-4, 8)$ **6a)** Yes **b)** Yes **c)** P'$(-10, 4)$, Q'$(-6, 1)$.
R'$(-2, -2)$ **7a)** $(1, 4)$, $(8, 8)$, $(3, -2)$ **b)** $(3, 1)$, $(10, 5)$ $(5, -5)$ **8a)** Yes
c) Yes **9a)** 112 units² **c)** P' **d)** 102 units² **10a)** 14 units²
b) 13.5 units² **c)** 33 units² **d)** 28 units² **11a)** $\frac{9\sqrt{2}}{2}$ units

c) $\frac{9\sqrt{2}}{2}$ units **12a)** $-\frac{3}{5}$ **b)** $3x + 5y - 24 = 0$ **c)** $-\frac{3}{5}$ **13a)** $(3, -8)$
b) $(1, -5)$ **c)** $(-1, -2)$ **d)** $-\frac{3}{2}$ **e)** $3x + 2y + 7 = 0$ **14)** $3x - y = 23$
15a) $3x - y = 6$ **b)** $2x - y = 4$ **c)** $x + 2y = 17$
16a) $4x - 3y = -13$ **b)** $2x - y = -19$ **c)** $2x - 2y = -17$
d) $x - 2y = -6$

CHAPTER 8

8.2 exercise/page 265

7a) 10 **b)** Yes **8a)** No. Sample not large enough **b)** S_2 **c)** 90 **9)** B
11a) C **b)** A: 70%, B: 64%, C: 64.9%

8.3 exercise/page 268

3a) ND **b)** D **c)** ND **d)** D **e)** D **f)** ND **g)** D **h)** ND **i)** D **4a)** B **b)** G
c) G **d)** B **e)** G **f)** B **7a)** 40 **b)** 20 **8a)** 33 **b)** 14 **c)** 3 **9)** R **10)** R
11) A **12)** A **13)** A **14)** R

8.5 exercise/page 274

10a) mean 10.4, median 10, mode 10 and 12 **b)** mean 63.3, median
63, mode none **c)** mean 6.3, median 6, mode 6 and 9 **2a)** not
necessarily **b)** yes **c)** not necessarily **3a)** M: mean 54.1, median 18;
N: mean 62.2, median 94 **4a)** P: mode 0, median 18; Q: mode 20,
median 18 **b)** Two values differ **c)** the mode **6a)** mean 26.8, median
27.5, mode 28 **d)** mean

8.7 exercise/page 279

1a) 0 **b)** 72 **2a)** 0 **b)** 118.9 **3)** 12 **4)** $-6, -1, 0, 0, 1, 6$ **5a)** 0 **b)** 0
6a) 74 **b)** 10.6 **8)** 12 **9)** $-6, -6, -4, 0, 4, 6, 6$ **10a)** 0 **b)** 0 **11a)** 176
b) 25.1 **12a)** Range P: 12, Range Q: 12 **b)** The same = 0 **d)** 3.3 **e)** 5
14a) 8 **c)** 2.6 **15a)** 20 **c)** 8.4 **17a)** 98 **b)** 9.4 **c)** No **18a)** 14 **b)** 11
c) 4.62 **d)** No **19a)** 8 **b)** 31.6 **c)** 2.9 **20c)** 30 **d)** 2.3 **21a)** 121.2
b) 8.2 **c)** 6 **22a)** 2.6 **b)** 8

8.8 exercise/page 282

1) 6.2 **2)** 1.1 **3)** 68% **4)** A = 2.9, B = 4, C = 8.4, D = 9.5 **5)** 1980
6) 690 **7)** 24 **8)** 24 **9)** No **10a)** 3.3 **b)** 8 **11a)** 6800 **b)** 500 **12a)** 24
b) 4 **13a)** 2.1 **b)** 12 **c)** 1 **14a)** 6.5 **b)** 16 **c)** 68.6 — 81.6
d) 3420 (50%)

8.9 exercise/page 285

1a) $\frac{1}{26}$ **b)** $\frac{3}{26}$ **c)** $\frac{5}{26}$ **d)** $\frac{8}{26}$ **2a)** $\frac{1}{6}$ **b)** $\frac{1}{2}$ **c)** $\frac{5}{6}$ **d)** $\frac{1}{2}$ **e)** 0 **3)** $\frac{21}{39}$
4a) $\frac{2}{3}$ **b)** $\frac{1}{3}$ **5a)** $\frac{7}{17}$ **b)** $\frac{10}{17}$ **6a)** 50 **b)** 0.96 **7a)** 0.49 **b)** 0.51 **8a)** 0.48

b) 0.52 **9a)** 0.5, 0.5 **10)** 0.54 **11)** 0.12 **12a)** $\frac{1}{4}$ **b)** $\frac{1}{4}$ **c)** $\frac{3}{4}$ **13a)** $\frac{1}{8}$ **b)** $\frac{3}{8}$ **c)** $\frac{3}{8}$ **d)** $\frac{1}{4}$ **14a)** $\frac{3}{8}$ **b)** $\frac{5}{8}$ **c)** $\frac{7}{8}$ $\frac{1}{2}$ **e)** $\frac{1}{2}$ **20a)** $\frac{35}{36}$ **b)** $\frac{5}{6}$ **c)** $\frac{17}{18}$ **21a)** $\frac{1}{2}$ **b)** $\frac{1}{2}$ **22a)** $\frac{5}{18}$ **b)** $\frac{1}{12}$ **23a)** $\frac{1}{52}$ **b)** $\frac{51}{52}$ **c)** $\frac{1}{13}$ **d)** $\frac{12}{13}$ **e)** $\frac{1}{26}$

8.10 exercise/page 288

7a) $\frac{1}{8}$ **b)** $\frac{3}{8}$ **c)** $\frac{3}{8}$ **8a)** $\frac{1}{48}$ **b)** $\frac{3}{16}$ **9a)** $\frac{5}{144}$ **b)** $\frac{1}{18}$ **c)** $\frac{5}{36}$ **10a)** $\frac{1}{2704}$ **b)** $\frac{1}{169}$ **c)** $\frac{1}{169}$ **d)** $\frac{12}{169}$ **e)** $\frac{1}{4}$ **11a)** $\frac{1}{40}$ **b)** $\frac{1}{39}$ **c)** $\frac{1}{1560}$ **12a)** $\frac{1}{50}$ **b)** $\frac{1}{49}$ **c)** $\frac{1}{117\,600}$ **13a)** $\frac{1}{13}$ **b)** $\frac{1}{650}$

8.11 exercise/page 291

1a) 0.094 **b)** 0.214 **2)** 0.009 **3a)** J **b)** 0.222 **5a)** 14.4 **b)** 38.3% **6a)** 12, 14 **b)** 66.7% **7a)** 14 **b)** 50% **8a)** 3.8 **b)** 68.3% **c)** 13 660 **9a)** 0.2 **b)** 0.583 **c)** 0.233 **d)** 0.55 **e)** 0.45 **f)** 0.683 **10a)** 2.6 − 2.7 cm **b)** 68.2% **11a)** 2.7 cm **b)** 0.5 cm **12)** $4140 **13a)** 0.068 **b)** 0.318 **c)** 0.074 **d)** 0.297 **14a)** 1.8 cm **b)** 50% **15a)** 2.3 cm **b)** 0.7 cm

CHAPTER 9

9.2 exercise/page 301

1) △ABC ≅ △RQP(ASA), △DEF ≅ △VTU(ASA), △GHI ≅ △UVL(ASA), △RAP ≅ △TMN(HS), △JKL ≅ △XWV(SSS), △MNP ≅ △SQR(ASA), △TAC ≅ △MST(SAS) **2a)** AC = FE **b)** PQ = UT or ∠R = ∠S or ∠P = ∠U **c)** ∠U = ∠V or ∠S = ∠X **d)** ∠J = ∠N, ∠L = ∠M **e)** RT = VT or ∠S = ∠U

9.4 exercise/page 306

21a) ∠ACB, ∠ABC **b)** ∠BAC, ∠BCA **c)** ∠CAB, ∠CBA **d)** ∠BCA, ∠BAC **e)** ∠ACB, ∠ABC **22a)** ∠DFE or ∠DEF **b)** ∠EDK or ∠EFH **c)** ∠FED or ∠FDE **d)** ∠EDK **e)** ∠HFE

9.5 exercise/page 310

1a) ∠1 and ∠5, ∠3 and ∠7, ∠2 and ∠6, ∠4 and ∠8, ∠9 and ∠13, ∠11 and ∠15, ∠10 and ∠14, ∠12 and ∠16, ∠7 and ∠15, ∠8 and ∠16, ∠5 and ∠13, ∠6 and ∠14, ∠3 and ∠11, ∠4 and ∠12, ∠1 and ∠9, ∠2 and ∠10. **b)** ∠3 and ∠6, ∠11 and ∠14, ∠4 and ∠5, ∠12 and ∠13, ∠9 and ∠4, ∠11 and ∠2, ∠13 and ∠8, ∠15 and ∠6. **c)** ∠3 and ∠5, ∠4 and ∠6, ∠11 and ∠13, ∠12 and ∠14, ∠2 and ∠9, ∠4 and ∠11, ∠6 and ∠13, ∠8 and ∠15. **2a)** ∠1 and ∠2, ∠1 and ∠3, ∠2 and ∠4, ∠3 and ∠4, ∠5 and ∠6, ∠5 and ∠7, ∠7 and ∠8, ∠8 and ∠6, ∠9 and ∠10, ∠9 and ∠11, ∠10 and ∠12, ∠11 and ∠12, ∠13 and ∠14, ∠13 and ∠15, ∠15 and ∠16, ∠14 and ∠16, ∠3 and ∠5, ∠4 and ∠6, ∠11 and ∠13, ∠12 and ∠14, ∠2 and ∠9, ∠4 and ∠11, ∠6 and ∠13, ∠8 and ∠15. **b)** ∠1 and ∠4, ∠2 and ∠3, ∠5 and ∠8, ∠6 and ∠7, ∠9 and ∠12, ∠10 and ∠11, ∠13 and ∠16, ∠14 and ∠15 **4a)** ∠DAC = ∠BCA, ∠BAC = ∠DCA, ∠BAD = ∠DCB, ∠ADC = ∠CBA. **b)** ∠PSQ = ∠RQS, ∠RSQ = ∠PQS, ∠QPS = ∠SRQ, ∠PSR = ∠RQP **5a)** $m = 96, n = 36$ **b)** $m = 28, n = 47$ **6)** $n = 144°, m = 36°$ **25a)** $x = 52.5$ **b)** $x = 54$ **c)** $x = 20$ **d)** $x = 17$ **e)** $x = 60$ **f)** $x = 140$ **26a)** 40° **b)** Isosceles **27)** ∠C = 40°, ∠A = 60°, ∠B = 80° **28a)** $x = 100, y = 160$ **b)** $x = 30, y = 110$ **c)** $x = 60, y = 20$ **d)** $x = y = 46$ **e)** $x = 70, y = 80$ **f)** $y = 25, x = 120$

9.6 exercise/page 314

1a) AB ∥ CD, GE ∥ HF **b)** AB ∥ DC, AD ∥ BC

9.9 exercise/page 321

6a) $x = 90, y = 35$ **b)** $k = 55$ **c)** $y = 60, z = 30$ **d)** $x = 60, y = 60, z = 40, n = 120$ **e)** $d = e = 90, b = 55, c = 40, a = 95$ **f)** $k = 60, n = 120, h = 90, v = 60, w = 30$ **g)** $a = b = 50, g = f = 40, d = 90, c = e = 45$ **h)** $c = 80, b = 30, n = 80, a = 50; m = 50$ **7a)** 3 **b)** $h = 4, k = 3$ **c)** 12 **11a)** 160° **b)** 40° **c)** 40°

9.11 exercise/page 324

12) (a) = (e), (b) = (c), (d) = (k), (f) = (j), (g) = (i), (h) = (l) **14a)** A(−1, 5), B(−4, 4) **b)** P(4, −2), Q(−1, −6), R(5, −7)

9.12 exercise/page 327

16a) (3, −2) **b)** (−1, −3) **c)** (−2, 5) **17a)** (−1, 5) **b)** (−3, −2) **d)** (4, −3)

9.13 exercise/page 330

11a) (−4, −3) **b)** (3, 5) **c)** (5, −4) **12a)** (−3, −5) **b)** (2, −3) **c)** 4, 3)

CHAPTER 10

10.2 exercise/page 338

1a) 11.8 cm² **b)** 6.7 cm² **c)** 50.0 cm² **d)** 26.7 cm² **2a)** 5.2 cm **b)** 3.7 cm **c)** 4.7 cm **d)** 6.2 cm **e)** 4.5 cm **3a)** 41.3 cm² **b)** 119.4 cm² **c)** 36 cm² **d)** 187.5 cm² **4a)** 22.7 cm² **b)** 72 cm² **c)** 90.4 cm² **d)** 48.1 cm² **5)** 6.7 cm **6)** 2.9 cm **7)** 14 cm **8)** 24 **11a)** 93.6 cm² **b)** 97.2 cm² **c)** 91.2 cm² **d)** 40.8 cm² **12)** $11a^2$

10.3 exercise/page 338

2a) ATT **b)** ATT **c)** ATT **d)** ATT **e)** Diagonal bisects the area of parallelogram **f)** Median EC bisects △ADC, median AD bisects △ABC. **g)** Diagonal bisects area of ∥gm APCB **h)** Median BE bisects △ABD, median CE bisects △ADC, median AD bisects △ABC **i)** See (h) **19a)** 2:5 **b)** 2:7 **c)** 5:7 **20a)** 1:1 **b)** 2:5 **c)** 1:2 **d)** 5:14 **e)** 1:7 **21a)** 1:8 **b)** 3:8

10.4 exercise/page 345
3a) 15 **b)** 3.9 **c)** 18.7 **d)** 3.5 **4a)** DE∥FG **b)** KJ∥PN

10.5 exercise/page 347
1a) Internal **b)** External **c)** Internal **d)** External **2a)** 1:2 **b)** 2:7 **c)** 3:2 **d)** 3:1 **e)** 3:4 **f)** 3:7 **g)** 2:9 **h)** 2:3 **i)** 7:3 **j)** 9:7 **k)** 1:2 **l)** 2:1 **3a)** 5:3 **b)** 2:3 **c)** 3:2 **d)** 5:2 **4)** RS = 43.75 cm, ST = 18.75 cm **5)** PN = 36 cm, NQ = 8 cm **b)** ED = 33.5 cm, DE = 12.5 cm **7)** AC = 16.8 m, CB = 11.2 m **8)** DE = 40 cm, EF = 24 cm, **9)** OZ = 66 cm, ZP = 22 cm **10a)** PR = 6.7 cm, RQ = 5.3 cm **b)** AC = 4.3 cm, CB = 10.7 cm **c)** PM = 6 cm, PN = 18 cm **d)** SV = 26.25 cm, VT = 11.25 cm **13a)** 2:1 **b)** 5:6 **c)** 1:1 **d)** 3:2 **15a)** 5:2 **b)** 5:7 **16a)** 3:10 **b)** 3:7 **c)** 7:10 **17a)** 4:7 **b)** 4:3 **c)** 3:7 **19a)** 4:7 **b)** 4:3 **c)** 3:1 **d)** 4:9 **20)** $\left(4, \frac{7}{3}\right)$ **21)** $(-31, -8)$ **22)**

10.6 exercise/page 350
3a) △RSV ~ △TPV **b)** $\frac{RS}{TP} = \frac{SV}{PV} = \frac{RV}{TV}$ **4a)** 9 **b)** 16 **c)** x = 10, y = 6 **d)** y = 3, x = 4 **e)** x = 20, y = 12 **f)** x = 10.5, y = 14 **5)** $\frac{44}{7}$ **6)** $\frac{15}{4}$ **7)** 12.5 **14)** 1.7 m, 2.5 m **15)** 26.2 m **17)** **18b)** **19)** 10.5 m **20)** 10.5 m **21)** 225 m **22)** 63.6 m **23)** 5.7 m **24)** 146.7 m **25)** 24 m

10.7 exercise/page 354
6a) A′(6, 9), B′(3, 15), C′(9, 12) **b)** A′(4, 6), B′(2, 10), C′(6, 8) **c)** A′(−2, −3), B′(−1, −5) C′(−3, −4) **9a)** 20 unit² **b)** 80 unit² **10a)** 9 units² **b)** 1 unit² **12)** 5x − 4y = 90 **13)** $k = -\frac{1}{2}$

10.8 exercise/page 356
1) (a), (b) **a)** (a), (b), (c), (d) **3)** b = 80°, a = 35°, c = 65° **4a)** x = 4, y = 18 **b)** x = 6, y = 4 **c)** x = 71°, y = 12 **d)** x = 4.5, y = 6 **e)** x = 9, y = 4.8 **5)** 100 cm, 120 cm, 140 cm **6)** 13.6 cm, 14.3 cm, 17.0 cm **7a)** x = 12 **b)** z = 15.5 **9)** 6.9

10.9 exercise/page 359
1a) $\frac{9}{16}$ **b)** $\frac{25}{4}$ **c)** 36:1 **2a)** 25 cm² **b)** 21 cm² **3)** 5 cm **4)** 35:25 **5)** 256:169 **6)** 8 **10a)** 1:4 **b)** 1:12 **11a)** $\frac{1}{4}$ **b)** $\frac{1}{3}$ **c)** $\frac{1}{12}$

10.10 exercise/page 361
2e) $\sqrt{13}, \sqrt{13}, 3\sqrt{13}, 2\sqrt{13}$ **3a)** [5, 2], $\sqrt{29}$ **b)** [−8, 0], 8 **c)** [−7, 3], $\sqrt{58}$ **d)** [−3, −2], $\sqrt{13}$ **4a)** [6, 4], $2\sqrt{13}$ **b)** [−6, −4], $2\sqrt{13}$ **c)** [−8, 5], $\sqrt{89}$ **d)** [−7, −1], $5\sqrt{2}$ **5a)** [1, 6] **b)** [1, 6] **c)** $\sqrt{37}$ **6b)** [3, 10], $\sqrt{109}$ **c)** [−5, 0], 5 **7a)** [21, −8] **b)** [−21, 8] **c)** [−24, −7] **8a)** m = 1, n = 8 **b)** m = 7, n = 18 **c)** m = 2, n = 6 **10a)** T **b)** F **c)** T **d)** T **e)** F **f)** T **g)** T **h)** F **12a)** $\overrightarrow{DC} = \vec{a}$ **b)** $-\overrightarrow{CD} = \vec{a}$ **c)** $\overrightarrow{AD} = \vec{b}$ **d)** $\overrightarrow{AC} = \vec{a} + \vec{b}$ **e)** $\overrightarrow{DB} = \vec{a} - \vec{b}$ **f)** $-\overrightarrow{BD} = \vec{a} - \vec{b}$ **13a)** $\overrightarrow{AC}$ **b)** $\overrightarrow{DC}$ **c)** $\overrightarrow{DA}$ **d)** $\overrightarrow{DA}$ **14a)** $\vec{b}$ **b)** $2\vec{b}$ **c)** $2\vec{a} + 2\vec{b}$ **15a)** $\overrightarrow{QR} = -\overrightarrow{PQ} + \overrightarrow{PR}$ **b)** $\overrightarrow{QS} = -\frac{1}{3}\overrightarrow{PQ} + \frac{1}{3}\overrightarrow{PR}$ **c)** $\overrightarrow{RS} = \frac{2}{3}\overrightarrow{PQ} - \frac{2}{3}\overrightarrow{PR}$ **d)** $\overrightarrow{PS} = \frac{1}{3}\overrightarrow{PR} + \frac{2}{3}\overrightarrow{PQ}$ **16a)** $\overrightarrow{OP} = \frac{2}{5}\overrightarrow{OA} + \frac{3}{5}\overrightarrow{OB}$ **b)** $\overrightarrow{OP} = \frac{4}{5}\overrightarrow{OA} + \frac{1}{5}\overrightarrow{OB}$ **c)** $\overrightarrow{OP} = \frac{4}{9}\overrightarrow{OA} + \frac{5}{9}\overrightarrow{OB}$ **17)** $\left(-1, \frac{37}{7}\right)$ **18a)** $\left(-\frac{2}{3}, 6\right)$ **b)** $\left(\frac{8}{3}, 9\right)$ **c)** $\left(2, \frac{42}{5}\right)$

Chapter 11

11.2 exercise/page 371
8a) $\sin \theta = \pm\frac{4}{5}$, $\cos \theta = \frac{3}{5}$, $\tan \theta = \pm\frac{4}{3}$ **b)** $\sin \theta = -\frac{4}{5}$, $\cos \theta = \pm\frac{3}{5}$, $\tan \theta = \pm\frac{4}{3}$ **c)** $\sin \theta = \pm\frac{2}{\sqrt{13}}$, $\cos \theta = \frac{3}{\sqrt{13}}$, $\tan \theta = \pm\frac{2}{3}$ **d)** $\sin \theta = \pm\frac{1}{2}$, $\cos \theta = -\frac{\sqrt{3}}{2}$, $\tan \theta = \pm\frac{1}{\sqrt{3}}$

9a) $(-\sqrt{3}, -1)$ **b)** $\sin \alpha = -\frac{1}{2}$, $\tan \alpha = \frac{1}{\sqrt{3}}$ **10a)** $(-8, 15)$ **b)** $\cos \theta = -\frac{8}{17}$, $\sec \theta = -\frac{17}{8}$, $\cot \theta = \frac{-8}{15}$ **11a)** $\cos B = -\frac{3}{5}$ **b)** $\tan B = -\frac{4}{3}$ **12a)** $-\frac{60}{169}$ **b)** $-\frac{5}{6}$ **c)** $\frac{144}{169}$ **d)** 1 **13)** $-\frac{16}{15}$ **14)** $\frac{26}{15}$

15a) second, third **c)** $\sin \theta = \pm\frac{24}{25}$ **16a)** $\pm\frac{15}{17}$ **b)** $\pm\frac{5}{13}$ **c)** $\pm\frac{24}{7}$ **d)** $\pm\sqrt{3}$ **17a)** I **b)** III **c)** II **d)** IV **e)** I **f)** I **18a)** I, II **b)** II, III **c)** I, III **d)** III, IV **e)** I, III **f)** I, IV **19a)** positive I, II; negative III, IV **b)** positive I, IV; negative II, III **c)** positive I, III; negative II, IV **d)** positive I, II; negative III, IV **e)** positive I, IV; negative II, III **f)** positive I, III; negative II, IV **g)** Quadrant I, $\sin \theta$, $\cos \theta$, $\tan \theta$ are all positive, II only $\sin \theta > 0$, III only $\tan \theta > 0$, IV only $\cos \theta > 0$

11.3 exercise/page 375
7a) $\frac{1}{\sqrt{2}}$ **b)** $\frac{1}{2}$ **c)** $\frac{\sqrt{3}}{2}$ **d)** $\frac{\sqrt{3}}{2}$ **e)** $\sqrt{2}$ **f)** $\sqrt{3}$ **g)** $\frac{1}{\sqrt{3}}$ **h)** $\frac{2}{\sqrt{3}}$ **i)** 1 **j)** $\frac{2}{\sqrt{3}}$ **k)** $\sqrt{2}$ **l)** $\frac{1}{\sqrt{3}}$

8b) $\sin 300° = -\frac{\sqrt{3}}{2}$, $\cos 300° = \frac{1}{2}$, $\tan 300° = -\sqrt{3}$

9b) $\sin(-225°) = \frac{1}{\sqrt{2}}$, $\cos(-225°) = -\frac{1}{\sqrt{2}}$, $\tan(-225°) = -1$
10a) $-\frac{\sqrt{3}}{2}$ **b)** $-\frac{\sqrt{3}}{2}$ **c)** same **11a)** $-\frac{1}{\sqrt{2}}$ **b)** $\frac{\sqrt{3}}{2}$ **c)** $\frac{1}{\sqrt{3}}$ **d)** -1
e) $\frac{2}{\sqrt{3}}$ **f)** $-\sqrt{3}$ **g)** $\sqrt{2}$ **h)** -1 **i)** $-\frac{2}{\sqrt{3}}$ **12a)** $\frac{1}{\sqrt{3}}$ **b)** $-\frac{1}{2}$ **c)** $-\frac{1}{\sqrt{3}}$
d) $-\sqrt{2}$ **e)** 1 **f)** $\frac{1}{2}$ **13a)** $1\frac{1}{2}$ **b)** 1 **c)** $\frac{\sqrt{3}}{2}$ **14a)** 0 **b)** $\frac{\sqrt{6}}{4}$ **c)** $\frac{1}{4}$
15a) $-\frac{1}{\sqrt{3}}$ **b)** $150°$ **16a)** $\frac{1}{2}$ **b)** $300°$ **17a)** $\pm\frac{\sqrt{3}}{2}$ **b)** $120°, 240°$
18a) $30°, 330°$ **b)** $210°, 330°$ **c)** $45°, 225°$ **d)** $225°, 315°$ **e)** $120°, 240°$
f) $225°, 315°$ **19)** $1\frac{1}{2}$ **20)** $\frac{2\sqrt{3}-3}{4}$

11.4 exercise/page 378
1a) 0.7314 **b)** 1.1106 **c)** 1.4142 **d)** 0.6428 **e)** 1.4663 **f)** 1.0000
g) 1.2361 **h)** 0.7431 **i)** 0.6157 **j)** 0.7986 **k)** 0.7813 **l)** 1.2868 **2a)** I
b) D **c)** I **d)** D **e)** I **f)** D **3a)** $49°$ **b)** $49°$ **c)** $50°$ **d)** $48°$ **e)** $52°$ **f)** $54°$
4a) $48°$ **b)** $47°$ **c)** $47°$ **d)** $52°$ **e)** $48°$ **f)** $51°$ **5a)** 0.9205 **b)** 0.7193
c) 5.2408 **d)** 1.1223 **e)** 0.3249 **f)** 6.3138 **g)** 0.7431 **h)** 1.0187
i) 0.8192 **6a)** $10°$ **b)** $18°$ **c)** $78°$ **d)** $65°$ **e)** $83°$ **f)** $25°$ **g)** $36°$ **h)** $25°$
7a) $59°$ **b)** $-304°$ **8a)** $20°$ **b)** $26°$ **c)** $75°$ **d)** $21°$ **e)** $73°$ **f)** $65°$
g) $10°$ **h)** $80°$ **9a)** $27°$ **b)** $34°$ **c)** $37°$ **d)** $56°$ **e)** $63°$ **f)** $74°$ **10a)** $30°$
b) $48°$ **c)** $56°$ **d)** $37°$ **e)** $24°$ **f)** $71°$ **11a)** I **b)** D **c)** I **d)** D **e)** D
f) D **13)** a, d **15a)** $10°$ **b)** $-74°$ **c)** $15°$ **16a)** $\theta = 40°$, $\alpha = 30°$
b) $\theta = 45°$, $\alpha = 30°$ **20a)** 0.7986 **b)** 0.0349 **c)** 0.7071
d) -0.7071 **e)** 1.8361 **f)** 2.9238 **g)** 1.1504 **h)** 7.1154 **i)** -0.3839
21a) $155°$ **b)** $209°$ **c)** $240°$ **d)** $275°$ **e)** $55°$ **f)** $211°$ **22a)** $105°$ **b)** $196°$
c) $169°$ **d)** $336°$ **e)** $248°$ **f)** $116°$ **23)** $245°, 295°$ **24a)** $110°, 250°$
b) $155°, 335°$ **25a)** $243°, 297°$ **b)** $22°, 338°$ **c)** $103°, 283°$ **d)** $116°$,
$244°$ **26a)** $256°, 284°$ **b)** $15°, 195°$ **c)** $58°, 122°$ **d)** $70°, 290°$ **e)** $149°$,
$329°$ **f)** $108°, 252°$

11.5 exercise/page 381
1a) $60°$ **b)** $90°$ **c)** $45°$ **d)** $30°$ **e)** $135°$ **f)** $-90°$ **g)** $-120°$ **h)** $150°$
i) $240°$ **j)** $-135°$ **k)** $-300°$ **l)** $270°$ **m)** $360°$ **n)** $-720°$ **o)** $540°$
p) $-270°$ **2a)** π **b)** 2π **c)** $\frac{\pi}{2}$ **d)** $\frac{\pi}{4}$ **e)** $-\frac{\pi}{3}$ **f)** $-\frac{5\pi}{6}$ **g)** $\frac{\pi}{6}$ **h)** $\frac{4\pi}{3}$
i) $\frac{-11\pi}{6}$ **j)** $\frac{3}{2}\pi$ **k)** $-\frac{\pi}{2}$ **l)** $\frac{2\pi}{3}$ **3a)** π **b)** $-\pi$ **c)** $\frac{\pi}{2}$ **d)** $-\frac{5\pi}{4}$ **e)** $\frac{7\pi}{6}$
f) $\frac{5\pi}{3}$ **5a)** II **b)** III **c)** III **d)** I **e)** IV **f)** IV **g)** II **h)** II **6a)** $\frac{\pi}{2}$ **b)** π
c) $\frac{\pi}{6}$ **d)** 1 **e)** $\frac{\pi}{2}$ **f)** 2.6 **7a)** $63°$ **b)** $52°$ **c)** $-17°$ **d)** $86°$ **e)** $-46°$
f) $23°$ **g)** $109°$ **h)** $132°$ **8a)** π **b)** $\frac{4\pi}{3}$ **c)** 4π **d)** 20π **9a)** $\frac{2\pi}{3}$
b) $-\frac{3\pi}{4}$ **c)** $\frac{11\pi}{6}$ **d)** $\frac{13\pi}{6}$ **e)** $-\frac{5\pi}{4}$ **f)** $\frac{7\pi}{4}$ **10a)** 1.0 **b)** 2.8 **c)** 4.3
d) 5.3 **e)** 4.1 **f)** 3.6 **11a)** $-\frac{5\pi}{3}$ **b)** $\frac{7\pi}{4}$ **c)** $-\frac{11\pi}{6}$ **d)** π **e)** $-\frac{5\pi}{4}$

f) $\frac{\pi}{2}$ **g)** $\frac{3\pi}{4}$ **h)** $-\frac{4\pi}{3}$ **12a)** 0.0175 **b)** 0.1396 **c)** 0.4363
d) 1.1345 **e)** 1.2217 **f)** 2.9671 **g)** -4.1015 **h)** -6.8068
13) $\frac{5\pi}{4}$ **14a)** $\frac{1}{2}$ **b)** 0.5000 **15a)** $114.6°$ **b)** 0.9093 **16a)** -0.4161
b) 1.5574 **c)** -1.0101 **d)** -0.6421 **e)** -0.9093 **f)** 2.0858
g) 0.8912 **h)** -0.5048 **17a)** $\frac{1}{\sqrt{2}}$ **b)** $\frac{1}{2}$ **c)** $\frac{1}{2}$ **d)** 1 **e)** $\sqrt{3}$ **f)** 2
18a) $\frac{1}{\sqrt{2}}$ **b)** $\frac{1}{\sqrt{2}}$ **c)** $-\sqrt{3}$ **d)** $\frac{2}{\sqrt{3}}$ **e)** $-\frac{2}{\sqrt{3}}$ **f)** $-\sqrt{3}$ **g)** $-\frac{1}{\sqrt{2}}$
h) $-\sqrt{2}$ **i)** $-\sqrt{3}$ **j)** $-\frac{1}{\sqrt{3}}$ **k)** 2 **l)** $-\frac{1}{2}$ **19a)** $\frac{\pi}{4}, \frac{3\pi}{4}$ **b)** $\frac{\pi}{6}, \frac{11\pi}{6}$
c) $\frac{\pi}{3}, \frac{5\pi}{3}$ **20a)** $\frac{5\pi}{6}, \frac{7\pi}{6}$ **b)** $\frac{3\pi}{4}, \frac{7\pi}{4}$ **c)** $\frac{4\pi}{3}, \frac{5\pi}{3}$

11.6 exercise/page 385
6a) $(1 - \cos\theta)(1 + \cos\theta)$ **b)** $(1 - \sin\theta)(1 + \sin\theta)$
c) $(\sin\theta - \cos\theta)(\sin\theta + \cos\theta)$ **d)** $\sin\alpha(1 - \sin\alpha)$
e) $(\tan\alpha - \cot\alpha)(\tan\alpha + \cot\alpha)$ **f)** $(\sec\theta - 1)(\sec\theta + 1)$ **7a)** $\cos\theta$
b) $\sin^2\theta + \cos^2\theta$ **c)** $\cos^2\theta$ **d)** $\sin\theta$ **e)** $-\frac{\cos^2\theta}{\sin^2\theta}$
f) $\frac{1}{\cos^2\theta}$ **8a)** LS $= \frac{1 - \sin^2\alpha}{\sin\alpha}$ **9a)** $1 - \sin^2\theta$

11.7 exercise/page 388
17a) $\frac{\sqrt{3}}{2}$ **b)** $\frac{1}{\sqrt{2}}$ **c)** 1 **d)** -1 **e)** -1 **f)** 0 **g)** $-\frac{1}{2}$ **h)** $\frac{1}{\sqrt{2}}$ **i)** $-\frac{\sqrt{3}}{2}$
18a) $\frac{1}{2}$ **b)** $\frac{\sqrt{3}}{2}$ **c)** 1 **d)** -1 **e)** $-\frac{1}{\sqrt{2}}$ **f)** $\frac{\sqrt{3}}{2}$ **19a)** $\frac{\sqrt{3}}{2}$ **b)** $\frac{1}{\sqrt{2}}$ **c)** 0
d) $\frac{\sqrt{3}}{2}$ **e)** $\frac{1}{\sqrt{2}}$ **f)** 0 **g)** 1 **h)** -1 **i)** $-\frac{1}{\sqrt{2}}$ **j)** $\frac{\sqrt{3}}{2}$ **k)** $\frac{1}{2}$ **l)** -1
m) $-\frac{1}{\sqrt{2}}$ **n)** $\frac{1}{2}$ **o)** $\frac{1}{\sqrt{2}}$ **20a)** $\sqrt{3}$ **b)** $-\sqrt{3}$ **c)** undefined **d)** $\frac{1}{\sqrt{3}}$ **e)** 1
f) undefined **21a)** 0 m **b)** 4 m **c)** 0 m **d)** -2 m **e)** 2 m **f)** $-2\frac{2}{3}$ m
22a) 12 h **b)** 0 h, 24 h **c)** 6 h, 18 h **d)** 3 h, 21 h **e)** 1 h 40 min
f) 9 h, 15 h **23b)** 12 h, 36 h **c)** 0 h, 24 h, 48 h **24b)** $23:00$

11.8 exercise/page 392
1a) $\csc\beta$ **b)** $\tan\beta$ **c)** $\cos\beta$ **d)** $\cot\beta$ **e)** $\sec\beta$ **f)** $\sin\beta$ **2a)** $\frac{4}{5}$ **b)** $\frac{3}{5}$
c) $\frac{3}{5}$ **d)** $\frac{3}{4}$ **3a)** $\frac{13}{5}$ **b)** $\frac{12}{5}$ **c)** $\frac{13}{12}$ **d)** $\frac{13}{5}$ **4a)** $\cot$ **b)** $\sec$ **c)** $\tan$ **d)** $\sin$
e) $\cos$ **f)** $\csc$ **5a)** 12.9 **b)** 14.0 **c)** 35.5 **d)** 10.9 **e)** 22.9 **f)** 140.3
6a) $51°$ **b)** $61°$ **c)** $44°$ **d)** $56°$ **e)** $37°$ **f)** $25°$ **7a)** 0.3333 **b)** 0.6521
c) 0.3659 **d)** 0.5217 **8a)** 6.8 **b)** 4.2 **c)** 300.9 **9a)** $\angle B \doteq 36°$
$\angle C \doteq 54°$ **b)** $\angle B \doteq 55°$ $\angle A \doteq 25°$ **c)** $\angle A \doteq 47°$ $\angle C \doteq 43°$
10a) $\angle B \doteq 37°$ **b)** RP $\doteq 9.3$ **c)** S $\doteq 9.4$ **d)** $r \doteq 7.3$ **11a)** $r \doteq 6.7$
b) $\angle P \doteq 42°$ **11)** $\sin R = 0.7444$, $\cos R = 0.6667$ $\tan R = 1.1167$,
$\csc R = 1.3432$, $\sec R = 1.5000$, $\cot R = 0.8955$ **12)** $\angle A = 55°$,

$b = 6.9$, $a = 9.8$ **13)** $\angle P = 12°$, $p = 2.83$, $q = 13.30$ **14)** $\angle P = 42°$, $\angle R = 48°$, $m = 19$ **15a)** $q = 10$, $\angle P = 37°$, $\angle R = 53°$ **b)** $\angle Q = 62°$, $t = 15.9$, $u = 7.4$ **c)** $\angle A = 60°$, $c = 2.4$, $a = 4.2$ **d)** $d = 20.7$, $\angle D = 56°$, $\angle F = 34°$ **e)** $\angle M = 52°$, $m = 153.6$, $r = 194.9$
16a) $a = 7.0$, $\angle A = 29°$, $\angle C = 61°$ **b)** $\angle D = 51°$, $f = 17.7$, $d = 21.8$ **c)** $\angle H = 47°$, $h = 153.3$, $i = 209.7$ **18)** $p = 10.8$ **19)** $RS = 39.7$

11.9 exercise/page 395

1) 177.9 m **2)** 46.3 m **3a)** 67.9 m² **b)** 20.9 cm² **4a)** 711 m **b)** 106 621 m² **5)** 40° **6)** 11.5 m, 12.2 m **7)** 55.4 m **8)** 6974.8 m **9)** 31.4 m **10)** 877.2 m **11)** 1166.1 m **12)** 2.5 m **13)** 22.4 m **14)** 2036.1 m **15)** 266.4 m **16)** **17a)** 1° **b)** 3° **18)** 128.6 m **19)** 60° **20)** 13.1 m

11.10 exercise/page 397

4) 12 **5)** 8 **6)** 0.375 **7a)** $b = 36.8$ **b)** $d = 34.4$ **c)** $s = 25.4$ **8a)** 60° **b)** 112° **c)** 55° **d)** 44° **9a)** $p = 17.3$, $q = 17.3$ **b)** $r = 32$, $s = 55.4$ **10)** 7.4 **13a)** $\angle A = 85°$, $a = 66.8$, $b = 34.6$ **b)** $\angle U = 56°$, $s = 46.3$, $t = 34.0$ **c)** $\angle P = 28°$, $p = 32.6$, $q = 38.9$ **d)** $\angle R = 75°$, $r = 30.4$, $t = 23.8$ **14a)** 50.2 **b)** 70° **c)** $p = 93.0$, $q = 56.7$ **15)** 16.2 **16)** $\angle P = 86°$, $q = 18.2$, $r = 17.0$ **17a)** 48°, 132° **b)** 23°, 157° **c)** 46°, 134° **d)** 60°, 120° **18)** 109°, 71° **19)** 70°, 70°, 10.1 or 110°, 30°, 5.4 **20)** no solution **21)** $\angle A = 47°$ or 133° **22a)** $\angle R = 73°$, $q = 119.2$, $r = 130.3$ **b)** $\angle B = 75°$, $c = 51.7$ $b = 50.9$ **c)** $\angle R = 75°$, $p = 12.5$, $q = 10.5$ **d)** $\angle D = 65°$, $d = 12.2$, $f = 10.5$ **e)** $\angle P = 56°$, $n = 5.5$, $p = 17.6$ **f)** $\angle B = 125°$, $b = 39.2$, $c = 10.8$ **23)** $\angle E = 135°$ or 45° **24a)** $\sqrt{2}$ **b)** $\dfrac{\sqrt{6}}{2}$ **25)** 49.0 **26)** 22.4 m

11.11 exercise/page 400

1) 2.4 km, 3.6 km **2)** 10.6 m, 7.8 m **3)** 155.7 m **4)** 10 m **5)** 39.0 m, 45.5 m **6)** 4.5 km **7)** 12.9 m, 14.6 m **8)** 10.2 m **9)** 323 km, 422.5 km, **10a)** $\angle B = 71°$, $\angle C = 71°$ **b)** 27.6 cm **11)** 22.7 m **12)** 1.36 A.U. **13)** 40.6 m **14)** 3.9 cm **15)** $x = 115.6$, $p = 49.7$ **16)** 354.5 m **17a)** $AT = 288.5$ m **b)** $AC = 215.2$ m **18)** 178.6 m

11.12 exercise/page 404

4a) $c = 5.2$ **b)** $r = 12.2$ **5a)** $\alpha \doteq 30°$ **b)** $\beta \doteq 117°$ **6a)** $c = 6.7$ **b)** $x = 102.6$ **7b)** $\angle E = 41°$ **8)** $a = 21$ **9)** $r = 14$ **10)** 0.2 **11)** 36 **12)** $\sqrt{37}$ **13a)** 6.5 **b)** 70° **c)** 82° **14a)** 5.03 **b)** 15.96 **15a)** 0.6625 **b)** 60° **16)** 60° **17)** $\angle A = 45°$, $\angle B = 56°$, $\angle C = 79°$, $a = 3.5$, $b = 4.1$, $c = 4.9$ **18)** $\angle P = 27°$, $\angle Q = 32°$, $\angle R = 121°$, $p = 6.8$, $q = 7.9$, $r = 12.81$ **19a)** $\angle T = 80°$, $\angle U = 43°$, $\angle V = 57°$, $t = 7$, $u = 4.85$, $v = 6$ **b)** $\angle K = 74°$, $\angle J = 43°$, $\angle F = 62°$, **c)** $\angle P = 19°$, $\angle H = 136°$, $\angle W = 25°$, $p = 4.9$, $w = 6.3$, $h = 10.4$ **20)** -0.0667 **21)** 12.6 **b)** 78° **22)** 26° **23)** 42 cm **24)** 19.1 cm **27)** 877.3 m **28)** 940.6 m **29)** 252 m **30)** 2.1 A.U. **31)** 1 km **32)** 16 km **33)** 44.6 m **34)** 75 m **35)** 7.3 km **36)** 22° **37)** 14 cm **38)** 32.7 km **39)** 14.8 cm **40a)** 13.3 km **b)** 53 km² **41)** 16 903.25 m² **42)** 171 m **43)** 157 km **44)** 311 km **45)** 22.8 km **46)** 8 km **47)** 105.2 km

11.13 exercise/page 408

1) 16 **2)** 187.2 **3)** 4.5 **4)** $10\sqrt{2}$ **5)** $\dfrac{1}{5}$ **6)** 0.0926 **7)** 306.4 m **8)** 69.3 m **9)** Distance from A: 137.2 km, Distance from B: 104 km **10)** 1814.2 m **11)** 6.1 m, 13.3 m **12)** $a = 62.1$ cm, $b = 18.6$ cm, $c = 69.3$ cm

11.14 exercise/page 408

1) 199.8 m **2)** 0.712 km **3)** 164.1 m **4)** 36 m

CHAPTER 12

12.2 exercise/page 414

1) (a), (c) **2a)** $y = 3x^2 - 6x + 3$ **b)** $y = -2x^2 - 4x - 2$ **c)** $y = -2x^2 + 12x - \dfrac{58}{3}$ **d)** $y = \dfrac{1}{2}x^2 - 4x + 13$
3a) domain: $\{x \mid 1 \leq x \leq 7\}$ range: $\{y \mid -4 \leq y \leq 5\}$ **b)** $(4, -4)$ **c)** $x = 4$ **d)** 2, 6 **e)** 2, 6 **f)** minimum **4b)** minimum: A, maximum: B **c)** A **5a)** U **b)** D **c)** U **6b)** $(0, -3)$, **c)** minimum $(0, -3)$ **7b)** Domain $\{x \mid x \in R\}$, Range $\{y \mid y \geq 2, y \in R\}$ vertex: $(0, 2)$, axis: $x = 0$
8b) Domain $\{x \mid x \in R\}$, Range $\{y \mid y \leq 4, y \in R\}$ vertex $(0, 4)$, axis $x = 0$ **c)** maximum **9b)** minimum -3, vertex $(0, -3)$
10b) maximum 3, vertex $(1, 3)$ **11a)** A, C **b)** $x = 0$ **c)** $(-1, 1)$, $(2, 7)$
12b) $(0, -9)$, $x = 0$ **c)** $(-3p)$, $(3, 0)$ **d)** 3, -3 **13)** -0.5, -1.2
14a) -1.4 or -7.6 **b)** 2.8 or -2.6 **15)** $a = \dfrac{5}{2}$, $b = -8$
16) $y = -x^2 + 1$ **17)** $a = -\dfrac{3}{5}$, $b = -5$ **18)** $a = \dfrac{1}{2}$, $b = \dfrac{3}{2}$, $c = 1$
19) $y = -3x^2 + 5x + 2$ **20)** $a = 4$, $b = -3$, $c = -7$
31a) $(x, y) \longrightarrow (x, y - 1)$ **b)** $(x, y) \longrightarrow (x, y + 4)$ **32b)** 12.5 m
33a) 8 m **b)** 8 m **34a)** 12 m **b)** 12 m **c)** 12.5 m **35)** 4 s **36a)** 1.8 m **b)** 1.1 m **37)** 1.2 s **38)** 3 s **39)** 8 min 20 s **40)** 3 s **41a)** 5.9 m **b)** Yes.

12.3 exercise/page 419

1a) 4 **b)** 4 **c)** 9 **d)** 16 **e)** $\dfrac{25}{4}$ **f)** $\dfrac{9}{4}$ **2a)** $p = -2$, $q = -4$ **b)** $p = 3$, $q = -9$ **c)** $p = \dfrac{9}{2}$, $q = -\dfrac{81}{4}$ **d)** $p = -\dfrac{5}{2}$, $q = -\dfrac{25}{4}$ **3a)** $(x - 1)^2 - 1$
b) $(x + 2)^2 - 4$ **c)** $(x - 1)^2 - 3$ **d)** $(x + 3)^2 - 1$ **e)** $2(x - 1)^2 - 1$
f) $-3(x - 1)^2 - 1$ **g)** $4(x - 1)^2 - 1$ **h)** $-6(x - 1)^2 + 4$ **4a)** $(x - 2)^2 + 2$
b) $-2(x - 1)^2 - 4$ **c)** $-2\left(x - \dfrac{3}{2}\right)^2 + \dfrac{35}{2}$ **d)** $-3\left(x - \dfrac{2}{3}\right)^2 + \dfrac{22}{3}$
11a) $x = -2$, $(-2, -1)$ **b)** $x = 3$, $(3, -4)$ **c)** $x = 1$, $(1, -4)$ **d)** $x = 1$,

(1, −1) **12a)** $x = −2, (−2, 2)$ **b)** $x = 1, (1, −3)$ **c)** $x = 1, (1, −1)$ **d)** $x = 1, (1, 1)$ **14a)** minimum −1 **b)** maximum −2 **c)** minimum −1 **d)** maximum −4 **15a)** $x = −1, (−1, −2)$ **b)** $x = 1, (1, 1)$ **c)** $x = 1, (1, 4)$ **d)** $x = 1, (1, 2)$ **16a)** $y = 3(x + 2)^2, (−2, 0)$ **b)** $y = −2\left(x − \frac{5}{4}\right)^2 + \frac{73}{8}, \left(\frac{5}{4}, −\frac{73}{8}\right)$ **c)** $y = 2\left(x − \frac{5}{4}\right)^2 − \frac{1}{8}, \left(\frac{5}{4}, −\frac{1}{8}\right)$ **d)** $y = −3\left(x + \frac{5}{6}\right)^2 + \frac{205}{12}, \left(−\frac{5}{6}, \frac{205}{12}\right)$ **18a)** $a < 0, q = 0$ **b)** $a > 0, q < 0$ **c)** $a < 0, p, q = 0$ **d)** $a > 0, p, q = 0$ **e)** $a < 0, q < 0$ **19a)** minimum $−\frac{5}{4}$ **b)** minimum 2 **c)** minimum −6 **d)** maximum $\frac{13}{12}$ **20a)** $x = \frac{3}{2}, \left(\frac{3}{2}, −\frac{21}{4}\right)$ **b)** $x = \frac{3}{4}, \left(\frac{3}{4}, \frac{7}{8}\right)$ **c)** $x = \frac{1}{3}, \left(\frac{1}{3}, \frac{2}{3}\right)$ **d)** $x = −\frac{1}{6}, \left(−\frac{1}{6}, \frac{61}{12}\right)$ **22a)** $y = x^2 + 6x + 14$ **b)** $y = 2x^2 + 12x + 23$ **c)** $y = −3x^2 − 18x − 22$ **d)** $y = x^2 + 2x + 6$ **e)** $y = x^2 + 14x + 51$ **23a)** $y = x^2 + 4x + 9$ **b)** The axis of the image is a translation **24a)** $y = 5(x + 3)^2 + 4$ **b)** A translation is an isometry **c)** $\{x \mid x \in R\}$ **d)** $\{y \mid y \geq 0, y \in R\}$ $\{y \mid y \geq 4, y \in R\}$

12.4 exercise/page 425

1a) Min $−\frac{3}{2}$ **b)** max $\frac{12}{5}$ **c)** max −25 **d)** min 16 **2a)** $13\frac{1}{3}$ **b)** $\frac{2}{3}$ **3a)** $3\frac{7}{8}$ **b)** $\frac{3}{4}$ **4a)** 1 **b)** $\frac{3}{4}$ **5a)** $\frac{3}{4}$ **b)** 1 **6a)** 144 **b)** 41 **c)** $17\frac{1}{8}$ **d)** $27\frac{1}{12}$ **7)** −4, 4 **8)** $−\frac{7}{2}, \frac{7}{2}$ **9)** $−\frac{13}{2}, \frac{13}{2}$ **10)** 40, 40 **11)** 7, 21 **12)** 50, 25 **13a)** 6.5 m **b)** 3 s **14a)** 32 m **b)** 4 s **15)** 9 × 18 km **16a)** 5 m **b)** 45 × 55 m **17)** 7.5 × 15 m **18)** $27.50 each **19)** 900 m² **20a)** 3 s **b)** 11.0 m **21)** 13, 14, 15 **22)** 11.0 m **23)** 46.1 m **24)** 10.9 m **25)** 60 s

12.5 exercise/page 428

1a) 3, −2 **b)** 5, −3 **c)** 7, 3 **d)** $\frac{1}{2}, −1$ **e)** $1, \frac{4}{3}$ **f)** $\frac{5}{2}, 3$ **2a)** $1 \pm \sqrt{3}$ **b)** $2 \pm \sqrt{5}$ **c)** −3, 1 **d)** $\frac{2 \pm \sqrt{5}}{3}$ **e)** $\frac{2 \pm \sqrt{6}}{4}$ **f)** $\frac{4 \pm 3\sqrt{10}}{12}$ **5a)** $\frac{2\sqrt{6}}{3}$ **b)** $\frac{3 \pm 4\sqrt{2}}{2}$ **c)** $2 \pm 2\sqrt{2}$ **d)** $−1 \pm \sqrt{6}$ **6a)** $5 \pm \sqrt{21}$ **7a)** $\frac{3 \pm \sqrt{5}}{2}$ **b)** $\frac{5 \pm \sqrt{17}}{2}$ **c)** $\frac{5 \pm \sqrt{17}}{4}$ **d)** 2 or $−\frac{1}{2}$ **8a)** $\frac{3 \pm \sqrt{21}}{6}$ **b)** $\frac{−5 \pm 3\sqrt{5}}{2}$ **9a)** $\frac{−5 \pm \sqrt{37}}{2}$ **b)** 3, −5 **c)** $\frac{1}{3}$ or −3 **d)** $\frac{−4 \pm \sqrt{14}}{2}$ **e)** 2 or $−\frac{1}{3}$ **f)** $\frac{2}{3}$ or 5 **g)** $\frac{1}{6}$ or −1 **h)** $\frac{3 \pm \sqrt{14}}{5}$ **10a)** $\frac{3 \pm \sqrt{5}}{2}$ **b)** x-intercepts **11)** $−4 \pm \sqrt{13}$ **12)** $x = −1$ or $\frac{2}{5}$ **13)** $x = \frac{5}{2}$ or −1 **15a)** 0.3, −3.3 **b)** 4.6, 0.44 **c)** −0.3, 1 **d)** ±1.6 **16a)** 3.6 or −1.6 **b)** −3 or 1 **c)** 2.7 or −0.2 **d)** −1.2 or 0.8 **17a)** (1, −7) **b)** (−1, 4) **c)** (1.3, −4.1) **d)** (−0.2, 3.1) **18a)** $k = 6$ or 8

b) $x = 4$ or 10 **c)** $x = 9$ or 5 **19a)** 8 **b)** 15 **c)** 30 **20)** $2 \pm \sqrt{19}$ **21a)** $1 \pm \sqrt{6}$ **b)** $2 \pm \sqrt{5}$ **22)** $m = \frac{4}{3}$ or −1 **23)** $m = 3$ or −1 **26a)** $−1 \pm \sqrt{1 − c}$ **b)** $\frac{−b \pm \sqrt{b^2 + 12}}{2}$ **c)** $\frac{−2 \pm \sqrt{4 − a}}{a}$ **d)** $\frac{−b \pm \sqrt{b^2 − 4c}}{2}$ **27a)** $a = 1, b = −2, c = −5$ **b)** $a = 1, b = −3, c = 1,$ **c)** $a = 2, b = −5, c = −1$ **d)** $a = 3, b = −2, c = 5$ **e)** $a = 5, b = −3, c = −8$ **f)** $a = 2, b = −4, c = −1$ **g)** $a = 3, b = −2, c = −3$ **h)** $a = 2, b = −3, c = −2$ **28a)** 5 or −7 **b)** −5 or 3 **c)** $\frac{1}{3}$ or 3 **d)** $\frac{1}{2}$ or −3 **29)** $\frac{3 \pm \sqrt{17}}{4}$ **30a)** $\frac{2 \pm \sqrt{6}}{2}$ **b)** $\frac{3 \pm 2\sqrt{6}}{3}$ **31a)** $\frac{5 \pm \sqrt{33}}{4}$ **b)** $\frac{3 \pm \sqrt{57}}{6}$ **32)** $\frac{1 \pm \sqrt{61}}{10}$ **33a)** 1.8 or −0.2 **b)** 1.9 or −0.4 **34)** −0.99, 2.51 **35a)** 3.52, 1.93 **b)** −1.75, 0.67

12.6 exercise/page 433

1) 8, 9 or −8, −9 **2)** 24.1 cm **3)** 10 cm, 24 cm, 26 cm **4)** Ordinary highway: 70 km/h, superhighway 90 km/h **5)** 8, 12 or −8, −12 **6)** 14, 15 or −14, −15 **7)** 3, 5, 7 or 1, 3, 5 **8)** base 8 cm, height 24 cm **9)** 5 m **10)** 7 m **11)** 30 cm, 40 cm **12)** 10 cm **13)** house: 12 m × 12 m, Garage 6 m × 6 m **14)** 4 cm × 4 cm, 6 cm × 6 cm **15)** 3.3 m **16)** 100 m **17)** Up 3.5 h, down 2.5 h, total 6 h **18)** 9/month **19a)** John 4 km/h, Petra 4.5 km/h **b)** John 45 min, Petra 40 min **20)** 1st day 80 km/h, 2nd day 70 km/h **21a)** Mark 4 km/h, Tina 13 km/h **b)** Mark 15 h, Tina 13 h **22)** 30 cm × 50 cm **23)** 8.8 cm

12.7 exercise/page 435

8a) 0; equal; **b)** 25; rational, unequal **c)** 9; rational, unequal **d)** −23; not real **e)** 12, −8; irrational **f)** −11; not real **9a)** B **b)** A **c)** B **d)** B **e)** B **f)** C **g)** B **h)** B **10)** $m > 2$ or $m < −2$ **11)** 1 **12)** $C < \frac{4}{3}$ **13)** $C < \frac{81}{4}$ **14)** 65 **15a)** $k < \frac{4}{5}$ **b)** $k > 2\sqrt{5}$ or $k < −2\sqrt{5}$ **16a)** $k > \frac{25}{12}$ **b)** $k > \frac{3}{2}$

12.8 exercise/page 437

1a) $y = 2x − 12$ **c)** $(5, −2), (2, −8)$ **2b)** $(3, −9), (1, −5)$ **3a)** $(−4, 4), (3, 18)$ **4)** $(0, 5), (−24, −43)$ **5)** $(−1, 0), (−7, 42)$ **6)** $(−4, −4), (−7, 14)$ **7)** $(1, −3), (3, 1)$ **8a)** $(0, 5), (−1, 4)$ **b)** $(4, 40), (−8, 16)$ **c)** $(3, 15), (−9, 63)$ **d)** $(1, 4), (−2, 7)$ **e)** $(3, 3), (−3, −3)$ **f)** $(4, 5), (5, 4)$ **9)** $(3, −9), (7, 7)$ **10)** $(4.39, 2.39), (−2.39, −4.39)$ **11)** $(2, −4), \left(\frac{−3}{2}, \frac{−9}{4}\right)$

13b) B **14b)** P belongs in A and C, Q belongs in A, R belongs in A,B **17a)** $y \geq (x − 2)^2 − 2$ **b)** $y < −(x + 3)^2 + 2$